Starch

Starch is the principal source of stored energy in plants, and its chemical composition varies depending on the botanical source of the starch. Starch plays a significant role in determining the structural characteristics of finished food products. *Starch: Structure, Properties, and Modifications for Food Applications* explores the comprehensive overview of the basic structure and properties of starch as well as the modification of starch with physical, chemical, and enzymatic methods. Each chapter presents an in-depth review of a specific research area, updated with current research. Chapters of this book provide comprehensive information regarding starch modification, which will help to design new, healthy starch-based food products.

KEY FEATURES

- This book will cover the functional characteristics of conventional and non-conventional starches.
- It covers the different methods of starch modification, including physical, chemical, and enzymatic methods.
- The latest information on the properties of modified starch is from different sources.
- This book will explore the current and emerging application trends of modified starches.

With contributions from esteemed researchers worldwide, this book serves as an invaluable resource for students, food technologists, researchers, and industry professionals seeking to deepen their understanding of modified starches and their diverse applications. We hope that the insights offered within these pages inspire new avenues of research and innovation, ultimately contributing to continued advancement in food technology.

Starch

Structure, Properties, and Modifications for Food Applications

Edited by
Sneh Punia Bangar, K.V. Sunooj and
Anil Kumar Siroha

CRC Press
Taylor & Francis Group
Boca Raton London New York

CRC Press is an imprint of the
Taylor & Francis Group, an **informa** business

First edition published 2025
by CRC Press
2385 NW Executive Center Drive, Suite 320, Boca Raton FL 33431

and by CRC Press
4 Park Square, Milton Park, Abingdon, Oxon, OX14 4RN

CRC Press is an imprint of Taylor & Francis Group, LLC

Library of Congress Cataloging-in-Publication Data
Names: Bangar, Sneh Punia, editor.
Title: Starch : structure, properties, and modifications for food applications / edited by
  Sneh Punia Bangar, K.V. Sunooj, and Anil Kumar Siroha.
Other titles: Starch (CRC Press)
Description: First edition. | Boca Raton, FL : CRC Press, 2025. | Includes bibliographical
  references and index.
Identifiers: LCCN 2024017730 | ISBN 9781032647166 (hbk) | ISBN 9781032655604
  (pbk) | ISBN 9781032655598 (ebk)
Subjects: LCSH: Starch.
Classification: LCC TP248.S7 S65 2025 | DDC 664/.2—dc23/eng/20240521
LC record available at https://lccn.loc.gov/2024017730

ISBN: 978-1-032-64716-6 (hbk)
ISBN: 978-1-032-65560-4 (pbk)
ISBN: 978-1-032-65559-8 (ebk)

DOI: 10.1201/9781032655598

Typeset in Times LT Std
by Apex CoVantage, LLC

Contents

7 Pulsed Electric Fields

Milad Tavassoli and Behnam Bahramian

8 Ionizing Radiation

*Mirela Braşoveanu, Betul Oskaybaş Emlek, Hassan Sabbaghi, Kevser Kahraman,
Serpil Özturk Muti, Farooq Sher, and Monica R. Nemţanu*

12 Acetylation/Esterification of Starch **281**
Ranjan Kaushik, Ankit Kumar, Neha Rani, Rekha Phogat,
and Rakesh Gehlot

13 Oxidation of Starch **314**
Plachikkattu Parambil Akhila, Sneh Punia Bangar, and K.V. Sunooj

Preface

Starch: Structure, Properties, and Modifications for Food Applications explores the intricate world of starch, providing a comprehensive overview of the basic structure, the diverse properties of starch, and the various modifications that make it indispensable to the food science and technology community. This book begins by emphasizing starch's properties and functions in food systems, which gives readers a solid understanding of its behavior and functionality. Detailed information about starch's physicochemical properties further enhances the reader's understanding of the ingredient, revealing its importance for numerous food applications. In the later chapters of this book, physical, chemical, enzymatic, and dual modifications of starch are extensively discussed so that specific food applications may be tailored according to their functional properties. From physical and chemical modifications to the incorporation of starch derivatives, this book presents a detailed analysis of the techniques and their impact on the performance of starch in food products. This book is designed to serve as a valuable resource for food scientists, researchers, industry professionals, and students seeking a comprehensive understanding of starch and its versatile applications in the food industry. We hope that the insights offered within these pages will inspire new avenues of research and innovation, ultimately contributing to continued advancement in food technology.

Chapter 1, contributed by Mounir, Ghandour, Shatta, and Farid, focuses on the properties of conventional and non-conventional starch and their applications. Heat moisture treatment of starches is elaborated upon in Chapter 2, and Siroha and his co-workers contribute to this chapter. Chapter 3, "Microwave Treatment", was authored by Navaf, Bangar, and Sunooj. This chapter highlights the history of the development of microwave heating and the advantages and disadvantages of microwave heating. Liu and her co-workers contributed to Chapter 4. The effect of high hydrostatic pressure on starch properties is addressed in Chapter 5. Dutta and Sit contributed to Chapter 6, which highlights the effect of ultrasound treatment on starch properties. Chapter 7 covers pulsed electric fields (PEFs) and was authored by Milad and Bahramian. In this chapter, the principles, fundamentals, and properties of PEF starch are elaborated upon. Braşoveanu and her co-workers contributed to Chapter 8. Non-thermal plasma modification of starch is explored in Chapter 9. Milling is a physical modification to alter starch properties using mechanical force. Aaliya, Bangar, and Sunooj authored Chapter 10. The effect of milling on starch properties, its application, and future recommendations are discussed in this chapter. Acid hydrolysis has emerged as a prominent method for chemical modification of starch. This method of chemical modification is discussed in Chapter 11, authored by Zambelli and his co-workers. Kaushik, Kumar, Rani, Phogat, and Gehlot authored the chapter on the acetylation/esterification of starch. Acetylated starch is derived from the esterification process of starch. Chapter 13 explores the methods and properties of starch modified by oxidation. In this chapter, physicochemical and functional properties, applications, and challenges of oxidative starches are discussed. Sultana and her co-workers authored the chapter "Starch Grafting". Chapter 15 elaborates on enzymatic modifications, properties, and applications. Enzymatic modification is a green and environmentally friendly method of starch modification. It can promote the structural transformation of natural starch to produce modified starch or derivatives with different physicochemical properties.

In closing, we extend our gratitude to the contributors whose expertise and dedication have made this endeavor possible. We express our sincere anticipation that *Starch: Structure, Properties, and Modifications for Food Applications* will prove an invaluable asset to all those with a vested interest in the multifaceted world of starch.

About the Editors

Sneh Punia Bangar, Ph.D., is a postdoctoral researcher in the Department of Food, Nutrition, and Packaging Sciences at Clemson University, United States. Her research interest includes the extraction and functional characterization of functional foods, starches, modified starches, biopolymers, and nanocomposite films/coatings. She has authored or co-authored more than 150 research/review articles, 12 authored/edited books, 35 book chapters, and 20 conference proceedings/seminars. Currently, she is the Associate Editor of the *Journal of Food Measurement and Characterization* (Springer), an editorial board member of the *International Journal of Food Science and Technology* (Wiley), Academic Editor of the *Journal of Food Quality* (Hindawi), and a section editor of *Current Food Science and Technology Reports* (Springer). Her significant contributions to academia and research were acknowledged with the Award of Honour in 2019 for her outstanding achievements. She is the recipient of the GSG Outstanding Graduate Research Assistant (2022) and CAFLS Outstanding Graduate Student Research Award (2023) at Clemson University, USA. Dr. Bangar has been listed (2023) in the World Ranking of Top 2% Scientists, published by Stanford University, USA. Also, she received the Professor Gurcharan Bains Award from the Association of Food Scientists & Technologists (India) in 2023 for her incredible contribution to food science and technology. She has an H-index of 50 with ~7500 citations per Google Scholar.

Dr. K.V. Sunooj, Ph.D., is currently working as an assistant professor in the Department of Food Science and Technology, Pondicherry University, Puducherry, R. Venkataraman Nagar, Kalapet. Dr. K.V. Sunooj has more than 16 years of experience teaching and conducting research in the area of food technology. He obtained a master's degree from the University of Calicut in 2003 and a Ph.D. from the Defence Food Research laboratory (Ministry of Defence, DRDO), Mysore (awarded by the University of Mysore). He started his career as a lecturer at the University of Calicut, Kerala, India. After completing his Ph.D., he joined as an assistant professor at the Department of Food Science and Technology, Pondicherry University, in 2009. Dr. Sunooj's teaching and research interests are in the areas of food material science, biopolymers, and non-conventional starch. He has more than 100 international publications, including book chapters. He is a member of several international associations; a member of the governing body and research council member of Amala Cancer Research Center, Kerala; and a member of the board of studies at the University of Calicut.

Anil Kumar Siroha, Ph.D., is presently working as an assistant professor (food science and technology) at Chaudhary Devi Lal University (CDLU), Sirsa (Haryana), India. He received his doctorate in FST from CDLU, Sirsa. His areas of interest include starch, starch modification, and the development of new products. He has authored or co-authored more than 30 research/review articles, 20 book chapters, 5 edited books, and 1 authored book. His significant contributions to research and publication were acknowledged with the Certificate of Felicitation in 2022 and Prof. Carl Hoseney Award in 2023. Dr. Siroha also serves as a reviewer for various national and international journals. He is an active member of the Association of Food Scientists and Technologists (India) (AFSTI) in Mysuru, India. Per Google Scholar, he has an H-index of 19 with ~1155 citations.

Contributors

Basheer Aaliya
Department of Food Science and Technology
Pondicherry University
Puducherry, India

Clement K. Ajani
African Research Center on Air Quality and
    Climate
College of Sustainable Agriculture and
    Environmental Science, University of
    Mohammed VI Polytechnic
Benguerir, Morocco
and
Organization of African Academic Doctors
    (OAAD)
Tatu, Nairobi, Kenya

Plachikkattu Parambil Akhila
Department of Food Science and Technology
Pondicherry University
Puducherry, India

Mfrekemfon G. Akpan
Department of Agricultural and Food Engineering
Faculty of Engineering, University of Uyo
Uyo, Akwa Ibom State, Nigeria

Behnam Bahramian
Department of Food Science and Technology,
    Faculty of Nutrition and Food Science
Tabriz University of Medical Sciences
Tabriz, Iran

Sneh Punia Bangar
Department of Food, Nutrition, and Packaging
    Sciences
Clemson University
Clemson, SC, USA

Andressa Barbosa Barroso
Federal University of Ceará, Food Engineering
    Department
Food Biomaterials Laboratory
Fortaleza, Ceará, Brazil

Edidiong J. Bassey
Organization of African Academic Doctors (OAAD)
Tatu, Nairobi, Kenya
and
School of Food Science and Engineering
South China University of Technology
Guangzhou, Guangdong, China

Barbara Biduski
Food Quality and Sensory Science Department
Teagasc Food Research Centre
Ashtown, Dublin, Ireland

Mirela Braşoveanu
Electron Accelerators Laboratory
National Institute for Laser, Plasma and Radiation
    Physics
Bucharest-Măgurele, Romania

Long Chen
School of Food Science and Technology
Jiangnan University
Wuxi, China
and
State Key Laboratory of Food Science and Resources
Jiangnan University
Wuxi, China

Wenjing Chen
College of Food Science
Fujian Agriculture and Forestry University
Fuzhou, China
and
Fujian Provincial Key Laboratory of Quality Science
    and Processing Technology in Special Starch
Fujian Agriculture and Forestry University
Fuzhou, China

Rosana Colussi
Center for Chemical, Pharmaceutical and Food
    Sciences
Federal University of Pelotas
Pelotas, Campus Universitário, s/n, Pelotas, RS,
    Brazil

Mariana Lopes dos Anjos
Food Engineering Department, Food Biomaterials
   Laboratory
Federal University of Ceará
Fortaleza, Ceará, Brazil

Ditimoni Dutta
Department of Food Engineering and
   Technology
Tezpur University
Assam, India

Betul Oskaybaş Emlek
Department of Food Engineering, Faculty of
   Engineering
Ömer Halisdemir University
Niğde, Turkey

Okon J. Esua
Department of Agricultural and Food
   Engineering, Faculty of Engineering
University of Uyo
Uyo, Akwa Ibom State, Nigeria
and
Organization of African Academic Doctors
   (OAAD)
Tatu, Nairobi, Kenya.

Eman Farid
Department of Food Science, Faculty of
   Agriculture
Zagazig University
Zagazig, Egypt

Sabrina Feksa Frasson
Department of Agroindustrial Science and
   Technology
Federal University of Pelotas
Pelotas, RS, Brazil

Rakesh Gehlot
Centre of Food Science and Technology
CCS HAU
Hisar, Haryana, India

Atef Ghandour
Agricultural Research Center ARC
Agricultural Engineering Research
   Institute
Dokki, Giza, Egypt

Zebin Guo
College of Food Science
Fujian Agriculture and Forestry University
Fuzhou, China
and
Fujian Provincial Key Laboratory of Quality Science
   and Processing Technology in Special Starch
Fujian Agriculture and Forestry University
Fuzhou, China

Guifang Huang
School of Food Science and Technology
Jiangnan University
Wuxi, China

Victory S. Igwe
Department of Food Science
Purdue University
West Lafayette, Indiana

Ru Jia
College of Food Science
Fujian Agriculture and Forestry University
Fuzhou, China
and
Fujian Provincial Key Laboratory of Quality Science
   and Processing Technology in Special Starch
Fujian Agriculture and Forestry University
Fuzhou, China

Simoneth Paulina Jimenez
Department of Food, Nutrition and Packaging
   Sciences
Clemson University
Clemson, SC, USA

Kevser Kahraman
Department of Materials Science and
   Nanotechnology Engineering
Abdullah Gül University
Kayseri, Turkey

Ranjan Kaushik
Centre of Food Science and Technology
CCS HAU
Hisar, Haryana, India

Ankit Kumar
Centre of Food Science and Technology
CCS HAU
Hisar, Haryana, India

Lu Liu
College of Food Science
Fujian Agriculture and Forestry University
Fuzhou, China
and
Fujian Provincial Key Laboratory of Quality
    Science and Processing Technology in Special
    Starch
Fujian Agriculture and Forestry University
Fuzhou, China

Ming Miao
School of Food Science and Technology
Jiangnan University
Wuxi, China

Sabah Mounir
Department of Food Science, Faculty of
    Agriculture
Zagazig University
Zagazig, Egypt

Serpil Özturk Muti
Department of Food Engineering
Sakarya University
Sakarya, Turkey

Muhammed Navaf
Department of Food Science and Technology
Pondicherry University
Puducherry, India

Monica R. Nemţanu
Electron Accelerators Laboratory, National
    Institute for Laser
Plasma and Radiation Physics
Bucharest-Măgurele, Romania

Perpetual O. Onyeaka
Organization of African Academic Doctors (OAAD)
Tatu, Nairobi, Kenya
and
College of Food Science
Southwest University
Chongqing, China

Rekha Phogat
Centre of Food Science and Technology
CCS HAU
Hisar, Haryana, India

Sukhvinder Singh Purewal
University Centre for Research and Development
    UCRD
Chandigarh University
Mohali, Punjab, India

Neha Rani
Centre of Food Science and Technology
CCS HAU
Hisar, Haryana, India

Hassan Sabbaghi
Department of Food Science and Technology,
    Faculty of Agriculture and Animal Science
University of Torbat-e Jam
Torbat-e Jam, Razavi Khorasan Province, Iran

Matheus Calixto Saraiva
Federal University of Ceará, Food Engineering
    Department
Food Biomaterials Laboratory
Fortaleza, Ceará, Brazil

Adel Shatta
Food Technology Department, Faculty of
    Agriculture
Suez Canal University
Ismailia, Egypt

Farooq Sher
Department of Engineering, School of Science
    and Technology
Nottingham Trent University
Nottingham, United Kingdom

Min Jae Shin
Department of Food, Nutrition and Packaging
    Sciences
Clemson University
Clemson, SC, USA

Charan Singh
College of Agriculture
Chaudhary Charan Singh Haryana Agricultural
    University
Bawal, Rewari, India

Jaspreet Singh
School of Food and Advanced Technology
Massey University

Palmerston North, New Zealand
and
Riddet Institute, Massey University
Palmerston North, New Zealand

Anil Kumar Siroha
College of Agriculture
Chaudhary Charan Singh Haryana Agricultural
  University
Bawal, Rewari, India

Nandan Sit
Department of Food Engineering and Technology
Tezpur University
Assam, India

Deandrae L. Smith
Department of Food Science
Purdue University
West Lafayette, Indiana

Afreen Sultana
Department of Food, Nutrition and Packaging
  Sciences
Clemson University
Clemson, SC, USA

K.V. Sunooj
Department of Food Science and Technology
Pondicherry University
Puducherry, India

Milad Tavassoli
Department of Food Science and Technology,
  Faculty of Nutrition and Food Science
Tabriz University of Medical Sciences
Tabriz, Iran

Rafael Audino Zambelli
Food Engineering Department, Food Biomaterials
  Laboratory
Federal University of Ceará
Fortaleza, Ceará, Brazil

Baodong Zheng
College of Food Science
Fujian Agriculture and Forestry University
Fuzhou, China
and
Fujian Provincial Key Laboratory of Quality Science
  and Processing Technology in Special Starch
Fujian Agriculture and Forestry University
Fuzhou, China

Starch
Properties and Functionality

1

Sabah Mounir, Atef Ghandour,
Adel Shatta, and Eman Farid

1.1 INTRODUCTION

Starch has a semi-crystalline structure, containing 70% amorphous regions (mainly water-soluble amylose) and 30% crystalline parts (mainly water-insoluble amylopectin). Starch is a polysaccharide consisting of glucose monomers in linear and branched chains. Amylose is a straight chain containing 100 or more D-glucose units linked by α-1,4-glycosidic bonds. It accounts for10–30% of starch (Alcázar-Alay & Meireles, 2015), while amylopectin is a branched form, consisting of about 20 glucose units linked by β-1,4-glycosidic bonds. It represents about 70–90% of starch (Durrani & Donald, 1995).

There are numerous applications of starch in different industries, such as the food and non-food industries. Indeed, the utilization of starch in different applications depends on its functional properties, where gelatinization and rheological behavior are the main functional properties that define its applications. In the food industry, starch is used as a thickening and binding agent in the preparation of pudding, soups, sauces, salad dressings, mayonnaise, and infant foods. Similarly, it has a potential application in the pharmaceutical and cosmetic industries (non-food sector); it is used in the manufacturing of tablets, capsules, and granules as a binder, disintegrant, diluent, lubricant, and glidant.

1.2 CONVENTIONAL AND NON-CONVENTIONAL SOURCES OF STARCH

There are various sources of starch, varying between conventional and non-conventional sources. The conventional sources that are widely and commercially available worldwide include cereal grains, such as wheat, maize, rice, and sorghum; tubers, like potato and sweet potato; and roots, such as cassava and arrowroot, as shown in Figure 1.1. Furthermore, there is a great interest in novel, non-conventional sources of starch to maintain sustainable production. Starch can be isolated from non-conventional sources like fruits (peach palm, green banana, and banana kachkal) (Oderinde et al., 2020; Patiño-Rodríguez et al., 2020, Guo et al., 2018; Fontes et al., 2017; Khawas & Deka, 2017); pseudocereal (Jan et al., 2017), including buckwheat, amaranth, quinoa, canihua, lotus seed, and faba beans (Li et al., 2019; Zheng et al., 2019; Ai et al., 2016); culms and grasses (Felisberto, Beraldo, et al., 2020, 2019); pulps (Felisberto, Costa, et al.,

DOI: 10.1201/9781032655598-1

1

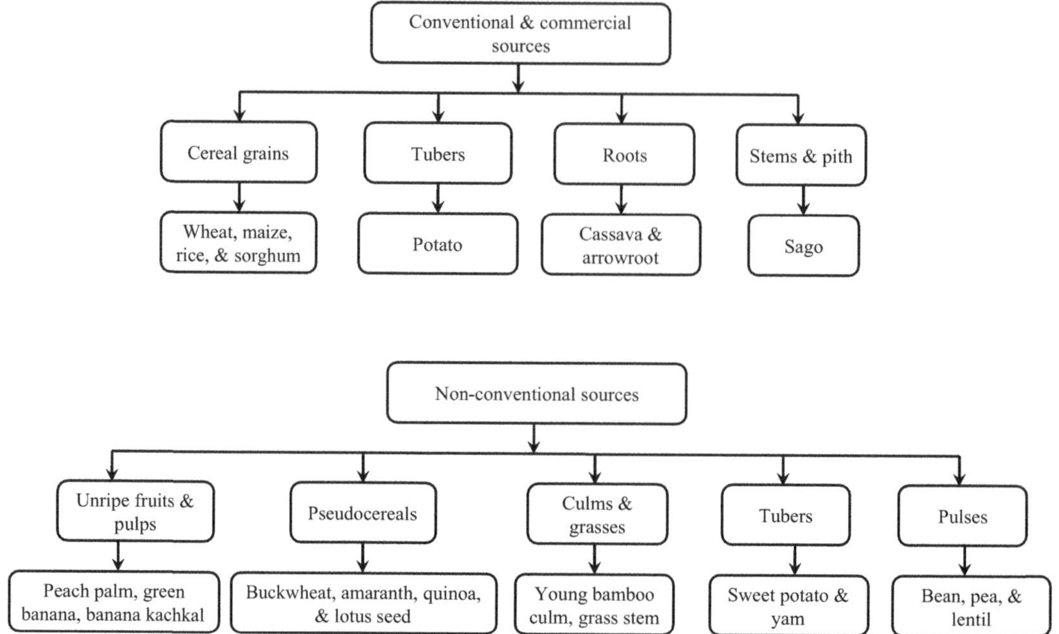

FIGURE 1.1 Conventional and non-conventional sources of starch.

2020; Fontes et al., 2017); and tubers (Shao et al., 2020; T. Li et al., 2018; Van Hung & Vo, 2017). Non-conventional starches may have similar or different characteristics to conventional and industrialized starches.

1.3 STARCH EXTRACTION METHODS

Starch from conventional and non-conventional sources can be extracted by several methods, which mainly depend on the botanical source of starch and its end use. These methods involve different unit operations, including washing, softening, grinding, sieving, tabling or centrifugation, dewatering, and drying. The most common extracting method for starch is the wet-milling, but this method has been enhanced over time. The following section summarizes the extraction methods of starch from its different common sources.

1.3.1 Extraction of Starch from Cereal Grains

1.3.1.1 Corn Starch

Corn starch is the main starch used worldwide, and the characteristics and applications of this type of starch are mostly dependent on the extraction technique. The most popular technique used for corn starch extraction is the wet-milling technique. Briefly, corn is steeped for 24 hrs in a solution of sulfurous acid (0.2–0.4%, w/v) as a reducing agent at a temperature of 50°C. This step is necessary and recommended before the wet-milling of corn grains because it makes the grains softer and activates the endogenous proteases (Eckhoff & Watson, 2009). As a result, the molecular structure of glutelin is weakened through

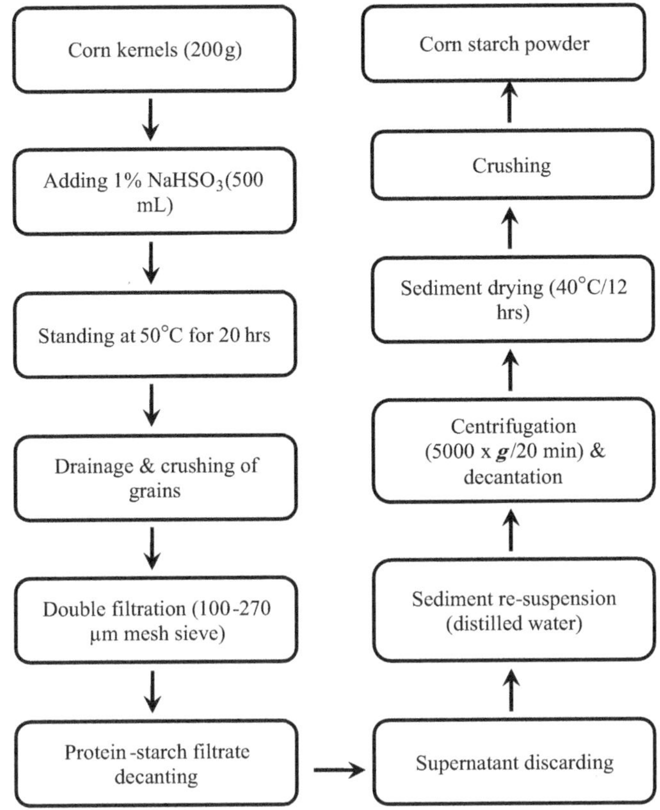

FIGURE 1.2 Main steps for extracting starch from corn kernels (Paraginski et al., 2019).

the breakdown of the inter- and intra-molecular disulfide bonds, losing the dense structure of glutelin and allowing the easy release of corn starch. Also, this step of soaking allows for the proper balance of water flow, pH, temperature, and sulfur dioxide concentration. Figure 1.2 illustrates the main steps for extracting corn starch.

Despite the higher extraction yield of the wet-milling technique, the sulfur pollution resulting from using sulfurous acid causes a very serious problem; therefore, Li et al. (2015a) proposed an eco-friendly process for isolating corn starch. This technique is based on the use of an acid solution coupled with a novel reducing agent, L-cysteine. This technique has shown a higher extraction yield at lower soaking temperatures, as well as improved physical, chemical, and functional characteristics of the extracted starch. Moreover, the obtained starch granule was smaller in size, had lower amylose content, and had higher solubility and swelling properties compared to those of commercial corn starch.

Another alternative technique, the alkali method, has also been used for isolating corn starch at lower concentrations. This technique has many features compared to wet-milling; soaking could be performed at lower temperatures in a short period of time. The alkali technique allows for improved extraction yield and purity of isolated starch by applying the optimum operating conditions in terms of alkali concentration, processing time, and temperature. However, alkaline conditions can promote the growth of putrefactive microorganisms; therefore, the pH of water must be adjusted to avoid such an effect. For instance, Uriarte-Aceves et al. (2018) isolated the corn starch by wet-milling under an adjusted pH of 3.25 using $NaHCO_3$ solution (0.2 mol/L). However, changes in the pH may disrupt the electrostatic interaction, causing the adsorption and desorption of proteins. Therefore, adding neutral salts like NaCl may alter this type of interaction (Li et al., 2015b).

1.3.1.2 Rice Starch

The production of rice starch is very expensive compared to other types of starch owing to the fact that the starch is closely linked with the protein molecules, and this entails some pre-treatments to assist starch extraction and recovery. Generally, the extraction process of rice starch involves the application of alkaline solvents, surfactants, or protein-hydrolyzing enzymes in order to eliminate the protein from rice flour (Choi et al., 2018). NaOH as an alkaline solvent and dodecylbenzene sulfonate and sodium lauryl sulfate as surfactants are frequently used to remove the protein during starch extraction (El Halal et al., 2019).

Alkaline extraction is an effective technique for isolating rice starch from glutelin, using the alkaline soaking in a NaOH solution at a concentration of 0.10–0.30%, w/v. This technique has shown higher extraction yield and purity of the starch. However, Cardoso et al. (2007) showed that the optimum concentration of NaOH in the alkaline extraction of rice starch varied from 0.15 to 0.18% (w/v), and a level higher than 0.24% caused changes in the structure of starch granules, as shown in Figure 1.3.

Enzymes, such as proteases, can also be applied in rice starch extraction to digest the proteins found in the rice flour; this process is known as enzymatic extraction. In this process, most proteins are separated by a laboratory centrifuge, where the protein-rich fraction is manually scraped, as well as the layer found at the top of the starch sediment (Choi et al., 2018). Puchongkavarin et al. (2005) extracted rice starch from soaked grains pre-treated by cellulase under slightly acidic conditions to promote the cell walls breaking down. Afterward, two proteolytic enzymes (papain and Corolase 7089) were added after pH adjustment (neutral conditions). The authors found that this technique resulted in starch with a higher residual protein and lower content of damaged starch. For this reason, alkaline extraction is an effective technique for isolating rice starch extraction, but this method produces chemical wastes that require water treatment.

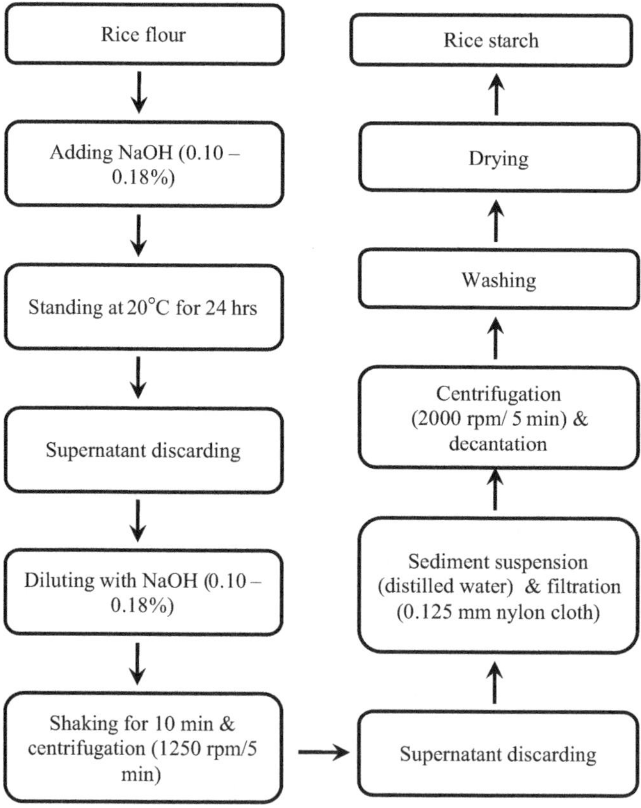

FIGURE 1.3   Main steps for rice starch extracting (Cardoso et al., 2007).

1.3.1.3 Wheat Starch

Wheat starch can be extracted from wheat flour or whole kernels by wet-milling (Liu & Ng, 2016). Generally, wheat starch is isolated from non-starch components by physical separation. Therefore, a number of processes for wheat wet-milling have been proposed, involving soaking, grinding, and separating different components of wheat grain. Yet the soaking conditions generally affect both extraction yield and starch quality; using SO_2 results in the highest extraction yield and starch purity, while acid soaking using lactic acid produces the lowest extraction yield and starch purity (Shevkani et al., 2017). Also, alkaline, sodium dodecyl sulfate, or salt solutions can be applied as solvents in the soaking step. Hu et al. (2017a) extracted the starch from waxy wheat flour by a wet-milling process using NaOH (0.03 mol/L), followed by shaking, centrifugation, washing, neutralization (HCl, 1 mol/L), hot air-drying, and milling. In addition, waxy wheat starch was isolated by Zhang et al. (2013), but in this method, the dough was prepared by mixing wheat flour with distilled water, and the remains were suspended in 2% sodium dodecyl sulfate. Figure 1.4 shows the main steps applied for isolating the wheat starch.

1.3.2 Extraction of Starch from Tubers

1.3.2.1 Potato Starch

Starch isolation from potatoes is easy and simple owing to their lower contents of protein and fat (< 4%), along with their unique tissue structure (Kringel et al., 2020). Briefly, the simplest method for potato starch isolating involves milling/grinding followed by decantation, centrifugation, successive washing

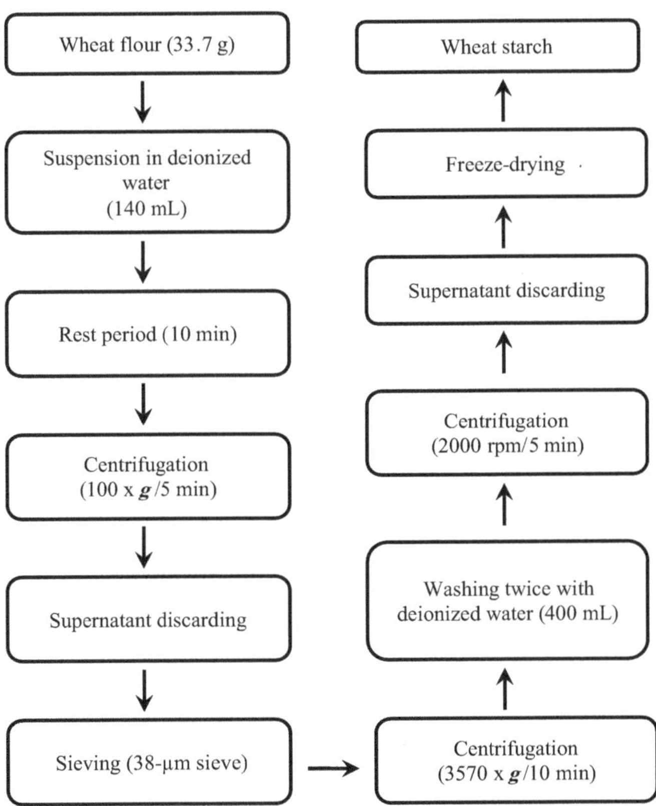

FIGURE 1.4   Main steps for wheat starch isolation (Pauly et al., 2012).

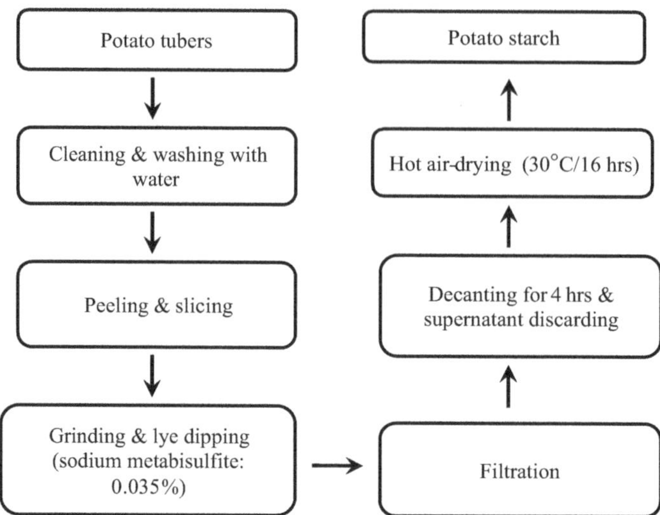

FIGURE 1.5   Main steps for potato starch isolation.

with distilled water, and then drying (Singh et al., 2006; Singh & Singh, 2001), as shown in Figure 1.5. After cleaning and peeling, the potatoes are ground and dipped in a sodium metabisulfite solution (0.035%). The grinding step is necessary to breakdown the cell walls, forming a mixture of broken cell walls with starch granules and other water-soluble components (salts, sugars, soluble proteins, and amino acids). The dipping in sodium metabisulfite solution prevents enzymatic browning, which results in the formation of compounds with a reddish-brown color, such as melanin. Also, enzymes such as cellulase (0.5% of potato meal) can be applied for starch isolated from potatoes in order to increase the starch yield (Hameed et al., 2020).

1.3.2.2 Sweet Potato Starch

Many processes have been reported in the literature for starch extraction from various colored sweet potatoes, including water extraction, NaOH extraction, sodium metabisulfite extraction, and sodium bisulfite extraction. NaOH is used to eliminate the protein, while sodium metabisulfite or sodium bisulfite is applied to prevent browning during the starch extraction (Xu, Guo et al., 2018). However, Trung et al. (2017) found that water extraction was effective for starch isolation from fresh sweet potatoes with high purity. This process involved only grinding, sieving, and washing with distilled water. Similar results were reported by Xu, Guo et al. (2018), who found that water extraction was an appropriate method for starch extracting from various colored sweet potatoes. On the other hand, starch whiteness is mostly affected by the presence of some pigments, such as β-carotene, which cannot be eliminated during the previously mentioned processes. Therefore, some other solvents may be used during the starch extraction (e.g., anhydrous ethanol solutions) in order to improve the starch purity and whiteness (Kim et al., 2013).

1.3.3 Extraction of Starch from Roots

1.3.3.1 Sago Starch

Sago starch is generally isolated from pith using water in order to form a slurry with pith pieces. The resulting slurry is sieved using a series of sieves with different meshes to eliminate any impurities present, followed by drying the obtained starch cake (Zhu, 2019), as shown in Figure 1.6. The traditional extraction

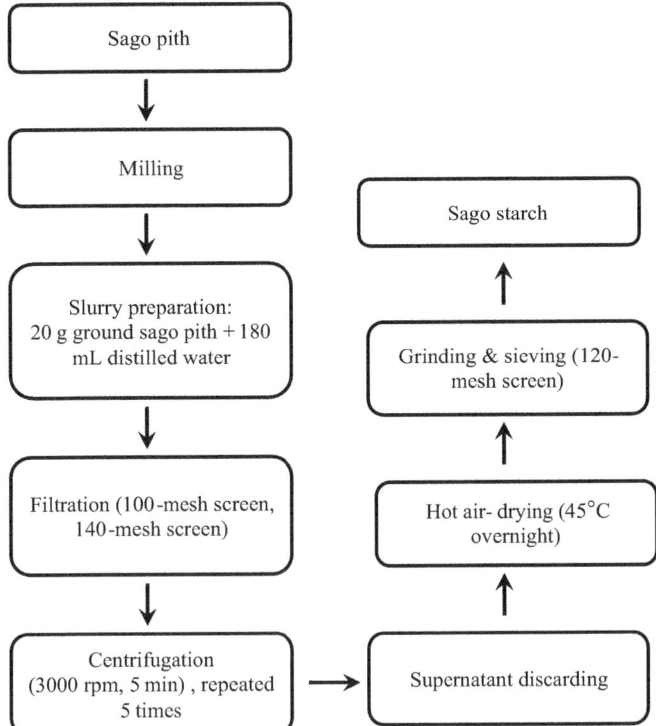

FIGURE 1.6   The traditional method for sago starch extracting (Pinyo et al., 2017).

technique produces a lower extraction yield and starch quality than that of processes with pretreatment, such as microwaves or ultrasound. This may be due to the fact that the starch granules are still entrapped in the fibrous matrix of the sago pith (Othman et al., 2015). Therefore, some laboratory pretreatment can be applied prior to the traditional extraction method in order to improve the sago starch yield and quality, including enzymes, microwaves, ultrasound, or a combination of pretreatments (Zhu, 2019; Pinyo et al., 2017). For instance, Pinyo et al. (2017) studied the effect of pretreatments, namely enzymatic, ultrasonic, microwave, and combined pretreatments; the authors found that the starch yield was 1.5, 63.7, 71.4, and 72.5%, respectively, for ultrasonic, microwave, enzymatic, and combined ultrasonic-enzymatic pretreatments. The authors also recommended ultrasound as a pretreatment for sago starch extraction by the traditional method with conditions of 10% (w/w) sago pith slurry, 160 W power input, and 7 min as a pretreatment time.

1.3.4 Extraction of Starch from Pseudocereal

1.3.4.1 Amaranth Starch

The isolation of starch from amaranth grains is a complicated process owing to the grain's high protein content and its small size (about 1 mm in diameter) (Condés et al., 2018). Alkaline soaking is often employed for amaranth starch isolation, using diluted solutions of NaOH, sodium metabisulfite, and other alkaline solutions to eliminate the protein (Loubes et al., 2012), with varying soaking temperatures and concentrations of chemicals applied. For instance, Calzetta et al. (2006) isolated amaranth starch using soaking in SO_2 solution (0.041%) at 53.9°C for 16 hrs, as shown in Figure 1.7. Yet these methods require a high concentration of dangerous chemicals; therefore, enzymes have been applied in order to reduce the

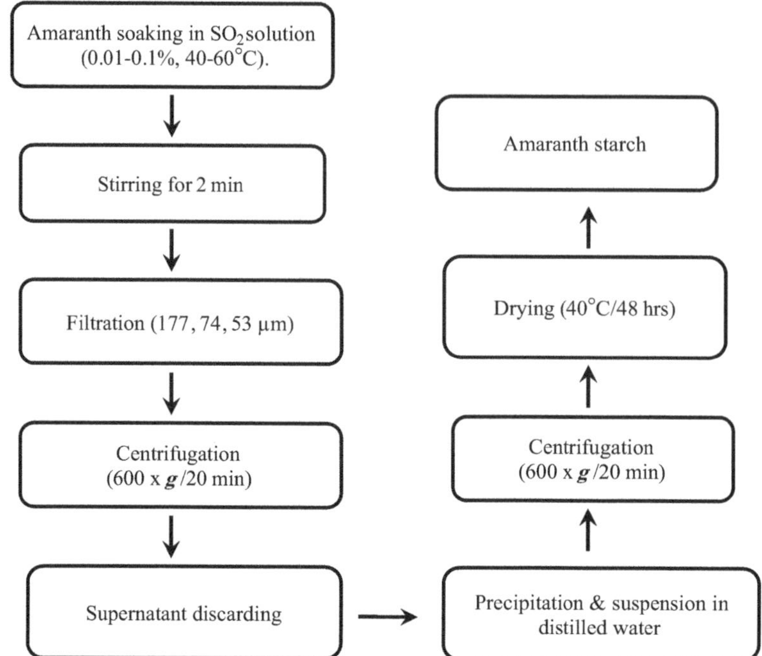

FIGURE 1.7   Main steps for amaranth starch extraction.

concentration of these chemicals. However, Bet et al. (2018) isolated amaranth starch with an efficient and eco-friendly method, an aqueous extraction. This method involves only water soaking, followed by mixing, filtration, decantation, centrifugation, and finally drying. Moreover, lye dipping can be coupled with aqueous extraction to eliminate the protein. For instance, Leal-Castañeda et al. (2018) extracted amaranth starch from seeds by soaking the seeds in distilled water at 4°C for 24 hrs, followed by alkaline dipping in a 0.3% NaOH solution in order to separate the protein. The isolated amaranth starch showed 86% purity and had 4.9% protein. The alkaline method could be coupled with enzymatic application; Villarreal et al. (2013) isolated amaranth starch from whole grains and whole flour by applying a NaOH solution (0.25%) coupled to protease (0.05%). This method involves successive soaking followed by wet-milling of the fibrous fraction, protease digestion, and finally multi-staged centrifugation.

1.4 MORPHOLOGY, STRUCTURE, AND CHEMICAL COMPOSITION OF STARCH

1.4.1 Morphology of Starch Granules

In order to understand the physicochemical characteristics and functionality of starch in different applications, it is important first to understand the structure and chemical composition of the starch granule. Starch granules are slightly dense and insoluble in cold water, with limited hydration (at temperatures below 50°C). The granules of starch are widely varied in shape, size, morphology, molecular structure, and chemical composition, depending on the source of starch (Smith, 2001). The size and shape of starch granules vary according to the starch source; starch granules have microscopic size with diameters

TABLE 1.1   Morphological Characteristics of Starch Granules from Different Sources (Modified from Tester & Karkalas, 2002)

STARCH	TYPE	SHAPE	SIZE (µM)	DISTRIBUTION
Wheat		Lenticular (A-type)	2–10	Spherical (B-type)
			15–35	
Barley		Lenticular (A-type)	15–25	
		Spherical (B-type)	2–5	
Rye		Lenticular (A-type)	10–40	Bimodal
		Spherical (B-type)	5–10	
Rice		Polyhedral	3–8 (single)	
			150 (compound)	
Maize (waxy and normal)	Cereal	Spherical/polyhedral	2–30	
Amylomaize		Irregular	2–30	
Millet		Polyhedral	4–12	
Oat		Polyhedral	3–10 (single)	
			80 (compound)	Unimodal
		Spherical (B-type)	5–10	
Sorghum		Spherical	5–20	
Triticale		Spherical	1–30	
Potato	Tuber	Lenticular	5–100	
Pea	Legume	Rentiform (single)	5–10	
Sago	Stem and pith	Oval	20–40	
Tapioca	Root	Spherical/lenticular	5–45	

varying from 0.1 to more than 200 µm (Gul et al., 2021), with different shapes, including spheres, ovals, platelets, irregular tubules, lenticulars, polygons, and ellipsoids, depending on the source of starch (Singh et al., 2003; Hoover, 2001). For instance, waxy maize and normal maize starches have spherical and polygonal shapes with a diameter of 5–20 µm, while potato starches exhibit large granules with spherical and oval shapes and a diameter of 16–75 µm. Also, the majority of pulse starches display many different shapes (kidney-like, spherical, round, elliptical, and irregular shapes) with a diameter of 5–70 µm (Hoover et al., 2010). However, high-amylose maize starches have spherical, polygonal, and elongated (filamentous) shapes. The diameter and length of elongated granules are in the range of 3–5 and more than 50 µm, respectively (Jiang, Horner, et al., 2010). Table 1.1 shows the morphological characteristics of different starch granules.

1.4.2 Structure and Chemical Composition of Starch

Starch granules principally consist of two polymers, amylose and amylopectin, which represent about 10–30% and 70–90%, respectively (Kaur et al., 2020). Starch polymers (amylose and amylopectin) are arranged in a semi-crystalline texture. The hilum is a central line (Maltese cross); the starch granule may contain one or more Maltese crosses, which reduces its birefringence ability (Jiang, Jane, et al., 2010). Amylose and amylopectin vary in their structure and molecular weight. Amylose is an amorphous fraction with a molecular weight ranging from 1×10^5 to 1×10^6 (Biliaderis, 1998); it is composed of linear chains of D-glucose units that are bound by α-1,4-glycosidic bonds. However, evidence has revealed that amylose is not a straight chain of glucose units; it is coiled, containing six glucose monomers/turn (Ogori & Taofeek, 2022; Gul et al., 2021). The content of amylose depends on the starch type; in waxy starches, amylose accounts for 0–8%, while in normal starches, it represents 20–30%, and high-amylose starches contain more than 40% (Li et al., 2008).

Amylopectin is a crystalline fraction with a higher molecular weight than that of amylose, ranging from 1×10^7 and 1×10^9 (Biliaderis, 1998); it consists of a branched chain of glucose units, which are mainly linked by α-1,4-glycosidic bonds (95%) and by α-1,6-glycosidic bonds (5%) at branch points. Amylopectin contains numerous thousands of glucose units, with branch points every 25 to 30 units, as shown in Figure 1.8. Amylopectin molecules form double helices to build the crystalline architecture of the starch granule. Amylose molecules are distributed between the amylopectin molecules. Therefore, the higher amount of amylopectin reflects the higher crystallinity of the starch granules. According to the sources of the starch, there are three different forms of crystalline structure, A-, B-, and C-type structures (Singh et al., 2003). In A-type structures, which are found in cereal starches, the glucose chains are short and bound by α-1,6-glycosidic bonds, with a polymerization degree of 6–15. However, in B-type structures (which are present in tuber starches and high-amylose starches), the amylopectin branches are more polymerized. These branches may be A-type or can form branched amylopectin molecules. This type of structure is classified into B_1, B_2, B_3, and B_4, with different degrees of polymerization, where the polymerization degree of B_1 and B_2 varies from 15 to 25 and from 40 to 50, respectively, while B_3 and B_4 have the highest degree of polymerization. The C-type is a combination of the A- and B-type structures and is composed of amylopectin molecules with non-reduced ends. This type is found in legumes, roots, stems, and some fruits (Sevenou et al., 2002; Cheetham & Tao, 1998).

Starch granules contain lipids in a small amount (< 1.5%). According to the starch sources, lipids differ in their chemical composition. For instance, cereal starches contain mainly phospholipids, and their content is positively correlated with the content of amylose (Morrison, 1995). The lipids in normal maize starch are generally composed of triglycerides, free fatty acids (FFAs), and a small amount of phospholipids. Conversely, normal rice starches contain a significant amount of phospholipids and some FFAs. Lipids can form a complex with amylose, known as the amylose-lipid complex. The formation of this complex promotes entanglements between amylose and amylopectin. Therefore, the swelling of the granule is limited during the heating process (Ai et al., 2013), leading to an increase in the pasting temperature, a reduction of the peak viscosity, and the formation of an opaque appearance of the paste. Other endogenous lipids also have considerable effects on the functional characteristics of starch (Debet & Gidley, 2006).

Starch granules also contain a small amount of protein, ranging between 0.06 and 0.4% (Debet & Gidley, 2006). These proteins are only present on the granule surface, and their removal is facilitated by washing them with a solution of sodium dodecyl sulfate (Bancel et al., 2010). Similar to the amylose-lipid complex, the surface proteins restrict the swelling of the granules of wheat and maize starches during pasting (Debet & Gidley, 2006). However, some proteins are also present within the starch granules (integral proteins), including biosynthetic and hydrolysis enzymes (Buléon et al., 2014; Bancel et al., 2010). For instance, some proteins found in the channels of the granules of waxy maize, normal maize, wheat, and sorghum can be removed by protease hydrolysis (Han et al., 2005). Integral proteins have a higher molecular weight than those of surface proteins, 50–150 versus 15–30 KDa, respectively.

Starch granules also contain some minerals (< 0.4%), including calcium, phosphorus, magnesium, sodium, and potassium. Among these minerals, phosphorus is the most important and is found in three forms: phosphate monoesters, phospholipids, and inorganic phosphates (Blennow et al., 2000). Phosphate monoesters are generally present in numerous starches, such as rice, potato, sweet potato, and arrowroot (Hoover, 2001), where potato starch contains a considerable amount of phosphate monoesters (< 0.1%) (Kasemsuwan & Jane, 1996). Phosphate monoesters are present in amylopectin molecules and bound to specific regions (e.g., O-6 of the D-glucopyranose unit) (Blennow et al., 2000). They carry two negative charges, which induce repulsion between adjacent amylopectin molecules. This results in distinctive functional characteristics of starch. Potato starch shows low gelatinization and pasting temperatures due to the repulsion induced by phosphate monoesters, thus destabilizing the crystalline texture.

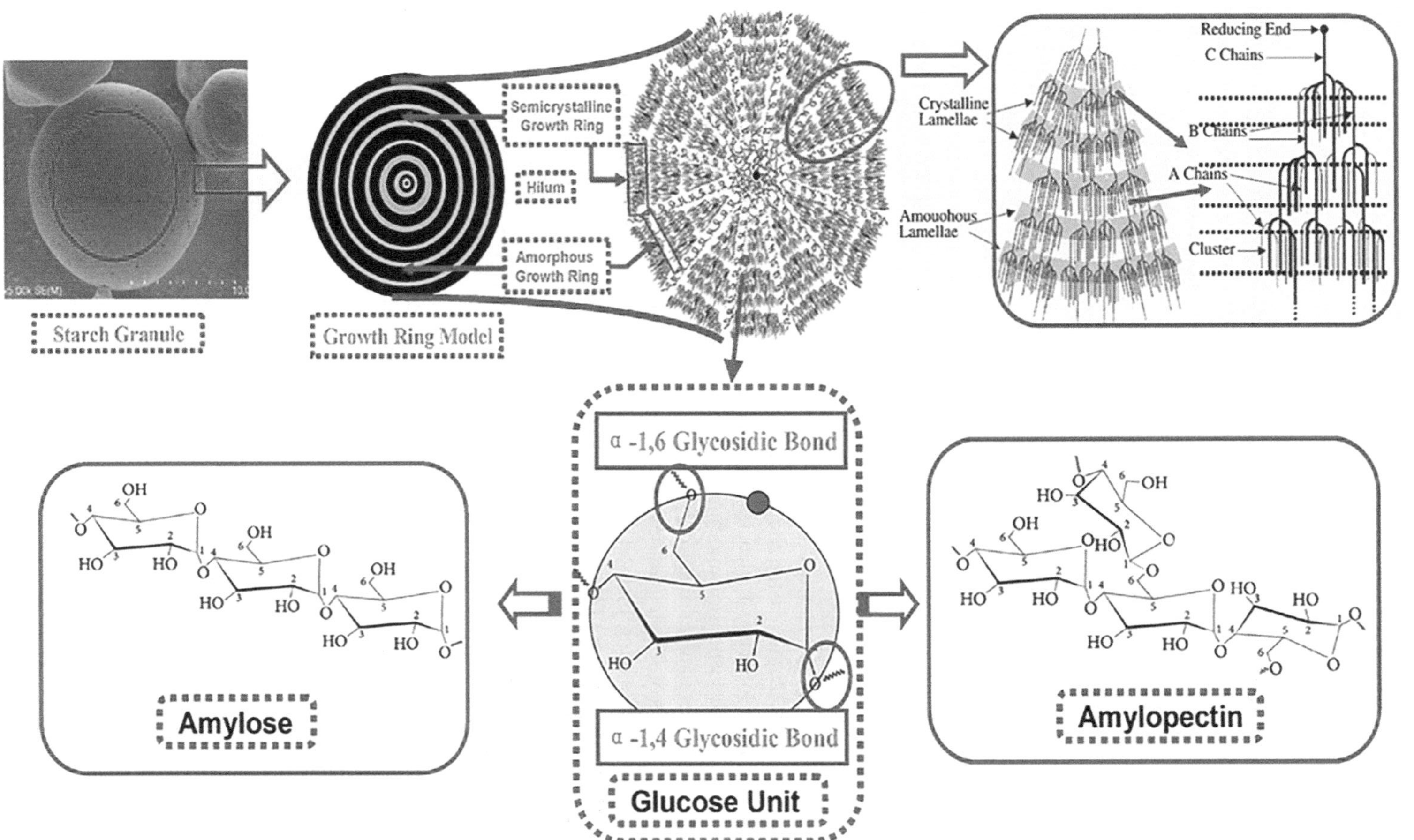

FIGURE 1.8   Schematic diagram of the starch structure (Liu et al., 2017).

1.5 PHYSICOCHEMICAL PROPERTIES OF STARCH

1.5.1 Swelling and Solubility

Generally, native starch is insoluble in cold water (at temperatures below 50°C) and absorbs a limited amount of water, which is mainly attributed to the hydrogen bonds and the crystalline structure. However, when starch is heated in the presence of excess water, it can absorb enough water and thus swell in water, increasing its volume several times. This process refers to starch swelling, which involves two stages: (1) hydration; and (2) the irreversible phase. The first stage is the hydration of the starch granules, which usually takes place below the gelatinization temperature of starch and slightly increases the granule diameter. In this stage, water molecules diffuse inside the starch granule and enter the amorphous regions, forming new hydrogen bonds with the starch polymers. This causes the starch granule to become soft and fragile, which in turn swells in water, leading to the amylose leaching out from the swollen granule and thus increasing the viscosity of the starch suspension. The second stage is the irreversible phase that occurs at temperatures above the gelatinization temperature, wherein an obvious increase in the granule diameter and the suspension viscosity is observed. The coupled action of heating and shearing breaks down the crystalline structure, causing a loss of birefringence. This increases the affinity between starch polymers and water molecules, enhancing water absorption and starch solubilization. Additional heating increases the swelling, and the granule starts to rupture. Also, the hydrogen bonds between starch polymers are broken, and water molecules form new hydrogen bonds with the exposed hydroxyl groups in amylose and amylopectin (Gul et al., 2021), increasing the swelling power of starch granules. Simultaneously, starch polymers (amylose and amylopectin) are leached from the swollen granules, causing a significant increase in the viscosity of the starch suspension and finally forming a starch paste (Vamadevan & Bertoft, 2020). Figure 1.9 illustrates the mechanism of starch swelling and rupture.

The swelling properties of starch granules are generally measured by using the hydrothermal method, which is a simple and time-saving method. Therefore, the conditions applied during the measurement could influence the swelling power of the granules. For instance, starch granules show low swelling capacity at low temperatures, which gradually increases with an increase in the test temperature (Zhang et al., 2020; Xu, Saleh, et al., 2018). This may be due to the gradual weakening of the interaction between the starch molecules with increasing temperature, resulting in an increase in the swelling degree. The

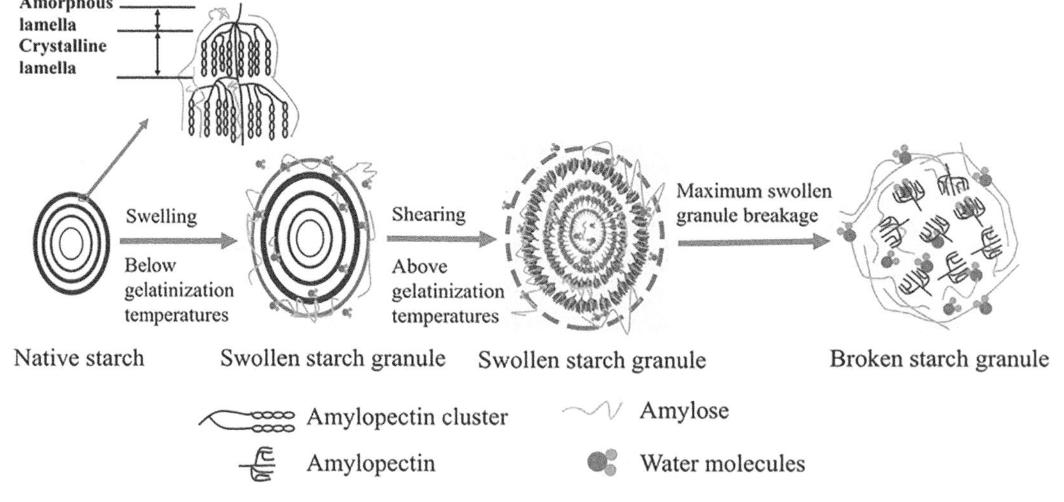

FIGURE 1.9 Mechanism of starch swelling and rupture (Jia et al., 2023).

optimum temperature for starch swelling ranges between 50 and 85°C. A similar trend is observed with the test time; an increase in the swelling degree is observed during a prolonged test time, which may be due to the increased interaction between water and starch granules over time.

On the other hand, the composition and structure of starch could affect the swelling properties of starch granules. Table 1.2 shows the swelling power and solubility of starches from different botanical

TABLE 1.2 Swelling Power and Solubility of Different Crystalline Starches from Several Botanical Sources

STARCH TYPE	CRYSTAL TYPE	AMYLOSE CONTENT (%)	TEST TEMPERATURE (°C)	SWELLING POWER (G/G)	SOLUBILITY (%)	REFERENCES
Normal wheat starch	A-type	20–30	80	7–12	3–15	Su et al. (2020);
Waxy wheat starch		6.40 ± 0.06		35–50	10–20	Zhang, et al. (2021)
Blue wheat starch	A-type	27.11 ± 0.18	100	13.5	7	Wang et al. (2022a)
Common buckwheat starch	A-type	28.05 ± 0.23	90	15–18.8	10–31.3	Goel et al. (2020); Sindhu et al. (2019)
Rice starch	A-type	19.5	75	17	ND	Vamadevan and
Waxy rice starch		1		23	ND	Bertoft, (2020)
Corn starch	A-type	22	80	8–15	ND	Li, Tian, et al.
Waxy corn starch		0		20–35	ND	(2016); Wang et al. (2022b)
Waxy maize starch	A-type	1.2	80	39.6	ND	Obadi et al. (2023)
Normal maize starch	A-type	29.9		15.9	8.2	
High-amylose maize starch	B-type/ C-type	65		10.9	7.5	
Normal maize starch	A-type	25.1	92.5	12.5–15	11.1	Liu et al. (2019b)
High-amylose maize starch	B-type	53.4		5–7	ND	
Potato starch	B-type	26.21 ± 0.16	90–95	32–35.99	8.03–14	Zhao et al. (2018)
Potato starch	B-type	17.9	75	57	ND	Vamadevan and
Waxy potato starch		0		80	ND	Bertoft (2020)
Sweet potato starch	C-type	ND	80	17.5–20	ND	Lv et al. (2018)
Sweet potato starch	A-type	ND	90	11–12	ND	Wang et al. (2022c)
Sweet potato starch	A-type	ND	90	26.61 ± 0.02	9.30 ± 0.05	Gou et al. (2019)
Barley starch	A-type	28.3	75	13–15	ND	Vamadevan and
Waxy barley starch		4.5		24	ND	Bertoft (2020)
Pearl millet starch	A-type	14.1 ± 0.5	90	14.0 ± 0.6	10.4 ± 0.4	Siroha et al. (2019)
Amaranth starch	A-type	ND	90	25.38 ± 1.4	87.45 ± 2.14	Sindhu et al. (2021)
Oat starch	A-type	22.30 ± 0.40	95	15–20	20–25	Falsafi et al. (2019)
Taro starch	A-type	13.9 ± 1.1	90	10.57 ± 1.02	8.27 ± 0.36	Deka and Sit (2016)
Cassava starch	A-type	27.3	90–95	8.42–17.5	12.17–12.5	Dudu et al. (2021); Pereira and Del Pino Beleia (2021)
Sago starch	C-type	25–30	80	10–15	0–2	Chan et al. (2010)
Acorn starch	C-type	20–48.9	90	24.49–25	0.09–13	Molavi et al. (2018); Chen et al. (2022)
Pea starch	C-type	34.08 ± 0.21	90	13.44 ± 0.31	17.27 ± 0.23	Han et al. (2021)
Quinoa starch	A-type	6.3–27	80	11.71 ± 0.07	3.94 ± 0.05	Zhou et al. (2021)
				20.89 ± 0.03	15.5 ± 0.4	Zhu and Zhang (2019)

ND: Not determined

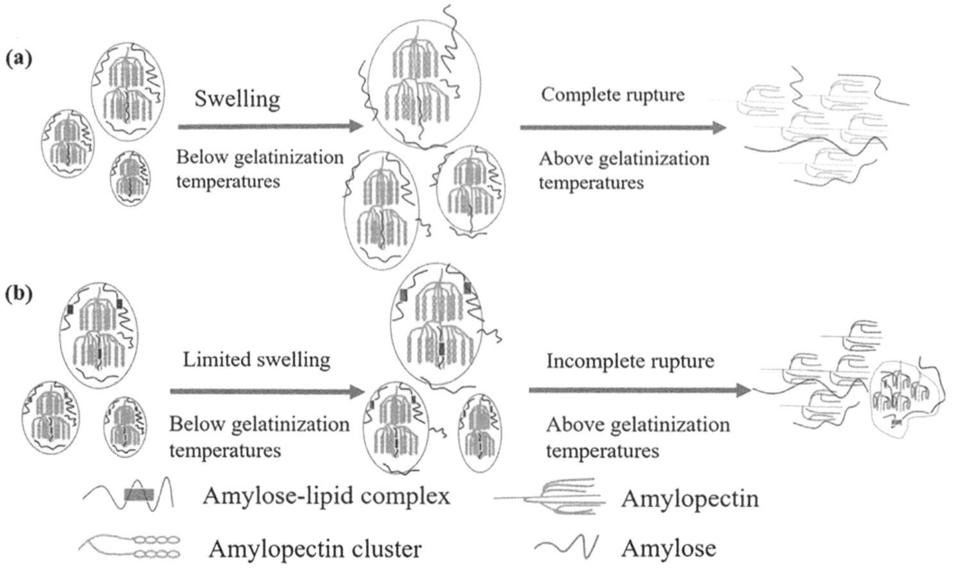

FIGURE 1.10   Schematic diagram representing the effect of endogenous lipids on the swelling power and solubility of starch: (a) in the absence of lipids; (b) in the presence of lipids (Jia et al., 2023).

sources. The swelling properties vary significantly according to the botanical source of starch, amylose content, and crystal type (A-, B-, or C-type). Amylopectin enhances the swelling properties of starch granules, while amylose may act as a diluent and an inhibitor of swelling, decreasing the swelling power. This may be attributed to the complexation with endogenous or exogenous lipids, which inhibits the hydration of amylose (Singh et al., 2003). Therefore, the swelling properties of starch granules are primarily related to the fine structure and proportion of starch molecules (amylose and amylopectin).

Furthermore, the presence of endogenous and exogenous lipids can alter the swelling power and solubility of starch granules, while the effect of exogenous lipids is slightly more significant. The presence of endogenous lipids causes a decrease in the swelling degree and solubility of starch granules, as shown in Figure 1.10. This can be explained by forming a complex with amylose (the amylose-lipid complex), which (1) inhibits the hydration of amylose, reducing the swelling of the starch granules (Singh et al., 2003); and (2) forms a hydrophobic film on the granule surface, preventing the interaction between starch and water (Liu et al., 2019a; Chang et al., 2013). Additionally, the type of lipids can also affect these properties; the swelling power of granules increases with increasing triglycerides and decreases with increasing monoglycerides (Navarro et al., 1996). Similar behavior has been reported by Lan et al. (2020), who found that the complexation of monoglyceride decreased the swelling power of potato starch. Therefore, the swelling power and solubility of starch can be improved by removing the endogenous lipids owing to the reduced amylose-lipid complex content (Hu et al., 2017b; Li, Gao, et al., 2016).

Similar to lipids, endogenous proteins can also affect the swelling power and the solubility of starch granules. The endogenous proteins decrease the swelling power and the solubility of starch granules. This may be due to the interaction between proteins and starch, which increases the rigidity and integrity of the starch granules, decreasing their swelling power and then their solubility (Jia et al., 2023), as shown in Figure 1.11. Removal of endogenous proteins can, however, improve the swelling degree and the solubility of starch (Sun et al., 2021; Zhan et al., 2020). After endogenous protein removal, the enhancement in the swelling powder and solubility of starch can be explained by several phenomena: (1) decreasing the complexation between starch and proteins, which in turn increases the starch swelling (Wang et al., 2020); (2) reducing the surface tension and internal tension of the granules: the interaction between endogenous proteins and starch increases the surface tension and internal tension of the granule, and removal of these

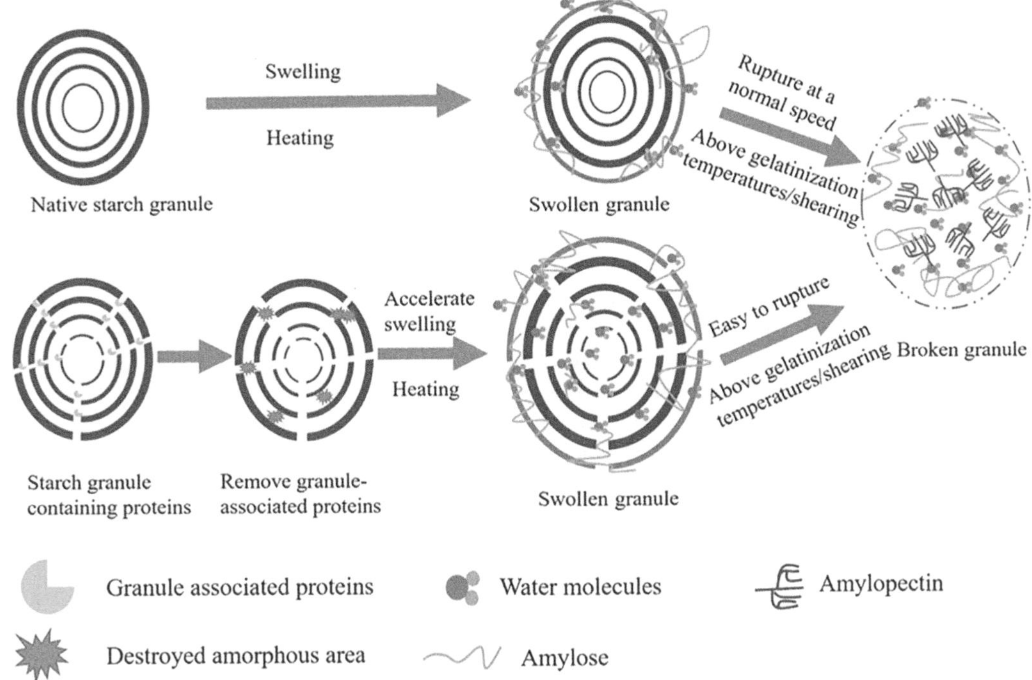

FIGURE 1.11   Schematic diagram of the swelling process of native starch with endogenous protein removed (Jia et al., 2023).

proteins reduces this tension, thereby improving granule swelling (Zhan et al., 2020); and (3) reducing the inhibitory effect of amylose, which acts as a diluent, improving the water absorption in the amorphous regions (Ding et al., 2021; Li, Wu, et al., 2016).

In contrast, the effect of exogenous proteins on the swelling power and solubility of the granule is inconsistent because they can form hydrophobic or hydrophilic films on the granule surface. The changes in both swelling power and solubility depend on the type of film formed on the granule surface. For example, globular proteins (casein and globulin) increase starch swelling. This may be attributed to the hydrophilicity of these proteins, which promotes the formation of a hydrophilic film on the granule surface, which in turn facilitates the absorption of water and then the granule swelling. Conversely, sodium caseinate can form a hydrophobic film on the granules' surface, thus reducing their swelling power (Wang et al., 2021).

Furthermore, starch modification can affect the swelling properties of starch granules by inducing different degrees of damage to the crystalline structure. The effect of starch modification on the swelling power and solubility of starch granules has been reported in many studies (Sindhu et al., 2021; Q. Wang et al., 2022a; Jia et al., 2021; Zhang et al., 2019; L. Wang et al., 2018; Yang et al., 2021; Bajaj et al., 2021; Sudheesh et al., 2020; W. Li et al., 2018). This effect depends on the botanical source of starch, the type and method of modification, and the conditions applied. A detailed discussion of the effect of starch modification on the swelling power and solubility of starch granules will be covered in the next chapters of this book.

1.5.2 Gelatinization and Retrogradation

Native starches are insoluble in cold water; therefore, gelatinization is indispensable to improve water absorption and their chemical and physical characteristics to meet and suit food applications. Also, the accessibility of the starch granule to enzymatic action is improved after gelatinization. Starch gelatinization can also affect the rheological and viscosity characteristics of the paste.

Gelatinization is known as a phase transition (from an amorphous state to a rubbery state) or a collapse of molecular order within the starch granule, involving a complete dissolution of starch polymers (amylose and amylopectin) and physical rupture of the starch granule. In other words, starch gelatinization is a two-step process, including (1) the swelling of starch granules as a result of the rupture of hydrogen bonds in the crystalline regions; and (2) the plasticizing effect of water, which allows hydration and swelling of the crystalline regions (Wani et al., 2012).

Starch gelatinization occurs by heating starch granules in excess water, which acts as a plasticizer (Oyeyinka & Oyeyinka, 2018; Hoover et al., 2010). During heating (above 52°C) in excess water, the starch granule undergoes a glass transition in the amorphous region before gelatinization, which is an irreversible change. The gelatinization temperature ranges between 55 and 80°C, depending on the starch type. Gelatinization is initiated when water diffuses within the granule to hydrate the amorphous regions and then spreads to the crystalline regions, where the granules are broken down with the absorption of water. This leads to a loss in their crystalline structure and birefringence, which allows their swelling and the leaching of amylose (Jiménez et al., 2012). As previously discussed, continued heating causes the melting of amylopectin crystallites to become amorphous, starch solubilization, and the leaching out of starch polymers (amylose and amylopectin), which increase the viscosity of the starch paste. The state and phase transition of starch when heated in excess water are shown in Figure 1.12.

The progress of starch gelatinization depends on some factors, including (1) the physicochemical characteristics of the starch granule (amylose-to-amylopectin ratio and fine structures and molecular weight distribution of amylose and amylopectin molecules); (2) the amount of available water; (3) the processing parameters (temperature, time, and shear); and (4) the presence of other ingredients, such as protein, lipid, and other polysaccharides (Schirmer et al., 2015; Hoover et al., 2010). Other solvents may be used for starch gelatinization, including dimethyl sulfoxide (DMSO), liquid ammonia, chloroacetic acid, formic acid, ethylene glycol, or 1,4-butanediol (Ai & Jane, 2015; Jiménez et al., 2012). In this case, the starch gelatinization is affected by some parameters, such as the type of solvent, the starch-to-solvent ratio, and the ability of the solvent to form hydrogen bonds with the starch molecules (Alcázar-Alay & Meireles, 2015; Jiménez et al., 2012).

The gelatinization properties can be determined by many methods, such as differential scanning calorimetry (DSC), nuclear magnetic resonance spectroscopy, polarized light microscopy, or

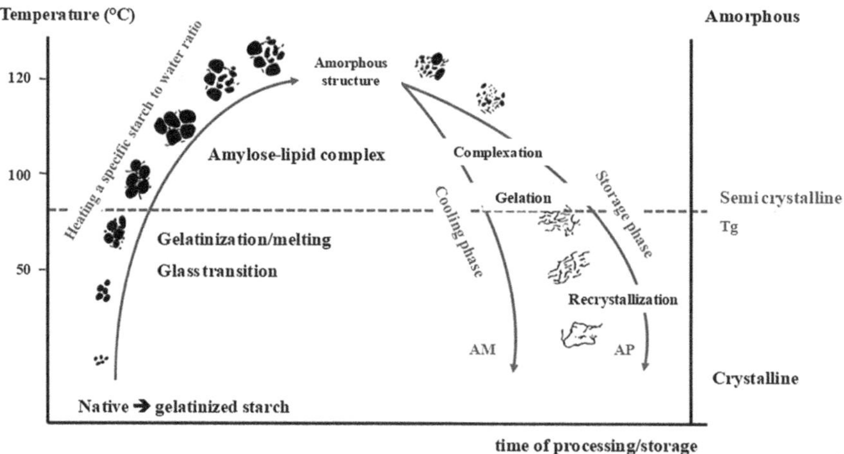

FIGURE 1.12 State and phase transition of starch when heated in excess water. Starch undergoes a phase transition from a crystalline to an amorphous state when heated and subsequent recrystallization when cooled and during storage. Tg: gelatinization temperature; AM: amylose; AP: amylopectin (Schirmer et al., 2015).

TABLE 1.3 Gelatinization Properties of Native Starches with Different Crystal Types (Jane et al., 1999)

STARCHTYPE	T_O (°C)	T_P (°C)	T_C (°C)	RANGE (°C)	ΔH (J/G)	BCL
A-type starch						
Wheat	57.1 ± 0.3	61.6 ± 0.2	66.2 ± 0.3	9.1	10.7 ± 0.2	22.7
Normal maize	64.1 ± 0.2	69.4 ± 0.1	74.9 ± 0.6	10.8	12.3	24.4
Waxy maize	64.2 ± 0.2	69.2	74.6 ±0.4	10.4	15.4 ± 0.0	23.5
du waxy maize	66.1 ± 0.5	74.2 ± 0.4	80.5 ± 0.2	14.4	15.6 ± 0.2	23.1
Normal rice	70.3 ± 0.2	76.2	80.2	9.9	13.2 ±0.6	22.7
Waxy rice	56.9 ± 0.3	63.2 ± 0.3	70.3 ± 0.7	13.4	15.4 ± 0.2	18.8
Sweet rice	58.6 ± 0.2	64.7	71.4 ± 0.5	12.8	13.4 ± 0.6	21.6
Waxy amaranth	66.7 ± 0.2	70.2 ± 0.2	75.2 ± 0.4	8.5	16.3 ± 0.2	21.8
Barley	56.3	59.5	62.9 ± 0.1	6.6	10 ± 0.3	22.1
Tapioca	64.3 ± 0.1	68.3 ± 0.2	74.4 ± 0.1	10.1	14.7 ± 0.7	27.6
Cattail millet	67.1	71.7	75.6	8.5	14.4 ± 0.3	21.5
Chinese taro	67.3 ± 0.1	72.9 ± 0.1	79.8 ± 0.2	12.5	15 ± 0.5	23.4
Mung bean	60.0 ± 0.4	65.3 ± 0.4	71.5 ± 0.4	11.5	11.4 ± 0.5	24.8
B-type starch						
ae Waxy maize	71.5 ± 0.2	81.0 ± 1.7	97.2 ± 0.8	25.7	22 ± 0.3	29.5
Amylomaize V	71.0 ± 0.4	81.3 ± 0.4	112.6 ± 1.2	41.6	19.5 ± 1.5	28.9
Amylomaize VII	70.6 ± 0.3	ND	129.4 ± 2.0	58.8	16.2 ± 0.8	30.7
Potato	58.2 ± 0.1	62.6 ± 0.1	67.7 ± 0.1	9.5	15.8 ± 1.2	29.4
Green leaf canna	59.3 ± 0.3	65.4 ± 0.4	80.3 ± 0.3	21	15.5 ± 0.4	28.9
C-type starch						
Lotus root	60.6	66.2	71.1 ± 0.2	10.5	13.5 ± 0.1	25.4
Green banana	68.6 ± 0.2	72.0 ± 0.2	76.1 ± 0.4	7.5	17.2 ± 0.1	26.7
Water chestnut	58.7 ± 0.5	70.1 ± 0.1	82.8 ± 0.2	24.1	13.6 ± 0.5	26.4

T_O: onset temperature; T_P: peak temperature; T_C: conclusion temperature; range of gelatinization: T_C– T_O; ΔH: enthalpy change; BCL: average branch chain length of amylopectin; ND: not detectable.

thermomechanical analysis. Yet DSC is the most commonly used method for determining the temperature of starch gelatinization (Gul et al., 2021). Four parameters can be determined by DSC, including onset temperature (T_O), peak temperature (T_P), conclusion temperature (T_C), and the enthalpy of gelatinization (Δ H). Table 1.3 shows the gelatinization properties of native starch from different botanical sources. Gelatinization temperature increases with increasing branch chain length (BCL) of amylopectin, reflecting the thermal stability of the crystallites formed of long-branch chains of amylopectin (Jane et al., 1999). Similarly, *ae* waxy and high-amylose maize starches, Amylomaize V and Amylomaize VII, show higher gelatinization temperatures. In contrast, waxy and sweet rice starches, which have a very short BCL of amylopectin, show much lower gelatinization temperatures. This may be attributed to the negative charges of phosphate monoesters, which destabilize the crystalline structure through repulsion. The high-amylose maize starches display significantly higher conclusion gelatinization temperatures than the *ae* waxy maize starches due to the thermal stability of the double-helical crystallites of amylose and intermediate components (Jiang, Campbell, et al., 2010).

Retrogradation, or setback, is the process that takes place after the cooling of gelatinized starch (the hot paste) and involves a re-association or recrystallization of the leached amylose (Gul et al., 2021). During retrogradation, amylose forms a double helix by association with other glucose units, releasing the water bound during the gelatinization and thus causing the syneresis phenomenon or gel hardness, whereas amylopectin re-crystallizes by association with its small chains because amylopectin is less sensitive to retrogradation than amylose (Singh et al., 2003), as shown in Figure 1.12. After retrogradation, starch shows lower gelatinization than the native starch owing to the weakened crystalline structure (Sasaki et al., 2000). This behavior determines the rheological and textural characteristics of starch-containing foods, such as bread staling during refrigerated storage (Gray & BeMiller, 2003). Retrogradation depends on the botanical starch source; starches from cereals generally retrograde more slowly, with a lesser extent of retrogradation than starches from tubers and roots. This may be attributed to many factors, such as the intrinsic structure of amylopectin, the amylose-to-amylopectin ratio, and lipid content (Gul et al., 2021). Indeed, starch retrogradation is also affected by the presence of proteins, lipids, other polysaccharides, and salts (Fu et al., 2015). For instance, proteins or lipids may form complexes with starch, which delay starch retrogradation during storage under refrigeration (Wu et al., 2010). Also, freeze–thaw cycles may accelerate starch retrogradation (Yu et al., 2010); when frozen foods are thawed, the bound water during gelatinization is easily separated from the gel matrix, resulting in syneresis, which causes changes in the quality attributes of frozen starch-based foods; therefore, freeze–thaw stability is an essential feature here.

1.5.3 Pasting Properties

The textural characteristics and stability of high-moisture starch-based foods are defined by the properties of starch granules, such as swelling, fragmentation degree, and retrogradation, during processing and retrogradation during the storage period. Thus, it is necessary to understand the mechanism of pasting and the formation of gel and paste. Indeed, it is difficult to distinguish between a paste and a gel; a paste is the hot system resulting from freshly cooked starch, while a gel signifies the system after cooling. Pasting is the process that occurs after or simultaneously with the gelatinization of starch by continuously heating in excess water. This involves continued swelling, rupture of swollen granules, and further leaching of dissolved starch molecules, forming a viscoelastic paste (BeMiller, 2011). A hot starch paste is a two-phase system composed of (1) a continuous phase, which contains dissolved and partly dissolved starch molecules (mainly amylose and low-molecular-weight amylopectin molecules); and (2) a discontinuous phase, which consists of swollen granules, granule fragments, and granule ghosts (swollen starch granules from which amylose and/or amylopectin are leached) (Wani et al., 2016).

The paste characteristics determine the suitability of starch for various food applications, which are defined and affected by the composition and nature of the continuous phase (e.g., the sizes and fine structures of amylose and amylopectin and their interactions), the nature and size of the particles in the discontinuous phase, and the interaction between the continuous and discontinuous phases (BeMiller, 2011). Viscosity is the most important characteristic of starch paste because it determines the starch applications in different industries (Gryszkin et al., 2014; Sarker et al., 2013). For instance, high paste viscosity is used for thickening applications in the food industry (Gul et al., 2021). The pasting characteristics can be determined by many techniques, including viscometers, rheometers (Park et al., 2007), the Brabender Visco-Amylograph (Lin et al., 2009), and the Rapid Visco-Analyzer.

Viscosity parameters are determined by using a Rapid Visco-Analyzer as a function of temperature and time during different periods of time, as shown in Figure 1.13. These periods are divided into (1) an initial period, where water is added to starch; (2) a controlled heating period, in which the temperature of the suspension is increased from ambient temperature to 95°C; (3) an isothermal period, where the temperature of the suspension is maintained at the maximum temperature (95°C) with shearing; (4) a cooling period, where the temperature of the suspension is dropped to about 50°C; and (5) a final holding period, where the temperature is maintained at 50°C (Balet et al., 2019). Table 1.4 summarizes the pasting properties of some native starches analyzed using a Rapid Visco-Analyzer with 8% (dry starch basis) starch content. Amylose is responsible for the higher values of the final viscosity and retrogradation. There is an inverse relation between amylose and peak viscosity and the gelatinization temperature; therefore, it is the primary polymer responsible for the higher values of final viscosity and retrogradation.

During the initial period, hydration of starch occurs where water diffuses within the starch granules, and the granules start to swell at temperatures below 50°C. The swelling of the starch increases with an increase in the heating temperature, resulting in the breaking down of the starch granules and the leaching

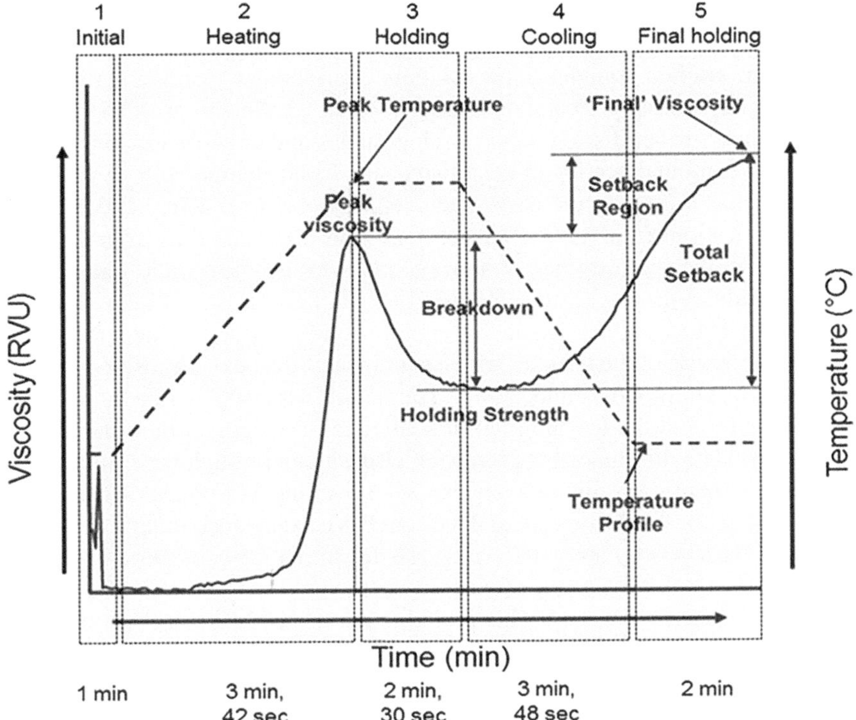

FIGURE 1.13 A typical RVA pasting profile of a cereal (the standard profile), indicating the major parameters measured (Balet et al., 2019).

TABLE 1.4 Pasting Properties of Some Native Starches Analyzed Using a Rapid Visco-Analyzer with 8% (Dry Starch Basis) Starch Content (Jane et al., 1999)

		VISCOSITY (RVU)			
STARCH TYPE	T_P (°C)	PEAK	HOT PASTE	FINAL	SETBACK
Wheat	88.6	104	75	154	79
Normal maize	82	152	95	169	74
Waxy maize	69.5	205	84	100	16
*du*waxy maize	75.7	109	77	99	22
*ae*waxy maize	83.2	162	150	190	40
Normal rice	79.9	113	96	160	64
Waxy rice	64.1	205	84	100	16
Sweet rice	64.6	219	100	128	28
Barley	91.2	88	58	116	58
Waxy amaranth	70.2	125	75	86	11
Cattail millet	74.2	201	80	208	128
Potato	63.5	702	165	231	66
Chinese taro	73.1	171	88	161	73
Lotus root	67.4	307	84	138	54
Tapioca	67.6	173	61	107	46
Green banana	74	250	194	272	78
Mung bean	73.8	186	161	363	202
Water chestnut	74.3	61	16	27	11

T_P: pasting temperature; RVU: rapid viscosity unit.

out of amylose from the swollen granules in the solution. Therefore, an increase in the suspension viscosity is observed during the heating period at 95°C due to starch gelatinization. Also, the applied mechanical shear stress can cause further disruption of starch granules and amylose leaching. The starch paste is then formed due to the continual heating of gelatinized starch. The temperature at which the viscosity of starch starts to increase is known as the pasting temperature (Liang & King, 2003), which is generally higher than the gelatinization temperature. So, starch pasting occurs directly after gelatinization, which describes the changes in the viscosity before, during, and after gelatinization. The pasting temperature is used to define the minimum temperature required for starch cooking (Sandhu et al., 2007). A high pasting temperature indicates high resistance of starch to swelling and disintegration. The maximum viscosity is attained at the end of the heating, which is known as the peak viscosity and reflects the water-binding capacity of the starch (Cozzolino, 2016). The paste's viscosity is rapidly decreased due to the rupture of starch granules and the leaching out of solubilized fractions in the solution; this viscosity is termed trough viscosity. The difference between peak viscosity and trough viscosity is called breakdown viscosity. As previously mentioned, the rate and extent of swelling and breakdown are dependent on the temperature difference, shear forces applied, and the starch type and concentration (Schirmer et al., 2013). In the cooling period, the viscosity starts to re-increase due to the retrogradation of starch after cooling. Amylopectin and amylose molecules realign and rearrange themselves to form a viscous gel after gelling, leading to a further increase in viscosity. This period is known as the setback (retrogradation), and the viscosity is termed cold paste viscosity (Alcázar-Alay & Meireles, 2015). The difference between the peak viscosity and the final viscosity is defined as the setback viscosity (Zhou et al., 2007), which describes the starch tendency to retrograde during storage or thawing. The temperature is kept constant during the final holding period, while the viscosity continues to increase until a plateau is attained, representing the final viscosity at the end of the pasting cycle, which reflects the starch capacity to form a viscous paste or gel after cooking and cooling.

Figure 1.14 shows the pasting properties of native starches analyzed by the Rapid Visco-Analyzer (8% starch content, dry starch basis). For example, waxy maize and waxy rice starches have lower pasting temperatures and a higher peak viscosity than their normal counterparts. An inverse relationship has been reported between the peak viscosity and the amylose content, where amylose, especially in the presence of lipids, limits the swelling of starch granules. The formation of the amylose-lipid complex increases the entanglements between amylose and amylopectin molecules, limiting the swelling of the starch granules (Ai et al., 2013). This leads to increasing the pasting temperature and reducing the peak viscosity. However, the normal maize and rice starches display greater setback viscosities than those of the waxy maize and rice starches. This may be attributed to the formation of a gel as a result of the interaction between amylose and other starch molecules during the cooling period, thereby increasing the setback viscosity.

1.5.4 Rheological Properties

The rheological properties reflect the behavior of starch when it is subjected to deformation and shearing forces. These properties play a substantial role in the different applications of starch (Berski et al., 2011), which include viscosity, texture, shear strength, and the tendency for retrogradation. Among the rheological properties, viscosity is the most important because starch is usually used as a thickening agent in food applications. The viscosity of the starch paste reflects the thickening capacity of starch for different applications as well as the stability of food products and consumer acceptability (Šubarić et al., 2014). The rheological properties of starch depend on several parameters, such as the starch source, the properties of the native starch, the processing conditions applied (e.g., temperature, time, and shearing), and the presence of non-starch components (e.g., lipids and proteins).

The rheological properties are determined by different techniques that include an amylograph, a viscoamylograph, or a dynamic rheometer (Schirmer et al., 2015). Some parameters are used to describe these characteristics, such as the storage modulus (G'), the loss modulus (G''), and the loss tangent (tan δ). The storage modulus (G') is a measure of energy stored in the material and recovered from it per cycle of

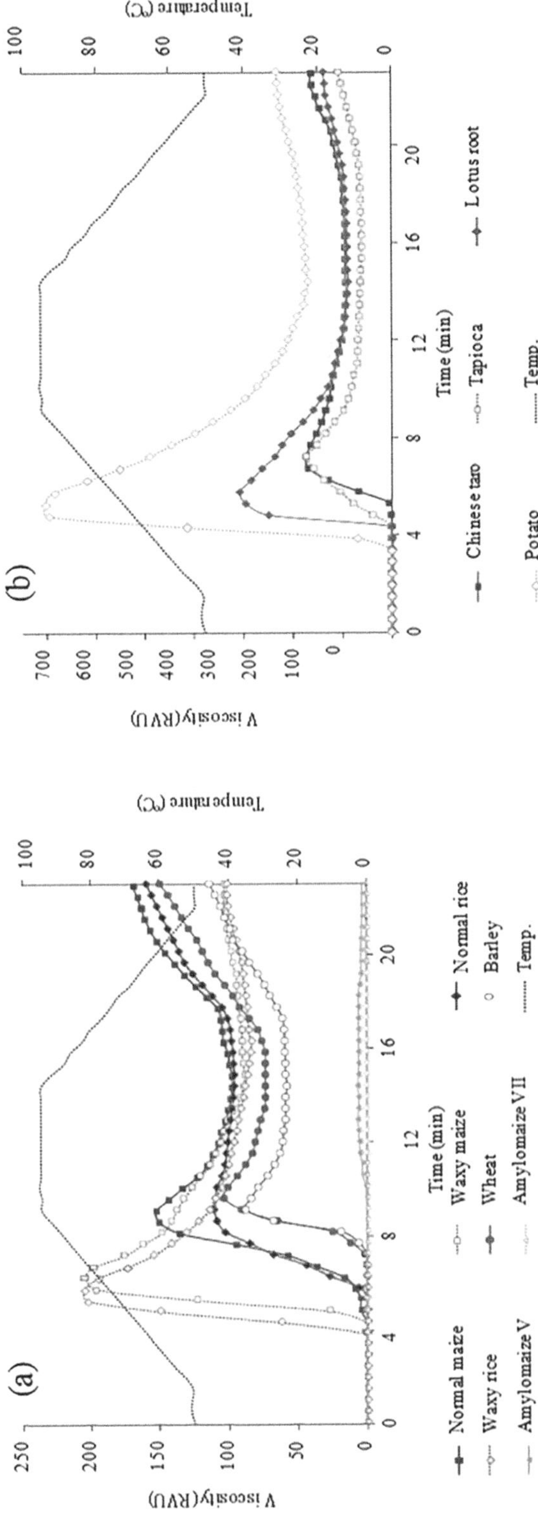

FIGURE 1.14 Pasting properties of native starches analyzed using a Rapid Visco-Analyzer: (a) cereal starches; (b) root and tuber starches (Ai & Jane, 2015).

deformation, reflecting the elastic characteristics. The loss modulus (G'') measures the energy dissipated or lost per cycle of deformation, indicating viscous behavior. The loss tangent, the ratio of the energy lost to the energy stored for each cycle (G''/G'), describes the behavior of the paste or gel, where high values (> 1) indicate liquid-like behavior, while low values (< 1) indicate solid-like behavior (Waterschoot et al., 2015; BeMiller & Whistler, 2009).

The suspension of native starch exhibits a low viscosity at room temperature, even at a high concentration of 35–40% (w/w). However, this viscosity is significantly developed after heating at temperatures higher than the gelatinization temperature; this process is known as starch pasting. Other developments in the viscosity occur after the formation of a viscous gel during the cooling period (as discussed in Section 1.5.3). Starch paste is characterized as a non-Newtonian fluid (Ai & Jane, 2015) and exhibits a complex viscoelastic behavior under heating conditions. During the heating of starch suspension, the storage modulus (G') and loss modulus (G'') of starch increase gradually at a certain temperature, indicating the swelling of the granule, the leaching out of solubilized molecules from the swollen granule, and the transformation of the suspension into a "sol" (Sodhi & Singh, 2003). With continual heating, the viscosity is greatly developed along with G' and G'', reaching the maximum value (peak viscosity) until a plateau (a constant value) is reached. The maximum viscosity and G' are related to the maximum swelling of the starch granules and the formation of a three-dimensional network. tan δ decreases simultaneously, also reflecting the formation of a three-dimensional network by the leached amylose. The maximum increase in the storage modulus (G') occurs at temperatures ranging between 60 and 80°C, while a plateau is reached at temperatures between 80 and 85°C (Ahmed et al., 2008). However, a further increase in the temperature (beyond gelatinization temperature) causes a sharp decrease in G' and an increase in tan δ, reflecting the breakdown of the gel structure, as shown in Figure 1.15, owing to the melting of the

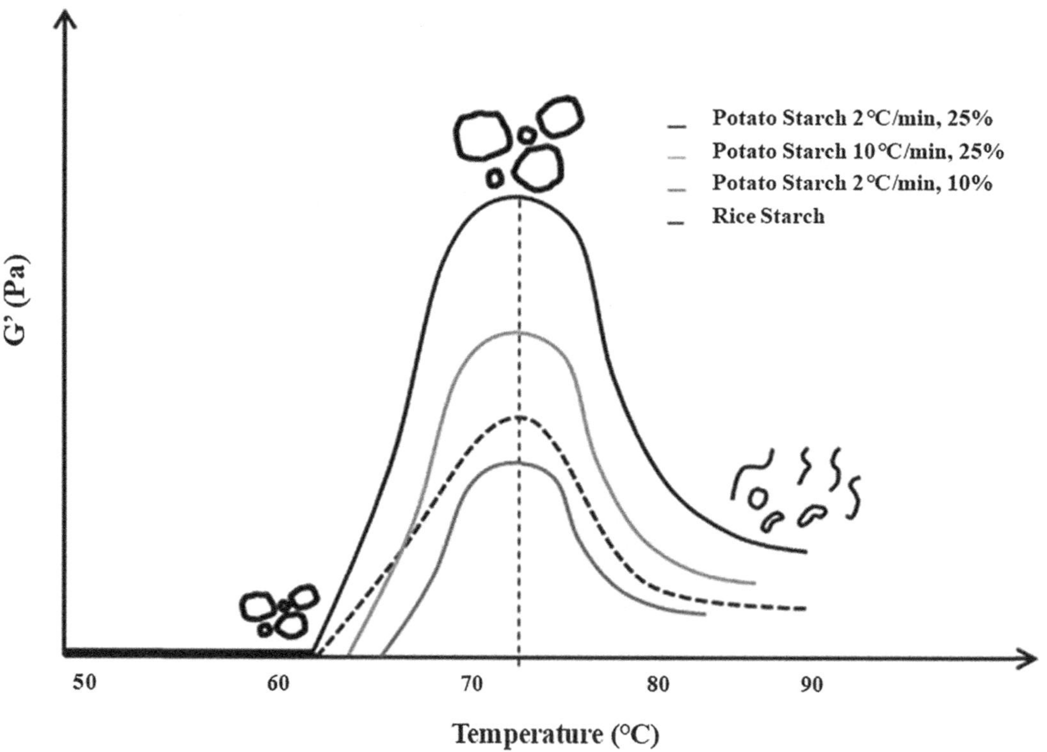

FIGURE 1.15 Changes in the viscoelastic properties of potato and rice starch suspensions under heating and further cooling; peak and plateau values can be observed at maximum starch swelling and then network breakdown and paste formation (Mandala, 2012).

crystalline regions remaining in the swollen starch granule or the disentanglement of the amylopectin molecules in the swollen particles that softens the particles (Tsai et al., 1997). Also, the loss of interactions between particles and the network may cause the network to collapse (Ahmed et al., 2008). The peak height at a given concentration indicates the ability of the granules to swell freely prior to their physical breakdown. Starches that are capable of swelling to a high degree are also less resistant to breakdown during cooking, showing a rapid drop in viscosity after reaching the maximum value (Singh et al., 2008).

1.5.5 Digestibility Properties

Starch digestion takes place in the human gastrointestinal tract; it is generally degraded into glucose through various sequential steps by the action of digestive hydrolytic enzymes. For instance, starch digestion first starts in the mouth with the action of salivary α-amylase, which plays a slight role in starch digestion and is then quickly broken down in the stomach under its acidic conditions. Most starch digestion occurs in the small intestine through the action of pancreatic α-amylase, which acts on the α-1,4-glycosidic bonds in amylose and amylopectin molecules (Quezada-Calvillo et al., 2007; Lehmann & Robin, 2007). Yet α-amylases have no specificity for α-1,6-glycosidic bonds in the amylopectin molecules. Therefore, their ability to break α-1,4-glycosidic bonds adjacent to the branching points is reduced owing to the steric hindrance (Singh et al., 2010). The products obtained from amylose hydrolysis contain oligosaccharides (e.g., maltose and maltotriose) and maltotetraose. Additionally, some microbial α-amylases can hydrolyze the amylose into maltotriose, maltohexose, and maltoheptose (Yook & Robyt, 2002). However, dextrins or branched oligosaccharides are the main products obtained from amylopectin hydrolysis. Also, the intestinal brush border enzymes can hydrolyze the obtained oligosaccharides (e.g., maltose, maltotriose, and dextrins) into glucose.

According to Englyst et al. (1992), starch can be classified into three types: (1) rapidly digestible starch (RDS); (2) slowly digestible starch (SDS); and (3) resistant starch (RS), depending on its digestibility in the small intestine. This classification is based on the in vitro digestion of starch by a simulation of stomach and intestinal conditions and the determination of glucose release at various times. Rapidly digestible starch contains the portion that can be quickly hydrolyzed and absorbed in the mouth and the small intestine. The glucose is released within 20 min, leading to an immediate increase in the glycemic index (GI). Slowly digestible starch contains the fraction that the enzymes in the small intestine can slowly and completely hydrolyze, and the glucose is released within a time range of 20–120 min. This type of starch plays a great role in long-term glycemic control. On the other hand, resistant starch contains the fraction that is resistant to digestion in the small intestine. The enzymes in the small intestine cannot hydrolyze it within 120 min, but it passes to the large intestine and can be fermented into short-chain fatty acids by the gut flora present in the colon. Figure 1.16 shows the process of starch digestion in the human gastrointestinal tract for the different types of starch and the approximate digestion time for each.

Indeed, the mechanism of starch swelling significantly affects its digestibility, involving changes in the starch digestibility, as shown in Figure 1.17. These changes may be attributed to many phenomena, including (1) reducing the pore size in the granules as a result of decreasing the swelling power of starch, which in turn inhibits the penetration of hydrolytic enzymes into the starch granules; (2) reducing the available specific surface area of granules to the hydrolytic enzymes due to the decreased swelling degree, limiting the starch accessibility; or (3) increasing the cracks on the surface of granules as a result of increased swelling power, which promotes the enzymatic attack (Cui et al., 2021). Starch digestibility is also affected by other factors, such as the source of starch and its morphological properties; food processing conditions (temperature and shearing); the textural and rheological properties of food; and the chemical composition of food, like the presence of proteins, lipids, and non-starch components and their interaction during food processing (Singh et al., 2010). For instance, small barley and wheat starch granules hydrolyze more quickly than large granules (Lindeboom et al., 2004). The presence of proteins and lipids may restrict the enzymatic hydrolysis of starch, which is attributed to the formation of complexes with amylose. These complexes limit the hydration and swelling of starch granules, thereby preventing

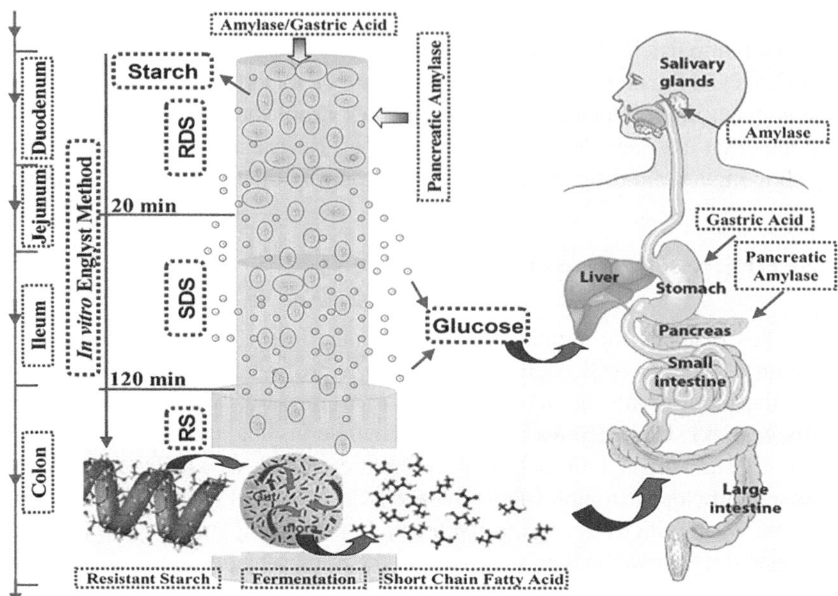

FIGURE1.16 Schematic diagram of the process of starch digestion in the human gastrointestinal tract, indicating the approximate time for starch digestion. RD: rapidly digestible starch; SDS: slowly digestible starch; RS: resistant starch (Liu et al., 2017).

the penetration of hydrolytic enzymes into the starch granules. On the other hand, food processing, like cooking at 100°C for some minutes, increases the rate of starch hydrolysis due to starch gelatinization, making it more susceptible to enzymatic attack. Similarly, the extrusion process improves the digestibility of starch due to the loss of its crystallinity, which is attributed to the increased shearing and kneading extent within the extruder barrel. Other processes, such as dehulling, soaking, and germination of cereals, can also improve starch digestibility owing to the elimination of phytic acid, tannins, and polyphenols. Roopa and Premavalli (2008) reported that the digestibility of finger millet starch increases by 35–40% during cooking, autoclaving, and puffing, followed by pressure cooking and germination. Conversely, some starch modifications may decrease the digestibility of starch, making it more resistant to enzymatic attacks. For example, chemically modified starches show greater resistance to α-amylase (Han & BeMiller, 2007).

1.6 STARCH FUNCTIONALITY AND APPLICATIONS

1.6.1 Starch Applications in the Food Industry

Starch plays an important role in the food industry, has many effects, and functions in a wide range of foods. Starch in food applications is used as a thickening, gelling, and emulsifying agent, and it can create new textures as well as enhancing the mouthfeel. Starch, in conjunction with other components, can impart or maintain flavor and nutrients (Cui, 2005). During food processing, starch undergoes many modifications, depending on the process and the processing parameters applied. These modifications include starch swelling, fragmentation, and solubilization at different levels. The distribution of starch between swollen granules, fragmented granules, and solubilized polymers affects the quality attributes of

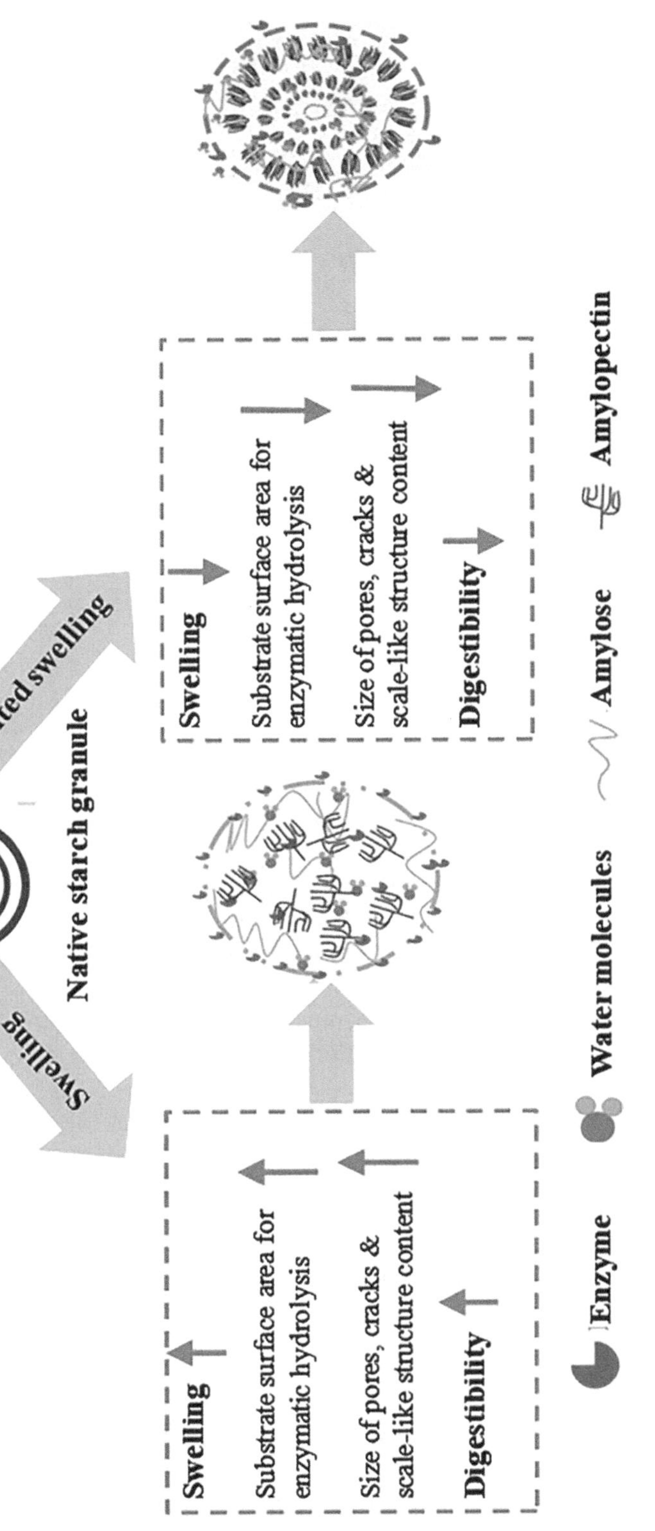

FIGURE1.17 Effect of the swelling mechanism on starch digestibility (Jia et al., 2023).

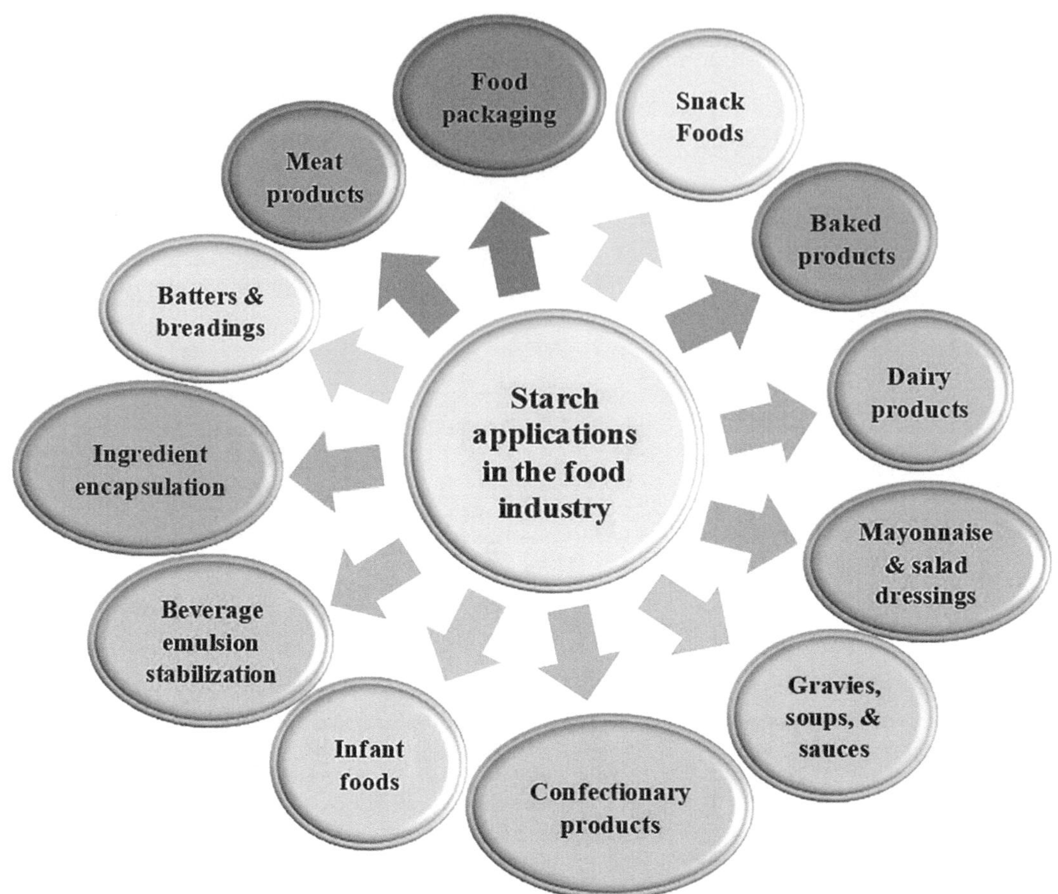

FIGURE 1.18   Different applications of starch in the food sector.

food containing starch, such as texture, structure, stability, and appearance. The different applications of starch in the food industry are addressed in the next section, as shown in Figure 1.18.

1.6.1.1 Snack Foods

Starch has a wide application in snack production, where it is a primary ingredient in the production of this type of food. Starch gives snack foods special textural characteristics, such as crispiness, expansion, and open-porous structure. It can also improve the oil-holding properties and eating quality (Cui, 2005). Starch plays a significant role in the production of extruded snacks and fried or baked snacks. In the extrusion process, the thermal and mechanical stresses produced in different zones within the extruder cause starch to breakdown to create a molten matrix that is necessary for pressure to develop within the extruder and achieve the desired expansion of the final product. In addition, the characteristics of amylose and amylopectin are other parameters affecting the expansion of extrudates. Starch with high-branched amylopectin produces dough with optimum viscosity, which is important to obtain the expected expansion and thus produce crispy and expanded snacks with an open-porous structure. Moreover, amylose gives strength to the dough, enhances its formation, and improves its cutting properties (Taggart, 2004). Modified starch can be used in the production of snack foods without affecting the crispness and crunchiness of the final products, in particular when used in the production of fiber-fortified snacks (Aprianita et al., 2014). In fried snacks, starch has been used to improve the texture of the final product by controlling oil absorption using high-amylose starches, thus enhancing its crispness.

1.6.1.2 Baked Products

Wheat flour is the main component in different baked products. Starch also plays a significant role in the production of baked products due to its intrinsic properties, like water absorption capacity, gelatinization, and retrogradation (Park et al., 2014). Starch gelatinization is a critical parameter for building structure and texture in baked foods. Since the wheat starch thickens only during baking, pregelatinized starches can be used to bind the limited available water more. This provides many advantages, such as a reduction in the dough's stickiness, enhanced handling and machinability properties, suspension of particulates (in muffin blends), increased cake volume, and increased water binding (Aprianita et al., 2014; Taggart, 2004). The main problem associated with backed products is staling resulting from starch retrogradation. Therefore, stabilized starches, mainly pregelatinized and hydroxypropylated starches, can be used to enhance the quality attributes of baked products, especially their freshness. This may be attributed to the high ability of these starches to bind water (Taggart, 2004). Pregelatinized starches are used for the preparation of cakes and cookies to improve their quality attributes, such as crumb softening and moisture retention for cakes and spreadability for cookies. Also, they may reduce the costs of heating and cooling filling. Highly cross-linked starches are also required for low-pH filling or where shear is applied. Highly stabilized waxy maize starch is used for fruit filling, while modified tapioca starch is preferred for cream filling. Furthermore, starch is used for the preparation of pies and tarts to stabilize and thicken fruit and cream filling, as well as to provide resistance to boil-out during baking and resistance to weeping during distribution (Mason, 2009). Rice starch is favored to produce freeze–thawed cakes (Jongsutjarittam & Charoenrein, 2013). On the other hand, native maize, waxy maize, or tapioca starches may be used as alternatives to wheat flour for gluten-free products for celiac sufferers (intolerant to wheat gluten) (Taggart, 2004).

1.6.1.3 Dairy Products

Modified starch is used in different dairy products, such as milk, yogurt, cheese, ice cream, pudding, sour cream, and buttermilk. Starch added to dairy products provides several functions, including improved viscosity and mouthfeel and increased stability (Mason, 2009). For instance, starch reduces syneresis and acts as a thickening agent, improving the mouthfeel and stability of yogurt and sour cream. Similar effects have been observed when starch is added to puddings. In addition, starch dressing enhances the curd properties of cottage cheese. Native starches can also be used for dairy products containing starch, such as pudding and other dairy desserts, to provide gelled texture; however, these products suffer from syneresis during long-period storage. For instance, wheat starch is used to provide thickening, adhesiveness, and gelling in ice cream and yogurt (Shevkani et al., 2017). Therefore, the selection of starch highly depends on the production method, equipment, processing parameters, required function, and product characteristics (Taggart, 2004). Homogenization and pasteurization, which are important processes in the dairy industry, may affect starch function. Also, the sequence of these operations may affect the starch functions; homogenization should be performed before pasteurization to avoid mechanical damage to the swollen starch granule, improving the viscosity. On the other hand, highly cross-linked, stabilized starches are required for dairy products that are homogenized after pasteurization in order to avoid viscosity loss. Highly cross-linked, stabilized waxy maize and tapioca starches are frequently used in the production of dairy products containing added starch (Mason, 2009). For instance, highly cross-linked, stabilized waxy maize and tapioca starches used in chilled desserts and UHT puddings improve viscosity, stability, and mouthfeel (a short, smooth structure). In addition, modified corn starch imparts the required texture (a short, firm texture) to some desserts.

1.6.1.4 Mayonnaise and Salad Dressings

The main function of starch in the production of mayonnaise and salad dressings is to thicken and stabilize the dispersed phase (Aprianita et al., 2014). The process of these products involves heat application and shear thinning, and they are mostly produced under acidic conditions (pH: 2.8). Thus, the starch's ability to tolerate acidity, heat, and shear is an important parameter in starch selecting for mayonnaise and salad dressings. Cross-linked and stabilized starches are frequently used in the manufacturing of these products (Kim et al., 2015). Moreover, hydroxypropylated starch can enhance the texture of low-fat dressings by developing a rich, creamy texture. This type of starch provides another effect: it can stabilize the products during freeze–thaw cycles, which is important and beneficial for products that need refrigerated storage after opening. The mouthfeel of low-fat salad dressing can be improved by enhancing the rheological and gelling properties using starch-based fat mimetics as vegetable oil replacers. Similarly, using amylose-containing starches combined with cross-linked and stabilized waxy maize starches can also enhance the gelling properties (Taggart, 2004). Lipophilic starches can also be used in the manufacturing of mayonnaise and salad dressings for emulsion stabilization, where they can interact with both oil and water, stabilizing the oil–water interface (Taggart, 2004).

1.6.1.5 Gravies, Soups, and Sauces

This group includes a wide range of products whose shelf stability varies from immediate consumption to a very long ambient shelf life. The shelf stability of this kind of food is ensured by its heating application (pasteurization or sterilization) and its pH. Starch selection for the production of these products mainly depends on the manufacturing process, which is also affected by heating and the pH of the product (Taggart, 2004). For instance, higher cross-linked, stabilized starches are suitable for high acidic foods (with pH < 4.5), which need a lower temperature or a shorter time to be sterilized, compared to neutral foods, providing process tolerance via strengthening and stabilizing the starch granules. Other parameters, including heat penetration, fill-viscosity, and shelf-life requirements, can also affect the suitability of starch for the production of gravies, soups, and sauces. The function of fill-viscosity starches, such as waxy maize starch, is to provide adequate viscosity during preheating to (1) ensure homogeneous suspension of particles; (2) provide uniform distribution of particles in the container; (3) avoid splashing; and (4) reduce unhygienic conditions around the filling line. Additionally, highly cross-linked and stabilized starches ensure the final viscosity, which helps heat penetrate during pasteurization or sterilization, maintaining the quality of the final product (Taggart, 2004).

Gravies, soups, and sauces are available in two forms: wet and dry mix. Therefore, other processing parameters should be taken into consideration during the production of wet products, such as the operating parameters of preheating and pasteurization or sterilization (temperature and time), cooling (method and rate), pumping mechanism, and shear during cooking or addition of difficult-to-disperse powders (Taggart, 2004). Therefore, the stabilized starch ensures the shelf-life of these products, which require high stability during freeze–thaw cycles. This effect is important for developing chilled, frozen, or ambient-stable products that need refrigerated storage after opening. Hydroxypropylated starches with high freeze–thaw stability are suitable for the production of these products (Aprianita et al., 2014) owing to their good compatibility with milk or milk protein, enhancing the texture and mouthfeel of cream-based sauces. On the other hand, dry mixes require high ease of dispersion in cold or hot water with rapid hydration to attain the expected viscosity at a suitable time. Therefore, pre-gelatinized starch is recommended for use in producing dry mixes, but it tends to lump in hot water, limiting the viscosity development. In order to overcome this problem, it is recommended to use either agglomerated, pre-gelatinized starches or highly stabilized hydroxypropylated starches, which are easily dispersed and quickly rehydrated, producing a rich, thick, and creamy product. Native wheat starch is also used to thicken and produce adhesiveness in gravies, soups, and dressings (Shevkani et al., 2017), while native maize starch is used to improve opacity and modify mouthfeel. However, it is recommended to use one part native and two parts modified starch for all types of gravy, soup, and sauce, achieving a high quality of the final products (Taggart, 2004).

1.6.1.6 Confectionary Products

Starches are used in the production of several confectionery products, including gums, pastes, molds, and dusting sweets, to prevent them from sticking together. These products vary from soft to hard gels and brittle to chewy textures. The starch function in these products is to build the structure of coatings and facilitate the shaping and designing of confectionaries. The starch selection mainly depends on some criteria, such as its ability to cook in sugar-rich media and its ease of handling during manufacturing (Aprianita et al., 2014; Taggart, 2004). For instance, dextrins are used with high-sugar solutions in order to produce stable, flexible coatings, as in jelly beans or shell-coated chocolates, owing to their excellent film-forming characteristics. This effect can be further enhanced by using starch in conjunction with other hydrocolloids. Meanwhile, starch molds are used to shape and design confectionery pastes. Also, these molds prevent moisture absorption since the moisture content of molding starch may alter the quality of final products. Very low moisture content (typically below 6%) increases the rate of moisture evaporation, producing confections with a hard crust, while a level higher than 9% makes the drying time longer, decreasing the production yield (Taggart, 2004).

1.6.1.7 Infant Foods

Infant foods are specially designed for weaned babies under the age of 12 months and young children aged from 1 to 3 years as a supplement to their diet or to help them adapt to common foods (European Commission, 2006). This type of food is generally limited to cereal-based foods (Pascari et al., 2019). In the preparation of these foods, enzymatic hydrolysis using amylases is an important step, which supports the partial cleavage of the starch present in the cereal grains in order to facilitate its bioavailability for the infant digestive system (Fernández-Artigas et al., 1999). Amylases, including α- and β-amylases, are used to hydrolyze starch into mono- and oligosaccharides and high-molecular-weight dextrins, respectively (Martínez & Gómez, 2016). Pregelatinized starches are generally used in the preparation of infant foods to thicken (or increase water retention without the need for heat) and afford short texture, thus enhancing the processing of these products and their quality (Ding et al., 2005). In order to prepare infant foods, starch pregelatinization involves mixing cereal flour with water (about 60–80% w/w) and then heating at 70°C, followed by the addition of α- and β-amylases after attaining gelatinization and stirring for a time ranging between 10 and 90 min (Pascari et al., 2019). Due to the wide range of infant foods and their complexity, the processing parameters may vary depending on the source of starch and the desired degree of hydrolysis (Pascari et al., 2019). Moreover, starches used in retorted infant foods (e.g., fruit-based and dessert-type products) should have temperature stability, resistance to low pH, stability in jars, and freeze–thaw stability (Mason, 2009). Therefore, cross-linked acetylated waxy maize and tapioca starches are suitable for the production of this type of infant food.

1.6.1.8 Beverage Emulsion Stabilization

A beverage emulsion is known as an oil-in-water emulsion (O/W), containing two phases: a dispersed oily phase and a continuous aqueous phase. In other words, it is a concentrate mixed with sugar and carbonated water to prepare soda and fruit drinks (Mason, 2009). The main function of O/W emulsion is to provide flavor and a cloudy appearance (opacity) to the final products. Conventionally, gum Arabic has been used as a stabilizer in the beverage industry to stabilize O/W emulsions by lowering the interface tension during the homogenization step. Recently, amphiphilic starches containing hydrophilic and lipophilic groups have been used to replace gum Arabic to stabilize these emulsions (Taggart, 2004) by preventing creaming, sedimentation, and loss of desired flavor and cloudy appearance in the concentrate and final product (Mason, 2009). Hydrophobic side-groups facilitate starch accumulation at the interface, whereas hydrophilic side-groups stick out into the aqueous phase, improving the oil droplet stability against aggregation (Tesch et al., 2002). On the other hand, it was observed that the addition of hydrocolloids such as xanthan gum may enhance the emulsion properties and the final product quality, where adding 3% xanthan gum to the emulsions stabilized with modified starch improved the emulsion's stability, decreased the overall size of the oil droplet, and increased the beverage's cloudy appearance (Taherian et al., 2007). Generally, a fine emulsion with an oil droplet size of 1 μm or less is required to obtain the desired stability and cloudy appearance (Mason, 2009). The storability of emulsions containing more than 0.02% starch nanoparticles has been improved and extended to more than 2 months without aggregation of oil droplets (Kim et al., 2015).

1.6.1.9 Food Ingredient Encapsulation

Encapsulation of food functional ingredients can be performed using different materials and techniques in order to control their release in the food and digestive systems (Zhengyu, 2018). These include phenolic compounds, vitamins, probiotics, flavorings, and oils. However, there is an increasing interest in using modified starch in the encapsulation of these ingredients. Therefore, encapsulation techniques, including spray-drying, freeze-drying, drum-drying, extrusion, fluidized-bed addition, emulsification, physical trapping, anti-solvent precipitation, and complexation, are developed (Zhu, 2017; Mason, 2009). The encapsulation process involves dispersing the oil in a solution containing about four times as much encapsulant, which is then homogenized and spray-dried. The modified starch stabilizes the emulsion through film formation and provides a dry matrix to occlude oil within, thereby reducing the loss of flavor during drying and storage as well as increasing the oxidative stability of oil during the storage period. Some studies reported the use of porous starches to encapsulate flavorings, polyphenols, vitamins, colorants, sweeteners, probiotics, oils, enzymes, and minerals (Benavent-Gil & Rosell, 2017; H. Li et al., 2016; Z. Li, Jiang et al., 2015; Belingheri et al., 2015a, 2015b; Wang et al., 2016). Additionally, hydrolyzed starch can replace caseinate in the preparation of non-dairy creamers. Also, amphiphilic starch applications are increasing to include high-load encapsulation of flavor oils (Taggart, 2004). Similarly, phenolic compounds and lipids (unsaturated fatty acids) can also be encapsulated using amylose from partially gelatinized starches (Zhu, 2017). On the other hand, diverse types of nanoparticles have been used to increase the absorption and bioavailability of food functional ingredients after their absorption (Z. Li, Jiang et al., 2015), where vitamin D_3 and curcumin can be loaded with starch nanoparticles. Similarly, phenolic compounds and lipids (unsaturated fatty acids) can also be encapsulated using amylose from partially gelatinized starches (Zhu, 2017).

1.6.1.10 Meat Products

Starch can be used in processed meat products to reduce shrinkage and cooking losses, thus increasing the cooking yield and improving the juiciness and texture of the final products. This is attributed to the ability of starch to retain water. This effect is important for achieving high-quality meat products and extending their shelf life. Modified starch can also be used in the production of emulsified meat products (i.e., bologna and frankfurters) and plays a crucial role in freeze–thaw cycles (Lawrie & Ledward, 2014) to ensure freeze–thaw stability. For instance, waxy maize or tapioca starches with a high degree of stabilization can improve water retention in chilled and frozen meat products (Taggart, 2004). Also, in reformed meat products, coarse, pregelatinized starch can bind moisture quickly, which enhances the dough viscosity, facilitates molding, and thus improves the final texture.

1.6.1.11 Batters and Breadings

Batters and breadings are generally used as a coating layer for a wide range of deep-fried products to provide the desired crispiness. These products include meat, chicken meat, seafood, and vegetables. Flour is the main ingredient in the batter mixes and breadings, but the quality variation of native flour may cause a substantial problem for food producers. Therefore, starch can be added to the batter mixes to remedy this problem. In the coating process, the core is first coated with a layer of predust flour, which allows adhesion to a liquid batter, and then coated with a layer of breadcrumbs. Starch added to the batter blends can provide an adequate viscosity, which controls the amount and thickness of the battered coating, to improve the visual, textural, and flavor characteristics and shelf life of the final products (Taggart, 2004). For instance, pregelatinized starch can be used to control the cold viscosity by providing shear and freeze–thaw stability for batter slurries, which are used for chilled and frozen food products or to be recycled. The textural characteristics of oven-baked meat can be improved by using high-amylose starches. However, these starches require high temperatures to attain the desired crispy texture; therefore, the use of modified amylose starches can produce a smooth, uniform coating at lower cooking temperatures as a result of lowering the gelatinization temperature (Taggart, 2004). Furthermore, these starches can also decrease oil uptake due to their excellent film-forming characteristics. On the other hand, dextrins are used to improve the color of vegetables' coatings. However, high levels of dextrins (about 30%) generate blistering effects, which in turn change the coating surface. Also, very high levels of dextrins and incompletely gelatinized high-amylose starches could produce stickiness. Stickiness can be prevented by lowering the amount of dextrins or increasing the amylose content in order to attain the expected crispiness (Taggart, 2004).

1.6.1.12 Food Packaging

Food packaging plays a significant role in maintaining food quality during its storage and distribution. Therefore, starch-based packaging materials have received significant attention owing to the abundance, biodegradability, renewability, thermoplastic characteristics, and low price of starch (Majid et al., 2018; López-Córdoba et al., 2017). Additionally, starch-based films are transparent, colorless, odorless, tasteless, biodegradable, biocompatible, and inexpensive; however, they tend to absorb moisture from a high relative humidity environment owing to their hydrophilic nature (Jiménez et al., 2012). Starch-based packaging films undergo retrogradation and crystallization through the storage period, causing alterations in their thermomechanical characteristics (Mukurumbira et al., 2017). Also, other drawbacks have been reported for developed starch-based films, including poor solubility, poor freeze–thaw stability, and uncontrollable paste consistency (Shah et al., 2016). For these reasons, modified starches have been used to enhance film characteristics, such as solubility, retrogradation resistance, thermal and freeze–thaw stability, and mechanical strength (Zhengyu, 2018; Winkler et al., 2014). Additionally, starch-based packaging films have been shown to have effective antimicrobial and antioxidant effects (Samsudin & Hani, 2018). On the other hand, starch nanoparticles could be used as an ingredient in barrier films for food packaging to enhance their properties, such as oxygen and water vapor permeability (Kim et al., 2015). A significant decrease in water vapor permeability was observed when 30–40% waxy maize was added to the film formulation (Kristo & Biliaderis, 2007). A similar trend was observed in cassava starch-based films containing 2.5% starch nanoparticles (Garcia et al., 2011).

1.6.2 Starch Applications in the Pharmaceutical Industry

Excipients are essential components of traditional and new drug products. According to the International Pharmaceutical Excipient Council, an excipient is defined as "any substance other than active drug or pro-drug that is included in the manufacturing process or is contained in finished pharmaceutical dosage forms". Excipients in drug production are classified according to their functions into binders, disintegrants, diluents, and lubricants (Desai et al., 2016; Hartesi et al., 2016). These excipients make the drugs more elegant, stable, active, and safe (Boonme et al., 2012). The wide use of starch in a native or modified form as a pharmaceutical excipient is basically due to its physical and functional characteristics. For instance, native starch has some important features that render it desirable for application as a pharmaceutical excipient, such as its whiteness, softness, and smoothness, ability to be molded, and gelling and viscosity properties (Ogunsona et al., 2018; Tester et al., 2004). Native starch has usually been applied in the production of capsules, tablets, and granules. Also, modified starch has some superior characteristics, making it more effective in both traditional and non-traditional drug delivery techniques, such as improved flow disintegration, direct compression, and its ability to form stable gels in both hot and cold

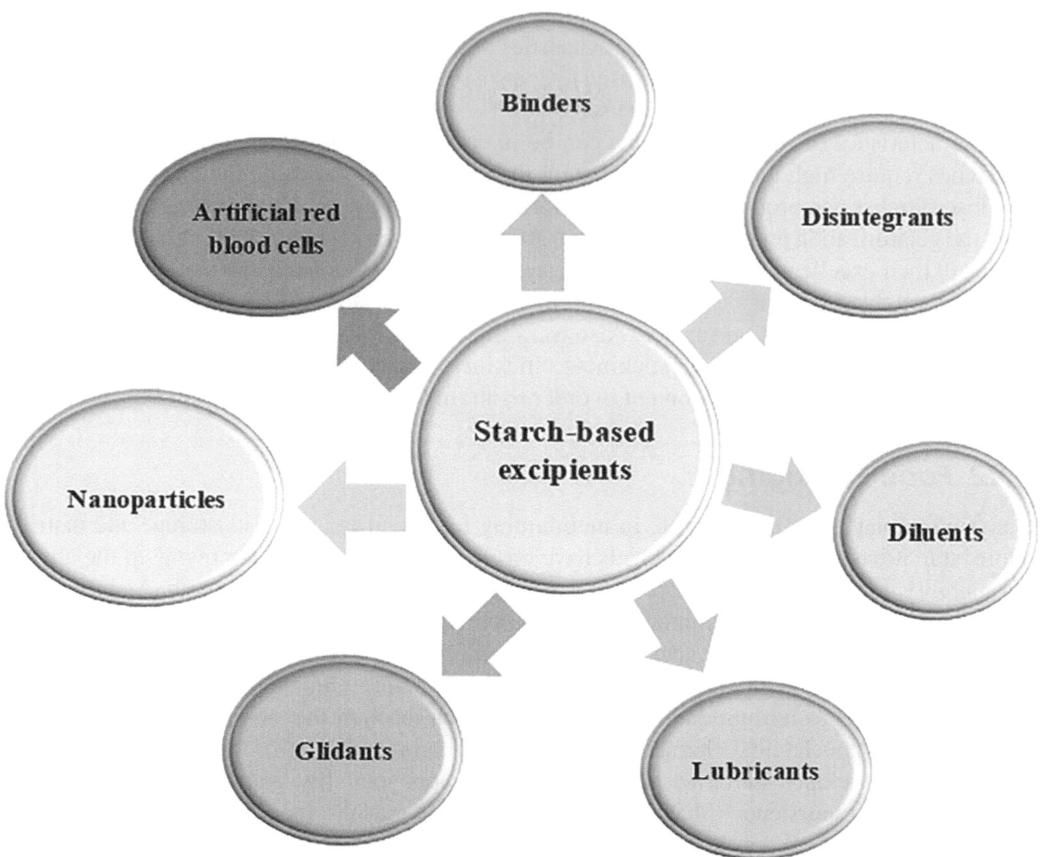

FIGURE 1.19 Various applications of starch in the pharmaceutical industry.

water (Reis et al., 2008). Starch-based excipients have many advantages in the drug industry, including reduced production costs, safety, and a higher-quality product. Starch-based excipients have been applied as drug carriers in controlled drug delivery systems (Szepes et al., 2008). For example, starch has been used as a binder in tablet and granule production; as a glidant, disintegrant, and diluent in traditional tablet or capsule formulations; as a carrier and lubricant in body and face cosmetic powders; and as a surface carrier for colors and flavors (Builders & Arhewoh, 2016; Builders et al., 2013a; Nattapulwat et al., 2008). The different applications of starch in the pharmaceutical industry are elaborated upon in the following section, as shown in Figure 1.19.

1.6.2.1 Starch-Based Binders

In solid oral dosage forms, the function of binders is to maintain the active drug component and other inactive materials together in a cohesive blend. Binders are classified into (1) dry binders, which can be added to the powder mix as a part of direct powder compression or after wet granulation step, such as cellulose and methyl cellulose; and (2) solution binders used in wet granulation technique after their dissolving in a solvent (water or alcohol), such as corn starch (Patil et al., 2010). The use of starch as a binder has some advantages, including uniform distribution, controlled drug release, and the compatibility of starch with the active drug component and other inactive materials in several traditional and new formulations (Builders & Arhewoh, 2016). Generally, starch is used in small amounts, varying from 3 to 20% of the total drug weight, depending on the concentration and type of other materials and the operative parameters applied in the production (Sasaki et al., 2000). As dry binders, starch-based binders have been shown to have excellent compression because starch produces a cohesive dry blend owing to its excellent adhesive characteristics, but it suffers from some plastic deformations during the compression. Therefore, starch has effectively been applied as a disintegrant and a binder in tablet and capsule manufacturing using wet granulation due to its partial solubility in cold water (Patil et al., 2010; Santander-Ortega et al., 2010). In the wet granulation technique, the starch gel acts as a glue to give the desired binding forces, holding together the powder particles and then forming the agglomerates. The main advantages of using native starch as a binder in wet granulation are (1) attractive appearance of granules; (2) improved density and flow of the powder and subsequently compaction; (3) homogeneity of ingredients; (4) excellent release of the active drug component; (5) providing desirable mechanical properties to the granule, which may be compressed in tablets or packed in capsule shells and sachets; and 6) reduced dustiness during tablet and capsule production (Builders et al., 2013b). Corn starch is mainly applied in the wet granulation technique, but other types of starch obtained from non-conventional sources (e.g., cocoyam starch) can also be used as a binder and replace the corn starch in the wet granulation technique (Uhumwangho et al., 2006).

1.6.2.2 Starch-Based Disintegrants

The function of a disintegrant is to accelerate active component release by disintegrating the tablets or the capsule shell, increasing its availability and facilitating its absorption by the body (Hartesi et al., 2016). Starch is one of the main disintegrants used in the manufacturing of tablets and capsules. Corn and potato starches are frequently used as disintegrants in the drug industry, and other types of starch, such as yam starch, can also be used as disintegrants (Nattapulwat et al., 2008).

The disintegrant working mechanism is based on breaking down the hydrogen bonds formed during compression between the starch and the drug components. During compaction, starch molecules bind the active drug component and other inactive components. On the other hand, when the drug is exposed to water, the starch granule absorbs water, and the hydrogen bonds breakdown, leading to starch swelling, which is elastically deformed afterward (Adedokun & Itiola, 2011). This process loosens the compacted tablet particles. Starch is used in small amounts, ranging from 3 to 15% of the drug weight (Hu et al., 2015; Sasaki et al., 2000). In general, a perfect disintegrant should be effective in small amounts to prevent or decrease its effect on tablet properties like friability, hardness, and compaction ability. Also, the effectiveness of starch as a disintegrant is also affected by the characteristics of the active drug component and other drug substrates. To obtain the optimum efficiency of starch as a disintegrant, half of the starch amount is incorporated into the granulation blend as an endo-disintegrant, and the other half is added directly to the dried granules as an exo-disintegrant (Builders & Arhewoh, 2016).

1.6.2.3 Starch-Based Diluents

Diluents are known as fillers and are incorporated at concentrations of 5–80% as an integral component in low-dose traditional formulations like tablets, capsules, or granules in order to facilitate handling and mixing, improve the content uniformity, and increase weight (Quodbach & Kleinebudde, 2015). However, the drug's characteristics, including the drug release and granule properties (cohesion, flowability, and compaction), may be affected by incorporating the diluents in great amounts (Zàmostný et al., 2012). Native starch is considered a natural diluent (Bolhuis & Waard, 2011), used for mixing and handling, facilitating, and the production of colorants and potent drugs. As an insoluble diluent, the use of native starch is attributed to some inherent characteristics, including the inability to interact with the active drug component and the ability to provide constant functional and physicochemical characteristics. Contrary to other diluents, when incorporated in a large amount, starch improves the drug's disintegration and interaction with water, maintaining the inherent solubility of the active drug component in the medium. However, the use of starch as a diluent can be determined by the formulation technique. For instance, it is not suitable for direct compression owing to its poor compaction ability, which results from low plasticity and tensile strength, stickiness to the punches and dies, and high moisture sensitivity (Zhang & Schwartz,

2000). Nevertheless, starch compaction ability may be enhanced through wet granulation techniques (Bayor et al., 2013).

1.6.2.4 Starch-Based Lubricant

Lubrication is necessary for pharmaceutical operations like mixing, tablet compaction, and capsule filling in order to reduce friction between the organic compounds and surfaces of production equipment, ensuring a continuous process (Li & Wu, 2014). Lubricants are incorporated in very small concentrations into the tablet and capsule formulations in order to enhance the powder processing and properties. Lubricants are usually added at concentrations in the range of 0.25–5% (Pagar et al., 2022). Starch is used in tablet and capsule production at a concentration ranging from 3 to 10% as a lubricant and anti-adherent material in order to enhance the powder flowability during the process by preventing stickiness between the formulation materials and the production equipment, such as punch faces, die walls, dosators, and tamping pins for encapsulates.

1.6.2.5 Starch-Based Glidant

Similarly, glidants are used to enhance powder flowability by reducing interparticle friction, cohesiveness, and surface charge, reducing the angle of repose. Glidants are commonly added as a dry powder just before direct compression, in combination with lubricants, because they are unable to reduce die wall friction. Starch is used as a glidant in conventional tablet and capsule manufacturing at concentrations of 2–10% (Bolhuis & Chow, 1996). Starches from different sources, such as starches from cassava, yam, and fonio, have been used as glidants in the manufacturing of tablets and capsules (Pagar et al., 2022). The glidant properties of native starch in pharmaceutical applications can be determined by measuring the flow rate and factor as well as the angle of repose. The particle size is another crucial factor affecting the glidant properties; the smaller the particle size, the more effective the glidant (Builders & Arhewoh, 2016; Bolhuis & Chow, 1996).

1.6.2.6 Other Pharmaceutical Applications

As previously shown, starch has been used as an excipient and special drug carrier in conventional and new drug formulations. Nowadays, starch has also been used to fabricate nanoparticles in order to deliver drugs to special sites, such as cancer cells (Ahmad et al., 2011), reducing the dose of harmful anticancer materials without affecting their therapeutic efficacy. Also, starch has been used to manufacture artificial red blood cells with great oxygen-carrying ability. For instance, potato starch has been used as artificial red blood cells by self-assembly, wherein hemoglobin was encapsulated with long-chain fatty acid-grafted potato starch (Xu et al., 2015).

1.7 STARCH MODIFICATIONS

Native starches have limited applications in the food industry and other sectors. This may be related to their lower solubility and swelling power, lower water-binding capacity, higher tendency for retrogradation, instability in pH, and poor resistance to severe processing conditions (high temperature and shear). Therefore, the modification of starch is indispensable to overcome these drawbacks and enhance the functionality and suitability of starch for different applications.

There are many different methods for starch modification, which depend on the required functional properties of starch in the different applications. These include physical, chemical, enzymatic, and dual techniques (hybrid techniques). Physical modification, which does not cause chemical alteration in the

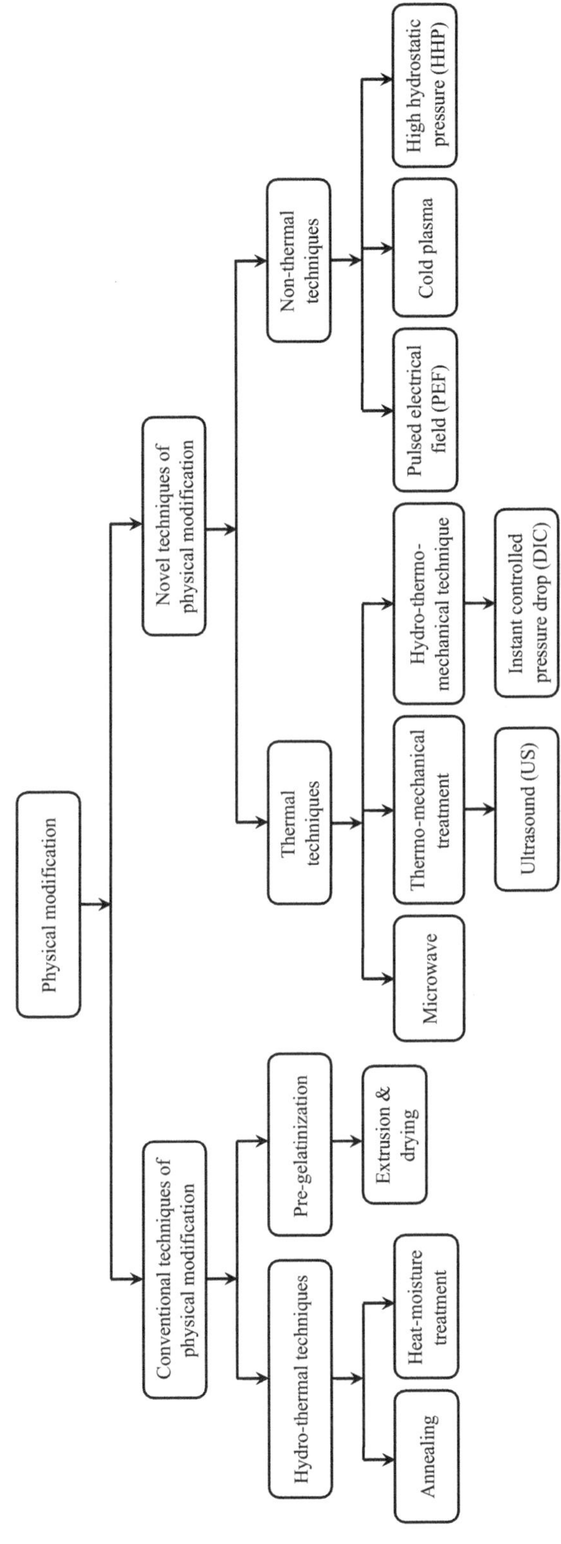

FIGURE 1.20 Conventional and innovative methods for physical starch modification.

starch except for the cleavage of some glycosidic bonds, includes conventional and novel techniques for starch modification. The conventional processes include pre-gelatinization or pre-cooking, like extrusion and drying, and hydro-thermal techniques, such as annealing and heat-moisture treatment. On the other hand, the novel processes are considered green and eco-friendly technologies to modify starches with high efficiency. These techniques involve the use of thermal and emerging non-thermal methods (Suri & Singh, 2023), as shown in Figure 1.20. Among the physical methods, pre-gelatinization is the most common technique used for starch modification.

In contrast to the physical modification of starch, chemical modification results in chemical changes in the starch by introducing functional groups in the starch molecules. These changes include replacing hydroxyl groups with other functional groups; grafting side chains with different compositions, lengths, and structures; or changing the distribution of groups or chains along the backbone (Fan & Picchioni, 2020). This type of modification includes oxidation, acetylation, succinylation, acid-thinning, and cross-linking. Although chemical modification methods are greatly effective in starch modification, they have some limitations with respect to costs, time, chemical residues, and environmental pollution. A detailed discussion of the different processes for starch modification and their effects on various starch properties will be covered in the next chapters of this book.

1.8 CONCLUSIONS AND FUTURE TRENDS

Starch granules have an extremely complex structure due to the variation in their chemical composition, the structure of starch polymers, and the mechanism of their association. Starches display different morphological, textural, physicochemical, and functional characteristics. These characteristics are significantly affected by the content of amylose and amylopectin and the presence of lipids, proteins, and phosphorous in the starch granule. Starch has several applications in the food industry; starch has very significant functionality in the preparation of various food products, such as snack foods, baked products, dairy products, confectionary products, meat products, infant foods, soups, and sauces and in the formation of biodegradable film. Starch can be used in the food industry to provide thickening, binding, stabilizing, adhesion, clouding, dusting, moisture retention, and texturizing properties. Similarly, starch exhibits great applications in the pharmaceutical industry, particularly in capsules and compressed tablets. It plays a significant role in controlling the release of the active ingredient of the drug. Different techniques can be applied to extract starch from various botanical sources, but wet-milling is the most common method used for isolating starch from different sources.

Although starch is well studied, there is a need for further studies to enhance starch characteristics and applications in different sectors, including (1) study and use of new sources of native starch that have desirable functional properties to extend the starch uses in the different food and pharmaceutical applications; (2) deep study and understanding of the interactions between starch and different components present in food and the effect of these interactions on the quality attributes of the final product; and (3) in-depth studies on the changes in the starch structure and functionality during food processing and storage.

REFERENCES

Adedokun, M. O., & Itiola, O. A. (2011). Disintegrant activities of natural and pregelatinized trifoliate yams, rice and corn starches in paracetamol tablets. *Journal of Applied Pharmaceutical Science, 1*(10), 200–206.

Ahmad, M. Z., Akhter, S., Ahmad, I., Rahman, M., Anwar, M., Jain, G. K., Ahmad, F. J., & Khar, R. K. (2011). Development of polysaccharide based colon targeted drug delivery system: Design and evaluation of Assam Bora rice starch based matrix tablet. *Current Drug Delivery, 5*, 575–581. https://doi.org/10.2174/156720111796642327

Ahmed, J., Ramaswamy, H. S., Ayad, A., & Alli, I. (2008). Thermal and dynamic rheology of insoluble starch from basmati rice. *Food Hydrocolloids*, 22(2), 278–287. https://doi.org/10.1016/j.foodhyd.2006.11.014

Ai, Y., Gong, L., Reed, M., Huang, J., Zhang, Y., & Jane, J.-L. (2016). Characterization of starch from bamboo seeds. *Starch*, 68(1–2), 131–139. https://doi.org/10.1002/star.201500206

Ai, Y., Hasjim, J., & Jane, J. (2013). Effects of lipids on enzymatic hydrolysis and physical properties of starch. *Carbohydrate Polymers*, 92(1), 120–127. https://doi.org/10.1016/j.carbpol.2012.08.092

Ai, Y., & Jane, J.-L. (2015). Gelatinization and rheological properties of starch. *Starch*, 67(3–4), 213–224. https://doi.org/10.1002/star.201400201

Alcázar-Alay, S. C., & Meireles, M. A. A. (2015). Physicochemical properties, modifications and applications of starches from different botanical sources. *Food Science & Technology*, 35(2), 215–236. http://dx.doi.org/10.1590/1678-457X.6749

Aprianita, A., Vasiljevic, T., Bannikova, A., & Kasapis, S. (2014). Physicochemical properties of flours and starches derived from traditional Indonesian tubers and roots. *Journal of Food Science & Technology*, 51(12), 3669–3679. https://doi.org/10.1007/s13197-012-0915-5

Bajaj, R., Singh, N., Ghumman, A., Kaur, A., & Mishra, H. N. (2021). Effect of high pressure treatment on structural, functional, and in-vitro digestibility of starches from tubers, cereals, and beans. *Starch*, 74(1–2), 2100096. https://doi.org/10.1002/star.202100096

Balet, S., Guelpa, A., Fox, G., & Manley, M. (2019). Rapid Visco Analyser (RVA) as a tool for measuring starch-related physiochemical properties in cereals: A review. *Food Analytical Methods*, 12, 2344–2360.

Bancel, E., Rogniaux, H., Debiton, C., Chambon, C., & Branlard, G. (2010). Extraction and proteome analysis of starch granule-associated proteins in mature wheat kernel (*Triticum aestivum* L.). *Journal of Proteome Research*, 9(6), 3299–3310. https://doi.org/10.1021/pr9010525

Bayor, M. T., Tuffour, E., & Lambon, P. S. (2013). Evaluation of starch from new sweet potato genotypes for use as a pharmaceutical diluent, binder or disintegrant. *Journal of Applied Pharmaceutical Science*, 3(8), S17–S23. https://doi.org/10.7324/JAPS.2013.38.S4

Belingheri, C., Ferrillo, A., & Vittadini, E. (2015b). Porous starch for flavor delivery in a tomato-based food application. *LWT-Food Science & Technology*, 60(1), 593–597. https://doi.org/10.1016/j.lwt.2014.09.047

Belingheri, C., Giussani, B., Rodriguez-Estrada, M. T., Ferrillo, A., & Vittadini, E. (2015a). Oxidative stability of high-oleic sunflower oil in a porous starch carrier. *Food Chemistry*, 166, 346–351. https://doi.org/10.1016/j.foodchem.2014.06.029

BeMiller, J. N. (2011). Pasting, paste, and gel properties of starch–hydrocolloid combinations. *Carbohydrate Polymers*, 86(2), 386–423. https://doi.org/10.1016/j.carbpol.2011.05.064

BeMiller, J. N., & Whistler, R. L. (2009). *Starch: Chemistry and technology*. Academic Press.

Benavent-Gil, Y., & Rosell, C. M. (2017). Comparison of porous starches obtained from different enzyme types and levels. *Carbohydrate Polymers*, 157, 533–540. https://doi.org/10.1016/j.carbpol.2016.10.047

Berski, W., Ptaszek, A., Ptaszek, P., Ziobro, R., Kowalski, G., Grzesik, M., & Achremowicz, B. (2011). Pasting and rheological properties of oat starch and its derivatives. *Carbohydrate Polymers*, 83(2), 665–671. http://dx.doi.org/10.1016/j.carbpol.2010.08.036

Bet, C. D., de Oliveira, C. S., Colman, T. A. D., Marinho, M. T., Lacerda, L. G., Ramos, A. P., & Schnitzler, E. (2018). Organic amaranth starch: A study of its technological properties after heat-moisture treatment. *Food Chemistry*, 30(264), 264, 435. https://doi.org/10.1016/j.foodchem.2018.05.021

Biliaderis, C. G. (1998). Structures and phase transitions of starch polymers. In R. H. Walter (Ed.), *Polysaccharide association structures in foods*. Marcel Dekker.

Blennow, A., Bay-Smidt, A. M., Olsen, C. E., & Moller, B. L. (2000). The distribution of covalently bound phosphate in the starch granule in relation to starch crystallinity. *International Journal of Biological Macromolecules*, 27(3), 211–218. https://doi.org/10.1016/S0141-8130(00)00121-5

Bolhuis, G. K., & Chow, Z. T. (1996). Materials for direct compaction. In G. Alderborn & C. Nystrum (Eds.), *Pharmaceutical powder compaction technology*. Mercel Dekker Inc.

Bolhuis, G. K., & Waard, H. (2011). Compaction properties of directly compressible materials. In M. Celik (Ed.), *Pharmaceutical powder compaction technology* (2nd ed.). CRC Press. https://doi.org/10.3109/9781420089189

Boonme, P., Pichayakorn, W., Prapruit, P., & Boromthanarat, S. (2012, October 29–31). Application of sago starch in cosmetic formulations. *Proceedings of 2nd ASEAN Sago Symposium 2012, UNIMAS, Kota Samarahan Advances in Sago Research and Development*, 53–60.

Builders, P. F., & Arhewoh, M. I. (2016). Pharmaceutical applications of native starch in conventional drug delivery—review. *Starch*, 68, 864–873. https://doi.org/10.1002/star.201500337

Builders, P. F., Anwunobi, A. P., Mbah, C. C., & Adikwu, M. U. (2013a). New direct compression excipient from tiger nut starch: Physicochemical and functional properties. *AAPS PharmSciTech*, 14(2), 818–827. https://doi.org/10.1208/s12249-013-9968-7

Builders, P. F., Mbah, C. C., Adama, K. K., & Momoh, M. A. (2013b). Effect of pH on the physicochemical and binder properties of tigernut starch. *Starch*, *66*(3–4), 281–293. https://doi.org/10.1002/star.201300014

Buléon, A., Cotte, M., Putaux, J.-L., D'Hulst, C., & Susini, J. (2014). Tracking sulfur and phosphorus within single starch granules using synchrotron X-ray micro fluorescence mapping. *Biochimica et Biophysica Acta (BBA)—General Subjects*, *1840*(1), 113–119. https://doi.org/10.1016/j.bbagen.2013.08.029

Calzetta Resio, A. N., Tolaba, M. P., & Suárez, C. (2006). Effect of steeping conditions on wet-milling attributes of amaranth. *International Journal of Food Science & Technology*, *41*(2), 70–76. http://dx.doi.org/10.1111/j.1365-2621.2006.01395.x

Cardoso, M. B., Putaux, J.-L., Samios, D., & Da Silveira, N. P. (2007). Influence of alkali concentration on the deproteinization and/or gelatinization of rice starch. *Carbohydrate Polymers*, *70*(2), 160–165. https://doi.org/10.1016/j.carbpol.2007.03.014

Chan, H.-T., Bhat, R., & Karim, A. A. (2010). Effects of sodium dodecyl sulphate and sonication treatment on physicochemical properties of starch. *Food Chemistry*, *120*, 703–709. https://doi.org/10.1016/j.foodchem.2009.10.066

Chang, F., He, X., & Huang, Q. (2013). Effect of lauric acid on the V-amylose complex distribution and properties of swelled normal corn starch granules. *Journal of Cereal Science*, *58*(1), 89–95. https://doi.org/10.1016/j.jcs.2013.03.016

Cheetham, N. W. H., & Tao, L. (1998). Variation in crystalline type with amylose content in maize starch granules: An X-ray powder diffraction study. *Carbohydrate Polymers*, *36*(4), 277–284. http://dx.doi.org/10.1016/S0144-8617(98)00007-1

Chen, P., Xie, Q. T., Wang, R. M., Wang, S. Y., Cheng, J. S., & Zhang, B. (2022). Effects of pullulanase enzymatic hydrolysis on the textural of acorn vermicelli and its influencing mechanism on the quality. *Food Research International*, *156*, 111294. https://doi.org/10.1016/j.foodres.2022.111294

Choi, J. M., Park, C.-S., Baik, M. Y., Kim, H. S., Choi, Y. S., Choi, H. W., & Seo, D. H. (2018). Enzymatic extraction of starch from broken rice using freeze–thaw infusion with food-grade protease. *Starch*, *70*(1–2), 1700007. https://doi.org/10.1002/star.201700007

Condés, C., Anon, M. C., Dufresne, A., & Mauri, A. N. (2018). Composite and nanocomposite films based on amaranth biopolymers. *Food Hydrocolloids*, *74*, 159–167. https://doi.org/10.1016/j.foodhyd.2017.07.013

Cozzolino, D. (2016). The use of the rapid viscoanalyser (RVA) in breeding and selection of cereals. *Journal of Cereal Science*, *70*, 282–290. https://doi.org/10.1016/j.jcs.2016.07.003

Cui, S. W. (2005). *Food carbohydrates: Chemistry, physical properties, and applications*. CRC Press.

Cui, W., Ma, Z., Li, X., & Hu, X. (2021). Structural rearrangement of native and processed pea starches following simulated digestion in vitro and fermentation characteristics of their resistant starch residues using human fecal inoculum. *International Journal of Biological Macromolecules*, *172*, 490–502. https://doi.org/10.1016/j.ijbiomac.2021.01.092

Debet, M. R., & Gidley, M. J. (2006). Three classes of starch granule swelling: Influence of surface proteins and lipids. *Carbohydrate Polymers*, *64*(3), 452–465. https://doi.org/10.1016/j.carbpol.2005.12.011

Deka, D., & Sit, N. (2016). Dual modification of taro starch by microwave and other heat moisture treatments. *International Journal of Biological Macromolecules*, *92*, 416–422. https://doi.org/10.1016/j.ijbiomac.2016.07.040

Desai, P. M., Liew, C. V., & Heng, P. W. S. (2016). Review of disintegrants and the disintegration phenomena. *Journal of Pharmaceutical Sciences*, *105*, 2545–2555. https://doi.org/10.1016/j.xphs.2015.12.019

Ding, Q.-B., Ainsworth, P., Tucker, G., & Marson, H. (2005). The effect of extrusion conditions on the physicochemical properties and sensory characteristics of rice-based expanded snacks. *Journal of Food Engineering*, *66*(3), 283–289. https://doi.org/10.1016/j.jfoodeng.2004.03.019

Ding, Y., Cheng, J., Lin, Q., Wang, Q., Wang, J., & Yu, G. (2021). Effects of endogenous proteins and lipids on structural, thermal, rheological, and pasting properties and digestibility of adlay seed (*Coix lacryma-jobi* L.) starch. *Food Hydrocolloids*, *111*, 106254. https://doi.org/10.1016/j.foodhyd.2020.106254

Dudu, O. E., Ma, Y., Olurin, T. O., Oyedeji, A. B., Oyeyinka, S. A., & Ogungbemi, J. W. (2021). Synergistic effect of hydrothermal and additive treatments on structural and functional characteristics of cassava starch. *Journal of Food Processing and Preservation*, *45*(11), e15904. https://doi.org/10.1111/jfpp.15904

Durrani, C. M., & Donald, A. M. (1995). Physical characterisation of amylopectin gels. *Polymer Gels & Networks*, *3*(1), 1–27. http://dx.doi.org/10.1016/0966-7822(94)00005-R

Eckhoff, S. R., & Watson, S. A. (2009). Corn and sorghum starches: Production. In J. BeMiller & R. L. Whistler (Eds.), *Starch: Chemistry and technology* (3rd ed., Ch. 9, pp. 374–439). Academic Press.

El Halal, S. L. M., Kringel, D. H., Zavareze, E. R., & Dias, A. R. G. (2019). Methods for extracting cereal starches from different sources: A review. *Starch*, *71*, 1900128. https://doi.org/10.1002/star.201900128

Englyst, H. N., Kingman, S. M., & Cummings, J. H. (1992). Classification and measurement of nutritionally important starch fractions. *European Journal of Clinical Nutrition*, *46*(2), S33–S50.

European Commission. (2006). Commission directive 2006/125/EC on processed cereal-based foods and baby foods for infants and young children. *Official Journal of the European Union, 399*, 16–35.

Falsafi, S. R., Maghsoudlou, Y., Rostamabadi, H., Rostamabadi, M. M., Hamedi, H., & Hosseini, S. M. H. (2019). Preparation of physically modified oat starch with different sonication treatments. *Food Hydrocolloids, 89*, 311–320. https://doi.org/10.1016/j.foodhyd.2018.10.046

Fan, Y., & Picchioni, F. (2020). Modification of starch: A review on the application of "green" solvents and controlled functionalization. *Carbohydrate Polymers, 241*, 116350. https://doi.org/10.1016/j.carbpol.2020.116350

Felisberto, M. H. F., Beraldo, A. L., Costa, M. S., Boas, F. V., Franco, C. M. L., & Clerici, M. T. P. S. (2019). Physicochemical and structural properties of starch from young bamboo culm of *Bambusa tuldoides*. *Food Hydrocolloids, 87*, 101–107. https://doi.org/10.1016/j.foodhyd.2018.07.032

Felisberto, M. H. F., Beraldo, A. L., Costa, M. S., Boas, F. V., Franco, C. M. L., & Clerici, M. T. P. S. (2020). Bambusa vulgaris starch: Characterization and technological properties. *Food Research International, 132*, 109102. https://doi.org/10.1016/j.foodres.2020.109102

Felisberto, M. H. F., Costa, M. S., Boas, F. V., Leivas, C. L., Franco, C. M. L., de Souza, S. M., Clerici, M. T. P. S., & Cordeiro L, M. C. (2020). Characterization and technological properties of peach palm (*Bactris gasipae*s var. *gasipaes*) fruit starch. *Food Research International, 136*, 109569. https://doi.org/10.1016/j.foodres.2020.109569

Fernández-Artigas, P., Guerra-Hernández, E., & García-Villanova, B. (1999). Browning indicators in model systems and baby cereals. *Journal of Agricultural & Food Chemistry, 47*(7), 2872–2878. https://doi.org/10.1021/jf9808729

Fontes, S. M., Cavalcanti, M. T., Candeia, R. A., & Almeida, E. L. (2017). Characterization and study of functional properties of banana starch green variety of Mysore (*Musa AAB—Mysore*). *Food Science & Technology, 37*(2), 224–231. https://doi.org/10.1590/1678-457X.18916

Fu, Z., Chen, J., Luo, S. J., Liu, C. M., & Liu, W. (2015). Effect of food additives on starch retrogradation: A review. *Starch, 67*(1–2), 69–78. http://dx.doi.org/10.1002/star.201300278

Garcia, N. L., Ribba, L., Dufresne, A., Aranguren, M., & Goyanes, S. (2011). Effect of glycerol on the morphology of nanocomposites made from thermoplastic starch and starch nanocrystals. *Carbohydrate Polymers, 84*(1), 203–210. https://doi.org/10.1016/j.carbpol.2010.11.024

Goel, C., Semwal, A. D., Khan, A., Kumar, S., & Sharma, G. K. (2020). Physical modification of starch: Changes in glycemic index, starch fractions, physicochemical and functional properties of heat-moisture treated buckwheat starch. *Journal of Food Science and Technology, 57*, 2941–2948. https://doi.org/10.1007/s13197-020-04326-4

Gou, M., Wu, H., Saleh, A. S. M., Jing, L., Liu, Y., Zhao, K., Su, C., Zhang, B., Jiang, H., & Li, W. (2019). Effects of repeated and continuous dry heat treatments on properties of sweet potato starch. *International Journal of Biological Macromolecules, 129*, 869–877. https://doi.org/10.1016/j.ijbiomac.2019.01.225

Gray, J. A., & BeMiller, J. N. (2003). Bread staling: Molecular basis and control. *Comprehensive Reviews in Food Science & Food Safety, 2*(1), 1–21. https://doi.org/10.1111/j.1541-4337.2003.tb00011.x

Gryszkin, A., Zieba, T., Kapelko, M., & Buczek, A. (2014). Effect of thermal modifications of potato starch on its selected properties. *Food Hydrocolloids, 40*, 122–127. http://dx.doi.org/10.1016/j. foodhyd.2014.02.010

Gul, K., Mir, N. A., Yousuf, B., Allai, F. M., & Sharma, S. (2021). Starch: An overview. In A. Gani & B. A. Ashwar (Eds.), *Food biopolymers: Structural, functional and nutraceutical properties*. Springer. https://doi. org/10.1007/978-3-030-27061-2_1

Guo, K., Lin, L., Fan, X., Zhang, L., & Wei, C. (2018). Comparison of structural and functional properties of starches from five fruit kernels. *Food Chemistry, 257*, 75–82. https://doi.org/10.1016/j.foodchem.2018.03.004

Hameed, M., Ahmad, S. W., Ahmad, S., Qutab, H. G., Dasih, M., & Imran, M. (2020). Enzymatic extraction of potato starch: A parametric optimization study using response surface methodology. *Polish Journal of Chemical Technology, 22*(3), 48–54. https://doi.org/10.2478/pjct-2020-0027

Han, J.-A., & BeMiller, J. N. (2007). Preparation and physical characteristics of slowly digesting modified food starches. *Carbohydrate Polymers, 67*(3), 366–374. https://doi.org/10.1016/j.carbpol.2006.06.011

Han, L., Cao, S., Yu, Y., Xu, X., Cao, X., & Chen, W. (2021). Modification in physicochemical, structural and digestive properties of pea starch during heat-moisture process assisted by pre- and post-treatment of ultrasound. *Food Chemistry, 360*, 129929. https://doi.org/10.1016/j.foodchem.2021.129929

Han, X. Z., Benmoussa, M., Gray, J. A., BeMiller, J. N., & Hamaker, B. R. (2005). Detection of proteins in starch granule channels. *Cereal Chemistry, 82*(4), 351–355. https://doi.org/10.1094/CC-82-0351

Hartesi, B., Sriwidodo, M., Abdassah, A., & Chaerunisaa, Y. (2016). Starch as pharmaceutical excipient. *International Journal Pharmaceutical Science Review and Research, 41*(2), 59–64.

Hoover, R. (2001). Composition, molecular structure, and physicochemical properties of tuber and root starches: A review. *Carbohydrate Polymers, 45*(3), 253–267. http://dx.doi.org/10.1016/S0144-8617(00)00260-5

Hoover, R., Hughes, T., Chung, H. J., & Liu, Q. (2010). Composition, molecular structure, properties, and modification of pulse starches: A review. *Food Research International, 43*(2), 399–413. https://doi.org/10.1016/j.foodres.2009.09.001

Hu, A., Jiao, S., Zheng, J., Li, L., Fan, Y., Chen, L., & Zhang, Z. (2015). Ultrasonic frequency effect on corn starch and its cavitation. *LWT–Food Science & Technology, 60*, 941–947. https://doi.org/10.1016/j.lwt.2014.10.048

Hu, P., Fan, X., Lin, L., Wang, J., Zhang, L., & Wei, C. (2017b). Effects of surface proteins and lipids on molecular structure, thermal properties, and enzymatic hydrolysis of rice starch. *Food Science & Technology, 38*(1), 84–90. https://doi.org/10.1590/1678-457X.35016

Hu, X. P., Zhang, B., Jin, Z. Y., Xu, X. M., & Chen, H. Q. (2017a). Effect of high hydrostatic pressure and retrogradation treatments on structural and physicochemical properties of waxy wheat starch. *Food Chemistry, 232*, 560–565. https://doi.org/10.1016/j.foodchem.2017.04.040

Jan, K. N., Panesar, P. S., Rana, J. C., & Singh, S. (2017). Structural, thermal and rheological properties of starches isolated from Indian quinoa varieties. *International Journal of Biological Macromolecules, 102*, 315–322. https://doi.org/10.1016/j.ijbiomac.2017.04.027

Jane, J., Chen, Y. Y., Lee, L. F., McPherson, A. E., Wong, K. S., Radosavljevic, M., & Kasemsuwan, T. (1999). Effects of amylopectin branch chain length and amylose content on the gelatinization and pasting properties of starch. *Cereal Chemistry, 76*(5), 629–637. https://doi.org/10.1094/CCHEM.1999.76.5.629

Jia, R., Cui, C., Gao, L., Qin, Y., Ji, N., Dai, L., Wang, Y., Xiong, L., Shi, R., & Sun, Q. (2023). A review of starch swelling behavior: Its mechanism, determination methods, influencing factors, and influence on food quality. *Carbohydrate Polymers, 321*, 121260. https://doi.org/10.1016/j.carbpol.2023.121260

Jia, S., Yu, B., Zhao, H., Tao, H., Liu, P., & Cui, B. (2021). Physicochemical properties and in vitro digestibility of dual-modified starch by cross-linking and annealing. *Starch, 74*(1–2), 2100102. https://doi.org/10.1002/star.202100102

Jiang, H., Campbell, M., Blanco, M., & Jane, J.-L. (2010). Characterization of maize amylose-extender (*ae*) mutant starches: Part II. Structures and properties of starch residues remaining after enzymatic hydrolysis at boiling-water temperature. *Carbohydrate Polymers, 80*(1), 1–12. https://doi.org/10.1016/j.carbpol.2009.10.060

Jiang, H., Horner, H. T., Pepper, T. M., Blanco, M., Campbell, M., & Jane, J.-L. (2010). Formation of elongated starch granules in high-amylose maize. *Carbohydrate Polymers, 80*(2), 533–538. https://doi.org/10.1016/j.carbpol.2009.12.016

Jiang, H., Jane, J.-L., Acevedo, D., Green, A., Shinn, G., Schrenker, D., Srichuwong, S., Campbell, M., & Wu, Y. (2010). Variations in starch physicochemical properties from a generation-means analysis study using amylomaize V and VII parents. *Journal of Agricultural & Food Chemistry, 58*(9), 5633–5639. https://doi.org/10.1021/jf904531d

Jiménez, A., Fabra, M. J., Talens, P., & Chiralt, A. (2012). Edible and biodegradable starch films: A review. *Food & Bioprocess Technology, 5*(6), 2058–2076. http://dx.doi.org/10.1007/s11947-012-0835-4

Jongsutjarittam, N., & Charoenrein, S. (2013). Influence of waxy rice flour substitution for wheat flour on characteristics of batter and freeze–thawed cake. *Carbohydrate Polymers, 97*(12), 306–314. https://doi.org/10.1016/j.carbpol.2013.04.087

Kasemsuwan, T., & Jane, J.-L. (1996). Quantitative method for the survey of starch phosphate derivatives and starch phospholipids by P-31 nuclear magnetic resonance spectroscopy. *Cereal Chemistry, 73*(6), 702–707.

Kaur, L., Dhull, S. B., Kumar, P., & Singh, A. (2020). Banana starch: Properties, description, and modified variations-A review. *International Journal of Biological Macromolecules, 165*, 2096–2102. https://doi.org/10.1016/j.ijbiomac.2020.10.058

Khawas, P., & Deka, S. C. (2017). Effect of modified resistant starch of culinary banana on physicochemical, functional, morphological, diffraction, and thermal properties. *International Journal of Food Properties, 20*(1), 133–150. https://doi.org/10.1080/10942912.2016.1147459

Kim, H. Y., Park, S. S., & Lim, S. T. (2015). Preparation, characterization and utilization of starch nanoparticles. *Colloids & Surfaces B: Biointerfaces, 126*, 607–620. https://doi.org/10.1016/j.colsurfb.2014.11.011

Kim, J., Ren, C., & Shin, M. (2013). Physicochemical properties of starch isolated from eight different varieties of Korean sweet potatoes. *Starch, 65*(11–12), 923–930. https://doi.org/10.1002/star.201200217

Kringel, D. H., El Halal, S. L. M., Zavareze, E. R., & Dias, A. R. G. (2020). Methods for the extraction of roots, tubers, pulses, pseudocereals, and other unconventional starches sources: A review. *Starch, 72*(11–12), 1900234. https://doi.org/10.1002/star.201900234

Kristo, E., & Biliaderis, C. G. (2007). Physical properties of starch nanocrystal-reinforced pullulan films. *Carbohydrate Polymers, 68*(1), 146–158. https://doi.org/10.1016/j.carbpol.2006.07.021

Lan, X., Wu, Q., Yang, D., Lin, J., Xu, S., Wu, J., Wang, Z., & Wang, S. (2020). Effect of amorphization methods on the properties and structures of potato starch-monoglyceride complex. *Starch, 72*(1–2), 1900138. https://doi.org/10.1002/star.201900138

Lawrie, R. A., & Ledward, D. (2014). *Lawrie's meat science* (7th ed.). Woodhead Publishing Limited.

Leal-Castañeda, E. J., García-Tejeda, Y., Hernández-Sánchez, H., Alamilla-Beltrán, L., Téllez-Medina, D. I., Calderón-Domínguez, G., García, H. S., & Gutiérrez-López, G. F. (2018). Pickering emulsions stabilized with native and lauroylated amaranth starch. *Food Hydrocolloids, 80*, 177–185. https://doi.org/10.1016/j.foodhyd.2018.01.043

Lehmann, U., & Robin, F. (2007). Slowly digestible starch—its structure and health implications: A review. *Trends in Food Science & Technology, 18*(7), 346–355. http://dx.doi.org/10.1016/j.tifs.2007.02.009

Li, H., Ho, V. T. T., Turner, M. S., & Dhital, S. (2016). Encapsulation of *Lactobacillus plantarum* in porous maize starch. *LWT–Food Science & Technology, 74*, 542–549. https://doi.org/10.1016/j.lwt.2016.08.019

Li, J., & Wu, Y. (2014). Lubricants in pharmaceutical solid dosage forms. *Lubricants, 2*(1), 21–43. https://doi.org/10.3390/lubricants2010021

Li, L., Jiang, H., Campbell, M., Blanco, M., & Jane, J.-L. (2008). Characterization of maize amylose-extender (*ae*) mutant starches. Part I: Relationship between resistant starch contents and molecular structures. *Carbohydrate Polymers, 74*(3), 396–404. https://doi.org/10.1016/j.carbpol.2008.03.012

Li, L., Yuan, T. Z., Setia, R., Raja, R. B., Zhang, B., & Ai, Y. (2019). Characteristics of pea, lentil and faba bean starches isolated from air-classified flours in comparison with commercial starches. *Food Chemistry, 276*, 599–607. https://doi.org/10.1016/j.foodchem.2018.10.064

Li, T., An, F., Teng, H., Huang, Q., Zeng, F., & Song, H. (2018). Comparison of structural features and in vitro digestibility of purple yam (*Dioscorea alata* L.) resistant starches by autoclaving and multi-enzyme hydrolysis. *Food Science & Biotechnology, 27*, 27–36. https://doi.org/10.1007/s10068-017-0206-z

Li, W., Gao, J., Saleh, A. S. M., Tian, X., Wang, P., Jiang, H., & Zhang, G. (2018). The modifications in physicochemical and functional properties of Proso millet starch after ultra-high pressure (UHP) process. *Starch, 70*, 5–6. https://doi.org/10.1002/star.201700235

Li, W., Gao, J., Wu, G., Zheng, J., Ouyang, S., Luo, Q., & Zhang, G. (2016). Physicochemical and structural properties of A- and B-starch isolated from normal and waxy wheat: Effects of lipids removal. *Food Hydrocolloids, 60*, 364–373. https://doi.org/10.1016/j.foodhyd.2016.04.011

Li, W., Tian, X., Wang, P., Saleh, A. S., Luo, Q., Zheng, J., Ouyang, S., & Zhang, G. (2016). Recrystallization characteristics of high hydrostatic pressure gelatinized normal and waxy corn starch. *International Journal of Biological Macromolecules, 83*, 171–177. https://doi.org/10.1016/j.ijbiomac.2015.11.057

Li, W., Wu, G., Luo, Q., Jiang, H., Zheng, J., Ouyang, S., & Zhang, G. (2016). Effects of removal of surface proteins on physicochemical and structural properties of A- and B-starch isolated from normal and waxy wheat. *Journal of Food Science & Technology, 53*(6), 2673–2685. https://doi.org/10.1007/s13197-016-2239-3

Li, X., Wang, C., Lu, F., Zhang, L., Yang, Q., & Li, X. (2015b). Effect of bonding forces on corn starch isolation. *Cereal Chemistry, 92*(4), 418–425. https://doi.org/10.1094/CCHEM-05-14-0092-R

Li, X., Wang, C., Lu, F., Zhang, L., Yang, Q., Mu, J., & Li, X. (2015a). Physicochemical properties of corn starch isolated by acid liquid and L-cysteine. *Food Hydrocolloids, 44*, 353–359. https://doi.org/10.1016/j.foodhyd.2014.09.003

Li, Z., Jiang, H., Xu, C., & Gu, L. (2015). A review: Using nanoparticles to enhance absorption and bioavailability of phenolic phytochemicals. *Food Hydrocolloids, 43*, 153–164. https://doi.org/10.1016/j.foodhyd.2014.05.010

Liang, X., & King, J. M. (2003). Pasting and crystalline property differences of commercial and isolated rice starch with added amino acids. *Journal of Food Science, 68*(3), 832–838. https://doi.org/10.1111/j.1365-2621.2003.tb08251.x

Lin, Q., Xiao, H., Zhao, J., Li, L., & Yu, F. (2009). Characterization of the pasting, flow and rheological properties of native and phosphorylated rice starches. *Starch, 61*(12), 709–715. https://doi.org/10.1002/star.200900184

Lindeboom, N., Chung, P. R., & Tyler, R. T. (2004). Analytical, biochemical and physicochemical aspects of starch granule size, with emphasis on small granule starches: A review. *Starch, 56*(3–4), 89–99. https://doi.org/10.1002/star.200300218

Liu, G., Gu, Z., Hong, Y., Cheng, L., & Li, C. (2017). Structure, functionality and applications of debranched starch: A review. *Trends in Food Science & Technology, 63*, 70–79. https://doi.org/10.1016/j.tifs.2017.03.004

Liu, P., Kang, X., Cui, B., Gao, W., Wu, Z., & Yu, B. (2019b). Effects of amylose content and enzymatic debranching on the properties of maize starch-glycerol monolaurate complexes. *Carbohydrate Polymers, 222*, 115000. https://doi.org/10.1016/j.carbpol.2019.115000

Liu, P., Kang, X., Cui, B., Wang, R., & Wu, Z. (2019a). Effects of glycerides with different molecular structures on the properties of maize starch and its film forming capacity. *Industrial Crops & Products, 129*, 512–517. https://doi.org/10.1016/j.indcrop.2018.12.039

Liu, Y., & Ng, P. K. W. (2016). Relationship between bran characteristics and bran starch of selected soft wheats grown in Michigan. *Food Chemistry, 197*(Part A), 427–435. https://doi.org/10.1016/j.foodchem.2015.10.112

López-Córdoba, A., Medina-Jaramillo, C., Piñeros-Hernandez, D., & Goyanes, S. (2017). Cassava starch films containing rosemary nanoparticles produced by solvent displacement method. *Food Hydrocolloids, 71*, 26–34. https://doi.org/10.1016/j.foodhyd.2017.04.028

Loubes, M. A., Calzetta Resio, A. N., Tolaba, M. P., & Suarez, C. (2012). Mechanical and thermal characteristics of amaranth starch isolated by acid wet-milling procedure. *LWT–Food Science & Technology, 46*(2), 519–524. https://doi.org/10.1016/j.lwt.2011.11.015

Lv, Q. Q., Li, G. Y., Xie, Q. T., Zhang, B., Li, X. M., Pan, Y., & Chen, H. Q. (2018). Evaluation studies on the combined effect of hydrothermal treatment and octenyl succinylation on the physic-chemical, structural and digestibility characteristics of sweet potato starch. *Food Chemistry, 256*, 413–418. https://doi.org/10.1016/j.foodchem.2018.02.147

Majid, I., Nayik, G. A., Dar, S. M., & Nanda, V. (2018). Novel food packaging technologies: Innovations and future prospective. *Journal of the Saudi Society of Agricultural Sciences, 17*(4), 454–462. https://doi.org/10.1016/j.jssas.2016.11.003

Mandala, L. G. (2012). Viscoelastic properties of starch and non-starch thickeners in simple mixtures or model food. In J. de Vicente (Ed.), *Viscoelasticity—from theory to biological applications* (pp. 217–236). IntechOpen. http://dx.doi.org/10.5772/50221

Martínez, M. M., & Gómez, M. (2016). Starch-active debranching and α-glucanotransferase enzymes. In R. C. Ray & Rosell, C. M. (Eds.), *Microbial enzyme technology in food applications*. CRC Press, Taylor and Francis Group. https://doi.org/10.1201/9781315368405-5

Mason, W. R. (2009). Starch use in foods. In J. Be Miller & R. Whistler (Eds.), *Starch: Chemistry and technology* (3rd ed., pp. 745–795). https://doi.org/10.1016/B978-0-12-746275-2.00020-3

Molavi, H., Razavi, S. M. A., & Farhoosh, R. (2018). Impact of hydrothermal modifications on the physicochemical, morphology, crystallinity, pasting and thermal properties of acorn starch. *Food Chemistry, 245*, 385–393. https://doi.org/10.1016/j.foodchem.2017.10.117

Morrison, W. R. (1995). Starch lipids and how they relate to starch granule structure and functionality. *Cereal Foods World, 40*, 437–446.

Mukurumbira, A. R., Mellem, J. J., & Amonsou, E. O. (2017). Effects of amadumbe starch nanocrystals on the physicochemical properties of starch biocomposite films. *Carbohydrate Polymers, 165*, 142–148. https://doi.org/10.1016/j.carbpol.2017.02.041

Nattapulwat, N., Purkkao, N., & Suwithayapanth, O. (2008). Evaluation of native and carboxymethyl yam (*Dioscorea esculenta*) starches as tablet disintegrants. *Silpakorn University Science & Technology Journal, 2*(2), 18–25. https://doi.org/10.14456/sustj.2008.7

Navarro, A. S., Martino, M. N., & Zaritzky, N. E. (1996). Modelling of rheological behaviour in starch-lipid systems. *Food Science and Technology, 29*(7), 632–639. https://doi.org/10.1006/fstl.1996.0096

Obadi, M., Qi, Y., & Xu, B. (2023). High-amylose maize starch: Structure, properties, modifications and industrial applications. *Carbohydrate Polymers, 299*, 120185. https://doi.org/10.1016/j.carbpol.2022.120185

Oderinde, A. A., Ibikunle, A. A., Bakre, L. G., & Babarinde, N. A. A. (2020). Modification of African breadfruit (*Treculia africana*, Decne) kernel starch: Physicochemical, morphological, pasting, and thermal properties. *International Journal of Biological Macromolecules, 153*, 79–87. https://doi.org/10.1016/j.ijbiomac.2020.02.293

Ogori, A. F., & Taofeek, A. (2022). Starch chemistry and application. *Journal of Food Technology & Preservation, 6*(6), 130. https://doi.org/10.35841/2591-796X-6.6.130

Ogunsona, E., Ojogbo, E., & Mekonnen, T. (2018). Advanced material applications of starch and its derivatives. *European Polymer Journal, 108*, 570–581. https://doi.org/10.1016/j.eurpolymj.2018.09.039.

Othman, Z., Hassan, O., & Hashim, K. (2015). Physicochemical and thermal properties of gamma-irradiated sago (*Metroxylon sagu*) starch. *Radiation Physics Chemistry, 109*, 48–53. https://doi.org/10.1016/j.radphyschem.2014.12.003

Oyeyinka, S. A., & Oyeyinka, A. T. (2018). A review on isolation, composition, physicochemical properties and modification of Bambara groundnut starch. *Food Hydrocolloids, 75*, 62–71. https://doi.org/10.1016/j.foodhyd.2017.09.012

Pagar, R., Pagar, Y., Pachorkar, P., Nikam, R., & Chaudhari, V. (2022). Starch: Natural sources and its pharmaceutical applications. *International Journal of Pharmaceutical Research & Applications, 7*(1), 545–554. https://doi.org/10.35629/7781-0701545554

Paraginski, R. T., Colussi, R., Dias, A. G., Zavareze, E., Elias, M. C., & Vanier, N. L. (2019). Physicochemical, pasting, crystallinity, and morphological properties of starches isolated from maize kernels exhibiting different types of defects. *Food Chemistry, 274*, 330–336. https://doi.org/10.1016/j.foodchem.2018.09.026

Park, I.-M., Ibáñez, A. M., Zhong, F., & Shoemaker, C. F. (2007). Gelatinization and pasting properties of waxy and non-waxy rice starches. *Starch, 59*(8), 388–396. https://doi.org/10.1002/star.200600570

Park, J. H., Kim, D. C., Lee, S. E., Kim, O. W., Kim, H., Lim, S. T., & Kim, S. S. (2014). Effects of rice flour size fractions on gluten free rice bread. *Food Science & Biotechnology, 23*(6), 1875–1883. https://doi.org/10.1007/s10068-014-0256-4

Pascari, X., Marín, S., Ramos, A. J., Molino, F., & Sanchis, V. (2019). Deoxynivalenol in cereal-based baby food production process: A review. *Food Control, 99*, 11–20. https://doi.org/10.1016/j.foodcont.2018.12.014

Patil, B. S., Soodam, S. R., Kulkarni, U., & Korwar, P. G. (2010). Evaluation of Moringa oleifera gum as a binder in tablet formulation. *International Journal of Research in Ayurveda & Pharmacy, 1*(2), 590–596.

Patiño-Rodríguez, O., Agama-Acevedo, E., Ramos-Lopez, G., & Bello-Pérez, L. A. (2020). Unripe mango kernel starch: Partial characterization. *Food Hydrocolloids, 101*, 105512. https://doi.org/10.1016/j.foodhyd.2019.105512

Pauly, A., Pareyt, B., De Brier, N., Fierens, E., & Delcour, J. A. (2012). Starch Isolation method impacts soft wheat (*Triticum aestivum* L. cv. Claire) starch puroindoline and lipid levels as well as its functional properties. *Journal of Cereal Science, 56*(2), 464–469. https://doi.org/10.1016/j.jcs.2012.06.003

Pereira, D. G., & Del Pino Beleia, A. (2021). Characterization of acid-thinned cassava starch and its technological properties in sugar solution. *LWT–Food Science & Technology, 151*, 112151. https://doi.org/10.1016/j.lwt.2021.112151

Pinyo, U., Luangpituksa, P., Suphantharika, M., Hansawasdi, C., & Wongsagonsu, R. (2017). Improvement of sago starch extraction process using various pretreatment techniques and their pretreatment combination. *Starch, 69*(9–10), 1700005. https://doi.org/10.1002/star.201700005

Puchongkavarin, H., Varavinit, S., & Bergthaller, W. (2005). Comparative study of pilot scale rice starch production by an alkaline and enzymatic process. *Starch, 57*(3–4), 134–144. https://doi.org/10.1002/star.200400279

Quezada-Calvillo, R., Robayo-Torres, C. C., Ao, Z., Hamaker, B. R., Quaroni, A., Brayer, G. D., & Nichols, B. L. (2007). Luminal substrate "brake" on mucosal maltase-glucoamylase activity regulates total rate of starch digestion to glucose. *Journal of Pediatric Gastroenterology & Nutrition, 45*(1), 32–43. https://doi.org/10.1097/mpg.0b013e31804216fc

Quodbach, J., & Kleinebudde, P. (2015). A critical review on tablet disintegration. *Pharmaceutical Development & Technology, 21*(6), 763–774. https://doi.org/10.3109/10837450.2015.1045618

Reis, A. V., Guilherme, M. R., Moia, T. A., Mattoso, L. H. C., Muniz, E. C., & Tambourgi, E. B. (2008). Synthesis and characterization of a starch-modified hydrogel as potential carrier for drug delivery system. *Journal of Polymer Science. Part A: Polymer Chemistry, 46*(7), 2567–2574. https://doi.org/10.1002/pola.22588

Roopa, S., & Premavalli, K. S. (2008). Effect of processing on starch fractions in different varieties of finger millet. *Food Chemistry, 106*(3), 875–882. https://doi.org/10.1016/j.foodchem.2006.08.035

Samsudin, H., & Hani, N. M. (2018). Use of starch in food packaging. In M. A. Villar, S. E. Barbosa, M. A. Gracía, L. A. Castillo, & O. V. López (Eds.), *Starch-based materials in food packaging* (Ch. 8, pp. 229–256). Academic Press.

Sandhu, K. S., Singh, N., & Singh Malhi, N. (2007). Some properties of corn grains and their flours I: Physicochemical, functional and chapatti–making properties of flours. *Food Chemistry, 101*(3), 938–946. https://doi.org/10.1016/j.foodchem.2006.02.040

Santander-Ortega, M. J., Stauner, T., Loretz, B., Ortega-Vinuesa, J. L., Bastos-González, D., Wenz, G., Schaefer, U. F., & Lehr, C. M. (2010). Nanoparticles made from novel starch derivatives for transdermal drug delivery. *Journal of Controlled Release, 141*(1), 85–92. https://doi.org/10.1016/j.jconrel.2009.08.012

Sarker, M. Z. I., Elgadir, M. A., Ferdosh, S., Akanda, M. J. H., Aditiawati, P., & Noda, T. (2013). Rheological behavior of starch-based biopolymer mixtures in selected processed foods. *Starch, 65*(1–2), 73–81. http://dx.doi.org/10.1002/star.201200072

Sasaki, T., Yasui, T., & Matsuki, J. (2000). Effect of amylose content on gelatinization, retrogradation, and pasting properties of starches from waxy and nonwaxy wheat and their seeds. *Cereal Chemistry, 77*(1), 58–63. https://doi.org/10.1094/CCHEM.2000.77.1.58

Schirmer, M., Höchstötter, A., Jekle, M., Arendt, E., & Becker, T. (2013). Physicochemical and morphological characterization of different starches with variable amylose/amylopectin ratio. *Food Hydrocolloids, 32*(1), 52–63. https://doi.org/10.1016/j.foodhyd.2012.11.032

Schirmer, M., Jekle, M., & Becker, T. (2015). Starch gelatinization and its complexity for analysis. *Starch, 67*(1–2), 30–41. http://dx.doi.org/10.1002/star.201400071

Sevenou, O., Hill, S. E., Farhat, I. A., & Mitchell, J. R. (2002). Organisation of the external region of the starch granule as determined by infrared spectroscopy. *International Journal of Biological Macromolecules, 31*(1–3), 79–85. https://doi.org/10.1016/S0141-8130(02)00067-3

Shah, U., Naqash, F., Gani, A., & Masoodi, F. A. (2016). Art and science behind modified starch edible films and coatings: A review. *Comprehensive Reviews in Food Science & Food Safety, 15*(3), 568–580. https://doi.org/10.1111/1541-4337.12197

Shao, Y., Mao, L., Guan, W., Wei, X., Yang, Y., Xu, F., Li, Y., & Jiang, Q. (2020). Physicochemical and structural properties of low-amylose Chinese yam (*Dioscorea opposita* Thunb.) starches. *International Journal of Biological Macromolecules, 164*, 427–433. https://doi.org/10.1016/j.ijbiomac.2020.07.054

Shevkani, K., Sing, N., Bajaj, R., & Kaur, A. (2017). Wheat starch production, structure, functionality and application: A review. *International Journal of Food Science & Technology*, *52*(1), 38–58. https://doi.org/10.1111/ijfs.13266

Sindhu, R., Devi, A., & Khatkar, B. S. (2019). Physicochemical, thermal and structural properties of heat moisture treated common buckwheat starches. *Journal of Food Science & Technology*, *56*(5), 2480–2489. https://doi.org/10.1007/s13197-019-03725-6

Sindhu, R., Devi, A., & Khatkar, B. S. (2021). Morphology, structure and functionality of acetylated, oxidized and heat moisture treated amaranth starches. *Food Hydrocolloids*, *118*, 106800. https://doi.org/10.1016/j.foodhyd.2021.106800

Singh, J., Dartois, A., & Kaur, L. (2010). Starch digestibility in food matrix: A review. *Trends in Food Science & Technology*, *21*(4), 168–180. https://doi.org/10.1016/j.tifs.2009.12.001

Singh, J., McCarthy, O. J., & Singh, H. (2006). Physico-chemical and morphological characteristics of New Zealand Taewa (*Maori potato*) starches. *Carbohydrate Polymers*, *64*(4), 569–581.https://doi.org/10.1016/j.carbpol.2005.11.013

Singh, J., & Singh, N. (2001). Studies on the morphological, thermal and rheological properties of starch separated from some Indian potato cultivars. *Food Chemistry*, *75*(1), 67–77. https://doi.org/10.1016/S0308-8146(01)00189-3

Singh, N., Isoro, N., Srichuwang, S., Noda, T., & Nishinan, K. (2008). Structural, thermal and viscoelastic properties of potato starches, *Food Hydrocolloids*, *22*(6), 979–988. https://doi.org/10.1016/j.foodhyd.2007.05.010

Singh, N., Singh, J., Kaur, L., Sodhi, N. S., & Gill, B. S. (2003). Morphological, thermal and rheological properties of starches from different botanical sources. *Food Chemistry*, *81*(2), 219–231. http://dx.doi.org/10.1016/S0308–8146(02)00416–8.

Siroha, A. K., Sandhu, K. S., Kaur, M., & Kaur, V. (2019). Physicochemical, rheological, morphological and in vitro digestibility properties of pearl millet starch modified at varying levels of acetylation. *International Journal of Biological Macromolecules*, *131*, 1077–1083. https://doi.org/10.1016/j.ijbiomac.2019.03.179

Smith, A. M. (2001). The biosynthesis of the starch granule. *Biomacromolecules*, *2*(2), 335–341. https://doi.org/10.1021/bm000133c

Sodhi, N. S., & Singh, N. (2003). Morphological, thermal and rheological properties of starches separated from rice cultivars grown in India. *Food Chemistry*, *80*(1), 99–108. https://doi.org/10.1016/S0308-8146(02)00246-7

Su, C., Saleh, A. S. M., Zhang, B., Zhao, K., Ge, X., Zhang, Q., & Li, W. (2020). Changes in structural, physico-chemical, and digestive properties of normal and waxy wheat starch during repeated and continuous annealing. *Carbohydrate Polymers*, *247*, 116675. https://doi.org/10.1016/j.carbpol.2020.116675

Šubarić, D., Ačkar, D., Babić, J., Sakač, N., & Jozinović, A. (2014). Modification of wheat starch with succinic acid/acetic anhydride and azelaic acid/acetic anhydride mixtures I. Thermophysical and pasting properties. *Journal of Food Science & Technology*, *51*, 2616–2623. http://dx.doi.org/10.1007/s13197-012-0790-0

Sudheesh, C., Sunooj, K. V., Navaf, M., Bhasha, S. A., George, J., Mounir, S., Kumar, S., & Sajeevkumar, V. A. (2020). Hydrothermal modifications of nonconventional Kithul (*Caryota urens*) starch: Physico-chemical, rheological properties and in vitro digestibility. *Journal of Food Science & Technology*, *57*, 2916–2925. https://doi.org/10.1007/s13197-020-04323-7

Sun, L., Xu, Z., Song, L., Ma, M., Zhang, C., Chen, X., Xu, X., Sui, Z., & Corke, H. (2021). Removal of starch granule associated proteins alters the physicochemical properties of annealed rice starches. *International Journal of Biological Macromolecules*, *185*, 412–418. https://doi.org/10.1016/j.ijbiomac.2021.06.082

Suri, S., & Singh, A. (2023). Modification of starch by novel and traditional ways: Influence on the structure and functional properties. *Sustainable Food Technology*, *1*, 348. https://doi.org/10.1039/d2fb00043a

Szepes, A., Makai, Z., Blümer, C. M., Mäder, K., Kása, P., Jr., & Szabó-Révész, P. (2008). Characterization and drug delivery behaviour of starch-based hydrogels prepared via isostatic ultrahigh pressure. *Carbohydrate Polymers*, *72*, 571–578. https://doi.org/10.1016/j.carbpol.2007.09.028

Taggart, P. (2004). Starch as an ingredient: Manufacture and applications. In A.-C. Eliasson (Ed.), *Starch in food: Structure, function and applications* (pp. 363–392). Woodhead Publishing Series in Food Science, Technology and Nutrition. CRC Press. https://doi.org/10.1533/9781855739093.3.363

Taherian, A. R., Fustier, P., & Ramaswamy, H. S (2007). Effect of added weighting agent and xanthan gum on stability and rheological properties of beverage emulsions formulated using modified starch. *Journal of Food Process Engineering*, *30*, 204–224. https://doi.org/10.1111/j.1745-4530.2007.00109.x

Tesch, S., Gerhards, C., & Schubert, H. (2002). Stabilization of emulsions by OSA starches *Journal of Food Engineering*, *54*(2), 167–174. https://doi.org/10.1016/S0260-8774(01)00206-0

Tester, R. F., & Karkalas, J. (2002). Starch. In A. Steinbüchel (Series Ed.), E. J. Vandamme, S. De Baets, & A. Steinbüchel (Vol. Eds.), *Biopolymers (Vol. 6), polysaccharides. II. Polysaccharides from eukaryotes* (pp. 381–438). Wiley-VCH.

Tester, R. F., Karkalas, J., & Qi, X. (2004). Starch-composition, fine structure and architecture. *Journal of Cereal Science*, *39*(2), 151–165. https://doi.org/10.1016/j.jcs.2003.12.001

Trung, P. T. B., Ngoc, L. B. B., Hoa, P. N., Tien, N. N. T., & Hung, P. V. (2017). Impact of heat moisture and annealing treatments on physicochemical properties and digestibility of starches from different colored sweet potato varieties. *International Journal of Biological Macromolecules*, *105*, 1071–1078. https://doi.org/10.1016/j.ijbiomac.2017.07.131

Tsai, M-L., Li, C-F., & Lii, C-Y. (1997). Effects of granular structure on the pasting behavior of starches. *Cereal Chemistry*, *74*(6), 750–757. https://doi.org/10.1094/CCHEM.1997.74.6.750

Uhumwangho, M. U., Okor, R. S., Eichie, F. E., & Abbah, C. M. (2006). Influence of some starch binders on the brittle fracture of paracetamol tables. *African Journal of Biotechnology*, *5*(20), 1950–1953.

Uriarte-Aceves, P. M., Milán-Carrillo, J., Cuevas-Rodríguez, E. O., Gutierrez-Dorado, R., Reyes-Moreno, C., & Milán-Noris, E. M. (2018). In vitro digestion properties of native isolated starches from Mexican blue maize (*Zeamays* L.) landrace. *LWT–Food Science & Technology*, *93*, 384–389. https://doi.org/10.1016/j.lwt.2018.03.015

Vamadevan, V., & Bertoft, E. (2020). Observations on the impact of amylopectin and amylose structure on the swelling of starch granules. *Food Hydrocolloids*, *103*, 105663. https://doi.org/10.1016/j.foodhyd.2020.105663

Van Hung, P., & Vo, T. N. D. (2017). Structure, physicochemical characteristics, and functional properties of starches isolated from yellow (*Curcuma longa*) and black (*Curcuma caesia*) turmeric rhizomes. *Starch*, *69*(5–6), 1600285. https://doi.org/10.1002/star.201600285

Villarreal, M. E., Ribotta, P. D., & Iturriaga, L. B. (2013). Comparing methods for extracting amaranthus starch and the properties of the isolated starches. *LWT–Food Science & Technology*, *51*(2), 441–447. https://doi.org/10.1016/j.lwt.2012.11.009

Wang, H., Lv, J., Jiang, S., Niu, B., Pang, M., & Jiang, S. (2016). Preparation and characterization of porous corn starch and its adsorption toward grape seed proanthocyanidins. *Starch*, *68*(11–12), 1254–1263. https://doi.org/10.1002/star.201600009

Wang, J., Zhao, S., Min, G., Qiao, D., Zhang, B., Niu, M., Jia, C., Xu, Y., & Lin, Q. (2021). Starch-protein interplay varies the multi-scale structures of starch undergoing thermal processing. *International Journal of Biological Macromolecules*, *175*, 179–187. https://doi.org/10.1016/j.ijbiomac.2021.02.020

Wang, L., Zhang, C., Chen, Z., Wang, X., Wang, K., Li, Y., Wang, R., Luo, X., Li, Y., & Li, J. (2018). Effect of annealing on the physico-chemical properties of rice starch and the quality of rice noodles. *Journal of Cereal Science*, *84*, 125–131. https://doi.org/10.1016/j.jcs.2018.10.004

Wang, Q., Li, L., Liu, C., & Zheng, X. (2022a). Heat-moisture modified blue wheat starch: Physicochemical properties modulated by its multi-scale structure. *Food Chemistry*, *386*, 132771. https://doi.org/10.1016/j.foodchem.2022.132771

Wang, S., Chao, C., Cai, J., Niu, B., Copeland, L., & Wang, S. (2020). Starch-lipid and starch-lipid-protein complexes: A comprehensive review. *Comprehensive Reviews in Food Science & Food Safety*, *19*(3), 1056–1079. https://doi.org/10.1111/1541-4337.12550

Wang, W., Hu, A., Li, J., Liu, G., Wang, M., & Zheng, J. (2022c). Comparison of physicochemical properties and digestibility of sweet potato starch after two modifications of microwave alone and microwave-assisted L-malic acid. *International Journal of Biological Macromolecules*, *210*, 614–621. https://doi.org/10.1016/j.ijbiomac.2022.04.215

Wang, W., Xue, L., Dong, Y., Xia, Z., Liu, X., Chen, G., & Du, X. (2022b). Application of multistage induced electric field for acid hydrolysis of starch in a continuous-flow reactor. *International Journal of Biological Macromolecules*, *221*, 703–713. https://doi.org/10.1016/j.ijbiomac.2022.09.057

Wani, A. A., Singh, P., Shah, M. A., Schweiggert-Weisz, U., Gul, K., & Wani, I. A. (2012). Rice starch diversity: Effects on structural, morphological, thermal and physicochemical properties: A review. *Comprehensive Reviews in Food Science & Food Safety*, *11*(5), 417–436. https://doi.org/10.1111/j.1541-4337.2012.00193.x

Wani, I. A., Sogi, D. S., Hamdani, A. M., Gani, A., Bhat, N. A., & Shah, A. (2016). Isolation, composition, and physicochemical properties of starch from legumes: A review. *Starch*, *68*(9–10), 834–845. https://doi.org/10.1002/star.201600007

Waterschoot, J., Gomand, S. V., Fierens, E., & Delcour, J. A. (2015). Starch blends and their physicochemical properties. *Starch*, *67*(1–2), 1–13. http://dx.doi.org/10.1002/star.201300214

Winkler, H., Vorwerg, W., & Rihm, R. (2014). Thermal and mechanical properties of fatty acid starch esters. *Carbohydrate Polymers*, *102*, 941–949. https://doi.org/10.1016/j.carbpol.2013.10.040

Wu, Y., Chen, Z., Li, X., & Wang, Z. (2010). Retrogradation properties of high amylose rice flour and rice starch by physical modification. *LWT–Food Science & Technology*, *43*(3), 492–497. https://doi.org/10.1016/j.lwt.2009.09.017

Xu, A., Guo, K., Liu, T., Bian, X., Zhang, L., & Wei, C. (2018). Effects of different isolation media on structural and functional properties of starches from root tubers of purple, yellow and white sweet potatoes. *Molecules*, *23*(9), 2134. https://doi.org/10.3390/molecules23092135

Xu, M., Saleh, A. S. M., Gong, B., Li, B., Jing, L., Gou, M., Jiang, H., & Li, W. (2018). The effect of repeated versus continuous annealing on structural, physicochemical, and digestive properties of potato starch. *Food Research International, 111*, 324–333. https://doi.org/10.1016/j.foodres.2018.05.052

Xu, R., Zhang, J., Zhou, P., Yang, R., Feng, X., & Xu, L. (2015). A novel artificial red blood cell substitute: Grafted starch encapsulated hemoglobin. *RSC Advances, 5*(54), 43845–43853. https://doi.org/10.1039/C5RA00772K

Yang, Z., Hao, H., Wu, Y., Liu, Y., & Ouyang, J. (2021). Influence of moisture and amylose on the physicochemical properties of rice starch during heat treatment. *International Journal of Biological Macromolecules, 168*, 656–662. https://doi.org/10.1016/j.ijbiomac.2020.11.122

Yook, C., & Robyt, J. F. (2002). Reactions of alpha amylases with starch granules in aqueous suspension giving products in solution and in a minimum amount of water giving products inside the granule. *Carbohydrate Research, 337*(12), 1113–1117. https://doi.org/10.1016/S0008-6215(02)00107-6

Yu, S., Ma, Y., & Sun, D-W. (2010). Effects of freezing rates on starch retrogradation and textural properties of cooked rice during storage. *LWT–Food Science & Technology, 43*(7), 1138–1143. https://doi.org/10.1016/j.lwt.2010.03.004

Zàmostný, P., Petrů, J., & Majerovà, D. (2012). Effect of maize starch excipient properties on drug release rate. *Procedia Engineering, 42*, 482–488. https://doi.org/10.1016/j.proeng.2012.07.439

Zhan, Q., Ye, X., Zhang, Y., Kong, X., Bao, J., Corke, H., & Sui, Z. (2020). Starch granule-associated proteins affect the physicochemical properties of rice starch. *Food Hydrocolloids, 101*, 105504. https://doi.org/10.1016/j.foodhyd.2019.105504

Zhang, B., Saleh, A. S. M., Su, C., Gong, B., Zhao, K., Zhang, G., Li, W., & Yan, W. (2020). The molecular structure, morphology, and physicochemical property and digestibility of potato starch after repeated and continuous heat-moisture treatment. *Journal of Food Science, 85*(12), 4215–4224. https://doi.org/10.1111/1750-3841.15528

Zhang, B., Wu, H., Gou, M., Xu, M., Liu, Y., Jing, L., Zhao, K., Jiang, H., & Li, W. (2019). The comparison of structural, physicochemical, and digestibility properties of repeatedly and continuously annealed sweet potato starch. *Journal of Food Science, 84*(8), 2050–2058. https://doi.org/10.1111/1750-3841.14711

Zhang, B., Zhang, Q., Wu, H., Su, C., Ge, X., Shen, H., Han, L., Yu, X., & Li, W. (2021). The influence of repeated versus continuous dry-heating on the performance of wheat starch with different amylose content. *LWT–Food Science & Technology, 136*(Part 2), 110380. https://doi.org/10.1016/j.lwt.2020.110380

Zhang, H., Zhang, W., Xu, C. X., & Zhou, X. (2013). Morphological features and physicochemical properties of waxy wheat starch. *International Journal of Biological Macromolecules, 62*, 304–309. https://doi.org/10.1016/j.ijbiomac.2013.09.030

Zhang, Y.-E., & Schwartz, J. B. (2000). Effect of diluents on tablet integrity and controlled drug release. *Drug Development & Industrial Pharmacy, 26*(7), 761–765. https://doi.org/10.1081/DDC-100101295

Zhao, K., Li, B., Xu, M., Jing, L., Gou, M., Yu, Z., & Li, W. (2018). Microwave pretreated esterification improved the substitution degree, structural and physicochemical properties of potato starch esters. *LWT–Food Science & Technology, 90*, 116–123. https://doi.org/10.1016/j.lwt.2017.12.021

Zheng, M., You, Q., Lin, Y., Lan, F., Luo, M., Zeng, H., Zheng, B., & Zhang, Y. (2019). Effect of guar gum on the physicochemical properties and in vitro digestibility of lotus seed starch. *Food Chemistry, 272*, 286–291. https://doi.org/10.1016/j.foodchem.2018.08.029

Zhengyu, J. (2018). *Functional starch and applications in food.* Springer Nature. https://doi.org/10.1007/978-981-13-1077-5

Zhou, Y., Cui, L., You, X., Jiang, Z., Qu, W., Liu, P., Ma, D., & Cui, Y. (2021). Effects of repeated and continuous dry heat treatments on the physicochemical and structural properties of quinoa starch. *Food Hydrocolloids, 113*, 106532. https://doi.org/10.1016/j.foodhyd.2020.106532

Zhou, Z. X., Robards, K., Helliwell, S., & Balanchard, C. (2007). Effect of the addition of fatty acids on rice starch properties. *Food Research International, 40*(2), 209–214. https://doi.org/10.1016/j.foodres.2006.10.006

Zhu, F. (2017). Encapsulation and delivery of food ingredients using starch based systems. *Food Chemistry, 229*, 542–552. https://doi.org/10.1016/j.foodchem.2017.02.101

Zhu, F. (2019). Recent advances in modifications and applications of sago starch. *Food Hydrocolloids, 96*, 412–423. https://doi.org/10.1016/j.foodhyd.2019.05.035

Zhu, F., & Zhang, Y. (2019). Effect of konjac glucomannan on physicochemical properties of quinoa and maize starches. *Cereal Chemistry, 96*(5), 878–884. https://doi.org/10.1002/cche.10188

Heat Moisture Treatment

2

Anil Kumar Siroha, Sneh Punia Bangar,
Sukhvinder Singh Purewal, and Charan Singh

2.1 INTRODUCTION

Starch is usually modified to increase its application in the food and non-food industries due to its instability during heating, shearing, acidity, and the ensuing storage (Lin et al., 2019). The process of changing starch structure due to various factors, such as operational, processing, and environmental, is known as modification. Changes might have a favorable or detrimental impact on the structure and functionality of starch molecules. Starches are modified to optimize and elevate particular functional attributes employing physical, chemical, and enzymatic techniques (Zia-ud-Din et al., 2017). There is a growing interest in enhancing the natural starch's qualities without having to undergo chemical alterations, as the industry is currently trending towards producing more natural food components. In order to improve starch adaptability, physical methods have received significant recognition because of their low cost, safety, and effective qualities. They are also considered a green alternative because they do not require chemical reagents and can achieve specific enhanced attributes for particular applications (Punia, 2020). There are various physical modification methods. These include pre-gelatinization, which is a thermal process, and heat moisture treatment (HMT) and annealing, both categorized as hydrothermal procedures. Furthermore, non-thermal modifications include high-pressure processing, micronization, ultrasonication, and pulse electric field methods (Alcázar-Alay & Meireles, 2015). This chapter aims to discuss physicochemical, pasting, morphological, and rheological properties and applications of HMT-modified starches.

2.2 HEAT MOISTURE TREATMENT

The first study of HMT for modifying corn starch dates back to 1944 (Sair & Fetzer, 1944). HMT, a physical method, is executed under limited moisture levels (10–30%) and elevated temperatures (90–120°C), with durations from 15 minutes to 16 hours (Maache-Rezzoug et al., 2008). The HMT should be applied at temperatures above the glass transition temperature (T_g) but below the gelatinization temperature (T_{gel})

DOI: 10.1201/9781032655598-2

(Jacobs & Delcour, 1998). Per Gunaratne and Hoover (2002), HMT facilitates the interaction of polymer chains by disrupting the crystalline structure and dissociating the double-helical configuration within the amorphous region, succeeding by reconfiguring the disrupted crystals.

Irrespective of the source of starch, HMT promotes gelatinization transition temperatures, broadens the range of temperatures for gelatinization, reduces granular swelling and amylose leaching, and enhances thermal stability, as demonstrated by da Rosa Zavareze et al. (2011). The characteristics of HMT starches are significantly influenced by two primary factors: the treatment conditions and the starch origin, as highlighted by Molavi et al. (2018). Parameters such as starch-to-moisture ratio, temperature, and duration of heating must be vigilantly monitored and managed during the HMT process to prevent starch gelatinization, as indicated by studies conducted by da Rosa Zavareze and Dias (2011) and Klein et al. (2013). The amylose content level and starch molecular weight influence the reassembly processes of starch during HMT, as elucidated by Li et al. (2019) and Yang et al. (2019). The modification mechanism involves enhancing the interactions among starch chains, commencing with the disruption of crystalline structures and subsequent dissociation of double-helix configurations and culminating in the reassembly of the disrupted crystals (Fonseca et al., 2021).

After HMT modification, notable alterations are evident in the digestibility of starches. Nutritional enhancements, including increased levels of slowly digestible starch (SDS) and resistant starch (RS), have been attained through chemical modifications such as starch ethers, starch esters, and crosslinked starches, as well as physical modifications like HMT and annealing (Singh et al., 2010). During the process, starch granules undergo partial gelatinization, prompting molecular chains to disassemble and rearrange, forming thermostable structures. Consequently, this retards starch gelatinization during cooking, diminishes pasting viscosity and swelling power, and mitigates starch digestion and retrogradation (Yang et al., 2019; Wang et al., 2021; Zhang et al., 2021). From 1981 onwards, studies focused on wheat and potato starches were conducted, revealing enhancements in the quality of cake and bread when utilizing modified potato starch (Donovan et al., 1983).

2.3 METHODS OF HEAT MOISTURE TREATMENT

Many researchers have employed different combinations of moisture, temperature, and time to modify starches, as summarized in Table 2.1.

2.4 PHYSICOCHEMICAL PROPERTIES

The critical structural characteristic of starches in food goes through many stages, from water binding to swelling to the disintegration of starch granules. The process of HMT significantly influences the properties and uses of starch. A decrease in amylose content for oat and pearl millet starches following HMT modification was observed (Kaur & Singh, 2019; Sandhu et al., 2020; Bangar et al., 2021). The alterations observed in HMT can be linked to the additional interactions among amylose-amylose and amylose-amylopectin chains, thereby modifying the crystalline structure of starch. Consequently, these modifications render amylose less soluble and less readily quantifiable (da Rosa Zavareze et al., 2010). Alternatively, an increase in lipid-complexed amylose chains also contributes to these changes (Chung et al., 2009). Chandla et al. (2017) observed an increase in the amylose content (3.47 to 3.73%) after the HMT modification of amaranth starch. The amylose content of starch showed no significant change during HMT (Singh et al., 2009; Sun et al., 2014; Rafiq et al., 2016).

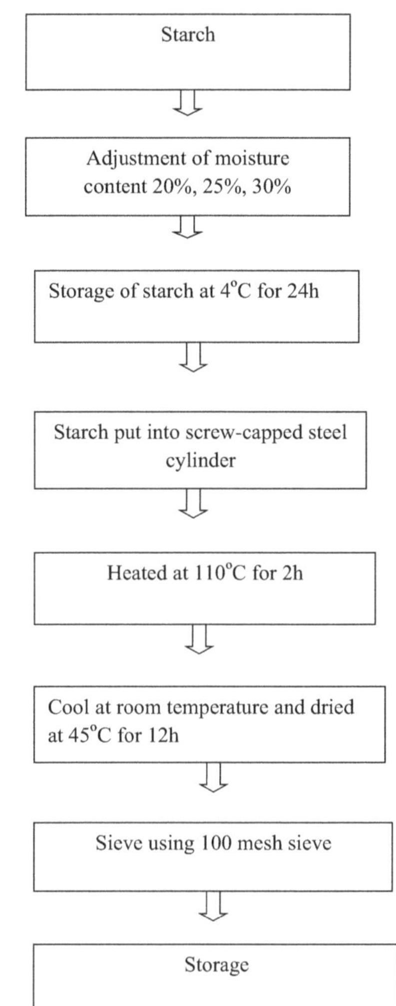

FIGURE 2.1   Method for heat moisture treatment (Liu et al., 2019).

The swelling power and solubility serve as indicators of the extent of interaction between starch chains found in both the amorphous and crystalline domains, and these measurements are impacted by the proportion of amylose to amylopectin (Blazek & Copeland, 2008). Swelling power demonstrates a direct correlation with the presence of amylopectin, while amylose content results in dilution and functions as an inhibitor to swelling (Singh et al., 2003). Compared to native starches, it was found that following HMT (oat and pearl millet), starches' swelling power was reduced (Kaur & Singh, 2019; Sandhu et al., 2020). Increased crystallinity, decreased hydration, enhanced interactions between amylose and amylopectin molecules, strengthened intramolecular bonds, the formation of amylose-lipid complexes, and modifications in the arrangements of the crystalline regions of starch have all been linked to this decrease in swelling power (Waduge et al., 2006; Jacobs et al., 1995). Chandla et al. (2017) documented an increase in the swelling power (10.10 to 10.81g/g) of amaranth starch subsequent to HMT. The increased swelling observed in HMT-treated amaranth starch could be attributed to an enlargement in particle size, an enhancement in granule hardness, and, subsequently, increased susceptibility of starch granules to interact with water.

TABLE 2.1 Heat Moisture Treatment Conditions for Starches

S. NO.	STARCH SOURCE	TEMPERATURE (°C)	TIME (H)	WATER CONTENT (%)	REFERENCE
1.	Rice	110	4	25	Chi et al. (2024)
2.	Corn, potato, rice, and tapioca	100	10	25	Kim et al. (2023)
3.	Wheat	110	4	15, 20, 25, 30	Wang et al. (2023)
4.	Corn	100	4	20, 25, 30, 35, 40	Zhong et al. (2023)
5.	Proso millet	110	4	25	Kumar et al. (2023)
6.	Potato	120	4	20	Yang et al. (2022)
7.	Wheat	100	3	15, 20, 25, 30, 35	Wang et al. (2022)
8.	Cowpea	100	2, 5 12	20	Acevedo et al. (2022)
9.	Corn	105	4	25	Zhang et al. (2021)
10.	Black rice	90	1, 2	25, 30	Wattananapakasem et al. (2021)
11.	Oat	100	16	30	Kaur and Singh (2019)
12.	Waxy maize	120	1, 3, 5	30	Park et al. (2018)
13.	Amaranth	110	2.5	28	Chandla et al. (2017)
14.	Horse chestnut	110	3	20, 25, 30	Rafiq et al. (2016)
15.	Rice	100	1, 4 or 8	17, 20	Chung et al. (2012)
16.	Rice	100	16	25	Jiranuntakul et al. (2011)
17.	Cocoyam tubers	100	16	18–27	Lawal (2005)

Starch solubility results from the leaching of amylose, which separates from and disperses out of granules during the swelling process. This leaching represents a transition from order to disorder within the starch granules that occurs when starch is heated with water (Tester & Morrison, 1990). Following HMT, the solubility of starches was observed to decrease, as indicated by studies conducted by Sandhu et al. (2020), Bangar et al. (2021), and Kim et al. (2023). The reduction in solubility can be attributed to the development of more structured amylopectin clusters and amylose-lipid complexes within starch granules. This structural rearrangement of starch granules leads to increased interactions, resulting in decreased solubility (Zavareze & Dias, 2011). It is reported that there is an increase in solubility power after HMT for pearl millet starch (Sandhu et al., 2020). Li et al. (2011) suggested that HMT starch might facilitate easier access to water in the amorphous regions of starch. Consequently, the starch chains that remain unassociated could readily dissolve into water, thereby increasing the overall solubility of the starch.

2.5 PASTING PROPERTIES

The pasting properties of starches provide important information about cooking behavior during successive heating and cooling cycles (Figure 2.2). HMT modification reduces peak viscosity, breakdown viscosity, and setback viscosity as compared to their native counterpart starch (Kaur & Singh, 2019; Sandhu et al., 2020; Bangar et al., 2021). It was observed that after HMT modification, peak, trough, breakdown, and setback viscosity were not observed (Srijunthongsiri et al., 2016). The pasting temperature of HMT starches was higher than that of native counterpart starch (Srijunthongsiri et al., 2016; Kaur & Singh, 2019; Bangar et al., 2021). The breakdown viscosity serves as an indicator of the stability of starch

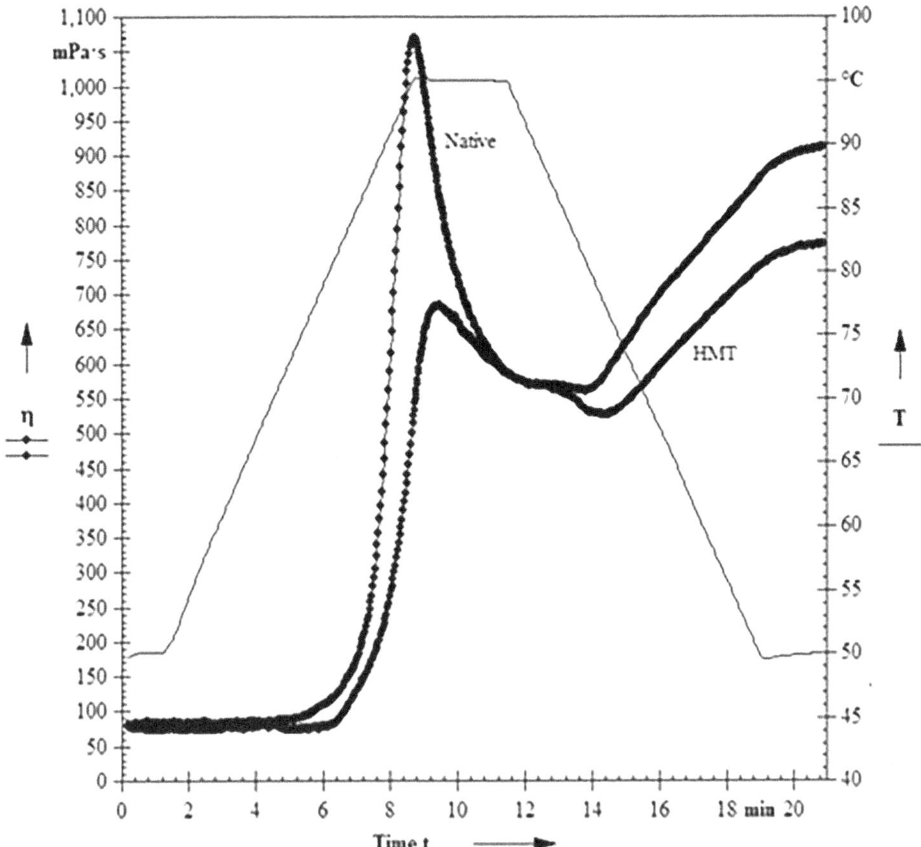

FIGURE 2.2 Pasting profile of HMT starch.

granules when subjected to continual heating and agitation. Yadav et al. (2013) documented a decrease in breakdown viscosity (347 to 57 cP) subsequent to the HMT modification of water chestnut starch. This reduction post-HMT signifies that starches exhibit enhanced stability under prolonged heating and agitation conditions, a finding corroborated by Adebowale et al. (2005). An increased pasting temperature was observed in comparison to native starch. They attributed the enhancements in starch pasting characteristics to molecular reorganization within the starch granule. The viscosity and pasting temperature exhibited higher values for HMT-treated black rice starch in comparison to native starch (Wattananapakasem et al., 2021). The alterations in pasting properties following HMT have been linked to reduced granular swelling and amylose leaching, an increase in granule rigidity, and intensified interactions among starch chains (Hoover & Vasanthan, 1994). The setback value serves as an indicator of the retrogradation tendency and is associated with the structural composition of amylose and amylopectin. This association arises from the tendency of smaller amylose molecules and low-chain amylopectin molecules to retrograde more rapidly (Akingbala & Rooney, 1987). Varatharajan et al. (2010) stated that setback decreased in normal potato starch but increased in waxy potato starch on HMT. Jayakody et al. (2007) demonstrated that the degree of setback is influenced by both the gelation of amylose and the presence of rigid swollen granules embedded within the leached amylose network. The absence of setback viscosity in HMT40 waxy and long-grain rice starch was noted. The setback viscosity may be contingent upon various factors, such as the extent of amylose leaching, granule size, and the length of amylose chains (Shih et al., 2007).

Jiranuntakul et al. (2011) determined the impact of HMT modification on waxy and normal corn, rice, and potato starches. They observed that, following treatment, all types except waxy potato starch exhibited increased pasting temperature (73.2–84.3°C) and reduced viscosity (ND to 113.6 RVU). The differences in viscosity values pre- and post-HMT were more evident in normal starches compared to waxy starches, whereas changes in pasting temperature showed the opposite trend (waxy > normal). HMT starches, due to their low viscosity values, can withstand processing conditions like crosslinked starches with the additional benefit of no chemical usage during their preparation (Molavi et al., 2018). Watcharatewinkul et al. (2009) evaluated the effect of moisture content during the treatment pasting profile. It was found that starch was treated with 15% moisture content, which resulted in lower paste viscosity, more stable viscosity (with no breakdown), and a noticeable decrease in setback value (45.1 RVU). This effect was more pronounced when starch was treated at 25% moisture content. After HMT, significant effects were observed on pasting properties, and breakdown and setback viscosity decreased, which showed the stability of starches during heating and cooling.

2.6 THERMAL PROPERTIES

The thermal properties of gelatinization, which were onset temperature (To), peak temperature (Tp), and maximum peak temperature (Tc), have been used as the thermal stability indices of the starch granules. Differential scanning calorimetry (DSC) is a method facilitating the investigation and understanding of starch structures. Additionally, it serves as a tool to assess starch behavior in industrial processing. Through DSC, the difference in enthalpy between a sample and a reference substance can be determined concerning temperature variations under controlled heating conditions (Schafranski et al., 2021).

It was observed that after HMT modification, To, Tp, and Tc increased (Li et al., 2011; Yadav et al., 2013; Rafiq et al., 2016) (Table 2.2). Yadav et al. (2013) observed a shift toward higher values in gelatinization temperatures (To, Tp, and Tc) following HMT modification. Simultaneously, a decrease in the

TABLE 2.2 Thermal Properties of Heat Moisture–Treated Starches

		NATIVE STARCH			HMT STARCH			
S. NO.	SOURCE	T_O (°C)	T_P (°C)	T_C (°C)	T_O (°C)	T_P (°C)	T_C (°C)	REFERENCE
1.	Wheat	55.7	61.9	78.4	57.1–65.4	62.8–78.4	78.7–83.5	Wang et al. (2023)
2.	Corn	65.01	73.26	81.49	68.82–99.85	73.64–103.79	78.92–108.94	Zhong et al. (2023)
3.	Proso millet	70.63	73.67	77.12	78.59	84.78	87.95	Kumar et al. (2023)
4.	Field bean starch	60.6	65.9	74.6	60.7–65.6	66.4–83.8	74.4–89.5	Piecyk and Domian (2021)
5.	Pearl millet	62.92	67.95	73.78	76.16	81.05	84.50	Bangar et al. (2021)
6.	Waxy maize	66.3	71.4	86.9	68.0–68.3	72.2–72.6	91.0–91.4	Park et al. (2018)
7.	Horse chestnut	53.35	58.81	63.57	61.06–76.27	65.17–80.47	70.28–83.56	Rafiq et al. (2016)
8.	Water chestnut	66.77	71.99	79.84	75.53–82.98	78.85–87.35	89.01–91.87	Yadav et al. (2013)
9.	Pinhão	60.77	66.65	74.22	62.03–73.03	68.13–76.78	74.41–83.84	Pinto et al. (2012)
10.	*Canna edulis* ker starch	59.70	64.18	71.86	61.20–65.20	65.35–68.65	74.56–79.76	Zhang et al. (2010)

gelatinization temperature range (Tc–To) was noted. The study found that higher moisture (25 and 30%) content during HMT correlated with a reduced gelatinization temperature range (13.07 to 8.89). This decrease in the range indicated improved uniformity, leading to enhanced perfection of crystallites and limited hydration of these crystalline structures (Jacobs et al., 1998; Tester et al., 2000). Hoover (2010) also stated that these increases are due to the structural change in the starch granule after HMT, such as the formation of amylose-amylose and amylose-lipid complexes, which inhibit the mobility of the starch chain within the amorphous lamella.

During HMT, simultaneous processes involving partial disorganization and gelatinization of starch granules, reordering of disintegrated molecules, and increases in associations among starch chains occurred. This resulted in a reduction of the content associated with the double-helical structure and an increase in the idealization of this structure. Consequently, this led to a decrease in ΔH (enthalpy change) and an elevation in gelatinization temperatures (Huang et al., 2016). During the hydrothermal treatment, short chains initially located within amorphous lamellae could transition towards the crystalline region, contributing to the formation of double helices and prompting the dense packing of crystallites (Piecyk & Domian, 2021). Wang et al. (2020) examined the impact of moisture content on the gelatinization characteristics of adlay seed starch. The study revealed that after HMT, To and Tp consistently increased, whereas Tc decreased, except in the case of starch treated at 30% moisture. The rise in To and Tp was associated with interactions occurring between amylose–amylose and/or amylose–amylopectin chains. After increasing the gelatinization temperature, the thermal stability of starches increased. Oliveira et al. (2018) analyzed potato and sweet potato starches subjected to HMT modification, observing elevated gelatinization temperatures alongside a reduction in gelatinization enthalpy.

2.7 MORPHOLOGICAL PROPERTIES

Scanning electron microscopy (SEM) serves as a crucial tool for understanding the granular structure of starches (Figure 2.3; Table 2.3). Attributes such as granule morphology, size distribution, and surface characteristics significantly influence numerous applications of starch in both food and non-food industries (da Rosa Zavareze et al., 2011). HMT-treated starches (corn, potato, rice, and tapioca) exhibited an absence of fractures or cracks on their granule surfaces (Kim et al., 2023). The shape and structural integrity of HMT-treated starch granules remained unchanged, although concavity was noted under SEM (Li et al., 2011). Yang et al. (2022) observed that the morphology of potato starch granules was not changed

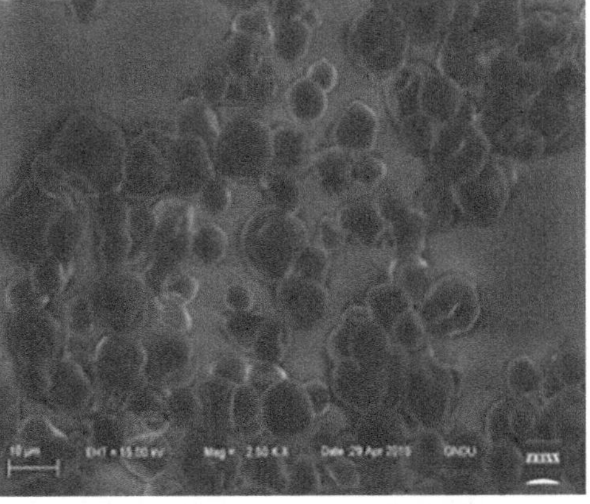

FIGURE 2.3   Scanning electron micrograph of HMT pearl millet starch.

TABLE 2.3 Morphological Properties of HMT Starches

S. NO.	MORPHOLOGICAL PROPERTIES	REFERENCE
1.	HMT starch did not show any fractures or cracks on the granule surface	Kim et al. (2023)
2.	The surface of starch granules became wrinkled, and some "pits" appeared on the surfaces	Zhong et al. (2023)
3.	Intact and slick, although some indentations appeared on their surface	Wang et al. (2022)
4.	HMT produced evident changes in the form of indentations at the center of some granules	Piecyk and Domian (2021)
5.	HMT treatment forms cavities and indentation	Bangar et al. (2021)
6.	Cracks were observed on the surface of highland barley starch treated with HMT	Liu et al. (2019)
7.	Exhibited smooth surface after HMT	Kaur and Singh (2019)
8.	Agglomerates were formed, mainly at the higher levels of moisture content	Bet et al. (2018)
9.	Showing some dents on the surface (HMT 30)	Rafiq et al. (2016)
10.	HMT induced some dents/holes on the surface of starch granules, and loss of physical integrity was also observed	Sharma et al. (2015b)
11.	After HMT, a slightly rough surface and aggregation of the granules were observed	Jiranuntakul et al. (2011)
12.	More aggregated and the surface of the granules more irregular after HMT	da Rosa Zavareze et al. (2010)

after HMT modification. However, HMT treatment with varying moisture levels (20 and 25%) notably influenced the shape and surface characteristics of corn starch granules. The findings indicated surface wrinkling and the emergence of "pits" on the starch granules following HMT treatment, particularly noticeable at moisture levels of 20 and 25% (Zhong et al., 2023). Similarly, in a previous study by Shi et al. (2018), they reported that large depressions were observed on the surface of corn starch treated with HMT (25% moisture, 120°C). Piecyk and Domian (2021) concluded that no alterations in granule morphology were evident in starches modified at 15% moisture content, irrespective of temperature variations. Conversely, higher moisture content (30%) during HMT resulted in noticeable changes, such as indentations at the center of certain granules, particularly prominent in starch modified at 120°C. Similar effects were also observed by Rafiq et al. (2016) in horse chestnut starch following HMT modification. The hydrothermal gelatinization of starch led to granule size increase, accompanied by surface cracks due to heat-induced damage (Kim et al., 2017). This phenomenon occurs from the disintegration of the double helix structure of amylopectin induced by excessive water and heat. Consequently, the granule undergoes swelling, leading to the leaching out of amylose, gradually expanding the granule's size until it ruptures (Song et al., 2015). Subsequent to HMT treatment, proso millet starch granules exhibited damage, presenting cracks and holes on their surfaces (Kumar et al., 2023). During HMT, the elevated temperatures led to moisture evaporation from the surface, resulting in granule aggregation and enlargement. Subsequent to the completion of the HMT process, the reduction in temperature prompted the outer layer of the granules to contract, leading to the formation of cracks and holes (Dong et al., 2021). Singh et al. (2011) determined that the impact of HMT on starch granules is dependent upon moisture levels. The study observed a bread dough-like structure when starch underwent modification at 40% moisture content. After modification, a reduction in granule size compared to their native state was evident, and this decrease in size was linked to the moisture content during the modification process. Kaur and Singh (2019) concluded that HMT did not significantly alter the surface properties of starch granules, although certain granules exhibited a few surface holes, potentially indicating surface degradation.

HMT modification resulted in the development of dents or holes on the surface of pearl millet starch granules (Sharma et al., 2015b). Additionally, HMT led to a loss of physical integrity with surface degradation, displaying characteristics akin to partial gelatinization in the modified samples. Barua et al. (2022) investigated the impact of different heating methods during HMT. They observed the formation of several aggregates along with surface damages such as scratches, pitting, and dents on the granular surfaces. The study highlighted that these effects were more pronounced in elephant foot yam starch treated with a hot air oven, followed by microwave and autoclave treatments. After HMT, the morphology of starches changed significantly. Higher moisture content affects the morphological properties during HMT.

2.8 DIGESTIBILITY

HMT-treated starches exhibited comparatively reduced levels of RDS in contrast to native starches, showing an increase in SDS and RS contents with the progression of modification (Kim et al., 2023). Post-HMT treatment, RDS and SDS content increased, while RS content decreased in comparison to the native starch (Kumar et al., 2023). In most gelatinized HMT starches, the SDS content decreased (1.4 to 0.1%), while the RS content did not significantly alter when compared to the native starch. Despite the modified physicochemical properties, these starches continue to serve as a substantial source of resistant starch (Piecyk & Domian, 2021).

A reduction in RDS content by 10.2% in gelatinized HMT corn starch, alongside an increase in SDS and RScontents by 2.5 and 7.7%, respectively (Chung et al., 2009). This phenomenon was attributed to heightened thermo-stability resulting from structural rearrangement during HMT. The improved starch portions post-gelatinization potentially limited starch chain accessibility to hydrolyzing enzymes. HMT facilitated improved chain mobility by altering the radial direction of amylopectin, weakening the crystalline area's structure via internal rearrangement, and enhancing starch integrity. Consequently, this facilitated easier access for α-amylase, albeit reducing the enzyme's sensitivity to hydrolysis, resulting in decreased RS and increased SDS (Huang et al., 2016). Significantly, a strong correlation exists between starch enzymatic digestion and its structures (Chi et al., 2024). The variations in in vitro digestibility among heat-moisture–treated starches verified the preparation of starches containing different structures (Wang et al., 2023).

The impact of moisture content during HMT modification on the digestibility of corn starch was determined by Zhong et al. (2023). Their findings indicated that an increase in moisture content led to a decrease in RDS content, coupled with an increase in SDS and RS contents. HMT reduced the native starch crystal content from 15.02 to 19.03%, consequently enhancing starch thermostability, resulting in a reduction in RDS content from 86.91 to 76.71% (Chi et al., 2024). This effect might be attributed to partial gelatinization and surface damage of the starch granules, facilitating the penetration of digestive enzymes into the granules and ultimately improving digestibility. Chung et al. (2009) stated that enzyme hydrolysis decreases or increases after HMT depending on the starch source, cultivar, and treatment conditions. After modification, there was an increase in RS and SDS content, along with a decrease in RDS content in comparison to their native starch counterparts (Sandhu et al., 2020). HMT parameters, including temperature, moisture content, and the botanical source of starch, have distinctive effects on SDS, RDS, and RS. Consequently, understanding these effects is crucial for the efficient utilization of HMT starches in applications that necessitate high levels of RS (Fonseca et al., 2021). RS has properties similar to dietary fibers, so it has many health benefits. RS has been found to be related to improvements in lipid and cholesterol metabolisms, prebiotic effects on colon microorganisms, and reduction in the risk of ulcerative colitis and colon cancer (Lopez-Rubio et al., 2008; Fuentes-Zaragoza et al., 2010).

2.9 X-RAY DIFFRACTION ANALYSIS

The starch granule represents a semi-crystalline system comprising crystalline and amorphous regions. Each distinct type of starch demonstrates a specific X-ray diffraction (XRD) pattern unique to its crystalline structure. Crystalline segments within starch exhibit well-defined peaks in XRD patterns, while the amorphous segments display broader and more diffuse peaks (Gernat et al., 1990). Based on diffraction patterns, starch is classified into A, B, C, and V types. The characteristic A-type X-ray pattern is commonly found in cereal starches and specific tubers. The B-type X-ray pattern is typical of root starches, tubers rich in amylose from starchy cereals, and certain fruits. Starches from fruits, legumes, and stems often exhibit the C-type diffraction pattern. Additionally, there exists another crystalline type, the V-type, resulting from the crystallization of amylose with lipids (Blazek & Gilbert, 2011; He & Wei, 2017; Oyeyinka & Oyeyinka, 2018). Kaur and Singh (2019) reported that the XRD type does not change, but the peak at 15° and 17.3° (2θ) widened for HMT starches as compared to native starches. The widening of the peak in HMT starches may be due to the formation of new crystallites (Sun et al., 2014), reorientation of crystallites, change in crystal size or a number of crystalline regions (Singh et al., 2018), or perfection of small crystalline regions of the starch granules (Sun et al., 2014) in HMT starches, which may lead to changed X-ray diffraction pattern. Kumar et al. (2023) also observed that after HMT modification, the crystalline pattern does not change. But additional peaks at 2θ values of 13.1° and 20.5° were observed. This clearly indicates a V-type XRD pattern (Jain et al., 2019), suggesting the formation of an amylose-lipid complex during HMT (Chen et al., 2015).

A reduction in relative crystallinity following HMT modification (Klein et al., 2013; Zhong et al., 2023). The application of thermal energy prompts the association of adjacent double helices and an increase in the number of hydrogen bonds, causing alterations in the X-ray intensities of starches (da Rosa Zavareze& Dias, 2011). Wang et al. (2023) noted that the crystalline pattern remained unchanged after HMT modification. However, they observed a significant decrease in starch crystallinity due to HMT, with moisture content playing a critical role in this reduction. The elevated levels of heat and moisture during HMT contributed to a decrease in crystallinity. These findings suggested the destruction of relative crystallinity within starch granules after HMT (Zou et al., 2020).

The transformation of diffraction patterns from B type to A type after HMT treatment was observed for potato starch (Hoover & Vasanthan, 1994). This implies that the impact of HMT might be dependent upon the source of the starch and its moisture content. Li et al. (2011) observed that after HMT, the crystalline pattern remained unaltered, while the diffraction intensity increased. Conversely, a reduction in X-ray intensity was noted in HMT-treated starches obtained from various legumes such as black bean, express field, pinto bean, and eston lentil (Hoover & Manuel, 1996).

The treatment led to an increase in crystallinity within waxy corn starch, from 41.4 to 44.3%. In contrast, there was no observable change in crystallinity for normal corn starch (32.6%) (Rocha et al., 2012). This increase in the crystallinity (42.8 to 45.3%) of the waxy corn starch may be attributed to a better organization of double helices and the larger size of crystallites after annealing (Lan et al., 2008). The relative crystallinity of native starch was 43%, which decreases significantly by as much as 35% for hot air oven-treated elephant foot yam starch but slightly by as much as to 41 and 40% in autoclave and microwave-treated starch during HMT, respectively (Barua et al., 2022).

2.10 RHEOLOGICAL PROPERTIES

The amount of energy stored in the material and recovered from it per cycle is called G', while the amount of energy dissipated or lost per cycle of sinusoidal deformation is described as G''. For all starches, G' and G'' increased progressively during heating, followed by a drop. Before TG' (temperature at which G'

is maximum) reached a maximum, both G' and G'' increased, which may be attributed to the swelling of granules of starch and leaching of chains of amylose, contributing to the formation of a composite network of solved materials supporting partially disintegrated starch granules (Arocas et al., 2009). G' and G'' overall decreased and increased, respectively, confirming the strength of starch hydrogels gradually decreased after HMT. It was observed that with the increase in the moisture content, a decrease in the G' values was observed (Wang et al., 2023). Sandhu et al. (2020) also observed the effect of HMT modification on pearl millet starches from different cultivars. After the modification of starches, a decrease of peak G' and breakdown in G' with an increase in TG' was observed as compared to their native counterpart starches. $\tan\delta$ (G''/G') of modified starches varied from 0.07 to 0.12, indicating their elastic behavior than viscous. The effect of HMT on pearl millet starches during the frequency sweep test was also observed. G' values decreased after the treatment. Bangar et al. (2021) also observed a decrease in G' values, while the reverse was observed for G'' and $\tan\delta$ during the frequency sweep test. Zhong et al. (2023) observed that the G' values of HMT-20 and 25% were much higher than those of the corn starch at the beginning of the angular frequency and showed a similar value at the end. G' showed lower values in HMT-30, 35, and 40% than that of corn starch, and the reduction was dependent on the increase of moisture. Sharma et al. (2015a) determined the effect of moisture content (20, 25, and 30%) on the HMT starch rheological properties. It was observed that G' and G'' values decreased after treatment. The effect of different heating methods (hot air oven, autoclave, and microwave) on rheological properties was examined by Barua et al. (2022), who observed that during rheological properties during heating and frequency sweep tests, G' values increased, except for the autoclave-treated sample.

The effect of HMT on steady shear properties of HMT starch from different pearl millet cultivars was observed. After HMT modification, a decrease in values of K and σo was observed as compared to their native counterpart starch pastes (Sandhu et al., 2020). The flow behavior index of modified starch was observed to be <1. The decrease in the values of K and σo may be due to the decrease in the swelling power of starch granules. Hoover and Manuel (1996) reported that the development of starch chain interactions and dissociation of a few double helices during HMT might be responsible for restricted starch swelling behavior. The flow behavior index of $n < 1$ demonstrated that the gel structure is broken down by the applied shear stress, which showed their shear-thinning behavior (Chan et al., 2011).

2.11 COMBINED MODIFICATIONS

HMT can be performed with various other modifications, including chemical, physical, and enzymatic methods. These combinations enhance material properties, thereby unlocking novel applications across diverse fields (Fonseca et al., 2021).

Ismailoglu and Basman conducted studies in 2015 and 2016 to determine the impact of distinct infrared treatments (at 550 or 730W) on corn and wheat starch in conjunction with varied moisture levels and durations during HMT. Their findings indicated that higher moisture content (30%) led to an elevation in gelatinization temperature and a reduction in enthalpy subsequent to treatment. Additionally, the relative crystallinity (RC) and viscosity also decreased after treatment, specifically under conditions of 730W and 30% moisture content. Conversely, for corn starch, an increase in RC was observed when treated with lower moisture content and infrared power (20% moisture, 550W).

The impact of HMT at various moisture levels (12%, 15%, 18%, 21%, and 24%) was observed and starches with 30% and 50% of their granular surface removed through chemical surface gelatinization to explore its internal structure (Bartz et al., 2017). At moisture contents of 12 and 15%, HMT decreased the gelatinization temperatures and relative crystallinity of the starches. Conversely, at moisture contents of 21 and 24%, both parameters increased. Additionally, the starches exhibited a more irregular surface morphology due to the removal of gelatinization layers. Chemical modification along with HMT was also reported to have an impact on thermal, pasting, and digestibility, as observed by Maior et al. (2021). It was observed that enthalpy, RC, and peak viscosity of acid-HMT starches decreased as compared to native counterpart starch. Gelatinization peaks were not detected when starch was modified with HMT-citric acid and HMT-lactic acid. It was observed that after HMT-acid modification, RDS (9.16 to 21.21%) and RS content increased (8.45 to 23.18%), whereas SDS content decreased (82.39 to 55.61%).

2.12 APPLICATIONS

Modified starches play a significant role in processed foods, offering diverse applications across various industries. As a result of the growth and improvement of the modified starch market, the ingredient industries have been developing starches with unique properties to supply the peculiarities of each segment of the market (Schafranski et al., 2021). HMT-modified starches exhibit increased thermal stability and resistance to shear, making them suitable for utilization in bread, frozen products, sauces, confectionery, canned goods, and pasta (Arns et al., 2015; Kaur & Singh, 2019).

2.12.1 Bakery Products

Lorenz and Kulp (1981) observed the bread and cake baking capabilities along with the thickening properties of wheat and potato starches following HMT treatment under different moisture levels (ranging from 18 to 27%) at 100°C for 16 hours. Their findings indicated a decrease in the bread baking quality and volume of bread made from wheat starch after HMT. Conversely, the baking potential, grain texture, and volume of bread derived from potato starch increased as a result of HMT treatment.

Chandla and colleagues (2017) prepared noodles using amaranth and corn starch treated with heat moisture. The study revealed that employing HMT led to improved texture and eating qualities in amaranth noodles. Noodles prepared from HMT amaranth showed better taste. Purwani and colleagues (2006) also produced noodles utilizing heat moisture-treated sago starch. Noodles derived from starch subjected to HMT exhibited higher firmness and elasticity, but they have lower stickiness compared to those of non-HMT. Additionally, these HMT-treated noodles demonstrated decreased cooking loss (4.15 to 2.86%) and rehydration weight. However, the HMT process extended the cooking duration and intensified the aroma in comparison to noodles prepared from amaranth and corn starch. Noodles fortified with starch, particularly those fortified with heat-moisture-treated starch possessing higher moisture content and displayed diminished water absorption index, hardness, springiness, and chewiness compared to noodles fortified with native starch (Wang et al., 2023). Yang et al. (2022) prepared gluten-free noodles utilizing potato starch. They employed HMT-modified starch mixed with sodium chloride as a binder paste. The resulting starch dough displayed increased hardness, while the starch noodles demonstrated reduced loss in solidity and breakage. Moreover, these noodles exhibited firmer texture and improved elasticity.

2.12.2 Sauce

Subroto and colleagues (2020) investigated the impact of HMT starch on the properties of tomato sauce. Their findings revealed that incorporating 1.5% HMT potato starch into the tomato sauce enhanced viscosity, color, taste, flavor, and overall acceptability. Furthermore, considering the heat stability of the tomato sauce, HMT-treated potato starch proved suitable for use in sterilized products.

2.12.3 Nanoparticles

HMT enhances the thermal characteristics and relative crystallinity of nanoparticles. Remarkably, HMT has gained widespread use in the modification of starch nanoparticles owing to its ability to increase both the thermal properties and crystallinity of these particles (Ji et al., 2019; Kumari et al., 2020).

2.12.4 Bioplastics

Starch-based bioplastics represent 85–90% of those available on the market and are produced with native or modified starch, isolated, or mixed with synthetic compounds (Ashok et al., 2018). The impact of HMT on the properties of bioplastics derived from green plantain banana starch. Among the various treatments, the PM04 modification (with moisture at 25% and temperature at 100°C) exhibited notable differences in mechanical properties compared to other modifications (Viana et al., 2022). This particular treatment resulted in significantly higher values for mechanical attributes such as elongation, tensile strength, force, and puncture deformation. Additionally, it displayed reduced water vapor permeability compared to bioplastics made from native starch. Bangar et al. (2021) studied the effect of HMT on pearl millet starch. It was observed that after treatment, the thickness of the film decreased, whereas the reverse was observed for water solubility and opacity of the film. Dutta and Sit (2022) evaluated that films produced from HMT starch have decreased solubility (13.57%), WVP (0.20 g mm m^{-2} h^{-1} kPa^{-1}), and enhanced mechanical properties (7.62 MPa tensile strength, elongation of 11.36%, and seal strength 4.06 MPa) compared to annealed and native starch films. Indrianti et al. (2018) showed that the thickness, tensile strength, and elongation of the modified (HMT) sweet potato starch edible films were higher, while the water vapor permeability and the solubility were lower than those prepared native sweet potato starch. The longer durations of HMT had no effects on the characteristics of the edible films.

2.12.5 Printing

Three-dimensional (3D) printing represents an innovative and transformative manufacturing technology that deposits materials layer by layer according to predetermined models (Liu et al., 2020). Its distinctive advantage lies in the capability for customized design and its user-friendly operation in contrast to conventional processing technologies (Zheng et al., 2023). HMT stands as an effective technique for altering starchy materials employed in 3D food printing, enhancing the stacking height, precision, and stability of the printed samples. This method offers considerable potential benefits to the food industry, providing a straightforward and efficient approach to enhance the effectiveness of starch-based 3D printing (Zheng et al., 2023).

2.13 CONCLUSION

In this chapter, the properties and applications of HMT starches are discussed. HMT modification is a physical type starch modification method. After this modification, the thermal stability of starches improved, which further enhanced their application. It was observed that moisture content, heating time, and temperature showed a significant effect. HMT increases the resistant starch content, which may reduce many diseases. HMT starch is used for bread, frozen products, sauces, confectionery, canned products, and pasta. In general, HMT has shown great potential for the physical modification of starches because it can provide desirable physicochemical changes, as well as increases in thermal stability and shear resistance. These characteristics are appreciated in the industrial sector. HMT is an alternative to be applied in producing bioplastics that demand greater tensile strength and less water vapor permeability, such as edible bio-plastics.

REFERENCES

Acevedo, B. A., Villanueva, M., Chaves, M. G., Avanza, M. V., & Ronda, F. (2022). Modification of structural and physicochemical properties of cowpea (Vigna unguiculata) starch by hydrothermal and ultrasound treatments. *Food Hydrocolloids, 124*, 107266.

Adebowale, K. O., Afolabi, T. A., & Olu-Owolabi, B. I. (2005). Hydrothermal treatments of Finger millet (*Eleusine coracana*) starch. *Food Hydrocolloids, 19*, 974–983.

Akingbala, J. O., & Rooney, L. W. (1987). Paste properties of sorghum flour and starches. *Journal of Food Processing and Preservation, 11*, 13–24.

Alcázar-Alay, S. C., & Meireles, M. A. A. (2015). Physicochemical properties, modifications and applications of starches from different botanical sources. *Food Science and Technology, 35*, 215–236.

Arns, B., Bartz, J., Radunz, M., do Evangelho, J. A., Pinto, V. Z., da Rosa Zavareze, E., & Dias, A. R. G. (2015). Impact of heat-moisture treatment on rice starch, applied directly in grain paddy rice or in isolated starch. *LWT-Food Science and Technology, 60*(2), 708–713.

Arocas, A., Sanz, T., & Fiszman, S. M. (2009). Influence of corn starch type in the rheological properties of a white sauce after heating and freezing. *Food Hydrocolloids, 23*(3), 901–907.

Ashok, A., Abhijith, R., & Rejeesh, C. R. (2018). Material characterization of starch derived bio degradable plastics and its mechanical property estimation. *Materials Today: Proceedings, 5*(1), 2163–2170.

Bangar, S. P., Nehra, M., Siroha, A. K., Petrů, M., Ilyas, R. A., Devi, U., & Devi, P. (2021). Development and characterization of physical modified pearl millet starch-based films. *Foods, 10*(7), 1609.

Bartz, J., da Rosa Zavareze, E., & Dias, A. R. G. (2017). Study of heat–moisture treatment of potato starch granules by chemical surface gelatinization. *Journal of the Science of Food and Agriculture, 97*(10), 3114–3123.

Barua, S., Hanewald, A., Bächle, M., Mezger, M., Srivastav, P. P., & Vilgis, T. A. (2022). Insights into the structural, thermal, crystalline and rheological behavior of various hydrothermally modified elephant foot yam (*Amorphophallus paeoniifolius*) starch. *Food Hydrocolloids, 129*, 107672.

Bet, C. D., de Oliveira, C. S., Colman, T. A. D., Marinho, M. T., Lacerda, L. G., Ramos, A. P., & Schnitzler, E. (2018). Organic amaranth starch: A study of its technological properties after heat-moisture treatment. *Food Chemistry, 264*, 435–442.

Blazek, J., & Copeland, L. (2008). Pasting and swelling properties of wheat flour and starch in relation to amylose content. *Carbohydrate Polymers, 71*, 380–387.

Blazek, J., & Gilbert, E. P. (2011). Application of small-angle X-ray and neutron scattering techniques to the characterisation of starch structure: A review. *Carbohydrate Polymers, 85*, 281–293.

Chan, H. T., Leh, C. P., Bhat, R., Senan, C., Williams, P. A., & Karim, A. A. (2011). Molecular structure, rheological and thermal characteristics of ozone-oxidized starch. *Food Chemistry, 126*, 1019–1024.

Chandla, N. K., Saxena, D. C., & Singh, S. (2017). Processing and evaluation of heat moisture treated (HMT) amaranth starch noodles; an inclusive comparison with corn starch noodles. *Journal of Cereal Science, 75*, 306–313.

Chen, X., He, X., Fu, X., & Huang, Q. (2015). In vitro digestion and physicochemical properties of wheat starch/flour modified by heat-moisture treatment. *Journal of Cereal Science, 63*, 109–115.

Chi, C., Ren, W., Yang, Y., Guo, X., Zhang, Y., Chen, B., He, Y., Chen, H., Zheng, X., & Wang, H. (2024). Starch ordered structures control starch reassembly behaviors during heat–moisture treatment for modulating its digestibility. *Food Chemistry, 430*, 136966.

Chung, H. J., Cho, A., & Lim, S. T. (2012). Effect of heat-moisture treatment for utilization of germinated brown rice in wheat noodle. *LWT-Food Science & Technology, 47*(2), 342–347.

Chung, H. J., Liu, Q., & Hoover, R. (2009). Impact of annealing and heat-moisture treatment on rapidly digestible, slowly digestible and resistant starch levels in native and gelatinized corn, pea and lentil starches. *Carbohydrate Polymers, 75*, 436–447.

da Rosa Zavareze, E., & Dias, A. R. G. (2011). Impact of heat-moisture treatment and annealing in starches: A review. *Carbohydrate Polymers, 83*(2), 317–328.

da Rosa Zavareze, E., Storck, C. R., de Castro, L. A. S., Schirmer, M. A., & Dias, A. R. G. (2010). Effect of heat-moisture treatment on rice starch of varying amylose content. *Food Chemistry, 121*(2), 358–365.

Dong, J., Huang, L., Chen, W., Zhu, Y., Dun, B., & Shen, R. (2021). Effect of heat-moisture treatments on digestibility and physicochemical property of whole quinoa flour. *Foods, 10*(12), 3042.

Donovan, J. W., Lorenz, K., & Kulp, K. (1983). Differential scanning calorimetry of heat-moisture treated wheat and potato starches. *Cereal Chemistry, 60*, 381–387.

Dutta, D., & Sit, N. (2022). Comparison of properties of films prepared from potato starch modified by annealing and heat–moisture treatment. *Starch/Stärke, 74*(11–12), 2200110.

Fonseca, L. M., El Halal, S. L. M., Dias, A. R. G., & da Rosa Zavareze, E. (2021). Physical modification of starch by heat-moisture treatment and annealing and their applications: A review. *Carbohydrate Polymers, 274*, 118665.

Fuentes-Zaragoza, E., Riquelme-Navarrete, M. J., Sánchez-Zapata, E., & Pérez-Ál-Varez, J. A. (2010). Resistant starch as functional ingredient: A review. *Food Research International, 43*, 931–942.

Gernat, C., Radosta, S., Damaschun, G., & Schierbaum, F. (1990). Supramolecular structure of legume starches revealed by X-ray scattering. *Starch/Starke, 42*, 175–178.

Gunaratne, A., & Hoover, R. (2002). Effect of heat-moisture treatment on the structure and physicochemical properties of tuber and root starches. *Carbohydrate Polymers, 49*, 425–437.

He, W., & Wei, C. (2017). Progress in C-type starches from different plant sources. *Food Hydrocolloids, 73*, 162–175.

Hoover, R. (2010). The impact of heat-moisture treatment on molecular structures and properties of starches isolated from different botanical sources. *Critical Reviews in Food Science and Nutrition, 50*, 835–847.

Hoover, R., & Manuel, H. (1996). Effect of heat-moisture treatment on structure and physicochemical properties of legume starches. *Food Research International, 29*, 731–750.

Hoover, R., & Vasanthan, T. (1994). Effect of heat-moisture treatment on the structure and physicochemical properties of cereal, legume, and tuber starches. *Carbohydrate Research, 252*, 33–53.

Huang, T. T., Zhou, D. N., Jin, Z. Y., Xu, X. M., & Chen, H. Q. (2016). Effect of repeated heat-moisture treatments on digestibility, physicochemical and structural properties of sweet potato starch. *Food Hydrocolloids, 54*, 202–210.

Indrianti, N., Pranoto, Y., & Abbas, A. (2018). Preparation and characterization of edible films made from modified sweet potato starch through heat moisture treatment. *Indonesian Journal of Chemistry, 18*(4), 679–687.

Ismailoglu, S. O., & Basman, A. (2015). Effects of infrared heat-moisture treatment on physicochemical properties of corn starch. *Starch-Stärke, 67*(5–6), 528–539.

Ismailoglu, S. O., & Basman, A. (2016). Physicochemical properties of infrared heat-moisture treated wheat starch. *Starch-Stärke, 68*(1–2), 67–75.

Jacobs, H., & Delcour, J. A. (1998). Hydrothermal modifications of granular starch, with retention of the granular structure: A review. *Journal of Agricultural and Food Chemistry, 46*(8), 2895–2905.

Jacobs, H., Eerlingen, R. C., Clauwaert, W., & Delcour, J. A. (1995). Influence of annealingon the pasting properties of starches from varying botanical sources. *Cereal Chemistry, 72*, 480–487.

Jacobs, H., Mischenko, N., Koch, M. H. J., Eerlingen, R. C., Delcour, J. A., & Reynaers, H. (1998). Evaluation of the impact of annealing on gelatinization at intermediate water content of wheat and potato starches: A differential scanning calorimetry and small angle X-ray scattering study. *Carbohydrate Research, 306*, 1–10.

Jain, S., Winuprasith, T., & Suphantharika, M. (2019). Design and synthesis of modified and resistant starch-based oil-in-water emulsions. *Food Hydrocolloids, 89*, 153–162.

Jayakody, L., Hoover, R., Liu, Q., & Donner, E. (2007). Studies on tuber starches. II. Molecular structure, composition and physicochemical properties of yam (*Dioscorea* Sp.) starches grown in Sri Lanka. *Carbohydrate Polymers, 69*, 148–163.

Ji, N., Ge, S., Li, M., Wang, Y., Xiong, L., Qiu, L., Bian, X., Sun, C., & Sun, Q. (2019). Effect of annealing on the structural and physicochemical properties of waxy rice starch nanoparticles. *Food Chemistry, 286*, 17–21.

Jiranuntakul, W., Puttanlek, C., Rungsardthong, V., Puncha-Arnon, S., & Uttapap, D. (2011). Microstructural and physicochemical properties of heat-moisture treated waxy and normal starches. *Journal of Food Engineering, 104*(2), 246–258.

Kaur, M., & Singh, S. (2019). Influence of heat-moisture treatment (HMT) on physicochemical and functional properties of starches from different Indian oat (*Avena sativa* L.) cultivars. *International Journal of Biological Macromolecules, 122*, 312–319.

Kim, H. Y., Oh, S. M., Bae, J. E., Yeom, J. H., Kim, B. Y., Kim, H. S., & Baik, M. Y. (2017). Preparation and characterization of amorphous granular potato starches (AGPS) and crosslinked amorphous granular potato starches (CLAGPS). *Carbohydrate Polymers, 178*, 41–47.

Kim, H. Y., Ye, S. J., & Baik, M. Y. (2023). Physicochemical properties of pressure moisture treated (PMT) and heat moisture treated (HMT) starches. *Innovative Food Science & Emerging Technologies, 87*, 103392.

Klein, B., Pinto, V. Z., Vanier, N. L., da Rosa Zavareze, E., Colussi, R., do Evangelho, J. A., Gutkoski, L. C., & Dias, A. R. G. (2013). Effect of single and dual heat–moisture treatments on properties of rice, cassava, and pinhao starches. *Carbohydrate Polymers, 98*(2), 1578–1584.

Kumar, S. R., Tangsrianugul, N., Sriprablom, J., Wongsagonsup, R., Wansuksri, R., &Suphantharika, M. (2023). Effect of heat-moisture treatment on the physicochemical properties and digestibility of proso millet flour and starch. *Carbohydrate Polymers, 307*, 120630.

Kumari, S., Yadav, B. S., & Yadav, R. B. (2020). Synthesis and modification approaches for starch nanoparticles for their emerging food industrial applications: A review. *Food Research International, 128*, Article 108765.

Lan, H., Hoover, R., Jayakody, L., Liu, Q., Donner, E., Baga, M., Asare, E. K., Hucl, P., & Chibbar, R. N. (2008). Impact of annealing on the molecular structure and physicochemical properties of normal, waxy and high amylose bread wheat starches. *Food Chemistry, 111*(3), 663–675.

Lawal, O. S. (2005). Studies on the hydrothermal modifications of new cocoyam (Xanthosoma sagittifolium) starch. *International Journal of Biological Macromolecules, 37*(5), 268–277.

Li, M., Zhang, B., Xie, Y., & Chen, H. (2019). Effects of debranching and repeated heat-moisture treatments on structure, physicochemical properties and in vitro digestibility of wheat starch. *Food Chemistry, 294*, 440–447.

Li, S., Ward, R., & Gao, Q. (2011). Effect of heat-moisture treatment on the formation and physicochemical properties of resistant starch from mung bean (*Phaseolus radiatus*) starch. *Food Hydrocolloids, 25,* 1702–1709.

Lin, C. L., Lin, J. H., Lin, J. J., & Chang, Y. H. (2019). Progressive alterations in crystalline structure of starches during heat-moisture treatment with varying iterations and holding times. *International Journal of Biological Macromolecules, 135,* 472–480.

Liu, K., Zhang, B., Chen, L., Li, X., & Zheng, B. (2019). Hierarchical structure and physicochemical properties of highland barley starch following heat moisture treatment. *Food Chemistry, 271,* 102–108.

Liu, Z., Chen, H., Zheng, B., Xie, F., & Chen, L. (2020). Understanding the structure and rheological properties of potato starch induced by hot-extrusion 3D printing. *Food Hydrocolloids, 105,* Article 105812.

Lopez-Rubio, A., Flanagan, B. M., Shrestha, A. K., Gidley, M. J., & Gilbert, E. P. (2008). Molecular rearrangement of starch during *in vitro* digestion: Toward a better understanding of enzyme resistant starch formation in processed starches. *Biomacromolecules, 9,* 1951–1958.

Lorenz, K., & Kulp, K. (1981). II. Functional properties and baking potential. *Cereal Chemistry, 58*(1), 49–52.

Maache-Rezzoug, Z., Zarguili, I., Loisel, C., Queveau, D., & Buleon, A. (2008). Structural modifications and thermal transitions of standard maize starch after DIC hydrothermal treatment. *Carbohydrate Polymers, 74,* 802–812.

Maior, L. D. O., de Almeida, V. S., Barretti, B. R. V., Ito, V. C., Beninca, C., Demiate, I. M., Schnitzler, E., Carvalho Filho, M. A. D. S., & Lacerda, L. G. (2021). Combination of organic acid and heat–moisture treatment: Impact on the thermal, structural, pasting properties and digestibility of maize starch. *Journal of Thermal Analysis and Calorimetry, 143,* 265–273.

Molavi, H., Razavi, S. M. A., & Farhoosh, R. (2018). Impact of hydrothermal modifications on the physicochemical, morphology, crystallinity, pasting and thermal properties of acorn starch. *Food Chemistry, 245,* 385–393.

Oliveira, C. S., Bet, C. D., Bisinella, R. Z. B., Waiga, L. H., Colman, T. A. D., & Schnitzler, E. (2018). Heat-moisture treatment (HMT) on blends from potato starch (PS) and sweet potato starch (SPS). *Journal of Thermal Analysis and Calorimetry, 133,* 1491–1498.

Oyeyinka, S. A., & Oyeyinka, A. T. (2018). A review on isolation, composition, physicochemical properties and modification of Bambara groundnut starch. *Food Hydrocolloids, 75,* 62–71.

Park, E. Y., Ma, J. G., Kim, J., Lee, D. H., Kim, S. Y., Kwon, D. J., & Kim, J. Y. (2018). Effect of dual modification of HMT and crosslinking on physicochemical properties and digestibility of waxy maize starch. *Food Hydrocolloids, 75,* 33–40.

Piecyk, M., & Domian, K. (2021). Effects of heat–moisture treatment conditions on the physicochemical properties and digestibility of field bean starch (*Vicia faba* var. minor). *International Journal of Biological Macromolecules, 182,* 425–433.

Pinto, V. Z., Vanier, N. L., Klein, B., Zavareze, E. D. R., Elias, M. C., Gutkoski, L. C., Helbig, E., & Dias, A. R. G. (2012). Physicochemical, crystallinity, pasting and thermal properties of heat-moisture-treated pinhão starch. *Starch-Stärke, 64*(11), 855–863.

Punia, S. (2020). Barley starch modifications: Physical, chemical and enzymatic—a review. *International Journal of Biological Macromolecules, 144,* 578–585.

Purwani, E. Y., Widaningrum, W., Thahir, R., & Muslich, M. (2006). Effect of heat moisture treatment of sago starch on its noodle quality. *Indonesian Journal of Agricultural Science, 7*(1), 8–14.

Rafiq, S. I., Singh, S., & Saxena, D. C. (2016). Effect of heat-moisture and acid treatment on physicochemical, pasting, thermal and morphological properties of Horse Chestnut (Aesculus indica) starch. *Food Hydrocolloids, 57,* 103–113.

Rocha, T. S., Felizardo, S. G., Jane, J. L., & Franco, C. M. (2012). Effect of annealing on the semicrystalline structure of normal and waxy corn starches. *Food Hydrocolloids, 29*(1), 93–99.

Sair, L., & Fetzer, W. R. (1944). Water sorption by cornstarch and commercial: Modifications of starches. *Industrial and Engineering Chemistry, 36*(4), 316–319.

Sandhu, K. S., Siroha, A. K., Punia, S., & Nehra, M. (2020). Effect of heat moisture treatment on rheological and *in vitro* digestibility properties of pearl millet starches. *Carbohydrate Polymer Technologies and Applications, 1,* 100002.

Schafranski, K., Ito, V. C., & Lacerda, L. G. (2021). Impacts and potential applications: A review of the modification of starches by heat-moisture treatment (HMT). *Food Hydrocolloids, 117,* 106690.

Sharma, M., Yadav, D. N., Singh, A. K., & Tomar, S. K. (2015a). Effect of heat-moisture treatment on resistant starch content as well as heat and shear stability of pearl millet starch. *Agricultural Research, 4*(4), 411–419.

Sharma, M., Yadav, D. N., Singh, A. K., & Tomar, S. K. (2015b). Rheological and functional properties of heat moisture treated pearl millet starch. *Journal of Food Science and Technology, 52,* 6502–6510.

Shi, M., Gao, Q., & Liu, Y. (2018). Corn, potato, and wrinkled pea starches with heat–moisture treatment: Structure and digestibility. *Cereal Chemistry, 95*(5), 603–614.

Shih, F., King, J., Dailek, A. J., & Ali, R. (2007). Physicochemical properties of rice starch modified by hydrothermal treatments. *Cereal Chemistry, 84*, 527–531.

Singh, G. D., Singh, S., Bawa, A. S., Riar, C. S., & Saxena, D. C. (2009). Influence of heat-moisture treatment and acid modifications on physicochemical, rheological, thermal and morphological characteristics of Indian water chestnut (*Trapanatans*) starch and its application in biodegradable films. *Starch/Starke, 61*, 503–513.

Singh, H., Chang, Y. H., Lin, J. H., Singh, N., & Singh, N. (2011). Influence of heat–moisture treatment and annealing on functional properties of sorghum starch. *Food Research International, 44*(9), 2949–2954.

Singh, J., Dartois, A., & Kaur, L. (2010). Starch digestibility in food matrix: A review. *Trends in Food Science & Technology, 21*(4), 168–180.

Singh, N., Singh, J., Kaur, L., Sodhi, N. S., & Gill, B. S. (2003). Morphological, thermal and rheological properties of starches from different botanical sources. *Food Chemistry, 81*, 219–231.

Singh, S., Thakur, S., Singh, M., Kumar, A., Kumar, A., Kumar, A., Punia, R., Kushwaha, J., Kumar, R., & Singh, H. (2018). Influence of different isolation methods on physicochemical and rheological properties of native and heat-moisture-treated chickpea starch. *Journal of Food Processing and Preservation, 42*(2), e13523.

Song, M. R., Choi, S. H., Kim, H. S., Kim, B. Y., & Baik, M. Y. (2015). Efficiency of high hydrostatic pressure in preparing amorphous granular starches. *Starch-Starke, 67*(9–10), 790–801.

Srijunthongsiri, S., Pradipasena, P., & Tulyathan, V. (2016). Influence of heat-moisture modification in the presence of calcium compound on physicochemical properties of pigeon pea [*Cajanus cajan* (L.) *Millsp.*] starch. *Food Hydrocolloids, 53*, 192–198.

Subroto, E., Indiarto, R., Marta, H., & Shalihah, S. (2020). Application of heat-moisture treatment potato starch to improve the heat stability of tomato sauce. In *IOP conference series: Earth and environmental science* (Vol. 443, No. 1, p. 012076). IOP Publishing.

Sun, Q., Han, Z., Wang, L., & Xiong, L. (2014). Physicochemical differences between sorghum starch and sorghum flour modified by heat-moisture treatment. *Food Chemistry, 145*, 756–764.

Tester, R. F., Debon, S. J. J., & Sommerville, M. D. (2000). Annealing of maize starch. *Carbohydrate Polymers, 42*, 287–299.

Tester, R. F., & Morrison, W. R. (1990). Swelling and gelatinization of cereal starches. I. Effects of amylopectin, amylose, and lipids. *Cereal Chemistry, 67*, 551–557.

Varatharajan, V., Hoover, R., Liu, Q., & Seetharaman, K. (2010). The impact of heat-moisture treatment on the molecular structure and physicochemical properties of normal and waxy potato starches. *Carbohydrate Polymers, 81*(2), 466–475.

Viana, E. B. M., Oliveira, N. L., Ribeiro, J. S., Almeida, M. F., Souza, C. C. E., Resende, J. V., Santos, L. S., & Veloso, C. M. (2022). Development of starch-based bioplastics of green plantain banana (*Musa paradisiaca* L.) modified with heat-moisture treatment (HMT). *Food Packaging and Shelf Life, 31*, 100776.

Waduge, R. N., Hoover, R., Vasanthan, T., Gao, J., & Li, J. (2006). Effect of annealing on the structure and physicochemical properties of barley starches of varying amylose content. *Food Research International, 39*, 59–77.

Wang, H., Ding, J., Xiao, N., Liu, X., Zhang, Y., & Zhang, H. (2020). Insights into the hierarchical structure and digestibility of starch in heat-moisture treated adlay seeds. *Food Chemistry, 318*, 126489.

Wang, H., Xu, K., Liu, X., Zhang, Y., Xie, X., & Zhang, H. (2021). Understanding the structural, pasting and digestion properties of starch isolated from frozen wheat dough. *Food Hydrocolloids, 111*, 106168.

Wang, H., Zhang, Y., Su, P., Pan, N., Liu, X., Zhang, Y., & Zhang, H. (2023). Insights into the aggregation structure and physicochemical properties of heat-moisture treated wheat starch and its associated effects on noodle quality. *Journal of Cereal Science, 112*, 103704.

Wang, Q., Li, L., Liu, C., & Zheng, X. (2022). Heat-moisture modified blue wheat starch: Physicochemical properties modulated by its multi-scale structure. *Food Chemistry, 386*, 132771.

Watcharatewinkul, Y., Puttanlek, C., Rungsardthong, V., & Uttapap, D. (2009). Pasting properties of a heat-moisture treated canna starch in relation to its structural characteristics. *Carbohydrate Polymers, 75*(3), 505–511.

Wattananapakasem, I., Penjumras, P., Malaithong, W., Nawong, S., Poomanee, W., & Kinoshita, H. (2021). Effect of heat–moisture treatment of germinated black rice on the physicochemical properties and its utilization by lactic acid bacteria. *Journal of Food Science and Technology, 58*, 4636–4645.

Yadav, B. S., Guleria, P., & Yadav, R. B. (2013). Hydrothermal modification of Indian water chestnut starch: Influence of heat-moisture treatment and annealing on the physicochemical, gelatinization and pasting characteristics. *LWT-Food Science and Technology, 53*(1), 211–217.

Yang, S., Dhital, S., Zhang, M. N., Wang, J., & Chen, Z. G. (2022). Structural, gelatinization, and rheological properties of heat-moisture treated potato starch with added salt and its application in potato starch noodles. *Food Hydrocolloids, 131*, 107802.

Yang, X., Chi, C., Liu, X., Zhang, Y., Zhang, H., & Wang, H. (2019). Understanding the structural and digestion changes of starch in heat-moisture treated polished rice grains with varying amylose content. *International Journal of Biological Macromolecules, 139*, 785–792.

Zavareze, E. D. R., & Dias, A. R. G. (2011). Impact of heat-moisture treatment and annealing in starches: A review. *Carbohydrate Polymers, 83*, 317–328.

Zhang, J., Wang, Z. W., & Yang, J. A. (2010). Physicochemical properties of Canna edulis Ker starch on heat-moisture treatment. *International Journal of Food Properties, 13*(6), 1266–1279.

Zheng, L., Zhang, Q., Yu, X., Luo, X., & Jiang, H. (2023). Effect of annealing and heat-moisture pretreatment on the quality of 3D-printed wheat starch gels. *Innovative Food Science & Emerging Technologies, 84*, 103274.

Zhang, Y., Zhao, X., Bao, X., Xiao, J., & Liu, H. (2021). Effects of pectin and heat-moisture treatment on structural characteristics and physicochemical properties of corn starch. *Food Hydrocolloids, 117*, 106664.

Zhong, Y., Yin, X., Yuan, Y., Kong, X., Chen, S., Ye, X., & Tian, J. (2023). Changes in physiochemical properties and in vitro digestion of corn starch prepared with heat-moisture treatment. *International Journal of Biological Macromolecules, 248*, 125912.

Zia-ud-Din, Xiong, H., & Fei, P. (2017). Physical and chemical modification of starches: A review. *Critical Reviews in Food Science and Nutrition, 57*(12), 2691–2705.

Zou, J., Xu, M., Tang, W., Wen, L., & Yang, B. (2020). Modification of structural, physicochemical and digestive properties of normal maize starch by thermal treatment. *Food Chemistry, 309*, 125733.

Microwave Treatment

3

Muhammed Navaf, Sneh Punia Bangar, and K.V. Sunooj

3.1 INTRODUCTION

Starch is one of the most prevalent carbohydrates on earth. It plays a vital role in our daily lives. It performs as a primary source of energy and is widely explored in different industries, including food, pharmaceuticals, textiles, and paper (Navaf, Sunooj, Aaliya, Akhila et al., 2022). Starch's unique properties, including its ability to form a gel or thicken solutions, make it an ingredient in many products. Even though starch has good functionality, sometimes it does not meet the specific requirements for diverse applications. Hence, there has been growing interest in modifying starch to enhance its functional attributes and expand its utility in recent years (Navaf & Sunooj, 2022). Starch modifications could be achieved by different methods, such as physical, chemical, enzymatic, and their combinations. However, due to increased environmental concerns and cost-effectiveness as compared to chemical modification, physical modifications have gained much popularity. Microwave treatment (MT) is classified under physical modification methods and is cheap and eco-friendly (Bemiller & Huber, 2015).

Microwave radiation can induce rapid heating in food systems. Under microwave conditions, the polar molecules of the system are oriented towards the electric field, which leads to the creation of molecular friction and thus generates heat. This rapid and efficient means of heating by microwave radiation can lead to significant changes in starch functionalities. Studies reported that the MT of starch induces the rearrangement of starch chains and depends on the starch type, microwave power, and treatment duration (Kong, 2019). This chapter contains a synopsis of the microwave modification of starch, its mechanism, and the impacts of microwave modification on different functional attributes of starches.

3.2 HISTORY AND DEVELOPMENT OF MICROWAVE HEATING

Microwaves are a form of electromagnetic radiation with a wavelength between 1 mm and 1 m and frequencies between 300 MHz and 300 GHz. The development of microwave heating was achieved by following the earlier applications of radio frequencies. Before World War II, the work on microwave heating was very limited. The discovery of microwave heating is credited to Percy Spencer, an American engineer, in 1945. While working on radar technology during World War II, he noticed that a candy bar in his pocket had melted after being exposed to microwave radiation emitted by a magnetron (a vacuum tube used in radar systems). Interestingly, Spencer began experimenting with other foods, leading to the development of the first microwave oven. By 1947, the first commercial microwave oven, known as the "Radarange", was introduced by Raytheon, the company Spencer worked for (Orsat et al., 2017).

3.3 FUNDAMENTALS OF STARCH MODIFICATION USING MICROWAVE HEATING

Microwave radiation with lower frequency exhibits better penetrability. Microwave fields are alternating magnetic fields in which the direction of the electric field determines the direction of polarity molecules from the original random thermal motion (Guo et al., 2017). Microwave heating produces micromovements and friction in molecules through high-frequency electric fields, explained by the principle of dielectric loss. On the other hand, MT also induces the molecular friction that arises from the interaction of different molecules present in the treating sample and the oscillating microwave radiation. It produces heat throughout the sample within a short period (Oyeyinka, Akintayo, et al., 2021). These micro-movements and frictions transform electromagnetic energy into thermal energy, resulting in instantaneous heating both internally and externally. The dielectric property of the food determines its response under microwave conditions. Further, the penetration depth and capacity to convert microwaves to heat also impact the food's response under microwaves. Dielectric properties are the characteristics that describe the storage and release of electrostatic energy in an electric field, usually articulated as a dielectric constant or a dielectric loss (Fan et al., 2013).

The temperature-time curve of starches under MT influences the moisture content, starch type, and microwave power. Starches without moisture content are not polarized under the microwave since the dielectric properties mainly rely on the free moisture content present in them. In general, a higher microwave power produces a higher temperature. Likewise, starch containing a low moisture content (1–5%) exhibits a quick rise in temperature; on the other hand, food with a higher moisture content (7–15%) exhibits a lower rise in temperature. Rather, a temperature plateau accompanied by an increasing interval length was observed for starches with high moisture content (>20%) due to isothermal transformation (Brasoveanu & Nemtanu, 2014).

The alteration in starch structure by MT occurs in several stages.

1. The dielectric relaxation of water occurs, which produces initial heating.
2. A quick increase in temperature results in the removal of moisture from the interior of granules.
3. This moisture loss creates internal pressure in the granules and leads to granule expansion starting from the center.
4. In the final stage, granule degradation will occur.

The heat produced during MT generates free radicles with the ability to cleave glycosidic linkages of large starch molecules and depolymerize them into smaller ones. However, some studies reported that MT induces intense movement of starch chains without destroying their intact granules (Oyeyinka, Akintayo, et al., 2021). Oyeyinka et al. (2019) observed significant alteration in starch structure and functional attributes even during a short period (<60 s) of treatment.

3.4 MICROWAVE-INDUCED CHANGES IN STARCH

3.4.1 Starch Composition

The MT of starch does not cause potential changes in the major components of starch. However, it may facilitate minor changes in components like protein and water. It has been reported that MT does not destroy the chemical bonds, groups, or their interaction, indicating the absence of chemical reactions (Wu, 2018). MT induces changes in physicochemical properties like evaporation of water molecules and reduction in protein and fiber content. The MT of taro starch (25% moisture content (mc)) at 180 W for 5 min significantly reduced its protein, fiber, ash, and amylose content (Deka & Sit, 2016). Similarly, 440 and 800 W MT of potato starch for 5 min significantly decreased the moisture and protein content. However, it did not cause any changes in its fat and amylose content (Przetaczek-Rożnowska et al., 2019).

3.4.2 Morphology

Depending on treatment conditions, MT has diverse effects on granule shape, size, and surface attributes. In general, a mild treatment condition did not cause a remarkable effect on the morphological attributes of microwave-treated starches. For example, the granule morphology of maize and potato starches remained unaffected after a low-power MT (2.06 W/g), whereas the granules were severely damaged by high-power microwave (6.63 W/g), and some hollows formed in the starch hilum (Xu et al., 2019). A mild MT of millet starch (20% mc for 20 s) did not cause any noticeable changes in granule morphology. However, a higher moisture content (20–30%) and treatment time (60 and 90 s) led to the destruction of the granules (Li, Hu, Zheng, et al., 2019). The outcome of different exposure times on the surface morphology of Bambara groundnut starch (30% mc) treated at 700 W was reported by Oyeyinka et al. (2019). The MT caused remarkable granule shape and size changes depending on treatment time. A 10-s treatment resulted in granule clump formation while retaining the granule integrity. Further, pore formation and changes in the shape and size of the granules were noticed for 30-s treated samples. A 60-s treatment led to the collapse of starch granules. A 5–10-s MT (700 W) of potato starch evidenced the formation of flaws or fractures in starch granules, whereas it did not affect granule integrity. However, a 15–20-s exposure resulted in severe deformation, breaking, and ruin of starch granules (Xie et al., 2013). At a power of 300 W, the microwave-treated potato starch maintained its granule integrity when treated for 1 min; a higher treatment time (3 min) resulted in a lower degree of disintegration in terms of cracks, deformed shape, and indentations (Kumar et al., 2020). The MT of sago starch (30% mc) at different durations resulted in irregularity in its granular shape (Figure 3.1) (Zailani et al., 2021). On the contrary, MT (30% mc, 1.2 kW) of starch with high amylose content retained its granule integrity even at an exposure of up to 4 min (Zhong et al., 2019).

Similarly, Li, Hu, Wang, et al. (2019) postulated the outcome of different mc on granule morphology of millet starch after MT at a power of 700 W for 60 s. Pores were formed on the granule surface at a mc of 30 and 35%, and noticeable cracks and deep cavities were formed at the center of the granules. However, at higher moisture contents of 40 and 45%, the native granules lost their intact shape, and large gel blocks were formed. It could be concluded that the MT at a higher mc could gelatinize the starch.

The MT of taro starch (25% mc) at 180 W for 5 min induced changes in granule shape and loss of physical integrity compared to the native taro starch (Deka & Sit, 2016). Han et al. (2021) noticed the effect of different microwave heating temperatures (100–160°C) by slow and rapid heating. The MT gradually changed the granule's surface roughness with increased temperature, and some granules broke at higher temperatures. Moreover, rapid heating exhibited more pronounced changes than slow microwave heating. The MT of cassava starches impacts its granule morphology and depends on the treatment power and duration. Oyeyinka, Akinware, et al. (2021) reported the impact of different treatment times and power (600 and 700 W) on the morphological attributes of cassava starch. At both treatment powers, the granule integrity was retained by exposure of 5–15 s, whereas a higher treatment time caused remarkable changes in granule morphology. A 30-s treatment by 600 and 700 W resulted in the formation of indentation and pores, respectively, on granule surfaces, and a 60-s treatment enlarged the size of pores formed.

The atomic force microscopy (AFM) analysis of cassava starch treated with microwave for different times (5–15 min) did not cause any characteristic changes in the granule surface, and it exhibited the depressions and protrusions seen in native granules (Colman et al., 2014). The polarized light micrograph of potato starch microwave treated 5 s slightly changed its Maltese cross, a 10-s microwave exposure led to the disappearance of birefringence of half of the granules, and 15-and 20-s exposures led to the complete disappearance of birefringence (Xie et al., 2013).

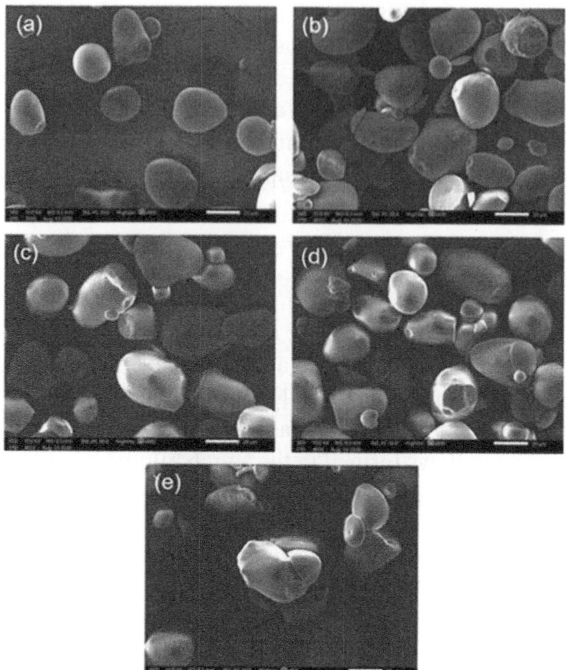

FIGURE 3.1 Micrograph of microwave-modified sago starch at different times: a) 0 min, b) 5 min, c) 10 min, d) 15 min, e) 20 min (Zailani et al., 2021).

3.4.3 Microwave-Induced Changes in Starch Structural Attributes

3.4.3.1 Lamellar and Crystalline Structures

The structure of starch is made up of layers of alternately amorphous and semicrystalline rings. The crystalline part of the structure is made up of double helices that contain chains of more than 10 glucose units long entwined and arranged in a double helix pattern. Starches exhibit four diverse types of crystalline patterns: A-, B-, C-, and V-type patterns. Dissimilarities have been reported in the crystalline pattern and relative crystallinity of starches after MT. In general, MT did not alter the crystalline pattern of starch; besides, it will either raise or lower the relative crystallinity (RC) of the starches, depending on starch and treatment parameters (Oyeyinka, Akintayo, et al., 2021). The MT of cassava starch for the different durations (0–15 min) did not shift its intact crystalline peaks. Also, a reduction in RC was noticed with an increase in exposure duration (Colman et al., 2014). Similarly, A 180 W MT for 5 min retained the A-type crystalline pattern of taro starch (25% mc), whereas it decreased the RC from 33.68 to 25.92% (Deka & Sit, 2016). A 1600-W microwave exposure of waxy maize starch also retained its intact A-type crystalline pattern. It decreased the RC with an increase in treatment duration (Yang et al., 2017). The authors also stated that a short treatment time (5 min) destroyed the amorphous and inferior crystalline regions, and prolonged treatment (10 and 20 min) damaged the entire crystalline region and exhibited a very low value of RC.

The MT of millet starch slurry at different concentrations of starch (10–30 w/w, dry weight basis) and treatment time (30–120 s) significantly reduced its RC (Li, Hu, Zheng, et al., 2019). The RC value was found to be reduced with an increase in starch concentration and prolonged treatment. At all starch concentrations, a 120 s MT starch did not exhibit any crystalline peak; thus, prolonged MT led to the destruction of the crystalline structure. Likewise, moisture content (30–50%) also caused noticeable changes in the crystalline attributes of millet starch (Li, Hu, Wang, et al., 2019). It was noticed that a 1000 W MT for 60 s significantly decreased the peak intensity in all the samples, and the crystalline peaks disappeared when the moisture content reached 40% and above. Furthermore, the intensity of peak and RC of a low-power (300 W) microwave-treated potato starch increased slightly after 1 min exposure. However, a further increase in treatment duration to 3 and 5 min decreased the RC (Kumar et al., 2020). Similarly, the RC of waxy maize starch slightly increased after 1 min MT and further decreased as the treatment duration increased (Zhong et al., 2019). The reduction in RC during prolonged treatment could be due to the following changes in granules:

1. Rearrangement of double helices to an amorphous domain.
2. At higher MT durations, the breakdown of water molecules generates hydroxyl radicals, further decreasing starch's crystallinity.

The impact of MT on crystalline attributes varies among different crystalline-type starches when treated under the same conditions. The B-type crystalline starches are more susceptible to MT than the A-type starches since they have a weaker structure. Xu et al. (2019) also postulated the impact of MT on different crystalline starches (A and B) (30% mc). A low (2.06 W/g) and high (6.63 W/g) power treatment for 5 min retained the crystalline pattern of A-type starch (maize), whereas a high power exposure of potato starch shifted its crystalline pattern from B-type to A-type. Further, the RC of maize starch decreased by 2.2% and 7.8%, and potato starches decreased by 2.6% and 21.0%, respectively, while exposed to a low-power and high-power microwave treatment. The transformation in the crystalline pattern from B type to A type by MT could be due to the following phenomena.

1. The removal of inter-helical water molecules of B-type starches.
2. The shift of double helices to the central cavity, where water molecules were previously present.

Contrary to this, a clear shift in crystalline peaks from A to B type and C to B type was reported by M. Wang et al. (2019) when corn starch and chestnut starch were exposed to MT for 90 s. However, the potato starch did not alter its intact B-type crystalline pattern. It can be concluded that microwave treatment's effect on crystalline patterns depends on the starch type and treatment parameters. The MT breaks down the α-1,6-glycosidic bonds of amylopectin, reducing the number of amylopectin chains, that is, the A-chain. In addition, it was reported that the breakdown of inter- and intramolecular hydrogen bonds, before the breakage of chemical bonds, also contributes to a reduction in diffraction peak strength and RC of starches after MT. Further, the gelatinization of starches by the production of a high temperature during MT also involves the reduction in their RC (Tao et al., 2020). During MT, the movement of double helices of starch occurs, which disrupts the crystallites and alters the crystalline orientation, owing to the reduction in RC (Ma et al., 2020). The enhanced reduction of RC at higher moisture content could be due to the high dielectric property of water, which lead to a more pronounced impact of MT on starch by causing a quick increase in heating and migration of water molecules towards the amorphous region (Li, Hu, Wang, et al., 2019)

3.4.3.2 Molecular Chain Structure

Starch molecules change their molecular structure through bond breaks, resulting in the release of shorter chains or by vibrations of glucose rings leading to the alteration in chain segment conformation and configuration (Tao et al., 2020). The unit chains of amylopectin consist of different fractions, that is, A, B, and a main chain C. The A chain (degree of polymerization (DP) 6–12) is the rearmost chain and is bonded to other unit chains using α-1,6-glycosidic interaction. At the same time, the B chain contains several other chains. The C chain has only one reducing residue per molecule. B1 (DP 13–24), B2 (DP 25–36), and B3

(DP > 36) chains are different sub-divisions of B chains and are short in length. As the A chains form double helices, B chains are involved in the development of the crystalline lamellae, and they are found in the amorphous regions in the branch clusters. Chen et al. (2021) noticed that the MT of hull-less barley starch with different power (640,720 and 800 W) and time (60,80 and 120 s) did not significantly impact its A and B chains. Microwave exposure might initially break the amorphous area and require higher power and extended duration to degrade the branches of the clusters further.

The MT of waxy maize starch (30% mc) at a power of 1600 W for 5 min increased the B1, B2, and B3 chains by 0.02%, 0.11%, and 0.11%, respectively; on the other hand, the amount of A chain lowered by 0.24% (Yang et al., 2017). Further, an increase in treatment time from 10 and 20 min noticeably decreased the amount of A chain by 6.23% and 13.23%, respectively. Long-term exposure to high-energy microwaves can break the crystalline regions of waxy maize starch. Similarly, the MT of potato starch with a power of 750 W significantly decreased its molecular weight. A 5–15 s treatment reduced the molecular weight at a slower rate, whereas prolonged treatment time abruptly reduced the molecular weight (Xie et al., 2013).

3.4.4 Functional Properties

3.4.4.1 Water and Oil Absorption Capacity

The water absorption capacity (WAC) is the measure of the amount of water in starch, which can be held by the unit dry weight of the starch. In general, MT increases the WAC of starches. During MT, the disruption and the development of cracks in the granule facilitate the passage of more water molecules to the granules. Moreover, the intergranular changes from crystalline to amorphous and the removal of water during treatment results in the development of severe strain in the granules, facilitating an easy binding of water with starch molecules. The alteration in the WAC of starches after MT depends on the microwave power and the MT duration. Kumar et al. (2020) postulated that an increase in MT duration from 1 to 5 min resulted in a 1.03- to 1.46-fold rise in WAC. Similarly, Mollekopf, Treppe, Fiala, et al. (2011) and Mollekopf, Treppe, Dixit, et al. (2011) postulated a rise in WAC of potato starch while the microwave power increased from 460 to 750 W and from 600 to 1500 W, respectively.

Like the WAC, the MT alters the oil absorption capacity (OAC) of treated starches. Kumar et al. (2020) noticed a reduction in OAC of microwave-treated potato starch as the exposure time was prolonged from 1 to 5 min. A similar reduction in OAC was noticed in finger millet starch at a power of 510 W for 10 min (Balakumaran et al., 2023). The decline in the amount of hydrophobic sites in starches during MT decreases their OAC. Contrary to this, the OAC of sago starch was found to increase when it was exposed to a microwave power of 900 W for 15 minutes (Zailani et al., 2022). Similarly, the OAC of microwave-treated lotus seed starch increased from 92.87 to 123.93% while exposed to microwave for 5 min (Nawaz et al., 2018). The development of fissures and irregularity on granules lead to the capillary action capturing the oil particles in the amorphous area of starch and, thus, results in increased OAC of microwave-treated starches.

3.4.4.2 Swelling Power and Solubility

The swelling power (SP) of starch in the presence of water includes the association of amorphous and crystalline phases of starches. The MT alters the crystalline and amorphous areas of starch, resulting in SP and solubility changes. In general, the MT of starches has been found to decrease their SP. This could be attributed to the rearrangement of starch chains, especially in crystalline regions, to a more randomly distributed appearance (Oyeyinka, Akintayo, et al., 2021). The studies show that the reduction in SP of microwave-treated starches is influenced by the starch type and treatment conditions like power and duration. For example, Ma et al. (2020) reported the outcome of different microwave power and treatment durations on the SP of waxy and non-waxy hull-less barley starch. It was postulated that the SP of both waxy and non-washed starch decreased with an increase in power (640 to 800 W) and treatment duration (60 to 180 s). Similarly, 700 W MT of Bambara ground nut starch increased the SP while increasing the treatment time from 10 to 60 s. Likewise, an increase in MT duration from 30 to 60 s decreased the SP of microwave-treated (480 W) potato starch. A reduction in the SP of normal and waxy maize starch was noticed after 20 minutes MT (Luo et al., 2006). Similarly, the MT of chestnut starch (30 W for 90 s) (M. Wang et al., 2020), lotus rhizome starch (180 W for 3), white finger millet (510 W for 10 min) (Balakumaran et al., 2023), and others were also found to be decreased. The rearrangement of starch chains in crystalline domine leads to a more random distribution of chains in the granules, which may collectively contribute to the alteration in the SP of starches after MT. On the other side, the SP of sago starch was reported to be increased by prolonging the MT duration from 5 to 20 min. Similarly, an increase in the SP of Indian horse chestnut starch was reported, with a prolonging in treatment duration from 15 to 45 s (Shah et al., 2016). The researchers stated that the higher SP of microwave-treated sago starch could be due to its higher WAC, which is the result of the formation of a hydrophilic hydroxyl group during microwave treatment.

Similar to the SP, the solubility of starches was also significantly altered by microwave treatments. The solubility of normal and waxy maize starches was reported to be decreased when exposed to microwave for 20 minutes (Luo et al., 2006). However, the solubility of Indian horse chestnut starch was found to be increased when the treatment duration was prolonged (Shah et al., 2016). Similarly, the solubility of starches from lotus rhizome (Chandak et al., 2022), sago (Zailani et al., 2022), barley (Ma et al., 2020), and others were found to be decreased after exposure to microwave radiation.

3.4.4.3 Gelatinization Properties

Gelatinization is the process of disordering the molecular structure of starch, resulting in permanent alteration in their functionalities. In its native form, starch can absorb roughly 30% (w/w) of moisture after being suspended in a high amount of water at certain conditions. The amorphous area of starch is easily accessible to water molecules. Granules of starch become amorphous and lose their crystalline structure when heated to a specific temperature with excess moisture content. This temperature-dependent transformation of starch structure is termed gelatinization, and the temperature is termed gelatinization temperature (GT) (Mason, 2009). Dissimilarities were reported in phase transition and gelatinization among starches from different sources. The amount of amylose and the amylopectin chain length are the features that influence the gelatinization properties like onset gelatinization temperature (T_o), peak gelatinization temperature (T_p), conclusion gelatinization temperature (T_c), and enthalpy of gelatinization (ΔH). Generally, starch with low amylose content exhibits higher gelatinization attributes (Oyeyinka, Akintayo, et al., 2021). Yang et al. (2017) postulated that the MT significantly increased the T_o, T_p, and T_c values and reduced the ΔH value of waxy maize starch as the MT time increased from 0 to 20 min at a power of 1600 W. They also stated that the GT is negatively correlated to the short-branched (DP 6–12) amylopectin proportion, destabilizing the crystalline structure. However, it is positively related to long-branched (DP > 36) chains contributing to longer double helices formation and needing higher energy for total destruction. The destruction of a few double helices that existed in both crystalline and non-crystalline domains during microwave exposure may reduce their ΔH value. A relatable change was also noticed in Bambara ground nut starch when it was exposed to a 700-W microwave for 0 to 60 s (Oyeyinka et al., 2019). The GT of microwave-treated samples was reported to be increased with an increase in moisture content. For example, MT (700 W) of millet starch at a mc of 5 to 30% for 30 s significantly increased their T_o, T_p, and T_c values and decreased the ΔH value. However, the endothermic peak disappeared after prolonged treatment (60 and 90 s), denoting a complete gelatinization of starch during prolonged treatment (Li, Hu, Zheng, et al., 2019). Similarly, Li, Hu, Wang, et al. (2019) also postulated an increase in the GT of millet starches after MT when the mc increased from 30–50%. The increased GT with increased mc could be attributed to the enhanced interaction of starch chains in the amorphous domain during the microwave treatment.

Likewise, the increase in treatment power also contributes to an increase in the GT and a reduction in ΔH values. A low-power (2.06 W/g) and high-power (6.63 W/g) treatment on thermal attributes of different crystalline starches were postulated by Xu et al. (2019). The MT potentially increased the T_o, T_p, and T_c values and decreased the ΔH value of potato and maize starches. The high-power MT led to a drastic increase in the GT of both potato and maize starches. A lower and high power treatment increased the T_p value of maize starch to 72.20 and 84.04°C, respectively, from 69.45°C. Likewise, a lower and high power treatment increased the T_p value of potato starch to 78.91 and 80.39°C, respectively, from 76.28°C.

The exposure of cassava starch to different microwave power for different time combinations increased cassava starch's GT (Oyeyinka, Akinware, et al., 2021). However, the increase caused by microwave power is not significant. The preferential destruction of relatively weaker crystallites and the existence of relatively stable crystallites by MT increases the T_o value. The rearrangement of dissociated starch chains and their increased association lead to the development of double helices and stable crystallites, which in turn results in increased T_c value of starches after MT (Xu et al., 2019). The DSC gelatinization curve of microwave-treated cassava starch at different treatment times is given in Figure 3.2. Contrary to this, a short (1 min) exposure of potato starch to a microwave (300 W) increased the T_o, T_p, and T_c values, whereas an increase in treatment time (5 min) reduced the T_p and T_c values (Kumar et al., 2020). The destruction of amylopectin chains during prolonged heating decreases the GT of treated starches.

3.4.4.4 Pasting Properties

In examining the appropriateness of starch for use as a functional ingredient in food and other products, its pasting attributes are taken into consideration. The selection of starch for thickening and binding will be based on their pasting characteristics. Granular starch dispersion has a high viscosity, which is an essential characteristic of its pasting. The high viscosity of the paste suggests its suitability as a thickening

TABLE 3.1 Gelatinization Attributes of Microwave-Treated Starches

STARCH	TREATMENT CONDITIONS	TIME	T_O (°C)	T_P (°C)	T_C (°C)	ΔH (JG^{-1})	REFERENCES
Waxy maize	1600 W, 30% moisture	5 min	71.20	81.00	87.70	11.60	Yang et al. (2017)
Waxy maize	1600 W, 30% moisture	10 min	72.90	82.40	91.30	11.10	Yang et al. (2017)
Waxy maize	1600 W, 30% moisture	20 min	75.10	83.90	93.69	11.20	Yang et al. (2017)
Millet	1000 W, 30% moisture	60 sec	72.65	80.29	84.1	6.75	Li, Hu, Wang, et al. (2019)
Millet	1000 W, 35% moisture	60 sec	72.92	80.41	85.32	4.82	Li, Hu, Wang, et al. (2019)
Millet	1000 W, 40% moisture	60 sec	72.77	80.72	86.07	4.38	Li, Hu, Wang, et al. (2019)
Millet	1000 W, 45% moisture	60 sec	73.29	80.72	89.1	3.72	Li, Hu, Wang, et al. (2019)
Millet	1000 W, 50% moisture	60 sec	73.33	80.76	89.27	2.7	Li, Hu, Wang, et al. (2019)
Maize	2.06 W/g, 30% moisture	5 min	73.13	78.91	85.50	10.78	Xu et al. (2019)
Maize	6.63 W/g, 30% moisture	5 min	77.58	80.39	85.57	7.66	Xu et al. (2019)
Potato	2.06 W/g, 30% moisture	5 min	67.11	72.20	81.00	11.69	Xu et al. (2019)
Potato	6.63 W/g, 30% moisture	5 min	81.72	84.04	88.65	2.93	Xu et al. (2019)
Potato	300 W	1 min	53.8	65.5	66.3	–	Kumar et al. (2020)
Potato	300 W	3 min	57.3	59.4	61.3	–	Kumar et al. (2020)
Potato	300 W	5 min	57.4	59.5	60.8	–	Kumar et al. (2020)
Bambara groundnut	700 W, 30% moisture	10 sec	75.0	79.0	85.0	11.0	Oyeyinka et al. (2019)
Bambara groundnut	700 W, 30% moisture	30 sec	76.0	81.0	87.0	10.3	Oyeyinka et al. (2019)
Bambara groundnut	700 W, 30% moisture	60 sec	76.2	84.2	92.0	10.0	Oyeyinka et al. (2019)

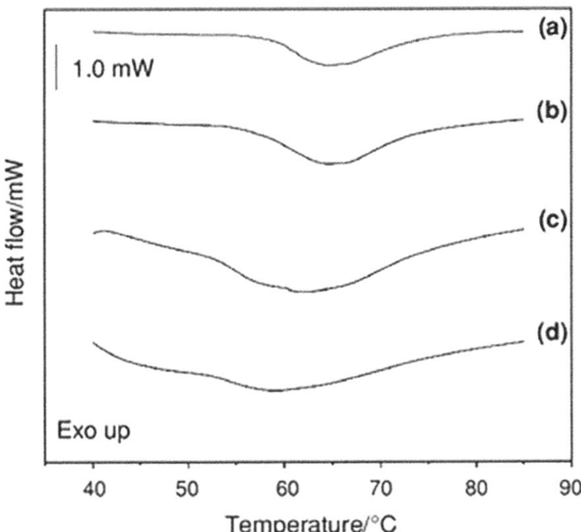

FIGURE 3.2 DSC gelatinization curve of native and microwave-treated cassava starches at different times: a) native, b) 5 minutes, c) 10 minutes, d) 15 minutes (Colman et al., 2014).

agent in diverse applications (Deka & Sit, 2016). Pasting is the process of modifying the starch structure, followed by gelatinization. In pasting, the viscosity of the sample is measured during and after gelatinization. When the starch is continuously heated, starch granules will swell, disrupt, and exudate molecular components, that is, leaching of amylose, which gives viscosity to the aqueous medium. The process of development of viscosity in the previous phenomena is termed pasting. The Rapid Visco-Analyzer (RVA) and Brabender viscograph are tools used to analyze the pasting behavior of starch gels.

Microwave treatments generally decrease the breakdown, trough, final, and setback viscosities. The extent of the impact of microwaves on starch pasting attributes depends on the treatment time and microwave power. An extended MT caused more reduction in pasting attributes. However, the impact on pasting temperature varies with starch sources. For example, the MT of taro starch (25% mc) decreased the pasting temperature (PT) and increased the pasting attributes like peak viscosity (PV), breakdown viscosity (BDV), setback viscosity (SBV), and final viscosity (FV) (Deka & Sit, 2016). On the contrary, MT of cassava starch (30% mc) with different power (600 and 700 W) and time (5–60 s) increased the PT and decreased the pasting attributes (Oyeyinka, Akinware, et al., 2021). A low power (300 W) for varying time (1–5 min) increased the PT and viscosity attributes of microwave-treated potato starch (Kumar et al., 2020). In contrast, MT of waxy and non-waxy hull-less bary starch for different power (640–800 W) and time (60–180 s) and Bambara groundnut starch at 700 W for different times (10–60 s) decreased the PT and pasting attributes. The MT of high amylose maize starch for 2–4 min increased the PV. A short period (1 minute) exhibited a lower PV (Zhong et al., 2019). The RVA profile of microwave-treated cassava starch showed that a short (5 min) microwave exposure potentially increased the PV, BDV, and FV. However, a further increase in treatment time considerably decreased the pasting attributes (Figure 3.3) (Colman et al., 2014).

The Brabender viscogram of microwave-treated starch is varied by several features, and an increase in PT and a reduction in PV were frequently observed. For example, the Brabender viscogram of microwave-treated waxy maize starch (30% mc) showed that an increase in treatment time from 5–20 min significantly increased the PT from 67.80 to 72.10°C. However, it decreased the PV, BDV, SBV, and FV values (Yang et al., 2017). Likewise, the Brabender viscogram of millet starch treated with different mc (10–30%) and treatment time (30–120 s) decreased the PT and viscosities (Li, Hu, Zheng, et al., 2019). A similar finding was observed when millet starch was treated with a mc of 30 to 50% (Li, Hu, Wang, et al., 2019).

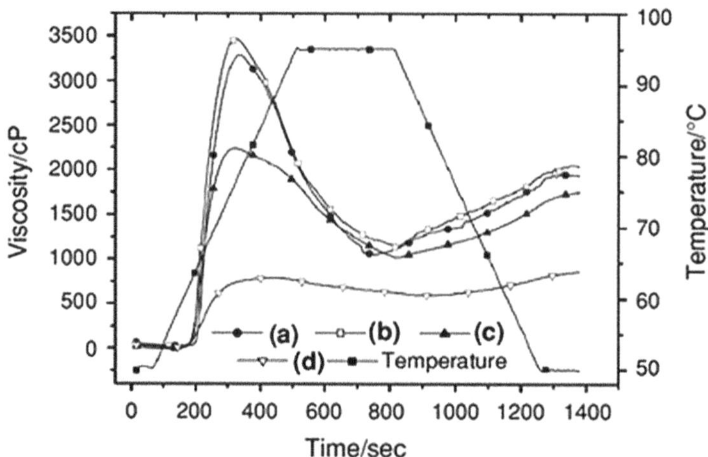

FIGURE 3.3   Pasting profile of native and microwave-treated cassava starches at different times: a) native, b) 5 minutes, c) 10 minutes, d) 15 minutes (Colman et al., 2014).

TABLE 3.2   Pasting Attributes of Microwave-Treated Starches

STARCH	TREATMENT CONDITIONS	TIME	PV	BDV	FV	SBV	REFERENCES
Indian horse chestnut	22% moisture	15 sec	5171 cP	3340 cP	3081 cP	1616 cP	Shah et al. (2016)
Indian horse chestnut	22% moisture	30 sec	3780 cP	2910 cP	2710 cP	1567 cP	Shah et al. (2016)
Indian horse chestnut	22% moisture	45 sec	3040 cP	2410 cP	1689 cP	1213 cP	Shah et al. (2016)
Millet	700 W, 30% moisture	60 sec	546.33 Bu	74 Bu	686.33 Bu	214 Bu	Li, Hu, Wang, et al., (2019)
Millet	700 W, 35% moisture	60 sec	537.33 Bu	76 Bu	683.67 Bu	222.33 Bu	Li, Hu, Wang, et al., (2019)
Millet	700 W, 40% moisture	60 sec	5.22.67 Bu	66.33 Bu	665.67 Bu	209.33 Bu	Li, Hu, Wang, et al., (2019)
Millet	700 W, 45% moisture	60 sec	506.00 Bu	54 Bu	661.33 Bu	209.33 Bu	Li, Hu, Wang, et al., (2019)
Millet	700 W, 50% moisture	60 sec	448.67 Bu	37.67 Bu	613 Bu	202 Bu	Li, Hu, Wang, et al., (2019)
Potato	300 W	1 min	7847cP	5234 cP	4509 cP	1989 cP	Kumar et al., (2020)
Potato	300 W	3 min	9947 cP	6938 cP	5569 cP	2559 cP	Kumar et al., (2020)
Potato	300 W	5 min	9428 cP	6458 cP	5705 cP	3723 cP	Kumar et al., (2020)
Bambara groundnut	700 W, 30% moisture	10 sec	420 RVU	221 RVU	271 RVU	62 RVU	Oyeyinka et al., (2019)
Bambara groundnut	700 W, 30% moisture	30 sec	415RVU	222 RVU	268 RVU	60 RVU	Oyeyinka et al., (2019)
Bambara groundnut	700 W, 30% moisture	60 sec	403 RVU	237 RVU	240 RVU	62 RVU	Oyeyinka et al., (2019)

The increased PT of starches after MT could be due to the enhanced interaction of starch chains in amorphous and crystalline domains during MT, making them highly stable. Hence, granule disintegration needs higher energy (Kumar et al., 2020). The increase in viscosities of starch after MT could be attributed to the breakup of starch granules and realignment of amylopectin chains, which gelatinize gradually as a function of the vibrational motion of polar groups during MT (Deka & Sit, 2016).

3.4.4.5 Retrogradation Properties

Starch retrogradation is the process of realignment and recrystallization of gelatinized material while undergoing cooling and storage, resulting in the release of water from the gel matrix, known as syneresis. Starch retrogradation is mainly studied in terms of syneresis or freeze–thaw stability. During retrogradation, the starch chains like amylose-amylose, amylose-amylopectin, and amylopectin-amylopectin interactions will happen, giving a more ordered structure to the starch gel. The retrogradation of starches significantly impacts the gel texture formation and resistant starch (RS) development. However, it may have certain negative impacts, like staling of bread. The MT of starches significantly impacts their retrogradation properties. Many studies reported that the MT of starches decreases its tendency to retrograde. For example, Yang et al. (2017) reported that the syneresis values of MT of waxy maize were reduced with an increase in treatment time from 5–20 min at a treatment power of 1600 W. Likewise, an increase in MT time from 15–45 s decreased the syneresis of microwave treated Indian horse chestnut starch (Shah et al., 2016). Luo et al. (2006) also postulated a reduction in syneresis and enhanced freeze–thaw stability of normal and waxy maize starches. On the other hand, an increase in the syneresis value was reported in microwave-treated sago starch (900 W for 15 and 20 min) (Zailani et al., 2021).

The structural breakdown and fragmentation of the starch chain under MT produces high amylose starch gels. These have a strong network structure, which reduces the amount of water expelled, thereby reducing the syneresis and retrogradation of modified starches. Likewise, the fragmentation of starch chin by MT forms small-sized starch chains that have higher water retention capacity and thus reduce the syneresis value (Shah et al., 2016).

3.4.4.6 Paste Clarity

It is important to take into account the paste clarity of starches when making food products such as drinks and puddings. A lack of clarity in the paste will result in a vague visual appearance, which may influence consumer interest in it (Navaf et al., 2020). Depending on the MT conditions and the starch origin, the MT of starches alters their light transmittance and paste clarity. Many studies show that MT potentially decreases starches' light transmittance and paste clarity. For example, the MT of sago starch at 900 W for 15 min noticeably decreased its light transmittance (Zailani et al., 2022). Zailani et al. (2021) also postulated a lowering in light transmittance of sago starch gel when it was treated for 5 to 20 min. Further, a 180 W for 3 min MT of lotus rhizome starch also showed a reduction in light transmittance (Chandak et al., 2022). On the other hand, Shah et al. (2016) noticed an increase in light transmittance and paste clarity of microwave-treated horse chestnut starch when it was exposed to a microwave for a period of 15–45 s. The enhanced transmittance of microwave-treated starch past could account for the hydrogen bonding between the OH group of starch and the water molecules, which increases paste clarity.

3.4.4.7 Rheological Attributes

The rheological attributes of a material explain the flow behavior and deformation of the material under stress. The rheological attributes of the starches depend on several factors like chemical structure and concentration of starches, pasting, and storage conditions. The rheological parameters storage modulus (G') and loss modulus (G''), denoting the elasticity and viscosity, respectively (Navaf, Sunooj, Aaliya, Sudheesh, et al., 2022), are determined to study the viscoelastic behavior of the gel. The rheological attributes of the starch could be explained by different means.

3.4.4.7.1 Steady Shear

With an increase in shear rate, the shear stress increases progressively and remains stable as more increases in shear rate occur. This kind of shear-thinning property is mainly observed in pseudo-plastic fluids. The starch paste commonly behaves like a pseudo-plastic and exhibits shear-thinning fluid attributes with a flow behavior index (n) less than 1. The n indicates the degree of shear thinning property, and the consistency index (K) is a rheological attribute that reflects the viscosity. In general, the MT of starches reduces its k value. Also, it increases the n value. During microwave treatment, amylopectin's molecular size is reduced, and molecules might become more mobile, leading to less intertwist between them and higher diffusion rates, which would lower its K value (Chen et al., 2021).

3.4.4.7.2 Dynamic Shear Properties

The temperature plays an important role in the rheological characteristics of the starch. During frequency sweep tests, it was seen that the G' and G" values of starch paste increase with an increase in angular frequency. A higher G' value over the G" value denotes the strong elastic nature of the gel (Navaf et al., 2023; Navaf, Sunooj, Krishna, et al., 2021). The dynamic shear analysis of microwave (300 W)-treated potato starch showed an increase in G' and G" with an increase in angular frequency and a higher value as compared to the native starch. Further, G' and G" were found to be increased with a prolong in treatment duration from 1 to 5 min (Kumar et al., 2020). Similarly, a 700 W MT of potato starch for 0–15 s increased both G' and G" values. However, a further prolonged treatment duration caused a reduction in it (Figure 3.4). The long MT leads to the breakdown of a high molecular weight fraction to a low molecular weight one, and it could not account for the development of a continuous gel network, ending in a reduction in rheological attributes (Xie et al., 2013; Navaf, Sunooj, Gyati, Aaliya, et al., 2022). Likewise, an increase in G' and G" values was noticed in microwave-treated hull-less barley starch when exposed to microwave power of 640, 720, and 800 W and duration of 60, 120, and 180 s (Chen et al., 2021).

3.4.4.8 Starch Digestibility

Starch is the prime carbohydrate source for daily human energy requirements. Hence, its digestibility is of paramount importance. Starch is classified into rapidly digestible starch (RDS), slowly digestible starch (SDS), and RS based on the kinetics of digestibility (Navaf, Sunooj, Aaliya, et al., 2021). RDS is the starch portion that may cause a quick rise in blood glucose content after ingestion, and SDS is the starch portion that gets completely digested in the small intestine at a lower rate. RS is the starch portion that cannot be digested in the small intestine and can ferment in the large intestine by microbes, forming short-chain fatty acids. It could significantly benefit human intestinal health (Oyeyinka, Akintayo, et al., 2021). In general, native starch consists of a higher proportion of RDS and belongs to high-glycemic foods. Several factors that influence the digestibility of starch are botanical origin, amylose content and chain length, degree of crystallinity, and molecular arrangement of amylopectin (Wu, 2018).

In addition, the type of modification also plays a vital place in determining the digestibility of starches. Microwave treatments mostly reduce the digestibility of starches, and it depends on treatment conditions like microwave power, moisture content, and treatment duration. For example, Xu et al. (2019) reported that low- and high-power MT of maize and potato starches exhibited different changes in their digestibility. In the case of maize starch, both low- and high-power MT decreased the amount of both RS and RDS and increased the amount of SDS. Also, in potato starch, low-power treatment reduced the amount of SDS and increased the amount of RS, whereas high-power treatment increased the amount of SDS and reduced the amount of RS as compared to their native counterparts.

Like microwave power, starch's mc also has a paramount impact on the digestibility of microwave-treated starches. Li, Hu, Wang, et al. (2019) reported the impact of mc on the digestibility of millet starches. The digestibility of millet starches was increased with an increase in mc from 30 to 50%. A mc of 30% increased the RDS and SDS content by 143.4 and 15.8%, respectively, and a mc of above 30% exhibited an RDS content of nearly 80%. Similar findings were reported in rice flour's in vitro starch

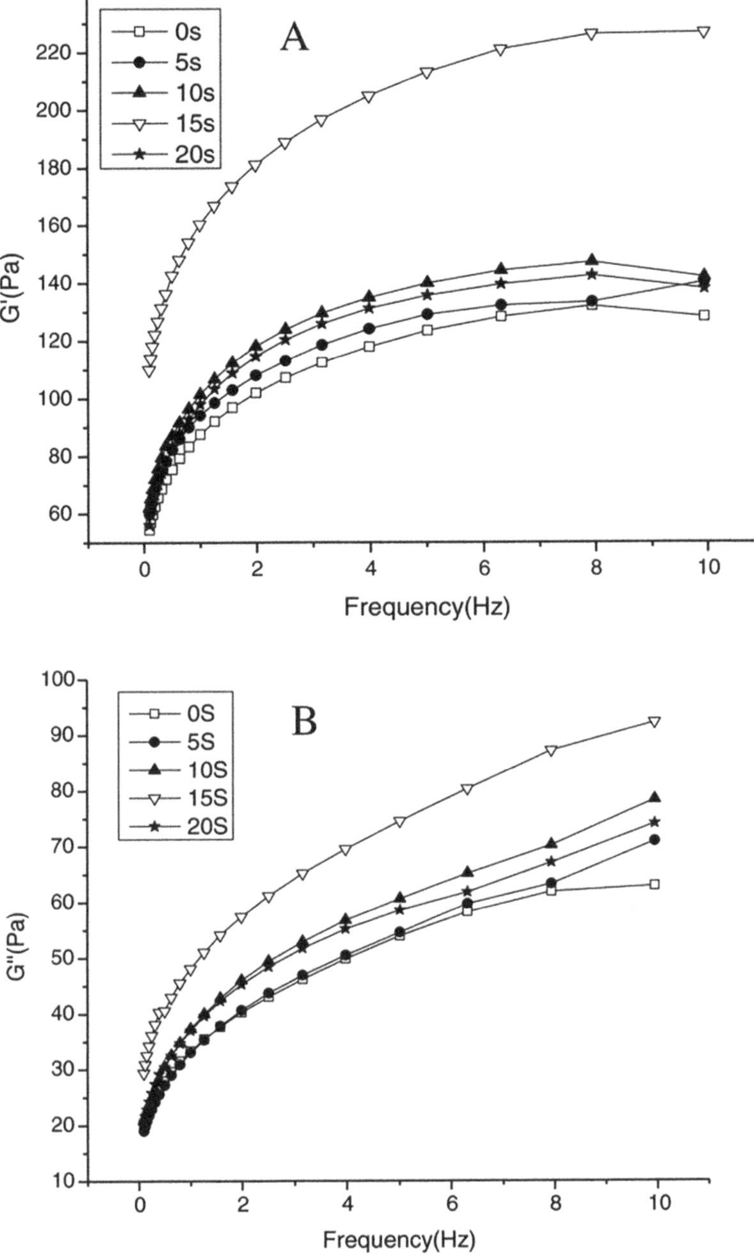

FIGURE 3.4   Rheology of native and microwave-treated potato starch at different treatment times: a) storage modulus, b) loss modulus (Xie et al., 2013).

digestibility when its moisture content increased from 3 to 30% (Solaesa et al., 2022). A lower moisture content (3 and 8) did not cause remarkable changes in the digestibility as compared to the native one. However, a higher moisture content of 20 and 30% increased the RDS content by 40 and 47% while decreasing the SDS content by 48 and 70%, respectively. Likewise, the RS content decreased to a lesser extent between 70 and 50%, at mc of 8, 13, and 15%.

The MT time also significantly impacts the digestibility of starches, depending on the moisture content and treatment power. In general, starch digestibility was increased with increased treatment time. For example, the MT of sweet potato starches (mc <7%) at a power of 480 W potentially increased their digestibility (W. Wang et al., 2022). The amount of RDS and SDS increased, and the amount of RS reduced when exposed for 30 and 60 s. Similarly, Zhong et al. (2019) postulated that a prolonged treatment time (2–4 min) significantly increased the digestibility of maize starch (mc 30%), whereas a lower treatment time (1 min) potentially decreased its digestibility. M. Wang et al. (2019) observed varying results when different crystalline starches (30% mc) were exposed at microwave for 90 s. It increased the digestibility of corn starch and reduced its RS content. However, the digestibility of potato and chestnut starch decreased, and their RS content increased.

The degradation of helical and crystalline structures by MT decreases the degree of branching and facilitates the entry of enzymes into the starch granules, exposing a higher number of sites for enzymatic action and increasing the digestibility. Moreover, the changes in the granule morphology also play a significant impact on starch digestibility. The appearance of cracks and holes in the granule surface also facilitates the diffusion of enzymes and increases digestibility (Xu et al., 2019). On the contrary, the decreased branching chain by MT facilitates the amylose-amylose interaction, and the retrogradation of starches promotes the development of RS (Wu, 2018).

3.5 MICROWAVE-ASSISTED CHEMICAL MODIFICATIONS

In addition to being used as a physical means of starch modification, microwaves are also used to assist different forms of chemical modifications, including esterification, cross-linking, and oxidation. Microwave-assisted chemical modifications enhance the efficacy of modifications since the presence of microwaves may facilitate several structural changes in starch granules, which facilitates better reactivity to the chemical agents.

3.5.1 Microwave-Assisted Esterification

In starch granules, MT may produce both thermal and non-thermal effects and thus can accelerate the rate of chemical modifications like esterification and reduce their reaction time (Hu et al., 2021). The higher temperature produced during MT helps to melt the esterifying agents and facilitates their diffusion into the granules without structural damage or gelatinization of starches (Lu et al., 2020).

The esterification of potato starch using acetic anhydride and glacial acetic acid after microwave pretreatment (600 W) remarkability altered their physicochemical and functional attributes as compared to the acetylation without pretreatment (Zhao et al., 2018). It was postulated that pre-MT of potato starch facilitates the rate of esterification and increases the gelatinization temperature range, SBV, and FV when compared with the untreated starch ester. The MT produces unstable crystals between starch chains, and it damages the partially weaker starch granules, increasing the gelatinization temperature range. Similarly, destroying weak amylopectin chains by microwave treatments produces small segments, which have a greater reassociation tendency and thus increase the setback and final viscosities.

The exposure time of starch to microwave plays a credential role in the degree of substitution (DS) or esterification. Zhang et al. (2014) postulated that the DS of potato starch esterified using vinyl acetate increased with a microwave exposure time of 3–12 min. Similarly, Lu et al. (2020) also postulated an increase in the DS of maize starch esterified with maleic anhydride with a prolonged treatment duration of 3 min. This could be due to the formation of more reactive sites with an increase in treatment duration. The microwave-assisted esterification of mung bean starch with vinyl acetate potentially increased its WAC and decreased the ΔH value with an increase in catalyst concentration (Bushra et al., 2013).

Likewise, microwave-assisted potato starch esterification with citric acid remarkably reduced the amylose content, degree of crystallinity, paste clarity, and digestibility. It increased the freeze–thaw stability (Hu et al., 2021).

3.5.2 Microwave-Assisted Oxidation

The oxidation of starches can be enhanced by using microwave-assisted modification. Microwave activation intensifies the oxidation and reduces the oxidation time and oxidizing agent concentration. Microwave-assisted oxidation was first performed by Lukasiewicz et al. (2011) for the oxidation of potato starch using hydrogen peroxide. Researchers observed that microwave-assisted oxidation resulted in a higher degree of oxidation and a higher depolymerization rate. Komulainen et al. (2013) noticed that the microwave-assisted oxidation of wheat starch by acidic bromate reduced the amount of oxidant from 1.05 to 0.1–0.25 equiv lowered their reaction time from 2–5.5 h to 10 min, and it gave higher yield of oxidized starch.

Likewise, the synergetic microwave-assisted oxidation of Adlay starch by ozone exhibited excellent physicochemical and functional attributes compared to individual treated starches (Subroto et al., 2022). The authors also reported that the synergetic modification by giving a 6-minute MT followed by oxidation for 20 minutes exhibited superior quality. It exhibited higher swelling ability and water absorption ability. However, an opposite trend was noticed in the pasting profile of synergetic modified starch. The PV and BV were found to be lower for microwave-assisted oxidized starch. The damage and reduction in granule size affect the granule integrity and reduce the PV of dual-modified starches. A low BV of dual-modified starch indicates the stability of the starch granules under shear force.

3.5.3 Microwave-Assisted Crosslinking

The MT of starches results in several changes in structural and functional attributes. Studies showed that microwave-assisted crosslinking of starch enhances its reaction efficacy, and the degree of crosslinking increases with prolonged MT duration. Microwave-induced crosslinking provides a convenient and efficient method for the synthesis of crosslinked starch that has a higher DS value in a short period (Gui-Jie et al., 2006). Researchers observed that the microwave-assisted crosslinking of corn starch with sodium trimetaphosphate considerably increased the DS value and dramatically reduced reaction time. A higher microwave power enhances reaction kinetics to achieve the desired level of crosslinking (Gui-Jie et al., 2006).

The microwave-assisted crosslinking of banana starch exhibited an increasing trend in the degree of crosslinking with a prolonged treatment duration (Surendra Babu et al., 2018). Researchers also reported that the dual-modified starch exhibited a lower retrogradation property and freeze–thaw stability. Likewise, microwave-assisted crosslinking of talipot starch exhibited reduced swelling power, pasting properties, and in vitro digestibility, whereas it improved solubility and paste clarity as compared to the starch crosslinked by conventional method (Aaliya et al., 2022). Similarly, microwave-assisted crosslinking of breadfruit starch exhibited higher RS content and swelling power, whereas MT of crosslinked starch reduced solubility, WAC, and OAC (Otemuyiwa & Aina, 2021).

3.5.4 Microwave-Assisted Carboxymethylation

The microwave is an efficient source of heat energy, allowing substances to be heated quickly and uniformly. Under microwave radiation, the molecules become more energetic and thus lead to an enhanced reaction rate under microwave-assisted reactions. Liu et al. (2012) reported that under microwave conditions, a 1–25 min reaction time is enough to attain a DS range from 0.06 to 0.32. Instead, it takes 150 min in conventional methods of carboxymethylation for the carboxymethylation of potato starch. Similarly, the microwave-assisted carboxymethylation of corn starch diminished the reaction duration and improved the physicochemical and functional attributes (Zhang et al., 2015).

3.6 APPLICATION OF MICROWAVE-TREATED STARCH

In general, MT of starch induces molecular rearrangement. This results in changes in its functional attributes like swelling ability, pasting and rheological attributes, and gelatinization properties. Hence, microwave-treated and microwave-assisted modified starches have broad applicability across various domains. MT lowers starches' SBV and retrogradation ability and could be explored as an ingredient in bread manufacturing that helps prevent the staling of bread (Roman et al., 2020). A study reported that the MT increased the water-holding capacity of the cassava starch. The gluten-free bread prepared from microwave-treated cassava starch and chickpea flour exhibited better quality attributes than the control bread. Bread prepared from microwave-treated ingredients exhibited lower baking loss and higher porosity (Garske et al., 2023). Similarly, the improved thermal properties of microwave-treated starch could be explored as an additive in high-temperature processing foods like extruded, canned, and baked products. Owing to the reduced digestibility of microwave-treated starch, it could be explored in low-calorie food products (Navaf, Sunooj, Aaliya, et al., 2021). The partial gelatinization of microwave-treated starch could promote its utility in many products since it eases cooking and reduces cooking time (Xie et al., 2013). The preheating of the dough using a microwave reduced the cooking time of the noodles (Xue et al., 2008). Further, microwave-modified starches could be explored in controlled drug release (Malafaya et al., 2001).

3.7 ADVANTAGES AND DISADVANTAGES OF MICROWAVE HEATING

Microwave radiation is advantageous for processing because it can heat materials rapidly, selectively, systematically, and environmentally cleanly. Moreover, microwave energy generates heat primarily in the load, not in the oven walls or the atmosphere, resulting in a lower operational cost. There are some disadvantages associated with microwave processing, such as non-uniform heating caused by too high energy transfer rates, the laborious optimization of processing parameters, and high-cost specialized equipment. It is difficult to heat transparent materials in the microwave frequency range due to the lack of dielectric data at a range of temperatures and the lack of data on the dielectric properties of microwave dielectrics. Microwave heating can result in uneven heating and "hot spots" or thermal runaway in some dielectric materials whose dielectric loss increases with temperature. Therefore, uniform temperature control during microwave heating is a difficult task that requires accurate temperature control (Brasoveanu & Nemtanu, 2014).

3.8 CONCLUSION

Microwave treatment is an efficient and rapid tool for starch modification, and it offers several benefits over conventional methods. Further, it offers a promising avenue for tailoring its properties to meet the demands of different industries. One of the key advantages of microwave modification is its ability to achieve the desired modification in a short period, potentially reducing the processing time. Microwave modification of starches potentially alters their physicochemical, morphological, structural, and functional attributes, depending on the starch type, microwave power, treatment duration, and moisture content. Understanding the mechanism of microwave action and the alteration in the functional attributes of microwave-treated starch paves the way for the innovative applications of microwave-treated starches in different fields.

REFERENCES

Aaliya, B., Sunooj, K. V., John, N. E., Navaf, M., Akhila, P. P., Sudheesh, C., Sabu, S., Sasidharan, A., Mir, S. A., & George, J. (2022). Impact of microwave irradiation on chemically modified talipot starches: A characterization study on heterogeneous dual modifications. *International Journal of Biological Macromolecules, 209,* 1943–1955. https://doi.org/10.1016/j.ijbiomac.2022.04.172

Balakumaran, M., Gokul Nath, K., Giridharan, B., Dhinesh, K., Dharunbalaji, A. K., Malini, B., & Sunil, C. K. (2023). White finger millet starch: Hydrothermal and microwave modification and its characterisation. *International Journal of Biological Macromolecules, 242,* 124619. https://doi.org/10.1016/j.ijbiomac.2023.124619

Bemiller, J. N., & Huber, K. C. (2015). Physical modification of food starch functionalities. *Annual Review of Food Science and Technology, 6,* 19–69. https://doi.org/10.1146/annurev-food-022814-015552

Brasoveanu, M., & Nemtanu, M. R. (2014). Behaviour of starch exposed to microwave radiation treatment. *Starch/Staerke, 66*(1–2), 3–14. https://doi.org/10.1002/star.201200191

Bushra, M., Xu, X. Y., & Pan, S. Y. (2013). Microwave assisted acetylation of mung bean starch and the catalytic activity of potassium carbonate in free-solvent reaction. *Starch/Staerke, 65*(3–4), 236–243. https://doi.org/10.1002/star.201200081

Chandak, A., Dhull, S. B., Chawla, P., Fogarasi, M., & Fogarasi, S. (2022). Effect of single and dual modifications on properties of lotus rhizome starch modified by microwave and γ-irradiation: A comparative study. *Foods, 11*(19), 2969. https://doi.org/10.3390/foods11192969

Chen, X., Liu, Y., Xu, Z., Zhang, C., Liu, X., Sui, Z., & Corke, H. (2021). Microwave irradiation alters the rheological properties and molecular structure of hull-less barley starch. *Food Hydrocolloids, 120,* 1–7. https://doi.org/10.1016/j.foodhyd.2021.106821

Colman, T. A. D., Demiate, I. M., & Schnitzler, E. (2014). The effect of microwave radiation on some thermal, rheological and structural properties of cassava starch. *Journal of Thermal Analysis and Calorimetry, 115*(3), 2245–2252. https://doi.org/10.1007/s10973-012-2866-5

Deka, D., & Sit, N. (2016). Dual modification of taro starch by microwave and other heat moisture treatments. *International Journal of Biological Macromolecules, 92,* 416–422. https://doi.org/10.1016/j.ijbiomac.2016.07.040

Fan, D., Ma, W., Wang, L., Huang, J., Zhang, F., Zhao, J., Zhang, H., & Chen, W. (2013). Determining the effects of microwave heating on the ordered structures of rice starch by NMR. *Carbohydrate Polymers, 92*(2), 1395–1401. https://doi.org/10.1016/j.carbpol.2012.09.072

Garske, R. P., Mercali, G. D., Thys, R. C. S., & Cladera-Olivera, F. (2023). Cassava starch and chickpea flour pretreated by microwave as a substitute for gluten-free bread additives. *Journal of Food Science and Technology, 60*(1), 53–63. https://doi.org/10.1007/s13197-022-05586-y

Gui-Jie, M., Peng, W., Xiang-Sheng, M., Xing, Z., & Tong, Z. (2006). Crosslinking of corn starch with sodium trimetaphosphate in solid state by microwave irradiation. *Journal of Applied Polymer Science, 102*(6), 5854–5860. https://doi.org/10.1002/app.24947

Guo, Q., Sun, D. W., Cheng, J. H., & Han, Z. (2017). Microwave processing techniques and their recent applications in the food industry. *Trends in Food Science and Technology, 67,* 236–247. https://doi.org/10.1016/j.tifs.2017.07.007

Han, Z., Li, Y., Luo, D. H., Zhao, Q., Cheng, J. H., & Wang, J. H. (2021). Structural variations of rice starch affected by constant power microwave treatment. *Food Chemistry, 359,* 129887. https://doi.org/10.1016/j.foodchem.2021.129887

Hu, A., Chen, X., Wang, J., Wang, X., Zheng, J., & Wang, L. (2021). Effects on the structure and properties of native corn starch modified by enzymatic debranching (ED), microwave assisted esterification with citric acid

(MCAE) and by the dual ED/MCAE treatment. *International Journal of Biological Macromolecules, 171,* 123–129. https://doi.org/10.1016/j.ijbiomac.2021.01.012

Komulainen, S., Diaz, E., Pursiainen, J., & Lajunen, M. (2013). Fast and efficient benign oxidation of native wheat starch by acidic bromate under microwave activation. *Carbohydrate Research, 367,* 58–62. https://doi.org/10.1016/j.carres.2012.11.018

Kong, X. (2019). Starches modified by nonconventional techniques and food applications. In *Starches for food application* (pp. 271–295). Elsevier. https://doi.org/10.1016/B978-0-12-809440-2.00007-1

Kumar, Y., Singh, L., Sharanagat, V. S., Patel, A., & Kumar, K. (2020). Effect of microwave treatment (low power and varying time) on potato starch: Microstructure, thermo-functional, pasting and rheological properties. *International Journal of Biological Macromolecules, 155,* 27–35. https://doi.org/10.1016/j.ijbiomac.2020.03.174

Li, Y., Hu, A., Wang, X., & Zheng, J. (2019). Physicochemical and in vitro digestion of millet starch: Effect of moisture content in microwave. *International Journal of Biological Macromolecules, 134,* 308–315. https://doi.org/10.1016/j.ijbiomac.2019.05.046

Li, Y., Hu, A., Zheng, J., & Wang, X. (2019). Comparative studies on structure and physiochemical changes of millet starch under microwave and ultrasound at the same power. *International Journal of Biological Macromolecules, 141,* 76–84. https://doi.org/10.1016/j.ijbiomac.2019.08.218

Liu, J., Ming, J., Li, W., & Zhao, G. (2012). Synthesis, characterisation and in vitro digestibility of carboxymethyl potato starch rapidly prepared with microwave-assistance. *Food Chemistry, 133*(4), 1196–1205. https://doi.org/10.1016/j.foodchem.2011.05.061

Lu, K., Zhu, J., Bao, X., Liu, H., Yu, L., & Chen, L. (2020). Effect of starch microstructure on microwave-assisted esterification. *International Journal of Biological Macromolecules, 164,* 2550–2557. https://doi.org/10.1016/j.ijbiomac.2020.08.099

Lukasiewicz, M., Bednarz, S., & Ptaszek, A. (2011). Environmental friendly polysaccharide modification—microwave-assisted oxidation of starch. *Starch/Staerke, 63*(5), 268–273. https://doi.org/10.1002/star.201000124

Luo, Z., He, X., Fu, X., Luo, F., & Gao, Q. (2006). Effect of microwave radiation on the physicochemical properties of normal maize, waxy maize and amylomaize V starches. *Starch/Staerke, 58*(9), 468–474. https://doi.org/10.1002/star.200600498

Ma, M., Zhang, Y., Chen, X., Li, H., Sui, Z., & Corke, H. (2020). Microwave irradiation differentially affect the physicochemical properties of waxy and non-waxy hull-less barley starch. *Journal of Cereal Science, 95,* 103072. https://doi.org/10.1016/j.jcs.2020.103072

Malafaya, P. B., Elvira, C., Gallardo, A., San Román, J., & Reis, R. L. (2001). Porous starch-based drug delivery systems processed by a microwave route. *Journal of Biomaterials Science, Polymer Edition, 12*(11), 1227–1241. https://doi.org/10.1163/156856201753395761

Mason, W. R. (2009). Starch use in foods. In J. BeMiller & R. Whistler (Eds.), *Starch chemistry and technology* (3rd ed., Vol. 1, Issue 1, pp. 745–795). Academic Press. https://doi.org/10.1016/B978-0-12-746275-2.00020-3

Mollekopf, N., Treppe, K., Dixit, O., Bauch, J., & Führlich, T. (2011). Simultaneous drying and modification of properties of starch using vacuum microwave treatment. *Drying Technology, 29*(5), 599–605. https://doi.org/10.1080/07373937.2010.519839

Mollekopf, N., Treppe, K., Fiala, P., & Dixit, O. (2011). Vacuum microwave treatment of potato starch and the resultant modification of properties. *Chemie-Ingenieur-Technik, 83*(3), 262–272. https://doi.org/10.1002/cite.201000105

Navaf, M., & Sunooj, K. V. (2022). Impact of different cross-linking agents on functional, rheological, and structural properties of talipot palm starch: A new source of stem starch. *Biology and Lifescience, 20,* 16. https://doi.org/10.3390/iecbm2022-13391

Navaf, M., Sunooj, K. V., Aaliya, B., Akhila, P. P., Sudheesh, C., Sinha, S. K., Murugesan, P., Sabu, S., Sasidharan, A., Mir, S. A., George, J., & Lackner, M. (2022). Impact of low-pressure argon plasma on structural, thermal, and rheological properties of Corypha umbraculifera L. starch: A non-conventional source of stem pith starch. *Starch—Stärke, 75,* 2200165. https://doi.org/10.1002/star.202200165

Navaf, M., Sunooj, K. V., Aaliya, B., Sudheesh, C., Akhila, P. P., Sabu, S., Sasidharan, A., & George, J. (2021). Talipot palm (Corypha umbraculifera L.) a nonconventional source of starch: Effect of citric acid on structural, rheological, thermal properties and in vitro digestibility. *International Journal of Biological Macromolecules, 182,* 554–563. https://doi.org/doi.org/10.1016/j.ijbiomac.2021.04.035

Navaf, M., Sunooj, K. V., Aaliya, B., Sudheesh, C., Akhila, P. P., Sabu, S., Sasidharan, A., & George, J. (2022). Impact of gamma irradiation on structural, thermal, and rheological properties of talipot palm (Corypha umbraculifera L.) starch: A stem starch. *Radiation Physics and Chemistry, 201,* 110459. https://doi.org/10.1016/j.radphyschem.2022.110459

Navaf, M., Sunooj, K. V., Aaliya, B., Sudheesh, C., & George, J. (2020). Physico-chemical, functional, morphological, thermal properties and digestibility of Talipot palm (Corypha umbraculifera L.) flour and starch grown in

Malabar region of South India. *Journal of Food Measurement and Characterization, 14*(3), 1601–1613. https://doi.org/10.1007/s11694-020-00408-1

Navaf, M., Sunooj, K. V., Gyati, R., Aaliya, B., Sudheesh, C., Akhila, P. P., Akhila, V., Sabu, S., Sasidharan, A., Mir, S. A., & George, J. (2022). Improving the structural, functional, and rheological properties of nonconventional stem pith starch from Corypha umbraculifera, by different chemical methods: A characterization study. *Journal of Food Measurement and Characterization, 17*, 1921–1931. https://doi.org/10.1007/s11694-022-01761-z

Navaf, M., Sunooj, K. V., Krishna, N. U., Aaliya, B., Sudheesh, C., Akhila, P. P., Sabu, S., Sasidharan, A., Mir, S. A., & George, J. (2021). Effect of different hydrothermal treatments on pasting, textural, and rheological properties of single and dual modified corypha Umbraculifera L. starch. *Starch/Starke, 74*, 2100236. https://doi.org/10.1002/star.202100236

Navaf, M., Sunooj, K. V., Saji, H., Aaliya, B., Akhila, P. P., Mir, S. A., Yadav, D. N., Lackner, M., George, J., & Nemţanu, M. R. (2023). Ultrasound and gamma-irradiation assisted development of starch-fatty acid complex: Impact of palmitic acid and stearic acid on structural, functional properties of Corypha umbraculifera L. starch. *Biomass Conversion and Biorefinery*, 1–13. https://doi.org/10.1007/s13399-023-04863-w

Nawaz, H., Shad, M. A., Saleem, S., Khan, M. U. A., Nishan, U., Rasheed, T., Bilal, M., & Iqbal, H. M. N. (2018). Characteristics of starch isolated from microwave heat treated lotus (Nelumbo nucifera) seed flour. *International Journal of Biological Macromolecules, 113*, 219–226. https://doi.org/10.1016/j.ijbiomac.2018.02.125

Orsat, V., Raghavan, G. S. V., & Krishnaswamy, K. (2017). Microwave technology for food processing: An overview of current and future applications. In *The microwave processing of foods* (2nd ed., pp. 100–116). Woodhead Publishing. https://doi.org/10.1016/B978-0-08-100528-6.00005-X

Otemuyiwa, I. O., & Aina, A. F. (2021). Physicochemical properties and in-vitro digestibility studies of microwave assisted chemically modified breadfruit (Artocarpus altilis) starch. *International Journal of Food Properties, 24*(1), 140–151. https://doi.org/10.1080/10942912.2020.1861007

Oyeyinka, S. A., Akintayo, O. A., Adebo, O. A., Kayitesi, E., & Njobeh, P. B. (2021). A review on the physicochemical properties of starches modified by microwave alone and in combination with other methods. *International Journal of Biological Macromolecules, 176*, 87–95. https://doi.org/10.1016/j.ijbiomac.2021.02.066

Oyeyinka, S. A., Akinware, R. O., Bankole, A. T., Njobeh, P. B., & Kayitesi, E. (2021). Influence of microwave heating and time on functional, pasting and thermal properties of cassava starch. *International Journal of Food Science and Technology, 56*(1), 215–223. https://doi.org/10.1111/ijfs.14621

Oyeyinka, S. A., Umaru, E., Olatunde, S. J., & Joseph, J. K. (2019). Effect of short microwave heating time on physicochemical and functional properties of Bambara groundnut starch. *Food Bioscience, 28*, 36–41. https://doi.org/10.1016/j.fbio.2019.01.005

Przetaczek-Rożnowska, I., Fortuna, T., Wodniak, M., Łabanowska, M., Pająk, P., & Królikowska, K. (2019). Properties of potato starch treated with microwave radiation and enriched with mineral additives. *International Journal of Biological Macromolecules, 124*, 229–234. https://doi.org/10.1016/j.ijbiomac.2018.11.153

Roman, L., Reguilon, M. P., Martinez, M. M., & Gomez, M. (2020). The effects of starch cross-linking, stabilization and pre-gelatinization at reducing gluten-free bread staling. *LWT-Food Science and Technology, 132*, 109908. https://doi.org/10.1016/j.lwt.2020.109908

Shah, U., Gani, A., Ashwar, B. A., Shah, A., Wani, I. A., & Masoodi, F. A. (2016). Effect of infrared and microwave radiations on properties of Indian Horse Chestnut starch. *International Journal of Biological Macromolecules, 84*, 166–173. https://doi.org/10.1016/j.ijbiomac.2015.12.020

Solaesa, Á. G., Villanueva, M., Vela, A. J., & Ronda, F. (2022). Impact of microwave radiation on in vitro starch digestibility, structural and thermal properties of rice flour: From dry to wet treatments. *International Journal of Biological Macromolecules, 222*, 1768–1777. https://doi.org/10.1016/j.ijbiomac.2022.09.262

Subroto, E., Filianty, F., Indiarto, R., & Andita Shafira, A. (2022). Physicochemical and functional properties of modified adlay starch (Coix lacryma-jobi) by microwave and ozonation. *International Journal of Food Properties, 25*(1), 1622–1634. https://doi.org/10.1080/10942912.2022.2096061

Surendra Babu, A., Mohan Naik, G. N., James, J., Aboobacker, A. B., Eldhose, A., & Jagan Mohan, R. (2018). A comparative study on dual modification of banana (Musa paradisiaca) starch by microwave irradiation and cross-linking. *Journal of Food Measurement and Characterization, 12*(3), 2209–2217. https://doi.org/10.1007/s11694-018-9837-x

Tao, Y., Yan, B., Fan, D., Zhang, N., Ma, S., Wang, L., Wu, Y., Wang, M., Zhao, J., & Zhang, H. (2020). Structural changes of starch subjected to microwave heating: A review from the perspective of dielectric properties. *Trends in Food Science and Technology, 99*(August 2019), 593–607. https://doi.org/10.1016/j.tifs.2020.02.020

Wang, M., Sun, M., Zhang, Y., Chen, Y., Wu, Y., & Ouyang, J. (2019, June). Effect of microwave irradiation-retrogradation treatment on the digestive and physicochemical properties of starches with different crystallinity. *Food Chemistry, 298*. https://doi.org/10.1016/j.foodchem.2019.125015

Wang, M., Wu, Y., Liu, Y., & Ouyang, J. (2020). Effect of ultrasonic and microwave dual-treatment on the physicochemical properties of chestnut starch. *Polymers, 12*(8), 1718. https://doi.org/10.3390/polym12081718

Wang, W., Hu, A., Li, J., Liu, G., & Wang, M. (2022). Comparison of physicochemical properties and digestibility of sweet potato starch after two modifications of microwave alone and microwave-assisted L-malic acid. *International Journal of Biological Macromolecules, 210*(29), 614–621.

Wu, K. (2018). Microwave treatment. In X. K. Z. Sui (Ed.), *Physical modifications of starch* (pp. 97–117). Springer Nature. https://doi.org/10.1007/978-981-13-0725-6_6

Xie, Y., Yan, M., Yuan, S., Sun, S., & Huo, Q. (2013). Effect of microwave treatment on the physicochemical properties of potato starch granules. *Chemistry Central Journal, 7*(1). https://doi.org/10.1186/1752-153X-7-113

Xu, X., Chen, Y., Luo, Z., & Lu, X. (2019). Different variations in structures of A- and B-type starches subjected to microwave treatment and their relationships with digestibility. *LWT—Food Science and Technology, 99*, 179–187. https://doi.org/10.1016/j.lwt.2018.09.072

Xue, C., Sakai, N., & Fukuoka, M. (2008). Use of microwave heating to control the degree of starch gelatinization in noodles. *Journal of Food Engineering, 87*(3), 357–362. https://doi.org/10.1016/j.jfoodeng.2007.12.017

Yang, Q., Qi, L., Luo, Z., Kong, X., Xiao, Z., Wang, P., & Peng, X. (2017). Effect of microwave irradiation on internal molecular structure and physical properties of waxy maize starch. *Food Hydrocolloids, 69*, 473–482. https://doi.org/10.1016/j.foodhyd.2017.03.011

Zailani, M. A., Kamilah, H., Husaini, A., Awang Seruji, A. Z. R., & Sarbini, S. R. (2022). Functional and digestibility properties of sago (Metroxylon sagu) starch modified by microwave heat treatment. *Food Hydrocolloids, 122*, 107042. https://doi.org/10.1016/j.foodhyd.2021.107042

Zailani, M. A., Kamilah, H., Husaini, A., & Sarbini, S. R. (2021). Physicochemical properties of microwave heated sago (Metroxylon sagu) starch. *CYTA—Journal of Food, 19*(1), 596–605. https://doi.org/10.1080/19476337.2021.1934550

Zhang, H., Wang, J. K., Liu, W. J., & Li, F. Y. (2015). Microwave-assisted synthesis, characterization, and textile sizing property of carboxymethyl corn starch. *Fibers and Polymers, 16*(11), 2308–2317. https://doi.org/10.1007/s12221-015-5321-y

Zhang, H., Wang, R., Wang, J., & Dong, Y. (2014). Microwave-assisted synthesis and characterization of acetylated corn starch. *Starch/Staerke, 66*(5–6), 515–523. https://doi.org/10.1002/star.201300165

Zhao, K., Li, B., Xu, M., Jing, L., Gou, M., Yu, Z., Zheng, J., & Li, W. (2018). Microwave pretreated esterification improved the substitution degree, structural and physicochemical properties of potato starch esters. *LWT—Food Science and Technology, 90*, 116–123. https://doi.org/10.1016/j.lwt.2017.12.021

Zhong, Y., Liang, W., Pu, H., Blennow, A., Liu, X., & Guo, D. (2019). Short-time microwave treatment affects the multi-scale structure and digestive properties of high-amylose maize starch. *International Journal of Biological Macromolecules, 137*, 870–877. https://doi.org/10.1016/j.ijbiomac.2019.07.025

Pre-Gelatinized Starch

4

Lu Liu, Wenjing Chen, Ru Jia, Baodong Zheng,
and Zebin Guo

ABBREVIATIONS

PGS   Pre-gelatinized starch
T_o Initial gelatinization temperature
T_p Peak gelatinization temperature
T_c Termination gelatinization temperature
ΔH Gelatinization enthalpy value

4.1 INTRODUCTION

Starch modification can improve the physical, chemical, and functional properties of natural starch itself and expand the application range of starch to meet the needs of production. At present, the modification technology of starch has been widely studied and can be summarized as physical modification, chemical modification, enzymatic modification, and combined modification technology. The physical, chemical, and functional properties of modified starch will change to some extent, and the selection of modification technology often needs to consider the production demand and the composition and structure characteristics of starch itself. Amylose and amylopectin are arranged to form a starch particle structure with a partial crystalline region, while natural starch has a microcrystalline bundle structure, so it cannot be dissolved in room-temperature water. Pre-gelatinized starch (PGS), also known as α-starch, refers to a class of physically modified starch that is gelatinized by a certain heating method and quickly dried at high-temperature conditions. It makes up for the shortcomings of natural starch insoluble in cold water in specific food processing and has good adhesion. It can improve the viscosity of the system and play a role in thickening and has a wide range of applications in staple foods, baked goods, meat products, snack foods, and other products (Roman et al., 2020; Fitriani et al., 2022; Sobini et al., 2022).

The essence of starch gelatinization is the breaking of hydrogen bonds between crystalline and amorphous starch molecules, that is, the melting of starch granule crystals. As shown in Figure 4.1, in this process, the amorphous part of starch particles first absorbs water and expands. Then, as the temperature rises under the action of an external heat source, water begins to enter the starch crystal gap. In this process, the water absorption capacity of starch is greatly improved, the phenomenon of starch birefringence is gradually blurred or even disappears, and the volume of starch particles increases to 50 to 100 times the original. Finally, the starch granules disintegrate, all the starch molecules enter the solution, and the starch is completely gelatinized (Roulet et al., 1988). Therefore, factors that can affect the amorphous and crystalline starch granules include the amylose/amylopectin content of the starch itself, starch crystal

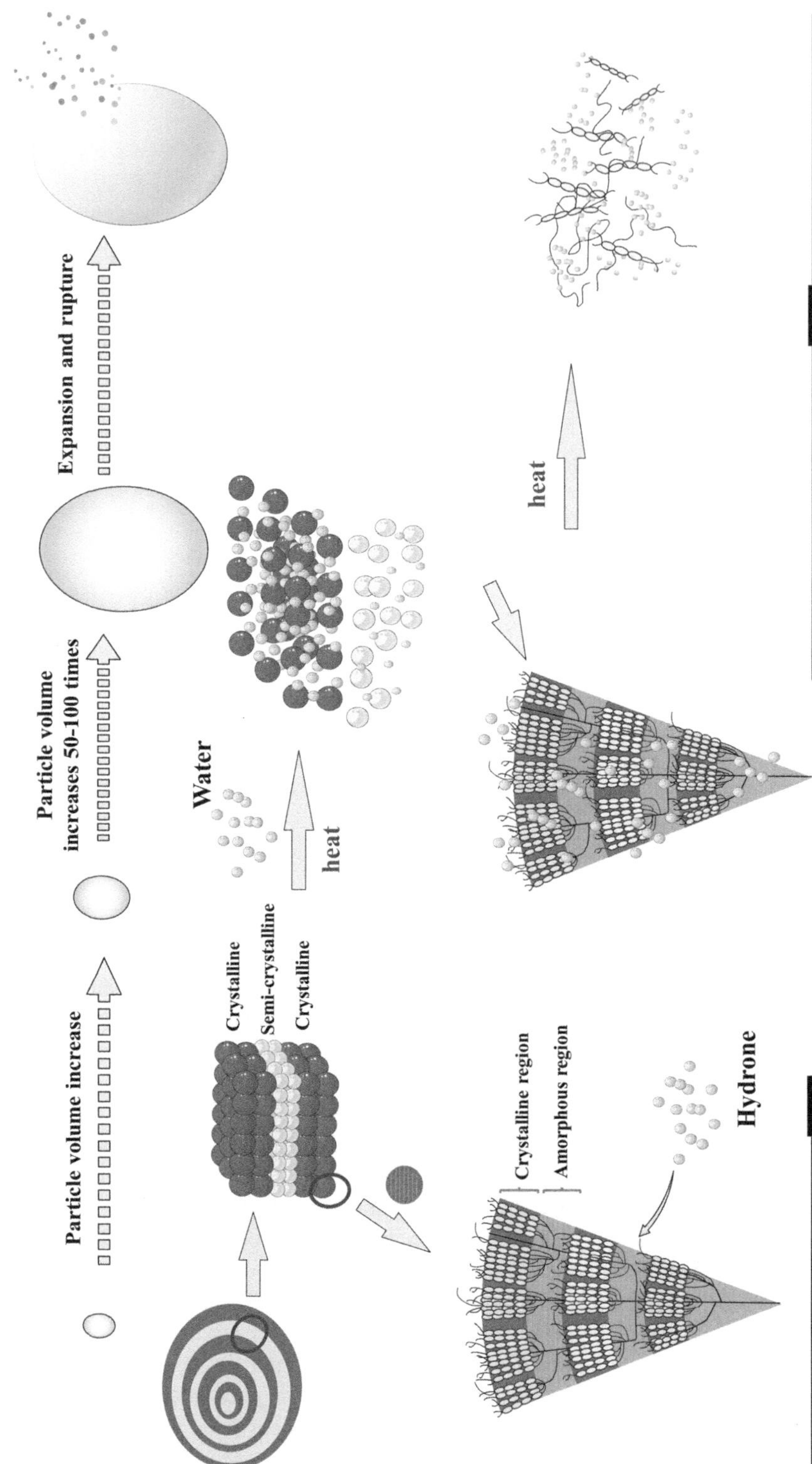

FIGURE 4.1   Starch gelatinization process.

form, and particle size, as well as moisture, pH, and salt ions. Even additional substances (lipids, proteins, non-starch polysaccharides, polyphenols, etc.) added to the starch can have an effect on the thermal properties of the starch (Table 4.1). In addition, common technologies used to produce PGS include spray drying, extrusion, and drum drying (Ma et al., 2022) (Figure 4.2). These different drying methods will also cause differences in the physical and chemical properties of PGS products. Therefore, to study PGS, we first need to fully understand the changes in starch gelatinization characteristics under specific processes and then optimize the production process of PGS under comprehensive considerations. During the gelatinization process, differential scanning calorimetry (DSC) is commonly used to measure gelatinization temperature and gelatinization enthalpy. They are two important parameters in the study of starch gelatinization. They are useful for identifying different types of starch and indicating starch gelatinization. The extent of the interaction of starch with water and the kinetics of the starch swelling stage before gelatinization is of great significance (Donmez et al., 2021). In addition, the enzymatic hydrolysis method, viscometry, birefringence method, near-infrared spectroscopy analysis technology, X-diffraction, and nuclear magnetic resonance spectroscopy technology can also be used to determine the gelatinization degree of starch (P. Guo et al., 2018).

In this chapter, the influencing factors of starch gelatinization were analyzed, the drying methods and structural characteristics of pre-gelatinization starch were systematically reviewed, and the application of pre-gelatinization starch in various fields was summarized and prospected in order to provide a theoretical basis for the production and application of pre-gelatinization starch.

4.2 FACTORS AFFECTING STARCH GELATINIZATION

4.2.1 Starch Properties

Starch from different plant sources and different origins and different extraction and processing methods of the same starch will cause differences in the amylose/amylopectin content, crystal type, particle size, molecular weight, and other structural and physical properties of starch particles, which will affect the product performance of PGS (Table 4.1).

The side chain of amylopectin is stably connected to the double helix of amylose by hydrogen bonds, but when the starch particles are combined with water under heating conditions, the hydrogen bonds break and are replaced by water. The hydration and swelling ability of starch granules depends on the ability of starch molecules to retain water through hydrogen bonds, which is affected by the amylose content and the side chains of amylopectin.

Short chains of amylopectin can bind to water molecules through hydrogen bonds, in contrast to amylose or long amylopectin side chains that interact through helical boundaries that do not bind to water molecules (Cornejo-Ramírez et al., 2018). Studies have shown that the gelatinization temperature of starch with high amylose content is high because the presence of spiral structure or complex molecules formed by amylose and other substances in starch will delay the swelling and crystallization of starch particles, so a higher gelatinization temperature is required (Tangsrianugul et al., 2019). Similarly, starch with higher amylopectin content and shorter branched chain has a lower gelatinization time, but it is not clear that the temperature and time of starch gelatinization are correlated with the amylopectin–amylopectin ratio because gelatinization time is only affected by the size of starch particles. In addition, a recent study showed that the gelatinization temperature of rice in a certain range decreased first and then increased with the amylose content. When the amylose content of rice starch was >25% or <15%, the gelatinization temperature was higher than that when the amylose content was 15%–25% (E. Li et al., 2022). This situation may be due to the fact that for the same starch, low amylose content will cause the amylose part to interfere with the double helix arrangement of the branched part, while amylose in the starch with high amylose content has a longer single-layer chain and a thicker crystal layer (C. Li, 2022).

TABLE 4.1 Factors Affecting Starch Gelatinization

FACTORS	STARCH SOURCE	EXPERIMENT CONTENT	RESULT	REFERENCES
Amylose/ amylopectin content	Rice	Determination of thermal properties of rice starch with different amylose content	When the amylose content is >25% or <15%, the gelatinization temperature is higher than that when the amylose content is 15%–25%.	(E. Li et al., 2022)
Granular pattern	Wheat, triticale, and barley	The structure and physicochemical properties of B type and A type starch were determined	B-type particles have higher gelatinization temperature, wider gelatinization temperature range, less peak value, disintegrability, and tempering viscosity.	(Ao & Jane, 2007)
Grain size	Corn	The effect of grain size distribution on gelatinization characteristics of whole grain corn flour was analyzed	When the average particle size is smaller, the particle size is smaller and the particle surface area is larger.	(Sun et al., 2022)
Moisture	Rice	The gelatinization characteristics of rice starch under different water conditions were studied	With the decrease in water content, the gelatinization temperature of starch increased, and a two-phase heat absorption peak appeared.	(E. Li et al., 2022)
Lipid	Lotus seed	Fatty acids with different chain length and saturation were added to amylose of lotus seed	The gelatinization temperature of the complex system gradually decreased with the decrease of the chain length of saturated fatty acids, and the gelatinization temperature decreased with the higher unsaturation of fatty acids.	(Z. Guo et al., 2018; Jia et al., 2018)
Protein	Cassava	Heat treatment of tapioca starch with natural gluten, gluten hydrolysate, or lemon-acid modified gluten	Natural gluten, glutenhydrolysate, and monic acid modified gluten increased the enthalpy of starch gelatinization.	(Wenchao Li et al., 2023)
	Lotus seed	Lotus seed protein was added to lotus seed starch	The initial temperature and peak temperature of starch gelatinization increased with the increase in protein content.	(S. Liu et al., 2022)
Non-starch polysaccharides	Corn	Add soluble dietary fiber to corn starch	The addition of flaxseed gum, xanthan gum, and konjac gum increased the viscosity of corn starch paste.	(X. Liu et al., 2019)
	Potato, corn, and lotus seed	Porphyra polysaccharides are added to three crystalline starches	The gelatinization temperature and enthalpy of the three starches increased significantly with the increase in polysaccharide concentration.	(Z. Wang et al., 2023)
Polyphenol	Rice	Green tea polyphenols were added to rice flour and rice starch	The peak temperature and gelatinization enthalpy of rice flour and rice starch decreased.	(Wu et al., 2020)
Salts	Corn	Different salts are added to the corn and waxy corn starch	The gelatinization temperature and enthalpy of starch increase.	(W. Wang et al., 2017)

According to the different X-ray diffraction characteristics, starch crystal morphology can be divided into type A (cereal starch), type B (potato tuber starch), type C (beans, root starch with the characteristics of both type A and type B), and type V (starch and lipids, polysaccharides, polyphenols, and other substances formed by the complex). The crystallinity of starch is also related to the content of amylose and amylopectin in starch particles. The amylose content of A-type starch particles is higher than that of B-type starch particles, and the crystallinity is lower than that of B-type starch particles (Ao & Jane, 2007), so A-type starch often requires A higher gelatinization temperature. The C-type starch amylose content is between A and B, so the initial gelatinization temperature (T_o) is theoretically higher than that of B-type starch but lower than that of A-type starch. The hydrophobic single helical structure of starch with a V-type structure inhibits the entry of water and can resist the degradation of the starch structure by thermal effect (Chao et al., 2018). Therefore, the gelatinization temperature required by V-type starch is higher than that of A, B, and C-type starch.

In terms of particle size, smaller particles have a larger surface area to bind and absorb water molecules, and during heating, such starch particles can rapidly absorb water and expand (Cornejo-Ramírez et al., 2018). Moreover, the pores and channels of starch particles enhance the ability to hydration, expansion, and gelatinization. Gelatinization capacity can be improved by reducing particle size through grinding, which can be used to prepare PGS with ultrafine particles (Sun et al., 2022). However, it should be noted that the smaller the starch particle size, the easier gelatinization is generally for the same starch; for different sources of starch, such as potato starch and corn starch, although the average particle size of potato starch is higher than that of corn starch, its gelatinization temperature is still lower than that of corn starch (Nithyadevi & Pandiyan, 2021), which is determined by the amylopectin content and crystal type of the starch itself.

4.2.2 Moisture Content during Processing

The gelatinization of starch is closely related to the water content in the suspension. In general, when the moisture content is too high (>65%), a single and symmetrical gelatinization peak will appear on the DSC map, but within a certain range (35%–60%), multiple gelatinization peaks can be observed, and when the moisture content drops below 35%, only one heat absorption may be observed (C. Li, 2022; E. Li et al., 2022; S. Wang & Copeland, 2012). The single and symmetrical pasting peak is produced by melting of a large amount of water and some special double helix structures (a double helix of amylopectin in the starch amorphous region, double helix formed between amylopectin and amylose intermediate chain, and entanglement between amylose short chain and amylopectin blockchain, etc.) (E. Li et al., 2022). The short chain of amylose participated in the formation of the first gelatinizing heat absorption peak when starch gelatinized with a certain moisture content, and the second gelatinizing heat absorption peak was formed by the interaction between the short straddle chain of amylose and the intermediate chain of amylose (E. Li et al., 2022).

4.2.3 Effect of Additives on Starch Gelatinization

4.2.3.1 Lipids

Lipids can change the gelatinization characteristics of starch. Amylose molecules can form a complex with a single or double helix structure with lipids in the amorphous lamella. The formation of amylose–lipid complex will hinder the water absorption and expansion of starch, reduce the solubility of starch in water, and lead to an increase in the gelatinization temperature of the starch system. At the same time, different types of lipids have different effects on starch gelatinization properties, and the common lipids that can form V-type complexes with starch include free fatty acids and monoglycerides (Chao et al., 2018). The results showed that when fatty acids with different chain lengths (palmitic acid, myristate acid, lauric acid) and different saturation (stearic acid, oleic acid, linoleic acid) were added to lotus amylose starch, it was found that the T_o, T_p, T_c, and ΔH of starch increased significantly. The formation of the amylose fatty acid complex decreased the solubility and swelling power of lotus amylose starch. The gelatinization temperature of the complex system gradually decreased with the decrease of the chain length of saturated fatty acids, indicating that the longer the carbon chain, the higher the temperature required for the gelatinization of the complex (Jia et al., 2018). For fatty acids with the same chain length, the higher the degree of unsaturation of the fatty acid, the lower the gelatinization temperature (Z. Guo et al., 2018).

4.2.3.2 Protein

Protein is one of the main nutrients of food, and the starch-based food system is inevitably affected by protein components in the gelatinization process, gelatinization process; on the one hand, the competition between starch and protein for water molecules leads to water redistribution (Lu et al., 2016). On the other hand, the increase in protein concentration leads to the formation of extensive network structures between protein molecules, which interferes with the hydration of starch, changes its thermal properties, and significantly reduces its ΔH (Yang et al., 2019). W. Li et al. (2023) showed that natural gluten, gluten hydrolysate, or lemon acid-modified gluten all increased the gelatinization enthalpy of cassava starch and enhanced the ordered structure of starch and starch-protein interaction. Narciso and Brennan (2018) showed that pea protein and whey protein may reduce the peak and final viscosity of rice starch by inhibiting amylose molecular rearrangement in rice starch gel. S. Liu et al. (2022) studied the interaction between lotus seed protein and amylose in lotus seed starch and confirmed that the initial and peak temperature (T_p) of lotus seed starch gelatinization would increase with the increase of protein content.

4.2.3.3 Non-Starch Polysaccharides

The effects of sugar on starch can be divided into monosaccharides, disaccharides, and non-starch polysaccharides. It has been reported that the interaction between starch and water is directly affected by the type of sugar added to the mixture (i.e., monosaccharides and disaccharides) (Donmez et al., 2021). On the one hand, hydrophilicity in the presence of small sugars reduces the reaction of water with starch molecules (Donmez et al., 2021). On the other hand, sugars can establish intermolecular interactions with starch through hydrogen bonding, thereby increasing the gelatinization temperature (Allan et al., 2018).

For non-starch polysaccharides, the addition of some soluble dietary fibers (such as chitosan, xanthan gum, konjac gum, etc.) can change the gelatinization characteristics of corn starch, and the addition of chitosan can reduce the viscosity of corn starch paste. The addition of flaxseed gum, xanthan gum, and konjac gum increased the viscosity of corn starch paste (X. Liu et al., 2019). Other studies have shown that the gelatinization temperature and ΔH of different crystalline types of starch (corn starch, potato starch and lotus seed starch) increase significantly under the action of *Porphyra haitanensis* polysaccharides due to the combination of hydroxyl group on the main and side chain of polysaccharide macromolecules and starch molecules (Z. Wang et al., 2023). These interactions increase the steric hindrance of water molecular diffusion, resulting in the obstruction of the binding of water molecules and starch molecules, and thus require higher temperatures to destroy the molecular structure of starch chains. In addition, non-starch polysaccharides with different structures have different effects on starch gelatinization. For example, water-soluble arabinoxylan has a significant inhibitory effect on amylose leaching, thus hindering starch gelatinization (Hou et al., 2020). Rhamnogalacturonan type I can adsorb on the surface of potato starch particles, inhibit particle expansion and dissolution of amylose, and ultimately delay the gelatinization of potato starch (Yin et al., 2021). Therefore, given the instability of natural starches during shearing and heating and their sensitivity to acids, non-starch polysaccharides are often added to starch-based foods to improve the physicochemical properties of starches (W. Liu et al., 2022).

4.2.3.4 Polyphenol

When starch and polyphenols are non-covalent, the combination of hydrogen bond and hydrophobic interaction significantly reduces the digestion and absorption properties of starch and changes the gelatinization, retrogenesis, and rheology properties of starch to varying degrees. During starch gelatinization, phenolic substances may enter the starch spiral cavity to form a V-type amylose complex with hydrophobicity (Zhu, 2015). The carbonyl and hydroxyl groups of polyphenols can also interact with hydroxyl groups contained in starch molecules, inducing starch molecule aggregation through hydrogen bonding and van der Waals forces (Chai et al., 2013). In the traditional heating process, the extracted amylose can interact with polyphenols to form complexes, and there are a small number of V-type complexes in this method, but a non-inclusive-type complex is dominant (Zhao et al., 2019). Green tea polyphenols decreased the T_p and ΔH of gelatinization of rice flour and rice starch by crosslinking with starch and promoting the hydration of starch particles themselves (Wu et al., 2020). In addition, green tea polyphenols can also significantly affect the gelatinization and thermodynamic properties of brown rice and white rice powder and have a good application prospect in inhibiting the rejuvenation of rice products and improving the freeze–thaw stability (Wu et al., 2020).

In addition to the aforementioned factors, the presence of inorganic salt ions can also inhibit starch gelatinization. For example, starch gelatinization will be affected by the electrostatic interaction between ions in salt and starch hydroxyl groups. Salinization or structural ions F^- and So_4^{2-} can reduce the swelling force, solubility, and transparency of starch. At the same time, the addition of inorganic salts makes the ionized salt ions combine with water molecules, so that starch gelatinization requires higher energy to overcome the hydrogen bond between water molecules and salt ions, resulting in an increase in gelatinization temperature and ΔH (W. Wang et al., 2017). Of course, excluding the difference in starch raw materials, the drying method has a significant impact on the structure and physical and chemical properties of the PGS produced.

4.3 MODIFICATION METHOD OF PRE-GELATINIZED STARCH

This section discusses the moddification methods of pre-gelatinized starches.

4.3.1 Spray Drying

Spray drying is the process of forming a spray state (fine dispersion state) through the atomizer of liquid materials. In the process of settling, the water is evaporated by the hot air stream, and the process of dehydration and drying is carried out. The powder or pellet products obtained after drying are separated from the air and collected together, and the two processes of spraying and drying are completed at the same time in this process (Figure 4.2A).

Spray drying is a rapid drying technology. In the process of spray drying, the starch suspension is sprayed into small droplets by a sprayer, and then the water in the starch solution evaporates in an instant in the high-temperature airflow. It is rapidly dried to form amorphous or semi-amorphous starch granule powder (T. P. R. D. Santos et al., 2018). The main production process is starch paste preparation → temperature gelatinization → steam atomization → drying → packaging. In the process of spray drying, it is necessary to adjust the spray angle, spray pressure, intake temperature, and other parameters of the equipment according to the nature and requirements of the starch solution in order to obtain the ideal drying effect. PGS prepared by spray drying underwent less shear and thermal aging, and the integrity of starch

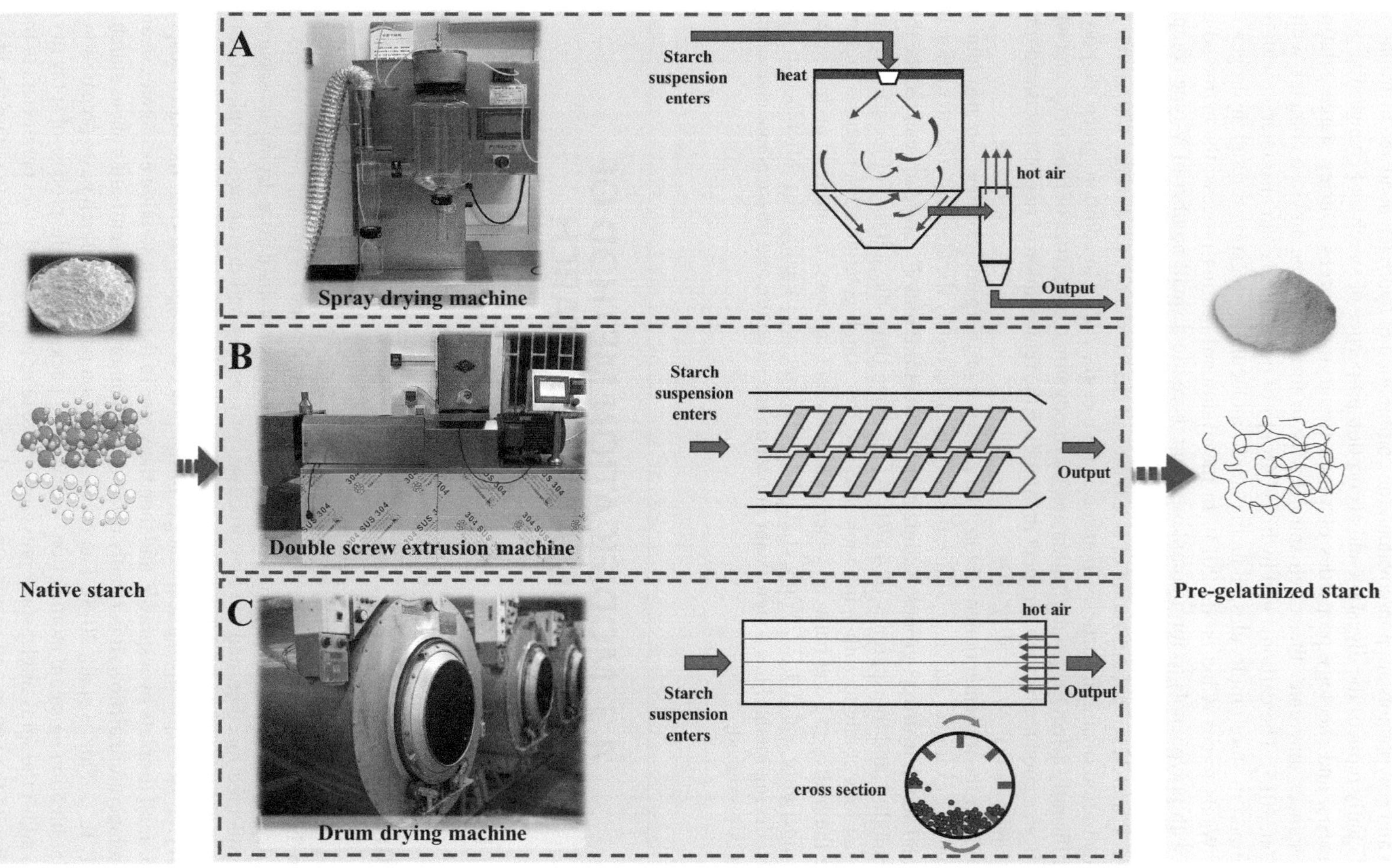

FIGURE 4.2 Pre-gelatinized starch drying equipment. (A) Spray drying; (B) drum drying; (c) microwave drying.

particles was maintained (Ma et al., 2022). The characteristic of this method is that the solubility of the product is improved compared with other methods, and the concentration of starch milk is lower, but the obtained product has low solubility, poor quality, large energy consumption, high cost, and harsh conditions. The product quality can be improved by optimizing the equipment parameters in the spray drying process or combining other methods.

4.3.2 Extrusion Method

The two screws of the twin screw extruder are toothed to each other, and the thread of one screw is inserted into the screw groove of the other screw so that the continuous screw groove is divided into a C-shaped chamber isolated from each other. When the screw rotates with the axial movement of the tooth part, the C-shaped chamber also moves along the axis. Every turn of the screw, the C-shaped chamber moves forward by a lead distance. The material in the C-shaped chamber, due to the thrust of the tooth thread, obstructs the tendency of the material to hold the screw and is pushed forward by the thread (Figure 4.2B). The twin-screw expansion process involves the uniform introduction of wet materials into the extruder cavity. Through the application of compression force, mechanical shear force, and friction force exerted by the extruder screw, the materials are thoroughly mixed, sheared, melted, cured, and sterilized within the extruder cavity. As a result of pressure differentials between the external cylinder body and atmospheric pressure, instantaneous expansion occurs.

The extrusion bulking method uses the heat generated by extrusion friction to gelatinize the starch. After explosive eruption at the top, the gelatinized starch is bulked and dried by instant decompression. Compared with traditional heating gelatinization, the gelatinization degree of starch by the extruding method is significantly improved, and the gelatinization degree is related to the temperature, pressure, shear force, and water content of extrusion (B. Wang et al., 2022; Yan et al., 2019). In the process of extrusion, starch is subjected to mechanical shear under high temperature and low moisture, resulting in the breakage of covalent bonds between starch components and the loss of particle integrity, partial decomposition and aggregation of starch groups, and changes in functional properties (T. P. R. D. Santos et al., 2018). The main production process is starch slurry preparation → melting → mixing → exhaust → decompression → high-pressure extrusion expansion → drying → crushing → screening → packaging. The advantages of this method are high yield, fast speed, and low cost, but the equipment is complicated, and the operation is difficult, which means it is easy to cause excessive gelatinization of starch and affect product quality.

4.3.3 Drum Drying

The drum drying method was invented by Wulkank Supf in Europe in 1980. Currently, the double-drum drying method is mainly used at home and abroad because double-drum drying has less impact on the mechanical shearing of starch granules and can produce a thinner film than single-drum drying (Ma et al., 2022). Compared with the starch produced by single-cylinder drying, the PGS produced by double-cylinder drying has a higher viscosity and less granular structure (Majzoobi et al., 2011). Drum drying PGS is processed by evenly distributing starch slurry on the surface of a steam-heated drum, allowing the water to quickly evaporate to form a starch film (Narciso & Brennan, 2018). Then, the film is scraped off with a scraper to obtain a brittle melt, which is then crushed to obtain dry particles of PGS (Gavrielidou et al., 2002). Its main production process is preparation of starch slurry → heating and gelatinization → drying → crushing → sieving → packaging. The PGS prepared by the drum drying method has no granular structure, low crystal structure, high cold water viscosity, good water absorption, and strong water solubility (Majzoobi et al., 2011). The disadvantage is that the production cost of PGS is high, and the energy consumption is high.

4.3.4 Microwave Drying

A drum dryer is a kind of rotary drying equipment with internal heating conduction. The wet material obtains heat transferred by heat conduction on the outer wall of the drum, removes water, and achieves the required wet water content. Heat is transmitted from the inner wall of the cylinder to the outer wall of the cylinder and then through the material film, which has high thermal efficiency and can be continuously operated, so it is widely used in the drying of liquid materials or ribbon materials (Figure 4.2C). Drum dryers can be divided into two forms: single-drum and double-drum dryers. In addition, they can also be divided into normal pressure and decompression, two forms according to the operating pressure.

The microwave method uses microwave heat to gelatinize, dry, and finally pulverize the starch slurry. The microwave method is a new method for preparing PGS. Microwave processing eliminates the influence of shear force on starch. One of the more advanced aspects is that the progress of preparing PGS can be calculated according to the Runge-Kuttagel method. Predicting the heat transfer of dough provides development space for the future application of standards for controlling the degree of pregelatinization in the food industry (Xue et al., 2010). At present, the preparation of PGS by microwave is still in the laboratory stage.

4.3.5 Other Methods

In addition to the previous methods, the combined application of some technologies with potential production applications can improve the product quality of PGS. For example, there are studies on the preparation of oat PGS based on spray drying combined with electron beam irradiation (Shen et al., 2023). After an electron beam irradiated spray drying starch, its solubility and dilatability increased significantly, but the gelatinization viscosity, gelatinization temperature, apparent viscosity, energy storage modulus, and loss modulus decreased. The depolymerization of starch molecules and breakage of amylopectin chains under electron beam irradiation may cause the result. Guan (2015) improved the spray method by combining the extrusion method to form the spray prefabricated wet powder method. The production process is to spray starch with water, introduce it into the twin-screw extruder, heat it to 120~140°C, press it to 3000–5000 KPa, press it out of the hole with a diameter of 0.2~1 mm, and dry it. After natural cooling and crushing, the product is mixed by atomizing fungicide spray, mixed by multi-angle variable pitch mixer and crushed, and dried by pipeline airflow. The method is suitable for potato flour and tapioca flour, which is characterized by low energy consumption, less water content in feed, moderate viscosity, and stable quality.

4.4 STRUCTURAL PROPERTIES OF PRE-GELATINIZED STARCH

4.4.1 Morphology of Pre-Gelatinized Starch

The morphology of starch particles after pre-gelatinization will change significantly in different degrees, depending on the processing technology used to produce pre-gelatinization and the degree of gelatinization (Fu et al., 2012; T. P. R. D. Santos et al., 2018). A scanning electron microscope (SEM) is common equipment for starch particle morphology characterization, which can directly show the appearance, structure, size, shape, and so on.

As shown in Figure 4.3 (Fu et al., 2012), the surface of the PGS particles after spray drying generally shows folding and shrinkage. With the increase of spray drying heating temperature, the depression and

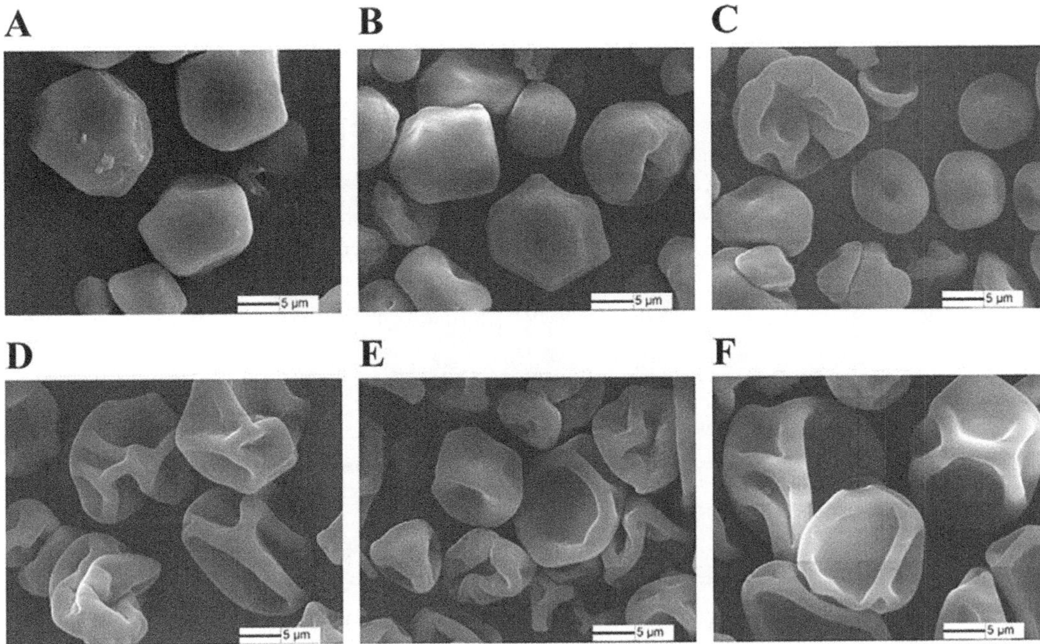

FIGURE 4.3   SEM microphotographs of partially gelatinized starch as affected by the gelatinization process (3000×): (A) native corn starch; (B) gelatinization at 64°C; (C) gelatinization at 66°C; (D) gelatinization at 68°C; (E) gelatinization at 69°C; and (F) gelatinization at 70°C.

shrinkage degree of the surface of the particles will increase. The reason may be that during spray drying, the outermost layer of water of some hydrated particles will quickly vaporize, and the surface will start to sag and shrink slightly. With the rise of gelatinization temperature, starch particles began to absorb a lot of water, and the expansion degree of starch particles increased. At this time, the internal structure of starch was destroyed, and the particles swelled. Subsequently, the volume shrinkage of these particles was obvious after spray drying (Fu et al., 2012). During this process, the proportion of small-size particles increases, which may be due to particle breakage or, particle folding, or shrinkage (T. P. R. D. Santos et al., 2018).

The extrusion process is a continuous process in which the combination of shearing forces, high temperature, and pressure can change the physical properties of the starch product. As shown in Figure 4.4 (Lazou & Krokida, 2010; Zhang et al., 2023), the surface of the PGS particles after extrusion and drying showed irregular or porous structures (Yan et al., 2019). (Yan et al. (2019) found that starch particles absorb water due to high shearing forces, temperature, and pressure in the extrusion process and then expand and eventually decompose to form a uniform gel. As a result, starch particles, after extrusion, become more irregular, and the granulation degree decreases. When starch is passed through the extruder mold, the melted extrude usually expands due to a drop in pressure and evaporation of water. Thus, during this process, starch particles are destroyed, and a porous honeycomb structure is formed. Different extrusion processes, such as extrusion temperature, feed water content, and feed speed, will result in different pore sizes and wall thickness. Lazou and Krokida (2010) found that hole wall thickness decreased with the rise of extrusion temperature but increased with the increase of feed speed. In addition, higher screw speeds contribute to the formation of larger holes due to higher screw speeds that increase the pressure in the extruder barrel. The complex process of extruding starch finally leads to the destruction of the particles, the disappearance of the crystal structure, and the formation of amorphous melt (Shrestha et al., 2010).

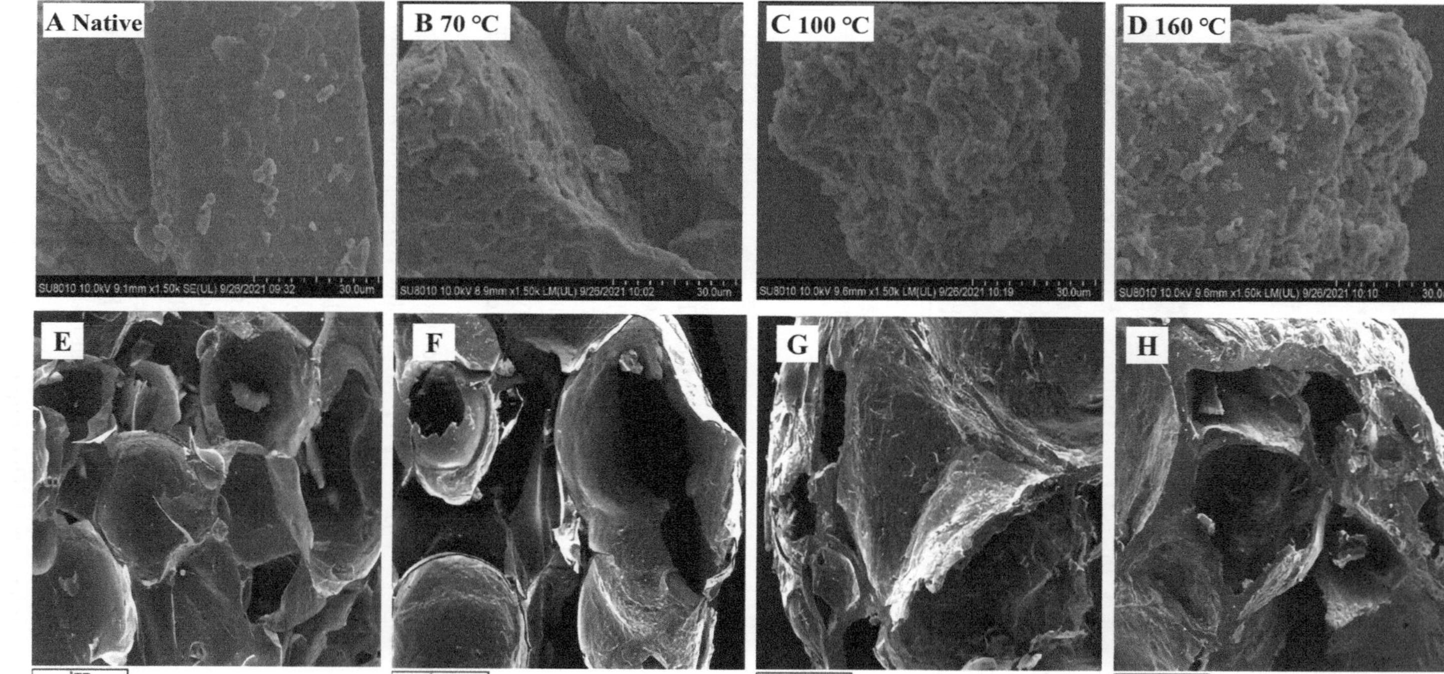

FIGURE 4.4   SEM images of native (A) and Tartary buckwheat powder pre-gelatinized at different temperatures (B–D). Macrostructure of corn–lentil extrudates as affected by feed rate (E–H): (E) lentil content = 10% and feed rate = 2.52 kg/h, (F) lentil content = 10% and feed rate = 4.68 kg/h, (G) lentil content = 50% and feed rate = 2.52 kg/h and (H) lentil content = 50% and feed rate = 4.68 kg/h.

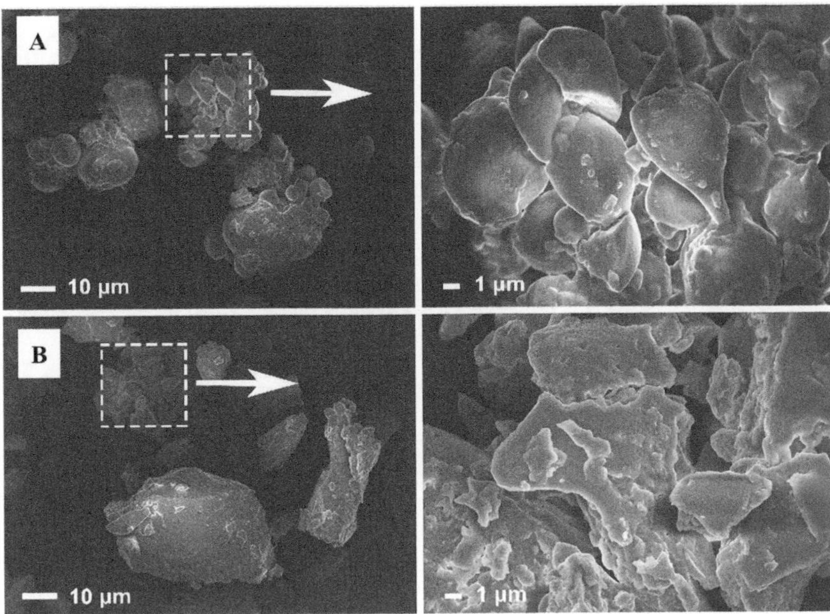

FIGURE 4.5 Scanning electron micrographs (magnified 1000 times) of original oat flour (A) and drum dried oat flour (B); arrows indicate the image on the right is a 5000 times magnified image of the dashed inset in the left imaged.

As shown in Figure 4.5 (C. He et al., 2020), the surface of the PGS after drum drying showed an irregular flake or porous structure. C. He et al. (2020) found that starch particles had a honeycomb structure, rough surface, and obvious edges during drum drying.

In addition to scanning electron microscopy, birefringence detected by a polarizing microscope can directly reflect the crystallization characteristics of starch particles. The apparent strength of birefringence means the thickness of the particles and the crystallinity of the grains. In the drying process, high temperature breaks the hydrogen bond between starch molecules and destroys the crystallinity of starch particles (Figure 4.6) (W. Li et al., 2014). Taking buckwheat starch particles as an example, the natural buckwheat starch particles were intact (Figure 4.6A) and showed strong birefringence, and a complete "cross" was formed in the center of the seed umbilical (Figure 4.6a). The micrograph of buckwheat starch after drying by roller showed an irregular shape and no complete starch particles (Figure 4.6A). All particles have lost the "Maltese cross", indicating that the starch crystallization zone has been destroyed and the starch has been completely gelatinized (Figure 4.6B) (W. Li et al., 2014).

4.4.2 Crystal Structure of Pre-Gelatinized Starch

The XRD diffraction peak of PGS after spray drying depends on raw materials and processing technology, but the crystallinity shows a decreasing trend (Fu et al., 2012; X. He et al., 2020; T. P. R. D. Santos et al., 2018). T. P. R. D. Santos et al. (2018) prepared pre-gelatinized starch by the spray drying method. They found that the crystal pattern of cassava starch and Peruvian carrot starch did not change, but the diffraction peak intensity decreased. A small peak appeared at $2\theta = 5.6°$, which may be related to its B-type diffraction pattern. Compared with A-type starch, B-type starch is less thermodynamically stable because B-type starch has a lower packing density. Fu et al. (2012) also found that with the rise of gelatinization temperature, the crystal pattern of corn starch after spray drying did not change, but the

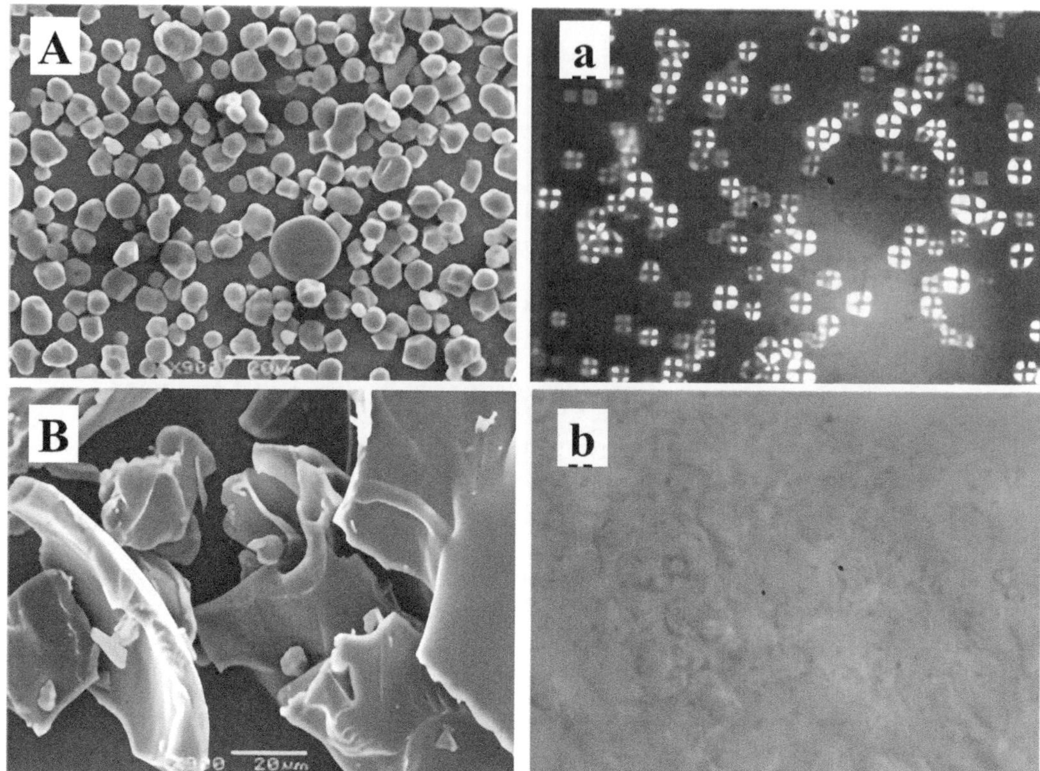

FIGURE 4.6 Scanning electron micrographs (left) and polarized light micrographs (×400) (right) of different modified buckwheat starch. (A, a) Native; (B, b) PGBS. Bars on the micrographs represent 20 µm.

peak value gradually decreased, and the crystallinity decreased. When the gelatinization temperature is higher than 70°C, the peaks at $2\theta = 15°$, $17°$, $18°$, and $23°$ disappear, and this observation indicates that the crystallized raw corn starch is mostly transformed into an amorphous state. X. He et al. (2020) found that the diffraction pattern of rice starch after spray drying was different from that of native starch, and all peaks almost disappeared, indicating that the crystalline structure of primary starch after spray drying treatment was completely destroyed.

As shown in Figure 4.7A (X. He et al., 2020), spray drying changed the diffraction pattern of oat starch, with a small peak at about 13°. In addition, peak strength decreased at 15°, 17°, 18°, and 23° but increased at 20°. A peak at 20° indicated the formation of amylose-lipid complexes. This phenomenon illustrated that the crystal structure of native starch is transformed into an amorphous form. The XRD peak strength of PGS prepared by the extrusion method decreased or disappeared with the increase in treatment conditions (T. P. R. D. Santos et al., 2018; Tao et al., 2021; Zhang et al., 2023); it was similar to the results obtained by spray drying. In Figure 4.7B (Zhang et al., 2023), the peak of pre-gelatinized Tartary buckwheat flours decreased obviously at $2\theta = 15.15°$, $17.26°$, $18.01°$, and $23.01°$ as the extrusion severity increased. Tao et al. (2021) found that wheat starch was treated by extrusion, and its diffraction peak intensity was weakened, but its A-type crystal structure was still maintained. As the extrusion temperature increased, the relative crystallinity of wheat starch apparently decreased from 27.83% of native starch to 13.71%. The loss of crystallinity may be related to the high temperature of the extrusion process; the heat degradation of starch leads to the formation of high-density branching points (α-1,6) of amylopectin. In the previously mentioned study by T. P. R. D. Santos et al. (2018), the diffraction peaks completely disappeared after the extrusion process, indicating that both starches were transformed into amorphous

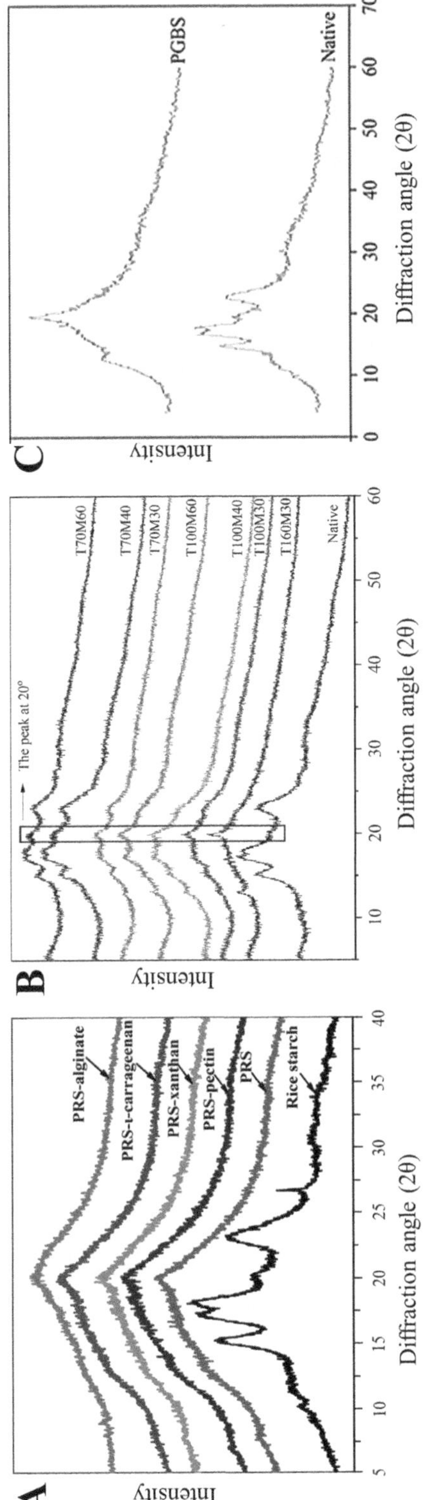

FIGURE 4.7 X-ray diffraction patterns. (A) Native rice starch, pre-gelatinized rice starch, pre-gelatinized rice starch (PRS) and pre-gelatinized rice starch-hydrocolloids (PRS-H). (B) Native and pregelatinized Tartary buckwheat flours. The peak at 20° represents the V-type amylose-lipid inclusion complexes. (C) Native and drum-dried buckwheat pre-gelatinized starch (PGBS) granules.

structures. In this process, the starch underwent a melting process under the action of shear force and temperature, forming a continuous amorphous mass, and the diffraction peak disappeared. With the increase in extrusion strength, the diffraction peak strength of PGS decreased significantly (Zhang et al., 2023). The crystallinity of extruded starch was also affected by amylose content. High amylose usually retains significant crystallinity, while low amylose loses the granular structure associated with crystal and underlying spiral melting, resulting in a more or less amorphous structure after extrusion (Shrestha et al., 2010).

In addition, the amylose–lipid complex was easily formed during the extrusion process, and the V-shaped diffraction peak was formed (Y. Liu et al., 2017). In some research, the crystal type of the PGS prepared by drum drying was damaged, and the crystallinity was significantly lower than that of the original starch (Majzoobi et al., 2011). (Nakorn et al., 2009) found that through drum drying three kinds of rice PGS (glutinous (PGS), jasmine (PJS) and Chiang (PCS)), the crystal order of PGS and PJS particles is completely lost, and the XRD diffraction pattern is completely amorphous. However, PCS with high amylose content had less structural damage. The X-ray diffraction pattern of PCS showed a peak at $2\theta = 20°$, indicating that PCS formed a V-shaped crystalline amylose-lipid complex structure. As shown in Figure 4.7C (W. Li et al., 2014), the drum-dried pre-gelatinized buckwheat starch showed very weak peaks around 13° and strong peaks around 23°, indicating that the initial A-type crystal was destroyed by drum-drying treatment. To sum up, the crystal peak intensity of pre-gelatinized starch will be greatly weakened or even disappear after different modification measures.

4.4.3 Interaction between Pre-Gelatinized Starch and Water

Swelling power indicates the ability of water to penetrate starch particles and can be used to evaluate the degree of interaction between starch chains in the crystalline and amorphous regions of starch particles. During gelatinization, the crystal structure of starch is destroyed due to the break of intermolecular and intramolecular hydrogen bonds, resulting in changes in the ability of starch to bind to water (Donmez et al., 2021; Roulet et al., 1988). Solubility is an indicator of starch change and molecular degradation and is used to measure the number of soluble components released after starch extrusion (Donmez et al., 2021). The swelling power and solubility of PGS after spray drying were higher than that of natural starch. In the process of pre-gelatinization, the crystal structure of starch particles is destroyed due to the breakdown of molecular hydrogen bonds. Water molecules readily diffuse to starch particles and form hydrogen bonds with exposed hydroxyl groups. At the same time, the fluidity of starch molecules and the leaching of amylose increase, which increases the solubility and swelling power of starch (X. He et al., 2020). The swelling power of PGS changes obviously with the increase in heating temperature. Fu et al. (2012) found that the dilatability of gelatinized corn starch at low temperatures (30°C) was higher than that of natural corn starch. However, when the heating temperature rises from 60°C to 70°C, the swelling power of natural corn starch increases rapidly.

When starch is gelatinized by extrusion, the crystal structure of starch is destroyed due to the break of inter- and intra-molecular hydrogen bonds so that more hydroxyl groups are exposed to form hydrogen bonds with water. In addition, compared with native starch, water molecules were more easily diffused into extruded starch to form amorphous regions (Shen et al., 2023). The damage degree of starch gelatinization induced by extrusion will increase the swelling power of starch (Maniglia et al., 2021; Yan et al., 2019).

The relationship between PGS and water after extrusion is affected by the treatment conditions. (Zhang et al., 2023) found that the water absorption index of pre-gelatinized Tartary buckwheat starch increased with the increase in temperature and the decrease of feed water content. Moreover, when the extrusion conditions (extrusion temperature, water content) are strong, the solubility is significantly higher than that of natural starch, and under mild extrusion conditions, the solubility of pre-gelatinized Tartary buckwheat starch is significantly reduced. Y. Liu et al. (2017) found in a study on the preparation of PGS by "improved extrusion cooking technology" that the PGS at low temperatures had higher water solubility and water absorption than natural starch, and the solubility gradually increased with the increase of water content.

The interaction between PGS and water prepared by drum drying is affected by the treatment conditions and raw materials. Some research has shown that the water absorption index, solubility, and dilatability of PGS after drum drying are significantly higher than those of native starch (Majzoobi et al., 2011; Nakorn et al., 2009; Wiriyawattana et al., 2018). With the rise of drum drying temperature, the water absorption index and solubility increased significantly, which may be related to the decomposition and degradation of starch macromolecules during heat treatment (W. Li et al., 2014). Obadi et al. (2020) also found that cassava starch has a high proportion of amylopectin and the longest average amylopectin chain length, so it has a high water-holding capacity. In addition, in the research of Wiriyawattana et al. (2018), it was found that in the range of 120–130°C, with the increase of drum drying temperature, the water absorption index and dilatability of starch significantly increased compared with the control sample, while the solubility decreased slightly. When the temperature rises again, the water absorption index and expansive force do not increase further.

4.4.4 Thermal Properties of Pre-Gelatinized Starch

The thermal properties of starch reflect the thermal stability and interaction between starch molecules (N. Wang et al., 2023). The thermogravimetric analyzer (TGA), differential thermal analyzer (DTA), and differential scanning calorimeter (DSC) are common analytical instruments for starch thermal properties. Thermogravimetric analysis is widely used to study all physical processes involving mass change. TGA is a branch of thermal analysis that studies the variation of a substance's weight with time or temperature. When the sample is in a controlled heating or cooling environment, a weight change curve is recorded. When the change in weight is recorded as a function of time, it is called an isothermal pattern. In scan mode, the weight change is recorded as a function of temperature (Loganathan et al., 2017). TGA is widely used to assess the stability of starch and starch/other component blends due to its simple and effective information provided by simple heat maps (Tian, Li, et al., 2011). DTA, similar to DSC, is a thermal analysis technique. In DTA, the starch sample in the research was subjected to the same thermal cycle as the inert reference (AL2O3), and any temperature difference between the sample and the reference was recorded. The differential temperature (DT) value was then used as a differential thermal analysis curve with temperature or time, which provides data on physical and chemical transformations such as crystallization, melting, and sublimation. DTA has been successfully used to study the stability and crystallization properties of natural and denatured starches due to their simplicity and the useful information provided by simple thermal images (Tian, Xu, et al., 2011). Because of its high thermal sensitivity, DSC rather than DTA is a suitable tool to study the structural transformation of polymers (Feist, 2015).

DSC is an effective method used in different areas of research, development, and quality inspection and testing. Thermal effects can be rapidly identified, and the associated temperature and characteristic calorific value can be determined using mass in the milligram range. DSC measurements can determine the heat capacity, transformation thermodynamic data, purity, and glass transition. DSC curves are used to identify substances. When the physical state of matter changes, such as melting or changing from one crystalline form to another, or a chemical reaction occurs, heat is either absorbed or released. DSC has great potential for studying the physical and chemical changes in foods. DSC is widely used to detect the endothermic or exothermic reaction when the sample is heated with the rise of temperature. When starch is mixed with water, water will form a chain-water crystal structure with the starch molecular chain; this structure will show an obvious endothermic melting peak with the increase of temperature. Different concentrations will lead to different melting temperatures and the size of heat absorption. DSC can calculate the area of the formed peak to judge the size of the heat absorption and understand the degree of gelatinization.

It has been found that the spray drying method can increase the gelatinization temperature and reduce the gelatinization enthalpy, but the gelatinization temperature range is reduced. During the pre-gelatinization process, water molecules penetrate into the starch particles, destroying the amylopectin chains that make up the crystalline zones. The increase in gelatinization temperature can be attributed to the rearrangement of crystal structure. The disturbance of amylopectin crystals causes the ΔH of starch, and the decrease in the ΔH of pasting may be caused by the disturbance of amylopectin crystals during heating (Shen et al., 2023; T. P. R. D. Santos et al., 2018). In addition, the gelatinization temperature and ΔH of some PGS prepared by spray drying together with other auxiliary methods have also been affected. Shen et al. (2023) treated oat starch by spray drying and electron beam irradiation and found that the gelatinization temperature decreased with the increase of irradiation dose, which meant that irradiation destroyed the crystal structure of part of the starch, but the change of irradiation dose had almost no effect on the gelatinization enthalpy. The thermal properties of green banana starch were studied after spray drying and ultrasonic co-treatment, and it was found that ultrasonic treatment reduced the energy required for the gelation of green banana starch (Izidoro et al., 2011).

The gelatinization enthalpy of PGS after extrusion treatment is significantly reduced or non-existent, which may be due to the destruction and loss of crystal structure (T. P. R. D. Santos et al., 2019, 2018; Zhang et al., 2023). Paes Rodrigues dos Santos et al. (2019) found that pre-gelatinized sweet potato starch prepared by extrusion treatment did not show curves in DSC analysis, and therefore, gelatinization temperature and ΔH values were not reported. This result confirms that the crystal structure of the extruded particles is completely lost. However, in the study of Yan et al. (2019), the gelatinization temperature and ΔH of corn starch combined with extrusion and hot and wet treatment increased compared with that of natural starch, indicating that the treated starch was highly gelatinized. The reason is that during the treatment process, the starch forms a crystalline zone, and at the same time, the rearrangement of the destroyed molecules occurs; the increase of correlation between starch chains leads to the increase of double helicity, which leads to an increase in ΔH.

With the increase of drum drying temperature, the gelatinization degree of starch increased gradually. When the temperature during the drum drying process is higher than the gelatinization temperature range of starch, no heat absorption peak can be found in the DSC analysis, indicating that the sample has been completely gelatinized (Wiriyawattana et al., 2018). Similar results were found by Taşkın and Savlak (2022) in the preparation of gluten-free fermented instant soup powders using mung beans and drum drying.

During the gelatinization process, the crystal structure of starch was destroyed due to the breaking of inter-and intra-molecular hydrogen bonds, causing the starch and the water binding ability to change. The solubility, swelling power, and water absorption index of pre-gelatinized starch increased with the increase of the treatment strength. After spray drying, the gelatinization temperature of pre-gelatinized starch increased, and the crystal structure was destroyed but it still had a better granule shape. After extrusion and drum drying, the morphology and crystalline structure of pre-gelatinized starch were seriously destroyed or even lost, and the gelatinization degree was greatly increased, or the endothermic peak disappeared. In short, pre-gelatinization can significantly alter the structure and functional properties of starch, which has a great impact on the quality of the starch products.

4.5 APPLICATION OF PRE-GELATINIZED STARCH

4.5.1 Application Status of Pre-Gelatinized Starch in Different Drying Methods

Pre-gelatinized starch prepared by different drying methods, such as screw extrusion, drum drying, and spray drying, may have different application scenarios because of the differences in structural characteristics. After extrusion treatment, its water-holding capacity is improved. Its elasticity and smoothness are increased, so the screw extrusion method is widely used in the preparation of bread (Ortolan et al., 2015), the preparation of rice flour (Lubowa et al., 2018), and the processing of noodles and for ingredients in fish feed. The spray drying method is often applied to the embedding of active substance microcapsules because of the short contact time with a high-temperature medium (Azhar et al., 2021; Gomes et al., 2019). The drum drying method has the characteristics of continuous production, simple operation, low energy consumption, high thermal efficiency, and stable drying quality of products and can be applied to the industry to obtain products with no granular structure, low crystal structure, high cold water viscosity, high water absorption, and high water solubility (Majzoobi et al., 2011).

4.5.2 Applications in Food Processing

PGS usually has high solubility in cold water and can provide ideal gelatinization and texture properties, so PGS is often used in the processing of noodles. Pre-gelatinization can improve the gel strength and viscosity of starch in cold water, thus improving the quality of noodles. Studies have shown that replacing part of rice flour with 5–15% (wt/wt) pre-pasted high amylose cornstarch to make rice flour can improve the texture and brightness of fresh (uncooked) noodles (Lubowa et al., 2018). Q. Li et al. (2020a) indicate that dough prepared by adding different concentrations (5–10%, w/w) of pre-gelatinized corn starch had greater tensile strength, and the addition of pre-gelatinized corn starch significantly improved the smoothness of noodles. Han et al. (2021) used the extrusion method to prepare pre-gelatinized buckwheat starch and then mixed it with starch in an appropriate proportion to make noodles. They found that pre-gelatinized starch with low relative molecular weight and high viscosity prepared by extrusion can form a gel network structure in gluten-free flour, which has a positive effect on the tensile properties of buckwheat noodles without gluten. In addition, drum-drying PGS has been reported to be superior to extrusion PGS in improving noodle processing quality (Ma et al., 2022).

In recent years, PGS has also commonly been used in the production of bread. As the basic skeleton of baking, it has good fresh-keeping properties and improves the taste of bread (Nasir et al., 2021). A study added a little pre-gelatinized tapioca starch to wheat flour to make French bread frozen dough, reducing the preparation and baking time of the frozen dough, thereby saving costs for the baking industry (Ortolan et al., 2015). In addition, pre-gelatinized wheat starch and pre-gelatinized tapioca starch had a cryoprotective effect on frozen dough, increasing the volume of dough after thawing and fermentation (Ortolan et al., 2015). In addition, the PGS also has anti-aging properties. The reason is that the PGS, due to its newly introduced groups, can effectively disperse the free water in the product, prevent dehydration between starch molecules due to hydrogen bonding, and thus have a certain inhibitory effect on the aging of the product. Studies have shown that adding 20% pre-gelatinized acetylated distarch phosphate to uncoated bread with a short shelf life can effectively delay the hardening of bread, play an anti-aging role, and extend the shelf life of uncoated bread (Roman et al., 2020).

PGS can also be added to soup as a food additive. Sobini et al. (2022) showed that the sensory properties of soup with Palmyrah PGS, including appearance, aroma, color, texture, taste, aftertaste, and overall acceptability, were improved to varying degrees. In addition, the pre-gelatinization of wheat flour using microwave, followed by cream mushroom soup preparation through infrared freeze drying, demonstrated that compared to single infrared freeze drying, microwave pre-gelatinization significantly enhanced the taste and viscosity of the rehydrated soup while reducing degradation tendencies (Q. Li et al., 2020b).

Compared with natural starch, part of the starch has absorbed water when the mixture is made with PGS, and the water escapes from the starch particles during baking, which has a better expansion effect, so PGS is also a good raw material for snack foods such as fried food, rice nuts, and crackers (Fitriani et al., 2022). Moreover, denatured PGS is sometimes used for better results.

4.5.3 Applications in Materials Science

Pre-gelatinized starch can also be used as raw material for further denaturation. The main principle is to form chains of hydrophilic and semi-rigid polymers with low starch content, and by introducing different functional groups, the required oligomer is grafted on the polymer in the form of branch chains by gelatinization (50–60°C), adjusting the ratio of hydrophilic to hydrophobic structure, and improving the molecular weight of the flocculant. Chen et al. (2022) prepared plywood from tapioca starch grafted with glycide methacrylate (GMA) and then crosslinked with sodium tripolyphosphate (STMP). By Fourier transform infrared spectroscopy (FTIR) analysis, it was found that GMA was successfully grafted onto the starch adhesive, and the hydrophobicity and shear strength of the adhesive were improved. Haq et al. (2022) grafted vinyl pyrrolidone onto carboxymethyl starch by free radical polymerization and synthesized carboxymethyl starch grafted polyethylpyrrolidone (CAR-St-g-PVP), which could be used as biodegradable adsorbents for wastewater treatment.

4.5.4 Applications in Biomedicine

As a physical denaturation of starch, pre-gummed starch has good fluidity, compressibility, and disintegrability and can be used as a disintegrator of tablets and capsules, as well as a binder and sustained-release agent of tablets (Alalaiwe et al., 2019). Kankate et al. (2020) compared the effects of natural starch and pre-gelatinized starch as sustained-release agents of drugs, respectively, and found that tablets prepared from natural starch would be released faster than PGS. Since pre-gummed starch has a low equilibrium relative humidity, it is preferred to enhance the stability of humidity-sensitive drugs by binding moisture. For example, PGS can be used as an excipient additive for pramexol dihydrochloric monohydrate tablets, which, after being added by internal precise methods, improves the stability of the tablets due to low water activity, resulting in improved efficacy and reduced side effects (Kim et al., 2014).

In addition, the bone hemostatic material (absorbable bone wax) developed in recent years, combined with the use of PGS, further improves the bone molding structure and reduces the adhesion of gloves, providing a better adhesion surface. This newly developed absorbable bone wax showed no systemic adverse reactions and reabsorption defects after two days of live testing, while commercial bone wax therapy showed inflammation and abnormal responses to damaged bones (Suwanprateeb et al., 2014).

4.5.5 Other Fields

The application of PGS in feed can effectively enhance bonding and improve palatability. When used in powdered feed, it acts as a cohesive agent during stirring, resulting in good viscoelasticity and water stability, promoting consumption while reducing loss. For instance, when cultivating American eels, the reasonable combination of various raw materials and additives, the appropriate proportion of PGS, and the formation of pellet feed can effectively alleviate the pollution of aquaculture water, reduce the frequency of water change, and improve the bonding effect of aquaculture feed, thus improving the feed utilization rate (Romano & Kumar, 2018). Kanmani et al. (2018) added PGS to tilapia feed, and the feed water stability, bulk density, and protein solubility were significantly improved, and the surface was smoother. Compared with tilapia on ordinary feed, tilapia's growth ability was significantly improved after eating feed containing PGS, which may be related to the easy absorption and utilization of starch by organisms. Romano et al. (2018) also showed that fish with PGS had less food intake than fish with other components, but growth was not affected.

The application of PGS in cosmetics can improve the texture and stability of cosmetics. Daudt et al. (2015) added Pinhão starch to the formula of the cosmetics, which improved the coating ability, stability, and viscosity of the cosmetics compared to the basic formula, and the formula did not cause any skin stimulation or skin pH change.

Starch molecular structure is very similar to the structure of fiber molecules in papermaking fiber raw materials, coupled with the advantages of wide source, low price, little environmental pollution, and so on; it is widely used in the papermaking industry. The most commonly used modified starch in the paper industry is cationic starch, and cationic starch is usually a PGS. In general, starch is often used as a surface sizing agent in the paper industry to achieve the desired smoothness, gloss, and print quality. Paper is mostly porous and hygroscopic, and water absorption will negatively affect its mechanical properties and increase the permeability of gas and water. Therefore, the use of cationic starch can improve its barrier properties without introducing toxic chemicals into the paper (A. D. A. D. Santos et al., 2023). The recycled pulp is reduced in mechanical strength through the recycling process, and the pulp fibers become stiff and expanded, and the surface area decreases. The addition of modified cationic starch can effectively enhance the mechanical properties of pulp fiber, which is conducive to its recycling and recycling production. For example, Salam et al. (2018) manufactured chitosan composite starch nanoparticles to enhance the physical properties of recycled pulp, as well as mechanical and chemical treatments can enhance the strength properties of OCC pulp boards. Marcello and Salam (2023) modified cationic starch with 2-chloroacetamide, then reacted it with formaldehyde to produce n-hydroxymethyl starch amide, and added 1% N-hydroxymethyl starch amide to the pulp, and the wet tensile index, dry tensile index, and dry break resistance index of the produced paper were significantly improved, which was conducive to the recycling of paper.

PGS is also often used in the processing of building materials. PGS can swell and dissolve in cold water to form a translucent paste with a certain viscosity. The paste silk is long and the bonding force is good. Because of its simple production, low price, cold water solubility, good bonding force, non-toxicity, and environmental protection, it is widely used in putty powder. Putty powder is widely used in the construction industry for leveling and bottoming interior walls. Among them, dry mix mortar putty is an auxiliary material of architectural decoration coating made of organic polymer or inorganic binder adding stone powder and other additives. PGS is one of the most commonly used organic polymer binders. Satisfactory results can be obtained by using PGS and CMC as binders of putty powder in dry climates in northern China. In addition, an appropriate amount of pre-gelated starch is added to gypsum powder to combine the adjacent gypsum crystals under the action of hydrogen bonding, making the structure of the gypsum crystals more complete, improving the mechanical strength of the gypsum, and thus achieving the purpose of reducing the cost and protecting the environment by reducing the amount of gypsum powder (J. Wang et al., 2023).

4.6 CONCLUSION

Pre-gelatinization is a common physical modification method of starch, which can improve the solubility and viscosity of starch in cold water, expand the application range of starch, and have good development prospects in the production of modified starch. In this chapter, the important factors affecting starch gelatinization are analyzed, the common drying methods (spray drying, extrusion, and drum drying) for industrial production of pre-gelatinization starch are summarized, and the structural characteristics of pre-gelatinization starch under different drying conditions are compared. In terms of particle integrity, the starch particles formed by spray drying are relatively complete and more convenient because the effect of shear force is weakened. However, in fact, for the starch characteristics required in the processing process, choosing the right drying method can maximize the production cost and improve the product quality.

PGS has a strong application market, and research on PGS still needs to be continuously explored, including different factors (such as starch source, pretreatment, additives, water activity, pH effect, etc.). The effect of PGS on the gelatinization process of raw starch, the optimization of PGS drying method, the interaction between PGS and other substances in the application process, and so on. Moreover, in the field of food nutrition, there is still a broad research space to analyze the relationship between the characteristics of PGS and the function of products, and to analyze the effect of PGS addition on the nutritional characteristics of food. These studies will facilitate enhanced utilization and widespread adoption of PGS across various domains.

REFERENCES

Alalaiwe, A., Fayed, M. H., Alshahrani, S. M., Alsulays, B. B., Alshetaili, A. S., Tawfeek, H. M., & Khafagy, E.-S. (2019). Application of design of experiment approach for investigating the effect of partially pre-gelatinized starch on critical quality attributes of rapid orally disintegrating tablets. *Journal of Drug Delivery Science and Technology, 49*, 227–234. https://doi.org/10.1016/j.jddst.2018.11.018

Allan, M. C., Rajwa, B., & Mauer, L. J. (2018). Effects of sugars and sugar alcohols on the gelatinization temperature of wheat starch. *Food Hydrocolloids, 84*, 593–607. https://doi.org/10.1016/j.foodhyd.2018.06.035

Ao, Z., & Jane, J.-L. (2007). Characterization and modeling of the A- and B-granule starches of wheat, triticale, and barley. *Carbohydrate Polymers, 67*(1), 46–55. https://doi.org/10.1016/j.carbpol.2006.04.013

Azhar, M. D., Hashib, S. A., Ibrahim, U. K., & Rahman, N. A. (2021). Development of carrier material for food applications in spray drying technology: An overview. *Materials Today: Proceedings, 47*, 1371–1375. https://doi.org/10.1016/j.matpr.2021.04.140

Chai, Y., Wang, M., & Zhang, G. (2013). Interaction between amylose and tea polyphenols modulates the postprandial glycemic response to high-amylose maize starch. *Journal of Agricultural and Food Chemistry, 61*(36), 8608–8615. https://doi.org/10.1021/jf402821r

Chao, C., Yu, J., Wang, S., Copeland, L., & Wang, S. (2018). Mechanisms underlying the formation of complexes between maize starch and lipids. *Journal of Agricultural and Food Chemistry, 66*(1), 272–278. https://doi.org/10.1021/acs.jafc.7b05025

Chen, X., Sun, C., Wang, Q., Tan, H., & Zhang, Y. (2022). Preparation of glycidyl methacrylate grafted starch adhesive to apply in high-performance and environment-friendly plywood. *International Journal of Biological Macromolecules, 194*, 954–961. https://doi.org/10.1016/j.ijbiomac.2021.11.152

Cornejo-Ramírez, Y. I., Martínez-Cruz, O., Toro-Sánchez, C. L. D., Wong-Corral, F. J., Borboa-Flores, J., & Cinco-Moroyoqui, F. J. (2018). The structural characteristics of starches and their functional properties. *CyTA—Journal of Food, 16*, 1003–1017. https://doi.org/10.1080/19476337.2018.1518343

Daudt, R. M., Back, P. I., Cardozo, N. S. M., Marczak, L. D. F., & Külkamp-Guerreiro, I. C. (2015). Pinhão starch and coat extract as new natural cosmetic ingredients: Topical formulation stability and sensory analysis. *Carbohydrate Polymers, 134*, 573–580. https://doi.org/10.1016/j.carbpol.2015.08.038

Donmez, D., Pinho, L., Patel, B., Desam, P., & Campanella, O. H. (2021). Characterization of starch–water interactions and their effects on two key functional properties: Starch gelatinization and retrogradation. *Current Opinion in Food Science, 39*, 103–109. https://doi.org/10.1016/j.cofs.2020.12.018

Feist, M. (2015). Thermal analysis: Basics, applications, and benefit. *ChemTexts, 1*(1), 8. https://doi.org/10.1007/s40828-015-0008-y

Fitriani, S., Riftyan, E., Saputra, E., & Wulandari, E. (2022). Utilization of pregelatinized sago starch in deep-fried snacks. *IOP Conference Series: Earth and Environmental Science, 1059*(1).

Fu, Z., Wang, L., Li, D., & Adhikari, B. (2012). Effects of partial gelatinization on structure and thermal properties of corn starch after spray drying. *Carbohydrate Polymers, 88*(4), 1319–1325. https://doi.org/10.1016/j.carbpol.2012.02.010

Gavrielidou, M. A., Valous, N. A., Karapantsios, T. D., & Raphaelides, S. N. (2002). Heat transport to a starch slurry gelatinizing between the drums of a double drum dryer. *Journal of Food Engineering, 54*(1), 45–58. https://doi.org/10.1016/S0260-8774(01)00184-4

Gomes, S., Finotelli, P. V., Sardela, V. F., Pereira, H. M. G., Santelli, R. E., Freire, A. S., & Torres, A. G. (2019). Microencapsulated Brazil nut (Bertholletia excelsa) cake extract powder as an added-value functional food ingredient. *LWT, 116*, 108495. https://doi.org/10.1016/j.lwt.2019.108495

Guan, Y. (2015). China patent no. CN103739721B.

Guo, P., Yu, J., Copeland, L., Wang, S., & Wang, S. (2018). Mechanisms of starch gelatinization during heating of wheat flour and its effect on in vitro starch digestibility. *Food Hydrocolloids, 82*, 370–378. https://doi.org/10.1016/j.foodhyd.2018.04.012

Guo, Z., Jia, X., Miao, S., Chen, B., Lu, X., & Zheng, B. (2018). Structural and thermal properties of amylose-fatty acid complexes prepared via high hydrostatic pressure. *Food Chemistry, 264*, 172–179.

Han, X., Xing, J., Han, C., Guo, X., & Zhu, K. (2021). The effects of extruded endogenous starch on the processing properties of gluten-free Tartary buckwheat noodles. *Carbohydrate Polymers, 267*, 118170. https://doi.org/10.1016/j.carbpol.2021.118170

Haq, F., Farid, A., Ullah, N., Kiran, M., Khan, R. U., Aziz, T., Mehmood, S., Haroon, M., Mubashir, M., Bokhari, A., Chuah, L. F., & Show, P. L. (2022). A study on the uptake of methylene blue by biodegradable and eco-friendly carboxylated starch grafted polyvinyl pyrrolidone. *Environmental Research, 215*, 114241. https://doi.org/10.1016/j.envres.2022.114241

He, C., Zheng, J., Liu, F., Woo, M. W., Xiong, H., & Zhao, Q. (2020). Fabrication and characterization of oat flour processed by different methods. *Journal of Cereal Science, 96*, 103123. https://doi.org/10.1016/j.jcs.2020.103123

He, X., Xia, W., Chen, R., Dai, T., Luo, S., Chen, J., & Liu, C. (2020). A new pre-gelatinized starch preparing by gelatinization and spray drying of rice starch with hydrocolloids. *Carbohydrate Polymers, 229*, 115485. https://doi.org/10.1016/j.carbpol.2019.115485

Hou, C., Zhao, X., Tian, M., Zhou, Y., Yang, R., Gu, Z., & Wang, P. (2020). Impact of water extractable arabinoxylan with different molecular weight on the gelatinization and retrogradation behavior of wheat starch. *Food Chemistry, 318*, 126477. https://doi.org/10.1016/j.foodchem.2020.126477

Izidoro, D. R., Sierakowski, M.-R., Haminiuk, C. W. I., Souza, C. F. D., & Scheer, A. D. P. (2011). Physical and chemical properties of ultrasonically, spray-dried green banana (Musa cavendish) starch. *Journal of Food Engineering, 104*(4), 639–648. https://doi.org/10.1016/j.jfoodeng.2011.02.002

Jia, X., Sun, S., Chen, B., Zheng, B., & Guo, Z. (2018). Understanding the crystal structure of lotus seed amylose–long-chain fatty acid complexes prepared by high hydrostatic pressure. *Food Research International, 111*, 334–341.

Kankate, D., Panpalia, S. G., Kumar, K. J., & Kennedy, J. F. (2020). Studies to predict the effect of pregelatinization on excipient property of maize and potato starch blends. *International Journal of Biological Macromolecules, 164*, 1206–1214. https://doi.org/10.1016/j.ijbiomac.2020.07.170

Kanmani, N., Romano, N., Ebrahimi, M., Nurul Amin, S. M., Kamarudin, M. S., Karami, A., & Kumar, V. (2018). Improvement of feed pellet characteristics by dietary pre-gelatinized starch and their subsequent effects on growth and physiology in tilapia. *Food Chemistry, 239*, 1037–1046. https://doi.org/10.1016/j.foodchem.2017.07.061

Kim, J.-U., Ha, J.-M., Rhee, Y.-S., Park, C.-W., Chi, S.-C., & Park, E.-S. (2014). Influence of pharmaceutical excipients on stability of pramipexole dihydrochloride monohydrate in tablets. *Journal of Pharmaceutical Investigation, 44*(3), 177–185. https://doi.org/10.1007/s40005-013-0113-0

Lazou, A., & Krokida, M. (2010). Structural and textural characterization of corn-lentil extruded snacks. *Journal of Food Engineering, 100*(3), 392–408. https://doi.org/10.1016/j.jfoodeng.2010.04.024

Li, C. (2022). Recent progress in understanding starch gelatinization—an important property determining food quality. *Carbohydrate Polymers, 293*, 119735. https://doi.org/10.1016/j.carbpol.2022.119735

Li, E., Cao, P., Cao, W., & Li, C. (2022). Relations between starch fine molecular structures with gelatinization property under different moisture content. *Carbohydrate Polymers, 278*, 118955. https://doi.org/10.1016/j.carbpol.2021.118955

Li, Q., Liu, S., Obadi, M., Jiang, Y., Zhao, F., Jiang, S., & Xu, B. (2020a). The impact of starch degradation induced by pre-gelatinization treatment on the quality of noodles. *Food Chemistry, 302*, 125267. https://doi.org/10.1016/j.foodchem.2019.125267

Li, Q., Liu, S., Obadi, M., Jiang, Y., Zhao, F., Jiang, S., & Xu, B. (2020b). A novel strategy for improving drying efficiency and quality of cream mushroom soup based on microwave pre-gelatinization and infrared freeze-drying. *Innovative Food Science & Emerging Technologies, 66*, 102516. https://doi.org/10.1016/j.ifset.2020.102516

Li, W., Cao, F., Fan, J., Ouyang, S., Luo, Q., Zheng, J., & Zhang, G. (2014). Physically modified common buckwheat starch and their physicochemical and structural properties. *Food Hydrocolloids, 40*, 237–244. https://doi.org/10.1016/j.foodhyd.2014.03.012

Li, W., Yan, J., Chang, X., Tian, Y., & Liu, S. (2023). Effects of different glutenin types on the physicochemical properties and in vitro digestion of cassava starch. *CyTA—Journal of Food, 21*(1), 394–403. https://doi.org/10.1080/19476337.2023.2209133

Liu, S., Chen, W., Zhang, C., Wu, T., Zheng, B., & Guo, Z. (2022). Structural, thermal and pasting properties of heat-treated lotus seed starch–protein mixtures. *Foods, 11*(19), 2933. https://doi.org/10.3390/foods11192933

Liu, W., Zhang, Y., Xu, Z., Pan, W., Shen, M., Han, J., Sun, X., Zhang, Y., Xie, J., Zhang, X., & Yu, L. (2022). Cross-linked corn bran arabinoxylan improves the pasting, rheological, gelling properties of corn starch and reduces its in vitro digestibility. *Food Hydrocolloids, 126*, 107440. https://doi.org/10.1016/j.foodhyd.2021.107440

Liu, X., Liu, S., Xi, H., Xu, J., Deng, D., & Huang, G. (2019). Effects of soluble dietary fiber on the crystallinity, pasting, rheological, and morphological properties of corn resistant starch. *LWT, 111*, 632–639. https://doi.org/10.1016/j.lwt.2019.01.059

Liu, Y., Chen, J., Luo, S., Li, C., Ye, J., Liu, C., & Gilbert, R. G. (2017). Physicochemical and structural properties of pregelatinized starch prepared by improved extrusion cooking technology. *Carbohydrate Polymers, 175*, 265–272. https://doi.org/10.1016/j.carbpol.2017.07.084

Loganathan, S., Valapa, R. B., Mishra, R. K., Pugazhenthi, G., & Thomas, S. (2017). Chapter 4—thermogravimetric analysis for characterization of nanomaterials. In S. Thomas, R. Thomas, A. K. Zachariah, & R. K. Mishra

(Eds.), *Thermal and rheological measurement techniques for nanomaterials characterization* (pp. 67–108). Elsevier.

Lu, Z.-H., Donner, E., Yada, R. Y., & Liu, Q. (2016). Physicochemical properties and in vitro starch digestibility of potato starch/protein blends. *Carbohydrate Polymers, 154*, 214–222. https://doi.org/10.1016/j.carbpol.2016.08.055

Lubowa, M., Yeoh, S. Y., & Easa, A. M. (2018). Textural and physical properties of retort processed rice noodles: Influence of chilling and partial substitution of rice flour with pregelatinized high-amylose maize starch. *Food Science and Technology International, 24*(6), 476–486. https://doi.org/10.1177/1082013218766984

Ma, H., Liu, M., Liang, Y., Zheng, X., Sun, L., Dang, W., Li, J., Li, L., & Liu, C. (2022). Research progress on properties of pre-gelatinized starch and its application in wheat flour products. *Grain & Oil Science and Technology, 5*(2), 87–97.

Majzoobi, M., Radi, M., Farahnaky, A., Jamalian, J., Karrila, T., & Mesbahi, G. (2011). Physicochemical properties of pre-gelatinized wheat starch produced by a twin drum drier. *Journal of Agricultural Science and Technology, 13*, 193–202.

Maniglia, B. C., Polachini, T. C., Norwood, E.-A., Le-Bail, P., & Le-Bail, A. (2021). Thermal technologies to enhance starch performance and starchy products. *Current Opinion in Food Science, 40*, 72–80. https://doi.org/10.1016/j.cofs.2021.01.005

Marcello, C., & Salam, A. (2023). A biobased synthesized N-hydroxymethyl starch-amide for enhancing the wet and dry strength of paper products. *Carbohydrate Polymers, 299*, 120194. https://doi.org/10.1016/j.carbpol.2022.120194

Nakorn, K. N., Tongdang, T., & Sirivongpaisal, P. (2009). Crystallinity and rheological properties of pregelatinized rice starches differing in amylose content. *Starch—Stärke, 61*(2), 101–108. https://doi.org/10.1002/star.200800008

Narciso, J. O., & Brennan, C. (2018). Whey and pea protein fortification of rice starches: Effects on protein and starch sigestibility and starch pasting properties. *Starch—Stärke, 70*(9–10), 1700315. https://doi.org/10.1002/star.201700315

Nasir, M., Ahmad, S., Usman, M., Farooq, U., Naz, A., Murtaza, M. A., Shehzad, Q., & din, G. M. U. (2021). Influence of pregelatinized starch on rheology of composite flour, in vitro enzyme digestibility and textural properties of millet-based chapatti. *Carbohydrate Polymer Technologies and Applications, 2*, 100108. https://doi.org/10.1016/j.carpta.2021.100108

Nithyadevi, K., & Pandiyan, V. (2021). Effect of infrared heating on physical, structural and pasting properties of starch. *Applied Physics A, 127*(12).

Obadi, M., Chen, Y., Qi, Y., Liu, S., & Xu, B. (2020). Effects of different pre-gelatinized starch on the processing quality of high value-added Tartary buckwheat noodles. *Journal of Food Measurement and Characterization, 14*(6), 3462–3472. https://doi.org/10.1007/s11694-020-00572-4

Ortolan, F., Brites, L. T. G., Montenegro, F. M., Schmiele, M., Steel, C. J., Clerici, M. T. P. S., Almeida, E. L., & Chang, Y. K. (2015). Effect of extruded wheat flour and pre-gelatinized cassava starch on process and quality parameters of French-type bread elaborated from frozen dough. *Food Research International, 76*, 402–409. https://doi.org/10.1016/j.foodres.2015.07.010

Roman, L., Reguilon, M. P., Martinez, M. M., & Gomez, M. (2020). The effects of starch cross-linking, stabilization and pre-gelatinization at reducing gluten-free bread staling. *LWT, 132*, 109908. https://doi.org/10.1016/j.lwt.2020.109908

Romano, N., Kanmani, N., Ebrahimi, M., Chong, C. M., Teh, J. C., Hoseinifar, S. H., Amin, S. N., Kamarudin, M. S., & Kumar, V. (2018). Combination of dietary pre-gelatinized starch and isomaltooligosaccharides improved pellet characteristics, subsequent feeding efficiencies and physiological status in African catfish, Clarias gariepinus, juveniles. *Aquaculture, 484*, 293–302. https://doi.org/10.1016/j.aquaculture.2017.09.022

Romano, N., & Kumar, V. (2018). Starch gelatinization on the physical characteristics of aquafeeds and subsequent implications to the productivity in farmed aquatic animals. *Reviews in Aquaculture, 11*(4), 1271–1284. https://doi.org/10.1111/raq.12291

Roulet, P., MacInnes, W. M., Würsch, P., Sanchez, R. M., & Raemy, A. (1988). A comparative study of the retrogradation kinetics of gelatinized wheat starch in gel and powder form using X-rays, differential scanning calorimetry and dynamic mechanical analysis. *Food Hydrocolloids, 2*(5), 381–396. https://doi.org/10.1016/S0268-005X(88)80003-1

Salam, A., Lucia, L., & Jameel, H. (2018). Starch derivatives that contribute significantly to the bonding and antibacterial character of recycled fibers. *ACS Omega, 3*(5), 5260–5265. https://doi.org/10.1021/acsomega.8b00307

Santos, A. D. A. D., Matos, L. C., Mendonça, M. C., Lago, R. C. D., Muguet, M. C. D. S., Damásio, R. A. P., Ponzecchi, A., Soares, J. R., Sanadi, A. R., & Tonoli, G. H. D. (2023). Evaluation of paper coated with cationic starch and carnauba wax mixtures regarding barrier properties. *Industrial Crops and Products, 203*, 117177. https://doi.org/10.1016/j.indcrop.2023.117177

Santos, T. P. R. D., Franco, C. M. L., Carmo, E. L. D., & Leonel, J.-L. J. M. (2019). Effect of spray-drying and extrusion on physicochemical characteristics of sweet potato starch. *Journal of Food Science and Technology, 56*(1), 376–383. https://doi.org/10.1007/s13197-018-3498-y

Santos, T. P. R. D., Franco, C. M. L., Demiate, I. M., Li, X.-H., Garcia, E. L., Jane, J.-L., & Leonel, M. (2018). Spray-drying and extrusion processes: Effects on morphology and physicochemical characteristics of starches isolated from Peruvian carrot and cassava. *International Journal of Biological Macromolecules, 118*, 1346–1353. https://doi.org/10.1016/j.ijbiomac.2018.06.070

Shen, H., Yu, J., Bai, J., Liu, Y., Ge, X., Li, W., & Zheng, J. (2023). A new pre-gelatinized starch preparing by spray drying and electron beam irradiation of oat starch. *Food Chemistry, 398*, 133938–133938.

Shrestha, A. K., Ng, C. S., Lopez-Rubio, A., Blazek, J., Gilbert, E. P., & Gidley, M. J. (2010). Enzyme resistance and structural organization in extruded high amylose maize starch. *Carbohydrate Polymers, 80*(3), 699–710. https://doi.org/10.1016/j.carbpol.2009.12.001

Sobini, N., Darsiga, S., Kananke, T. C., & Srivijeindran, S. (2022). Characterization of modified palmyrah tuber starch by pre-gelatinization, acid and dextrinization processes and its applicability. *Food Chemistry Advances, 1*, 100143. https://doi.org/10.1016/j.focha.2022.100143

Sun, L., Hu, X., Zhang, Y., Li, J., & Wang, B. (2022). Effects of different particle size distribution on pasting characteristics of whole kernel corn flour. *Journal of the Chinese Cereals and Oils Association, 37*(5), 32–38. https://kns.cnki.net/kcms/detail/11.2864.TS.20210928.0350.006.html

Suwanprateeb, J., Kiertkrittikhoon, S., Kintarak, J., Suvannapruk, W., Thammarakcharoen, F., & Rukskul, P. (2014). In vivo assessment of new resorbable PEG-PPG-PEG copolymer/starch bone wax in bone healing and tissue reaction of bone defect in rabbit model. *Journal of Materials Science. Materials in Medicine, 25*(9), 2131–2139. https://doi.org/10.1007/s10856-014-5249-6

Tangsrianugul, N., Wongsagonsup, R., & Suphantharika, M. (2019). Physicochemical and rheological properties of flour and starch from Thai pigmented rice cultivars. *International Journal of Biological Macromolecules, 137*, 666–675. https://doi.org/10.1016/j.ijbiomac.2019.06.196

Tao, H., Zhu, X., Nan, B., Jiang, R., & Wang, H. (2021). Effect of extruded starches on the structure, farinograph characteristics and baking behavior of wheat dough. *Food Chemistry, 348*, 129017. https://doi.org/10.1016/j.foodchem.2021.129017

Taşkın, B., & Savlak, N. (2022). Functional, chemical, and sensorial properties of gluten-free fermented instant soup powders developed by use of mung bean and drum drying process. *Food Bioscience, 47*, 101677. https://doi.org/10.1016/j.fbio.2022.101677

Tian, Y., Li, Y., Xu, X., & Jin, Z. (2011). Starch retrogradation studied by thermogravimetric analysis (TGA). *Carbohydrate Polymers, 84*(3), 1165–1168. https://doi.org/10.1016/j.carbpol.2011.01.006

Tian, Y., Xu, X., Xie, Z., Zhao, J., & Jin, Z. (2011). Starch retrogradation determined by differential thermal analysis (DTA). *Food Hydrocolloids, 25*(6), 1637–1639. https://doi.org/10.1016/j.foodhyd.2011.01.015

Wang, B., Dong, Y., Fang, Y., Gao, W., Kang, X., Liu, P., Yan, S., Cui, B., & El-Aty, A. M. A. (2022). Effects of different moisture contents on the structure and properties of corn starch during extrusion. *Food Chemistry, 368*, 130804. https://doi.org/10.1016/j.foodchem.2021.130804

Wang, J., Xu, Q., Qin, J., Wu, Q., Zhu, H., Lu, B., & Zhu, Z. (2023). Performance enhancement and mechanism of pregelatinized starch-binding gypsum at high water-gypsum ratios. *Construction and Building Materials, 393*, 132142. https://doi.org/10.1016/j.conbuildmat.2023.132142

Wang, N., You, Y., Liao, X., Zhang, F., Kan, J., & Zheng, J. (2023). Ultrasonic modification of lotus starch based on multi-scale structure: Pasting, rheological, and thermal properties. *LWT, 184*, 115030. https://doi.org/10.1016/j.lwt.2023.115030

Wang, S., & Copeland, L. (2012). Phase transitions of pea starch over a wide range of water content. *Journal of Agricultural and Food Chemistry, 60*(25), 6439–6446. https://doi.org/10.1021/jf3011992

Wang, W., Zhou, H., Yang, H., Zhao, S., Liu, Y., & Liu, R. (2017). Effects of salts on the gelatinization and retrogradation properties of maize starch and waxy maize starch. *Food Chemistry, 214*, 319–327. https://doi.org/10.1016/j.foodchem.2016.07.040

Wang, Z., Zhong, Z., Zheng, B., Zhang, Y., & Zeng, H. (2023). Effects of Porphyra haitanensis polysaccharides on gelatinization and gelatinization kinetics of starches with different crystal types. *International Journal of Biological Macromolecules, 242*(Pt 3), 125117. https://doi.org/10.1016/j.ijbiomac.2023.125117

Wiriyawattana, P., Suwonsichon, S., & Suwonsichon, T. (2018). Effects of drum drying on physical and antioxidant properties of riceberry flour. *Agriculture and Natural Resources, 52*(5), 445–450. https://doi.org/10.1016/j.anres.2018.11.008

Wu, Y., Niu, M., & Xu, H. (2020). Pasting behaviors, gel rheological properties, and freeze–thaw stability of rice flour and starch modified by green tea polyphenols. *LWT, 118*, 108796–108796.

Xue, C., Fukuoka, M., & Sakai, N. (2010). Prediction of the degree of starch gelatinization in wheat flour dough during microwave heating. *Journal of Food Engineering, 97*(1), 40–45. https://doi.org/10.1016/j.jfoodeng.2009.09.013

Yan, X., Wu, Z., Li, M., Yin, F., Ren, K., & Tao, H. (2019). The combined effects of extrusion and heat-moisture treatment on the physicochemical properties and digestibility of corn starch. *International Journal of Biological Macromolecules, 134*, 1108–1112. https://doi.org/10.1016/j.ijbiomac.2019.05.112

Yang, C., Zhong, F., Goff, H. D., & Li, Y. (2019). Study on starch-protein interactions and their effects on physicochemical and digestible properties of the blends. *Food Chemistry, 280*, 51–58. https://doi.org/10.1016/j.foodchem.2018.12.028

Yin, X., Zheng, Y., Kong, X., Cao, S., Chen, S., Liu, D., Ye, X., & Tian, J. (2021). RG-I pectin affects the physicochemical properties and digestibility of potato starch. *Food Hydrocolloids, 117*, 106687. https://doi.org/10.1016/j.foodhyd.2021.106687

Zhang, Z., Zhu, M., Xing, B., Liang, Y., Zou, L., Li, M., Fan, X., Ren, G., Zhang, L., & Qin, P. (2023). Effects of extrusion on structural properties, physicochemical properties and in vitro starch digestibility of Tartary buckwheat flour. *Food Hydrocolloids, 135*, 108197. https://doi.org/10.1016/j.foodhyd.2022.108197

Zhao, B., Sun, S., Lin, H., Chen, L., Qin, S., Wu, W., Zheng, B., & Guo, Z. (2019). Physicochemical properties and digestion of the lotus seed starch-green tea polyphenol complex under ultrasound-microwave synergistic interaction. *Ultrasonics Sonochemistry, 52*, 50–61. https://doi.org/10.1016/j.ultsonch.2018.11.001

Zhu, F. (2015). Interactions between starch and phenolic compound. *Trends in Food Science & Technology, 43*(2), 129–143. https://doi.org/10.1016/j.tifs.2015.02.003

High Hydrostatic Pressure on Starch Modification

Process Conditions, Equipment Settings, and Characteristics of Modified Starches

Rosana Colussi, Sabrina Feksa Frasson, Barbara Biduski, and Jaspreet Singh

5.1 INTRODUCTION

Starch is a naturally occurring storage polysaccharide found abundantly in seeds, roots, and tubers, considered the second-largest biomass on earth (Suri & Singh, 2023). Starch's low cost, high availability, and biodegradability make it a widely studied polymer across diverse industry segments. However, in its native form, starch does not always present the desired characteristics for specific applications. Starches can be modified by physiological, enzymatic, chemical, and physical processes to overcome these limitations. Physical modification of starches involves treatments that preserve the D-glucopyranosyl units within starch molecules. These treatments produce changes only in the packaging arrangements of the molecules, within the granules, and in the overall structures of starch granules. These changes promote significant impacts on thermal and viscoamylographic properties and starch digestibility, offering the advantage of being environmentally friendly (Bemiller & Huber, 2015; Heydari et al., 2021b; Kim et al., 2022; Wu et al., 2022; Ye & Baik, 2023).

Thermal treatments are still the most common physical modification; however, non-thermal treatments are increasingly attracting interest due to the rapid modification and different effects caused to the starch granule. Contemporary research extensively explores non-thermal treatments such as ultrasound, pulsed electric field, radiation, cold plasma, and high hydrostatic pressure (HHP) (Suri & Singh, 2023; Wang et al., 2023b; Ye & Baik, 2023).

DOI: 10.1201/9781032655598-5

Widely recognized for its environmentally friendly and green technology, HHP is a technique that over the years has aroused interest, primarily in enzymatic and microorganism inactivation, as well as for improving the texture and cold sterilization of foods (Rostamabadi et al., 2023). The HHP market has been growing over the years, and according to Future Market Insights, the global high-pressure processing equipment market size is valued at US\$ 333.9 million in 2023 and projected to reach US\$ 1.1 billion by 2033. In recent years, significant technological improvements have enhanced the efficiency and effectiveness of the equipment. These ongoing innovations are expected to contribute to further market expansion, particularly in food safety and starch modification.

When applied in starch, high hydrostatic pressure promotes cold gelatinization, a mechanism different from heat-induced gelatinization. As a result, this process imparts different characteristics to modified starches. The effects caused by this modification depend on multiple process-related factors, including starch concentration, pressurization time, number of pressurization and depressurization cycles, and temperature. Additionally, starch-related factors, such as source, amylose content, relative crystallinity, morphology, and granule size, also play a crucial role (Pei-Ling et al., 2010).

For instance, a study carried out by Almeida et al. (2023) evaluated the technological properties of rice starch (20%, w/v) modified with pressures of 200, 400, and 600 MPa for 30 minutes at 30°C. The authors found that pressures above 400 MPa led to a granular disorder, reducing the energy required for gelatinization. Additionally, the diffraction pattern changed from type A to type V. Furthermore, the starches modified with 600 MPa showed improvements in water, oil, and milk absorption values, crucial data for widespread applications within the food industry. Furthermore, HHP-induced modifications in starch digestibility were investigated by Colussi et al. (2018) through the modified potato starch using HHP (400 and 600 MPa for 3 and 6 cycles of 10 minutes each). Their findings highlight that HHP, associated with retrogradation, allows the production of starch with different functional characteristics. After cooking, these starches exhibit less hydrolysis during in vitro gastrointestinal digestion, resulting in a slower glucose release. The HHP-modified starches can be used to develop products that have delayed digestion and low glycemic index.

It is evident that there is no universal standard for all types of starch, as each source responds differently under the same process conditions. Therefore, it is important to analyze how different starch sources respond to high hydrostatic pressure. Consequently, starches with distinct characteristics can be used for various food industry segments. With that in mind, this chapter will address the factors influencing the morphological, physical, chemical, thermal, and digestibility characteristics of starches modified through high hydrostatic pressure.

5.2 THE HHP SETUP AND MECHANISM

High-pressure processing, also known as high hydrostatic pressure and ultrahigh-pressure processing (UHP), is an emerging technology that operates based on three fundamental principles. The first one is the isostatic principle, stating that pressure uniformly acts in all directions, regardless of material size or shape, ensuring instantaneous and homogeneous impact. The second principle relies on Le Chatelier's principle, wherein pressure adjustments in response to volume changes drive a system toward a new equilibrium state for any given phenomenon (Balasubramaniam et al., 2015). The third microscopic ordering principle implies that the degree of molecular ordering of a substance increases when pressure increases at a constant temperature (Elamin et al., 2015; Jaeger et al., 2012). High hydrostatic pressure operates within a pressure range of 100 to 800 MPa and can be applied either with or without the presence of heat. In real-world food processing scenarios, the synergistic impact of both thermal and pressure effects can induce diverse physical, chemical, or biological changes in food products (Balasubramaniam et al., 2015).

Among the various non-thermal methods used in the food industry, HHP is already well established in developed countries and has gained broad acceptance in the commercial processing of diverse foods, including fruits, vegetables, dairy, and meat products. Renowned for its green technique and environmentally friendly nature, the HPP can be a sustainable technique that produces minimal loss or harm to the environment. The use of this technique can ensure microbial safety, enzyme inactivation, and shelf-life stability efficiently and effectively (Sehrawat et al., 2021).

A typical high hydrostatic pressure system consists of a high-pressure vessel and its closure and pressure-generating, temperature, and pressure-control devices. Water is the primary medium used for pressure transfer, and pressure is applied uniformly and instantly throughout the sample, regardless of direct contact with the pressure medium (Castro et al., 2020). The HHP selective affects non-covalent bonds, as these are sensitive to pressure. In contrast, pressure does not impact the components responsible for retaining color characteristics, bioactive compounds, and low molecular weight volatile and non-volatile compounds with covalent bonds. Thus, after processing, food preserves its natural characteristics (Sehrawat et al., 2021; Zhang et al., 2022a).

The use of high pressure in starch modification is a recent technique and has been extensively investigated over the last few years (Castro et al., 2023; Kim et al., 2023; Li et al., 2015; Obadi et al., 2023; Pulgarín et al., 2023; Seo et al., 2023; Yang et al., 2016; Zamani & Razavi, 2023; Zeng et al., 2018; Zhang et al., 2022a, 2023; Zhi-Guang et al., 2020; Zhou et al., 2022) and the gelatinization process induced by HHP reviewed by many researchers (Castro et al., 2020; Chen et al., 2023; Dominguez-Ayala et al., 2022; Fan et al., 2023; Nasehi & Javaheri, 2012; Pei-Ling et al., 2010; Wang et al., 2023b). The existing studies exploring the impact of HHP on starch can be categorized in three ways (Katopo et al., 2002). The first category involves pressure applications that did not reach a level sufficient for starch gelatinization. The second category includes studies where pressure was applied to mostly dry starch samples. The third category involves ultrahigh-pressure treatments, characterized by investigations conducted with excess water and pressures over 400 MPa.

During the HHP process, reversible hydration of the amorphous phase of the granule occurs, followed by irreversible distortion of the crystalline phase. Consequently, the secondary and tertiary structures break down while the covalent bonds remain intact. Depending on the intensity, this can lead to granular destruction. This process is defined as pressure-induced gelatinization. The packing of the chain in the starch granule structure is decisive for the granule resistance to HHP (Knorr et al., 2006; Pei-Ling et al., 2010). Unlike heat gelatinization, HHP-modified granules maintain their structure intact or partially intact, and amylose solubilization is low (Knorr et al., 2006). Figure 5.1 presents a scheme proposed by Buckow et al. (2007), which compares the gelatinization process by heat and HHP. Despite the many studies available, the mechanism of gelatinization due to high pressure is still not well understood. Therefore, evaluating and understanding the effects of HHP on starches from different sources can lead to better utilization of various starch resources.

As mentioned earlier, water content, temperature, pressure, and pressurization time are the main factors that need to be adjusted for starch modification by HHP. The concentration of starch to be processed is a key factor since the intensity of gelatinization is proportional to the reduction in starch content. The main medium used to transmit pressure is water. Heydari et al. (2021a) investigated different concentrations (10%, 15%, and 20% w/w) of starch from waxy corn, normal corn, and wheat starch submitted to 600 MPa for 20 minutes at a maximum temperature of 34°C. The increase in starch concentration considerably increased water absorption capacity, oil absorption capacity, and zeta potential values and pronouncedly decreased peak viscosities, breakdown, and retrogradation. Kawai et al. (2007) carried out a very in-depth study using potato starch. They evaluated the effects of pressure (600, 800, and 1000 MPa), pressurization time (18 and 66 hours), and starch concentration (10 to 70% w/v). Differential scanning calorimetry (DSC) analysis found that the mixture with 10% starch did not present any peak due to its complete gelatinization under pressure. The mixture with 40% (w/v), treated with HHP, presented a broad endothermic peak as a consequence of re-gelatinization. This demonstrates the heterogeneity of the crystals, indicating that, with the use of pressure, part of the crystalline structure was affected by HHP, but not in its entirety.

During the HHP process, increasing pressure generally raises the process temperature. In most studies, the system is adjusted to ensure that the temperature does not exceed 40°C. This precaution is taken

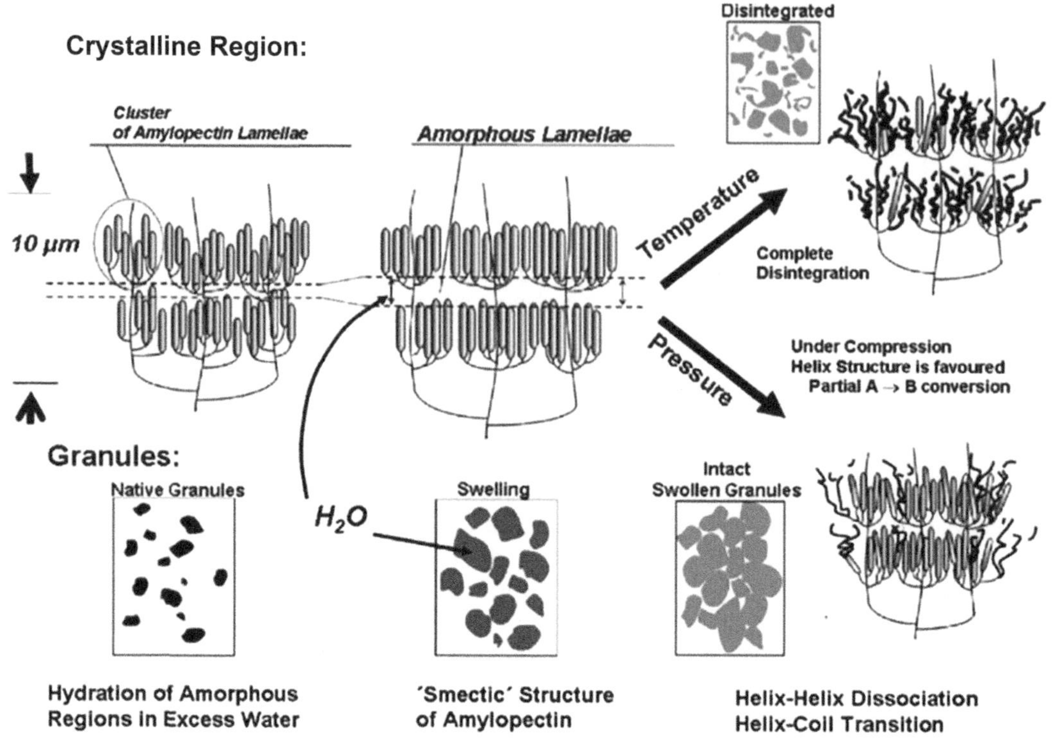

FIGURE 5.1 Comparative diagram between the heat gelatinization and cold gelatinization processes using HHP. In the presence of excess water, the amorphous regions of the growth ring become hydrated, and in the crystalline domain, amorphous lamellae are formed. The granule swells. The smectic crystal structure decomposes through helix-helix dissociation, followed by a helical coil transition when the gelatinization temperature is exceeded. It is suggested that under pressure, the disintegration of the macromolecule is incomplete, as the stabilization of hydrogen bonds under pressure favors the helix conformation (Buckow et al., 2007).

because pressure, when combined with heat, can lead to the gelatinization of starch, which requires careful management of both variables. Kawai et al. (2012) evaluated the treatment temperature (20–70°C) on potato starch with water mixtures (10–70%, w/w) submitted to HHP (400–1000 MPa). Both increasing temperature and reducing starch concentration favored gelatinization, and the data were summarized in a series of state diagrams. It is important to highlight that each starch possesses unique characteristics; therefore, starches from different sources may exhibit different behaviors even when subjected to the same conditions.

Pressure is the most commonly employed parameter in various studies. The effects on starch granules become noticeable at pressures above 400 MPa, and typically, the higher the pressure, the greater the impact on the starch granules. Zeng et al. (2018) investigated the effects of pressures ranging from 100 to 600 MPa for 20 minutes in a waxy rice starch slurry (10%, w/w). The authors found that HHP can modify the morphology, semi-crystalline structure, and in vitro digestibility. With increasing pressure, both the relative crystallinity and resistant starch content (RS) decreased. On the other hand, rapidly digestible starch (RDS) and slowly digestible starch (SDS) reached their highest levels at 600 and 400 MPa, respectively. The digestibility of the starch appears to be influenced by the applied pressure.

As the pressure treatment time increases, starch granules form a gelatinous connection zone, leading to an increase in particle size. Crystalline destruction occurs, and gelatinization takes place in starch treated with pressures above 500 MPa, causing a transformation in the crystal structure to a V-type. With an extended pressure treatment time, the starch becomes less stable, exhibiting reduced hardness and increased susceptibility to gelatinization. Consequently, the hydrolysis rate and the content of rapidly digestible starch increase (Zhang et al., 2022a).

5.3 EFFECT OF HIGH PRESSURE ON STARCH

5.3.1 Morphology

Starch is produced by green plants and serves as an energy storage compound, synthesized in a granular form. Biosynthesis of starch granules occurs in the amyloplast and is initiated at the hilum, where the granule grows by apposition. The granules are densely packed and semi-crystalline; therefore, starch granules are not soluble in water at room temperature (BeMiller & Whistler, 2009). Starches from different botanical sources present differentiated morphology, sizes, shapes, and diameters. The shape can be oval, spherical, polygonal, lenticular, or elongated, and the size can range from 1 μm to 100 μm in diameter. When submitted to HHP, the extension of the gelatinization will depend on the size of the granules and their composition.

When heated with water, starch granules swell and lose their lamellar ring structure and crystallinity. However, when subjected to high pressure, the shape and structure of the granules can be maintained. The response of each source to the same pressure varies. In Figure 5.2, images were selected from several

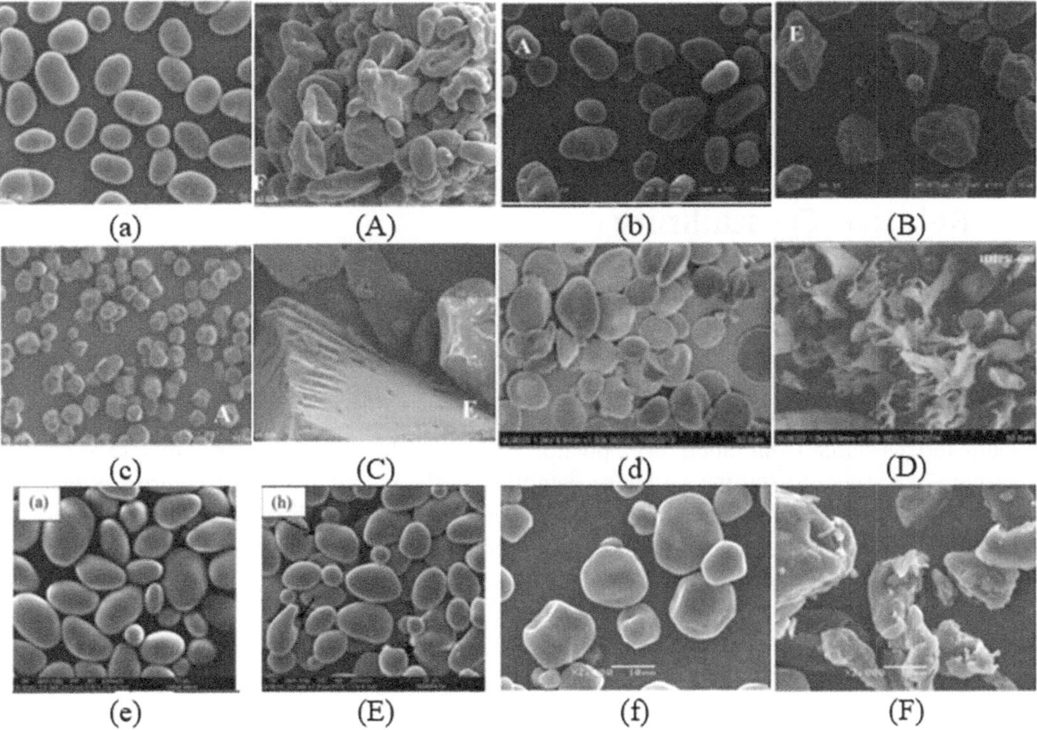

FIGURE 5.2 Compiling of scanning electron micrographs of native starches from mung bean (a), pea (b), millet (c), waxy wheat (d), potato (e), and sorghum (f) and HHP treated with 600 Mpa: mung bean (A), pea (B), millet (C), waxy wheat (D), potato (E), and sorghum (F). Compiled from Colussi et al. (2018); Hu et al. (2017); Li et al. (2011, 2018); Liu et al. (2016a, 2018).

studies that applied 600 MPa to modify starch. In general, starches with type B crystallinity exhibit greater resistance to changes in granule morphology (Figure 5.2E) than starches with a type A crystallinity pattern (Figure 5.2C, 5.2D, and 5.2F). However, starches with type C crystallinity patterns show intermediate resistance (Figure 5.2A and 5.2B). Following the modification, an increase in size and surface alteration are noticeable, resulting in a rougher and uneven texture, primarily due to the leaching of amylose.

The size of the granule is also a factor of great relevance, as the smaller the granule size, the greater the contact surface and, consequently, the higher the exposure to high pressure. It was observed that cereal starch granules (with a type A crystallinity pattern), varying in size, are affected to different extents. Millet starch granules have been observed to completely lost their structure (Li et al., 2018), while waxy wheat starch granules partially maintained their integrity (Hu et al., 2017). This difference is attributed to wheat having a bimodal granule size (large lenticular and small rounded granules), which exhibited distinct resistances when subjected to the same pressure.

According to Pei-Ling et al. (2010), the surface of starch granules is more resistant to HHP, as it features a tightly compacted layer that prevents the granule from losing its integrity. The internal part comprises a gel-like network, possibly arising from the hydration of the amorphous phase and/or fusion of the crystalline structure—the most organized (dense) part of the granule.

The gelatinization can be confirmed by the loss of birefringence. Some studies as Li et al. (2015) in which modified red adzuki bean starch suspensions (20%, w/w) at different pressures (0.1, 150, 300, 450, and 600 MPa) for 15 min verified that applying pressures varying from 150 to 450 MPa do not present relevant changes number and clarity of "Maltese cross". On the other hand, at pressures above 600 MPa, all starch granules lost the Maltese cross. In their study, Song et al. (2013) prepared amorphous granular starch. The researchers used six commercial starches (corn, waxy rice, tapioca, potato, sweet potato, and non-waxy rice starch) and subjected them to 550 MPa for 30 minutes. They observed that waxy rice starches gelatinized under these conditions have lost their granular form. Potato and sweet potato starches did not completely lose their crystalline and granular structure, expanding their potential applications in the food industry. The researchers also mentioned that they plan to conduct quantitative comparisons of morphological properties for specific purposes in future studies.

5.3.2 Relative Crystallinity

Amylose and amylopectin constitute the amorphous and crystalline regions within starch granules, forming a supramolecular structure that encompasses granule morphology, lamellar structure, fractal patterns, and crystalline arrangement. Starch exhibits several polymorphs distinguished by distinct crystalline structures, categorized as A-type, B-type, C-type, and V-type. The A-type crystalline structure is commonly associated with starch derived from cereals like corn, rice, and wheat, while the B-type pattern characterizes starch from tubers (e.g., potato), fruits (e.g., banana), and high-amylose corn starches. Furthermore, a C-type crystalline structure combines both A- and B-type structures, often found in legume starch (e.g., pea starch) as well as certain root and seed starches. Last, the V-type crystalline structure involves amylose single-helices co-crystallized with compounds like iodine, dimethyl sulfoxide (DMSO), alcohols, or fatty acids (Liu et al., 2018; Yang et al., 2016).

The crystallinity of starch is maintained through the organization of amylopectin double helices within a unit cell. These helices consist of two polyglycoside chains. In A-type starches, the closely arranged double helices form a monoclinic system, with each crystal cell containing eight water molecules interacting closely with the helices. Conversely, the arrangement of double helices in B-type starches results in a relatively loose structure, forming a hexagonal crystal. Each hexagonal arrangement comprising six double helices within the crystal cell accommodates 36 water molecules situated within the cavity surrounded by the helices (Feng et al., 2020). The dense packing characteristic of the A-type crystalline structure prevents certain chemical reactions (e.g., acid hydrolysis) from taking place, whereas the B-type structure is commonly found in high-amylose starches.

The x-ray diffraction pattern of starch can undergo a shift in types depending on the intensity of the HHP treatment. Typically, elevated pressure leads to weaker diffraction peaks due to the disruption of starch structure crystals caused by partial or complete gelatinization. Various studies have documented instances where C-type starches, like mung bean starch and lotus seed starch, transformed into B-type patterns following high-pressure treatment (Li et al., 2011); similarly, certain pressurized A-type starches, such as rice starch and maize starch, were observed to transform into B-type starches in other studies (Katopo et al., 2002).

It has been suggested that B-type starches show more resilience against HHP treatment in comparison to A-type starches. Despite this, even high amylose corn starches of the B-type can experience complete disruption of their crystalline structure under high-pressure applications. In addition to this alteration in crystallinity, it is expected that other larger-scale supramolecular arrangements, such as lamellar characteristics and fractal structures, would also change due to HHP treatment (Yang et al., 2016).

For instance, pea starch was treated with various pressure conditions, and up to 450 MPa, no significant alterations were observed in the X-ray diffraction, only changes in peak intensities. However, when subjected to 600 MPa treatment, specific alterations occurred: the peak intensity at 17.7° disappeared, while those at 15.3° and 23.3° decreased. Consequently, the X-ray diffraction pattern of pea starch shifted from a C-type to a B-type pattern (Liu et al., 2018). The authors conclude that the internal crystalline structure of pea starch underwent significant disruption due to HHP treatments.

HHP was used to modify mung bean starches, revealing no alterations in the X-ray diffraction type up to pressure levels of 480 MPa for 30 minutes (Figure 5.3). However, there were slight changes in peak intensities, indicative of reduced crystallinity during pressurization. However, when subjected to 600 MPa for 30 minutes, the X-ray diffraction patterns of mung bean starch showcased a transformation from the C-type toward a B-like pattern (Li et al., 2011). All pressurized starches, including maize, rice, waxy corn, wheat, and taro starches, exhibited a transformation from the A-type X-ray diffraction pattern toward a weaker B-type pattern. Conversely, like potato starch, starches displaying a B-type X-ray diffraction pattern retained their original X-ray pattern after HHP treatment (Katopo et al., 2002). Sorghum starch only changed the X-ray diffraction pattern at 600 MPa. The authors suggest that as sorghum starch presents an A-type structure, the amylopectin structure is flexible and easily destroyed, resulting in the rearrangement of double helices and changes in crystalline structure (Liu et al., 2016a).

The high pressure could potentially create an opportunity for starch granules to interact with activated water molecules, resulting in a subtle trend toward altering the X-ray pattern (Wang et al., 2008).

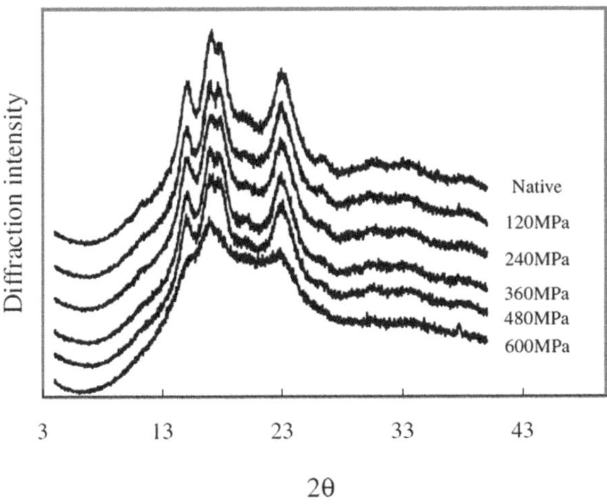

FIGURE 5.3 X-ray powder diffraction patterns of mung bean starch treated at different pressure levels (Li et al., 2011).

Small-angle X-ray scattering (SAXS) allows the detection of the lamellar structure within the semicrystalline growth rings when starch is dispersed in water. Studies employing SAXS on native starches derived from various plant sources have unveiled a consistent decline in the intensity of the scattering maximum, corresponding to the lamellar structure, as the amylose content increases (Yang et al., 2016).

5.3.3 Pasting and Rheology Properties (Rapid Visco Analyzer, Rheology, Swelling Power, and Solubility)

The structural and organizational alterations induced by HHP result in changes in the water solubility and swelling power, subsequently altering the viscosity profile of starch. Starch swelling power (SP) and water solubility (WS) serve as crucial indicators for the industry and provide valuable insights into starch behavior within food production systems (Clerici et al., 2018). Both SP and WS are influenced by the test temperature, typically showing an increase as the test temperature rises, often ranging from 50 to 95°C. However, following HHP treatment, the swelling power at higher temperatures tends to decrease, particularly with higher HHP intensities. This observed effect has been noted across various starch sources, including corn, potato (Rahman et al., 2020), pea (Liu et al., 2018), and quinoa (Ahmed et al., 2018).

Modified starch generally exhibits changes in gelatinization properties, affecting the viscosity profile and rheology characteristics. The Rapid Visco Analyzer (RVA) is a user-friendly and efficient tool for visualizing modification in starch macromolecules due to its reliability, repeatability, and versatility (Balet et al., 2019). The RVA analysis simulates the cooking process by subjecting the starch or flour to excess water with a specific heating and cooling cycle. Typically, the sample is dispersed in water at a concentration of about 10% w/w (starch/water). During the heating stage, the starch granules absorb water, leading to an increase in viscosity. The hydration causes a small amylose chain to leach out of the granule, followed by a granule fragmentation that causes a drop in viscosity. Subsequently, the cooling stage causes the viscosity to rise once again due to the reassociation of starch macromolecules and the retrogradation phase, known as setback.

When starch is modified by a physical process, the rearrangement or association of the starch chain occurs, impacting how these macromolecules interact and absorb water during heating. This altered structure can also affect the macromolecule tendency to retrograde, influencing the final viscosity or paste formation. Using RVA is essential to predict how modified starch will perform within a food system. There is a strong correlation between the pasting properties measured via RVA and the textural properties as well as the sensory characteristics of food products (Liu et al., 2019).

The influence of HHP on starch pasting properties is dependent on several factors, such as starch source, amylose content, and intensity of applied treatment. Moreover, the form in which the treatment is applied, whether on raw material, dried powder or isolated starch, also impacts the modification effect due to the interaction between starch and non-starch constituents. For instance, the presence of polyphenols has been reported to decrease the pH in flour-water systems, changing the gelatinization and retrogradation properties of starch (Zhu, 2015). Additionally, dietary fiber can compete for water binding, which can limit starch hydration.

Typically, when applying low pressure (about 200 MPa), minimal changes occur in the starch pasting properties, often resulting in results similar to the native starch (Zhu & Li, 2019). Increasing the pressure to above 500 MPa led to more intense alteration by inducing starch gelatinization. As a result, there is a decrease in swelling powder and, consequently, an increase in solubility. Table 5.1 summarizes the effect of HHP on the pasting properties of the modified starches.

In addition to the pressure intensity, the number of cycles applied can significantly influence the starch pasting properties. For instance, potato starch was treated at 400 and 600 MPa for 3 and 6 cycles of 10 min each (Colussi et al., 2017). The increase in the cycle number resulted in higher peak viscosity and breakdown. However, increasing the pressure to 600 MPa did not affect the starch properties. It is essential to highlight that each starch responds uniquely to HHP processing due to its composition and

TABLE 5.1 Pasting Properties of HHP-Modified Starches

INGREDIENT	PRESSURE (MPA)	SWELLING POWER (G/G) AT 95°C	SOLU-BILITY	PEAK VISCOSITY	BREAKDOWN	FINAL VISCOSITY	GELATINI-ZATION	REFERENCE
Potato starch	400	NA	NA	↑	↑	↓ or ≈	No	(Colussi et al., 2018)
	600			↓	↓	↓ or ≈	No	
Quinoa flour	200	↑ or ≈	↑ or ≈	↓ or ≈	↓	↓ or ≈	No	(Zhu & Li, 2019)
	600	↓	↑	↓	↓	↓	Yes	
Maize	200	↑ or ≈	or ≈	≈	≈	≈	No	(Li & Zhu, 2018)
	600	↓	↓	↓	↓	↓	Partially	
Lentil starch	400	NA	NA	≈	≈	≈	No	(Ahmed et al., 2016)
	600			↓	↓	↓	Yes	
Mung bean starch	240	NA	NA	↑	↑	↑	NA	(Jiang et al., 2015a)
	600			↓	↓	↓		
Rice starch	240	NA	NA	↓	↑	↑	NA	(Jiang et al., 2015b)
	600			↑	↓	↑		
Mango kernel starch	300	NA	NA	↑	↓	↓	No	(Kaur et al., 2019)
	600			↓	No	↓	Partially	
Japonica rice starch	200	NA	NA	↑	↑	↑	No	(Almeida et al., 2023)
	400			↑	↑	↑	No	
	600			↓	↓	↓	Yes	
Red adzuki bean	150	↓	≈	↓	↓	↑	No	(Li et al., 2015)
	300	↓	↓	↓	↓	↑	No	
	450	↓	↑	↓	↓	↑	No	
	600	↓	↑	↓	↓	↓	Yes	
Corn starch	500	↓	≈	↓	↓	↓	Partially	(Kim et al., 2023)
Potato starch	500	↓	≈	↓	≈	≈	No	
Rice starch	500	↓	↓	↓	↓	↓	Partially	
Tapioca	500	↓	↓	↓	↓	↓	Partially	
Pearl millet starch	200	↓	↓	≈	↓	≈	No	(Mirzababaee et al., 2022)
	400	↓	↓	≈	↓	≈	No	
	600	↓	↓	↓	↓	↓	Partially	
Oat starch	500	NA	NA	↓	↓ or ≈	↓	Partially	(Zhang et al., 2022b)
Maize starch (waxy)	600	NA	NA	↓	↓	↓	Yes	(Lin et al., 2023)
Maize starch (Hylon V high amylose)	600	NA	NA	↑	↑	↑	No	
Maize starch (Hylon VII high amylose)	600	NA	NA	↑	↑	↑	No	

NA: not analyzed; ↑ increased; ↓ decreased; ≈ similar to native.

crystallinity pattern (Colussi et al., 2017). As reported by Katopo et al. (2002), starches exhibiting B-type X-ray diffraction patterns are more resistant to high-pressure treatment. Consequently, starches derived from cereals source, mostly displaying an A-type pattern, tend to be more susceptible to alterations in viscosity profile due to HHP.

Beside the viscosity profile measured by RVA, the rheology characteristics of starch suspension or paste can be assessed through frequency sweep or apparent viscosity tests using rheometer equipment. The dynamic oscillation properties provide structural insights into starches by differentiating between the elastic and viscous contributions in measured stress concerning frequency. Storage modulus (G') and loss modulus (G") measurements show the starch behavior in relation to elasticity and viscosity, respectively. These tests are conducted within the linear viscosity region, preserving the sample structure due to the use of small strain. The frequency sweep test is used to estimate the long-term storage stability of gel systems (Mezger, 2020). In general, G' (storage modules) values are higher than G" (loss modules), displaying less frequency dependence, indicative of their gel-like behaviors. Several studies have reported increased G after HHP treatment across various material sources. Examples include lentil starch (Ahmed et al., 2016), mung bean starch (Jiang et al., 2015a), quinoa and maize flour (Li & Zhu, 2018), sweet potato starch (Chen et al., 2021), mango kernel starch (Kaur et al., 2019), tapioca starch (Ahmed et al., 2014), and rice starch (Jiang et al., 2015b).

The G' is directly associated with the cross-link density within a gel network, signifying that the higher G' values, the stronger the anti-disintegration ability and gel hardness of the starch gel (Chen et al., 2021). Conversely, G" indicates the energy required for the starch gel to resist flow. The increase in G' following HHP treatment indicates the rigidity and viscoelasticity of the gels, especially with the increasing pressure levels, in comparison to the native counterparts. Moreover, the complex viscosity commonly increases with the pressure treatment, indicating an increase in the viscoelasticity and changing the starch properties from viscoelastic-solid to fluid (Castro et al., 2020).

The changes in the starch functional properties mentioned in this section can be attributed to the rearrangement of starch molecules that can inhibit hydration and swelling. These rearrangements affect the crystalline structure of amylopectin; such structures are required to be intact in order to entrap the water (Wang & Copeland, 2015). Moreover, it is suggested that the gelatinization of starch by HHP is likely to be caused by water being forced into the starch granules and increasing their degree of hydration.

5.3.4 Thermal Properties

Thermal analysis can be used to confirm potential gelatinization of starch granules. The absence of endothermic peaks indicates complete gelatinization of the starch. Additionally, a reduction in peak temperature measured by DSC signifies loss of molecular order and damage to the crystalline structure. The melting temperature of starch granules relies on the structural arrangement of amylopectin clusters and their diverse polymorphic structures. Hence, the varying responses of different starches to high-pressure treatment may be attributed to differences in the inner structure of their granules (Li et al., 2011). Typically, pressures up to 500 MPa induce partial gelatinization, decreasing endotherm peak intensity and reducing enthalpy. However, pressures exceeding 500 MPa have been reported to disrupt the crystalline structure entirely, leading to the absence of peaks in DSC analysis (Figure 5.4).

Quinoa starch underwent gelatinization at 500 and 600 MPa, with pressures below 400 MPa showing minimal impact on its properties. At 600 MPa, complete gelatinization occurred in quinoa starch, evidenced by the absence of an endothermic peak in the DSC analysis of the treated sample (Li & Zhu, 2018). Similar behavior was reported in mung bean starch (Li et al., 2011), lentil starch (Ahmed et al., 2016), tapioca starch (Ahmed et al., 2014), and lotus seed starch (Guo et al., 2015). In contrast, sweet potato starch treated at 500 MPa displayed no significant differences in thermal properties compared to untreated starch (Rahman et al., 2020). This observation led to the hypothesis that sweet potato possesses a highly organized structure that withstands breakdown at 500 MPa. Cui and Zhu (2019) studied the impact of high pressure on different sweet potato types—orange, purple, and red sweet potatoes—and noted slight

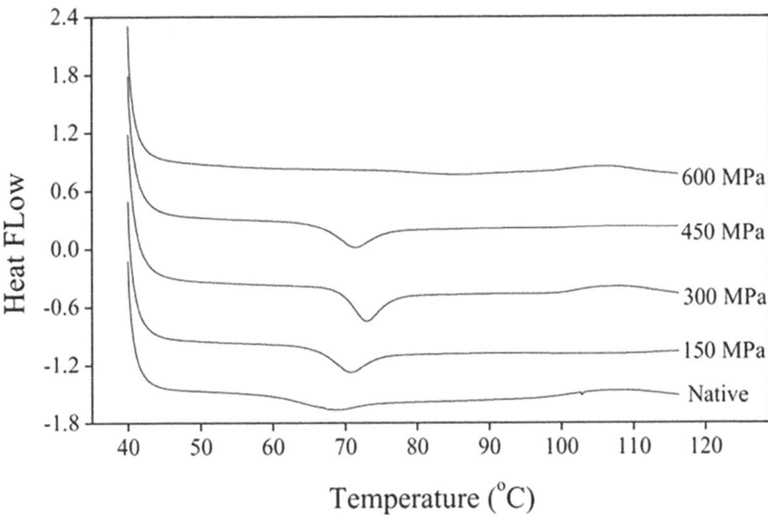

FIGURE 5.4 Thermal analysis of red adzuki bean starch granules before and after high-hydrostatic pressure treatment using DSC (Li et al., 2015).

changes in thermal properties at pressures up to 500 MPa. However, only purple and red sweet potatoes were completely gelatinized at 600 MPa, as indicated by the absence of an endothermic peak. Conversely, orange sweet potato exhibited decreased enthalpy and increased gelatinization temperature, suggesting partial gelatinization. Notably, all three varieties displayed a C-type polymorph, but purple and red sweet potatoes amylopectin contained more fingerprint A-chains than orange sweet potato amylopectin. Unable to form double helices, these chains might induce structural defects in starch granules, rendering purple and red sweet potato starches more sensitive to HHP (Li & Zhu, 2018). In addition, sorghum starch also did not gelatinize completely at 600 MPa conditions (Liu et al., 2016b). HHP sorghum starch exhibits a gradual reduction in the onset gelatinization temperature with the increase of pressure treatment. Lower energy was required for its gelatinization as the enthalpy reduced from 22 to 10 J/g.

The enthalpy (ΔH) correlates with the energy needed to break intermolecular bonds within starch macromolecules. A reduction in ΔH indicates a transition in molecular structure from an ordered to a disordered state. Extending the duration of HHP slightly decreases ΔH in shorter periods (20 minutes). However, with increasing HHP duration, a significant reduction in ΔH occurs, indicating a disruption in the organized structure (Wang et al., 2023a). The significant reduction in enthalpy is a result of the disruption of double helices and, hence, low relative crystallinity. Consequently, the loss of this structured arrangement suggests the onset of the gelatinization process.

In summary, high-pressure processing (HPP) has diverse effects on the thermal properties and gelatinization of starches from different origins. The differences in thermal behaviors observed among these starch sources under HPP highlight the intricate interplay between starch structure, pressure intensity, and gelatinization tendencies, offering valuable insights into the varied susceptibilities of starches to HPP-induced modifications.

5.3.5 Retrogradation

After gelatinization, as a result of reduced time and temperature (mainly through refrigeration or freezing), starch chains tend to interact more strongly with each other. This interaction forces water to leave, a phenomenon known as syneresis. Retrogradation or recrystallization occurs when amylose and amylopectin chains aggregate to form crystalline double helices stabilized by hydrogen bonds. These helices form highly stable three-dimensional crystalline structures during cooling, exhibiting X-ray diffraction pattern B (BeMiller & Whistler, 2009). Starch retrogradation can be divided into short-term and long-term retrogradation. Amylose is associated with short-term retrogradation, while long-term retrogradation is primarily linked to amylopectin.

Thermal processing is the most common technique for starch gelatinization, with research primarily focusing on the retrogradation mechanism of thermally gelatinized starch. Nevertheless, as an alternative to conventional thermal processing, high-pressure technology has been widely employed to gelatinize or physically modify various types of starch suspensions (Kawai et al., 2012).

At sufficiently high pressures, starch gelatinization can be achieved at room temperature. The modification by HHP is capable of maintaining the granular structure, leading to a distinct retrogradation process for the granules, which usually happens more slowly (Hu et al., 2011). The preservation of the granular structure with partially and/or completely gelatinized granules inside holds great potential for industrial applications. Additionally, it implies a more compact structure with lower enzymatic susceptibility (Chen et al., 2023), which consequently has lower digestibility (Colussi et al., 2018).

The relative crystallinity, enthalpy, and hardness of starch gels after high-pressure modification are higher than those of heat-treated starches, indicating that high-pressure pretreatment promotes starch retrogradation (Chen et al., 2023; Hu et al., 2011; Kawai et al., 2012; Okur et al., 2021; Zhang et al., 2022b). Despite several recent studies, the mechanism of retrogradation caused by HHP is still not well explained.

According to Kawai et al. (2012), starch retrogradation begins immediately after depressurization. Zhang et al. (2022b) conducted a study where oat starch (15%, w/v) was modified using pressures of 100 to 600 MPa for 15 minutes. Subsequently, retrogradation was allowed to occur for 0, 1, 3, 5, 7, 14, 21, and 35 days at 4°C. The effects of retrogradation were analyzed using FTIR spectroscopy, XRD analysis, and ^{13}C CP/MAS NMR spectroscopy to evaluate its impact. The R1045/1022 and R955/1022 values of oat starch treated with 500 MPa increased as the storage time progressed, indicating the reforming of the ordered structure and the double helix structure during storage. The R1045/1022 value may indicate the degree of starch retrogradation, where a higher value corresponds to a greater rate of recrystallization of the molecules, signifying rapid retrogradation. XRD analysis confirmed that starch molecules recrystallize during retrogradation. In comparison to native starch, the HHP-treated starches exhibited lower relative crystallinity during retrogradation, suggesting a slower molecular rearrangement of starch chains and a delayed retrogradation during storage. Notably, the starch treated with 500 MPa displayed diffraction peaks near 2θ of 20°, indicative of a type V crystalline pattern. Over the retrogradation period, the sample's diffraction peaks became sharper and higher, signifying an increase in the crystal structure. A new diffraction peak emerged near 2θ of 17°, highlighting a B- and V-type diffraction pattern in long-term retrogradation.

Colussi et al. (2018) verified that the microstructural characteristics of potato starch, modified by 6 cycles of 600 MPa with retrogradation (4°C for 7 days), presented a more compact network with much smaller spaces than those in native starch and HHP-treated starch. The authors assume that the density of the structure in the retrograded sample would make it difficult for digestive enzymes to access it compared to other treatments. The lower digestibility of retrograded gels confirmed this fact. Additionally, the authors report that the changes starch undergoes during gelatinization and retrogradation are the main determinants of its functional properties during processing and digestion. These properties, in turn, determine the quality, acceptability, nutritional value, and shelf life of the final food product.

5.3.6 Digestibility In Vitro

The effect of HHP on starch digestibility has been widely investigated, and the data obtained are presented in different ways. The most commonly used method is to express the levels of rapidly digestible starch (digested within 20 minutes in the small intestine), slowly digestible starch (digested between 20 minutes and 120 minutes in the small intestine), and resistant starch (not digested in the small intestine and acting as colon fiber), as described by Goni et al. (1997). Another commonly used approach is the percentage of hydrolysis, representing the amount of starch converted into glucose during gastrointestinal digestion. In this process, digestion is simulated in the mouth (through chewing and salivary juice), in the stomach (via peristaltic movements and gastric juice), and in the small intestine (through peristaltic movements and intestinal juice). The data were obtained to yield a curve, allowing for the prediction of the sample's digestion behavior. Furthermore, based on the area under the curve, it is possible to quantify the estimated glycemic index (IGe) (Dartois et al., 2010; Goebel et al., 2019; Minekus et al., 2014).

As well as the parameters mentioned in these topics, the impact of HHP on starch digestibility will depend on the intensity of pressure. Table 5.2 summarizes the effect of HHP on starch granules and the technique used for this quantification.

When modifying starch using HHP, it is possible to adjust the system so that the starch granules maintain their integrity during the pressurization process. When there is no disruption of the external structure of the granule, starch digestibility becomes slower, and when associated with retrogradation, it becomes even slower (Colussi et al., 2018; Tian et al., 2014). The surface properties of starch granules, supermolecular rearrangement, and changes in relative crystallinity during HHP treatments could be the main factors affecting the variation in digestibility (Zeng et al., 2018).

Zhang et al. (2022a) set the pressure at 500 MPa and used different holding times (5, 10, 15, 20, 25, and 30 min) in the modification of oat starch. The authors verified that the 5- to 10-minute holding time reduces the starch hydrolysis and RDS content while the SDS and RS increased. On the other hand, holding time higher than 20 minutes increases starch hydrolysis and reduces SDS and RS (Liu et al., 2016c). On another hand, for sorghum starch, the increase in pressure (100 to 600 MPa) reduced the degree of starch hydrolysis. The application of 600 MPa resulted in a lower RDS (20.3%), along with a higher SDS (45.1%) and RS (13.4%). Based on these results, it is possible to verify that only by adjusting the holding time in the modification of starch by HHP can the hydrolysis rate be changed to values higher or lower than that of native starch. Such modification can expand the range of applications in the food industry, targeting modified starches in specific sectors such as foods for athletes or people who require a slow and continuous energy rate.

Zhang et al. (2022a) further explain that with longer holding times, the pressure damages the starch granule, disrupting the crystalline region and disintegrating the double-helix structure. This, in turn, exposes the groups and makes the starch more sensitive to enzymes, leading to a higher rate of hydrolysis. As the hydrolysis rate increases, the starch becomes more digestible. Furthermore, an increase in treatment time results in larger starch particles, creating a larger reaction area, higher hydrolysis rates, and greater starch digestibility.

Tian et al. (2014) evaluated the effect of HHP on the slowly digestible properties of waxy and non-waxy rice starches. The starches were modified at 600 MPa for 30 minutes at a temperature of 30°C, followed by storage at 4°C for 0, 1, 3, 7, and 14 days. As the main results, the authors found that starches modified by HPP had higher SDS values. They attribute this increase to (1) intact granules, as they exhibit a reduction in susceptibility to amylolytic enzymes; (2) the formation of amylose-lipid complex in non-waxy starch; and (3) the formation of imperfect crystallites that hinder the access of amylolytic enzymes. When the starches were subjected to retrogradation, it was observed that those gelatinized by HHP were retrograded at a higher rate than those gelatinized by heat. However, retrogradation increased the SDS content until the seventh day, after which the values remained stable.

Aiming for an even greater reduction in the rate of starch hydrolysis, some studies employed double modification and achieved significant results. Colussi et al. (2020) investigated the effects of double modification through heat moisture treatment (HMT) and high hydrostatic pressure, assessing their impact on the digestibility of modified starches. The authors applied different humidity levels in the HMT modification (15%, 20%, and 25%) and subsequently modified the starch by HHP (starch concentration 33.3%, w/v, 600 MPa for 6 cycles of 10 minutes at 21°C). After undergoing 90-minute digestion in a simulated small intestine, around 96% of the untreated starch was hydrolyzed, while the digestibility of HMT-treated starch with 25% moisture stood at approximately 84%. The double-modified starch exhibited a reduced rate of glucose release, correlating with the increased moisture content in the HMT. This implies that when consumed as part of a meal, modified starch can extend the feeling of satiety, presenting an alternative for preparing foods with a low glycemic index.

TABLE 5.2 Effect of HHP on Starch Digestibility

SOURCE/ SAMPLE	HHP CONDITIONS	METHODOLOGY	EFFECTS ON DIGESTIBILITY	REFERENCES
Cassava flour	Starch concentration:10, 20, and 30% (wdb/v) Pressure: 300, 400, 500, and 600 MPa Temperature: room Holding time: 10 and 30 min	(Minekus et al., 2014)	—A pressure of 600 MPa for 10 minutes promoted a higher RDS. —The flour concentration influenced an increase in SDS concentration. —Pressures greater than 500 MPa promoted a significant decrease in RS content, resulting in a consequent increase in SDS and RDS.	(Conde et al., 2022)
Non-waxy rice starch and waxy rice starch	Starch concentration: non-informed. Pressure: 600 MPa Temperature: 30°C Holding time: 30 min Retrogradation: 4°C for 0, 1, 3, 7, and 14 days	(Englyst et al., 1992)	—HHP-treated samples exhibited a higher SDS concentration than starch gelatinized by heat. —The highest percentage of SDS was mainly attributed to the intact starch granules retained by the HHP treatment. —Retrogradation was greater in starches subjected to HHP, which may justify the higher SDS. —Retrogradation increased the SDS content until the seventh day, after which the values remained stable.	(Tian et al., 2014)
Oat starch	Starch concentration: 15% (w/v) Pressure: 500 MPa Temperature: room Holding time: 5, 10, 15, 20, 25, and 30 min	(Zhang et al., 2022b)	—The use of 500 MPa for 5 to 10 minutes reduced the rate of starch hydrolysis, decreased the RDS content, and increased the SDS and RS contents. —Holding time greater than 20 min increases starch hydrolysis and reduces the SDS and RS. —Longer pressure treatment promotes higher starch digestibility.	(Zhang et al., 2022a)
Corn, potato, rice, and tapioca starch	Starch concentration: 25% (w/v) Pressure: 350, 450, and 550 MPa Temperature: room Holding time: 10 min	(McCleary & Monaghan, 2002)	—Modified starches exhibited a decrease in rapidly digestible starch content and an increase in slowly digestible starch and resistant starch content. —The increases in SDS and RDS contents, as well as the decrease in RS content, were proportional to the pressure level in all types of starch. —Potato starch demonstrated the smallest decrease in RDS content and the smallest increase in SDS and RS contents per pressure modification time (PMT), possibly due to the pressure resistance of type B starch. —The greatest increase in SDS and RS content and the greatest decrease in RDS content in HHP treated were observed in tapioca starch, suggesting that the digestibility was influenced by the type of starch and pressure level.	Kim et al., 2023)
Sorghum starch	Starch concentration: 20% (w/v) Pressure: 120, 240, 360, 480, and 600 MPa Temperature: room Holding time: 20 min	(Liu et al., 2015)	—Elevating pressure levels led to a reduction in the degree of hydrolysis of sorghum starch. —After HHP, RDS decreased, while SDS and RS levels increased. —The application of 600 MPa resulted in lower RDS (20.3%), along with higher SDS (45.1%) and RS (13.4%).	(Liu et al., 2016a)

Starch	Conditions	Reference (method)	Observations	Reference
Waxy rice starch	Starch concentration: 10% (w/w) Pressure: 100, 200, 300, 400, 500, and 600 MPa Temperature: 25°C Holding time: 20 min	(Englyst et al., 1992)	—There was no significant difference in RDS, SDS, and RS content after treatments at 100–200 MPa. —At pressure levels ranging from 300 to 500 MPa, HHP treatments led to an increase in RDS (33.5–39.2%) and SDS (49.9–53.6%), accompanied by a significant decrease in RS (14.6–10.3%). —High-pressure treatment at 600 MPa increased RDS and decreased SDS and RS. —The optimal condition for enhancing SDS in waxy rice starch is 400 MPa.	(Zeng et al., 2018)
Buckwheat starch	Starch concentration: 20% (w/v) Pressure: 120, 240, 360, 480, and 600 MPa Temperature: room Holding time: 20 min	(Liu et al., 2015)	—The degree of hydrolysis in HHP-modified samples decreased with increasing pressure. —After HHP treatment, RDS content decreased, while SDS and RS levels increased. —The highest impact was obtained with 600 MPa with the lowest RDS (29.0%) and the highest SDS (49.1%) and RS (8.2%).	(Liu et al., 2016c)
Tartary buckwheat starch	Starch concentration: 20% (w/v) Pressure: 120, 240, 360, 480, and 600 MPa Temperature: room Holding time: 20 min	(Goñi et al., 1997)	—The degree of hydrolysis of the HHP-modified samples decreased with increasing pressure. —After HHP treatment, RDS content decreased markedly, while SDS and RS levels increased. —HHP treatment decreased the degree of TBS hydrolysis and improved its potential health benefits by reducing RDS content and increasing SDS and RS levels.	(Liu et al., 2016c)
Potato starch	Starch concentration: 25% (w/v) Pressure: 400 and 600 MPa; 3 and 6 cycles Temperature: 21°C Holding time: 10 min Retrogradation: 4°C for 7 days	(Dartois et al., 2010)	—Three cycles of 400 MPa was not enough to change digestibility. —Six cycles at 400 MPa significantly reduced the starch digestibility. —Six cycles at 600 MPa in association with retrogradation promoted lower hydrolysis during gastro-small intestinal digestion in vitro, as well as slower glucose release.	(Colussi et al., 2018)
Waxy wheat starch	Starch concentration: 10% (w/v) Pressure: 300, 400, 500, 600 MPa Temperature: 20°C Holding time: 30 min Retrogradation: 4 °C for 4 days	(Englyst et al., 1992)	—The content of RDS and SDS increased, while RS decreased proportionally to the increase in pressure. —The highest SDS values were obtained at the highest pressure applied.	(Hu et al., 2017)
Pea starch	Starch concentration: 15% (w/v) Pressure: 150, 300, 450, and 600 MPa Temperature: 30°C Holding time: 25 min	(Englyst et al., 1992; Minekus et al., 2014)	—The use of HHP promoted a slower hydrolysis rate of pea starch. —With increasing pressure, the degree of hydrolysis of pea starches decreased, the amount of RDS and SDS decreased, and RS levels increased.	(Liu et al., 2018)

RDS: rapidly digestible starch; SDS: slowly digestible starch; RS: resistant starch; n.i.: non-informed.

5.4 APPLICATIONS AND FINAL REMARKS

HHP offers novel opportunities for utilizing starch in food systems. Overall, the application of high pressure to modify food functionality remains to be used on different food products, especially including a range of cereal/grain foods, which are staple foods. Novel properties of grain products may be obtained using HHP processing (Zhu & Li, 2019). HHP has been applied to replace the thermal treatments commonly used by the food industry, focused on satisfying the consumers who demand minimally processed and no-preservative foods. HHP can be used beyond food safety; this emerging technology is capable of altering the macronutrient structure and, thereby, their properties. The use of emerging technology to modify starch has been the focus of recent research. The application of pressurized starch is a new ingredient that might be able to replace additives and provide a clean-label product.

HHP starches have been applied as fat replacers, particularly in food emulsion systems. Waxy corn starch (Heydari & Razavi, 2021); wheat starch (Heydari, Razavi, & Farahnaky, 2021); and starch nanocrystals from chestnut, corn starch, and potato sources (Lin et al., 2023) have been studied recently. Heydari et al. (2021a) used HHP wheat starch as a fat replacer in an oil-in-water emulsion. Their findings affirm that replacing fat with HHP wheat starch, particularly at higher concentrations (15–20%), facilitated suitable homogeneity and stability. It resulted in a lower creaming index, acceptable consistency and hardness, favorable viscoelastic properties, and appropriate yield stress in the prepared emulsions. The same behavior was reported for corn and waxy corn starch treated with HHP (Heydari & Razavi, 2021). Consequently, HHP starches can applied as a fat replacer and stabilizer, offering reduced-calorie benefits in low-fat emulsions.

Modified flour and starch applied in food formulation can provide an improvement in the final quality. For instance, high-pressure treated rice flour and corn starch were applied to gluten-free bread and resulted in a delay in the staling process of the bread by reducing the crumb hardness, resulting in softer bread (Cappa et al., 2016). The effect of pressured oat flour in gluten-free bread was dependent on the pressures applied. Generally, pressures higher than 350 MPa resulted in a loss in bread quality by reducing the specific volume, creating uneven gas cell distribution, regardless of the level of inclusion. However, at the 10% level of incorporation of oat flour treated at 200 MPa, an increase in the volume was recorded (Hüttner et al., 2010).

HHP was utilized in the modification of starch for hydrogel applications (Larrea-Wachtendorff et al., 2021). Starches from corn, rice, tapioca, and wheat underwent treatment at 600 MPa for 5 and 15 minutes. The resulting HPP hydrogels derived from corn, rice, and wheat starch displayed a "soft" gel structure, resembling creams due to their excellent spreadability. On the contrary, tapioca starch-formed HPP hydrogels are characterized by a gummy or rubbery compact structure, displaying resistance to flow.

The application of high pressure to modify the functionality of foods is already a reality (Heydari & Razavi, 2021; Cappa et al., 2016), especially in cereals and grains such as bakery products (Cappa et al., 2016) and noodles (Lee & Koo, 2019) which are basic foods serving a significant portion of the market, including those catering to people with celiac disease. Despite this, it is evident that most studies apply high pressure to isolated starch. However, new studies verifying the interaction of starch with other food constituents such as proteins, lipids, and bioactive compounds, still need to be conducted. Furthermore, research on the application of starches modified by high pressure should be further explored not only in the food industry but also in the pharmaceutical and chemical industries.

REFERENCES

Ahmed, J., Singh, A., Ramaswamy, H. S., Pandey, P. K., & Raghavan, G. S. V. (2014). Effect of high-pressure on calorimetric, rheological and dielectric properties of selected starch dispersions. *Carbohydrate Polymers*, *103*(1), 12–21. https://doi.org/10.1016/j.carbpol.2013.12.014

Ahmed, J., Thomas, L., Arfat, Y. A., & Joseph, A. (2018). Rheological, structural and functional properties of high-pressure treated quinoa starch in dispersions. *Carbohydrate Polymers, 197*, 649–657. https://doi.org/10.1016/j.carbpol.2018.05.081

Ahmed, J., Thomas, L., Taher, A., & Joseph, A. (2016). Impact of high pressure treatment on functional, rheological, pasting, and structural properties of lentil starch dispersions. *Carbohydrate Polymers, 152*, 639–647. https://doi.org/10.1016/j.carbpol.2016.07.008

Almeida, R. L. J., Santos, N. C., Feitoza, J. V. F., Ribeiro, V. H. A., Silva, V. M. A., Figueiredo, M. J., Ribeiro, C. A. C., Muniz, C. E. S., Eduardo, R. S., Cavalcante, J. A., & Mota, M. M. A. (2023). Evaluation of the technological properties of rice starch modified by high hydrostatic pressure (HHP). *Innovative Food Science & Emerging Technologies, 83*, 103241. https://doi.org/10.1016/j.ifset.2022.103241

Balasubramaniam, V. M. B., Martínez-Monteagudo, S. I., & Gupta, R. (2015). Principles and application of high pressure-based technologies in the food industry. In *Annual review of food science and technology* (Vol. 6, pp. 435–462). Annual Reviews Inc. https://doi.org/10.1146/annurev-food-022814-015539

Balet, S., Guelpa, A., Fox, G., & Manley, M. (2019). Rapid Visco Analyser (RVA) as a tool for measuring starch-related physiochemical properties in cereals: A review. *Food Analytical Methods, 12*(10), 2344–2360. https://doi.org/10.1007/s12161-019-01581-w

Bemiller, J. N., & Huber, K. C. (2015). Physical modification of food starch functionalities. *Annual Review of Food Science and Technology, 6*, 19–69. https://doi.org/10.1146/annurev-food-022814-015552

BeMiller, J., & Whistler, R. (2009). *Starch: Chemistry and technology* (3rd ed., Vol. 1). Elsevier.

Buckow, R., Heinz, V., & Knorr, D. (2007). High pressure phase transition kinetics of maize starch. *Journal of Food Engineering, 81*(2), 469–475. https://doi.org/10.1016/j.jfoodeng.2006.11.027

Cappa, C., Barbosa-Cánovas, G. V., Lucisano, M., & Mariotti, M. (2016). Effect of high pressure processing on the baking aptitude of corn starch and rice flour. *LWT, 73*, 20–27. https://doi.org/10.1016/j.lwt.2016.05.028

Castro, L. M. G., Alexandre, E. M. C., Saraiva, J. A., & Pintado, M. (2020). Impact of high pressure on starch properties: A review. In *Food hydrocolloids* (Vol. 106). Elsevier B.V. https://doi.org/10.1016/j.foodhyd.2020.105877

Castro, L. M. G., Caço, A. I., Pereira, C. F., Sousa, S. C., Brassesco, M. E., Machado, M., Ramos, Ó. L., Alexandre, E. M. C., Saraiva, J. A., & Pintado, M. (2023). Modification of acorn starch structure and properties by high hydrostatic pressure. *Gels, 9*(9), 757. https://doi.org/10.3390/gels9090757

Chen, L., Dai, Y., Hou, H., Wang, W., Ding, X., Zhang, H., Li, X., & Dong, H. (2021). Effect of high pressure microfluidization on the morphology, structure and rheology of sweet potato starch. *Food Hydrocolloids, 115*(61), 106606. https://doi.org/10.1016/j.foodhyd.2021.106606

Chen, Z., Yang, Q., Yang, Y., & Zhong, H. (2023). The effects of high-pressure treatment on the structure, physicochemical properties and digestive property of starch—a review. *International Journal of Biological Macromolecules, 244*. Elsevier B.V. https://doi.org/10.1016/j.ijbiomac.2023.125376

Clerici, M. T. P. S., Sampaio, U. M., & Schmiele, M. (2018). Identification and analysis of starch. In *Starches for food application: Chemical, technological and health properties*. Academic Press. https://doi.org/10.1016/B978-0-12-809440-2.00002-2

Colussi, R., Kaur, L., Zavareze, E. da R., Dias, A. R. G., Stewart, R. B., & Singh, J. (2017). High pressure processing and retrogradation of potato starch: Influence on functional properties and gastro-small intestinal digestion in vitro. *Food Hydrocolloids, 75*, 131–137. https://doi.org/10.1016/j.foodhyd.2017.09.004

Colussi, R., Kaur, L., Zavareze, E. da R., Dias, A. R. G., Stewart, R. B., & Singh, J. (2018). High pressure processing and retrogradation of potato starch: Influence on functional properties and gastro-small intestinal digestion in vitro. *Food Hydrocolloids, 75*, 131–137. https://doi.org/10.1016/j.foodhyd.2017.09.004

Colussi, R., Kringel, D., Kaur, L., da Rosa Zavareze, E., Dias, A. R. G., & Singh, J. (2020). Dual modification of potato starch: Effects of heat-moisture and high pressure treatments on starch structure and functionalities. *Food Chemistry, 318*. https://doi.org/10.1016/j.foodchem.2020.126475

Conde, L. A., Kebede, B., Leong, S. Y., & Oey, I. (2022). Effect of high hydrostatic pressure processing on starch properties of cassava flour. *Applied Sciences (Switzerland), 12*(19). https://doi.org/10.3390/app121910043

Cui, R., & Zhu, F. (2019, March). Physicochemical properties and bioactive compounds of different varieties of sweet potato flour treated with high hydrostatic pressure. *Food Chemistry, 299*, 125129. https://doi.org/10.1016/j.foodchem.2019.125129

Dartois, A., Singh, J., Kaur, L., & Singh, H. (2010). Influence of guar gum on the in vitro starch digestibility-rheological and Microstructural characteristics. *Food Biophysics, 5*(3), 149–160. https://doi.org/10.1007/s11483-010-9155-2

Dominguez-Ayala, J. E., Soler, A., Mendez-Montealvo, G., & Velazquez, G. (2022). Supramolecular structure and techno-functional properties of starch modified by high hydrostatic pressure (HHP): A review. *Carbohydrate Polymers, 291*, 119609. https://doi.org/10.1016/j.carbpol.2022.119609

Englyst, A. N., Kingman, S. M., & Cummings, J. H. (1992). Classification and measurement of nutritionally important starch fractions. *European Journal of Clinical Nutrition, 46*(2), 33–50.

Elamin, W. M., Endan, J. B., Yosuf, Y. A., Shamsudin, R., & Ahmedov, A. (2015). High pressure processing technology and equipment evolution: A review. *Journal of Engineering Science and Technology Review*, *8*(5). www.jestr.org

Fan, L., Ye, Q., Lu, W., Chen, D., Zhang, C., Xiao, L., Meng, X., Lee, Y.-C., Wang, H.-M. D., & Xiao, C. (2023). The properties and preparation of functional starch: A review. *Food Reviews International*, *39*(7), 3984–4008. https://doi.org/10.1080/87559129.2021.2015375

Feng, W., Ma, S., & Wang, X. (2020). Recent advances in quality deterioration and improvement of starch in frozen dough. *Grain & Oil Science and Technology*, *3*(4), 154–163. https://doi.org/10.1016/j.gaost.2020.07.002

Goebel, J. T. S., Kaur, L., Colussi, R., Elias, M. C., & Singh, J. (2019). Microstructure of indica and japonica rice influences their starch digestibility: A study using a human digestion simulator. *Food Hydrocolloids*, *94*, 191–198. https://doi.org/10.1016/j.foodhyd.2019.02.038

Goni, I., Garcia-Alonso, A., & Saura-Calixto, F. (1997). A starch hydrolysis procedure to estimate glycemic index. *Nutrition Research*, *17*(3), 427–437.

Guo, Z., Zeng, S., Lu, X., Zhou, M., Zheng, M., & Zheng, B. (2015). Structural and physicochemical properties of lotus seed starch treated with ultra-high pressure. *Food Chemistry*, *186*, 223–230. https://doi.org/10.1016/j.foodchem.2015.03.069

Heydari, A., & Razavi, S. M. A. (2021). Evaluating high pressure-treated corn and waxy corn starches as novel fat replacers in model low-fat O/W emulsions: A physical and rheological study. *International Journal of Biological Macromolecules*, *184*, 393–404. https://doi.org/10.1016/j.ijbiomac.2021.06.052

Heydari, A., Razavi, S. M. A., & Farahnaky, A. (2021a). Effect of high pressure-treated wheat starch as a fat replacer on the physical and rheological properties of reduced-fat O/W emulsions. *Innovative Food Science and Emerging Technologies*, *70*, 102702. https://doi.org/10.1016/j.ifset.2021.102702

Heydari, A., Razavi, S. M. A., Hesarinejad, M. A., & Farahnaky, A. (2021b). New insights into physical, morphological, thermal, and pasting properties of HHP-treated starches: Effect of starch type and industry-scale concentration. *Starch—Stärke*, *73*(7–8). https://doi.org/10.1002/star.202000179

Hu, X., Xu, X., Jin, Z., Tian, Y., Bai, Y., & Xie, Z. (2011). Retrogradation properties of rice starch gelatinized by heat and high hydrostatic pressure (HHP). *Journal of Food Engineering*, *106*(3), 262–266. https://doi.org/10.1016/j.jfoodeng.2011.05.021

Hu, X. P., Zhang, B., Jin, Z. Y., Xu, X. M., & Chen, H. Q. (2017). Effect of high hydrostatic pressure and retrogradation treatments on structural and physicochemical properties of waxy wheat starch. *Food Chemistry*, *232*, 560–565. https://doi.org/10.1016/j.foodchem.2017.04.040

Hüttner, E. K., Bello, F. D., & Arendt, E. K. (2010). Fundamental study on the effect of hydrostatic pressure treatment on the bread-making performance of oat flour. *European Food Research and Technology*, *230*(6), 827–835. https://doi.org/10.1007/s00217-010-1228-4

Jaeger, H., Reineke, K., Schoessler, K., & Knorr, D. (2012). Effects of emerging processing technologies on food material properties. In *Food materials science and engineering* (pp. 222–262). Wiley. https://doi.org/10.1002/9781118373903.ch9

Jiang, B., Li, W., Hu, X., Wu, J., & Shen, Q. (2015a). Rheology of mung bean starch treated by high hydrostatic pressure. *International Journal of Food Properties*, *18*(1), 81–92. https://doi.org/10.1080/10942912.2013.819363

Jiang, B., Li, W., Shen, Q., Hu, X., & Wu, J. (2015b). Effects of high hydrostatic pressure on rheological properties of rice starch. *International Journal of Food Properties*, *18*(6), 1334–1344. https://doi.org/10.1080/10942912.2012.709209

Katopo, H., Song, Y., & Jane, J. L. (2002). Effect and mechanism of ultrahigh hydrostatic pressure on the structure and properties of starches. *Carbohydrate Polymers*, *47*(3), 233–244. https://doi.org/10.1016/S0144-8617(01)00168-0

Kaur, M., Punia, S., Sandhu, K. S., & Ahmed, J. (2019). Impact of high pressure processing on the rheological, thermal and morphological characteristics of mango kernel starch. *International Journal of Biological Macromolecules*, *140*, 149–155. https://doi.org/10.1016/j.ijbiomac.2019.08.132

Kawai, K., Fukami, K., & Yamamoto, K. (2007). Effects of treatment pressure, holding time, and starch content on gelatinization and retrogradation properties of potato starch-water mixtures treated with high hydrostatic pressure. *Carbohydrate Polymers*, *69*(3), 590–596. https://doi.org/10.1016/j.carbpol.2007.01.015

Kawai, K., Fukami, K., & Yamamoto, K. (2012). Effect of temperature on gelatinization and retrogradation in high hydrostatic pressure treatment of potato starch-water mixtures. *Carbohydrate Polymers*, *87*(1), 314–321. https://doi.org/10.1016/j.carbpol.2011.07.046

Kim, H. Y., Ye, S. J., & Baik, M. Y. (2022). Pressure moisture treatment (PMT) of starch, a new physical modification method. *Food Hydrocolloids*, *134*. https://doi.org/10.1016/j.foodhyd.2022.108051

Kim, H. Y., Ye, S. J., & Baik, M. Y. (2023). Physicochemical properties of pressure moisture treated (PMT) and heat moisture treated (HMT) starches. *Innovative Food Science & Emerging Technologies*, *87*, 103392. https://doi.org/10.1016/j.ifset.2023.103392

Knorr, D., Heinz, V., & Buckow, R. (2006). High pressure application for food biopolymers. In *Biochimica et Biophysica Acta—Proteins and Proteomics, 1764*(3), 619–631. https://doi.org/10.1016/j.bbapap.2006.01.017

Larrea-Wachtendorff, D., Sousa, I., & Ferrari, G. (2021). Starch-based hydrogels produced by high-pressure processing (HPP): Effect of the starch source and processing time. *Food Engineering Reviews, 13*(3), 622–633. https://doi.org/10.1007/s12393-020-09264-7

Lee, N., & Koo, J. (2019). Effects of high hydrostatic pressure on quality changes of blends with low-protein wheat and oat flour and derivative foods. *Food Chemistry, 271*, 685–690. https://doi.org/10.1016/j.foodchem.2018.07.171

Li, G., & Zhu, F. (2018). Effect of high pressure on rheological and thermal properties of quinoa and maize starches. *Food Chemistry, 241*, 380–386. https://doi.org/10.1016/j.foodchem.2017.08.088

Li, W., Gao, J., Saleh, A. S. M., Tian, X., Wang, P., Jiang, H., & Zhang, G. (2018). The modifications in physicochemical and functional properties of proso millet starch after ultra-high pressure (UHP) process. *Starch/Staerke, 70*(5–6). https://doi.org/10.1002/star.201700235

Li, W., Tian, X., Liu, L., Wang, P., Wu, G., Zheng, J., Ouyang, S., Luo, Q., & Zhang, G. (2015). High pressure induced gelatinization of red adzuki bean starch and its effects on starch physicochemical and structural properties. *Food Hydrocolloids, 45*, 132–139. https://doi.org/10.1016/j.foodhyd.2014.11.013

Li, W., Zhang, F., Liu, P., Bai, Y., Gao, L., & Shen, Q. (2011). Effect of high hydrostatic pressure on physicochemical, thermal and morphological properties of mung bean (Vigna radiata L.) starch. *Journal of Food Engineering, 103*(4), 388–393. https://doi.org/10.1016/j.jfoodeng.2010.11.008

Lin, D., Zhao, J., Fan, H., Qin, W., & Wu, Z. (2023). Enhancing starch nanocrystal production and evaluating their efficacy as fat replacers in ice cream: Investigating the influence of high pressure and ultrasonication. *International Journal of Biological Macromolecules, 251*, 126385. https://doi.org/10.1016/j.ijbiomac.2023.126385

Liu, H., Fan, H., Cao, R., Blanchard, C., & Wang, M. (2016a). Physicochemical properties and in vitro digestibility of sorghum starch altered by high hydrostatic pressure. *International Journal of Biological Macromolecules, 92*, 753–760. https://doi.org/10.1016/j.ijbiomac.2016.07.088

Liu, H., Guo, X., Li, W., Wang, X., Lv, M., Peng, Q., & Wang, M. (2015). Changes in physicochemical properties and in vitro digestibility of common buckwheat starch by heat-moisture treatment and annealing. *Carbohydrate Polymers, 132*, 237–244. https://doi.org/10.1016/j.carbpol.2015.06.071

Liu, H., Lv, M., Wang, L., Li, Y., Fan, H., & Wang, M. (2016b). Comparative study: How annealing and heat-moisture treatment affect the digestibility, textural, and physicochemical properties of maize starch. *Starch—Stärke, 68*(11–12), 1158–1168. https://doi.org/10.1002/star.201500268

Liu, H., Wang, L., Cao, R., Fan, H., & Wang, M. (2016c). In vitro digestibility and changes in physicochemical and structural properties of common buckwheat starch affected by high hydrostatic pressure. *Carbohydrate Polymers, 144*, 1–8. https://doi.org/10.1016/j.carbpol.2016.02.028

Liu, M., Wu, N. N., Yu, G. P., Zhai, X. T., Chen, X., Zhang, M., Tian, X. H., Liu, Y. X., Wang, L. P., & Tan, B. (2018). Physicochemical properties, structural properties, and in vitro digestibility of pea starch treated with high hydrostatic pressure. *Starch/Staerke, 70*(1–2), 1–9. https://doi.org/10.1002/star.201700082

Liu, S., Yuan, T. Z., Wang, X., Reimer, M., Isaak, C., & Ai, Y. (2019). Behaviors of starches evaluated at high heating temperatures using a new model of Rapid Visco Analyzer – RVA 4800. *Food Hydrocolloids, 94*, 217–228. https://doi.org/10.1016/j.foodhyd.2019.03.015

McCleary, B. V., & Monaghan, D. A. (2002). Measurement of resistant starch. *Journal of AOAC International, 85*(3), 665–675. https://doi.org/10.1093/jaoac/85.3.665

Mezger, T. G. (2020). *The rheology handbook: For users of rotational and oscillatory* (5th revise ed.). European Coatings.

Minekus, M., Alminger, M., Alvito, P., Ballance, S., Bohn, T., Bourlieu, C., Carrière, F., Boutrou, R., Corredig, M., Dupont, D., Dufour, C., Egger, L., Golding, M., Karakaya, S., Kirkhus, B., Le Feunteun, S., Lesmes, U., MacIerzanka, A., MacKie, A., . . . Brodkorb, A. (2014). A standardised static in vitro digestion method suitable for food-an international consensus. *Food and Function, 5*(6), 1113–1124. https://doi.org/10.1039/c3fo60702j

Mirzababaee, S. M., Ozmen, D., Hesarinejad, M. A., Toker, O. S., & Yeganehzad, S. (2022). A study on the structural, physicochemical, rheological and thermal properties of high hydrostatic pressurized pearl millet starch. *International Journal of Biological Macromolecules, 223*, 511–523. https://doi.org/10.1016/j.ijbiomac.2022.11.044

Nasehi, B., & Javaheri, S. (2012). Application of high hydrostatic pressure in modifying functional properties of starches: A review. *Middle-East Journal of Scientific Research, 11*(7), 856–861.

Obadi, M., Qi, Y., & Xu, B. (2023). High-amylose maize starch: Structure, properties, modifications and industrial applications. *Carbohydrate Polymers, 299*, 120185. https://doi.org/10.1016/j.carbpol.2022.120185

Okur, I., Sezer, P., Oztop, M. H., & Alpas, H. (2021). Recent advances in gelatinisation and retrogradation of starch by high hydrostatic pressure. *International Journal of Food Science and Technology, 56*(9), 4367–4375. John Wiley and Sons Inc. https://doi.org/10.1111/ijfs.15174

Pei-Ling, L., Xiao-Song, H., & Qun, S. (2010). Effect of high hydrostatic pressure on starches: A review. *Starch/Staerke*, *62*(12), 615–628. https://doi.org/10.1002/star.201000001

Pulgarín, O., Larrea-Wachtendorff, D., & Ferrari, G. (2023). Effects of the amylose/amylopectin content and storage conditions on corn starch hydrogels produced by high-pressure processing (HPP). *Gels*, *9*(2). https://doi.org/10.3390/gels9020087

Rahman, M. H., Mu, T. H., Zhang, M., Ma, M. M., & Sun, H. N. (2020). Comparative study of the effects of high hydrostatic pressure on physicochemical, thermal, and structural properties of maize, potato, and sweet potato starches. *Journal of Food Processing and Preservation*, *44*(11), 1–11. https://doi.org/10.1111/jfpp.14852

Rostamabadi, H., Nowacka, M., Colussi, R., Frasson, S. F., Demirkesen, I., Mert, B., Singha, P., Singh, S. K., & Falsafi, S. R. (2023). Impact of emerging non-thermal processing treatments on major food macromolecules: Starch, protein, and lipid. *Trends in Food Science & Technology*, *141*, 104208. https://doi.org/10.1016/j.tifs.2023.104208

Sehrawat, R., Kaur, B. P., Nema, P. K., Tewari, S., & Kumar, L. (2021). Microbial inactivation by high pressure processing: Principle, mechanism and factors responsible. *Food Science and Biotechnology*, *30*(1), 19–35. https://doi.org/10.1007/s10068-020-00831-6

Seo, J. H., Jo, Y. J., Lee, Y. R., Lee, J., & Jeong, H. S. (2023). Physicochemical properties of soft and hard-type rice flour according to moisture content and high hydrostatic pressure treatment. *Foods*, *12*(1). https://doi.org/10.3390/foods12010227

Song, C.-G., Baik, M.-Y., & Kim, B.-Y. (2013). Rheological properties of native maize, waxy maize, and acetylated maize starches, and applications in the development of food products. *Journal of the Korean Society for Applied Biological Chemistry*, *56*(1), 63–68. https://doi.org/10.1007/s13765-012-2142-1

Suri, S., & Singh, A. (2023). Modification of starch by novel and traditional ways: Influence on the structure and functional properties. *Sustainable Food Technology*, *1*(3), 348–362. Royal Society of Chemistry. https://doi.org/10.1039/d2fb00043a

Tian, Y., Li, D., Zhao, J., Xu, X., & Jin, Z. (2014). Effect of high hydrostatic pressure (HHP) on slowly digestible properties of rice starches. *Food Chemistry*, *152*, 225–229. https://doi.org/10.1016/j.foodchem.2013.11.162

Wang, B., Li, D., Wang, L. J., Chiu, Y. L., Chen, X. D., & Mao, Z. H. (2008). Effect of high-pressure homogenization on the structure and thermal properties of maize starch. *Journal of Food Engineering*, *87*(3), 436–444. https://doi.org/10.1016/j.jfoodeng.2007.12.027

Wang, N., Dong, Y., Dai, Y., Zhang, H., Hou, H., Wang, W., Ding, X., Zhang, H., & Li, C. (2023a). Influences of high hydrostatic pressure on structures and properties of mung bean starch and quality of cationic starch. *Food Research International*, *165*, 112532. https://doi.org/10.1016/j.foodres.2023.112532

Wang, N., Li, C., Miao, D., Hou, H., Dai, Y., Zhang, Y., & Wang, B. (2023b). The effect of non-thermal physical modification on the structure, properties and chemical activity of starch: A review. *International Journal of Biological Macromolecules*, *251*, 126200. https://doi.org/10.1016/j.ijbiomac.2023.126200

Wang, S., & Copeland, L. (2015). Effect of acid hydrolysis on starch structure and functionality: A review. *Critical Reviews in Food Science and Nutrition*, *55*(8), 1081–1097. https://doi.org/10.1080/10408398.2012.684551

Wu, Z., Qiao, D., Zhao, S., Lin, Q., Zhang, B., & Xie, F. (2022). Nonthermal physical modification of starch: An overview of recent research into structure and property alterations. *International Journal of Biological Macromolecules*, *203*, 153–175. https://doi.org/10.1016/j.ijbiomac.2022.01.103

Yang, Z., Swedlund, P., Hemar, Y., Mo, G., Wei, Y., Li, Z., & Wu, Z. (2016). Effect of high hydrostatic pressure on the supramolecular structure of corn starch with different amylose contents. *International Journal of Biological Macromolecules*, *85*, 604–614. https://doi.org/10.1016/j.ijbiomac.2016.01.018

Ye, S.-J., & Baik, M.-Y. (2023). Characteristics of physically modified starches. *Food Science and Biotechnology*, *32*(7), 875–883. https://doi.org/10.1007/s10068-023-01284-3

Zamani, Z., & Razavi, S. M. A. (2023). Steady shear rheological properties, microstructure and stability of water in water emulsions made with basil seed gum and waxy corn starch or high pressure-treated waxy corn starch. *LWT*, *174*, 114453. https://doi.org/10.1016/j.lwt.2023.114453

Zeng, F., Li, T., Gao, Q., Liu, B., & Yu, S. (2018). Physicochemical properties and in vitro digestibility of high hydrostatic pressure treated waxy rice starch. *International Journal of Biological Macromolecules*, *120*, 1030–1038. https://doi.org/10.1016/j.ijbiomac.2018.08.121

Zhang, J., Zhang, M., Bai, X., Zhang, Y., & Wang, C. (2022a). The impact of high hydrostatic pressure treatment time on the structure, gelatinization and thermal properties and in vitro digestibility of oat starch. *Grain and Oil Science and Technology*, *5*(1), 1–12. https://doi.org/10.1016/j.gaost.2022.01.002

Zhang, J., Zhang, M., Zhang, Y., Bai, X., & Wang, C. (2022b). Effects of high hydrostatic pressure on the structure and retrogradation inhibition of oat starch. *International Journal of Food Science and Technology*, *57*(4), 2113–2125. https://doi.org/10.1111/ijfs.15642

Zhang, X., Wang, C., Liu, Z., Xue, Y., Zhao, Q., & Shen, Q. (2023). Four stages of multi-scale structural changes in rice starch during the entire high hydrostatic pressure treatment. *Food Hydrocolloids, 134.* https://doi.org/10.1016/j.foodhyd.2022.108012

Zhi-Guang, C., Jun-Rong, H., Hua-Yin, P., Qi, Y., & Chen-Lu, F. (2020). The effects of HHP (high hydrostatic pressure) on the interchain interaction and the conformation of amylopectin and double-amylose molecules. *International Journal of Biological Macromolecules, 155,* 91–102. https://doi.org/10.1016/j.ijbiomac.2020.03.190

Zhou, X., Chen, J., Wang, S., & Zhou, Y. (2022). Effect of high hydrostatic pressure treatment on the formation and in vitro digestion of Tartary buckwheat starch/flavonoid complexes. *Food Chemistry, 382.* https://doi.org/10.1016/j.foodchem.2022.132324

Zhu, F. (2015). Interactions between starch and phenolic compound. *Trends in Food Science and Technology, 43*(2), 129–143. https://doi.org/10.1016/j.tifs.2015.02.003

Zhu, F., & Li, H. (2019). Effect of high hydrostatic pressure on physicochemical properties of quinoa flour. *LWT, 114,* 108367. https://doi.org/10.1016/j.lwt.2019.108367

Ultrasound Treatment

6

Ditimoni Dutta and Nandan Sit

ABBREVIATIONS

BD	Breakdown viscosity
D.W.	Distilled water
FV	Final viscosity
MW	Microwave
PS	Potato starch
PT	Pasting temperature
PV	Peak viscosity
RVA	Rapid visco-analyzer
SB	Setback viscosity
US	Ultrasound

6.1 INTRODUCTION

Ultrasonic treatment applied to starch systems represents a promising technique gaining recognition across diverse sectors, including food processing, pharmaceuticals, and biofuel production. This method utilizes ultrasonic waves to modify the structure and attributes of starch, ultimately increasing its functionality and performance (Li et al., 2023). The term "ultrasound" denotes mechanical waves with frequencies exceeding the human auditory range (>16 kHz), classified into three frequency ranges: power ultrasound (6–100 kHz), high-frequency ultrasound (100 kHz–1 MHz), and diagnostic ultrasound (1–10 MHz) (Gallego-Juárez, 2017). The impact of ultrasonic treatment stems from acoustic cavitation, a rapid process involving the creation, expansion, and subsequent collapse of bubbles within a liquid. This collapse generates intense heat (reaching temperatures up to 5000 K) and pressure (up to 20 MPa) within a brief duration (Monroy et al., 2017). The application of ultrasonic treatment effectively disrupts hydrogen bonds within starch granules, leading to their disintegration and the formation of a more homogeneous starch structure (Zhang et al., 2022). In ultrasonic treatment (UT), ultrasound energy is conveyed to starch granules through cavitation, signifying the generation, enlargement, and swift collapse of microbubbles (Mallakpour, 2018). This process exposes a greater number of hydroxyl groups, fostering interaction with water molecules and resulting in enhanced solubility and gelatinization traits. Furthermore, ultrasonic treatment contributes to heightened starch crystallinity and alterations in its crystal structure (Monroy et al., 2018). These structural modifications significantly impact the physicochemical properties of starch, including improved thermal stability, reduced viscosity, and enhanced retrogradation (Acevedo et al., 2022). Additionally, this treatment results in the physical degradation of starch granules and a reduction in the size of starch polymer molecules.

Starch, a widely available biopolymer derived from agriculture, serves as a fundamental raw material across various industries due to its versatility, cost-effectiveness, ease of modification, and biodegradability

DOI: 10.1201/9781032655598-6

(Li et al., 2023). Despite its advantageous attributes, native starch faces limitations such as low solubility, limited freeze–thaw stability, weak resistance to pressure, thermal and shear conditions, resistance to enzymatic hydrolysis, and a high tendency toward retrogradation and syneresis ((Zarski et al., 2021). To address these drawbacks, starch undergoes modification via diverse techniques aimed at altering its properties and improving functionality (Dutta & Sit, 2022). Modified starch offers a range of advantages, including improved functionality, enhanced texture, increased stability, greater versatility, sustainability, and cost-effectiveness (Monroy et al., 2018). These attributes position modified starches as valuable ingredients widely applicable across industries. The primary modification methods encompass physical, chemical, and enzymatic processes (Kumari & Sit, 2023). Physical modification, an environmentally friendly approach, eliminates the need for hazardous chemicals and stands as a safe and straightforward process to execute (Mhaske et al., 2023). Ultrasonic treatment in starch systems involves the application of high-frequency sound waves (typically above 20 kHz) to modify starch-based materials. It holds promise for altering starch properties, enhancing functionality, and optimizing various industrial processes. Thus, the present chapter aims to comprehensively explore the influence of ultrasound technology on starch, delving into its structural aspects, inherent properties, and the diverse modifications achievable through ultrasound treatment. The chapter aims to elucidate how ultrasound alters the molecular structure, physicochemical attributes, and functional characteristics of starch, specifically focusing on its relevance and applicability. Additionally, it seeks to highlight the potential advantages, limitations, and optimization strategies associated with ultrasound treatment as a tool for enhancing starch properties, thereby addressing its significance in optimizing starch for various purposes.

Here's an overview of ultrasonic treatment in starch systems.

6.2 STARCH MODIFICATION

Modification using chemical treatment can leave behind residues and cause environmental problems. Physical modification can be thereby employed to avoid the negative effects of using hazardous chemicals in the final product (Satmalawati et al., 2020). Starch comprises about 20% to 25% amylose and 75% to 80% amylopectin. Amylose, a linear polysaccharide, is composed of D-glucose units connected by α-1,4 glycosidic bonds. In contrast, amylopectin, which is branched, consists of D-glucose units linked by α-1,4 glycosidic bonds for the main chain and α-1,6 glycosidic bonds for branching (Priyadarshi et al., 2024). Because of its physicochemical qualities, it is significant in the food sector. It is also very significant nutritionally since it provides the majority of the carbs in the diet (Li et al. 2018). The form, size, and composition of the starch are determined by the plant's origin (Joshi et al., 2013). According to Bitik et al. (2019), the lentil starch granule has a mean particle diameter range of 15.4–17.9 m with a surface oval to spherical shape. Chickpea starch, on the other hand, has a mean particle diameter ranging from 11.0–14.4 m with oval to spherical forms (Singh et al.). Ultrasonic treatment primarily promotes the movement of starch molecular chains via mechanical, cavitation, and thermal effects; alters the original hydrogen bond and double helix structure in starch granules; and regulates starch physicochemical properties via influence on the surface and internal structure of starch granules (Li et al., 2022; Acevedo et al., 2022). Utilizing ultrasonication is an effective method for modifying starch, enhancing stability, processability, and resistant starch content. It allows for precise control over viscosity and gelatinization time, enabling adjustments to meet specific requirements (Bitik et al., 2019).

6.3 ULTRASONIC TREATMENT

For ultrasonic treatment, starch is suspended in an ultrasonic homogenizer or ultrasonic bath. An ice bath is used to cool the sample vessel. The starch sample is heated to 60–80°C for 10, 20, and 30 min depending upon the treatment conditions. The ultrasound action and interval times were both set to 2 s (Hu et al.,

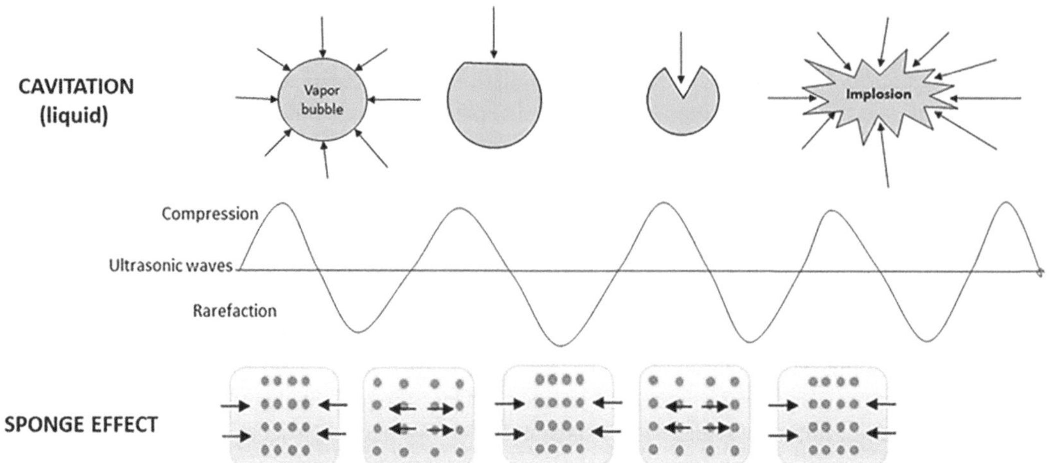

FIGURE 6.1 Operational mechanism of ultrasound.

2019). Following ultrasonic treatment, the starch is rinsed with distilled water and dried at 40°C. Sonication has recently been identified as a potential and practicable approach for improving functionality in the food and beverage industries (Acevedo et al., 2022). This technology is classified as either high-frequency low-intensity ultrasound (HF-LIU; 100 kHz^{-1} MHz, power 1 W/cm^2) or low-frequency high-intensity ultrasound (LF-HIU; 16–100 kHz, power 10–1000 W/cm^2). Compared to traditional thermal methods, the HIU process (in probe and bath sonicators) has several advantages, including higher production yield at a faster processing rate, higher quality and functionality of processed foods, better preservation of heat-labile food constituents, more cost-effective technology with less energy consumption, easier implementation and operation, lower equipment contamination, and more aseptic processing conditions (Gharibzahedi et al., 2020). Figure 6.1 illustrates the operational mechanism of ultrasound, showcasing how ultrasound waves interact with materials, in this case, potentially in the context of food processing or scientific applications.

Listed in the following are the broad uses of ultrasound in the realm of food technology.

6.3.1 Gelatinization

Ultrasonication can induce gelatinization of starch granules by disrupting the crystalline structure. This can improve starch solubility and create modified starches with altered functional properties. Ultrasound treatment was shown to alter the structured area within starch granules before causing reversible hydration of the less organized phase, breaking down the starch, and reducing the enthalpy of gelatinization (Hu et al., 2013).

6.3.2 Retrogradation Inhibition

Ultrasonic treatment can inhibit retrogradation, the process where gelatinized starch molecules re-associate and form a more crystalline structure. This can enhance the stability of starch-based products (Acevedo et al., 2022).

6.3.3 Particle Size Reduction

Ultrasonication can break down the granular structure of starch, resulting in a decrease in particle size. This attribute proves advantageous in multiple industries, including food and pharmaceuticals, where finer particle sizes are favored (Rahaman et al., 2021; Monroy et al., 2018).

6.3.4 Enhanced Enzymatic Reactions and Accessibility

Cavitation, caused by ultrasound, primarily triggers chemical reactions. The collapse of cavitation bubbles creates active radicals like radical dotOH, radical dotO, and radical dotHO2, which then target the starch (Rahaman et al., 2021) Additionally, ultrasonic treatment can increase the accessibility of starch granules to enzymes, facilitating enzymatic reactions. This is particularly relevant in processes like starch hydrolysis for the production of syrups and other starch derivatives (Chan et al., 2021)

6.4 EFFECT OF ULTRASONIC TREATMENT ON STARCH PROPERTIES

Figure 6.2 shows an illustration depicting how ultrasound impacts the structural and functional aspects of starch molecules and granules. Studies have shown that ultrasonic treatment can accelerate chemical reactions in starch including hydrolysis of starch, disruption of crystalline structure, enhanced enzymatic activity. These processes result in significant changes such as increased solubility, reduction in granule size, enhanced gelatinization, increased viscosity, altered thermal properties, and modification of functional properties in its structure and properties. Table 6.1 highlights the versatility of ultrasound treatment in modifying the properties of starches, making them suitable for various applications in the food, pharmaceutical, and other industries.

The impact of using ultrasonic pretreatment on diverse attributes of starches is described in the following.

6.4.1 Starch Structure

Ultrasound pretreatment can have various effects on the structure of sweet potato starch, depending on the ultrasonic condition (300 W for 15, 20, 25, and 30 min) with a 13-mm ultrasonic probe. Ultrasonic

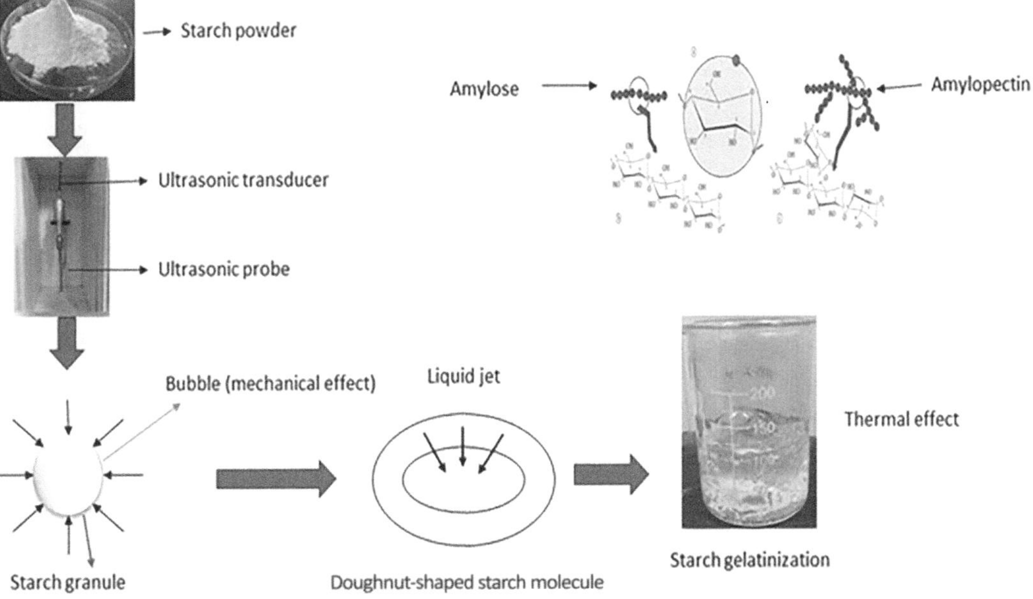

FIGURE 6.2   How ultrasound impacts starches.

TABLE 6.1 Effects Observed in Diverse Starches Treated with Ultrasound

STARCH	CONDITIONS	MAJOR FINDINGS	REFERENCES
Normal, waxy, and high-amylose corn starches	Starch suspension dissolved in 30% D.W. was placed inside a beaker. UT was conducted with stirring for 30 min at 30°C	• Pores and fissures on the surfaces. The ultrasonic treatment did not alter the crystalline structure of the three corn starches. • Increased in gelatinization transition temperatures, swelling power, and solubility, while decreasing their enthalpy and gelatinization temperature range. • The apparent viscosity of the samples remained unchanged, except for a noticeable drop in the high-amylose sample.	Han et al. (2021a)
Potato starch	The samples, comprising a starch suspension in D.W. (10%), were subjected to varying power levels (60, 105, 155 W) at a frequency of 20 kHz for 30 min	• Incisions and channels were observed on the surface of the starch granules. • Incisions became more pronounced with enhanced ultrasound power resulting in erosions. • Changes in the granule cluster structure leading to molecular disorder.	Wang et al. (2022)
Corn starch	Samples consisting of 100 g of starch suspension in D.W. (5%) were subjected to treatment for 40 min at 30°C, employing various ultrasonic frequencies (20 kHz, 25 kHz, and a combination of 20 kHz + 25 kHz)	• The apparent viscosity of the treated samples decreased. • The sonicated granules exhibited depressions on their surfaces, particularly utilizing the dual-frequency US (20 kHz + 25 kHz).	Hu et al. (2014)
Cassava starch	Starch suspension in D.W. (5%, w/v) underwent US employing 40% amplitude for varying durations (5, 10, and 20 min), along with 60% amplitude for 20 min	• At 60% amplitude, complete starch gelatinization occurred, while at 40%, partial gelatinization was observed. • Surface damages on starch granules, elevating surface roughness and facilitating the formation of cavities, channels, and cracks, thus broadening the granule size distribution. • ATR-FTIR spectra and X-ray diffraction results indicated structural disruptions in the granules' crystalline regions resulting in reduced degrees of crystallinity.	Monroy et al. (2018)
Oat starch	100 mL of starch suspension in D.W (5%) was treated. The bath operated at an intensity of 5 W/cm^2 for durations of 10 and 20 min, maintaining a controlled temperature of 25°C	• No alterations in oat starch granule surface, size, amylose content, thermal properties, paste viscosity, clarity, gel hardness, starch solubility, and swelling power under the investigated conditions. • Syneresis of the sonicated samples remained unchanged even after 5 days of storage. • Water and oil absorption increased following a 20-mi sonication period.	Falsafi et al. (2019)

Quinoa and maize starch	A 7.5% w/w starch suspension was prepared. The output amplitude of the probe was set at 50% and 20 kHz, and the pulse rate was 3 s of operational time with a 1-s pause. The intensity was 27.3 W/cm^2. The starch sample was treated for 2, 4, 8, 14, and 22 h	• Surface damage to the starch granules. • Increased gel hardness. • A decrease of relative crystallinity. • Enhanced formation of single helical structures.	Wei et al. (2023)
Rice starch	Ultrasonic probe (30 W) for 1, 3, and 5 min sonication time	• Reduction in both apparent and inherent viscosities. • A more translucent rice paste. • –No alterations observed in the starch degree of polymerization (DP).	Bonto et al. (2021)
Sorghum starch	Aqueous suspensions (30% w/w) sorghum starch were sonicated at pH 5.5 using a 24 kHz Sonication was performed at 30, 60, and 100% amplitudes for 10, 20, and 30 min	• Enhancement of physicochemical, starch gel network, and rheological properties of the sorghum starch. • However, severe ultrasonic power led to destruction of the starch granule structure and weakened the gelling ability of the starch, thus reducing pseudoplastic behavior.	Jafari and Koocheki (2024)

treatment caused a transformation in the surface of corn starch, rendering it rougher. This alteration led to the nearly complete disappearance of the distinctive pores typically found on the native corn surface (Wu et al., 2022). Sweet potato starch in its native form showed various shapes like polygons, ovals, and circles, with fairly smooth surfaces without any visible pores or cracks (Wang et al., 2020). Following ultrasonication, pores, fractures, and indentations became noticeable on the surface of the ultrasonicated sweet potato starch granules (Wang et al., 2020; Cui et al., 2018). Moreover, these alterations became increasingly pronounced with longer sonication periods, ndicating that the impact of ultrasonic treatment primarily affected the surface characteristics of the starch granules rather than their actual size (Zhang et al., 2021a). The rationale behind this lies in the intense shear force, mechanical stress, and pressure gradients that arise from the rapid creation and collapse of cavitation bubbles during ultrasonication. This phenomenon leads to the degradation of polymers and the disruption of the starch's structural aggregation (Cui et al., 2018). Additionally, the cavitation process causing water molecule dissociation results in the formation of free radicals like hydroxide (-OH) and hydrogen (-H) radicals (Yang et al., 2019). These free radicals are believed to trigger the cleavage of starch molecular chains, consequently disturbing the internal molecular structure and clustering, thereby compromising the integrity and firmness of starch granules (Zhu, 2015). These findings align with reports by Zheng et al. (2013), Sujka et al. (2013), and Sujka et al. (2017) which highlight that cracks, pores, and indentations observed on the particle surface are likely outcomes of the inherent cavitation effects induced by ultrasonic treatment. Furthermore, SEM analysis showed significant surface damage, characterized by scratches, on both B-type potato starch and C-type pea starch, a contrast to the surface of A-type corn starch (Ouyang et al., 2021). Additionally, according to reports of Zhang et al. (2021b); Hu et al. (2019), ultrasound pretreatment has the potential to influence the morphological, physicochemical, functional, and rheological characteristics of starch. Thus, the application of ultrasound can lead to alterations in the formation of cavities and fractures on the surface of starch granules. As a result, ultrasound is recognized for its distinctive impact on the degradation and modification of polymers, as well as on reaction or production rates (Carmona-García et al., 2016).

6.4.2 Enzyme Hydrolysis

The breakdown of starch through enzymatic hydrolysis holds significance in the food industry. However, ultrasound exposure, particularly under conditions of elevated temperatures, high ultrasonic power, and prolonged treatment, resulted in the deactivation of enzymes involved in this process (Wang et al., 2020). Ultrasound pretreatment has been shown to enhance the degree of hydrolysis (DE) values in the processing of starch. Compared to native corn starch, starches treated with ultrasound reach the desired DE value more rapidly during liquefaction. The saccharification process also experiences significant improvements in DE values, with ultrasound-treated starch exhibiting increased saccharification values as treatment time extends. It's noteworthy that ultrasound pretreatment accelerates both liquefaction and saccharification of starch-based feedstock, and different ultrasound conditions result in varying DE values. Additionally, ultrasound has been observed to boost the hydrolysis of starch amylase within a specific strength range, evidenced by higher rate constants compared to untreated starch (Li et al., 2018). The level of wheat starch hydrolysis experienced an increase with higher ultrasonic power levels initially. However, a subsequent rise in ultrasonic power led to a decline in the degree of hydrolysis (Karwasra et al., 2020). According to Yu et al. (2013), the degradation of polysaccharides intensifies with increasing power until reaching a certain limit, known as the "upper threshold". Therefore, the reduction in hydrolysis degree observed at 360 W in ultrasonic-assisted enzymatic hydrolysis isn't due to the limitations of starch degradation caused by ultrasound but rather the deactivation of glucoamylase at higher ultrasonic power levels (Carmona-Garcia et al., 2016).

6.4.3 Hydrolysis Rate

The hydrolysis rate of starch increased steadily when the ultrasonic treatment period was increased. Starch hydrolysis rose from 0.5% after 10 minutes to 4% after 30 minutes. The ultrasound acted on the liquid medium to generate mechanical vibrations, which resulted in a thermal, mechanical, and cavitation action, resulting in the fracture of the starch chains like potato, corn and wheat (Cai et al., 2022). The higher the degree of damage, the longer the ultrasonic time; thus, the hydrolysis rate of starch was enhanced (Li et al., 2023).

6.4.4 Morphological Properties

The impact of ultrasonic treatment on native starches provides evident results, leading to significant changes in their morphology. The treatment alters the structure of starch particles, with the level of alteration dependent on the higher intensity of the ultrasonic treatment in wheat (Rahaman et al., 2021). Utilizing ultrasonic treatment before chemical or enzymatic modification enhances the fissures, cracks, and pores on the starch surface, facilitating better penetration of chemicals or enzymes and thereby increasing the efficacy of starch modification. Such modified starches find application in stabilizing chilled food (Kumari & Sit, 2023). Cassava and corn starches underwent alteration through brief exposure to high-power ultrasound at 40 kHz. The outcomes indicated reduced relative crystallinity and enthalpy, along with surface cracks in both modified starches, while the starch size remained unchanged (Kumari & Sit, 2023). The damage caused by short-term ultrasonic treatment to potato starch granules appeared relatively minor, exhibiting only shallow surface damage. However, prolonged treatment, especially within a 30-minute timeframe, resulted in more pronounced surface damage, with rough edges, corners, and visible depressions and pores (Li et al., 2023). High-power ultrasonication up to 155 W induces rupture and mechanical damage to potato starch granules due to the collapse of cavitation bubbles, generating high-pressure gradients and localized liquid velocities that exert shear forces capable of breaking polymer chains and damaging granules (Kaur & Gill, 2019; Zhu et al., 2012). Observations of various starches revealed differences in their surface characteristics. Wheat and barley starches displayed smoother surfaces compared to rice and maize starches. Post-ultrasonication, wheat and maize starches exhibited more noticeable modifications, such as microscopic fractures in wheat starch and pit formation in maize starch, suggesting a lower granular structural integrity in maize compared to rice, wheat, and barley starches (Li et al., 2019). The microstructure of starch was found to be linked to the intensity of ultrasonic treatment and the botanical origin of the starch (Kumari & Sit, 2023). The formation of pores was attributed to mechanical damage caused by cavitation bubble collapse during ultrasonic treatment, as observed in cereals, legumes (Monroy et al., 2018), and tuber starch granules, resulting in cracks and depressions (Acevedo et al., 2022). The surfaces of granules exhibited smoothing and fissures, potentially induced by ultrasonic radiation, in line with findings by Bitik et al. (2019) attributing visible fissures to ultrasonic granule disintegration. Studies on kiwi starch showed that low-intensity ultrasonic treatment led to surface roughness and corrosion after 30 min, while increased ultrasonic power caused visible depressions and partial disintegration. The coarsening of starch granules was noted with higher intensity, demonstrating a deepening degree of corrosion and disintegration as ultrasonic time increased (Wang et al., 2022). Generally, size of the granules was not affected by ultrasound treatment, but roughness on the surface of the granules was noticed which increased with treatment time. The frictional/shear force produced by shock waves and due to the formation of high-pressure gradients in the surrounding area induces structural changes in starches (Manzoor et al., 2019). Several factors influence how ultrasound impacts the morphology of starch granules, such as the starch slurry's concentration, system temperature, ultrasound frequency, intensity, power, treatment duration, and whether it involves single or dual frequencies (Zhu, 2015). Following ultrasound treatment, the surfaces of potato starch exhibited cracks, pores, and even granule breakdown (Acevedo et al., 2022). Some of the visible morphological changes and findings upon ultrasound treatment in various starches are given in Table 6.2.

6.4.5 X-Ray Diffraction

X-ray diffraction (XRD) is employed to evaluate the crystallinity of starch granules that varies to extent, depending on starch type and experimental set up. Carmona-Garcıa et al. (2016) studied the X-ray

TABLE 6.2   Changes in Morphological Properties of Ultrasound-Treated Starches

SOURCE	MAJOR FINDINGS UPON ULTRASONIC TREATMENT	REFERENCES
Sorghum starch	Physical damage to the granule surface, resulting in some cracks and holes	Jafari & Koocheki (2024)
Cassava starch	Surface roughness of the starch granules	Monroy et al. (2018)
Pearl millet starch	Granules with irregular shapes and surfaces with pores, roughness, craters, and depressions	Sharma et al. (2023)
Pea starch	Starch granules are characterized with uneven shapes and surfaces exhibiting pores, rough textures, and depressions	Ding et al. (2021)
Potato starch	Deep cavities and damage into starch granules	Abedi et al. (2019)
Sweet potato and wheat starches	Formation of holes and fissures into the granules surface	Cui and Zhu (2020)
Finger millet starch	Breakdown of polymeric chains in starch, resulting in a reduction of particle size and the formation of a homogeneous size distribution	Yadav et al. (2021)
Tapioca starch	Formation of holes, dents, notches, and grooves on starch granules along with breakdown of intra- and inter-granule forces in starch granules	Pourmohammadi and Abedi (2020)
Rice starch	Cracks and visible pores were observed on the surface of starch granules	Yang et al. (2019)

diffraction patterns of native and modified plantain and taro starch samples sonicated at 80 W. They found that the X-ray diffraction intensity of plantain and taro starches decreased more as the processing time increased during 80 W ultrasonic treatment. Regular maize and potato starches underwent modification through ultrasound treatment. The findings demonstrated a decrease in the relative crystallinity of the modified starches. This decline could be attributed to the cavitation effect of ultrasound, causing the internal bonds to loosen and damaging the structure of the starch granules (Kumari & Sit, 2023). Zhu (2015) observed varied impacts of sonication on different starch types. Wide-angle X-ray diffraction analysis (WAXS) revealed that ultrasonication had minimal impact on the polymorph type of maize starch (A-type), potato starch (B-type), and maize starches with varying amylose contents (0–50%) (A- and B-type). Ouyang et al.'s (2021) findings indicated that B-type starch, when sonicated, displayed a surface that was more porous and coarser in comparison to A-type or C-type starches. Wang et al. (2023) similarly discovered that employing high-power ultrasonic treatment led to the disruption of the crystalline region within starch granules, causing a reduction in RC (relative crystallinity). These are in comparison with previous studies (Monroy et al., 2018; Manzoor et al., 2019). Several results showed that sonicated B-type starch had a more porous and coarser surface than A- or C-type starch. A-type starch had higher resistant starch content, retrogradation propensity, and thermal stability, whereas B and C-type starches had lower resistant starch content but higher amylose concentration (Hu et al., 2019). According to the findings of this study, the crystallinity pattern of starch influenced the physicochemical and microstructure changes that occurred during sonication. Following sonication, there's a decrease in the relative crystallinity of starches. The destabilization of the layered structure within starch could account for the decline in relative crystallinity observed in the ultrasound-treated samples (Rahaman et al., 2021). X-ray diffraction patterns of sweet potato starches before and after ultrasonication with different sonication time (15, 20, 25, and 30 min) were studied by Wang et al. (2020). Despite sweet potato and millet starches having different crystalline types, both exhibited similar relative crystallinity and amylose content. Ultrasound did not alter the crystalline types in either starch, but it caused more significant damage to the internal structure of the crystalline region in potato starch compared to millet starch (Hu et al., 2019). The possible reason might be due to despite potato starch being C-type and millet starch A-type, both starches showed similar relative crystallinity and amylose content in this study. Ultrasound treatment induced slight changes in diffraction patterns for both starches, with gradual decreases in peak intensities (5.7°, 21.9°, and 24°) observed specifically in potato starch. The alteration in relative crystallinity differed notably between the two starches, influenced by their distinct packing of crystalline and amorphous regions (Hu et al., 2019). Water bath sonication for 10 and 20 min, along with probe sonication at 39 W/cm^2 for 10 and 20 min, as well as at 48 W/cm^2 for 10 min, did not cause any alterations in the oat starch's degree of crystallinity. However, probe sonication at 48 W/cm^2 for 20 min and 63 W/cm^2 for 10 and 20 min led to a decrease in the degree of crystallinity. These reductions observed at higher sonication intensities are likely result from the breakdown of less stable crystalline structures during sonication (Falsafi et al., 2019). Likewise, studies on sonicated cassava (Monroy et al., 2018), sweet potato (Zheng et al., 2013), and potato (Zhu et al., 2012) starches consistently reported similar reductions in the degree of crystallinity.

6.4.6 Fourier Transform–Infrared Spectroscopy

Fourier transform–infrared spectroscopy (FT-IR) analysis indicates that the functional groups within starch remain intact following ultrasonic treatment. However, upon ultrasonic treatment, the crystal structure gets disrupted, leading to a decrease in the starch's crystalline index (Bitik et al., 2019). The alterations observed after ultrasonication suggest that ultrasound had a more pronounced impact on the short-range structure of potato starch compared to millet starch. Additionally, ultrasound caused greater damage to the double helix structures in potato starch (Bai et al., 2017). The decrease in short-range ordered degree of sweet potato starches suggested that ultrasound exposure could potentially disrupt both the amorphous and crystalline sections of starch granules, causing a disorganized arrangement or unwinding of helical structures. The FT-IR findings demonstrated that this effect became more pronounced with longer sonication periods (up to 30 min) (Wang et al., 2020). A comparable reduction in short-range ordered degree was observed in cassava starch subjected to ultrasound treatment (Monroy et al., 2018). However, the FT-IR spectroscopy showed no discernible distinctions in peaks, and there was neither the appearance of new absorption peaks nor the disappearance of existing ones upon ultrasonicated for treated starches of potato and millet starches. There were no new chemical bonds formed or lost during the ultrasound treatment (40 kHz and 40 kHz + 80 kHz) frequency sonication, supporting the notion that the effects of ultrasound were purely physical in nature (Hu et al., 2019). Figure 6.3 represents some FT-IR spectra of ultrasound-treated sweet potato starches.

6.4.7 Transmittance

Following ultrasonic treatment, there's an elevation in starch transmittance. Starch subjected to dual-frequency ultrasound for 30 min demonstrates the highest transmittance, showcasing a 5.1% increase compared to the original/native waxy corn starch (Xu et al., 2021). The rapid collapse of cavitation bubbles during ultrasound treatment leads to the creation of strong pressure differences and rapid movements of the surrounding liquid. These actions generate powerful shearing forces that can cause indentations on the starch granule. (Bitik et al., 2019). This process aids in enhancing water penetration into the starch granules and hence leads to improvement in light transmittance. Moreover, ultrasonic treatment can fracture amylopectin chains by interfering with their covalent bonds, resulting in heightened transmittance (Cui et al., 2018). Through the ultrasound process, there was an enhancement in the clarity of starch pastes, displaying higher transmittance compared to the control. In comparison to ultrasonicated sorghum starches (starch suspension), pearl millet starches exhibited even higher transmittance, leading to improved paste clarity. This is attributed to the disruption of swollen starch granules, possibly resulting from a lack of granular structure or minimal association of chains. These factors contribute to making the starch paste highly transparent (Sharma et al., 2023). The results obtained align with those documented

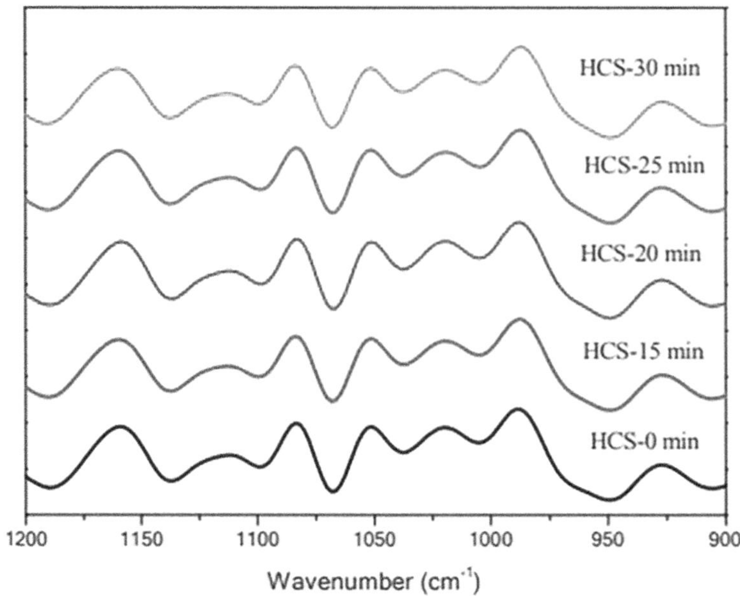

FIGURE 6.3 FT-IR spectroscopy of ultrasound-treated sweet potato starch (Wang et al., 2020).

by Yang et al. (2019) and Kumari and Sit (2023) in their studies on rice, potato, and mung bean starch, as well as with the findings reported by Sujka and Jamroz (2013) in their investigations on cereal starches.

6.4.8 Pasting Properties

Ultrasound treatment leads to significant changes in peak viscosity (PV), trough viscosity (TV), breakdown viscosity (BV), final viscosity (FV), and setback viscosity (SV). The pasting temperatures of sonicated starches rose because of the partial breakdown of the amorphous regions, leading to a heightened ratio of crystalline to amorphous content. When comparing with the control samples, ultrasound pretreatment (NU-10–30, NU-20–30, NU-30–60, and NU-30–100) results in substantial decreases in the PV (235–205 BU), PT (94.4–92°C), BV (30–28 BU), FV (250–230 BU), and SV (53–35 BU) values of sorghum starch (Jafari & Koocheki, 2024). The reduction in PV can be attributed to decreased granule integrity and rigidity, potentially due to glycosidic bond cleavage. The pasting temperatures of sonicated starches rose because of the partial breakdown of the amorphous regions, leading to a heightened ratio of crystalline to amorphous content. Comparable outcomes have been noted in studies involving sonicated corn starch as well (Li et al., 2018). The peak viscosity of potato starch (1900 BU) was notably higher than that of millet starch (384 BU). When starches underwent sonication at 30% amplitude for 10 and 20 min, the peak viscosity rose to 249 and 240 BU, respectively (Jafari & Koocheki, 2024). Ultrasound treatment (40 + 80 kHz) did not alter the pasting behavior of either starch type (potato and millet) compared to their native forms. The peak viscosity of potato starch decreased noticeably from 1900 to 1786 BU, while millet starch showed no change (Wang et al., 2020), showing that the millet starch was more resistant to ultrasound than potato starch. Samples subjected to gentle sonication conditions (10 min–30% amp and 20 min–30% amp) displayed elevated peak (237–240 BU), setback (58–65 BU), and final viscosities (259–265 BU). The substantial swelling power exhibited by the granules of ultrasonicated starches led to a higher volume fraction occupied, intensifying the inter-granular friction and subsequently elevating both peak and final

TABLE 6.3 Pasting Characteristics of Various Starches Treated with Various Ultrasound Frequencies, as Observed by Brabender Viscometer

STARCH	GELATINIZATION TEMPERATURE (°C)	PV (CP)	BD (CP)	FV	SETBACK (CP)	REFERENCES
NS	69.8	447	110		31	Yang et al. (2019)
25 kHz	70.5	424	96		13	
80 kHz	70.6	403	85		28	
25 + 80 kHz	70.5	384	96		50	
PS		1696–1764.68	930–1083	1250–1264.15	504.29–534.45	Hu et al. (2019)
Millet starch		369.38–384.37	56.12–64.64	510.04–526.76	192.26–213.20	Hu et al. (2019)
Sweet potato starch	69.0	630.5–633.2	151.0–159.0	675.5–687.0	186.5–218.5	Wang et al. (2020)
Rice starch	82.67	2357–2440	1021–1101	2363–2498	1055–1149	Yang et al. (2019)
Sorghum starch		205–249	27–35	228–279	45–65	Jafari and Koocheki, (2024)

viscosities (Jafari & Koocheki, 2024). Ultrasound-induced macromolecular chain rupture and crystalline structure destruction are suggested as reasons for the overall viscosity decrease in starch paste (Li et al., 2018). Lower BV values of ultrasonicated treated cereal starches (15 and 30 min) indicated increases resistance to shear thinning during cooking (Kaur et al., 2019). Final and setback viscosities are linked to the polymerization of leached amylose and long linear amylopectin. While FV and SV decrease insignificantly, the overall viscosity reduction is associated with starch chain degradation, leading to shorter molecular chains and lower viscosity during gelatinization (Jin et al., 2020). The rise in peak viscosity and final viscosity, along with the reduction in breakdown viscosity, could be attributed to the starch paste's increased resistance to shearing post-ultrasonication. This resistance led to the formation of a more rigid gel during the cooling process (Dhull et al., 2021). Following ultrasound treatment, starches exhibit greater resilience against high temperatures and intense mechanical forces compared to their native counterparts as discussed. Table 6.3, which highlights characteristic values observed in the Brabender curves of starch treated with different ultrasound frequencies for a duration of 60 minutes, likely provides insights into the effects of varying ultrasound frequencies on starch behavior during heating and cooling. These insights are depicted through Brabender viscometry.

Figure 6.4 illustrates the RVA viscosity as a function of pasting time for both non-sonicated (depicted by open symbols) and sonicated (depicted by solid symbols) waxy rice starch dispersions. The dashed line in the graph represents the heating profile according to Zuo et al. (2009). Additionally, pasting profiles for plantain and taro starches are presented, where the starches were sonicated for specific durations: 20 min (PSS-20) and 50 min (PSS-50) for plantain starch, native taro starch (NTS), 20 min sonicated taro starch (TSS-20), and 50 min sonicated taro starch (TSS-50), per Carmona-Gracia et al. (2016). Furthermore, RVA profiles of native and pregelatinized starches obtained through ultrasonication under various conditions are included. These conditions involve different starch types: native wheat starch (NWS), native tapioca starch (NTS), ultrasonic pregelatinized wheat starch using a 20-mm-diameter probe (UPWS2), ultrasonic pregelatinized tapioca starch using a 20-mm-diameter probe (UPTS2), ultrasonic pregelatinized wheat starch using a 100-mm-diameter probe (UPWS10), and ultrasonic pregelatinized tapioca starch using a 100-mm-diameter probe (UPTS10), as reported by Abedi et al. (2019). Additionally, Figure 6.4 represents pasting profiles of native rice starch and rice starch treated with different ultrasonic power levels, as reported by Yang et al. (2019).

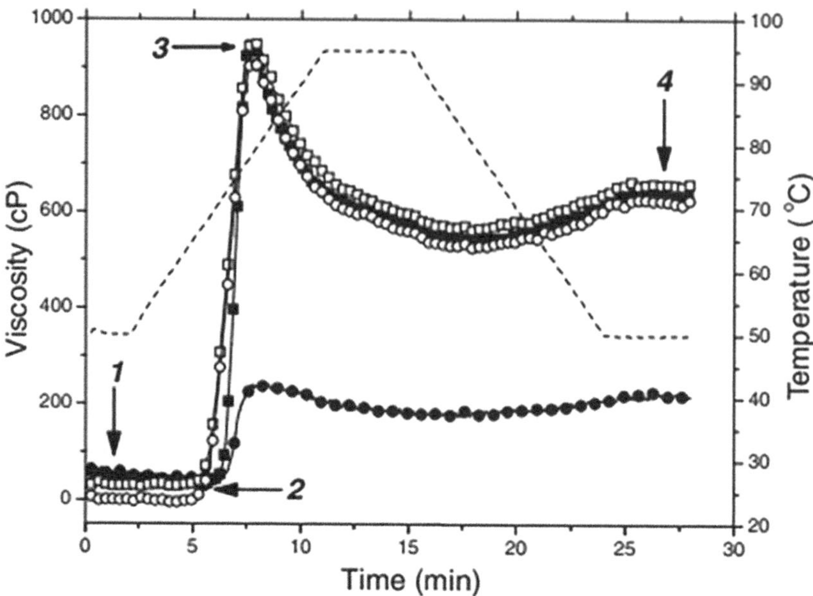

FIGURE 6.4   Pasting profiles of non-sonicated (open symbols) and sonicated (solid symbols) waxy rice starch dispersions. Suspensions were heated at 25°C □,■ or 63°C ○,● for 30 min. Sonication conditions are: power intensity of 0.18 W/cm² and a frequency of 211 kHz. Dashed line is the heating profile (Zuo et al., 2009).

6.5 EFFECT OF ULTRASONICATION ON THE THERMO-STRUCTURAL PROPERTIES OF STARCHES

6.5.1 Thermal Properties

The change in thermal properties due to ultrasound treatment in starches may be due to the higher degree of crystallinity that provides more thermal stability of granules against gelatinization. The pores and cracks developed in the starch due to ultrasonication made it easier for water to penetrate the granules that disrupt the crystalline regions of the starch (Hu et al., 2019). As compared to the amorphous region, the crystalline region. Sonication disrupted the anti-ultrasonic (crystalline) region in potato starch, whereas in millet starch, ultrasound affected only the areas more responsive to its effects while leaving other regions intact (Hu et al., 2019). The enthalpy of gelatinization (ΔH) represents the energy needed for the unfolding of double helices situated within both amorphous and crystalline regions. The gelatinization temperatures of both native and modified sorghum starches spanned from 64.7 to 82.3°C, with gelatinization enthalpy ranging between 9.7 and 14.2 (J/g). Following sonication under mild conditions (10–30 min), the gelatinization temperatures (onset, peak, and end temperatures) and enthalpy values of the starch exhibited a shift towards higher values. This shift can be attributed to the breakdown of less robust crystalline structures, particularly less organized double-helices, leading to a delayed gelatinization process (Jafari & Koocheki, 2024; Falsafi et al., 2019). The application of high-intensity ultrasound can potentially improve the uniformity of the corn starch granule structure by disrupting weaker and less structured crystalline fractions. This process may lead to the development of a more well-ordered crystalline structure with narrower ranges of gelatinization temperatures (Amini et al., 2015). However, the ΔH value for native normal maize starch was 2.73 J/g (Table 6.4). Treatment with either single ultrasound resulted in a reduction in the ΔH

TABLE 6.4 Thermal Properties of Various Native and Ultrasonicated Starches

SAMPLE	NATIVE				ULTRASONICATED				REFERENCE
	T_O (ONSET TEMPERATURE)°C	T_P (PEAK TEMPERATURE)°C	T_C (CONCLUSION TEMPERATURE)°C	ΔH (J/G)	T_O (ONSET TEMPERATURE)°C	T_P (PEAK TEMPERATURE)°C	T_C (CONCLUSION TEMPERATURE)°C	ΔH (J/G)	
Corn starch	71.58	77.01	87.9	12.0	71.98	77.72	87.99	10.08	Rahaman et al. (2021)
Cassava starch	66.8	74.31	86.77	12.65	66.35	74.12	84.55	10.32	Rahaman et al. (2021)
Potato starch	66.47	70.08	75.83	1.87	67.20	70.70	75.09	1.65	Hu et al., 2019
Millet starch	69.18	77.33	82.83	8.93	67.86	76.40	82.33	10.22	Hu et al., 2019
Sorghum starch	68.1	72.5	77.8	12.7	67.7	72.1	76.7	10.3–13.4	Jafari and Koocheki (2024)
Oat starch	57.35	64.45	69.31	9	57.45	64.71	69.24	7.39	Falsafi et al. (2019)
Pea starch	57.52	66.20	65.36	6.48	57.75	65.48	73.72	5.64	Zhang et al. (2021b)
Sweet potato starch	64.94	69.99	83.52	18.38	64.26	69.26	80.15	14.93	Wang et al. (2019)
Maize starch	67.9	73.3	78.9	2.73	67.7	72.1	76.5	1.54	Wang et al. (2022)
Rice starch	62.94	69.82	78.69	8.70	62.62–63.11	69.19–69.86	75.57–78.25	7.21–9.58	Yang et al. (2019)

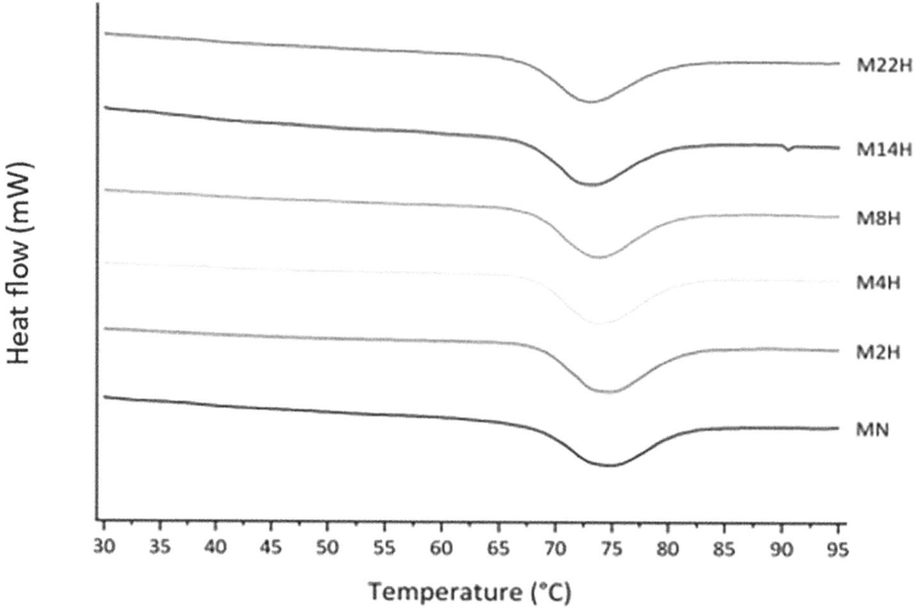

FIGURE 6.5   DSC profiles of native and ultrasound-treated maize starch (Wei et al., 2023).

of maize starch. Wang et al. (2019) suggested that this decrease in ΔH was attributed to the disruption of the double-helix structure. In comparison to the ΔH of native potato starch (1.87 J/g), ultrasound treatment exhibited a decrease. This suggests that ultrasound treatment facilitated the destruction of the crystalline area in potato starch (Hu et al., 2019). The decrease in ΔH was also observed in ultrasound-treated tapioca starch (Monroy et al., 2018) and quinoa starch (Zhu & Li, 2019). Table 6.4 showcases the thermal properties of various native ultrasonicated starches likely outlining the effects of ultrasound treatment of different starch samples. Additionally, DSC profile of native and ultrasound-treated maize starch is shown in Figure 6.5.

6.5.2 Microstructure

Numerous studies indicate that ultrasound treatment creates frictional forces, leading to structural alterations in starch granules. This occurs through the collapse of cavitation bubbles, resulting in high-pressure gradients and rapid liquid velocities around them. Consequently, shear forces are formed, significant enough to cause polymer degradation, resulting in small surface fissures and depressions on the granules (Sujka & Jamroz, 2013; Monroy et al., 2018). Ultrasonication increased the number of pores and cracks on the starch granule surface. The modification of the microstructure heavily depends on the size of starch granules, as larger granules possess a greater ability to trap kinetic energy (Carmona-García et al., 2016). Certain starch granules display minor visible cracks in their central areas. Following ultrasound treatment, particularly in water, some granules exhibited small surface fissures and depressions. The impact of ultrasounds on starch was notably more pronounced in the case of potato and wheat starches. Following sonication in water, the surface of potato starch granules exhibited numerous cracks and scratches. Ultrasonication led to a transformation in the natural morphology of rice starch granule surfaces, exhibiting the emergence of notches, grooves, and cracks. The concentration of free radicals, including OH radical and H radical, rose in the ultrasound environment (water). This increase was a result of the intense effects of ultrasound irradiation when applied to potato samples. These radicals could potentially accelerate the breakdown of clusters and the disturbance of internal structure, adversely affecting the smooth surfaces and flat edges of potato starch samples (Zhang et al., 2021a). These alterations in the microstructure could likely be attributed to the cavitation effect generated by ultrasound and the subsequent soaking process (Zia-ud-Din et al., 2017). Hj. Latip et al. (2021) research indicated that ultrasound treatment could create pores or channels within corn starch but did not cause damage to its crystalline structure. Ultrasonication led to a transformation in the natural morphology of rice starch granule surfaces, exhibiting the emergence of notches, grooves, and cracks. These alterations in the microstructure could likely be attributed to the cavitation effect generated by ultrasound and the subsequent soaking process (Zia-ud-Din et al., 2017). The combination of the cavitation and mechanical impact from ultrasound treatment led to noticeable depressions and minor disruptions on the surface of rice starch. The application of ultrasound caused significant shrinkage and substantial collapse on the starch surface (Li et al., 2022). The employment of sonication led to the generation of numerous bubbles, resulting in the creation of an uneven mass and a starch surface characterized by numerous microscopic holes. The rupture of these bubbles caused an immediate increase in pressure, leading to the disruption of native starch and the emergence of numerous tiny holes. Simultaneously, it is hypothesized that the structural changes were induced by the impact of high-power ultrasound in partially gelatinizing resistant starch. This process weakened amylose crystallization, enhancing surface accessibility through hydrogen bonding. Smooth and compact starch reticular structures were formed due to expanded channels, allowing easy entry of different molecules into starch granules and their subsequent binding. SEM micrographs indicate that ultrasound was responsible for the development of rice starch complexes while causing minimal damage to the starch surface (Raza et al., 2023). Another study indicated that ultrasound treatment could create pores or channels within corn starch but did not cause damage to its crystalline structure (Hi et al., 2021). The 20-kHz ultrasound treatment caused surface roughening of the UFPS, resulting in some cracks and scratches. Elevating the ultrasonic frequency to 25 and 28 kHz revealed an increased presence of cracks, scratches, and even granule breakdown on the UFPS surface. This observation suggests that higher ultrasonic frequencies (25 and 28 kHz) induce more significant starch breakdown compared to the effects produced by 20-kHz ultrasound (Xiao et al., 2021). Using two simultaneous ultrasound beams enhanced the vibration amplitude and generated increased ultrasonic cavitation compared to the effects observed with single-frequency ultrasound within the same timeframe (Hu et al., 2015). Scanning electron microscopy (SEM) images of native and ultrasonicated (UT) starches offering visual insights into the structural differences and morphological alterations induced by ultrasound treatment as reported by various researchers Li et al. (2019) reported the observed formation of fissures and pores, as well as the degradation of the starch granule under applied ultrasonic conditions. Additionally, Khurshida et al. (2021) investigated surface erosion (Figure 6.6), Wei et al. (2023) studied the dispersion of surface roughness, and surface damage aggregates were explored by Wei et al. (2023) in relation to the same subject.

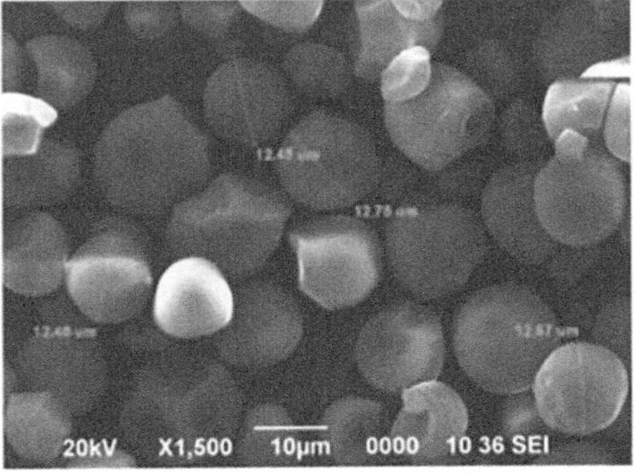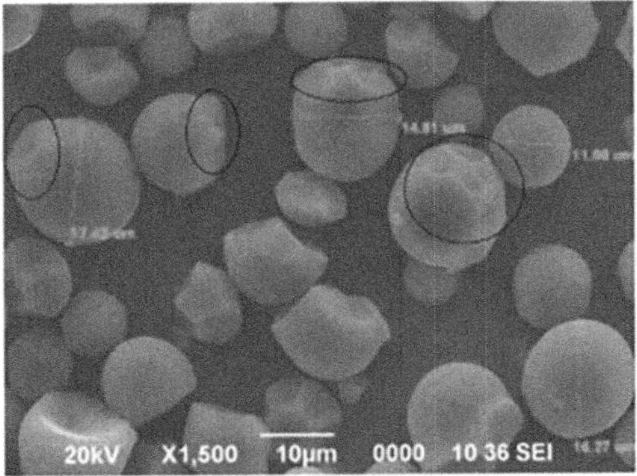

FIGURE 6.6 Scanning electron micrographs of native and ultrasonicated cassava starch (Khurshida et al., 2021).

6.6 EFFECT OF ULTRASONICATION ON THE RHEOLOGICAL PROPERTIES

6.6.1 Viscosity

The rheological properties of starches exhibited alterations after undergoing sonication. Ultrasonication has a multifaceted impact on the viscosity of starch solutions, with effects depending on factors like ultrasonication parameters and starch characteristics. Generally, ultrasonication (100% amplitude for 15, 30, and 45 min) induces shear thinning behavior in corn starch solutions, causing a reduction in viscosity under shear forces (Acevedo et al., 2022). As the duration of ultrasonic treatment extends to 30 min in cereal starches, both G′ (storage modulus) and G″ (loss modulus) values decrease. Consequently, the straightening of amylose molecules takes place, reducing the shear action within the fluid layers and thereby contributing to the decrease in viscosity (Kaur & Gill, 2019). The application of ultrasound waves disrupts the molecular structure of starch, potentially leading to a breakdown of intermolecular forces and a subsequent decrease in viscosity. Additionally, ultrasonication can promote gelatinization, enhancing the viscosity by breaking down the granular structure of starch molecules (Monroy et al., 2018). The process may also result in a reduction in particle size, influencing the rheological properties of the starch and contributing to changes in viscosity. Temperature effects, stemming from localized heating during ultrasonication, further complicate the picture, as they can impact the degree of gelatinization and molecular mobility, influencing overall viscosity. The complex interplay of these factors underscores the need for careful optimization of ultrasonication conditions for specific applications involving starch, such as in the food and pharmaceutical industries. The rheological characteristics of starch play a pivotal role in determining the quality of starch-based products as they directly impact the mechanical and technological properties of starch. In starch modification, reducing viscosity is a key factor (Bitik et al., 2019). Ultrasonication leads to the breakdown of starch molecules by breaking the glycosidic bonds in starch granules, resulting in decreased starch viscosity due to the loss in molecular weight. The outcomes indicated a decrease in the viscosity of the paste (Satmalawati et al., 2020). Typically, as indicated by Yu et al. (2013), the viscosity of starch tends to decrease with longer durations and higher levels of ultrasonic power. Using an ultrasound probe operating at 170 W power and 20-kHz frequency for 30 min significantly ($P < 0.05$) decreased the viscosity of starch paste. This reduction was linked to the breakdown of starch granules (as evident in SEM images) and the rupture of starch molecules due to intense shear forces, thereby weakening the starch structure and leading to decreased paste viscosity (Falsafi et al., 2019).

6.6.2 Steady Shear

In general, ultrasonic frequency amplified the shear-thinning characteristics, leading to a more pseudo-plastic nature of the starch gel. Moreover, the consistency index (K), indicating viscosity magnitude also increases for ultrasonicated starches (Xiao et al., 2021). The strength of starch gel is related to the loss tangent (tan δ). Lower tan δ values indicate higher gel strength. The results showed slight variations in tan δ values, except for starches treated for 30 min, which showed lower tan δ values compared to the control and starches ultrasonicated for 15 min. Therefore, it may be concluded that cereal starches treated for 30 min or higher duration of ultrasonication showed higher gel strength behavior (Kaur & Gill, 2019). Amini et al. (2015) discovered that the tan (δ) value of sonicated corn starch at 65°C exceeded that of the control sample and samples subjected to sonication at lower temperatures. This observation suggests that ultrasound led to the degradation of starch components by breaking polymer chains due to shear forces. The values of n for ultrasonicated treated corn starch, potato starch and pea starch were lower than those of native CS, PtS, and PS, respectively, indicating that ultrasonicated treated corn starch, potato starch and pea starch exhibited a greater degree of shear-thinning compared to their native counterparts. Furthermore, with increasing ultrasonic frequency from 20 to 28 kHz, the n values of ultrasonicated treated corn starch, potato starch and pea starch decreased. In general, ultrasonic frequency enhanced the shear-thinning behavior, making the starch gel more pseudo-plastic (Xiao et al., 2021).

6.6.3 Dynamic Rheological Properties

With the increase of ultrasonic frequency from 20 to 28 kHz, there was a corresponding increase in both the G′ (storage modulus) and G″ (loss modulus) values. Furthermore, the rise in G′ and G″ values contributed to an augmentation in the hydrogel network strength of ultrasonicated corn, potato, and pea starches, consequently creating a more organized and solidified gel. Based on the flow behavior parameters, it's apparent that corn starch samples displayed shear-thinning behavior (n < 1). The samples stronger pseudo-plasticity (Amini et al., 2015). The application of ultrasound (20% amp, 20 and 50 min) to plantain starch resulted in reduced storage (G′) and loss (G″) moduli when compared to the native counterpart. Moreover, this reduction was more noticeable with longer treatment durations (Carmona-Garcia et al., 2016). In summary, sonication notably weakened the entangled starch paste network without altering the flow behavior. However, when combined with acid hydrolysis, most of the amylose molecules degraded, leading to a less organized network structure with reduced resistance to deformation. This phenomenon suggests that ultrasonic frequency could potentially enhance the retrogradation and gelation capabilities of the starch gel (Sudheesh et al., 2019). Dynamic rheological assessments revealed that ultrasound treated starches displayed characteristics resembling those of solid substances, suggesting their potential use to increase the solid components in food without necessitating excessive thickening (Xiao et al., 2021). Based on the study of Kaur and Gill (2019), it was observed that barley starch exhibited the highest G′ (4930 Pa) and G″ (711 Pa) values at 90°C, followed by maize, rice, and wheat starches. The dominance of solid/elastic behavior was indicated by higher G′ values compared to G″ (Li et al., 2018). Ultrasound treatment facilitated starch granule degradation through cavitational forces, enhancing granule permeability to water during the heating phase (Kaur & Gill, 2019). The impact of ultrasonication on plantain starch (large granule size) was more pronounced than on taro starch (small granule size) due to the imperfect arrangement of starch components caused by the sonic energy trapped by small granules being insufficient to induce compact packing. This suggests that large granule-size starch treated with ultrasound is more affected, resulting in a more compact network due to interactions among linear starch chains produced by the treatment. Changes in the storage modulus of taro starch were observed for longer sonication times up to 50 min, indicating that the reduction of G′ storage modulus is due to energy accumulation within the granule microstructure (Carmona-Garcia et al., 2016). The starch gels' network structures were strong, as indicated by the low tan δ values observed in Figure 6.7. The values for ultrasonicated corn, potato, and pea starches were lower than 0.1 (Figure 6.7), which suggested that ultrasonic frequency affected the elastic properties more than the viscous properties (Wen et al., 2020). The decreasing tan δ values demonstrated that the degree of molecular cross-linking polymerization in the gel system had increased, making it more rigid in a three-dimensional network. Previous studies found that a small tan δ indicated that deformation was recoverable and the starch gel was rigid. Therefore, ultrasonic frequency showed potential for controlling the rheology of corn, potato, and pea starches (Xiao et al., 2021). Figure 6.7 illustrates various rheological curves of native and ultrasonicated starches reported by various researchers, including: (1) wheat starch (WSS-15), rice starch (RSS-15), maize starch (MSS-15), and barley starch (BSS-15), as well as wheat starch (WSS-30), rice starch (RSS-30), maize starch (MSS-30), and barley starch (BSS-30) (Kaur & Gill, 2019); (2) frequency sweep profiles of starch pastes, encompassing native plantain starch (NPS), plantain starch sonicated for 20 min (PSS-20), plantain starch sonicated for 50 min (PSS-50), native taro starch (NTS), taro starch sonicated for 20 min (TSS-20), and taro starch sonicated for 50 min (TSS-50) (Carmona-Garcia et al., 2016); and (3) storage modulus (G′), loss modulus (G″), and tan δ of native starch, ultrasonicated corn starch (UFCS), ultrasonicated potato starch (UFPtS), and ultrasonicated pea starch (UFPS); A–C represent G′, A1–C1 represent G″, and A2–C2 represent tan δ (Xiao et al., 2021).

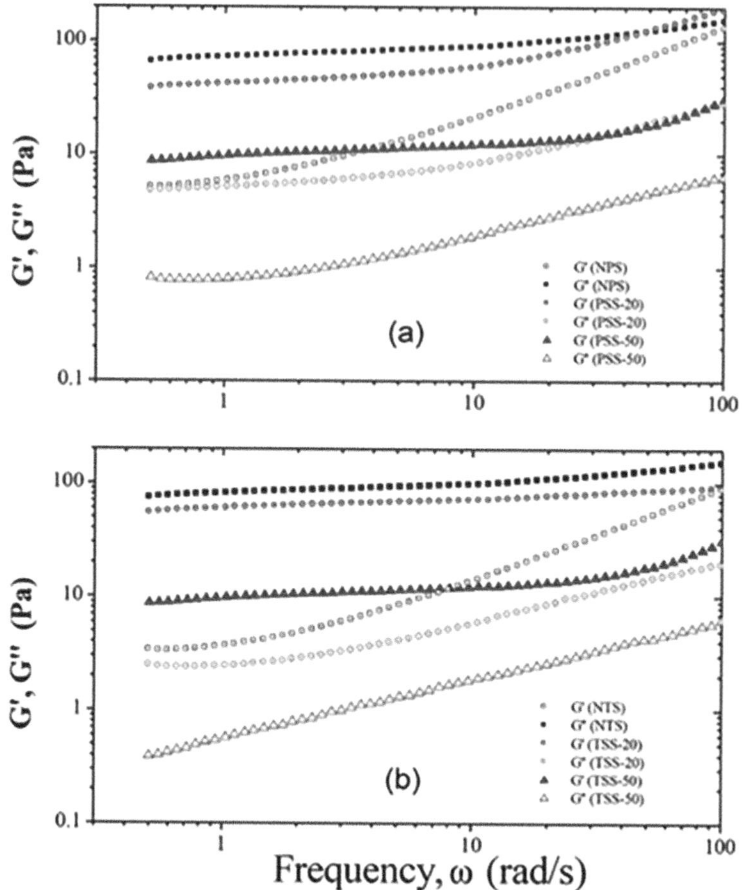

FIGURE 6.7 Rheological curves of some native and ultrasonicated starches reported by various researchers (Carmona-García et al., 2016).

6.7 EFFECT OF ULTRASONICATION ON PHYSICO-CHEMICAL AND FUNCTIONAL PROPERTIES

6.7.1 Amylose Content

The effect of ultrasonication on starches' amylose content relies on numerous factors, including ultrasonication parameters, specific starch types, and processing conditions. Ultrasonication exhibits the potential to disrupt starch granules by inducing cavitation, shear forces, and microstreaming (Chan et al., 2021). These mechanical forces could impact starch granule integrity, potentially altering amylose content (Acevedo et al., 2022). A higher amylose content is associated with a detrimental impact on glycemic index (GI). Following ultrasound (US) treatment for an extended duration and at increased power levels, the GI of rice with lower amylose content rose from 68.52 to 76.72. This suggests that ultrasound facilitated the breakdown of the rice grain surface and disruption of starch crystalline structure, allowing greater accessibility for enzymatic activity and leading to an elevation in glycemic index (Kunyanee & Luangsakul, 2020). Ultrasound extraction may lead to the partial depolymerization of both amylose and amylopectin, resulting in an augmentation of linear chains and an increase in *Radix puerariae* starch amylose content (Li et al., 2019). A distinct observation was reported by Li et al. (2023), indicating that the amylose content of rice starch remained unaltered even under high-intensity ultrasound extraction. This divergence could be attributed to variations in the molecular structure of starch in different materials, as well as differences in process parameters such as the additional quantity of water and the concentration of the extract solution. The amylose content of both native and ultrasound-treated sweet potato starch ranged from 25.1 to 28.75%. Ultrasonicated starches exhibited higher amylose content compared to native starch, with a more pronounced effect observed as sonication time increased from 15 to 30 minutes. Ultrasonic treatment enhances amylose content by inducing molecular scission of chains and depolymerization of amylopectin. This process contributes to the strengthening of linear fractions. The results suggest that ultrasound may promote increased swelling capacity and solubility of starch granules (Wang et al., 2020).

6.7.2 Particle Size

Ultrasonication significantly influences the particle size of starch by promoting mechanical and structural changes. The application of ultrasound waves induces cavitation, creating microbubbles that collapse near starch particles (Monroy et al., 2018). This phenomenon leads to intense shear forces and shockwaves, resulting in the fragmentation and reduction of starch particle size (Manzoor et al., 2019). The disruptive action of ultrasonication on the granular structure of starch enhances the accessibility of water to the internal regions of starch particles, facilitating their breakdown into smaller fragments. The degree of particle size reduction is dependent on ultrasonication parameters such as frequency, intensity, and duration. Generally, higher ultrasonication intensity and longer exposure times tend to produce smaller starch particles (Mallakpour, 2018). The controlled manipulation of particle size through ultrasonication has implications for various industrial applications, including food processing and pharmaceutical formulations, where the modification of starch characteristics is crucial for achieving desired product properties (Jin, 2018). The effect of ultrasonication on particle size stands as a well-established phenomenon spanning various fields, encompassing chemistry, materials science, and biotechnology. Ultrasonication entails the application of high-frequency sound waves (ultrasound) to a medium, instigating cavitation and mechanical forces capable of disintegrating larger particles into smaller ones. This process generates cavitation bubbles within a liquid medium, rapidly forming and collapsing, thereby exerting intense local forces that fragment particles. Consequently, ultrasonication improves particle dispersibility in a liquid, preventing agglomeration and ensuring a more even distribution of particle sizes (Xiao et al., 2021). The intensity and frequency of ultrasound treatment impact the particle size distribution due to the abrasive nature of ultrasonic energy (Carmona-Gracia et al., 2016). Karwasra et al. (2020) researched how ultrasonication affected the functional and structural properties of starches from Indian wheat (*Triticum aestivum* L.) cultivars. A-type granules dominated the total volume of starches (67.70 to 79.89%), followed by B-type (17.20 to 26.54%) and C-type granules (2.76 to 5.97%). Most A-type granules increased with ultrasonication treatments (US-15 and US-30), except for a couple of cultivars. B-type granules generally decreased with US-30, except for a few cultivars, and C-type granules reduced in volume after both ultrasonication treatments. The impact of ultrasound varies based on the starch's type and source (Bitik et al., 2019). Larger starch granules tend to capture ultrasound energy more easily due to their increased surface area (Carmona-García et al., 2016). Additionally, the composition of amylose and amylopectin plays a crucial role in the size reduction caused by ultrasonication (Jin, 2018). Microfluidization, as indicated in Liu et al.'s study (2009), also reduces particle size effectively in starches. Both ultrasound and microfluidization treatments significantly decreased the span value of native chickpea starch (p < 0.05). However, only ultrasound treatment showed effectiveness in reducing the span value of native lentil starch. The impact of treatment duration differed notably between maize and quinoa starches (Wei et al., 2023). Initially, maize starch showed a 2.9 μm increase in particle size within the first 4 h of treatment, followed by a consistent decrease. Xiao et al. (2021) also noted a swelling-induced particle size increase due to ultrasonication, which reversed with longer treatment times (14 and 22 h) as starch damage became dominant.

Morphological changes (SEM results) can also be aligned with the findings of particle size distribution of starches. Quinoa starch's smaller granule size appeared to offer some resistance to mechanical shearing from cavitation bubble collapse, resulting in less influence from ultrasonication on morphology and particle size compared to larger granule starches (Wei et al., 2023). The reduction in particle size via ultrasonication leads to an increased overall surface area, a beneficial aspect in applications such as catalysis and drug delivery that require increased surface area. Moreover, ultrasonication can expedite specific chemical reactions by fostering the collision and interaction of reactant particles, owing to enhanced particle mobility in the liquid medium. After ultrasonic treatment (40 + 80 kHz and 60°C), the average size of potato starch was reduced by 30.1%, but millet only decreased by 7.93%, which indicated that ultrasound had a great influence on potato starch than millet starch (Hu et al., 2019). In polymerization processes, ultrasonication plays a role in controlling the size of polymer particles. The cavitation and shear forces can influence nucleation and growth of polymer chains, thereby impacting the ultimate particle size. Furthermore, ultrasonication is applied in sono-crystallization processes, where it induces nucleation and crystal growth, resulting in smaller and more uniform crystal sizes (Thirunavookarasu et al., 2023).

6.7.3 Solubility

Solubility rises as the amplitude and duration of the process increased. This boost in solubility is linked to starch granule depolymerization and structural weakening caused by the oxidation process (Schmiele et al., 2019). Ultrasonicated starch tends to exhibit an amorphous nature (Satmalawati et al., 2020). The notable rise in solubility is a result of the breakdown of starch granules and molecules due to the application of ultrasonic treatment. The ultrasonic treatment leads to a considerable increase in mechanical damage to starch granules, particularly in the amorphous area. This heightened hydrophilicity of starch post-treatment enables greater water infiltration into the granules, thereby elevating starch solubility (Luo & Lu, 2010). Following ultrasonic treatment, starch granules, particularly in the amorphous region, experience heightened mechanical damage. Moreover, the considerable rise in starch hydrophilicity post-treatment enables enhanced water penetration into the starch granules, thereby increasing the starch's solubility.

6.7.4 Swelling Power

Swelling power and solubility serve as indicators of how starch chains interact within crystalline and non-crystalline areas. The extent of this interaction is influenced by factors like the amylose-amylopectin ratio; phosphorus content; and characteristics of amylose and amylopectin such as their molecular weight, branching, length, and branch structure (Sujka, 2017). Ultrasound likely caused disruption to the granules and fractured the crystalline molecular structure of starch. Consequently, water molecules bonded with the available hydroxyl groups in amylose and amylopectin via hydrogen bonds, leading to enhanced water absorption and retention in both types of starches. This effect has been verified in wheat, rice, waxy, and regular maize starches (Sujka & Jamroz, 2013).

6.7.5 Water and Oil Holding Capacity

Sonicated starch demonstrated greater water and oil absorption capacity compared to the original/native starch. The enhanced water and oil absorption capacity of sonicated starch could be attributed to a higher void space created within the starch due to intense sonication, encouraging smaller granules to attract water (Xiao et al., 2021). Additionally, sonication causes the breakdown of starch granules into smaller particles, leading to a larger surface area capable of absorbing water and oil (Rahaman et al., 2021). As sonication time increases, the oil absorption capacity rises, potentially because pores form on the starch granule surface, enhancing the capillary action that draws in oil molecules (Sujka et al., 2017). When starch granules experience heightened water absorption, it leads to a less robust gel network and reduced viscosity. The decreased swelling power and peak viscosity observed in potato starch indicate that the damage caused by ultrasound was more severe in potato starch compared to millet starches under similar conditions (Hu et al., 2019).

6.7.6 Gel Strength

Ultrasonication has a notable effect on the gel strength of starch by influencing the gelatinization process and modifying the molecular structure of starch molecules. When ultrasound waves are applied to starch solutions, they can induce localized heating and intense mechanical forces, promoting the disruption of starch granules and the subsequent gelatinization of amylose and amylopectin (Dhull et al., 2021). This gelatinization process contributes to the formation of a starch gel, and the application of ultrasonication can lead to changes in the gel structure. Depending on the ultrasonication parameters, such as intensity and duration, the gel strength may be influenced by factors like starch granule size, degree of gelatinization, and the resulting molecular interactions within the gel network (Xiao et al., 2021). Generally, ultrasonication can contribute to the enhancement of gel strength by promoting a more homogeneous gel structure and influencing the molecular arrangement of starch polymers (Acevedo et al., 2022). This is particularly relevant in food and industrial applications where starch gels are utilized for their thickening and gelling properties, and the controlled modulation of gel strength is essential for achieving desired product characteristics (Dhull et al., 2021).

6.7.7 Freeze–Thaw Stability

Ultrasonic treatment positively impacts the freeze–thaw stability of both cereal and tuber starches, although specific effects may vary between these two types of starches. When applied to cereal starches (e.g., corn, wheat), ultrasonication promotes more uniform dispersion, reducing the likelihood of ice crystal formation and improving stability during freeze–thaw cycles. The disruptive action of ultrasound on starch granules may affect retrogradation kinetics, minimizing undesirable texture changes during freezing and thawing (Dhull et al., 2021). Similarly, for tuber starches (e.g., potato, tapioca), ultrasonication enhances particle dispersion and may contribute to a more stable starch matrix, reducing the potential for syneresis or water separation during freeze–thaw cycles. The technique's ability to modify starch properties, including particle size and structural integrity, makes ultrasonication a promising tool for optimizing freeze–thaw stability in a variety of starch-based products, spanning both cereal and tuber starch sources, particularly in the food industry where maintaining product quality after freezing is critical (Acevedo et al., 2022).

6.8 EFFECT ON DUAL AND SINGLE MODIFICATION OF ULTRASONICATION ON STARCHES

The cavitation effect strongly depends on operation parameters, especially frequency. Some reports found that increasing ultrasonic frequency can get better performance (Hu et al., 2013). Long exposure to high temperature ultrasonication (U50T30) resulted in increased presence of small fragments and creases on the surface of the starch granules (Acevedo et al., 2022). Conversely, low-temperature ultrasonication notably exhibited more pores and profound cracks compared to the effects observed with high-temperature ultrasonication in case of maize starch (Chan et al., 2021). This suggests that the primary mechanism behind the damage to the surface of the starch granule caused by low-temperature ultrasonication is attributed to the shear force generated by cavitation and the assault of free radicals, as indicated by Kardos and Luche in 2001. Bai et al. (2017) devised a light microscopy method to measure the impact of high-power low-frequency ultrasound (0–160 W, 20 kHz) on potato starch. Their findings revealed a direct correlation between the rise in defects and the intensity of sonication. In contrast, Du et al. (2022) suggested that sonication at 360 kHz effectively broke down elongated chitosan structures. Furthermore, researchers found that combining two low-frequency ultrasound waves could amplify the cavitation effect without increasing ultrasonic intensity or duration. Zhang et al. (2021b) observed a higher extraction rate when employing dual-frequency ultrasound at 20 and 40 kHz, compared to treatments using only 20 or 40 kHz ultrasound waves. Zare et al. (2023) reported a substantial increase in luminescence intensity when using dual-frequency ultrasound compared to single-frequency ultrasound. Kabawa et al. (2023) demonstrated that appropriate ultrasound treatment could alter the enzyme configuration without changing its conformation. Sarkodie et al. (2023) concluded that combining ultrasound with α-amylase during the desizing process improved efficiency by approximately 10% compared to using only α-amylase. Hu et al. (2013) suggested that ultrasound not only increased the interactions between dye and enzyme but also enhanced enzyme activity. Finally, Zhu (2015) found that combining neutral protease with sonication was the most effective method for isolating starch from damaged corn flour, resulting in low residual protein and damaged starch. Reports suggested that the use of dual-frequency ultrasound resulted in more prominent indentations and cavities on starch surfaces, leading to a more significant decrease in both crystalline index and peak viscosity compared to single-frequency ultrasound treatments (Zheng et al., 2013). Relative to starch treated with single-frequency ultrasound, the surface of starch subjected to dual-frequency ultrasound exhibits more prominent dents and holes. With longer durations of ultrasonic treatment, there's a noticeable rise in starch solubility. Specifically, when employing dual-frequency ultrasound, starch treated for 60 minutes demonstrates the highest solubility, showcasing a 2.69% increase compared to the original/native starch. Nevertheless, prolonged exposure to ultrasonic treatment can result in a reduction in transmittance (Hj et al., 2021). The structural alterations and characteristics of starch treated with dual-frequency ultrasonic treatment show more pronounced differences compared to those treated with single-frequency ultrasound. However, there remains insufficient exploration regarding the impact of ultrasound treatments on starch (Luo & Lu, 2010). Optimizing parameters such as ultrasound frequency, power, temperature, processing duration, and other variables is crucial to counter or minimize the adverse effects stemming from the generation of free radicals during cavitation bubble implosion (Zheng et al., 2013). Research showed that ultrasonic treatment significantly harms the crystal structure, leading to an increase in the amorphous structure and a reduction in certain absorption peak intensities. Starch suspensions treated with dual-frequency ultrasound exhibit more evident damage to their crystal structures compared to those treated with single-frequency ultrasounds (Bai et al., 2017). The dual modification technique, referred to as synergistic treatment, involves altering natural starch using two distinct methods. Utilizing both ultrasonic and high-pressure treatments significantly influenced the structure and physicochemical attributes of pea starch, potentially enhancing its characteristics. Table 6.5, focusing on the comparison of the effects of ultrasound treatment on starches with other modification methods, provides a comprehensive overview of how ultrasound treatment stacks up against various conventional or alternative methods used to modify starch properties.

6.9 ULTRASONICATION PROCESS PARAMETERS

During the process of ultrasonic modification, factors such as ultrasonic parameters and starch properties wield significant influence on the modification outcome. These factors encompass the energy input, treatment duration, gas atmosphere, system temperature, and more (Hu et al., 2019). Modifying starch by ultrasound treatment: The impact of ultrasounds on starch granules is influenced by several factors, including sonication power and frequency, treatment temperature, and duration, as well as the properties of starch dispersions, such as concentration and the botanical source of the starch (Hu et al., 2019). Another effect is cavitation. The swift collapse of bubbles could also generate shear forces capable of fracturing polymer chains, such as those found in starch (Hj. Latip et al., 2021). Additionally, solvent molecules might

TABLE 6.5   Comparison of the Effects of Ultrasound Treatment on Starches with Other Modification Methods

STARCH TYPE	MODIFICATION COMBINATION	RESULTS	REFERENCE
Potato starch	US+ electric field	Dual modification showed a low level of conclusion temperature and resistant starch. The findings indicated that both single and dual modifications disrupted both the amorphous and crystalline regions of the starch. Ultrasonic treatment emerged as a promising physical method for starch applications.	Cao and Gao (2020)
Rice starch	Ultrasound-assisted cellulase enzymatic treatment	Ultrasound-assisted enzymatic treatment improved the comprehensive quality of brown rice noodles, making it more suitable for the market and human health requirements.	Geng et al. (2021)
Millet starch	Ultrasound+ annealing	The altered starches exhibited elevated amylose levels compared to the native starch, with the dual-modified starches displaying the highest amylose content. The alteration led to reduced water absorption capacity (WAC) and increased relative crystallinity, syneresis, and hardness in the starch gels. Additionally, the gelatinization temperature of the altered starches was higher compared to the native ones. SEM analysis of the altered starches revealed surface pore development, providing evidence of structural changes during the modification process.	Amarnath et al. (2023)
Lentil starch	Sonication and γ-irradiation	Swelling and syneresis decreased in sonicated and irradiated lentil starch along with pasting properties.	Majeed et al. (2017)
Pea starch	Ultrasound and heat-moisture	The starch experienced a reduction in its crystalline nature, molecular weight, and capacity to swell.	Han et al. (2021b)
Corn starch	US+MW and MW+US	US modification duration enhances the modification efficiency. The most significant alteration in starch granules was noted with 360 W microwave power and 30 to 40 min UT.	Yılmaz and Tugrul (2023)

separate to create radicals that can prompt polymer breakdown. The efficacy of sonication is influenced by the starch content within the slurry. An increase in starch content led to a reduced level of damage to the starch granules (Bitik et al., 2019). The cavitation phenomenon is the primary mechanism of ultrasonic structural alteration of dietary proteins. Under cavitational conditions, the rapid formation and collapse of gas bubbles caused by localized pressure differences over a few microseconds results in the release of significant hydrodynamic shear forces and elevated heat fluxes at the bubble collapse zone (O'sullivan et al., 2017). In the extraction, processing, and alteration of LPs, the employment of probe and bath sonication has some advantages and disadvantages when compared to other thermal and non-thermal technologies.

6.9.1 Frequency and Intensity

Understanding how changes in frequency and intensity influence cavitation and mechanical effects is crucial for harnessing ultrasound's potential for starch modification in food processing, pharmaceuticals, and other industries. Frequency alterations influence the degree of starch modification. Specific frequencies might enhance or reduce the effectiveness of processes like particle size reduction, enzymatic modification, or structural changes in starch granules. Intensity variations can significantly impact the degree and extent of starch modification (Estivi et al., 2022). Higher intensities tend to induce more significant changes in starch properties, including gelatinization behavior, granule size, or functional characteristics (Vela et al., 2021).

6.9.2 Amplitude

Amplitude plays a pivotal role in determining the intensity of mechanical effects in ultrasound processing of starches. Optimizing this parameter involves striking a balance between desired treatment outcomes and potential risks associated with higher amplitudes, ensuring the efficient and controlled application of ultrasound technology (Wei et al., 2023). Higher amplitudes generate stronger acoustic waves, intensifying cavitation and thereby enhancing mechanical effects like disruption, fragmentation, or mixing within the treated medium. This increased amplitude also yields more potent acoustic forces, impacting particle size reduction, emulsification, or extraction efficiency across diverse applications. Inhomogeneous amplitudes may cause irregular treatment or insufficient penetration of ultrasound waves, thereby affecting overall effectiveness. Understanding the maximum achievable amplitude by ultrasound equipment is crucial; this limit governs treatment intensity (Hu et al., 2019). This comprehension of equipment capabilities is vital for optimizing the amplitude parameter while respecting system limitations and preventing damage to equipment or materials.

6.9.3 Temperature, Pressure, and Viscosity

Temperature, pressure, and viscosity stand as critical parameters in ultrasound processing, exerting significant influence across various applications. Temperature intricately impacts the pace of chemical reactions, solubility, and material viscosity (Astráin-Redín et al., 2019). Within ultrasound-assisted processes, temperature acts upon cavitation, either augmenting or diminishing bubble formation. Precise temperature control holds paramount importance in optimizing ultrasound effects; elevated temperatures possess the potential to amplify cavitation, yet excessively high temperatures can result in unwelcome thermal effects or material degradation (Chavan et al., 2022). Pressure serves as a determinant in shaping the cavitation environment, directly impacting the size and intensity of cavitation bubbles, thereby influencing the mechanical outcomes during ultrasound treatment. Viscosity significantly contributes to the dynamics of cavitation bubbles, with higher-viscosity materials potentially impeding bubble collapse or altering the distribution of ultrasound energy within the medium. The fine-tuning of these parameters requires a delicate equilibrium. Tailoring temperature, pressure, and viscosity parameters to align with specific applications stands pivotal in maximizing the efficacy of ultrasound processes, all while mitigating potential adverse effects on the materials undergoing treatment (Silva, 2020).

6.10 APPLICATIONS

6.10.1 Food Industry

6.10.1.1 Thickening and Gelling Agent

Ultrasound-treated starch can be used as a thickening and gelling agent in food products such as sauces, gravies, and soups. The modified starch properties enhance its functionality in terms of texture and stability (Xiao et al., 2021). Corn starch pastes exhibited shear-thinning behavior, evident from G′ surpassing G″, a typical trait of pseudoplastic fluids. This signifies the starches' highly elastic nature rather than a viscous one. Ultrasonicated corn starches present an opportunity to attain elevated apparent viscosity, a valuable attribute in the food industry (Kaur & Gill, 2019).

6.10.1.2 Bakery Products

Ultrasound-treated starch can improve the texture and shelf life of bakery products, such as bread and cakes, by influencing starch gelatinization and retrogradation (Bitik et al., 2019). Ultrasonic treatment affected wheat starch-monoglyceride complex (WSG) structure. Shi et al. (2021) reported that high-intensity (375W) ultrasonic-treated wheat starch yielded maximum crystallinity, RS, and CI. Steamed bread with 5–10% WSG had good quality without harming its properties.

6.10.1.3 Confectionery

In confectionery products like gummies and candies, ultrasound-treated starch can be utilized to achieve desired textures and prevent crystallization (Xiao et al., 2021). In a previous study by Aslam et al. (2022), ultrasound was recognized as a technology that accelerates heat and mass transfer, ensuring superior quality in the final product while using significantly less time and energy compared to traditional methods.

6.10.2 Edible Film and Coating

Starch-based films and coatings, modified through ultrasound treatment, find applications in the food industry for packaging, preserving freshness, and improving the shelf life of fruits and vegetables (Han et al., 2021a). Sonication techniques notably affected the amorphous portion of the starch, resulting in the breakdown of amylose into smaller molecules and an increased formation of surface cracks in the starch. Such modified starch varieties can be employed to create clearer, smoother gels, and for developing films (Kumari & Sit, 2023).

6.10.3 Pharmaceuticals

6.10.3.1 Drug Delivery Systems

Ultrasound-treated starch can be employed in pharmaceuticals to develop drug delivery systems. Starch nanoparticles produced with ultrasound can enhance the solubility and bioavailability of drugs (Xiao et al., 2021). Wang et al. (2020) effectively produced highly lipophilic modified potato starch using ultrasound-freeze thaw cycle treatment. This enhanced the starch's oil adsorption capacity and heat resistance, making it more resistant to digestion. This suggests its potential use as resistant starch in food processing. This starch could serve as an encapsulating agent for various substances, including bioactive compounds and drugs like doxorubicin hydrochloride, potentially creating a novel and straightforward colon-targeted drug delivery system.

6.10.3.2 Excipient in Tablet Formulation

Ultrasound-treated starch can be used as an excipient in tablet formulations, contributing to the disintegration and release properties of the tablets (Han et al., 2021a). In a study, de Lima et al. (2023) formulated innovative use of oral films based on pre-gelatinized starch for delivering cinnamon essential oil, a source rich in total phenolic compounds. Ultrasound was assessed for its efficacy in integrating cinnamon essential oil (CEO) into the film solution. All CEO-added formulations led to a breakdown of the matrix's cell wall, resulting in enhanced release of emulsion droplets.

6.10.4 Biodegradable Materials

6.10.4.1 Biodegradable Plastics

Ultrasound-treated starch nanoparticles can be incorporated into biodegradable plastics, offering a sustainable alternative to traditional plastics (Dutta & Sit, 2023). Kumari and Sit (2023) highlighted that ultrasound-treated starch allows customization of starch, influencing the yield of starch nanoparticles and its film-forming characteristics. Also, resulting starch-and protein-based bio-nanocomposites, after an hour of ultrasound treatment, contribute to emerging biodegradable materials. These biodegradable plastics dissolve in warm water, offering a potential eco-friendly alternative for various applications (Dutta & Sit, 2023).

6.10.4.2 Biodegradable Packaging

Starch-based materials modified with ultrasound can be used in the development of environmentally friendly and biodegradable packaging solutions (Dutta & Sit, 2024). Falsafi and co-workers (2019) utilized ultrasound to modify oat starch, observing enhanced amylose content, swelling power, solubility, and water and oil holding capacity. This treatment also reduced starch retrogradation, making it advantageous for food packaging. The process of sonication is believed to relax the polymeric structure, facilitating the release of active substances, potentially accounting for the improved antimicrobial properties in the field of active packaging (Zhang et al., 2019). The films with 40% ultrasonic amplitude improved the properties of sweet potato starch-based films by reporting highest tensile strength values and lowest water vapor permeability (Liu et al., 2018). Further, the optimal characteristics of sago starch film, including the highest thermal resistance, lowest water vapor permeability, greatest transparency, and minimal moisture absorption, were achieved by subjecting it to ultrasonication for a duration of 10 min at 600 W (Abral et al., 2019). In the process of preparing the film, ultrasonication was utilized to enhance the compatibility of the film-forming components, leading to improvements in tensile strength, light transmittance, and barrier properties of the resulting film. Furthermore, ultrasonication contributed to enhanced antibacterial activity in the film. The composite film, incorporating water hyacinth nanocellulose-filled bengkuang starch and ginger nano-fibers through ultrasonic treatment, demonstrates promising potential for applications in the food packaging industry (Mi et al., 2021).

6.10.5 Textile Industry

6.10.5.1 Sizing Agent in Textiles

Ultrasound-treated starch can be applied as a sizing agent in the textile industry to improve the strength and adhesion of fibers during the weaving process. Tabasum et al. (2019) conducted a review focusing on blending corn starch with various materials: natural and synthetic polymers alongside inorganic nanoparticles. The resulting starch film was employed as a glass fiber sizing agent, enhancing its adhesion to the fibers and reducing stress on weaving machines. Ultrasound-assisted modification increased adhesion and hardness of the gel. Additionally, Wang et al. (2020) produced emulsions utilizing ultrasound-treated NaCas and WPI, which improved the thermal stability of sisal fiber.

6.10.5.2 Finishing Agent

Modified starch can be used as a finishing agent to impart desirable properties such as softness and wrinkle resistance to textiles. Rahaman et al. (2021) explored the application of various radiation methods, including gamma, ultrasound, plasma, and ultraviolet, for modifying different textile materials in the realm of textile processing and finishing. These methods were utilized to impart specific properties to textiles, such as achieving wrinkle resistance or a durable press finish.

6.10.6 Paper and Board Industry

Ultrasound-treated starch is employed as a key component in the formulation of adhesives and binders for the paper and board industry (Zhu et al., 2017). Starch products exhibit variations in structure and composition. Sonicated modified starch, acting as a cobinder alongside synthetic binders, influences mechanical, thermal, adhesive, and barrier properties (Tabasum et al., 2019). A comparison was made between native potato starch and ultrasound-treated potato starch, evaluating their respective characteristics (Zhang et al., 2019). These applications demonstrate the versatility of ultrasound-treated starch in various industries, offering improved properties and performance compared to untreated starch. The specific benefits depend on the processing parameters used during ultrasound treatment and the intended application.

6.11 CONCLUSION

In conclusion, ultrasonic treatment offers a versatile and promising approach for modifying starch properties and enhancing various processes. Variations in molecular structure and physical properties induced by ultrasound differed among different starch types. The process results in structural disorganization and significant changes in physicochemical properties. Ongoing research explores new applications and optimizes existing processes for industrial use. Pea starch underwent notable alterations under ultrasonic treatment, where the particles' surfaces became sunken due to ultrasound-induced hydrolysis. This allowed increased water absorption, expansion, and aggregation, enlarging particle size. Crystallinity reduced slightly with increased treatment time, impacting viscosity parameters but maintaining consistent pasting temperatures. Moreover, optimized ultrasonic treatment strengthened starch gel. While studies have examined advantages and drawbacks, fine-tuning specific ultrasonic parameters like frequency, power, temperature, and time per food item is crucial to assess their degradation impact confidently. Ultrasound emerges as an eco-friendly method, facilitating structural and physical changes in starch for potential food applications. Primarily affecting starch granule surfaces and inducing molecular breakdown, ultrasound significantly alters starch properties. However, conflicting findings often stem from variations in processing parameters and starch origin. These studies underscore ultrasound's effectiveness in altering starch and flour samples, albeit with outcomes being dependent on equipment, processing conditions, and starch source, suggesting inconclusive findings across literature.

REFERENCES

Abedi, E., Pourmohammadi, K., Jahromi, M., Niakousari, M., & Torri, L. (2019). The effect of ultrasonic probe size for effective ultrasound-assisted pregelatinized starch. *Food and Bioprocess Technology, 12,* 1852–1862. https://link.springer.com/article/10.1007/s11947-019-02347-2

Abral, H., Basri, A., Muhammad, F., Fernando, Y., Hafizulhaq, F., Mahardika, M., Sugiarti, E., Sapuan, S. M., Ilyas, R. A., & Stephane, I. (2019). A simple method for improving the properties of the sago starch films prepared by using ultrasonication treatment. *Food Hydrocolloids, 93,* 276–283. https://doi.org/10.1016/j.foodhyd.2019.02.012

Acevedo, B. A., Villanueva, M., Chaves, M. G., Avanza, M. V., & Ronda, F. (2022). Modification of structural and physicochemical properties of cowpea (Vigna unguiculata) starch by hydrothermal and ultrasound treatments. *Food Hydrocolloids, 124,* 107266. https://doi.org/10.1016/j.foodhyd.2021.107266.

Amarnath, M. S., Muhammed, A., Antony, A. K., Malini, B., & Sunil, C. K. (2023). White finger millet starch: Physical modification (annealing and ultrasound), and its impact on physicochemical, functional, thermal and structural properties. *Food and Humanity, 1,* 599–606.

Amini, A. M., Razavi, S. M. A., & Mortazavi, S. A. (2015). Morphological, physicochemical, and viscoelastic properties of sonicated corn starch. *Carbohydrate Polymers, 122,* 282–292. https://doi.org/10.1016/j.carbpol.2015.01.020.

Aslam, R., Alam, M. S., Kaur, J., Panayampadan, A. S., Dar, O. I., Kothakota, A., & Pandiselvam, R. (2022). Understanding the effects of ultrasound processing on texture and rheological properties of food. *Journal of Texture Studies*, *53*(6), 775–799. https://doi.org/10.1111/jtxs.12644

Astráin-Redín, L., Raso, J., Condón, S., Cebrián, G., & Álvarez, I. (2019). Application of high-power ultrasound in the food industry. In *Sonochemical reactions*. IntechOpen. https://doi.org/10.5772/intechopen.90444

Bai, W., Hébraud, P., Ashokkumar, M., & Hemar, Y. (2017). Investigation on the pitting of potato starch granules during high frequency ultrasound treatment. *Ultrasonics Sonochemistry*, *35*, 547–555. https://doi.org/10.1016/j.ultsonch.2016.05.022

Bitik, A., Sumnu, G., & Oztop, M. (2019). Physicochemical and structural characterization of microfluidized and sonicated legume starches. *Food and Bioprocess Technology*, *12*, 1144–1156. https://link.springer.com/article/10.1007/s11947-019-02264-4

Bonto, A. P., Tiozon Jr, R. N., Sreenivasulu, N., & Camacho, D. H. (2021). Impact of ultrasonic treatment on rice starch and grain functional properties: A review. *Ultrasonics Sonochemistry*, *71*, 105383. https://doi.org/10.1016/j.ultsonch.2020.105383

Cai, B., Mazahreh, J., Ma, Q., Wang, F., & Hu, X. (2022). Ultrasound-assisted fabrication of biopolymer materials: A review. *International Journal of Biological Macromolecules*, *209*, 1613–1628. https://doi.org/10.1016/j.ijbiomac.2022.04.055

Cao, M., & Gao, Q. (2020). Effect of dual modification with ultrasonic and electric field on potato starch. *International Journal of Biological Macromolecules*, *150*, 637–643. https://doi.org/10.1016/j.ijbiomac.2020.02.008

Carmona-García, R., Bello-Pérez, L. A., Aguirre-Cruz, A., Aparicio-Saguilán, A., Hernández-Torres, J., & Alvarez-Ramirez, J. (2016). Effect of ultrasonic treatment on the morphological, physicochemical, functional, and rheological properties of starches with different granule size. *Starch-Stärke*, *68*(9–10), 972–979. https://doi.org/10.1002/star.201600019

Chan, C. H., Wu, R. G., & Shao, Y. Y. (2021). The effects of ultrasonic treatment on physicochemical properties and in vitro digestibility of semigelatinized high amylose maize starch. *Food Hydrocolloids*, *119*, 106831. https://doi.org/10.1016/j.foodhyd.2021.106831

Chavan, P., Sharma, P., Sharma, S. R., Mittal, T. C., & Jaiswal, A. K. (2022). Application of high-intensity ultrasound to improve food processing efficiency: A review. *Foods*, *11*(1), 122. https://doi.org/10.3390/foods11010122

Cui, R., & Zhu, F. (2020). Effect of ultrasound on structural and physicochemical properties of sweet potato and wheat flours. *Ultrasonics Sonochemistry*, *66*, 105118. https://doi.org/10.1016/j.ultsonch.2020.105118

Cui, S., Li, M., Zhang, S., Liu, J., Sun, Q., & Xiong, L. (2018). Physicochemical properties of maize and sweet potato starches in the presence of cellulose nanocrystals. *Food Hydrocolloids*, *77*, 220–227. https://doi.org/10.1016/j.foodhyd.2017.09.037

de Lima, A. R., da Rosa, A. C. S., Stevanato, N., Raspe, D. T., dos Santos Garcia, V. A., & da Silva, C. (2023). Ultrasonic incorporation of cinnamon essential oil into pre-gelatinized cassava starch oral films. *Uningá Review*, *38*, eURJ4448–eURJ4448. https://doi.org/10.46311/2178-2571.38.eURJ4448

Dhull, S. B., Punia, S., Kumar, M., Singh, S., & Singh, P. (2021). Effect of different modifications (physical and chemical) on morphological, pasting, and rheological properties of black rice (Oryza sativa L. Indica) starch: A comparative study. *Starch-Stärke*, *73*(1–2), 2000098. https://doi.org/10.1002/star.202000098

Ding, Y., Xiao, Y., Ouyang, Q., Luo, F., & Lin, Q. (2021). Modulating the in vitro digestibility of chemically modified starch ingredient by a non-thermal processing technology of ultrasonic treatment. *Ultrasonics Sonochemistry*, *70*, 105350. https://doi.org/10.1016/j.ultsonch.2020.105350

Du, B., Jeepipalli, S. P., & Xu, B. (2022). Critical review on alterations in physiochemical properties and molecular structure of natural polysaccharides upon ultrasonication. *Ultrasonics Sonochemistry*, 106170. https://doi.org/10.1016/j.ultsonch.2022.106170

Dutta, D., & Sit, N. (2022). Comparison of properties of films prepared from potato starch modified by annealing and heat-moisture treatment. *Starch-Stärke*, *74*(11–12), 2200110. https://doi.org/10.1002/star.202200110

Dutta, D., & Sit, N. (2023). Comparison of properties of films prepared from casein modified by ultrasound and autoclave treatment. *Journal of Food Measurement and Characterization*, *17*(5), 5426–5439. http://dx.doi.org/10.1007/s11694-023-02037-w

Dutta, D., & Sit, N. (2024). Preparation and characterization of potato starch-based composite films reinforced by modified banana fibers and its application in packaging of grapes. *International Journal of Biological Macromolecules*, *254*, 127791. https://doi.org/10.1016/j.ijbiomac.2023.127791

Estivi, L., Brandolini, A., Condezo-Hoyos, L., & Hidalgo, A. (2022). Impact of low-frequency ultrasound technology on physical, chemical and technological properties of cereals and pseudocereals. *Ultrasonics Sonochemistry*, *86*, 106044. https://doi.org/10.1016/j.ultsonch.2022.106044

Falsafi, S. R., Maghsoudlou, Y., Rostamabadi, H., Rostamabadi, M. M., Hamedi, H., & Hosseini, S. M. H. (2019). Preparation of physically modified oat starch with different sonication treatments. *Food Hydrocolloids*, *89*, 311–320. https://doi.org/10.1016/j.foodhyd.2018.10.046

Gallego-Juárez, J. A. (2017). Basic principles of ultrasound, in M. Villamiel, A. Montilla, J. V. García-Pérez, J. A. Cárcel, J. Benedito (Eds). *Ultrasound in Food Processing: Recent Advances*, Wiley-Blackwell, Chichester, 2017, pp. 4–26.

Geng, D. H., Lin, Z., Liu, L., Qin, W., Wang, A., Wang, F., & Tong, L. T. (2021). Effects of ultrasound-assisted cellulase enzymatic treatment on the textural properties and in vitro starch digestibility of brown rice noodles. *LWT*, *146*, 111543. https://doi.org/10.1016/j.lwt.2021.111543

Gharibzahedi, S. M. T., & Smith, B. (2020). The functional modification of legume proteins by ultrasonication: A review. *Trends in Food Science & Technology*, *98*, 107–116. https://doi.org/10.1016/j.tifs.2020.02.002

Han, K. T., Kim, H. R., Moon, T. W., & Choi, S. J. (2021a). Isothermal and temperature-cycling retrogradation of high-amylose corn starch: Impact of sonication on its structural and retrogradation properties. *Ultrasonics Sonochemistry*, *76*, 105650. https://doi.org/10.1016/j.ultsonch.2021.105650

Han, L., Cao, S., Yu, Y., Xu, X., Cao, X., & Chen, W. (2021b). Modification in physicochemical, structural and digestive properties of pea starch during heat-moisture process assisted by pre-and post-treatment of ultrasound. *Food Chemistry*, *360*, 129929. https://doi.org/10.1016/j.foodchem.2021.129929

Hj. Latip, D. N., Samsudin, H., Utra, U., & Alias, A. K. (2021). Modification methods toward the production of porous starch: A review. *Critical Reviews in Food Science and Nutrition*, *61*(17), 2841–2862. https://doi.org/10.1080/10408398.2020.1789064

Hu, A., Jiao, S., Zheng, J., Li, L., Fan, Y., Chen, L., & Zhang, Z. (2015). Ultrasonic frequency effect on corn starch and its cavitation. *LWT-Food Science and Technology*, *60*(2), 941–947. https://doi.org/10.1016/j.lwt.2014.10.048

Hu, A., Li, L., Zheng, J., Lu, J., Meng, X., Liu, Y., & Rehman, R. U. (2014). Different-frequency ultrasonic effects on properties and structure of corn starch. *Journal of the Science of Food and Agriculture*, *94*(14), 2929–2934. https://doi.org/10.1002/jsfa.6636

Hu, A., Li, Y., & Zheng, J. (2019). Dual-frequency ultrasonic effect on the structure and properties of starch with different size. *LWT*, *106*, 254–262. https://doi.org/10.1016/j.lwt.2019.02.040

Hu, A., Lu, J., Zheng, J., Sun, J., Yang, L., Zhang, X., Zhang, Y., & Lin, Q. (2013). Ultrasonically aided enzymatical effects on the properties and structure of mung bean starch. *Innovative Food Science & Emerging Technologies*, *20*, 146–151. https://doi.org/10.1016/j.ifset.2013.08.005

Jafari, M., & Koocheki, A. (2024). Impact of ultrasound treatment on the physicochemical and rheological properties of acid hydrolyzed sorghum starch. *International Journal of Biological Macromolecules*, *256*, 128521. https://doi.org/10.1016/j.ijbiomac.2023.128521

Jin, J., Lin, H., Yagoub, A. E. A., Xiong, S., Xu, L., & Udenigwe, C. C. (2020). Effects of high-power ultrasound on the enzymolysis and structures of sweet potato starch. *Journal of the Science of Food and Agriculture*, *100*(8), 3498–3506. https://doi.org/10.1002/jsfa.10390

Jin, Z. (Ed.). (2018). *Functional starch and applications in food*. Springer. http://dx.doi.org/10.1007/978-981-13-1077-5

Joshi, M., Aldred, P., McKnight, S., Panozzo, J. F., Kasapis, S., Adhikari, R., & Adhikari, B. (2013). Physicochemical and functional characteristics of lentil starch. *Carbohydrate Polymers*, *92*(2), 1484–1496. https://doi.org/10.1016/j.carbpol.2012.10.035

Kabawa, B., Sampers, I., & Raes, K. (2023). Effect of ultrasonic treatment on enzymes: Decoupling the relation between the ultrasonic driven conformational change and enzyme activity. *Ultrasonics Sonochemistry*, 106720. https://doi.org/10.1016/j.ultsonch.2023.106720

Karwasra, B. L., Kaur, M., & Gill, B. S. (2020). Impact of ultrasonication on functional and structural properties of Indian wheat (Triticum aestivum L.) cultivar starches. *International Journal of Biological Macromolecules*, *164*, 1858–1866. https://doi.org/10.1016/j.ijbiomac.2020.08.013

Kaur, H., & Gill, B. S. (2019). Effect of high-intensity ultrasound treatment on nutritional, rheological and structural properties of starches obtained from different cereals. *International Journal of Biological Macromolecules*, *126*, 367–375. https://doi.org/10.1016/j.ijbiomac.2018.12.149

Khurshida, S., Das, M. J., Deka, S. C., & Sit, N. (2021). Effect of dual modification sequence on physicochemical, pasting, rheological and digestibility properties of cassava starch modified by acetic acid and ultrasound. *International Journal of Biological Macromolecules*, *188*, 649–656. https://doi.org/10.1016/j.ijbiomac.2021.08.062

Kumari, B., & Sit, N. (2023). Comprehensive review on single and dual modification of starch: Methods, properties and applications. *International Journal of Biological Macromolecules*, 126952. https://doi.org/10.1016/j.ijbiomac.2023.126952

Kunyanee, K., & Luangsakul, N. (2020). The effects of ultrasound–assisted recrystallization followed by chilling to produce the lower glycemic index of rice with different amylose content. *Food Chemistry*, *323*, 126843. https://doi.org/10.1016/j.foodchem.2020.126843

Li, G., Ge, X., Guo, C., & Liu, B. (2023). Effect of ultrasonic treatment on structure and physicochemical properties of pea starch. *Foods*, *12*(13), 2620. https://doi.org/10.3390/foods12132620

Li, M., Li, J., & Zhu, C. (2018). Effect of ultrasound pretreatment on enzymolysis and physicochemical properties of corn starch. *International Journal of Biological Macromolecules, 111,* 848–856. https://doi.org/10.1016/j.ijbiomac.2017.12.156

Li, S., Li, Q., Zhu, F., Song, H., Wang, C., & Guan, X. (2022). Effect of vacuum combined ultrasound treatment on the fine structure and physiochemical properties of rice starch. *Food Hydrocolloids, 124,* 107198. https://doi.org/10.1016/j.foodhyd.2021.107198

Li, Y., Hu, A., Zheng, J., & Wang, X. (2019). Comparative studies on structure and physiochemical changes of millet starch under microwave and ultrasound at the same power. *International Journal of Biological Macromolecules, 141,* 76–84. https://doi.org/10.1016/j.ijbiomac.2019.08.218

Liu, P., Wang, R., Kang, X., Cui, B., & Yu, B. (2018). Effects of ultrasonic treatment on amylose-lipid complex formation and properties of sweet potato starch-based films. *Ultrasonics Sonochemistry, 44,* 215–222. https://doi.org/10.1016/j.ultsonch.2018.02.029

Liu, W., Liu, J., Xie, M., Liu, C., Liu, W., & Wan, J. (2009). Characterization and high-pressure microfluidization-induced activation of polyphenoloxidase from Chinese pear (Pyrus pyrifolia Nakai). *Journal of Agricultural and Food Chemistry, 57*(12), 5376–5380. https://doi.org/10.1021/jf9006642

Luo, Z., & Lu, J. (2010). Effect of ultrasonic treatment on the thermal properties of maize starches. *Modern Food Science and Technology, 26*(7), 666–755. http://dx.doi.org/10.1002/star.200800014

Majeed, T., Wani, I. A., & Hussain, P. R. (2017). Effect of dual modification of sonication and γ-irradiation on physicochemical and functional properties of lentil (Lens culinaris L.) starch. *International Journal of Biological Macromolecules, 101,* 358–365. https://doi.org/10.1016/j.ijbiomac.2017.03.110

Mallakpour, S. (2018). Ultrasonic-assisted fabrication of starch/MWCNT-glucose nanocomposites for drug delivery. *Ultrasonics Sonochemistry, 40,* 402–409. https://doi.org/10.1016/j.ultsonch.2017.07.033

Manzoor, M. F., Zeng, X. A., Rahaman, A., Siddeeg, A., Aadil, R. M., Ahmed, Z., Li, J., & Niu, D. (2019). Combined impact of pulsed electric field and ultrasound on bioactive compounds and FT-IR analysis of almond extract. *Journal of Food Science and Technology, 56,* 2355–2364. http://dx.doi.org/10.1007/s13197-019-03627-7

Mhaske, V. P., Jilkar, S., & Yadav, M. D. (2023). Minireview on layered transition metal oxides synthesis using coprecipitation for sodium ion batteries cathode material: Advances and perspectives. *Energy & Fuels, 37*(21), 16221–16244. http://dx.doi.org/10.1021/acs.energyfuels.3c02861

Mi, T., Zhang, X., Liu, P., Gao, W., Li, J., Xu, N., Yuan, C., & Cui, B. (2023). Ultrasonication effects on physicochemical properties of biopolymer-based films: A comprehensive review. *Critical Reviews in Food Science and Nutrition, 63*(21), 5044–5062. https://doi.org/10.1080/10408398.2021.2012420

Monroy, Y., Rivero, S., & García, M. A. (2018). Microstructural and techno-functional properties of cassava starch modified by ultrasound. *Ultrasonics Sonochemistry, 42,* 795–804. https://doi.org/10.1016/j.ultsonch.2017.12.048

O'sullivan, J. J., Park, M., Beevers, J., Greenwood, R. W., & Norton, I. T. (2017). Applications of ultrasound for the functional modification of proteins and nanoemulsion formation: A review. *Food Hydrocolloids, 71,* 299–310. https://doi.org/10.1016/j.ultsonch.2018.05.037

Ouyang, Q., Wang, X., Xiao, Y., Luo, F., Lin, Q., & Ding, Y. (2021). Structural changes of A-, B-and C-type starches of corn, potato and pea as influenced by sonication temperature and their relationships with digestibility. *Food Chemistry, 358,* 129858. https://doi.org/10.1016/j.foodchem.2021.129858

Pourmohammadi, K., & Abedi, E. (2020). The effect of pre and post-ultrasonication on the aggregation structure and physicochemical characteristics of tapioca starch containing sucrose, isomalt and maltodextrin. *International Journal of Biological Macromolecules, 163,* 485–496. https://doi.org/10.1016/j.ijbiomac.2020.06.205

Priyadarshi, R., Ghosh, T., Roy, S., & Rhim, J. W. (2024). Biopolymer-based antimicrobial nanocomposite materials for food packaging and preservation. In *Food packaging and preservation* (pp. 33–52). Academic Press. https://doi.org/10.1016/B978-0-323-90044-7.00003-3

Rahaman, A., Kumari, A., Zeng, X. A., Farooq, M. A., Siddique, R., Khalifa, I., Siddeeg, A., Ali, M., & Manzoor, M. F. (2021). Ultrasound based modification and structural-functional analysis of corn and cassava starch. *Ultrasonics Sonochemistry, 80,* 105795. https://doi.org/10.1016/j.ultsonch.2021.105795

Raza, H., Li, S., Zhou, Q., He, J., Cheng, K. W., Dai, S., & Wang, M. (2023). Effects of ultrasound-induced V-type rice starch-tannic acid interactions on starch in vitro digestion and multiscale structural properties. *International Journal of Biological Macromolecules, 246,* 125619. https://doi.org/10.1016/j.ijbiomac.2023.125619

Sarkodie, B., Feng, Q., Xu, C., & Xu, Z. (2023). Desizability and biodegradability of textile warp sizing materials and their mechanism: A review. *Journal of Polymers and the Environment,* 1–21. http://dx.doi.org/10.1007/s10924-023-02801-5

Satmalawati, E. M., Pranoto, Y., Marseno, D. W., & Marsono, Y. (2020, December). The physicochemical characteristics of cassava starch modified by ultrasonication. In *IOP conference series: Materials science and engineering* (Vol. 980, No. 1, p. 012030). IOP Publishing. http://dx.doi.org/10.1088/1757-899X/980/1/012030

Schmiele, M., Sampaio, U. M., Gomes, P. T. G., & Clerici, M. T. P. S. (2019). Physical modifications of starch. In *Starches for food application* (pp. 223–269). Academic Press. https://doi.org/10.1016/B978-0-12-809440-2.00006-X

Sharma, S., Thakur, K., Sharma, R., & Bobade, H. (2023). Molecular morphology & interactions, functional properties, rheology and in vitro digestibility of ultrasonically modified pearl millet and sorghum starches. *International Journal of Biological Macromolecules*, *253*, 127476. https://doi.org/10.1016/j.ijbiomac.2023.127476

Shi, M., Wang, F., Lan, P., Zhang, Y., Zhang, M., Yan, Y., & Liu, Y. (2021). Effect of ultrasonic intensity on structure and properties of wheat starch-monoglyceride complex and its influence on quality of norther-style Chinese steamed bread. *LWT*, *138*, 110677. https://doi.org/10.1016/j.lwt.2020.110677

Silva, F. V. (2020). Ultrasound assisted thermal inactivation of spores in foods: Pathogenic and spoilage bacteria, molds and yeasts. *Trends in Food Science & Technology*, *105*, 402–415. https://doi.org/10.1016/j.tifs.2020.09.020

Sudheesh, C., Sunooj, K. V., George, J., Kumar, S., & Sajeevkumar, V. A. (2019). Impact of γ– irradiation on the physico-chemical, rheological properties and in vitro digestibility of kithul (Caryota urens) starch; a new source of nonconventional stem starch. *Radiation Physics and Chemistry*, *162*, 54–65. https://doi.org/10.1016/j.radphyschem.2019.04.031

Sujka, M. (2017). Ultrasonic modification of starch–Impact on granules porosity. *Ultrasonics Sonochemistry*, *37*, 424–429. https://doi.org/10.1016/j.ultsonch.2017.02.001

Sujka, M., & Jamroz, J. (2013). Ultrasound-treated starch: SEM and TEM imaging, and functional behaviour. *Food Hydrocolloids*, *31*(2), 413–419. https://doi.org/10.1016/j.foodhyd.2012.11.027

Tabasum, S., Younas, M., Zaeem, M. A., Majeed, I., Majeed, M., Noreen, A., Iqbal, M. N., & Zia, K. M. (2019). A review on blending of corn starch with natural and synthetic polymers, and inorganic nanoparticles with mathematical modeling. *International Journal of Biological Macromolecules*, *122*, 969–996. https://doi.org/10.1016/j.ijbiomac.2018.10.092

Thirunavookarasu, N., Kumar, S., Shetty, P., Shanmugam, A., & Rawson, A. (2023). Impact of ultrasound treatment on the structural modifications and functionality of carbohydrates–A review. *Carbohydrate Research*, 109017. https://doi.org/10.1016/j.carres.2023.109017

Vela, A. J., Villanueva, M., Solaesa, Á. G., & Ronda, F. (2021). Impact of high-intensity ultrasound waves on structural, functional, thermal and rheological properties of rice flour and its biopolymers structural features. *Food Hydrocolloids*, *113*, 106480. https://doi.org/10.1016/j.foodhyd.2020.106480

Wang, H., Xu, K., Ma, Y., Liang, Y., Zhang, H., & Chen, L. (2020). Impact of ultrasonication on the aggregation structure and physicochemical characteristics of sweet potato starch. *Ultrasonics Sonochemistry*, *63*, 104868. https://doi.org/10.1016/j.ultsonch.2019.104868

Wang, J., Lv, X., Lan, T., Lei, Y., Suo, J., Zhao, Q., Lei, J., Sun, X., & Ma, T. (2022). Modification in structural, physicochemical, functional, and in vitro digestive properties of kiwi starch by high-power ultrasound treatment. *Ultrasonics Sonochemistry*, *86*, 106004. https://doi.org/10.1016/j.ultsonch.2022.106004

Wei, Y., Li, G., & Zhu, F. (2023). Impact of long-term ultrasound treatment on structural and physicochemical properties of starches differing in granule size. *Carbohydrate Polymers*, *320*, 121195. https://doi.org/10.1016/j.carbpol.2023.121195

Wen, Y., Yao, T., Xu, Y., Corke, H., & Sui, Z. (2020). Pasting, thermal and rheological properties of octenylsuccinylate modified starches from diverse small granule starches differing in amylose content. *Journal of Cereal Science*, *95*, 103030. https://doi.org/10.1016/j.jcs.2020.103030

Wu, Z., Qiao, D., Zhao, S., Lin, Q., Zhang, B., & Xie, F. (2022). Nonthermal physical modification of starch: An overview of recent research into structure and property alterations. *International Journal of Biological Macromolecules*, *203*, 153–175. https://doi.org/10.1016/j.ijbiomac.2022.01.103

Xiao, Y., Wu, X., Zhang, B., Luo, F., Lin, Q., & Ding, Y. (2021). Understanding the aggregation structure, digestive and rheological properties of corn, potato, and pea starches modified by ultrasonic frequency. *International Journal of Biological Macromolecules*, *189*, 1008–1019. https://doi.org/10.1016/j.ijbiomac.2021.08.163

Xu, B., Ren, A., Chen, J., Li, H., Wei, B., Wang, J., Asam, S.M.R., Bhandari, B., Zhou, C., & Ma, H. (2021). Effect of multi-mode dual-frequency ultrasound irradiation on the degradation of waxy corn starch in a gelatinized state. *Food Hydrocolloids*, *113*, 106440.

Yadav, S., Mishra, S., & Pradhan, R. C. (2021). Ultrasound-assisted hydration of finger millet (Eleusine Coracana) and its effects on starch isolates and antinutrients. *Ultrasonics Sonochemistry*, *73*, 105542. https://doi.org/10.1016/j.ultsonch.2021.105542

Yang, Y., Li, W., Shi, W., Zhang, W., & A. El-Emam, M. (2019). Numerical investigation of a high-pressure submerged jet using a cavitation model considering effects of shear stress. *Processes*, *7*(8), 541. https://doi.org/10.3390/pr7080541

Yılmaz, A., & Tugrul, N. (2023). Effect of ultrasound-microwave and microwave-ultrasound treatment on physicochemical properties of corn starch. *Ultrasonics Sonochemistry*, *98*, 106516. https://doi.org/10.1016/j.ultsonch.2023.106516

Yu, S., Zhang, Y., Ge, Y., Zhang, Y., Sun, T., Jiao, Y., & Zheng, X. Q. (2013). Effects of ultrasound processing on the thermal and retrogradation properties of nonwaxy rice starch. *Journal of Food Process Engineering, 36*(6), 793–802. https://doi.org/10.1111/jfpe.12048

Zare, M., Bussemaker, M. J., Serna-Galvis, E. A., Torres-Palma, R. A., & Lee, J. (2023). Impact of sonication power on the degradation of paracetamol under single-and dual-frequency ultrasound. *Ultrasonics Sonochemistry, 99*, 106564. https://doi.org/10.1016/j.ultsonch.2023.106564

Zarski, A., Bajer, K., & Kapuśniak, J. (2021). Review of the most important methods of improving the processing properties of starch toward non-food applications. *Polymers, 13*(5), 832. http://dx.doi.org/10.3390/polym13050832

Zhang, R., Huang, Y., Sun, C., Xiaozhen, L., Bentian, X., & Wang, Z. (2019). Study on ultrasonic techniques for enhancing the separation process of membrane. *Ultrasonics Sonochemistry, 55*, 341–347. https://doi.org/10.1016/j.ultsonch.2018.12.041

Zhang, X., Mi, T., Gao, W., Wu, Z., Yuan, C., Cui, B., Dai, W., & Liu, P. (2022). Ultrasonication effects on physicochemical properties of starch -lipid complex. *Food Chemistry, 388*, 133054. https://doi.org/10.1016/j.foodchem.2022.133054

Zhang, J., Yu, P., Fan, L., & Sun, Y. (2021a). Effects of ultrasound treatment on the starch properties and oil absorption of potato chips. *Ultrasonics Sonochemistry, 70*, 105347. https://doi.org/10.1016/j.ultsonch.2020.105347

Zhang, L., Wang, X., Hu, Y., Fakayode, O. A., Ma, H., Zhou, C., Hu, Z., Xia, A., & Li, Q. (2021b). Dual-frequency multi-angle ultrasonic processing technology and its real-time monitoring on physicochemical properties of raw soymilk and soybean protein. *Ultrasonics Sonochemistry, 80*, 105803. https://doi.org/10.1016/j.ultsonch.2021.105803

Zheng, J., Li, Q., Hu, A., Yang, L., Lu, J., Zhang, X., & Lin, Q. (2013). Dual-frequency ultrasound effect on structure and properties of sweet potato starch. *Starch-Stärke, 65*(7–8), 621–627. https://doi.org/10.1002/star.201200197

Zhu, F., & Li, H. (2019). Modification of quinoa flour functionality using ultrasound. *Ultrasonics Sonochemistry, 52*, 305–310. https://doi.org/10.1016/j.ultsonch.2018.11.027

Zhu, B., Liu, J., & Gao, W. (2017). Process optimization of ultrasound-assisted alcoholic-alkaline treatment for granular cold water swelling starches. *Ultrasonics Sonochemistry, 38*, 579–584.

Zhu, F. (2015). Composition, structure, physicochemical properties, and modifications of cassava starch. *Carbohydrate Polymers, 122*, 456–480. https://doi.org/10.1016/j.carbpol.2014.10.063

Zhu, J., Li, L., Chen, L., & Li, X. (2012). Study on supramolecular structural changes of ultrasonic treated potato starch granules. *Food Hydrocolloids, 29*(1), 116–122. https://doi.org/10.1016/j.foodhyd.2012.02.004

Zia-ud-Din, Xiong, H., & Fei, P. (2017). Physical and chemical modification of starches: A review. *Critical Reviews in Food Science and Nutrition, 57*(12), 2691–2705. https://doi.org/10.1080/10408398.2015.1087379

Zuo, J. Y., Knoerzer, K., Mawson, R., Kentish, S., & Ashokkumar, M. (2009). The pasting properties of sonicated waxy rice starch suspensions. *Ultrasonics Sonochemistry, 16*(4), 462–468. https://doi.org/10.1016/j.ultsonch.2009.01.002

Pulsed Electric Fields

7

Milad Tavassoli and Behnam Bahramian

7.1 INTRODUCTION

Starch is a polysaccharide compound that exists as a natural biopolymer in potatoes, rice, barley, corn, and so on as an energy source. It is a versatile industrial constituent that finds widespread usage in a variety of industries, including food, feed, chemicals, petrochemicals, adhesives, paints, paper, textiles, and pharmaceuticals (Han, Zeng, Yu, et al., 2009; Y. Li, Wang, Han, et al., 2023). In order to attain new features and applications and hence improve starch performance, it is imperative to develop and improve starch modification methods continuously. (Castro et al., 2023; Hong, Chen, et al., 2016).

Since native starches are inert, resistant to enzymes, and lack the functional properties necessary to be successfully used in the food industry, some starches are rarely used in their original form, because they are weakly soluble in water, have a strong tendency to retrograde, are highly unstable in gels and pastes, and have poor thermal stability (Singh et al., 2022; Wu et al., 2019). In addition to improving its properties such as clarity and sheen, texture, film formation, adhesion, freezing-thawing stability, gelling tendencies, and syneresis and decreasing retrogradation, starch modification has been done for several years (Abduh, Leong, Agyei, & Oey, 2019; Zeng, Gao, Han, Zeng, & 2016). Starches can be modified chemically (by oxidation, acetylation, succinylation, and phosphorylation), enzymatically (by enzymes like cellulase, papain, xylanase, and pronase), or physically (by nonthermal, pregelatinization, and hydrothermal methods) (Maniglia et al., 2021). Starch is modified chemically by adding new functional groups to its backbone in order to achieve desired characteristics. As a result, even well-established technologies began to be reassessed; the classic illustration of this is the use of conventional chemical procedures (Hong et al., 2018; Qiu et al., 2021).

In addition to improving the paste's adherence, smoothness, gloss, and clarity, the alteration of starch also lessens the paste's propensity to gel, retrogradation, and syneresis. As a result, starch modifications are frequently used in the food processing sectors to get around these restrictions. Starch granule stability by modification slows down the rate of retrogradation, increases solubility, and boosts the gel texture and pasting qualities during processing (Castro et al., 2023; Zhu, 2018). In spite of this, native starch has a number of negative impacts, such as decreased solubility and decreased paste clarity. Moreover, under some circumstances, employing native starch is thought to be detrimental due to starch retrogradation and recrystallization, two significant factors (Castro et al., 2023; Chen et al., 2021). Therefore, the proper use of starch is limited by the unexpected processing characteristics of native starches. In order to combat these negative effects of starches, a number of safe therapeutic techniques, including enzymatic, chemical, and physical approaches, were developed. With the use of these approaches, the starch polymer may be modified, changing its structural and physicochemical characteristics and raising its value in both the food and non-food sectors (Almeida et al., 2023; Singh et al., 2022; Zhu, 2018).

This chapter provides an overview of recent studies on the use of pulsed electric field (PEF) technology to modify starch. The precise methods through which PEF modifies starch are yet to be established.

DOI: 10.1201/9781032655598-7

According to the literature, PEF treatment alters the in vitro digestibility of starch (increasing the level of rapidly digestible starch and decreasing the level of slowly digestible starch), damages the starch surface, and improves emulsion stability (Hong, Zeng, et al., 2016; Y. Li et al., 2023). The exact processing parameters used, such as the electric field intensity and specific energy input, determine the effectiveness and degree of PEF on starch characteristics. Additionally, the application of PEF was investigated as a support for other processes like esterification, enzymolysis, acetylation, and thermal processing to produce starches with new functionalities like higher porosity, a higher degree of starch gelatinization, capacity for rehydration, ability to form gels with greater stability under low-temperature conditions in the frozen food industry, improvement of noodle color, and faster water absorption rate (Almeida, Santos, Feitoza, et al., 2022). Despite the fact that PEF treatment is regarded as a non-thermal method, the sample is heated via ohmic heating. To prevent heat impacts, the temperature of the starch suspension must be carefully regulated. In this approach, a powerful temperature-controlling device is needed for the widespread deployment of PEF (Taha et al., 2022).

The main emphasis of this chapter's description of current PEF advancements is the extraction and modification of starch. Process characteristics, purification chamber designs, and pulsed power generator applications for starch extraction are reviewed in detail. The advantages and disadvantages of PEF's application for starch extraction and modification are discussed. It also offers certain rules and concepts for understanding process performance and makes recommendations for the development of these electrical systems in the future.

7.2 OVERVIEW OF PEF

7.2.1 Principles and Fundamentals of PEF for Starch

PEF-based processing is an environmentally benign method that works well for a variety of food processing tasks, including dehydration, freezing, recovering bioactive chemicals, and inactivating microorganisms and/or enzymes (Bhat et al., 2019; Dunn, 2019). The use of PEF technology for the processing of various liquid, semi-liquid, solid, and muscle meals has been the subject of several investigations. PEF technology has improved the yield and efficiency of extraction processes, allowing for the extraction of juice from apples or grapes, sugar from beets, and bioactive substances. In a similar vein, PEF pretreatment had a notable impact on process performance metrics when French fries were produced commercially (Kumari et al., 2018). PEF treatment improves tenderness and lessens shear force in beef muscles. PEF has an impact on drying kinetics as well; a shorter drying period helps preserve biocompounds in dried samples (Raso et al., 2022). As a result, PEF-based processing helps this industry advance significantly. As a nonthermal food processing technique, PEF uses electric pulses that last between nanoseconds and milliseconds with an electrical field intensity in the range of 10–80 kV/cm. This highly adaptable method has demonstrated the ability to physically alter the inherent characteristics of starch derived from various sources. Furthermore, this method may be used instead of traditional thermal processing to inactivate pathogenic and spoilage organisms at lower temperatures while preserving the sensory and nutritional qualities of food products when the intensity of the electric fields used is higher than 20 kV/cm (El Kantar et al., 2018; Gómez et al., 2019).

The treatment chamber and pulse generator are the two main parts of a PEF system. To be used by a PEF system, high-voltage alternating current is transformed by a pulse generator into direct high-voltage energy, or electrical current that flows in a single direction. After being transformed, this energy is kept in reserve in the condenser, which is used to quickly release electrical energy between the two electrodes in the treatment chamber to create electrical pulses (Gómez et al., 2019; Stewart et al., 2019). The switch, which is responsible for controlling the discharges, is the most important part because it must quickly turn the circuit on or off at high voltages and currents. A pulse transformer is used to set up the condenser when the voltage is inadequate (Panja, 2018). Electrical pulses and electrical pulse fields develop as a result of the creation of an electric potential differential by the discharges. Additionally, the material to be treated is positioned in the space between the electrodes of the treatment chamber (Kaufmann & Pitt, 2018).

The schematic of PEF operation for starch extraction and modification is shown in Figure 7.1. The PEF system generates pulses that can be either unipolar or bipolar, depending on whether one or two pairs of electrodes are used during treatment. They can also be classified based on their shape, such as squarewave, which occurs when the applied voltage is kept constant at its maximum value for a specific amount of time (known as the pulse width), after which it abruptly decreases or exponential, in which the maximum voltage applied initially decays exponentially over time (Arshad et al., 2020; Khan et al., 2018). The time it takes for the voltage to drop to 37% in this instance is the definition of the pulse width. Square wave pulses are more favorable than exponential ones because of this difference in geometry. Bipolar square wave pulses are the most often utilized in food preparation because of this. The shape of the chamber will vary based on how the electrodes are arranged. The electrodes are on opposing sides of the parallel plate chamber in a parallel arrangement. Because of this, the product flow has a direction perpendicular to the electric field, and the electric field can be uniform. Due to the chamber length reduces impedance—a measurement of the resistance the circuit has to the current when voltage is applied—it should be longer than the distance between the electrodes in order to provide a homogeneous electric field that is acceptable (Bhat et al., 2019; Cserhalmi et al., 2006). At high currents, however, electrode corrosion in the electrode-flow interface might result from the huge electrode surface and low electrical resistance. To stop current leakage, it is beneficial to operate symmetrically to the ground. Two cylindrical electrodes, the positive pole encircled by the negative, make up the coaxial arrangement. The direction of the product flow is perpendicular to the electric field, just like in the parallel setup. Although it is simple to construct and permits a higher degree of homogeneity, the electric field is not constant down the column. By adjusting the diameters, this may be standardized, but doing so may increase the electrodes' surface area and reduce the impedance, therefore this design is only appropriate for low-conductivity loads (Y. Li, Wang, Han, et al., 2023; Maniglia et al., 2021; Min et al., 2003). The electrodes in the collinear layout are next to one another and are spaced apart by an insulating substance; the chamber is tubular. This is highly desirable for food processing, as it allows fluids to flow more dynamically, is simple to clean, and has a high resistance because of its smaller cross-section area (Duque et al., 2020; Gómez et al., 2019; Y. Li, Wang, Han, et al., 2023). The reactivity of the electrodes is limited since more co-linear unities may be linked and operated at a lower current as compared to the parallel design. Nonetheless, the chamber's temperature and electric field distribution are not uniform. Treatments can be carried out continuously or in batches. Lower quantities of solid and semi-solid samples, which are regarded as static, can be used with batch treatments. Because they allow for stricter control over the parameters, they are more frequently encountered in experimental investigations and offer more benefits in the laboratory setting (Singh et al., 2022; Wu et al., 2019). On the other hand, agitation-free systems may lead to certain volumes not being handled appropriately. The lack of agitation in continuous treatments may be avoided by using numerous treatment zones inline or flow channels. Continuous treatments are more suited for processing liquids and are readily integrated into industrial processes. While coaxial and collinear chambers are more frequently employed in continuous systems—where the sample is pumped at a given flow rate, and the pulses are administered at a defined frequency—parallel chambers are typically utilized in batch systems (Bhat et al., 2019; Taha et al., 2022).

PEF is caused by a phenomenon known as electro-pulsation, which is the exposure of cells to electric pulses that change their membranes and increase their conductivity and/or permeability. A cell's ability to sense an external electric pulsed field causes a variation in the basal transmembrane potential, which is the difference between the intra- and extracellular electric voltages under normal physiological conditions (Duque et al., 2020; Hong, Zeng, et al., 2016). The effects of this variation depend on the length and intensity of the electric field. The most prevalent impact is caused by water molecules creating unstable metastable hydrophilic holes in bi-phospholipid membranes, which increases permeability for molecules without transmembrane transport pathways (Chen et al., 2023; Gómez et al., 2019; Panja, 2018). Nevertheless, the membrane's permeability and conductivity only significantly rise upon reaching

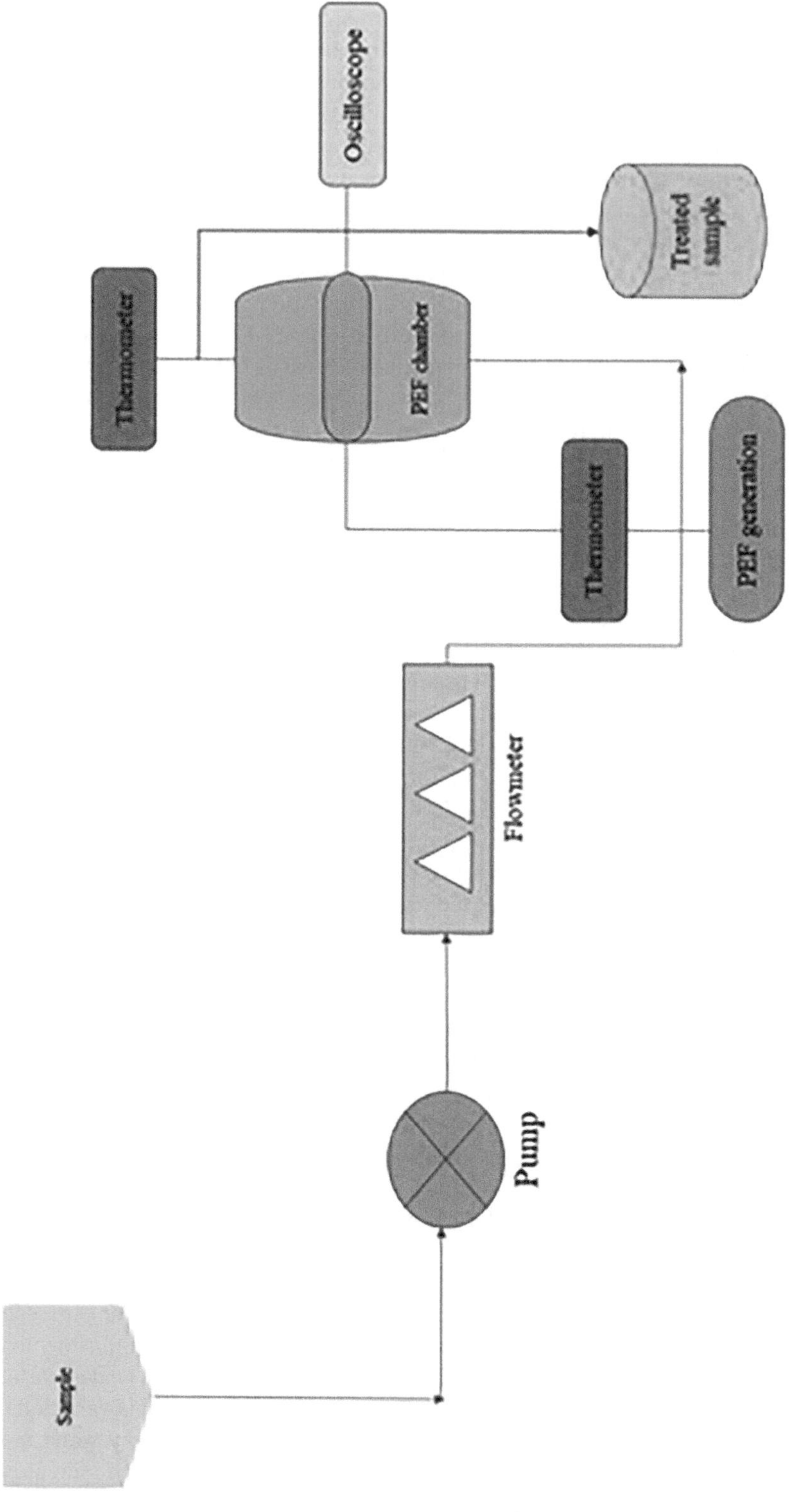

FIGURE 7.1   Schematic of PEF performance for starch modification.

a minimum transmembrane potential value, which is a non-universal value that depends on a number of variables. Permeability and conductivity modifications are preserved as long as this value is upheld. Both conductivity and permeability decline to a stable and measurable level when the electric field is withdrawn, enabling the diffusion of ions and small molecules. The transmembrane potential is also smaller than the lowest previously obtained (Almeida, Santos, Padilha, et al., 2022; Q. Li et al., 2019). Even after the membrane has sealed again, changes in the physiological functions of the cell and its responses to external stimuli may still be seen before the cell reverts to its original condition. If no harm is done and cellular viability is maintained, the membrane then gradually returns to its original state (Almeida, Santos, Feitoza, et al., 2022; Han, Zeng, Zhang, & Yu, 2009; Panja, 2018).

7.2.2 Performance and Effective Factors on Starch Extraction by PEF

PEF treatment is one of several physical modification methods, including cold plasma, ultrasonication, γ-irradiation, high hydrostatic pressure (HPP), and high-pressure homogenization (HPH) (Chen et al., 2023; Y. Li, Wang, Wang, et al., 2023; Maniglia et al., 2021; Zhu, 2018). In PEF treatment, starch suspension is subjected to a high-intensity pulse electric field strength of 10–80 kV/cm for a relatively brief period of time (typically µs to ms), which modifies the structural conformations of starch molecules and, therefore impacts their functional properties. When compared to other physical modification methods, PEF modification of starch offers some advantages, including lower energy usage and shorter processing times. Researchers have previously utilized this method to modify a wide range of starches, including maize starch, potato starch, wheat starch, cassava starch, pea starch, and waxy rice starch (Eliasson, 2017; Maniglia et al., 2021; Zeng et al., 2016).

High-voltage repeating pulses must be generated and transmitted by an effective PEF processing system. Precise regulation of both voltage amplitude and pulse duration is also essential. High voltage electric field pulses can be produced in a variety of methods, from straightforward circuits to intricate networks (El Kantar et al., 2018; Kumari et al., 2018; Wu et al., 2019). This is also determined by the type of waveform required and the availability of the switches, inductors, transformers, and capacitors. The PEF processing system's treatment chamber serves as its central component. Since the treatment chamber design affects the PEF process efficiency, which is mostly dependent upon peak field intensity and treatment uniformity. Treatment chambers are primarily divided into two batches and continuous depending on the kind of food (solid or liquid) and the amount of liquid processing (Y. Li, Wang, et al., 2023; Qiu et al., 2021). There are three major types of treatment chambers based on electrode configuration: parallel, coaxial, and co-linear. Voltage, temperature, treatment chamber current, product flow rate, and other processing variables are measured, controlled, and tracked by the control and monitoring system. It is the main component of the system that aids in the continuous observation of the PEF process by providing all the information and charts needed to assess the consistency of the field strength and pulse waveform with regard to their form, frequency, and duration of treatment during the course of the procedure (Raso et al., 2022; Taha et al., 2022). The minimum necessary and adequate intensity is required for PEF-induced alteration of starch, which alters the structural and functional characteristics of starch. The phrase "critical PEF treatment intensity" (Ec) refers to this lowest intensity, which might include temperature, treatment duration, frequency, pulse width, energy, and electric field strength (Arshad et al., 2020; Y. Li, Wang, et al., 2023).

Food PEF processing typically uses rectangular, exponential pulse waveforms that combine short and long pulse durations in both monopolar and bipolar modes. It has been shown that rectangular pulses with rapidly rising and falling edges are the most efficient due to their minimal temperature increase and largest effective area (Bhat et al., 2019; Taha et al., 2022). However, it is challenging and costly to produce an exact rectangular waveform. The exponential decay waveform, on the other hand, is more straightforward to create. On the other hand, the high-intensity electric field in the exponential decay waveform consists of a lengthy tail of the low-intensity field and a tiny usable peak. Sample resistivity and discharge circuit have an impact on pulse width. There is an inverse connection between the electric field strength and the outcomes; that is, a narrow pulse width and a high field intensity can provide results that are comparable to each other (Chen et al., 2023, 2021; Luo et al., 2023). Furthermore, increasing pulse width or frequency led to an increase in treatment intensity at a constant electric field strength. Applications involving starch modification usually have a pulse width of 1–50 µs. Treatment duration and pulse width are influenced by pulse repetition rate, commonly referred to as pulse frequency. Switches using solid state power are advised for higher frequencies. In a high-voltage pulse generator, however, running at higher frequencies results in larger switching losses (Castro et al., 2023; Raso et al., 2022). Starch modification is often carried out in the frequency range of 100–1000 Hz. The flow velocity of the starch suspension is crucial for PEF processing operations that are ongoing. It specifies the length of time spent within the treatment chamber overall. As a result, it significantly affects how treatment chambers and pulsed power supply are designed (Almeida et al., 2023; Maniglia et al., 2021).

7.3 PROCESSING AND MODIFY STARCH PROPERTIES BY PEF

7.3.1 Granule Morphology

Q. Li et al. (2019) investigated the effect of a pulsed electric field on starch morphology. They showed that potato (type B), wheat (type A), and pea (type C) starch grains were not damaged in pulsed electric field treatment (2.86 to 8.57 kV/cm) (Figure 7.2) (Q. Li et al., 2019). However, in this context, in 2019, researchers observed sunken areas in the rice starch grain that were exposed to the pulsed electric field, and fracture occurred at the electric field intensity of 8.57 kV/cm (the highest intensity) (Wu et al., 2019). The results of the damage caused to the starch grains in the low electric field strength can be related to

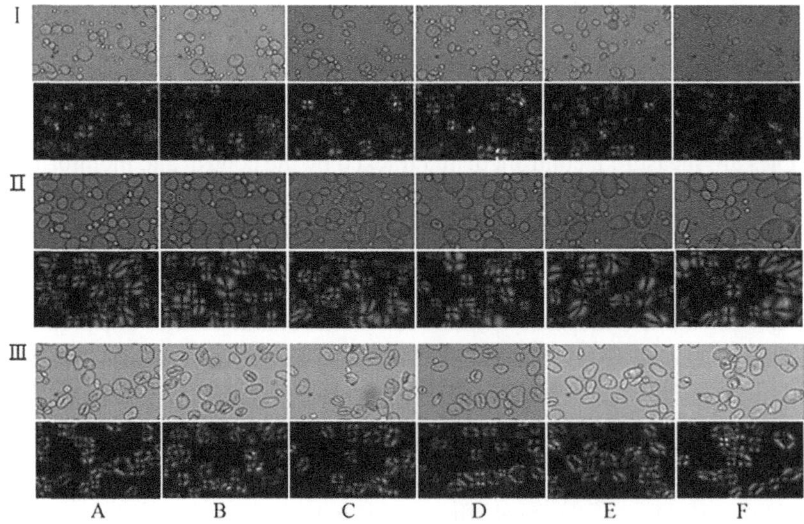

FIGURE 7.2 Micrographs of PEF-treated samples in polarized light (bottom rows) and normal light (top rows) for wheat starch (I), potato starch (II), and pea starch (III) samples. Column A refers to the control ($0\,kVcm^{-1}$) and columns B, C, D, E, F refer to the PEF intensity of 2.86, 4.29, 5.71, 7.14, and $8.57\,kV\,cm^{-1}$, respectively.

FIGURE 7.3 SEM of waxy rice starch treated with different electric field strengths: (A) native. (B) 30 kV cm⁻¹. (C) 40 kV cm⁻¹. (D) 50 kV cm⁻¹.

the origin of the starch plant. These damages are evident when a higher order of electric field intensity is used independent of starch polymorphism. In another similar study, which subjected waxy rice starch to a pulsed electric field, it was shown that with the increase in the intensity of the electric field, the amount of change in the appearance of the surface increased. In the outer region of the starch granule, there is a dense structure that is somewhat resistant to external stress and pressure. When the granules are exposed to high electric field intensity, the protection of the outer packaging is destroyed, which is why the starch granules tend to accumulate between the particles (Figure 7.3) (Zeng et al., 2016).

7.3.2 Particle Size

Several studies have investigated the effect of PEF on particle size (Table 7.1). For example, a 2009 study by Han et al. showed that when corn starch grains were treated with 50 kV/cm², the diameter of corn starch grains increased from 24.22 to 58.81. The results indicate that the pulsed electric field has damaged the outer part of the starch granule, and the inner part of the starch granule can absorb water and swell after the treatment with the electric field. As a result, it strengthens electrostatic and van der Waals forces between starch granules, and starch grains accumulate (Han, Zeng, Zhang, et al., 2009). In another similar study, the authors reported that the particle size of thermally processed oat flour treated with a pulsed electric field (electric field strengths 2 and 4.4 kV/cm combined with specific energy inputs between 48 and 484 kJ/kg) was much larger than that of raw oat flour treated with a pulsed electric field.

TABLE 7.1 Effect of Pulsed Electric Field on Starch Particle Size

SAMPLE	ELECTRIC FIELD STRENGTHS (KV/CM)	SPECIFIC ENERGY INPUT (KJ/KG)	$D_{4,3}$ (µM)	$D_{3,2}$ (µM)	D(0.1) (µM)	D(0.5) (µM)	D(0.9) (µM)	SSA (M²/G)	REF.
Native sample	-----	-------	37.93	16.67	16.12	35.76	63.68	0.36	(Han, Zeng, Yu, et al., 2009)
Potato starch	30	-------	56.11	16.80	16.93	38.14	89.74	0.36	
	40	-------	85.16	22.95	19.16	46.17	231.41	0.26	
	50	-------	113.84	24.91	21.79	51.14	341.96	0.24	
Native sample	-----	-------	15.22	7.37	7.40	14.89	24.22	0.81	(Han, Zeng, Zhang, et al., 2009)
Corn starch	30	-------	23.65	9.46	8.18	18.54	44.29	0.63	
	40	-------	27.74	10.15	9.28	22.77	53.80	0.59	
	50	-------	29.68	10.47	9.50	23.60	58.81	0.57	
Native sample	-----	-------	----	-----	7–8	23–33	331–451	-----	(Duque et al., 2020)
Oat flour (raw)	2.2	53	----	-----	6–7	23–30	64–298	-----	
	2.1	249	----	-----	7–8	22–31	43–370	-----	
	2.1	484	----	-----	24–30	93–108	244–403	-----	
	4.4	51	----	-----	7–8	29–34	234–365	-----	
	4.3	220	----	-----	7–8	32–34	341–358	-----	
	4.1	441	----	-----	22–27	85–123	295–412	-----	
Native sample	-----	------	----	-----	12–14	82–257	565–1299	-----	(Duque et al., 2020)
Oat flour	2.2	49	----	-----	14–15	150–192	889–914	-----	
(thermal	2.1	233	----	-----	16–20	106–238	871–955	-----	
treated)	2.1	434	----	-----	33–46	146–213	213–931	-----	
	4.4	48	----	-----	13–15	151–213	904–939	-----	
	4.3	200	----	-----	11–18	91–206	639–918	-----	
	4.1	418	----	-----	30–57	102–255	226–925	-----	

Heat treatment causes gelatinization of starch, which includes aggregation of grains leaching of amylose. The authors also reported that oat starch subjected to pulsed electric field treatment changed the oat protein's secondary structure, indicating that the changes that occur on these oat proteins can be partially responsible for the aggregation (Duque et al., 2020). Y. Li, Wang, et al. (2023) investigated the effect of a pulsed electric field on the particle size of dialdehyde starch. They reported that the particle size increased with the help of pulsed electric field and oxidation. Because the pulsed electric field can cause damage to the starch grains and facilitate chemical operations, and these factors increase the particles (Y. Li, Wang, et al., 2023).

7.3.3 Birefringence

The amylopectin of starch grains in its crystalline regions contains a radial arrangement from helium to the starch surface. Li et al. investigated the effect of a pulsed electric field (intensity 2.86 to 8.57 kV/cm) for potato, pea, and potato starches. The researchers found that under polarized light for electric intensities of 2.86 to 5.71 kV/cm, there was no significant difference in birefringence, which indicates that the arrangement of amylopectin and amylose was not disturbed (Q. Li et al., 2019a). Abduh et al. (2019) observed no change in potato starch birefringence under pulsed electric field treatment (specific input energy 50 and 150 kJ/kg and electric field intensity 0.5 to 1.1 kV/cm) (Abduh et al., 2019). In a study, researchers applied high electric field intensity (7.14 and 8.57 kV/cm) on potato, pea and wheat starch. The researchers showed that the Maltese cross disappeared for potato and wheat starch, but there was no change for pea starch. This indicates that potato and wheat starches are more susceptible to pulse electric field treatment than peas. The pulsed electric field can change the arrangement of starch amylopectin in the crystalline region, indicating that the malt cross-links are destroyed, and birefringence occurs (Q. Li et al., 2019). In another study conducted by Wu et al. (2019), who investigated the effect of a pulsed electric field on rice starch, they showed that because the diameter of the rice starch grains was small, there were no changes in Maltese crosses.

7.3.4 X-Ray Diffractometry

A and B starches are more sensitive to pulsed electric fields compared to the C type (Almeida et al., 2023). The only method that can determine the crystalline quantity of starch grains is X-ray diffraction (XRD). Table 7.2 shows the effect of the pulsed electric field on crystallinity of starches. It has been shown that the electric field intensity of 2.86 to 8.57 has no change in the diffraction peak in potato, wheat, and pea starch, indicating that the changes in the starch morphology, which are treated with a pulsed electric field, are small (Figure 7.4). The crystallinity of starches treated with 2.86 to 5.71 kV/cm was slightly higher than starches treated with 7.14 and 8.57 kV/cm. These findings indicate that during the treatment with a pulsed electric field with low intensity, the amylose chains are organized, which should increase the crystallinity as a result. While at a higher electric field intensity (approximately greater than 7.14 kV/cm), the hydrogen bond in the starch grains of the amylopectin chain is disrupted in the crystalline regions (Q. Li et al., 2019). When the hydrogen bonds are disturbed, the crystallinity is reduced, which causes the elimination of the Maltese crosses (birefringence) (Cornejo-Ramírez et al., 2018). These findings are similar to the results of Wu et al. in 2019, who treated rice starch with a pulsed electric field with an intensity of 2.86 to 8.57 kV/cm. In another study, researchers reported that crystallinity decreased when electric field intensity from 30 to 50 kV/cm was applied to waxy rice starch. The authors also reported that the diffraction peaks at 15.3, 17.1, 18.2, and 23.5° reduced with increasing electric field intensity (Zeng et al., 2016). These findings were in agreement with the results for corn (Han, Zeng, Zhang, et al., 2009), tapioca (Han, Zeng, et al., 2012), and potato starch (Han, Zeng, Yu, et al., 2009). It has been proven that by increasing

TABLE 7.2   The Effect of Pulsed Electric Field on the Crystallinity of Different Starches

STARCH	ELECTRIC FIELD STRENGTHS (KV/CM)	CRYSTALLINITY (%)	REF.
Native sample	---------	27.10	(Han, Zeng, Zhang, et al., 2009)
Corn	30	24.60	
	40	22.20	
	50	19.91	
Native sample		29.1	(Wu et al., 2019a)
Rice	2.86	30.0	
	5.71	30.6	
	8.57	28.7	
Native sample		31.8	(Zeng et al., 2016)
Waxy rice	30	35.1	
	40	32.4	
	50	28.7	
Native sample		29.4	(Q. Li et al., 2019)
Pea	2.86	29.8	
	4.29	30.4	
	5.71	30.2	
	7.14	29.4	
	8.57	29.5	
Native sample		25.6	(Q. Li et al., 2019)
Wheat	2.86	26.5	
	4.29	25.9	
	5.71	26.3	
	7.14	25.4	
	8.57	25.2	
Native sample		25.2	(Q. Li et al., 2019)
Potato	2.86	25.7	
	4.29	25.5	
	5.71	25.3	
	7.14	24.3	
	8.57	24.1	

the intensity of the pulsed electric field, more energy is provided for this treatment to disrupt the non-covalent bonds between the chains of starch grains, and in addition, it strengthens the interaction between the hydroxyl group of molecular chains of starch grains and water molecules. This causes the starch to transition from crystalline to non-crystalline (Han, Yu, et al., 2012). Disruption and changes in the starch amylopectin crystal can cause a reduction in starch crystallinity (Bertoft, 2017).

7.3.5 Molecular Weight

The pulsed electric field is able to affect the molecular weight of starch (Table 7.3). Han, Yu, et al. (2012) investigated the molecular weight of corn starch treated with an electric field. They showed that the molecular weight of corn starch decreased with increasing electric field intensity (30 to 50 kV/cm) and

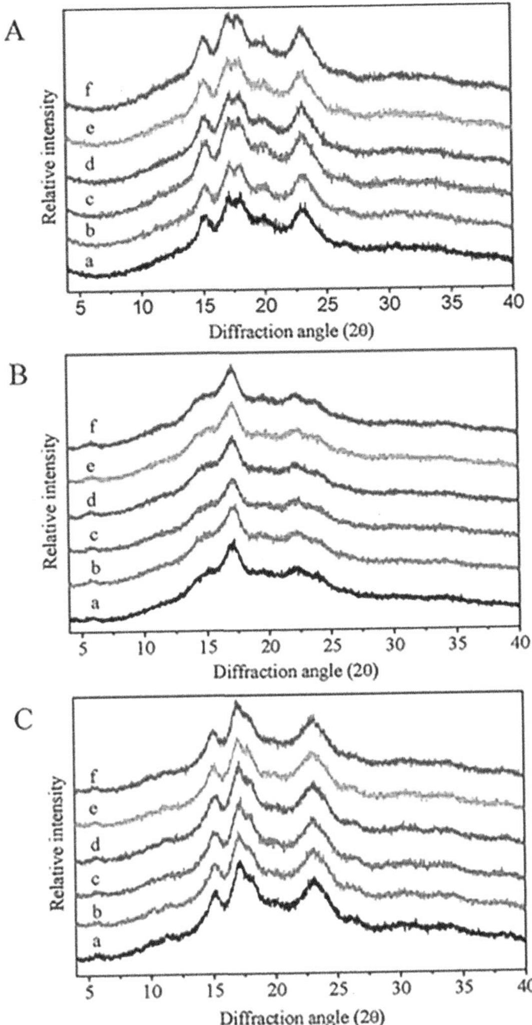

FIGURE 7.4 XRD patterns of wheat starch (A), potato starch (B), and pea starch (C) treated with pulsed electric field.

time (424 to 1272 microseconds). In addition, the authors proved that the molecular weight decreased further with the increase of electric field intensity. Also, the authors reported that the effect of the electric field intensity factor on molecular weight is more significant than time. The authors attributed the decrease in molecular weight to the instability of amylopectin (Han, Yu, et al., 2012). Zeng et al. proved otherwise in 2016 as they reported that the molecular weight of electric field-treated waxy rice starch was not significant. According to Q. Li et al., 2019, one of the reasons for starch digestion capacity may be the same changes that occur in molecular weight. Wu et al. (2019) showed that when the electric field intensity increased from 5.71 kV/cm, the molecular weight of long and short amylopectin chains increased. Q. Li et al. attributed the reason for such a change in molecular weight to the probability of breaking the molecular chain. The authors also reported that the ratio of the amylose chain length to the amylopectin chain length ratio was less than 1, and the amount of amylose was insignificant with the control (Wu et al., 2019). These results show that the pulsed electric field affects the amylopectin chain.

TABLE 7.3   The Effect of Pulsed Electric Field on the Molecular Weight of Different Starches

STARCH	ELECTRIC FIELD STRENGTHS (KV/CM)	MW (10⁷ G/ MOL)	MN (10⁷ G/ MOL)	MW/ MN	REF.
Native sample		1.56	---------	———	(Wu et al., 2019)
Rice	2.86	1.50	---------	———	
	5.71	1.54	---------	———	
	8.57	1.84	---------	———	
Native sample		8.13	4.79	2.46	(Q. Li et al., 2019)
Pea	2.86	2.97	1.16	2.56	
	4.29	2.96	1.24	2.39	
	5.71	3.62	1.04	3.60	
	7.14	3.60	1.03	3.30	
	8.57	6.93	1.81	3.83	
Native sample		6.60	3.83	1.72	(Q. Li et al., 2019)
Wheat	2.86	8.91	5.89	1.51	
	4.29	10.31	5.62	1.84	
	5.71	8.44	4.31	1.96	
	7.14	9.06	5.04	1.80	
	8.57	6.88	3.05	2.25	
Native sample		8.54	8.46	1.01	(Q. Li et al., 2019)
Potato	2.86	8.68	8.15	1.07	
	4.29	8.53	8.25	1.03	
	5.71	7.86	5.11	1.54	
	7.14	7.78	5.57	1.40	
	8.57	7.11	4.47	1.59	
Native sample		10.38	---------	———	(Zeng et al., 2016)
Waxy rice	30	10.15	---------	———	
	40	9.39	---------	———	
	50	8.90	---------	———	

7.3.6 Differential Scanning Calorimetry and Gelatinization Temperature

Table 7.4 shows the studies related to the effect of a pulsed electric field on the enthalpy and gelatinization temperature of starches. In the 2012 study, Han et al. showed that the enthalpy and gelatinization temperature of corn starch decreased with increasing time (424 to 1272 microseconds) and electric field strength (30 to 50 kV/cm) due to the decrease in molecular weight and breaking of the amylopectin chain. According to Han et al., these results show an interaction effect between time and electric field strength. Zeng et al. (2016), who investigated the enthalpy and gelatinization temperature of waxy rice starch, showed that the enthalpy and gelatinization temperature decreased with the increase in electric field intensity (30 to 50 kV cm). In another study, the gelatinization enthalpy and gelatinization temperature of potato starch reduced with the increase of the electric field intensity (from 30 to 50 kV/cm) (Han, Zeng, Yu, et al., 2009). Researchers reported that the degree of crystallinity and the degree of perfection of starch crystals play a role in the process of gelatinization enthalpy and gelatinization temperature, respectively. Studies have proven that potato starch crystallinity decreases with increasing electric field intensity in pulsed electric field treatments. Therefore, scanning calorimetry thermal treatment makes

TABLE 7.4   Results of the Effect of Pulsed Electric Field on Enthalpy Temperature and Gelatinization of Starch

STARCH	ELECTRIC FIELD STRENGTHS (KV/CM)	SPECIFIC ENERGY INPUT (KJ/KG)	T (µS)	TO (°C)	TP (°C)	TC (°C)	ΔT (°C)	ΔH (J/G)	REF.
Native sample			----	57.69	60.96	67.08	9.39	37.32	(Abduh et al., 2019)
Potato (isolated)	0.5	50	----	57.59	60.83	66.56	8.97	38.55	
	0.7	50	----	57.94	61.09	66.74	8.80	40.08	
	0.9	50	----	57.97	60.95	66.50	8.53	40.98	
	1.1	50	----	58.23	61.21	66.57	8.34	38.89	
	0.7	150	----	58.34	61.20	66.33	7.98	38.50	
	0.9	150	----	58.23	61.19	66.51	8.27	39.43	

Native sample			----	57.69	60.96	67.08	9.39	37.32	(S. B. Abduh et al., 2019)
Potato (leached)	0.5	50	----	57.59	60.83	66.56	8.97	38.55	
	0.7	50	----	57.94	61.09	66.74	8.80	40.08	
	0.9	50	----	57.97	60.95	66.50	8.53	40.98	
	1.1	50	----	58.23	61.21	66.57	8.34	38.89	
	0.7	150	----	58.34	61.20	66.33	7.98	38.50	
	0.9	150	----	58.23	61.19	66.51	8.27	39.43	
Native sample		----	----	68.78	68.78	78.14	16.06	16.06	(Han, Zeng, et al., 2012)
Tapioca	30	----	214	62.73	68.12	75.15	8.87	8.87	
	40	----	214	62.21	66.63	72.31	4.82	4.82	
	50	----	214	61.76	65.94	72.61	3.80	3.80	

Native sample			----	68.93	73.53	79.70	9.20	1.61	(Han, Zeng, Yu, et al., 2009)
Corn	30	----		68.60	73.38	78.49	9.56	1.22	
	40	----		67.46	72.21	77.68	9.50	1.14	
	50	----		67.34	71.51	77.40	8.34	0.95	

the molecules in the crystalline regions react more easily with the water molecules, and as a result, the enthalpy of gelatinization decreases (as stated in Table 7.4) (Eliasson, 2017; Wang et al., 2008). These results are consistent with those the researchers investigated for corn (Han, Zeng, Zhang, et al., 2009) and tapioca (Han, Zeng, et al., 2012). Starches treated with a pulsed electric field have broken hydrogen bonds, meaning less energy is needed to break the remaining pions. As previously shown, it is evident by the decrease in enthalpy and gelatinization temperature, which corresponds to the temperature at which starch gelatinizes (Jayaram, 2000; Schirmer et al., 2015). Abduh et al. investigated potatoes treated with a pulsed electric field (electric field intensity 0.5 to 1.1 kV/cm and input energy 50 and 150 kJ/kg). They reported that potatoes with higher specific energy had a narrower gelatinization zone than those with lower specific energy (2019).

7.3.7 Pasting Properties

One of the pasting properties of starch grains is the peak viscosity, which is related to the water absorption of starch grains. Breakdown viscosity, which is another property of pasting starch grains, describes

TABLE 7.5 The Effect of Electric Field on the Pasting Properties of Starch

STARCH	ELECTRIC FIELD STRENGTHS (KV/CM)	PEAK (BU)	SH (BU)	SC (BU)	EC (BU)	FV (BU)	BD (BU)	SB (BU)	REF.
Native sample		632.2	----	----	583.0	805.0	49.2	222.0	(Wu et al., 2019)
Rice	2.86	637.0			587.0	803.3	50.0	216.3	
	5.71	639.0			589.0	805.0	50.0	216.7	
	8.57	629.3			581.0	799.7	48.3	218.7	
Native sample		2961	1060	524	913	907	2437	----	(Han, Zeng, Yu, et al., 2009)
Potato	30	2771	1074	555	955	922	2216	----	
	40	2705	1063	524	917	862	2437	----	
	50	2641	1022	523	910	877	2119	----	
Native sample		982	496	279	557	505	703	----	(Han, Zeng, et al., 2012)
Tapioca	30	921	517	295	572	512	626	----	
	40	889	505	281	546	489	608	----	
	50	820	469	260	493	444	560	----	
Native sample		335	320	253	568	537	82	----	(Han, Zeng, Zhang, et al., 2009)
Corn	30	291	280	220	528	482	71	----	
	40	282	271	215	469	442	67	----	
	50	250	243	201	470	426	49	----	

the thermal stability of starch. Among the other factors of sticking properties of starch grains, regression viscosity, and final viscosity can be mentioned. The final viscosity is obtained in the last stage after cooling, and the setback viscosity of starch grains is obtained from the retrogradation degree. Various studies investigated the effect of the electric field on the pasting properties of starch (Table 7.5). In 2019, Wu et al. treated rice starch with a pulsed electric field (electric field intensity 2.86 to 8.57 kV/cm), which showed that this treatment does not affect the pasting temperature and breakdown viscosities. In a similar study, Han et al., who used an electric field intensity of 30 to 50 kV/cm for tapioca starch, showed that the viscosity peak decreased with increasing electric field intensity, indicating that the granules were less swollen before bursting. The breakdown viscosity also decreased with the increase of the electric field intensity from 30 to 50 kV/cm, which indicates that the hot paste has become more stable. In addition, the authors reported that the final viscosity and setback also decreased with increasing electric field intensity (according to Table 7.5), indicating that the sample was less prone to retrogression (Han, Zeng, et al., 2012). These findings are consistent with the results for potato (Han, Zeng, Yu, et al., 2009) and corn starch (Han, Zeng, Zhang, et al., 2009). Another study investigated the effect of pulsed electric field treatment on thermally processed and raw oat flour. The researchers reported that increasing the specific energy input caused a decrease in the pasting temperatures and viscosity. This reduction in pasting temperature compared to the control indicates that the oat flour swells earlier than the control. Pulsed electric field treatment did not significantly change thermally processed flour's peak viscosity but decreased in raw oat flour. The results indicate that raw oat flour is more sensitive than thermally processed flour because it has undergone thermal pretreatment, which may cause gelatinization and aggregation before treatment. The breakdown viscosity of both flours decreased after the pulsed electric field treatment. The improvement of paste stability is attributed to the reduction of breakdown viscosity. In addition, among these two flours, only raw oat flour had a low setback viscosity, indicating that this flour's tendency to retrograde is less (Duque et al., 2020).

7.3.8 Small-Angle X-Ray Scattering

Starch layers can be studied and characterized using small-angle X-ray scattering (SAXS). Starch grains are formed by amorphous rings (with semi-crystalline rings). Amorphous rings are composed of amylopectin and amylose, and semi-crystalline rings are composed of amorphous and crystalline regions (Z. Yang et al., 2017). To describe the amount of disorder, the fractal structure (D) in starch grains is used, and a mass dimension (Dm) is considered. The starch layer can be studied using SAXS (Table 7.6). Li et al. (2019) investigated the effect of a pulsed electric field on the thickness of semi-crystalline lamellae of potato, wheat, and pea starch. The authors reported that no significant difference was observed for wheat starch subjected to the pulsed electric field, indicating that the pulsed electric field did not alter the thickness of the wheat starch. However, a significant difference was observed for potato and pea starch in treatments of 5.71 and 2.86 kV/cm, respectively. The researchers also reported that the scattering peak position of potato starch increased but decreased for pea (according to Table 7.6) (Q. Li et al., 2019). Another study reported that the lamella repetition distance of waxy rice starch treated with a pulsed electric field increased with increasing electric field intensity (Zeng et al., 2016). Contrary to these results, Wu et al.'s (2019) study reported that the repetition distance of rice starch lamellas treated with a pulsed

TABLE 7.6   SAXS Results of Pulsed Electric Field–Treated Starches

STARCH	ELECTRIC FIELD STRENGTHS (KV/CM)	Q (NM⁻¹)	D (NM)	REF.
Native sample		-----	9.206	(Wu et al., 2019)
Rice	2.86	-----	9.392	
	5.71	-----	9.237	
	8.57	-----	9.015	
Native sample		0.7066	8.89	(Zeng et al., 2016)
Waxy rice	30	0.7035	8.93	
	40	0.6570	9.56	
	50	0.6520	9.63	
Native sample		0.612	10.267	(Q. Li et al., 2019)
Pea	2.86	0.597	10.525	
	4.29	0.612	10.267	
	5.71	0.612	10.267	
	7.14	0.612	10.267	
	8.57	0.612	10.267	
Native sample		0.669	9.392	(Q. Li et al., 2019)
Potato	2.86	0.669	9.392	
	4.29	0.669	9.392	
	5.71	0.683	9.199	
	7.14	0.669	9.392	
	8.57	0.669	9.392	
Native sample		0.597	10.525	(Q. Li et al., 2019)
Wheat	2.86	0.612	10.267	
	4.29	0.612	10.267	
	5.71	0.597	10.525	
	7.14	0.597	10.525	
	8.57	0.583	10.777	

electric field decreased with increasing electric field intensity. These results show that the pulsed electric field had a different effect on the repetition distance of starch lamellae.

7.3.9 FTIR and NMR

Nuclear magnetic resonance (NMR) and Fourier transform infrared spectrum (FTIR) results of starches treated with a pulsed electric field are reported in Table 7.7. Bands ~1047 and ~1022 cm^{-1} in FTIR can be used to detect the amorphous and crystallinity region of starch grains. An intensity ratio of $A_{1047}/_{1022}$ is used to provide information about double helix packing and crystallinity of short-range molecular order. NMR is used to obtain direct quantitative information on the ratio of short-range double helices (based on C_1 and C_4 positions) (Warren et al., 2016). For example, in the study of Han et al. in 2012, where corn starch was treated with a pulsed electric field with an intensity of 50 kV/cm and using NMR (^{1}H and ^{13}C spectra), they showed that this intensity did not significantly change the chemical structure of starch. In another similar study, which investigated the infrared spectrum of pea, potato, and wheat starch under the treatment of 2.86 to 8.75 kV/cm, they showed that the $A_{1047}/_{1022}$ intensity of potato starch is more variable than that of wheat and pea starch. In addition, ^{13}C nuclear magnetic resonance spectra show that the ordered structure of potato and wheat starch is reduced by 1.6% and 0.7%, respectively, compared to native starch (intensity 8.75 kV/cm). However, an increase of 4.5% was observed for pea starch. This decrease indicates the breaking of the hydrogen bond; in other words, the crystal structure of starch is disturbed (Figure 7.5) (Q. Li et al., 2019). In Wu et al.'s (2019) study, the intensity of $A_{1047}/_{1022}$ decreased in

TABLE 7.7   NMR and FTIR Results of Starches Treated with Pulsed Electric Field

STARCH	ELECTRIC FIELD STRENGTHS (KV/CM)	NMR (STRUCTURE) ORDERED (%)	DISORDERED (%)	FTIR $A_{1047/1022}$	REF.
Native sample		52.4	47.6	0.619	(Wu et al.,
Rice	2.86	52.4	47.6	0.620	2019)
	5.71	54.1	49.1	0.625	
	8.57	52.1	47.9	0.607	
Native sample		45.2	54.8	0.648	(Duque et al.,
Pea	2.86	46.8	53.2	0.702	2020)
	4.29	46.9	53.1	0.677	
	5.71	46.9	53.1	0.690	
	7.14	45.6	54.4	0.686	
	8.57	47.0	53.0	0.677	
Native sample		50.4	49.6	0.591	(Duque et al.,
Wheat	2.86	52.6	47.4	0.604	2020)
	4.29	50.9	49.1	0.608	
	5.71	51.2	48.8	0.567	
	7.14	49.8	50.2	0.543	
	8.57	49.7	50.3	0.539	
Native sample		45.1	54.9	0.792	(Duque et al.,
Potato	2.86	44.8	55.2	0.848	2020)
	4.29	52.8	47.2	0.927	
	5.71	43.2	56.8	0.790	
	7.14	42.4	57.6	0.772	
	8.57	43.5	56.5	0.757	

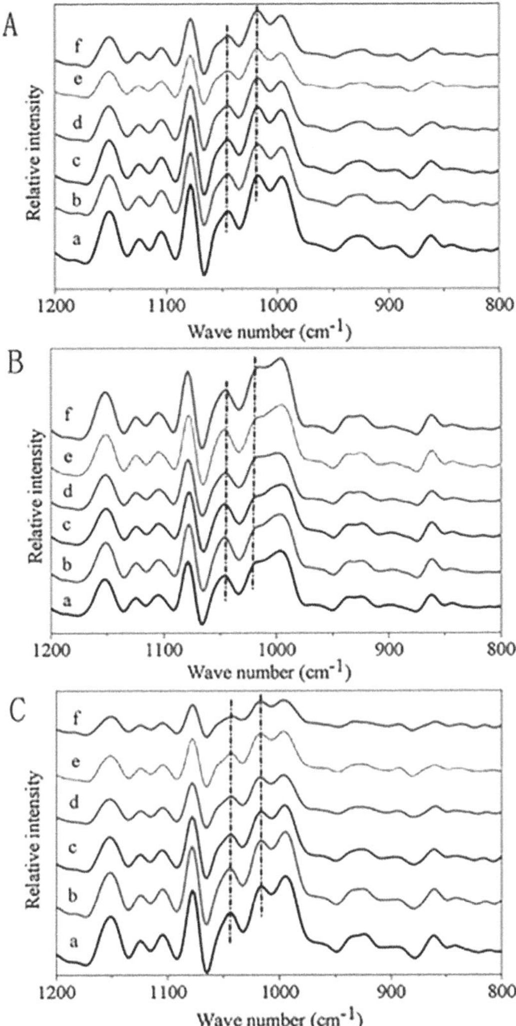

FIGURE 7.5 FTIR spectra for wheat starch (A), potato starch (B), and pea starch (C) samples treated with PEF. a: control (0 kV cm⁻¹); b, c, d, e, and f refer to PEF intensity of 2.86, 4.29, 5.71, 7.14, and 8.57 kV cm⁻¹, respectively.

waxy rice after treatment with a pulsed electric field with an intensity of 8.75 kV/cm, which shows that the pulsed electric field causes changes in the structure of starch polymorphisms. In another study, researchers showed that the intensity of $A_{1047}/_{1022}$ for raw oat flour was reduced after pulsed electric field treatment at 441 kJ/kg and 4.1 kV/cm compared to the control sample. This indicates that the pulsed electric field causes the starch structure to change and disturbs the crystallinity. However, no change was obtained for thermally processed oat flour (Duque et al., 2020).

7.3.10 In-Vitro Digestibility

Based on this factor, starch grains are divided into slowly digestible starch (SDS), rapidly digestible starch (RDS), and resistant starch (RS). RDS, which undergoes digestion in less than 20 minutes; SDS, which requires more than 20 minutes for digestion; and RS, which evades hydrolysis in the stomach and small intestine. Resistant starches are impervious to digestive enzymes in the stomach and small intestine and proceed to the large intestine, where they undergo fermentation by the gut microbiota. One of the essential factors for starch is in vitro digestion because it determines the glycemic response in in-vivo conditions. Studies show that the Englyst method is the most reliable and widely used to perform this test (Englyst et al., 1992). In Table 7.8, several studies on the effect of pulsed electric field on the in-vitro digestibility of starches have been investigated. For example, in 2019, researchers investigated the effect of a pulsed electric field (2.86 to 8.57 kV/cm) on starch digestion (pea, potatoes, and wheat). The researchers showed that starch treated with a pulsed electric field caused a decrease in slow-digesting starch and an increase in fast-digesting starch, but resistant starch did not change compared to native starch (Table 7.8) (Q. Li et al., 2019). In this context, Wu et al. reported similar results on rice with a pulsed electric field (2.86 to 8.57 kV/cm) (2019). These results are consistent with the results of Zeng et al., who found that the native rice starch was subjected to pulsed electric field treatment despite the reduction of resistant starch (2016). Different polymorphisms cannot explain these results. Because the pulsed electric field causes morphological changes and damage to starch grains, the digestible enzyme will have easier access to new bonds or more glycosidic bonds in an area that is not accessible at first. It has already been proven

TABLE 7.8　Effect of Pulsed Electric Field on In-Vitro Digestibility of Starches

STARCH	ELECTRIC FIELD STRENGTHS (KV/CM)	RDS (%)	SDS (%)	RS (%)	REF.
Native sample		42.47	20.09	37.44	(Wu et al., 2019)
Rice	2.86	42.76	19.23	38.02	
	5.71	43.33	18.94	37.73	
	8.57	46.20	15.78	38.02	
Native sample		50.88	37.29	11.73	(Q. Li et al., 2019)
Pea	2.86	54.55	33.13	12.32	
	4.29	55.29	31.82	12.89	
	5.71	58.04	29.83	12.13	
	7.14	58.40	29.29	12.31	
	8.57	59.40	28.36	12.24	
Native sample		37.91	39.46	22.63	(Q. Li et al., 2019)
Wheat	2.86	39.68	34.65	25.47	
	4.29	39.96	34.84	25.60	
	5.71	41.57	35.58	22.85	
	7.14	42.68	35.98	21.36	
	8.57	44.16	35.37	20.47	
Native sample		24.78	45.68	29.54	(Q. Li et al., 2019)
Potato	2.86	29.62	39.59	30.79	
	4.29	31.28	38.76	29.68	
	5.71	32.67	38.62	30.10	
	7.14	35.95	37.79	29.54	
	8.57	35.95	39.59	27.46	
Native sample		32.4	45.5	22.1	(Zeng et al., 2016)
Waxy rice	30	37.4	42.0	20.7	
	40	43.1	39.4	17.6	
	50	50.4	35.2	14.4	

that the pulsed electric field can reduce the crystallinity of starch and cause changes in the morphology of starch grains, which makes starch more sensitive to enzymatic activity. This indicates that many glycosidic bonds should be converted into more hydrolyzable starch (in other words, an increase of rapidly hydrolyzed starch), lowering the content of slowly digestible starch (F. Zhu, 2018).

7.4 APPLICATIONS, ADVANTAGES, AND LIMITATIONS OF PEF IN STARCH

Recent research suggests that PEF treatments may have an impact on the following aspects of starch: conformation, size, microstructure, solubility, viscoelastic characteristics, swelling effect, digestibility in vitro, thermal stability, and structural transition. PEF has the benefit of causing the changes in the physicochemical characteristics of starch with less energy for a brief amount of time when compared to other starch modification techniques (Abduh et al., 2019; Yang et al., 2018). Emerging technologies can manufacture modified starches with diverse functionalities, which might be used to improve 3D printing, among other things. The latter is a cutting-edge technique that may create materials with a complicated internal structure and a bespoke, individualized pattern.

Applications for starch-based materials in 3D printing have been researched for a variety of applications, including food and medical. Though several studies have been conducted to produce modified starches for 3D printing, none of them have examined the application of PEF-treated starch thus far (Koski & Bose, 2019; Mantihal et al., 2020).

In this regard, with an emphasis on the use of 3D printing, Maniglia and colleagues assessed the effect of PEF on the composition, characteristics, and abilities of cassava and wheat starches. Three distinct combinations of field strength and total specific energy input (T1: 15 kV/cm; 25 kJ/kg; T2: 25 kV/cm; 25 kJ/kg; and T3: 25 kV/cm; 50 kJ/kg) were used to PEF-treat aqueous starch solutions. For cassava starch, all three conditions produced the same results (no damage to the granule surface, a decrease in peak apparent viscosity, and firmer gels); however, T3 enhanced the effects on wheat starch (fractures on the granule surface, a decrease in peak apparent viscosity, firmer gels). After selection for further analysis, the T3 condition showed no changes in functional groups but depolymerization, a decrease in relative crystallinity, and a decrease in gelatinization enthalpy. While PEF-treated cassava starch performed similarly to native starch, PEF-treated wheat starch produced 3D printing with a smoother surface and distinct texture. Consequently, PEF influences each source differently, which might improve 3D printing applications (Maniglia et al., 2021).

PEF methodology has been applied to chemically change starches in addition to physically modifying them, and it has been further compared with conventional chemical acetylation procedures. Some advantages and disadvantages of PEF method for starch extraction and modification are shown in Figure 7.1. The conventional PEF acetylation of starch lowers expenses, conserves reagents, shortens the modification time, and enhances reaction efficiency by allowing for higher degrees of acetylation. The literature claims that acetic anhydride must be used as an acetylating agent in classical acetylation, which raises expenses and poses environmental hazards. As for the modification phase itself, it calls for extra caution that isn't involved in physical modification, including adjusting pH to favor acetylation conditions. NaOH is used to make this correction, implying the need for an additional chemical reagent (Almeida, Santos, Feitoza, et al., 2022; Maniglia et al., 2021; Raso et al., 2022). Furthermore, conventional acetylation takes a lot longer than physical alteration. Following modification, the acetylation reaction must be stopped with ethanol, and any leftover acetic anhydride from the acetylation process must be washed out of the starches. Physical modification just needs the use of water as a solvent, but chemical acetylation involves the use of several chemical solvents up until this point. Following that, the starches are sieved, dried, and stored. PEF technology is, therefore, quicker, safer, greener, and more ecologically friendly. The only physical adjustment required is to filter the modified, cooled starches and then dry, screen, and store them

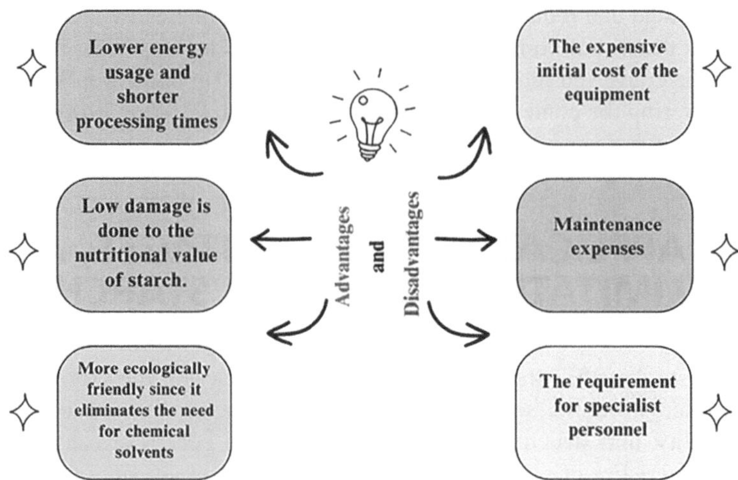

FIGURE 7.6   Advantages and disadvantages of using PEF for starch extraction and modification.

(Arshad et al., 2020; Prabhu et al., 2019; Stewart et al., 2019). In general, physical modification is superior to chemical methods in the following ways: (1) simplifies or reduces the number of steps required, which reduces the time required to prepare the starch suspension and store the modified starch; (2) significantly shortens the modification time; (3) uses only water instead of chemical solvents, resulting in less waste being produced (greener and more environmentally friendly); (4) is easier to use; (5) makes it easier to control experimental conditions because guidelines for the application of PEF in food and biotechnological processes have already been created, providing the parameters that must be controlled; and (6) allows for batch or continuous mode (Castro et al., 2023; Maniglia et al., 2021).

Ohmic heating results from the conversion of electrical energy into heat, even though PEF processing is non-thermal. Additional conformational changes in the starches caused by this heating result in minor swelling, mild gelatinization, or weak development of a gel-like structure. According to Han and coworkers (2009), corn and potato starch particles underwent PEF treatment and developed a gel-like shape, although the authors of both cases acknowledged utilizing a water bath to regulate the temperature below 50°C during PEF therapy. Peak temperature could go unnoticed, nevertheless, due to the process's speed. Thus, these and other PEF therapy effects may be explained by heating. Their heating significantly influenced the alteration of starch granules. Since boiling starch in aqueous solutions causes molecular chains and helical structures to move more, its crystalline structures become weaker (Han, Zeng, Zhang, et al., 2009; Han, Zeng, Yu, et al., 2009). Moreover, PEF has several drawbacks, including: (1) the expensive initial cost of the equipment; (2) maintenance expenses; and (3) the requirement for specialist personnel. But after a few years, when the equipment has fully depreciated, the initial expenditure proves more beneficial (Castro et al., 2023; Maniglia et al., 2021).

7.5 CONCLUSION AND FUTURE PERSPECTIVES

This chapter's findings have convincingly shown how to change the structural and functional characteristics of starches through the application of pulsed electric field processing. Based on both physical and chemical prospects, PEF treatment-induced electrochemical and electrolytic processes, as well as starch polarization effects, start changes in structural features such as morphology, crystalline structure, lamellar structure, and molecular order structure. Then these structural alterations have an impact on the starch's digestibility, pasting qualities, freeze–thaw stability, and thermal characteristics. These variations were impacted by the PEF process parameters, which included electric field intensity; treatment duration; pulse number; temperature; and intrinsic material attributes such as concentration, conductivity, and pH. The aforementioned investigations were carried out in labs. There are currently no established industrial facilities for PEF-induced starch changes. Technical viability research and cost assessment are therefore crucial for moving batch processes to continuous and pilot scale. In addition, PEF treatment results in a decrease in relative crystallinity; alters the lamellar repeating distance of starch based on the starch's botanical origin; and lowers viscosity, pasting temperature, gelatinization temperatures, and enthalpies. In terms of in vitro digestibility, it appears to retain the amount of resistant starch while increasing the amount of quickly digested starch and decreasing the amount of slowly digested starch. These may be connected to modifications in the starch chains, and more research into the intricate structure of amylopectin may yield fresh information to clarify the fluctuations in starch digestibility. The reduced digestibility of PEF-treated starch under in-vitro human simulated digestive circumstances appears to be advantageous for adding these starches to humans' diets. In comparison to conventional acetylation, PEF modification technology is safer and more ecologically friendly since it eliminates the need for chemical solvents. It also has a reduced processing cost.

Furthermore, PEF has demonstrated promise in helping to extract starch from algae. However, further research is required to assess this potential in other starch-rich matrices, including fruits, vegetables, roots and tubers, and cereals. Additionally, optimal operating conditions are needed to maximize protein removal and maximize starch purity. Further study might compare the effects of PEF-modified starch with those of other currently used technologies.

REFERENCES

Abduh, S. B., Leong, S. Y., Agyei, D., & Oey, I. (2019). Understanding the properties of starch in potatoes (Solanum tuberosum var. Agria) after being treated with pulsed electric field processing. *Foods*, 8(5), 159. https://doi.org/10.3390/foods8050159

Almeida, R. L. J., Santos, N. C., Feitoza, J. V. F., de Alcântara Ribeiro, V. H., de Alcântara Silva, V. M., de Figueiredo, M. J., Ribeiro, C. A. C., Galdino, P. O., Queiroga, A. H. F., & de Sousa Muniz, C. E. (2022). The impact of the pulsed electric field on the structural, morphological, functional, textural, and rheological properties of red rice starch (Oryza sativa). *Journal of Food Process Engineering*, 45(11), e14145.

Almeida, R. L. J., Santos, N. C., Muniz, C. E. S., da Silva Eduardo, R., de Almeida Silva, R., Ribeiro, C. A. C., da Costa, G. A., de Figueiredo, M. J., Galdino, P. O., & Dos Santos, E. S. (2023). Red rice starch modification-Combination of the non-thermal method with a pulsed electric field (PEF) and enzymatic method using α-amylase. *International Journal of Biological Macromolecules*, 253, 127030.

Almeida, R. L. J., Santos, N. C., Padilha, C. E., de Almeida Mota, M. M., de Alcântara Silva, V. M., André, A. M. M. C. N., & dos Santos, E. S. (2022). Application of pulsed electric field and drying temperature response on the thermodynamic and thermal properties of red rice starch (Oryza Sativa L.). *Journal of Food Process Engineering*, 45(2), e13947.

Arshad, R. N., Abdul-Malek, Z., Munir, A., Buntat, Z., Ahmad, M. H., Jusoh, Y. M., Bekhit, A. E. D., Roobab, U., Manzoor, M. F., & Aadil, R. M. (2020). Electrical systems for pulsed electric field applications in the food industry: An engineering perspective. *Trends in Food Science & Technology*, 104, 1–13.

Bertoft, E. (2017). Understanding starch structure: Recent progress. *Agronomy*, 7(3), 56.

Bhat, Z. F., Morton, J. D., Mason, S. L., & Bekhit, A. E.-D. A. (2019). Current and future prospects for the use of pulsed electric field in the meat industry. *Critical Reviews in Food Science and Nutrition*, 59(10), 1660–1674.

Castro, L. M. G., Alexandre, E. M. C., Saraiva, J. A., & Pintado, M. (2023). Starch extraction and modification by pulsed electric fields. *Food Reviews International*, 39(4), 2161–2182. https://doi.org/10.1080/87559129.2021.1945620

Chen, B.-R., Teng, Y.-X., Wang, L.-H., Xu, F.-Y., Li, Y., Wen, Q.-H., Wang, R., Li, J., Wang, Z., & Zeng, X.-A. (2023). Pulsed electric field-assisted esterification improves the freeze–thaw stability of corn starch gel by changing its molecular structure. *International Journal of Biological Macromolecules*, 231, 123085.

Chen, B.-R., Wen, Q.-H., Zeng, X.-A., Abdul, R., Roobab, U., & Xu, F.-Y. (2021). Pulsed electric field assisted modification of octenyl succinylated potato starch and its influence on pasting properties. *Carbohydrate Polymers*, *254*, 117294.

Cornejo-Ramírez, Y. I., Martínez-Cruz, O., Del Toro-Sánchez, C. L., Wong-Corral, F. J., Borboa-Flores, J., & Cinco-Moroyoqui, F. J. (2018). The structural characteristics of starches and their functional properties. *CyTA-Journal of Food*, *16*(1), 1003–1017.

Cserhalmi, Z., Sass-Kiss, A., Tóth-Markus, M., & Lechner, N. (2006). Study of pulsed electric field treated citrus juices. *Innovative Food Science & Emerging Technologies*, *7*(1–2), 49–54.

Dunn, J. (2019). *Pulsed electric field processing: An overview* (pp. 1–30). CRC Press.

Duque, S. M. M., Leong, S. Y., Agyei, D., Singh, J., Larsen, N., & Oey, I. (2020). Understanding the impact of pulsed electric fields treatment on the thermal and pasting properties of raw and thermally processed oat flours. *Food Research International*, *129*, 108839.

Eliasson, A.-C. (2017). Starch: Physicochemical and functional aspects. In *Carbohydrates in food* (pp. 501–600). CRC Press.

El Kantar, S., Boussetta, N., Lebovka, N., Foucart, F., Rajha, H. N., Maroun, R. G., Louka, N., & Vorobiev, E. (2018). Pulsed electric field treatment of citrus fruits: Improvement of juice and polyphenols extraction. *Innovative Food Science & Emerging Technologies*, *46*, 153–161.

Englyst, H. N., Kingman, S., & Cummings, J. (1992). Classification and measurement of nutritionally important starch fractions. *European Journal of Clinical Nutrition*, *46*, S33–S50.

Gómez, B., Munekata, P. E., Gavahian, M., Barba, F. J., Martí-Quijal, F. J., Bolumar, T., Campagnol, P. C. B., Tomasevic, I., & Lorenzo, J. M. (2019). Application of pulsed electric fields in meat and fish processing industries: An overview. *Food Research International*, *123*, 95–105.

Han, Z., Yu, Q., Zeng, X. A., Luo, D. H., Yu, S. J., Zhang, B. S., & Chen, X. D. (2012). Studies on the microstructure and thermal properties of pulsed electric fields (PEF)-treated maize starch. *International Journal of Food Engineering*, *8*(1).

Han, Z., Zeng, X. A., Fu, N., Yu, S. J., Chen, X. D., & Kennedy, J. F. (2012). Effects of pulsed electric field treatments on some properties of tapioca starch. *Carbohydrate Polymers*, *89*(4), 1012–1017.

Han, Z., Zeng, X.-A., Zhang, B.-S., & Yu, S.-J. (2009). Effects of pulsed electric fields (PEF) treatment on the properties of corn starch. *Journal of Food Engineering*, *93*(3), 318–323.

Han, Z., Zeng, X. A., Yu, S. J., Zhang, B. S., & Chen, X. D. (2009). Effects of pulsed electric fields (PEF) treatment on physicochemical properties of potato starch. *Innovative Food Science & Emerging Technologies*, *10*(4), 481–485.

Hong, J., Chen, R., Zeng, X.-A., & Han, Z. (2016). Effect of pulsed electric fields assisted acetylation on morphological, structural and functional characteristics of potato starch. *Food Chemistry*, *192*, 15–24.

Hong, J., Zeng, X.-A., Buckow, R., Han, Z., & Wang, M.-S. (2016). Nanostructure, morphology and functionality of cassava starch after pulsed electric fields assisted acetylation. *Food Hydrocolloids*, *54*, 139–150.

Hong, J., Zeng, X.-A., Han, Z., & Brennan, C. S. (2018). Effect of pulsed electric fields treatment on the nanostructure of esterified potato starch and their potential glycemic digestibility. *Innovative Food Science & Emerging Technologies*, *45*, 438–446.

Jayaram, S. H. (2000). Sterilization of liquid foods by pulsed electric fields. *IEEE Electrical Insulation Magazine*, *16*(6), 17–25.

Kaufmann, M., & Pitt, T. (2018). Pulsed-field gel electrophoresis of bacterial DNA. In *Methods in practical laboratory bacteriology* (pp. 83–92). CRC Press.

Khan, M. K., Ahmad, K., Hassan, S., Imran, M., Ahmad, N., & Xu, C. (2018). Effect of novel technologies on polyphenols during food processing. *Innovative Food Science & Emerging Technologies*, *45*, 361–381.

Koski, C., & Bose, S. (2019). Effects of amylose content on the mechanical properties of starch-hydroxyapatite 3D printed bone scaffolds. *Additive Manufacturing*, *30*, 100817. https://doi.org/10.1016/j.addma.2019.100817

Kumari, B., Tiwari, B. K., Hossain, M. B., Brunton, N. P., & Rai, D. K. (2018). Recent advances on application of ultrasound and pulsed electric field technologies in the extraction of bioactives from agro-industrial byproducts. *Food and Bioprocess Technology*, *11*, 223–241.

Li, Q., Wu, Q.-Y., Jiang, W., Qian, J.-Y., Zhang, L., Wu, M., Rao, S. Q., & Wu, C.-S. (2019). Effect of pulsed electric field on structural properties and digestibility of starches with different crystalline type in solid state. *Carbohydrate Polymers*, *207*, 362–370.

Li, Y., Wang, J.-H., Han, Y., Yue, F.-H., Zeng, X.-A., Chen, B.-R., Zeng, M. Q., Woo, M. W., & Han, Z. (2023). The effects of pulsed electric fields treatment on the structure and physicochemical properties of dialdehyde starch. *Food Chemistry*, *408*, 135231.

Li, Y., Wang, J.-H., Wang, E.-C., Tang, Z.-S., Han, Y., Luo, X.-E., Zeng, X. A., Woo, M. W., & Han, Z. (2023). The microstructure and thermal properties of pulsed electric field pretreated oxidized starch. *International Journal of Biological Macromolecules*, *235*, 123721.

Luo, X.-E., Wang, R.-Y., Wang, J.-H., Li, Y., Luo, H.-N., Zeng, X.-A., Woo, M. W., & Han, Z. (2023). Combining pulsed electric field and cross-linking to enhance the structural and physicochemical properties of corn porous starch. *Food Chemistry, 418*, 135971.

Maniglia, B. C., Pataro, G., Ferrari, G., Augusto, P. E. D., Le-Bail, P., & Le-Bail, A. (2021). Pulsed electric fields (PEF) treatment to enhance starch 3D printing application: Effect on structure, properties, and functionality of wheat and cassava starches. *Innovative Food Science & Emerging Technologies, 68*, 102602.

Mantihal, S., Kobun, R., & Lee, B.-B. (2020). 3D food printing of as the new way of preparing food: A review. *International Journal of Gastronomy and Food Science, 22*, 100260. https://doi.org/10.1016/j.ijgfs.2020.100260

Min, S., Jin, Z., Min, S., Yeom, H., & Zhang, Q. (2003). Commercial-scale pulsed electric field processing of orange juice. *Journal of Food Science, 68*(4), 1265–1271.

Panja, P. (2018). Green extraction methods of food polyphenols from vegetable materials. *Current Opinion in Food Science, 23*, 173–182.

Prabhu, M. S., Levkov, K., Livney, Y. D., Israel, A., & Golberg, A. (2019). High-voltage pulsed electric field preprocessing enhances extraction of starch, proteins, and ash from marine macroalgae Ulva ohnoi. *ACS Sustainable Chemistry & Engineering, 7*(20), 17453–17463. https://doi.org/10.1021/acssuschemeng.9b04669

Qiu, S., Abbaspourrad, A., & Padilla-Zakour, O. I. (2021). Changes in the glutinous rice grain and physicochemical properties of its starch upon moderate treatment with pulsed electric field. *Foods, 10*(2), 395.

Raso, J., Heinz, V., Alvarez, I., & Toepfl, S. (2022). *Pulsed electric fields technology for the food industry*. Springer.

Schirmer, M., Jekle, M., & Becker, T. (2015). Starch gelatinization and its complexity for analysis. *Starch-Stärke, 67*(1–2), 30–41.

Singh, H., Blennow, A., Gupta, A. D., Kaur, P., Dhillon, B., Sodhi, N. S., & Dubey, P. K. (2022). Pulsed light, pulsed electric field and cold plasma modification of starches: Technological advancements & effects on functional properties. *Journal of Food Measurement and Characterization, 16*(5), 4092–4109.

Stewart, M. T., Haines, D. E., Verma, A., Kirchhof, N., Barka, N., Grassl, E., & Howard, B. (2019). Intracardiac pulsed field ablation: Proof of feasibility in a chronic porcine model. *Heart Rhythm, 16*(5), 754–764.

Taha, A., Casanova, F., Šimonis, P., Jonikaitė-Švėgždienė, J., Jurkūnas, M., Gomaa, M. A., & Stirkė, A. (2022). Pulsed electric field-assisted glycation of bovine serum albumin/starch conjugates improved their emulsifying properties. *Innovative Food Science & Emerging Technologies, 82*, 103190.

Wang, B., Li, D., Wang, L.-J., Chiu, Y. L., Chen, X. D., & Mao, Z.-H. (2008). Effect of high-pressure homogenization on the structure and thermal properties of maize starch. *Journal of Food Engineering, 87*(3), 436–444.

Warren, F. J., Gidley, M. J., & Flanagan, B. M. (2016). Infrared spectroscopy as a tool to characterise starch ordered structure—a joint FTIR–ATR, NMR, XRD and DSC study. *Carbohydrate Polymers, 139*, 35–42.

Wu, C., Wu, Q.-Y., Wu, M., Jiang, W., Qian, J.-Y., Rao, S.-Q., Zhang, L., Li, Q., & Zhang, C. (2019). Effect of pulsed electric field on properties and multi-scale structure of japonica rice starch. *LWT, 116*, 108515.

Yang, F., Zhang, M., Bhandari, B., & Liu, Y. (2018). Investigation on lemon juice gel as food material for 3D printing and optimization of printing parameters. *LWT, 87*, 67–76. https://doi.org/10.1016/j.lwt.2017.08.054

Yang, Z., Chaib, S., Gu, Q., & Hemar, Y. (2017). Impact of pressure on physicochemical properties of starch dispersions. *Food Hydrocolloids, 68*, 164–177.

Zeng, F., Gao, Q.-Y., Han, Z., Zeng, X.-A., & Yu, S.-J. (2016). Structural properties and digestibility of pulsed electric field treated waxy rice starch. *Food Chemistry, 194*, 1313–1319.

Zhu, F. (2018). Modifications of starch by electric field based techniques. *Trends in Food Science & Technology, 75*, 158–169.

Ionizing Radiation

8

Mirela Braşoveanu, Betul Oskaybaş Emlek,
Hassan Sabbaghi, Kevser Kahraman, Serpil Özturk
Muti, Farooq Sher, and Monica R. Nemţanu

8.1 INTRODUCTION

Starch is one of the most plentiful and versatile carbohydrates, being of utmost importance in countless food applications. It is one of the most widespread and essential ingredients in the food sector, found in the composition of diverse products that are consumed daily. A polysaccharide known for its versatility, starch is composed of glucose molecules joined by α-1,4 and α-1,6 glycosidic linkages. It can participate in the transformation of texture, structure, and mouthfeel in various food products. From thickening sauces to providing a crunchy texture in fried snacks and imparting a creamy texture to puddings and custards, starch plays a critical role in shaping eating experiences.

However, the utilization of native starch comes with its set of challenges and limitations as a result of its functional properties (Bangar et al., 2022; Chakraborty et al., 2022; Compart et al., 2023; Maniglia et al., 2021), which can vary significantly depending on its botanical source and the processing methods employed. In food applications, specific limitations include restricted resistance to high temperatures, susceptibility to retrogradation leading to reduced shelf life and syneresis, sensory attribute issues, and limited compatibility with certain processing techniques and ingredients. The process itself is resource intensive due to the necessary extraction and purification procedures. Furthermore, challenges arise in the form of variations in functional properties due to botanical sources, shear sensitivity causing potential breakdown and loss of thickening ability, the need for precise control over viscosity, syneresis affecting product quality, the assessment of gelatinization temperatures suitable for specific processes, and concerns related to allergens.

The food industry is continuously looking for new methods to improve starch qualities and overcome the limitations of its native form. Starch modification is a practice that aims to alter the starch functions, thereby improving its performance in various applications. Different chemical, physical, enzymatic, and genetic methods can be used to modify the molecular and crystalline structure of starch, thereby increasing its usefulness in industrial applications (Maniglia et al., 2021). With the increasing demand for safer and more sustainable starch modification methods, ionizing radiation processing has emerged as a progressive and environmentally friendly alternative. Thus, from the perspective of starch, ionizing radiation offers unique benefits in terms of functionality, safety, and sustainability.

In food processing and preservation, the utilization of ionizing radiation is widely acknowledged and subject to stringent regulations, with applications ranging from dehydrated condiments to moist food such as meat (Danyo et al., 2024). Therefore, ionizing radiation-based techniques, such as gamma ray, X-ray, and electron beam irradiation, have found extensive applications in the food sector, including microbial decontamination, elimination of insects and parasites, control of sprouting in tubers and bulbs, inhibition

DOI: 10.1201/9781032655598-8

of enzyme activity, and maintaining the stability of stored products (Mshelia et al., 2023). These techniques necessitate minimal sample preparation, induce negligible temperature elevation, and generate no chemical residues or byproducts in the processed products, demonstrating rapidity, efficiency, and environmental friendliness (Braşoveanu & Nemţanu, 2018; Rostamabadi et al., 2023).

The role of ionizing radiation in starch modification is primarily associated with its ability to induce the formation of free radicals within the starch matrix (Polesi et al., 2016; Rostamabadi et al., 2023). These free radicals initiate reactions that can lead to alterations in the size and structure of starch components, particularly amylose and amylopectin. The consequent changes in the functional and physicochemical properties of starch are significantly influenced by the processing parameters, such as irradiation dose, irradiation dose rate, sample moisture content, and gas atmosphere type (Braşoveanu & Nemţanu, 2018). Additionally, compliance with regulatory guidelines and safety standards is essential when applying ionizing radiation to starch-based food processing. The recommended irradiation dose for starch modification in food applications depends on the desired modifications and the specific requirements of the food product. According to Codex Alimentarius (CAC, 2003), the highest absorbed dose applied to food should not exceed 10 kGy, except in cases where it is essential to fulfilling a valid technological objective.

Within the realm of starch modification for food applications, the main focus of this chapter is a summary of recent significant findings related to transformations in the features of starch originating in diverse botanical sources when subjected to ionizing radiation processing. Moreover, aspects encompassing analytical methods for characterizing irradiated starch, the reaction mechanism involved, and considerations on safety and regulations associated with the utilization of ionizing radiation for treating starch in the context of food applications are discussed.

8.2 IONIZING RADIATION-BASED TECHNIQUES

Radiation represents a form of energy that manifests as waves or particles, transferring from one point to another. Irradiation, the process by which this energy is directed at a material for a specific purpose, can induce various effects, both physical, such as color alteration and temperature elevation, or chemical, through ionizing and exciting atoms and molecules within the material. Electromagnetic radiation and high-energy electrons (a form of corpuscular radiation) stand as common radiation types employed in the processing of bio-based materials (Lima et al., 2018; Bisht et al., 2021).

8.2.1 Radiation Sources

The electromagnetic spectrum distinguished between non-ionizing (e.g., ultraviolet and visible light, infrared radiation, microwaves) and ionizing radiation (e.g., X-rays, gamma radiation), which exhibit distinct characteristics such as wavelength, energy, penetration depth, or effects on the biological matter (Lima et al., 2018; Bisht et al., 2021). Ionizing radiation, exemplified by X-ray and gamma-ray as forms of electromagnetic radiation, along with high-energy electron beams recognized as β particles, are the favored modalities for modifying starch macromolecules (Kumari & Sit, 2023). Gamma rays originate from radioisotopes like cobalt-60 (^{60}Co) and cesium-137 (^{137}Cs), whereas electron beams and X-rays are generated using electron accelerators and electron beam devices integrated into e$^-$/X converters (Chmielewski, 2023). Nevertheless, these various forms of radiation fundamentally interact similarly with materials subjected to irradiation. Their principal distinctions concerning applicability pertain to their penetration capability and dose rate. For instance, gamma radiation proves advantageous for processing comparatively thick or dense materials owing to its superior penetrating capabilities when contrasted with electron beams, which possess limited penetration depth but significantly higher dose rates (Rostamabadi et al., 2023).

8.2.2 Radiation–Matter Interaction

In the field of radiation physics, the fundamental concept of the interaction of ionizing radiation with substance includes a wide range of phenomena occurring at the atomic and molecular levels. Understanding how ionizing radiation interacts with materials is essential for both generating its advantages and mitigating its risks.

According to Braşoveanu and Nemţanu (2018), the fundamental mechanisms involved in the transfer of energy via γ-rays or X-rays predominantly encompasses the photoelectric effect, Compton scattering, and the production of electron–positron pairs, inducing the emission of rapid electrons, which dissipate energy by events similar to electron beams. These processes give rise to two principal effects: the ionization and excitation of atoms and molecules within the material. Ionization represents the primary event that occurs when a neutral atom or molecule acquires a charge, thereby forming an ion. When an ion is produced as a result of an electron loss or capture, it includes an unpaired electron, effectively constituting a highly reactive chemical species known as a free radical. The expelled electron can further ionize additional atoms and molecules through successive collisions and ionization events. Excitation, another main process, occurs as a charged particle of high energy goes through atoms, giving energy to electrons within them but not driving them out, thereby resulting in the excitation of the atom. Therefore, the primary species formed are ions, highly reactive free radicals, and molecules in an excited state.

The secondary effects encompass various reactions involving the primary species that result in the end products (Braşoveanu & Nemţanu, 2018). These events involve the splitting of an excited molecule, leading to the formation of two radicals or two distinct molecular entities. The free radicals actively engage in radical-radical recombination processes by which they recombine with one another to either regenerate the original molecular structure or form new ones. Additionally, these free radicals may undergo recombination with another fresh molecule, extracting an atom of hydrogen and thereby creating a novel free radical along with an entirely fresh molecular structure.

8.2.3 Radiation Quantities and Units

Two essential dosimetry parameters are the absorbed dose and the absorbed dose rate (ICRU, 1980). These radiation metrics stand as pivotal physical quantities within the realm of dosimetry, essential for optimizing and regulating the irradiation process (IAEA, 1987, 2002).

Absorbed dose, denoted as D, represents the quantity of energy absorbed per unit mass of irradiated material at a specific location within the area of concern (IAEA, 2002). The SI unit for absorbed dose is the *gray* (Gy), where 1 gray is equivalent to 1 joule of energy absorbed per kilogram of the absorbing material. The absorbed dose is crucial in assessing the radiation impact on treated material.

The rate of absorbed dose change over time is termed the absorbed dose rate (Braşoveanu & Nemţanu, 2018), denoted as $\dot{D}$, and it is expressed in grays per unit of time. The dose rate is essential in determining the radiation intensity at any given moment.

In general, the relationship between absorbed dose and dose rate in material processing by applying ionizing radiation is interlinked (Adlienė & Adlytė, 2017). The absorbed dose directly impacts the resulting modifications in material properties, while the dose rate determines the speed and uniformity of these modifications. Adjusting the dose rate can influence the time required for irradiation while ensuring the desired absorbed dose for the specific modification objectives.

8.2.4 Advantages and Disadvantages

The use of ionizing radiation in technological processing comes with its set of advantages and disadvantages. The specific advantages and disadvantages can vary based on the type of radiation used (e.g., X-ray,

TABLE 8.1   Advantages and Disadvantages of Ionizing Radiation Processing (Braşoveanu & Nemţanu, 2018; Fan & Niemira, 2020; Pi et al., 2021)

RADIATION TYPE	ADVANTAGES	DISADVANTAGES
X-rays	Good penetration ability, suitable for bulk operations Lower energy, reducing safety concerns Radiation can be turned on and off	Limited penetration power compared to gamma rays Can be absorbed by dense materials, limiting applicability Extended processing time Expensive costs
Gamma rays	High penetration power, suitable for dense materials Efficient for large-scale processing Long half-life of radiation sources Continual emitting of rays at a predictable rate	Low dose rate Extended processing time Utilization of radioactive sources, which requires more shielding for workers and environmental protection than X-rays and electron beams Continuous emission of rays Higher equipment and facility costs compared to X-rays
Electron beams	Applicability in high-flow and high-dose irradiation Short time of processing Precise energy control Radiation can be switched on and off The most cost-efficient form among the ionizing radiations	Low penetrability High initial investment cost

gamma ray, and electron beam). Table 8.1 provides several of the technological advantages and disadvantages of ionizing radiation processing.

8.3 IMPACT ON STARCH STRUCTURE

This section presents considerations on the alteration of granule morphology, crystalline structure, and molecular characteristics of starch originating from various botanical sources induced by ionizing radiation.

8.3.1 Morphological Properties

Due to its higher molecular size, starch is more sensitive to structural breakdown triggered by ionizing radiation than other food matrix components (Rostamabadi et al., 2023). Several key aspects, including the source of the starch, the level of irradiation (specifically the dose and duration), the sample moisture content, and the crystalline pattern, influence the morphological behavior of starch under irradiation. However, it may be difficult to observe the effects of irradiation on granule morphology when just a partial breakdown of glycosidic bonds occurs (Zhiguang et al., 2023). In addition, irradiation and starch characteristics contribute to the inability to observe the influence of radiation on morphological changes.

Research on the alterations in morphological properties after irradiation has revealed the significance of irradiation dose as a crucial parameter affecting starch morphology. Numerous scholarly studies using

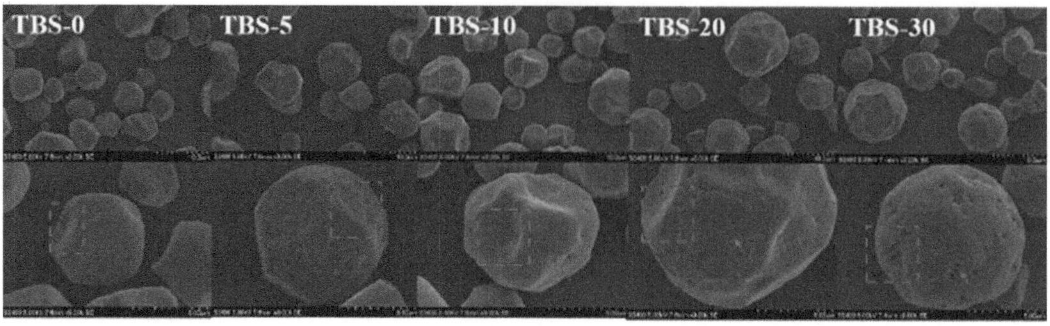

FIGURE 8.1 Particle morphology of EBI treated TBS with varied absorbed doses. A column represented a sample. Top line: observation at 3000 times; Bottom line: observation at 9000 times. (Reprinted from Huang et al. (2023), with permission from Elsevier).

scanning electron microscopy (SEM) have demonstrated that low-dose ionizing radiation does not cause noticeable effects on the granule morphology. Irradiation doses below 10 kGy did not significantly change the morphology of the starch granules extracted from potatoes (< 5 kGy) (Liang et al., 2022; Lei et al., 2023), oats (< 8 kGy) (Shen et al., 2023), quinoa (< 8 kGy) (Du et al., 2020), black rice (< 10 kGy) (Dhull et al., 2021), and rice (< 10 kGy) (Pan et al., 2020). In contrast, several investigations have found that the structural properties of starch undergo considerable transformations after being subjected to higher doses of irradiation. Irradiating talipot palm starch at doses of 50, 80, and 100 kGy, for instance, led to the loss of smoothness in certain granule surfaces (Navaf et al., 2022). These changes are assigned to granule disintegration caused by the effect of highly energetic irradiation (Aaliya et al., 2021). Punia et al. (2020) investigated dents and cavities that are formed on the granule surface with a few granules even rupturing at a high dose of irradiation of 20 kGy. In another study (Huang et al., 2023) it was shown that the holes and wrinkles occurred on Tartary buckwheat starch (TBS) granule surface after electron beam irradiation (EBI) (Figure 8.1). The degree of depression increased, and the granules became rougher at 20 and 30 kGy irradiation doses, respectively.

In another study, Govindaraju et al. (2022) indicated that the smaller-sized starch granule is more susceptible to irradiation compared to large starch granules. The optical images demonstrated that the potato starch granules exhibited fewer morphological alterations compared to the corn and rice starch granules after irradiation (Govindaraju et al., 2022). According to Chung and Liu (2010), B-type starch exhibits a higher susceptibility to irradiation-induced modifications in comparison to C-type starch. Furthermore, Lee et al. (2006) observed that the impact of irradiation on granule size is contingent upon the moisture level of the sample. In addition to that, polarized light was utilized to look into the morphological properties of the irradiated starch. When examined microscopically under polarized light, the starch granule exhibits a classic Maltese cross model due to the birefringence of its crystalline areas, which characterizes the radial orientation of the macromolecules (El-Esawi, 2019). It was reported by Liang et al. (2022) that irradiation doses did not affect the polarized cross's position, brightness, and blackness, indicating the preservation of the starch crystalline structure. Instead, Lei et al. (2023) investigated the starch morphological properties using polarized light, and it was found that the irradiation led to the disruption of the original semi-crystalline structure of larger granules of potato starch, and the crystalline arrangement was identified as disordered.

8.3.2 Crystalline Properties

The relationship between irradiation and starch crystallinity has been widely studied so far. The majority of the investigations have consistently demonstrated that the ionizing radiation treatment can lead to

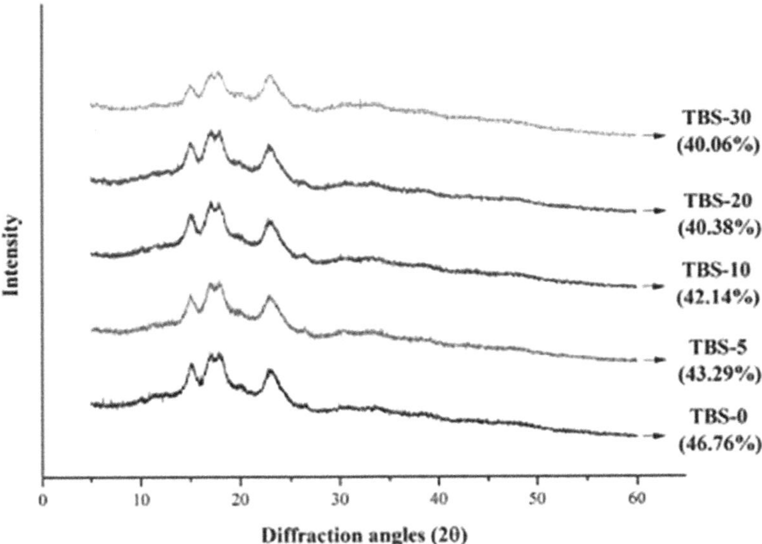

FIGURE 8.2   XRD pattern of EBI treated TBS with varied absorbed doses. Crystallinity was displayed in parentheses. (Reprinted from Huang et al. (2023), with permission from Elsevier).

a reduction in X-ray peak and relative crystallinity while the starch crystal configuration remains unaffected (Figure 8.2). Lei et al. (2023) demonstrated that the application of ionizing radiation treatment had no effects on the crystallization B-type of potato starch, while the relative crystallinity decreased. Shen et al. (2023) also mentioned that the position of oat starch X-ray peaks was not influenced by irradiation. Furthermore, it was discovered a direct relationship between the irradiation dose and oat starch crystallinity, with a resultant decrease. This drop may be caused by the degradation of amylopectin generated by the processing with radiation. The research conducted by Shen et al. (2023) highlighted that irradiation induces starch breakdown, leading to the generation of a greater number of short chains. Consequently, this phenomenon was associated with a decline in relative crystallinity. Comparable results were observed in starches from maize and potato (Di et al., 2023; Li et al., 2023; Tappiban et al., 2022), waxy maize (Zhou et al., 2020), rice (Pan et al., 2020; Zheng et al., 2023), Tartary buckwheat (Huang et al., 2023), cassava (Nontamas et al., 2022), talipot palm (Aaliya et al., 2021; Navaf et al., 2022), black rice (Dhull et al., 2021), quinoa (Du et al., 2020), lotus seed (Punia et al., 2020), kithul (Sudheesh et al., 2019), and buckwheat and oat (Dar et al., 2018). As opposed to this, Liang et al. (2022) have found that a rise in the irradiation dose of the electron beam causes an increase in the relative crystallinity of potato starch. Furthermore, it was indicated that the application of numerous irradiations facilitated the attainment of these outcomes. According to Liang et al. (2022), this observed inconsistency result can be attributed to starch diversity and different treatment conditions. Another explanation for this difference is the formation of more short chains and tiny fragments, particularly evident with elevated doses and multiple irradiations. Consequently, it was anticipated that these smaller molecules may undergo re-association, resulting in the formation of crystalline structural domains and thereby causing a rise in relative crystallinity (Liang et al., 2022). In another study, Li et al. (2023) applied electron beam irradiation to high amylose maize starch up to 20 kGy dose. The X-ray diffraction pattern exhibited features reminiscent of the native starch, and there was no consistent correlation between the irradiation dose level and the relative crystallinity. The authors suggested that the arrangement of molecules in high-amylose starch might exhibit a higher resistance to irradiation compared to low-amylose starch.

8.3.3 Molecular Properties

Irradiation is considered safe and suitable for foods up to a defined irradiation dose. However, irradiation can cause modifications in the molecular arrangements of protein, starch, and lipid composition of foods (Rostamabadi et al., 2023). Fourier-transform infrared spectrometry (FTIR) has found widespread application in evaluating the impact of irradiation on starch molecular structure (Zhiguang et al., 2023). FTIR spectra are used to illustrate the distinctive bands that correspond to the stretching, bending, and deformation properties of the primary functional groups present inside starch granules (Aaliya et al., 2021). The FTIR spectra of native starch should exhibit five distinct absorption bands. According to Braşoveanu and Nemţanu (2018), the absorption bands observed in the infrared spectrum can be assigned to specific molecular vibrations. The range of 3000–3700 cm^{-1} is assigned to the stretching vibrations of O-H bonds. The 2800–3000 cm^{-1} range is indicative of stretching vibrations of C-H bonds. Within 1550–1800 cm^{-1}, the vibrations of O-H bonds in water molecules that are bound to other molecules can be identified. The range of 800–1550 cm^{-1} is known as the fingerprint region, which contains a variety of molecular vibrations that are characteristic of the specific compound being analyzed. Finally, the absorption bands below 800 cm^{-1} are specifically associated with the pyranose ring of the glycosidic unit.

Most studies reported that the FTIR spectrum characteristics of irradiated starch exhibited similarities to those of native starch. Spectrograms following irradiation treatment showed a lack of bands associated with functional groups. For instance, according to the findings by Liang et al. (2022), Navaf et al. (2022), Du et al. (2020), and Punia et al. (2020), ionizing radiation was observed not to affect the characteristic peaks of starch and did not introduce new functional groups. However, the absorbance peak intensity may change with irradiation. The research conducted by Lei et al. (2023) revealed that the spectrum patterns of electron beam-irradiated potato starch exhibited similarities to the control group. Tran et al. (2022) also reported that there were no statistically significant alterations observed in the FTIR spectra when comparing the irradiated native starches. Furthermore, the band at 1156 cm^{-1}, attributed to glycosidic linkages, displayed a notable decrease in intensity. This decrease pointed towards the breakdown of starch molecules during the irradiation process (Tran et al., 2022). Lei et al. (2023) found an intensity increase at the wavelength of 2930 cm^{-1} for starch following irradiation. This increase suggested a breakdown of starch, leading to an elevated amylose-to-amylopectin ratio and the liberation of hydroxyl groups, C-O-C groups, glycosidic linkage, and other related compounds. Furthermore, Punia et al. (2020) documented an augmentation in the liberation of hydroxyl groups upon irradiation, leading to an elevation in the peak observed at 3300 cm^{-1} with increasing irradiation dose. In another study, the absorbance ratio at 1047 cm^{-1}/1022 cm^{-1}, which reflects the changes in starch crystalline and amorphous region characteristics, reduced with the rise in irradiation dose. Such change occurred by deterioration in the hydrogen bonding of starch with irradiation that involves the disruption of the double helix within the crystal zone (Zheng et al., 2023). In a similar vein, Navaf et al. (2022) observed that the heightened intensity of the peak at 1018 cm^{-1} representing the amorphous region occurred after exposure to large doses of irradiation. This finding suggested a reduction in the organized structure of starch chains due to irradiation. According to Sudheesh et al. (2019), it was observed that the strength of peaks at 3468 cm^{-1} and 2927 cm^{-1} associated with OH and CH_2 groups, respectively, in kithul starch reduced as the irradiation doses increased (Figure 8.3). Furthermore, using FTIR analysis, it was confirmed that the ionizing radiation treatment resulted in the degradation of glycosidic bonds coupled with a reduction in the level of short-range crystalline organization, specifically in the form of double helices (Punia et al., 2020). Conversely, particular research has observed the lack of a certain absorbance peak. According to Govindaraju et al. (2022), the O-H vibrational band at 3000 cm^{-1} completely disappeared after irradiation, showing a significant alteration in the starch structure. Based on the previously mentioned examples, it is evident that the O-H and C-H bonds were the most significantly impacted bonds by the irradiation. Specifically, the inter- and intramolecular hydrogen bonds within the starch structure exhibited a high degree of sensitivity to irradiation (Braşoveanu & Nemţanu, 2018).

Amylopectin's molecular weight is higher, and it has branched chains, while amylose possesses a lower molecular weight with linear chains (Shen et al., 2023). The assessment of starch's molecular weight

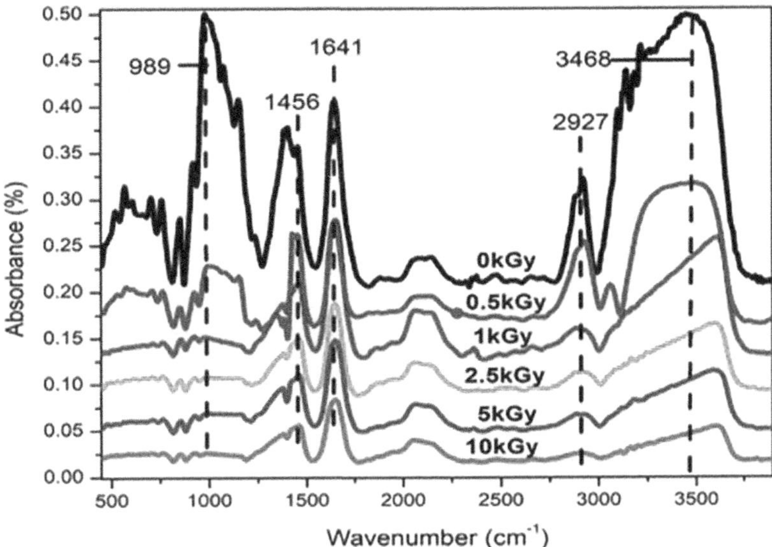

FIGURE 8.3   FT-IR spectra of native and irradiated kithul starches (reprinted from Sudheesh et al. (2019), with permission from Elsevier).

and the measurement of its polymer chain length offer significant insights into the irradiation influence on the starch's molecular composition. Zhou et al. (2020) reported that the molecular weight of waxy maize starch exhibited a decrease with the rising irradiation dose. Their empirical findings indicate that an escalation in irradiation dose is associated with a decrease in the size of amylopectin molecules. Consequently, this process led to debranching accompanied by an augmentation in the creation of short linear-chain molecules. A similar decrease in molecular weight was observed by Zheng et al. (2023) and attributed to the breaking of starch chains into shorter fragments with irradiation dose. Shen et al. (2023) also reported a correlation between the decrease in the molecular weight of amylopectin and the concurrent increase in the molecular weight of amylose. These observations for potato starch (Liang et al., 2022) and cassava starch (Tran et al., 2022) have been corroborated by other studies.

The length of the starch chain, specifically amylopectin, is generally classified into four distinct groups. These categories include branched starch chains with a single branch point in the A chain, as well as those with multiple branch points in the B chain. The latter group is further divided based on the degree of polymerization (DP) into B1, B2, and B3. The chain lengths for the A, B1, B2, and B3 types are DP of 6–12, 13–24, 25–36, and 37, respectively (Zhou et al., 2020). Zhou et al. (2020) observed that when the irradiation dose increased up to 10 kGy, there was a slight rise in A chains (DP6–12). However, a subsequent rise in the irradiation dose led to a significant reduction in starch-branched chains. Particularly at doses below 10 kGy, irradiation had a greater impact on the molecular structure of starch as opposed to the length distribution of branched chains in contrast to native waxy maize starch. The study suggested that an irradiation dose lower than 10 kGy cleaves more α-1,6-glucosidic bonds than α-1,4-glucosidic bonds, leading to the generation of additional linear branched chains (Zhou et al., 2020). Zheng et al. (2023) also noted a rise in the A chain of the starch and a decrease in B chains (B1, B2, B3) following the increasing irradiation dose to 6 kGy. This observed situation is likely a consequence of alterations in starch distribution, with a shift from long-chain to short-chain structures following exposure to elevated levels of irradiation (Zheng et al., 2023). Similarly, observations by Shen et al. (2023) indicate that irradiation causes the transformation of long starch chains into more linear, shorter chains in oat starch. Emphasizing the role of the irradiation dose is essential, as it is a major determinant in shaping the impact of irradiation on molecular structure. As revealed in the study conducted by Pan et al. (2020), the distribution of starch chain length did not exhibit a substantial alteration when subjected to low levels of irradiation (<4 kGy).

The absence of change is likely due to the restricted capacity of low-energy electrons to cleave the glycosidic bonds present in starch molecules. It was also stated that the stability of the crystalline regions of starches is greater than that of the amorphous parts. This finding suggests that short chains situated within the crystalline layers may possess a stronger capacity to withstand irradiation (Pan et al., 2020).

8.4 IMPACT ON FUNCTIONAL PROPERTIES

The subsequent sections detail the influence of ionizing radiation on pasting properties, rheological characteristics, and gelatinization behavior of starches originating from different botanical sources.

8.4.1 Pasting Properties

Pasting features of starch are associated with the alteration of swollen granules under the influence of heat in the presence of excess water and constant shear rate. These properties are integral to the starch cooking and processing dynamics. The application of ionizing radiation can induce substantial changes in the starch pasting properties (Tappiban et al., 2022). For instance, Punia et al. (2020) outlined that the pasting viscosity values (peak, trough, final, and setback viscosities) of lotus seed starch irradiated up to 20 kGy remarkably decreased compared with its native form. The decrease was mostly relevant to starch degradation upon gamma radiation. In other words, the ionizing radiation treatment induces the disruption of glycosidic bonds, leading to amylose-amylopectin degradation, thus, starch swelling power and interaction between starch and water decreases, and expectedly, pasting viscosity values reduce (Lei et al., 2023). Lower viscosities than native counterparts were also identified in irradiated starches from potatoes (Lei et al., 2023; Liang et al., 2022; Tappiban et al., 2022), corn (Tappiban et al., 2022), rice (Dhull et al., 2021; Pan et al., 2020; Zheng et al., 2023), talipot palm (Aaliya et al., 2021; Navaf et al., 2022), and kithul (Sudheesh et al., 2019) with increasing the irradiation dose from 1 to 100 kGy. For example, for rice starch irradiated with accelerated electrons of 5 MeV at a dose rate of 2 kGy/s (Figure 8.4), increasing the irradiation dose from 0 kGy to 10 kGy resulted in a decrease in peak, hold, and final viscosity values from 5189.5, 2538.5, and 3666.0 cP to 236.0, 186.0, and 93.0 cP, respectively (Pan et al., 2020).

There is a direct relationship between the pasting properties of the starch and the textural features, contributing to the overall quality of the final starch-based product (Rostamabadi et al., 2023). Therefore, the obtained lower pasting viscosity properties following irradiation could be seen as an advantage for easy cooking and reduced retrogradation (Punia et al., 2020).

8.4.2 Rheological Characteristics

The rheological properties of starch also play a substantial role in determining its application area. Starch exhibits various deformation and flow behaviors in response to applied stress, which are referred to as its rheological properties (Ai & Jane, 2015). These properties, such as the consistency index and flow behavior index, generally reduce because irradiation may cause the structural disintegration of starch molecules (Rostamabadi et al., 2023). According to Dhull et al. (2021), gamma irradiation at 20 kGy with a dose rate of 2.5 kGy/h led to a reduction of approximately 21% in the consistency index of black rice starch. Similarly, as per the findings of Laxminarayana et al. (2022), the viscosity of native starch solutions exhibited non-Newtonian characteristics, but irradiation induced a Newtonian behavior in the solutions due to damage to chemical structure upon irradiation. Furthermore, the rheological parameters, specifically the storage modulus (G') and loss modulus (G''), function as indicators of the energy stored within the material and the energy dissipated as heat, respectively. G' is used to represent the elastic behavior, while G'' is

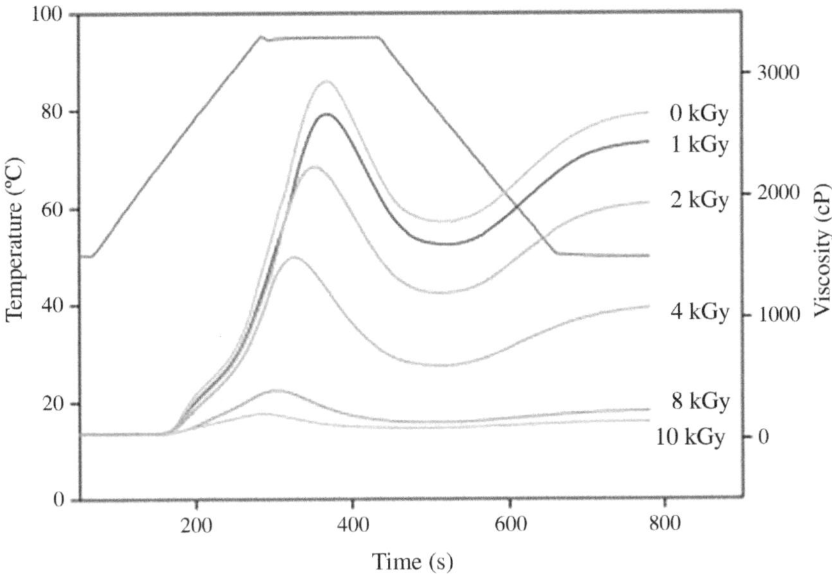

FIGURE 8.4 Pasting profiles of the starches isolated from irradiated-rice grains at various doses (reprinted from Pan et al. (2020), with permission from Elsevier).

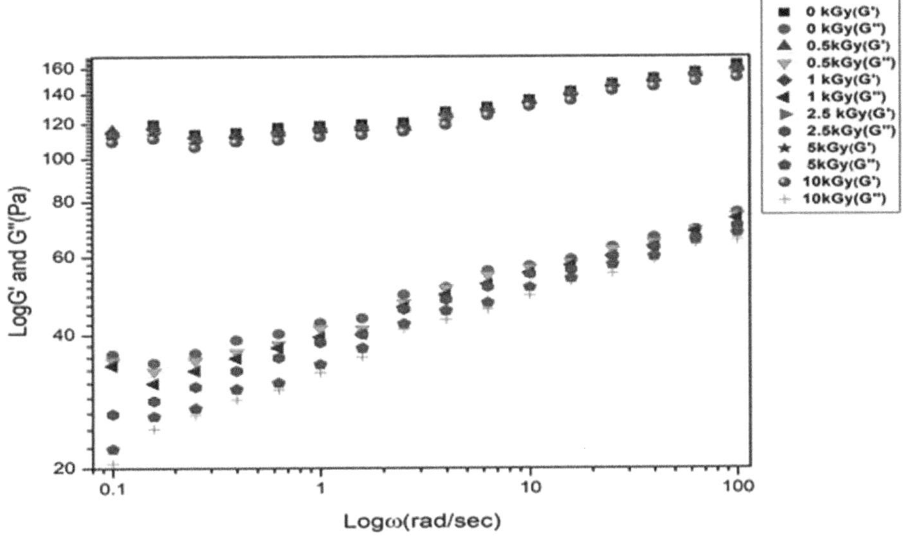

FIGURE 8.5 Rheological properties of native and irradiated kithul starches (G') storage modulus (G'') loss modulus (reprinted from Sudheesh et al. (2019), with permission from Elsevier).

used to indicate the viscous behavior of the gel (Navaf et al., 2022). For starches from talipot palm (Aaliya et al., 2021; Navaf et al., 2022) and rice (Dhull et al., 2021), it was observed a reduction in storage and loss moduli after irradiation. Depolymerization and disintegration of the macromolecular structure after irradiation were proposed as the underlying mechanisms for this phenomenon (Navaf et al., 2022). The case of kithul starch yielded a similar observation (Figure 8.5) (Sudheesh et al., 2019).

8.4.3 Gelatinization Behavior

Starch exhibits limited solubility in cold water. However, when exposed to temperatures exceeding 52°C, it undergoes a transformation referred to as "gelatinization". The gelatinization property, a critical functional property of starch, is generally studied using differential scanning calorimetry (DSC) (Bashir & Aggarwal, 2019). DSC thermograms allow for the examination of transition temperatures such as onset temperature (T_o), peak temperature (T_p), conclusion temperature (T_c), and transition enthalpy (ΔH) (Punia et al., 2020). Ionizing radiation processing of starch generally causes structural reconfiguration in starch macromolecules, thus leading to changes in gelatinization temperatures and process enthalpy (Braşoveanu & Nemţanu, 2018). Punia et al. (2020) reported that the starch gelatinization temperatures were reduced following gamma-irradiation treatment. The decrease in gelatinization temperatures was explained by the rapid degradation of intermolecular starch bonds induced by heat and irradiation. Tappiban et al. (2022) showed a comparable outcome, wherein the irradiation dose exhibited a modest decrease in the gelatinization temperatures (T_o, T_p, and T_c values) of electron beam-irradiated potato starch in comparison with those of the native starch. The decrease showed that the long amylopectin chains degraded following irradiation; therefore, less energy was required for starch gelatinization, and thus, the peak temperature decreased from 70.96°C for the native form to 69.99°C for irradiated starch at 10 kGy with a dose rate of 2 kGy/s. A similar decrease in the temperatures of gelatinization, T_o and T_p, occurred in electron beam-irradiated potato starch with increasing irradiation doses up to 24 kGy and irradiation frequency up to 8 times (Liang et al., 2022). Prior research has proven that irradiating certain starches, such as talipot palm starch (Aaliya et al., 2021; Navaf et al., 2022), waxy maize starch (Zhou et al., 2020), and kithul starch (Sudheesh et al., 2019), results in a significant reduction in gelatinization parameters.

Conversely, it was found that slight increases in T_o, T_p, and T_c values were produced in irradiated corn starch (Tappiban et al., 2022) and caused by alterations in the arrangement of starch crystallites with irradiation. A greater gelatinization temperature signifies an increased requirement of energy for the unfolding and melting of the double helices. Similar enhancements in T_o, T_p, and T_c of tapioca starch following ionizing radiation treatment were observed by Kanatt (2020). Similarly, according to a recent study (Lei et al., 2023), the T_o, T_p, T_c, and ΔH for potato starch rose following irradiation processing. It was indicated that the observed rise in T_c and T_o provided evidence of the uniformity and stability of the crystallites in irradiated starch. Furthermore, it was reported that the increase in gelatinization temperatures can also be a consequence of the degradation of comparatively feeble crystalline structures, thereby enhancing the stability of the crystalline structures. Lei et al. (2023) noted, nevertheless, that at lower doses (<10 kGy), the ΔH of both electron beam and X-ray irradiated potato starch was greater than that of native starch. Moreover, as the dose increased up to 30 kGy, the ΔH decreased, which was correlated with a decline in crystallinity. Similarly, Liang et al. (2022) reported that the ΔH of electron beam-irradiated potato starch elevated up to a certain irradiation dose (24 kGy). Sudheesh et al. (2019) reported that T_o, T_p, T_c, and ΔH of kithul starch significantly increased with gamma irradiation at 5 and 10 kGy at a dose rate of 2 kGy/h (Figure 8.6). In contrast, those values decreased following irradiation up to 2.5 kGy. This was primarily the result of the destruction of weak crystalline regions; however, the crystalline structure remained after irradiation treatment 5 and 10 kGy, and thus, starch became very strong, resulting in greater stability (Sudheesh et al., 2019). Instead, Pan et al. (2020) stated that rice starch thermal properties were unaffected at electron beam irradiation at low doses (<1 kGy); however, it was shown that T_o, T_p, T_c, and ΔH decreased from 57.8°C to 56.6°C, 63.9°C to 62.3°C, 70.2°C to 68.8°C, and 7.7 J/g to 6.3 J/g, respectively, at increasing irradiation doses from 1 to 10 kGy at a dose rate of 2 kGy/s. It was indicated that irradiation at high doses causes the destruction of the starch crystalline structure (Pan et al., 2020). Furthermore, another study showed that the T_o and T_c for quinoa starch were not impacted following radiation treatment, while gelatinization enthalpy decreased. The alterations in enthalpy could potentially be associated with variations in the degree of crystallinity and granule size of starch, as suggested by Du et al. (2020).

As seen in the studies given previously, different results obtained from the gelatinization properties of starch demonstrate that there is a strong correlation between gelatinization properties and crystalline structure (Liang et al., 2022; Tappiban et al., 2022). Another study conducted by Govindaraju et al. (2022) verified that the gelatinization behavior of starch depends on the starch type.

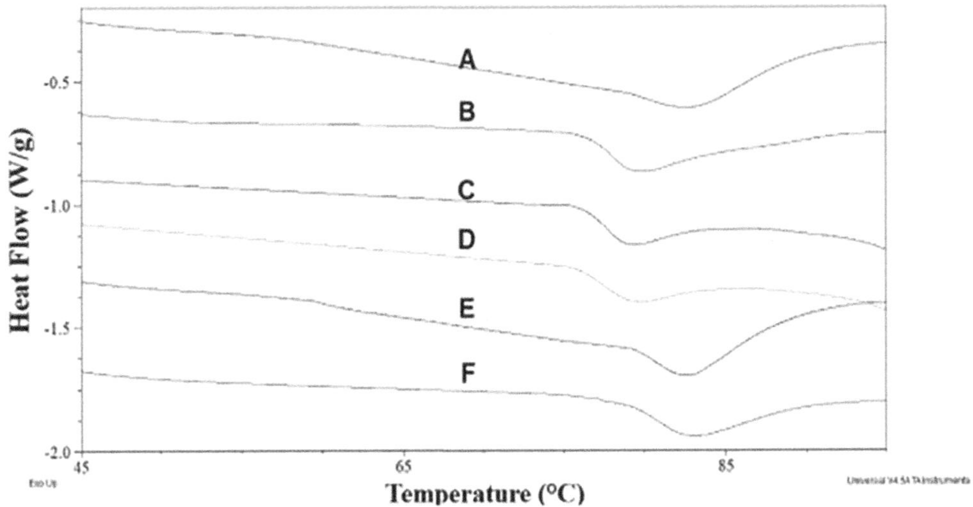

FIGURE 8.6 DSC thermogram of native and irradiated kithul starches (A) 0 kGy, (B) 0.5 kGy, (C) 1 kGy, (D) 2.5 kGy, (E) 5 kGy, (F) 10 kGy (reprinted from Sudheesh et al. (2019), with permission from Elsevier).

8.4.4 Swelling Power and Solubility

Both swelling power and solubility offer insights into the degree of interactions among starch chains inside the amorphous and crystalline regions of the granule (Rostamabadi et al., 2023). Ionizing radiation treatment generally causes a reduction in swelling power. This is due to the decrease in the amylopectin fraction as well as the denaturation of the protein matrix, both of which limit water passage into the starch matrix (Punia et al., 2020). It is worth noting that swelling power is directly related to the presence of amylopectin long branch chains in starch. Following irradiation treatment, these long branch chains undergo degradation and are converted into shorter branch chains. Consequently, the starch molecule's ability to retain water is diminished (Lei et al., 2023; Navaf et al., 2022).

Recent studies have found that various types of starches suffered a decrease in swelling power after radiation treatment. Such a decrease has recently been reported for maize starch (Di et al., 2023), potato starch (Lei et al., 2023; Liang et al., 2022), talipot palm starch (Aaliya et al., 2021; Navaf et al., 2022), tapioca starch (Kanatt, 2020), lotus seed starch (Punia et al., 2020), and kithul starch (Sudheesh et al., 2019). The utilization of irradiated starch as a food additive is supported by its potential to enhance the durability of starch products by a reduction in the starch swelling power, hence minimizing rupture during the cooking process (Lei et al., 2023). On the other hand, it was stated that the irradiation dose affects the water swelling capacity of starch (Tran et al., 2022). Thus, Tran et al. (2022) observed that the swelling capacity of cassava starch increased compared to the native sample when irradiated with gamma rays at a low dose of 5 kGy, whereas that value decreased at a higher irradiation dose (10 kGy). According to Tran et al. (2022), the higher swelling capacity was related to the higher degradation of amylose in the amorphous region at a low irradiation dose, while the disruption of amylopectin was responsible for the lower swelling capacity at a high irradiation dose.

Conversely, numerous investigations have documented the increase in solubility as a consequence of the increase of irradiation dose for starches extracted from maize (Di et al., 2023), potato (Lei et al., 2023; Liang et al., 2022), rice (Zheng et al., 2023), cassava (Tran et al., 2022), talipot palm (Aaliya et al., 2021; Navaf et al., 2022), tapioca (Kanatt, 2020), lotus seed (Punia et al., 2020), and kithul starch (Sudheesh et al., 2019). The observed increase in starch solubility as a consequence of irradiation dose is likely due to the starch depolymerization caused by irradiation. This depolymerization leads to a reduction in interchain hydrogen bonding, which in turn enhances the polarity of the starch molecules and promotes the formation of simple sugars. These simple sugars, due to their increased affinity for water, exhibit a greater solubility compared to intact starch molecules (Punia et al., 2020). This phenomenon can be described in the following manner: under the influence of irradiation, glycosidic bonds are cleaved, and long chains are fragmented, resulting in depolymerization. Consequently, shorter chains, which possess a higher propensity for hydration compared to the predominant branches, are formed. As a result, more hydrogen bonds are created with water, which enhances the starch solubility (Liang et al., 2022).

The water absorption capacity (WAC) is connected with the capability of starch molecules to associate with water molecules, and it holds significance concerning the textural and functional attributes of starch (Navaf et al., 2022). The irradiation treatment led to an increase in the WAC of starches from talipot palm from 1.70% for the native form to 1.83, 2.06, 2.17, and 2.35% for samples irradiated at 20, 50, 80, and 100 kGy, respectively (Navaf et al., 2022). Kanatt (2020) reported that the WAC of tapioca starch samples were 1.159, 1.189, and 1.228 g/g after being subjected to gamma irradiation at 5, 10, and 20 kGy, respectively, while the native counterpart was 1.093 g/g. The elevation in WAC after irradiation is attributable to the enzymatic breakdown of starch into monosaccharides and oligosaccharides, thereby liberating a substantial quantity of hydrogen molecules that subsequently form bonds with water molecules (Navaf et al., 2022).

Like WAC, oil absorption capacity (OAC) refers to the ability of starch molecules to capture and retain oil. The non-polar nature of oil particles is evident, and their integration into the starch molecule predominantly takes place via physical entrapment (Navaf et al., 2022). According to the findings of Navaf et al. (2022), irradiation processing increased the OAC of talipot palm starch. The OAC of native talipot palm starch (1.41%) increased to 1.57% as a result of 100 kGy gamma irradiation at a dose rate of 2.5 kGy/h, at the temperature of 25°C in air. The observed increase can be ascribed to the structural damage acquired by the starch granules due to the treatment. A comparable finding was documented in a study conducted on tapioca starch (Kanatt, 2020), where the OAC increased by 37% after gamma irradiation at 20 kGy with a dose rate of 1.8 kGy/h.

The physicochemical property of light transmittance offers valuable insights into the behavior of the starch paste as it undergoes the passage of light through the sample. A higher light transmittance is indicative of increased transparency inside the paste (Bashir & Aggarwal, 2019). In almost all studies, it was found that the ionizing radiation application can significantly increase the light transmittance of starches from rice (Zheng et al., 2023), talipot palm (Aaliya et al., 2021; Navaf et al., 2022), tapioca (Kanatt, 2020), lotus seed (Punia et al., 2020), and kithul (Sudheesh et al., 2019). The rise in transmittance can be correlated with a decrease in amylose content following irradiation (Zhiguang et al., 2023). An additional factor contributing to the observed enhancement in light transmittance of irradiated starch is the creation of smaller molecules as a consequence of starch molecule disintegration during the irradiation process (Navaf et al., 2022). Additionally, during the process of irradiation, carboxyl groups are generated, and thus, hydrogen bonds with water molecules are formed. As a consequence, there is an increase in light transmission and an improvement in the paste clarity (Navaf et al., 2022).

The term "freeze–thaw stability" is also one of the starch physicochemical properties, adverse physical alterations that transpire throughout the process of freeze–thawing (Lei et al., 2023). The application of ionizing radiation can induce substantial alterations in the starch freeze–thaw stability. Lei et al. (2023) conducted an investigation whereby they noticed that the freeze–thaw stability of potato starch was enhanced through both electron beam and X-ray irradiation application. For example, when subjected to an irradiation dose of 30 kGy, the electron beam-treated starch exhibited syneresis of 3.03%, 9.77%, and 12.86% in the first, second, and third cycles, respectively. Similarly, X-ray-treated starch had syneresis of 4.74%, 22.84%, and 29.44%. All these values were significantly less than those observed in the native counterpart. This circumstance was ascribed to the heightened interaction between amylose and amylose-amylopectin (Lei et al., 2023). A comparable enhancement in freeze–thaw stability was noted in kithul starch that underwent gamma irradiation at 10 kGy (Sudheesh et al., 2019), as well as in gamma-irradiated tapioca starch up to 20 kGy (Kanatt, 2020). The utilization of irradiated starch as an ingredient for frozen meals is a viable recommendation owing to its enhanced freeze–thaw stability, as demonstrated by Lei et al. (2023).

8.4.5 Digestive Properties

The assessment of starch digestibility characteristics holds significance in light of the increased interest in developing food items with reduced and delayed starch digestion qualities. These features are believed to contribute to the prevention of certain diseases, including colon cancer, type 2 diabetes, and cardiovascular issues (Bello-Perez et al., 2020). Starch digestibility property can be influenced by changes in its chemical structure and content. Numerous studies have demonstrated the efficacy of ionizing radiation in enhancing the in vitro digestibility of starch (Bashir & Aggarwal, 2019). For instance, according to Liang et al. (2022), the slowly digestible starch (SDS) and resistant starch (RS) of potato starch were enhanced by irradiation processing. Thus, the SDS content of native starch increased from 7.67% to 9.74% after electron beam irradiation at 24 kGy. Also, RS content increased to 55.3%, while native starch had an RS content of 49.85%. Starch digestibility properties are strongly tied to its structure, with SDS being linked to the amorphous region and RS being affected by the ordered crystalline structure. It was stated that the irradiation treatment produces more short chains, particularly A chains. These chains are joined to branched chains that are scattered throughout the crystal flakes to create aggregated formations. Furthermore, these aggregated structures in starch improved starch molecular chain alignment, which serves as a barrier to amylase, and thus, amylase hydrolysis decreases. Moreover, the use of radiation treatment facilitates the generation of carboxyl groups, hence potentially enhancing the capacity to withstand the role of digestive enzymes in the process of starch hydrolysis (Liang et al., 2022). A similar rise in RS and SDS contents following electron beam irradiation up to 8 kGy was also reported in quinoa starch (Du et al., 2020).

However, another study demonstrates that the application of ionizing radiation increased the digestibility of various starch varieties, such as potato, corn, and rice (Govindaraju et al., 2022). This enhancement was evident in the increased hydrolysis values, escalating from 1.66%, 5.7%, and 7.8% in their native forms to 10.66%, 14.58%, and 17.03%, respectively, following gamma irradiation at 10 kGy with a dose rate of 2.2 kGy/h. Similarly, Sudheesh et al. (2019) found that the kithul starch RS content decreased by approximately 16% while RDS (rapidly digestible starch) content increased by approximately 70%, respectively, following gamma irradiation at 10 kGy with a dose rate of 2 kGy/h. This situation was explained as the formation of fractures and fissures as a result of depolymerization following irradiation, which can make the interior of granules more accessible to digestive enzymes. In addition, an increase in amylose content and relative crystallinity has been found to decrease the digestion of starch. According to the statement, the utilization of free radicals to mediate molecular depolymerization resulted in a decrease in both the amylose content and the relative crystallinity of starch. Another possible explanation for the increase in starch digestibility following irradiation could be the generation of compounds of reduced molecular weight and a drop in the chain length of the polymer. Additionally, in another study, it was also stated that carboxyl group formation induced by irradiation at high doses led to a decrease in the digestibility of rice starch (Pan et al., 2020). A reduction of approximately 15% in the contents of RS and SDS was observed in rice starch following electron beam irradiation at 6 kGy (Zheng et al., 2023). Similarly, a comparable decrease was noted in talipot palm starch after gamma irradiation at 10 kGy. (Aaliya et al., 2021).

8.5 REACTION MECHANISM INDUCED IN STARCH BY IONIZING RADIATION

Whether discussing gamma radiation and X-rays or electron beams, electrons play an essential role in mediating the energy release of radiation to starch. Gamma radiation and X-rays impart energy through the photoelectric effect, the Compton effect, and pair generation, resulting in the release of high-speed electrons, which subsequently dissipate energy using similar mechanisms as those observed in the electron beam (Braşoveanu & Nemţanu, 2018). Consequently, irradiation creates a cloud of energetic electrons in the starch, initiating *direct effects* on the polymer chain, including electron removal (ionization), elevation of molecular energy to a higher level (excitation), and dissociation. These processes give rise to carbon-originated radicals (Nemţanu & Braşoveanu, 2017), further participating in various recombination processes as side effects (Braşoveanu & Nemţanu, 2022). According to Vazirov et al. (2023), a hydroxyl radical forms, featuring an unpaired electron located at the C_1 atom of the glucose unit.

Alternatively, the presence of water in starch molecules induces *indirect effects* through highly reactive chemical species resulting from water radiolysis. As a result of water molecule radiolysis within native starch, hydrogen and hydroxyl radicals initiate an event on macromolecules. This leads to the formation of macroradicals participating in free radical reactions, subsequently resulting in processes such as cross-linking, chain scission, and even the opening of the glucose ring. Degradation and crosslinking processes are competitive reactions depending mainly on input processing parameters, including irradiation dose and dose rate. Generally, crosslinking dominates when starch is subjected to ionizing radiation at doses up to 5 kGy and at lower dose rates, reaching up to 1 kGy/h (Chung et al., 2015; Polesi et al., 2016). Nevertheless, polysaccharides, including starch exposed to ionizing radiation, are known to be more susceptible to degradation under ionizing radiation (Nemţanu & Braşoveanu, 2021).

If radiation processing of starch occurs under an oxygen atmosphere, then carbon-centered radicals react rapidly with oxygen, leading to the cleavage of glycoside bonds and their conversion into peroxyl radicals (Nemţanu & Braşoveanu, 2017; Korkmaz & Polat, 2000). As per Sharpatyi (2003), irradiation under oxygen-rich conditions allows the involvement of any carbon atom ($C_1 \ldots C_6$) of the glucose unit to participate in the creation of peroxide radicals resulting from primary radicals and their transformation through isomerization, involving the breakdown of C–C bonds in glucose rings as well as O–O bonds in peroxide radicals. Conversely, it was suggested that the C_1–H and C_4–H bonds of the pyranose ring are particularly vulnerable to hydrogen abstraction being the weakest (Ershov, 1998). Furthermore, when radiation acts on poly-α-D-glucopyranoses, including starch, it predominantly results in the production of terminal macroradicals featuring unpaired electrons at C_1 and C_4, accompanied by concomitant decomposition of the pyranose ring due to the breakdown of the C–H bond (Ershov & Isakova, 1987). Simultaneously, primary macroradicals have the potential to experience rearrangement, fragmentation, or elimination processes, producing secondary radicals with the ability to abstract hydrogen.

A potential mechanism of reactions for starch subjected to ionizing radiation is proposed as follows:

$$\text{Starch} \xrightarrow{\text{ionization}} \underset{\text{ion}}{\text{Starch}^{\cdot+}} + \underset{\text{free electron}}{e^-} \xrightarrow{\text{charge recombination}} \underset{\text{excited molecule}}{\text{Starch*}}$$

$$\text{Starch} \xrightarrow{\text{excitation}} \text{Starch*}$$

$$\text{Starch*} \xrightarrow{\text{dissociation}} \begin{cases} \underset{\text{free radicals}}{R\cdot + R'\cdot} \\ \underset{\text{new molecules}}{M + M'} \end{cases}$$

$$R\cdot + R'\cdot \longrightarrow \begin{cases} \xrightarrow{\text{radical recombination}} \text{Starch} \\ \xrightarrow{\text{recombination}} M + M' \\ \xrightarrow[\text{abstraction}]{+\,MH} \underset{\substack{\text{new} \quad \text{free} \\ \text{molecule} \;\; \text{radical}}}{RH + M\cdot} \end{cases}$$

$$H_2O \xrightarrow{\text{water radiolysis}} e_{aq}^-, HO\cdot, H\cdot, H^+, \ldots\ldots$$

$$\text{Starch} + \cdot OH \longrightarrow \underset{\text{free radical}}{\text{Starch}\cdot} + H_2O$$

$$\text{Starch}\cdot + \text{Starch}\cdot \xrightarrow{\text{cross-linking}} \text{Starch-Starch} + H_2O$$

$$\text{Starch}\cdot + O_2 \longrightarrow \underset{\text{radical peroxyl}}{\text{Starch}-O-O\cdot} \longrightarrow \underset{\text{new molecules}}{ROH, ROOH, ROR, ROOR}$$

8.6 ANALYTICAL METHODS

The characterization of irradiated starch necessitates a multidimensional approach employing a range of analytical techniques. These methodologies facilitate the elucidation of molecular structure, physicochemical changes, and alterations in functionality following irradiation, thereby providing insights into the modified properties of the starch. Several characterization methods can be used to analyze and evaluate irradiated starch. The most used analytical methods for characterizing starches modified by ionizing radiation encompass chemical and imaging techniques based on various types of microscopies such as optical microscopy, confocal laser scanning microscopy (CLSM), polarization light microscopy (PLM), transmission and scanning electron microscopies (TEM/SEM). Additionally, X-ray powder diffractometry (XRD), spectroscopic techniques such as Fourier transform infrared spectroscopy, visible spectroscopy, electron spin resonance (ESR) spectroscopy, along with chromatographic methods such as high-performance size-exclusion chromatography (HPSEC), gel permeation chromatography (GPC), and anion exchange chromatography (HPAEC) are commonly employed. Furthermore, thermal techniques involving differential scanning calorimetry and thermogravimetric analysis (TGA), and rheological methods are pivotal in this context. Table 8.2 provides a summary of the diverse techniques utilized to characterize irradiated starch, each having mentioned its specific application in studying the alterations induced by irradiation.

Microscopy techniques are critical in the irradiated starch examination, enabling the identification of alterations in granule morphology, size distribution, and level of aggregation. Ionizing radiation, characterized by its energetic and penetrating nature, can induce effects throughout the sample's entire volume. Consequently, radiation-induced alterations in the starch granules can occur in both the inner and peripheral areas. Optical and electron microscopy methods, including TEM and SEM, offer direct insights into the morphology (size distribution and shape) and the degree of aggregation within irradiated starch granules (Barroso & del Mastro, 2019; Laxminarayana et al., 2022). TEM allows for detailed internal imaging, providing intricate insights into the internal structure of the granules. Conversely, SEM facilitates high-resolution imaging, enabling the observation of surface or near-surface morphological changes in the starch granules. It should be noted that for SEM analysis, a conductive material layer on the sample is often required to prevent the accumulation of static electric charge during measurements. PLM serves to depict the level of anisotropy and its organizational aspects within the starch granules, while CLSM is used for characterizing the arrangement of amylose and amylopectin, as well as identifying the localization of reducing ends (Liang et al., 2022; Wang et al., 2023).

Scattering techniques such as XRD are used to investigate the crystallinity and crystallinity degree. XRD stands as the predominant technique due to its capability to unveil the crystallographic patterns and determine the starch relative crystallinity (Liang et al., 2022; Sunder et al., 2022). Ionizing radiation treatments influence both the crystalline and amorphous regions within the starch granule, causing alterations in crystallinity degree following the region most affected by the irradiation dose (Braşoveanu & Nemţanu, 2018).

Spectroscopic methods are utilized to analyze and characterize irradiated starch, enabling the investigation of its structural, compositional, and functional alterations induced by the irradiation process. FTIR spectroscopy stands as a valuable and user-friendly technique that offers insights into the changes within both the crystalline and amorphous regions, while also identifying alteration in the short-range ordered structure, such as the single and double-helical order (Liang et al., 2022). The short-range molecular order can be assessed by analyzing the absorption bands at 1046 cm^{-1} (crystalline structure) and 1022 cm^{-1} (amorphous region) along with their ratio (van Soest et al., 1995). Visible spectroscopy, known for its rapidity and cost-effectiveness, serves as a commonly employed method for determining various characteristics including amylose content (Atrous et al., 2017; Laxminarayana et al., 2022; Lei et al., 2023; Navaf et al., 2022), paste clarity or opacity (Navaf et al., 2022; Nemţanu & Braşoveanu, 2010), and colorimetric attributes (Huang et al., 2023; Navaf et al., 2022; Trinh & Nguyen, 2020). ESR is employed for the monitoring of free radical formation (Atrous et al., 2015; Braşoveanu et al., 2013; Maghraby et al., 2023; Vazirov et al., 2023).

TABLE 8.2 Characterization Methods of Starches Modified with Ionizing Radiation

TYPE OF METHOD	CHARACTERIZATION TECHNIQUE	TYPE OF INVESTIGATION	STUDIED PROPERTIES	REFERENCE(S)
Imaging	Optical microscopy, CLSM, PLM, TEM, SEM	Structural, morphological	Crystallinity, size, shape, surface and internal features	Barroso & del Mastro, 2019; Laxminarayana et al., 2022; Liang et al., 2022; Wang et al., 2023
Diffraction	XRD	Structural	Crystalline structure/phase structure (crystallinity type, relative crystallinity)	Liang et al., 2022; Sunder et al., 2022
Spectroscopy	FTIR	Structural, physicochemical	Spectral features, chemical structure (functional groups and bonds), short-range molecular order	Du et al., 2020; Liang et al., 2022
	ESR	Structural	Free radicals	Atrous et al., 2015; Braşoveanu et al., 2013; Maghraby et al., 2023; Vazirov et al., 2023
	Visible	Physicochemical	Apparent amylose content, paste clarity/opacity, color	Huang et al., 2023; Lei et al., 2023; Navaf et al., 2022; Trinh & Nguyen, 2020
Chromatography	HPSEC, GPC, HPAEC	Structural	Molecular weights and molecular weight distribution, branch chain length distribution	Nemţanu et al., 2010; Nemtanu & Brasoveanu, 2012; Zhou et al., 2020
Thermal analysis	DSC, TGA	Thermal, structural	Thermal properties (gelatinization and melting behavior), thermal stability (degradation at high temperatures), crystallinity level	Du et al., 2020; Punia et al., 2020; Tran et al., 2022
Rheology	Pasting assessment, steady-shear rheology, dynamic rheology	Physicochemical	Pasting profile and pasting parameters, flow behavior, apparent viscosity, dynamic viscosity, elasticity (storage modulus, G'), and plasticity (loss modulus, G'')	Braşoveanu & Nemţanu, 2020; Du et al., 2020; Laxminarayana et al., 2022; Navaf et al., 2022; Wang et al., 2023
Chemical	Chemical methods	Chemical	Proximate composition, acidity, carboxyl and carbonyl contents, water solubility index and swelling power or volume, water and oil absorption capacities, syneresis, freeze–thaw stability	Laxminarayana et al., 2022; Lei et al., 2023; Navaf et al., 2022; Trinh & Nguyen, 2020; Wang et al., 2023
Enzymatic	In vitro digestibility assay	Enzymatic	Rapidly digestible starch, slowly digestible starch, and resistant starch contents	Du et al., 2020; Liang et al., 2022; Trinh & Nguyen, 2020; Wang et al., 2023

Chromatographic methods are also applicable for analyzing alterations in the starch structure following irradiation. HPSEC and GPC provide insights into the transformations in molecular weights concerning the irradiation dose (Nemţanu et al., 2010; Nemtanu & Brasoveanu, 2012, 2017; Zhou et al., 2020), while HPAEC is employed for evaluating the distribution of branch chain lengths (Zhou et al., 2020). The quantification of the radiation-induced degradation process, which relies on the determined molecular weights, can be performed by computing the degradation index, degradation rate constant, and half-value dose (Nemţanu et al., 2010; Nemtanu & Brasoveanu, 2012, 2017).

Techniques for thermal analysis are frequently employed to determine how radiation affects the functional characteristics of starch, such as gelatinization. The irradiation of starch typically results in significant alterations in gelatinization temperatures and enthalpy, owing to structural rearrangement inside the starch. Consequently, the gelatinization of starch is examined and tracked using DSC, a highly valuable tool offering a numerical assessment of gelatinization enthalpy along with the identification of the temperature range in which gelatinization occurs (Braşoveanu & Nemţanu, 2018; Du et al., 2020; Navaf et al., 2022; Punia et al., 2020). Additionally, TGA enables the pyrolysis process investigation to determine the thermal robustness of starch following irradiation (Tran et al., 2022).

The rheological evaluation is crucial in characterizing irradiated starches since ionizing radiation processing can induce substantial alterations to the pasting and rheological features of starch, notably reducing its viscosity and modifying its flow behavior (Braşoveanu & Nemţanu, 2018). Most studies (Barroso & del Mastro, 2019; Lei et al., 2023; Navaf et al., 2022; Wang et al., 2023) primarily focus on assessing the pasting behavior of irradiated starches. However, the flow behavior, apparent viscosity, as well as the elastic and viscous nature are evaluated through steady-shear rheoviscosimetric measurements (Atrous et al., 2017; Laxminarayana et al., 2022; Tran et al., 2022) and dynamic rheology (Du et al., 2020; Navaf et al., 2022), respectively. Typically, ionizing radiation leads to starch degradation, resulting in a noticeable exponential decrease in viscosity with the radiation dose (Braşoveanu & Nemţanu, 2020; Nemtanu & Brasoveanu, 2012). In this case, by evaluating and modeling starch viscosity concerning the irradiation dose, it becomes possible to ascertain the *characteristic irradiation dose* wherein the viscosity drops by *e* times for starches tailored under ionizing radiation (Braşoveanu & Nemţanu, 2020; Nemtanu & Brasoveanu, 2012). This irradiation dose is regarded as a material constant, reflecting the starch's sensitivity to irradiation (Nemtanu & Brasoveanu, 2012), a sensitivity that illustrates the intricate nature of starch structure as well as the diverse influences of its properties on the interaction with radiation (Braşoveanu & Nemţanu, 2020). Moreover, understanding the magnitude of this characteristic irradiation dose allows the determination of the necessary irradiation dose to achieve a particular viscosity in practical applications, simplifying the process of experimental design. Additionally, Braşoveanu and Nemţanu (2020) provided an algorithm consisting of six steps for the practical utilization of a mathematical model that uses the characteristic irradiation dose. This model aids in identifying the required irradiation dose to attain the targeted viscosity value for a specific technological application.

Chemical methods are typically employed to assess the composition and various properties of starch, including granule water solubility, swelling power or volume, acidity, carboxyl and carbonyl content, and water and oil absorption capacities, as well as syneresis and freeze–thaw stability (Laxminarayana et al., 2022; Lei et al., 2023; Navaf et al., 2022; Trinh & Nguyen, 2020; Wang et al., 2023).

Assessing the in vitro digestibility of modified starch matters greatly in appraising its nutritional attributes. The responsiveness of starches to digestion enzymes is intricately linked to the multiscale granule structure, encompassing the granule microstructure, as well as the crystalline and amorphous regions (Du et al., 2020; Liang et al., 2022; Trinh & Nguyen, 2020; Wang et al., 2023). Therefore, understanding how irradiation alters the ratios of RDS, SDS, and RS in starch is crucial for evaluating the nutritional impact of irradiated starches.

In conclusion, the range of characterization methods presented previously can provide a multifaceted understanding of the structural, compositional, and functional alterations induced by ionizing radiation in starch.

8.7 SAFETY AND REGULATORY CONSIDERATIONS

From the perspective of potential utilization, ensuring safety and compliance with regulatory standards for the modification of starch through ionizing irradiation (gamma rays, electron beam, and X-rays) to enhance its functional properties, both in food and non-food applications, is undoubtedly of significant interest. This interest largely stems from concerns expressed by scientists and consumers who harbor reservations and uncertainties regarding the utilization of irradiation and the potential level of induced chemical alterations.

In the food sector, even today, there remains a consumer sentiment that favors the benefits of food processing through irradiation over its disadvantages. These drawbacks can be significantly mitigated by employing appropriate irradiation parameters and minimizing the risk of irradiation leakage (Arvanitoyannis, 2010), ensuring equipment maintenance, and offering adequate personnel training. Conversely, effective consumer information and education about irradiation in general, and specifically about food irradiation, could lead to adequate knowledge of the irradiation. This could foster acceptance that irradiated ingredients, including starch, are both safe and nutritious.

Food irradiation, according to the World Health Organization (WHO) and the Food and Agriculture Organization (FAO), is a safe method for processing food and human-consumed products as long as the appropriate radiation dose is followed (Maherani et al., 2016). Concerning the legal recommendations concerning the consumption of specific raw substances in the food industry, a considerable Code of Federal Regulations (CFRs) containing relevant data has been supplied to industrial producers (Sabbaghi, 2023). Generally, the impact on macronutrient quality, particularly carbohydrates, due to radiation doses up to 10 kGy is minimal (Arapcheska et al., 2020; WHO, 1999). Additionally, various macronutrients found in foods, carbohydrates exhibit lower sensitivity to radiation, and when applied under controlled conditions, the radiation processes yield comparable or lesser damage to the nutritional integrity of the food in comparison to traditional methods such as cooking, drying, canning, and pasteurizing (Arapcheska et al., 2020; Lima et al., 2018; WHO, 1994). Nonetheless, at the current industrial level, modified starch can safely be utilized as a constituent in items designed for producing, manufacturing, packing, processing, preparing, treating, packaging, transporting, or storing food, following the guidelines specified by the FDA (21 CFR 178.3520). Therefore, among the industrial starch-modified is the starch irradiated under (1) radiation from a sealed cobalt-60 source, with a maximum absorbed dose not exceeding 50 kGy, or (2) an electron beam source with a maximum energy of 7 MeV, where the maximum absorbed dose does not exceed 50 kGy. The irradiation is used to generate free radicals in starch required for subsequent graft polymerization with the reactants such as acrylamide and [2-(methacryloyloxy) ethyl]trimethylammonium methyl sulfate. The resultant industrial starch-modified is exclusively intended for use as a retention resource and dry strength agent employed before the sheet-forming process in the production of paper and paperboard meant to contact food.

On the other hand, the association between radiation and radioactivity is prevalent among most individuals. Hence, an important aspect of the safety of irradiated starch lies in the assurance that the final product does not retain radioactive properties. Theoretically, certain chemical elements could potentially become activated and, consequently, become radioactive following irradiation. However, according to Nemtanu and Brasoveanu (2011), it is implausible for irradiated food, starch included, to become radioactive due to the level of radiation energy used and the chemical elements present in foods. It is essential to emphasize that ionizing radiation sources such as gamma rays emitted by radioactive isotopes like cobalt-60 or cesium-137, electron beams, and X-rays lack the energy to induce radioactivity but possess adequate energy to displace electrons from atoms, forming ions or free radicals (Smith & Pillai, 2004) Moreover, there are established strict guidelines on acceptable radiation levels in food products as well as labeling requirements. Compliance with these limits during the irradiation process is crucial to ensure the safety of the modified starch. Moving forward, we will briefly outline some of the key regulations concerning food irradiation that may also cover starch irradiation for food applications due to the absence of specific regulations solely for starch.

At the beginning of the 1980s, The Codex Alimentarius Commission (CAC, 1983) put forth a General Standard for Irradiated Foods and a Recommended International Code of Practice for the Operation of Radiation Facilities Used for the Treatment of Foods. These guidelines encompass aspects of treatment, handling, and distribution of irradiated food products, providing a solid framework for establishing specific protocols necessary for the practical use of commercial irradiation. Presently, Codex Alimentarius (CAC, 2003) specifies that the highest absorbed dose applied to food should not exceed 10 kGy, except in cases where it is essential to fulfilling a valid technological objective. However, the Food and Agriculture Organization, International Atomic Energy Agency (IAEA), and World Health Organization (1999) jointly determined the food irradiated to a level necessary to accomplish the desired technical goal is considered secure for consumption and suitable in terms of nutrition.

Considering the Codex General Standard of Irradiated Foods, various nations globally have adopted their specific policies regarding food irradiation. For instance, within the European Union (EU), food processing through ionizing radiation is monitored by two distinct directives: 1999/2/EC and 1999/3/EC. Directive 1999/2/EC pertains to the manufacture, marketing, and importation of foods and food ingredients, hereafter called "foodstuffs", treated with ionizing radiation. Meanwhile, Directive 1999/3/EC established the creation of an initial positive list within the Community, outlining food and food ingredients, allowing for ionizing radiation treatment alongside the approved maximum doses for what was intended. In the United States, the Food and Drug Administration (1986) is principally in charge of monitoring food irradiation due to its classification as a "food additive" even though there is no physical addition to the food. Additionally, other federal agencies, such as the United States Department of Agriculture (USDA), the Nuclear Regulatory Commission (NRC), and the United States Department of Transportation (DOT), regulate various aspects of food irradiation.

In areas where irradiation is permitted, there is a necessity for regulations to authorize the operation of the facility, handling of radioactive materials, and the entire process. These regulations are designed to guarantee safety from radiation, the integrity of the environment, and overall wellness and security during the operation of the facility. Moreover, they also encompass provisions for the safe disposal of any hazardous materials generated after the operation.

Labeling is one of the critical aspects of regulatory compliance. Clear and accurate labeling is generally necessary to inform consumers about the irradiation process. This transparency ensures that individuals can make informed choices while purchasing irradiated items, including radiation-modified starches. The Codex General Standard for Irradiated Foods mandates labeling for irradiated food and ingredients (CAC, 2003), yet there exist significant variations among countries in their interpretations of this requirement. For instance, the EU requests that any irradiated food or irradiated food ingredient, even if it is in trace amounts in a non-irradiated food, must be labeled with the words "irradiated" or "treated with ionising radiation" (Directive 1999/2/EC). Moreover, in the USA, the radura symbol (Figure 8.7) is mandatory to indicate food that has undergone irradiation, along with the requirement to include "treated with radiation" or "treated by irradiation" on the packaging, unless "irradiated" is already part of the product name

FIGURE 8.7 Radura, the symbol of food treated with ionizing radiation.

(FDA, 21CFR179.26). However, in the United States, there is no obligation to label irradiated individual ingredients if the whole food item containing them has not undergone irradiation (FDA, 21CFR179.26).

To control the correct implementation of this labeling requirement and correct identification of foods that may have been treated with ionizing radiation, several physical, chemical, and biological analytical tools have been suggested. An optimal detection methodology should be: (1) specifically designed to identify irradiation and remain unaffected by other treatments or storage conditions; (2) both precise and reproducible; (3) capable of detecting irradiation even at doses below the minimum typically applied to food; (4) adaptable for use across a variety of products; (5) fast, simple, easy, and cheap to execute; and (6) capable of estimating the applied irradiation dose (Parlato et al., 2014; Stevenson & Stewart, 1995).

Regarding irradiated food containing carbohydrates, including starch, electron spin resonance (ESR) spectroscopy provides a physical means to detect free radicals within the substance. The amount of radiation applied to the material directly correlates with the number of free radicals found there (Maghraby et al., 2023). Currently, in the European Union, ESR spectroscopy stands as one of the established tools for detecting irradiation in food that contains bone, cellulose, and crystalline sugar (European Commission, 1996, 2000, 2001). This approach depends on identifying the paramagnetic species that are produced as a result of the irradiation. These established European Standards have also been adopted by the Codex Alimentarius Commission as a General Method, referenced within the Codex General Standard for Irradiated Foods under the "Post-irradiation verification" section (CAC, 2003). Nonetheless, the primary challenge related to this method stems from the unstable nature of the quite low signals typical of radiation (Sanyal et al., 2009). ESR studies on starch and starch-containing products have revealed the potential of ESR spectroscopic techniques to identify irradiated starch and determine radiation doses accurately. For gamma-irradiated starch within the 0.5–10 kGy range, variations in the ESR signal were found to be directly proportional to the irradiation doses, exhibiting ±7.7% relative to the mean value throughout 38 days post-irradiation (Maghraby et al., 2023). A similar correlation between the ESR signal and irradiation dose was observed in wheat seeds exposed to a 0.5 MeV electron beam within the range of 1–20 kGy, indicating a reliable relationship during the initial 48 hours post-irradiation (Vazirov et al., 2023). However, the persistent signal from radiation treatment diminished over 23 days due to radical recombination, impeding the determination of irradiation for doses less than 1.1 kGy. Moreover, in wheat flour irradiated with a 6 MeV electron beam up to 25 kGy, a decay pattern of radicals was observed, characterized by a swift exponential loss of short-lived radicals within the first 38 hours before a progressive decrease of long-lived radicals (Braşoveanu et al., 2013). Consequently, the ESR technique's efficacy in identifying irradiated starch is strongly related to the durability of generated free radicals, enabling their detection within a limited time following irradiation, potentially less than the goods' shelf life.

It can be concluded from the evidence presented in this section that using irradiated starch as a food ingredient should not raise public health concerns. When adhering to all existing regulations and standards, the irradiation of starch and starch-based products is regarded as a safe and efficient process that holds potential benefits within the food industry.

8.8 CONCLUSION AND FUTURE PERSPECTIVES

Ionizing radiation for starch modification has been studied for several decades but with a deeper focus and performant methodologies in the last few years. The substantial body of results published recently demonstrates that ionizing radiation, including gamma radiation, X-ray, and electron beam, can induce alterations in both the structure and functions of starch, primarily attributable to the degradation phenomenon. Starch modification by physical techniques based on ionizing radiation is a rapid, effective, cost-efficient, and eco-friendly method, offering tailored adjustments to starch properties for various applications. The effects induced in starch by ionizing radiation are predominantly associated with the degradation phenomenon, which results in a molecular weight reduction and double helix and crystalline structure alterations. The reconfiguration of starch molecules induced by irradiation leads to a general reduction in crystallinity degree, accompanied by changes in spectral bands and thermal parameters. The concurrent decrease in viscosity and swelling power, accompanied by an increase in water solubility with rising irradiation doses, positions the irradiated starches to fulfill particular criteria in diverse scenarios. Notably, this process facilitates the production of resistant starch, offering distinct health benefits.

It is crucial to emphasize that the interplay of parameters linked to both starch and processing input variables significantly influences the behavior of starch under ionizing radiation. Key factors impacting the starch's responsiveness to ionizing radiation encompass its botanical origin, moisture percentage, density, and penetration depth. Furthermore, the processing input parameters that are essential contributors to the irradiation process include the radiation source, process variables (energy, irradiation dose, and dose rate), system geometry, and the presence of oxygen. Consequently, considering variations in ionizing radiation processing methodologies and in the composition and structure of starches, precaution is advised when comparing attributes across irradiated starches.

Future research endeavors could systematically investigate the physical, chemical, and structural features inherent to every starch variety, accounting for varying moisture and oxygen contents during irradiation processing. One of the top priorities is related to the still incompletely characterized internal structure of irradiated starch. Additionally, a comprehensive exploration of thermal properties is warranted, given the current variability in findings that have been reported with no reliable connections to other investigated starch structural characteristics. Also, further studies should delve into the deep evaluation of molecular weights and their distribution. An in-depth understanding of molecular weight dynamics can furnish valuable insights, contributing to the ability to predict distinct functionalities in radiation-modified starches. This can also assist in the optimization of processing parameters for desired functional attributes in various applications that necessitate continued exploration. Furthermore, research efforts should focus on unraveling the full potential and usability of radiation-modified starch in both newly developed and existing formulations, assessing its impact on product quality and functionality. This comprehensive approach will undoubtedly advance the understanding and utilization of ionizing radiation processing in starch modification for diverse practical applications in the food sector.

ACKNOWLEDGMENTS

This work was supported by (1) the Romanian Ministry of Research, Innovation and Digitalization under Romanian National Core Program LAPLAS VII—contract no. 30N/2023 and (2) a grant of the Romanian Ministry of Research, Innovation, and Digitalization, CNCS—UEFISCDI, project number PN-III-P4-PCE-2021–1778, within PNCDI III.

REFERENCES

Aaliya, B., Sunooj, K. V., Navaf, M., Akhila, P. P., Sudheesh, C., Sabu, S., Sasidharan, A., Mir, S. A., George, J., & Khaneghah, A. M. (2021). Effect of low dose γ-irradiation on the structural and functional properties, and in vitro digestibility of ultrasonicated stem starch from *Corypha umbraculifera* L. *Applied Food Research, 1*, 100013. https://doi.org/10.1016/j.afres.2021.100013

Adlienė, D., & Adlytė, R. (2017). Dosimetry principles, dose measurements and radiation protection. In Y. Sun & A. G. Chmielewski (Eds.), *Applications of ionizing radiation in materials processing* (Vol. 1, pp. 55–80). Institute of Nuclear Chemistry and Technology.

Ai, Y., & Jane, J. L. (2015). Gelatinization and rheological properties of starch. *Starch-Stärke, 67*, 213–224. https://doi.org/10.1002/star.201400201

Arapcheska, M., Spasevska, H., & Ginovska, M. (2020). Effect of irradiation on food safety and quality. *Current Trends in Natural Sciences, 9*, 100–106. https://doi.org/10.47068/ctns.2020.v9i18.014

Arvanitoyannis, I. S. (2010). *Irradiation of food commodities: Techniques, applications, detection, legislation, safety and consumer opinion* (1st ed.). Academic Press.

Atrous, H., Benbettaieb, N., Chouaibi, M., Attia, H., & Ghorbel, D. (2017). Changes in wheat and potato starches induced by gamma irradiation: A comparative macro and microscopic study. *International Journal of Food Properties, 20*, 1532–1546. https://doi.org/10.1080/10942912.2016.1213740

Atrous, H., Benbettaieb, N., Hosni, F., Danthine, S., Blecker, C., Attia, H., & Ghorbel, D. (2015). Effect of γ-radiation on free radicals formation, structural changes and functional properties of wheat starch. *International Journal of Biological Macromolecules, 80*, 64–76. https://doi.org/10.1016/j.ijbiomac.2015.06.014

Bangar, S. P., Ashogbon, A. O., Singh, A., Chaudhary, V., & Whiteside, W. S. (2022). Enzymatic modification of starch: A green approach for starch applications. *Carbohydrate Polymers, 287*, 119265. https://doi.org/10.1016/j.carbpol.2022.119265

Barroso, A. G., & del Mastro, N. L. (2019). Physicochemical characterization of irradiated arrowroot starch. *Radiation Physics and Chemistry, 158*, 194–198. https://doi.org/10.1016/j.radphyschem.2019.02.020

Bashir, K., & Aggarwal, M. (2019). Physicochemical, structural and functional properties of native and irradiated starch: A review. *Journal of Food Science and Technology, 56*, 513–523. https://doi.org/10.1007/s13197-018-3530-2

Bello-Perez, L. A., Flores-Silva, P. C., Agama-Acevedo, E., & Tovar, J. (2020). Starch digestibility: Past, present, and future. *Journal of the Science of Food and Agriculture, 100*, 5009–5016. https://doi.org/10.1002/jsfa.8955

Bisht, B., Bhatnagar, P., Gururani, P., Kumar, V., Tomar, M. S., Sinhmar, R., Rathi, N., & Kumar., S. (2021). Food irradiation: Effect of ionizing and non-ionizing radiations on preservation of fruits and vegetables–a review. *Trends in Food Science & Technology, 114*, 372–385. https://doi.org/10.1016/j.tifs.2021.06.002

Braşoveanu, M., Crăciun, G., Mănăilă, E., Ighigeanu, D., Nemţanu, M. R., & Grecu, M. N. (2013). Evolution of the levels of free radicals generated on wheat flour and wheat bran by electron beam. *Cereal Chemistry, 90*(5), 469–473. https://doi.org/10.1094/CCHEM-10-12-0131-R

Braşoveanu, M., & Nemţanu, M. R. (2018). Aspects on starches modified by ionizing radiation processing. In E. F. Huicochea & R. Rendon (Eds.), *Applications of modified starches* (pp. 49–68). IntechOpen. https://doi.org/10.5772/intechopen.71626

Braşoveanu, M., & Nemţanu, M. R. (2020). Pasting properties modeling and comparative analysis of starch exposed to ionizing radiation. *Radiation Physics and Chemistry, 168*, 108492. https://doi.org/10.1016/j.radphyschem.2019.108492

Braşoveanu, M., & Nemţanu, M. R. (2022). Dual modification of starch by physical methods based on corona electrical discharges and ionizing radiation: Synergistic impact on rheological behavior. *Foods, 16*, 2479. https://doi.org/10.3390/foods11162479

CAC (Codex Alimentarius Commission). (1983). *Codex general standard for irradiated foods* (CODEX STAN 106-1983). Joint FAO/WHO Food Standards Programme. United Nations Food and Agriculture Organization/World Health Organization/Codex Alimentarius Commission.

CAC (Codex Alimentarius Commission). (2003). *Revised codex general standard for irradiated foods* (CODEX STAN 106-1983, Rev. 1-2003). Codex Alimentarius Commission.

Chakraborty, I., N, P., Mal, S. S., Paul, U. C., Rahman, M. H., & Mazumder, N. (2022). An insight into the gelatinization properties influencing the modified starches used in food industry: A review. *Food and Bioprocess Technology, 15*, 1195–1223. https://doi.org/10.1007/s11947-022-02761-z

Chmielewski, A. G. (2023). Radiation technologies: The future is today. *Radiation Physics and Chemistry, 111233*. https://doi.org/10.1016/j.radphyschem.2023.111233

Chung, H. J., & Liu, Q. (2010). Molecular structure and physicochemical properties of potato and bean starches as affected by gamma-irradiation. *International Journal of Biological Macromolecules, 47*, 214–222. https://doi.org/10.1016/j.ijbiomac.2010.04.019

Chung, K. H., Othman, Z., & Lee, J. S. (2015). Gamma irradiation of corn starches with different amylose-to-amylopectin ratio. *Journal of Food Science and Technology, 52*, 218–6229. https://doi.org/10.1007/s13197-014-1700-4

Compart, J., Singh, A., Fettke, J., & Apriyanto, A. (2023). Customizing starch properties: A review of starch modifications and their applications. *Polymers, 15*, 3491. https://doi.org/10.3390/polym15163491

Danyo, E. K., Ivantsova, M. N., & Selezneva, I. S. (2024). Ionizing radiation effects on microorganisms and its applications in the food industry. *Foods and Raw Materials, 12*, 1–12. https://doi.org/10.21603/2308-4057-2024-1-583

Dar, M. Z., Deepika, K., Jan, K., Swer, T. L., Kumar, P., Verma, R., Verma, K., Prakash, K. S., Jan, S., & Bashir, K. (2018). Modification of structure and physicochemical properties of buckwheat and oat starch by γ-irradiation. *International Journal of Biological Macromolecules, 108*, 1348–1356. https://doi.org/10.1016/j.ijbiomac.2017.11.067

Dhull, S. B., Punia, S., Kumar, M., Singh, S., & Singh, P. (2021). Effect of different modifications (physical and chemical) on morphological, pasting, and rheological properties of black rice (*Oryza sativa* L. Indica) starch: A comparative study. *Starch-Stärke, 73*, 2000098. https://doi.org/10.1002/star.202000098

Di, Y., Na, R., Xia, H., Wang, Y., & Li, F. (2023). Irradiation effects on characteristics and ethanol fermentation of maize starch. *International Journal of Biological Macromolecules, 246*, 125602. https://doi.org/10.1016/j.ijbiomac.2023.125602

Directive 1999/2/EC of the European Parliament and of the Council of February 22, 1999 on the approximation of the laws of the Member States concerning foods and food ingredients treated with ionising radiation, Official Journal of the European Communities L 66/16, 1999.

Directive 1999/3/EC of the European Parliament and of the Council of February 22, 1999 on the establishment of a Community list of foods and food ingredients treated with ionising radiation, Official Journal of the European Communities L 66/24, 1999.

Du, Z., Li, Y., Luo, X., Xing, J., Zhang, Q., Wang, R., Wang, L., & Chen, Z. (2020). Effects of electron beam irradiation on the physicochemical properties of quinoa and starch microstructure. *Starch-Stärke, 72*, 1900178. https://doi.org/10.1002/star.201900178

El-Esawi, M. A. (Ed.). (2019). *Legume crops: Characterization and breeding for improved food security*. BoD-Books on Demand.

Ershov, B. G. (1998). Radiation-chemical degradation of cellulose and other polysaccharides. *Russian Chemistry Reviews, 67*, 315–334. https://doi.org/10.1070/RC1998v067n04ABEH000379

Ershov, B. G., & Isakova, O. V. (1987). Nature of the radicals arising upon the γ-irradiation of starch and maltose. *Bulletin of the Academy of Sciences of the USSR, Division of Chemical Science* (English Translation), *36*, 2166–2168.

European Commission. (1996). *Foodstuffs—detection of irradiated food containing bone—method by ESR spectroscopy* (EN 1786). European Commission.

European Commission. (2000). *Foodstuffs—detection of irradiated food containing cellulose by ESR spectroscopy* (EN 1787). European Commission.

European Commission. (2001). *Foodstuffs—detection of irradiated food containing crystalline sugar by ESR spectroscopy* (EN 13708). European Commission.

Fan, X., & Niemira, B. A. (2020). Gamma ray, electron beam, and X-ray irradiation. In A. Demirici, H. Feng, & K. Krishnamurthy (Eds.), *Food safety engineering*. Food Engineering Series (pp. 471–492). Springer. https://doi.org/10.1007/978-3-030-42660-6_18

FDA (Food and Drug Administration). (1986). Irradiation in the production, processing and handling of food: Final rule. *Federal Register, 51*, 13376–13399.

FDA (Food and Drug Administration). *Code of federal regulations* (21 CFR 179.26). https://www.accessdata.fda.gov/scripts/cdrh/cfdocs/cfcfr/cfrsearch.cfm?fr=179.26

FDA (Food and Drug Administration). *Industrial starch-modified* (21 CFR 178.3520). www.ecfr.gov/current/title-21/section-178.3520

Govindaraju, I., Sunder, M., Chakraborty, I., Mumbrekar, K. D., Mal, S. S., & Mazumder, N. (2022). Investigation of physico-chemical properties of native and gamma irradiated starches. *Materials Today: Proceedings, 55*, 12–16. https://doi.org/10.1016/j.matpr.2021.11.641

Huang, Y., Wu, X., Zhang, H., Liu, Q., Wen, M., Yu, J., Li, P., He, C., & Wang, M. (2023). Electron beam irradiation as a physical modification method to deform the properties and structures of Tartary buckwheat starch: A perspective of granule and crystal. *Radiation Physics and Chemistry, 202*, 110517. https://doi.org/10.1016/j.radphyschem.2022.110517

IAEA. (1987). *Absorbed dose determination in photon and electron beams* (Technical Reports Series No. 277). IAEA.

IAEA. (2002). *Dosimetry for food irradiation* (Technical Reports Series No. 409). IAEA.

ICRU. (1980). *Radiation quantities and units* (Report 33). ICRU.

Kanatt, S. R. (2020). Irradiation as a tool for modifying tapioca starch and development of an active food packaging film with irradiated starch. *Radiation Physics and Chemistry, 173*, 108873. https://doi.org/10.1016/j.radphyschem.2020.108873

Korkmaz, M., & Polat, M. (2000). Free radical kinetics of irradiated durum wheat. *Radiation Physics and Chemistry, 58*, 169–179. https://doi.org/10.1016/S0969-806X(99)00359-X

Kumari, B., & Sit, N. (2023). Comprehensive review on single and dual modification of starch: Methods, properties and applications. *International Journal of Biological Macromolecules, 126952*. https://doi.org/10.1016/j.ijbiomac.2023.126952

Laxminarayana, D., Raju, A., Rao, B. S., & Rao, N. R. (2022). A comparitive study on morphology, solution properties of electron beam and gamma irradited edible starch. *Materials Today: Proceedings, 64*, 261–266. https://doi.org/10.1016/j.matpr.2022.04.506

Lee, Y.-J., Kim, S.-Y., Lim, S.-T., Han, S.-M., Kim, H.-M., & Kang, I.-J. (2006). Physicochemical properties of gamma-irradiated corn starch. *Preventive Nutrition and Food Science, 11*, 146–154. https://doi.org/10.3746/jfn.2006.11.2.146

Lei, X., Yu, J., Hu, Y., Bai, J., Feng, S., & Ren, Y. (2023). Comparative investigation of the effects of electron beam and X-ray irradiation on potato starch: Structure and functional properties. *International Journal of Biological Macromolecules, 236*, 123909. https://doi.org/10.1016/j.ijbiomac.2023.123909

Li, H. T., Zhang, W., Pan, W., Chen, Y., Bao, Y., & Bui, A. T. (2023). Altered leaching composition of maize starch granules by irradiative depolymerization: The key role of degraded molecular structure. *International Journal of Biological Macromolecules, 253*, 126756. https://doi.org/10.1016/j.ijbiomac.2023.126756

Liang, W., Zhao, W., Liu, X., Zheng, J., Sun, Z., Ge, X., Shen, H., Ospankulova, G., Muratkhan, M., & Li, W. (2022). Understanding how electron beam irradiation doses and frequencies modify the multiscale structure, physicochemical properties, and in vitro digestibility of potato starch. *Food Research International, 162*, 111947. https://doi.org/10.1016/j.foodres.2022.111947

Lima, F., Vieira, K., Santos, M., & Mendes de Souza, P. (2018). Effects of radiation technologies on food nutritional quality. In A. V. Díaz & R. M. García-Gimeno (Eds.), *Descriptive food science*. IntechOpen. https://doi.org/10.5772/intechopen.80437

Maghraby, A. M., Salama, E., Anwar, S. A., & El-Sayed, M. (2023). EPR study of irradiated starch: Induced radicals and dose assessment. *Radiation Effects and Defects in Solids*, 1–10. https://doi.org/10.1080/10420150.2023.2265016

Maherani, B., Hossain, F., Criado, P., Ben-Fadhel, Y., Salmieri, S., & Lacroix, M. (2016). World market development and consumer acceptance of irradiation technology. *Foods, 5*, 79. https://doi.org/10.3390/foods5040079

Maniglia, B. C., Castanha, N., Le-Bail, P., Le-Bail, A., & Augusto, P. E. (2021). Starch modification through environmentally friendly alternatives: A review. *Critical Reviews in Food Science and Nutrition, 61*(15), 2482–2505. https://doi.org/10.1080/10408398.2020.1778633

Mshelia, R. D. Z., Dibal, N. I., & Chiroma, S. M. (2023). Food irradiation: An effective but under-utilized technique for food preservations. *Journal of Food Science and Technology, 60*, 2517–2525. https://doi.org/10.1007/s13197-022-05564-4

Navaf, M., Sunooj, K. V., Aaliya, B., Sudheesh, C., Akhila, P. P., Sabu, S., Sasidharan, A., & George, J. (2022). Impact of gamma irradiation on structural, thermal, and rheological properties of talipot palm (*Corypha umbraculifera* L.) starch: A stem starch. *Radiation Physics and Chemistry, 201*, 110459. https://doi.org/10.1016/j.radphyschem.2022.110459

Nemţanu, M. R., & Braşoveanu, M. (2010). Functional properties of some non-conventional treated starches. In M. Eknashar (Ed.), *Biopolymers* (pp. 319–344). Scyio.

Nemtanu, M. R., & Brasoveanu, M. (2011). Irradiation as a method to assure food quality and safety. In D. A. Medina & A. M. Laine (Eds.), *Food quality: Control, analysis and consumer concerns* (pp. 319–344). Nova Science Publishers, Inc.

Nemtanu, M. R., & Braşoveanu, M. (2012). Radio-sensitivity of some starches treated with accelerated electron beam. *Starch/Starke, 64*, 435–440. https://doi.org/10.1002/star.201100111

Nemţanu, M. R., & Braşoveanu, M. (2017). Degradation of amylose by ionizing radiation processing. *Starch/Starke, 69*, 1600027. https://doi.org/10.1002/star.201600027

Nemţanu, M. R., & Braşoveanu, M. (2021). Exposure of starch to combined physical treatments based on corona electrical discharges and ionizing radiation: Impact on physicochemical properties. *Radiation Physics and Chemistry, 184*, Article 109480. https://doi.org/10.1016/j.radphyschem.2021.109480

Nemţanu, M. R., Braşoveanu, M., & Iovu, H. (2010). Degradation rate of some electron beam irradiated starches. *UPB Scientific Bulletin, Series B, 72*, 69–74.

Nontamas, P., Phatthanakun, R., Chio-Srichan, S., Soontaranon, S., Sorndech, W., & Tongta, S. (2022). Physicochemical properties and digestibility of native and citrate starches change in different ways by synchrotron radiation. *International Journal of Biological Macromolecules, 207*, 475–483. https://doi.org/10.1016/j.ijbiomac.2022.03.033

Pan, L., Xing, J., Zhang, H., Luo, X., & Chen, Z. (2020). Electron beam irradiation as a tool for rice grain storage and its effects on the physicochemical properties of rice starch. *International Journal of Biological Macromolecules, 164*, 2915–2921. https://doi.org/10.1016/j.ijbiomac.2020.07.211

Parlato, A., Giacomarra, M., Galati, A., & Crescimanno, M. (2014). ISO 14470: 2011 and EU legislative background on food irradiation technology: The Italian attitude. *Trends in Food Science & Technology, 38*, 60–74. https://doi.org/10.1016/j.tifs.2014.04.001

Pi, X., Yang, Y., Sun, Y., Wang, X., Wan, Y., Fu, G., Li, X., & Cheng, J. (2021). Food irradiation: A promising technology to produce hypoallergenic food with high quality. *Critical Reviews in Food Science and Nutrition, 62*, 6698–6713. https://doi.org/10.1080/10408398.2021.1904822

Polesi, L. F., Sarmento, S. B. S., de Moraes, J., Franco, C. M. L., & Canniati-Brazaca, S. G. (2016). Physicochemical and structural characteristics of rice starch modified by irradiation. *Food Chemistry, 191,* 59–66. https://doi.org/10.1016/j.foodchem.2015.03.055

Punia, S., Dhull, S. B., Kunner, P., & Rohilla, S. (2020). Effect of γ-radiation on physico-chemical, morphological and thermal characteristics of lotus seed (*Nelumbo nucifera*) starch. *International Journal of Biological Macromolecules, 157,* 584–590. https://doi.org/10.1016/j.ijbiomac.2020.04.181

Rostamabadi, H., Demirkesen, I., Taze, B. H., Karaca, A. C., Habib, M., Jan, K., Bashir, K., Nemtanu, M. R., Colussi, R., & Falsafi, S. R. (2023). Ionizing and nonionizing radiations can change physicochemical, techno-functional, and nutritional attributes of starch. *Food Chemistry: X, 100771.* https://doi.org/10.1016/j.fochx.2023.100771

Sabbaghi, H. (2023). Perspective chapter: Cellulose in food production—principles and innovations. In R. B. Jeyakumar, K. Sankarapandian, & Y. K. Ravi (Eds.), *Cellulose—fundamentals and conversion into biofuel and useful chemicals.* IntechOpen. http://dx.doi.org/10.5772/intechopen.109204

Sanyal, B., Chawla, S. P., & Sharma, A. (2009). An improved method to identify irradiated rice by EPR spectroscopy and thermoluminescence measurements. *Food Chemistry, 116,* 526–534. https://doi.org/10.1016/j.foodchem.2009.02.067

Sharpatyi, V. A. (2003). Radiation chemistry of polysaccharides: 1. Mechanisms of carbon monoxide and formic acid formation. *High Energy Chemistry, 37,* 369–372. https://doi.org/10.1023/B:HIEC.0000003393.88711.e2

Shen, H., Yu, J., Bai, J., Liu, Y., Ge, X., Li, W., & Zheng, J. (2023). A new pre-gelatinized starch preparing by spray drying and electron beam irradiation of oat starch. *Food Chemistry, 398,* 133938. https://doi.org/10.1016/j.foodchem.2022.133938

Smith, J. S., & Pillai, S. (2004). Irradiation and food safety, *Food Technology, 58,* 48–55.

Sudheesh, C., Sunooj, K. V., George, J., Kumar, S., & Sajeevkumar, V. A. (2019). Impact of γ– irradiation on the physico-chemical, rheological properties and in vitro digestibility of kithul (*Caryota urens*) starch; a new source of nonconventional stem starch. *Radiation Physics and Chemistry, 162,* 54–65. https://doi.org/10.1016/j.radphyschem.2019.04.031

Sunder, M., Mumbrekar, K. D., & Mazumder, N. (2022). Gamma radiation as a modifier of starch–physicochemical perspective. *Current Research in Food Science, 5,* 141–149. https://doi.org/10.1016/j.crfs.2022.01.001

Stevenson, M. H., & Stewart, E. M. (1995). Identification of irradiated food: The current status. *Radiation Physics and Chemistry, 46,* 653–658. https://doi.org/10.1016/0969-806X(95)00236-Q

Tappiban, P., Zhao, J., Zhang, Y., Gao, Y., Zhang, L., & Bao, J. (2022). Effects of single and dual modifications through electron beam irradiation and hydroxypropylation on physicochemical properties of potato and corn starches. *International Journal of Biological Macromolecules, 220,* 1579–1588. https://doi.org/10.1016/j.ijbiomac.2022.09.091

Tran, M. Q., Nguyen, V. B., & Tran, X. A. (2022). Gamma radiation modification of cassava starch and its characterization. *Polymer Engineering & Science, 62,* 1197–1204. https://doi.org/10.1002/pen.25919

Trinh, K. S., & Nguyen, T. L. (2020). Electron beam irradiated maize starch: Changes in structural, physicochemical properties, and digestibility. *International Journal of Advanced and Applied Sciences, 7,* 119–124. https://doi.org/10.21833/ijaas.2020.03.013

van Soest, J. J. G., Tournois, H., de Wit, D., & Vliegenthart, J. F. G. (1995). Short-range structure in (partially) crystalline potato starch determined with attenuated total reflectance Fourier-transform IR spectroscopy. *Carbohydrate Research, 279,* 201–214. https://doi.org/10.1016/0008-6215(95)00270-7

Vazirov, R. A., Narkhova, A. A., Vazirova, E. N., & Sokovnin, S. Y. (2023). Electron paramagnetic resonance signal in wheat seeds irradiated with low-energy electron beam. *Radiation Physics and Chemistry, 208,* 110934. https://doi.org/10.1016/j.radphyschem.2023.110934

Wang, G., Wang, D., & Huang, M. (2023). Effect of 10 MeV electron beam irradiation on the structure and functional properties of wheat starch. *Food Biophysics, 1–8.* https://doi.org/10.1007/s11483-023-09787-6

WHO (World Health Organization). (1994). *Safety and nutritional adequacy of irradiated food.* WHO.

WHO (World Health Organization). (1999). *High-dose irradiation: Wholesomeness of food irradiated with dose above 10 kGy.* Report of a Joint FAO/IAEA/WHO Study Group (Technical Report Series 890). WHO.

Zheng, J., Zhu, W., Liu, X., Zhao, W., Liang, W., Zheng, Y., & Li, W. (2023). Effects of electron beam irradiation pretreatment on the substitution degree, multiscale structure and physicochemical properties of OSA-esterified rice starch. *Food Bioscience, 56,* 103136. https://doi.org/10.1016/j.fbio.2023.103136

Zhiguang, C., Rui, Z., Qi, Y., & Haixia, Z. (2023). The effects of gamma irradiation treatment on starch structure and properties: A review. *International Journal of Food Science & Technology, 58,* 4519–4528. https://doi.org/10.1111/ijfs.16575

Zhou, X., Ye, X., He, J., Wang, R., & Jin, Z. (2020). Effects of electron beam irradiation on the properties of waxy maize starch and its films. *International Journal of Biological Macromolecules, 151,* 239–246. https://doi.org/10.1016/j.ijbiomac.2020.01.287

Nonthermal Plasma

9

Okon J. Esua, Clement K. Ajani, Edidiong J. Bassey, Mfrekemfon G. Akpan, Perpetual O. Onyeaka, Victory S. Igwe, and Deandrae L. Smith

9.1 INTRODUCTION

The green-production and energy-saving requirements of modern food processing have necessitated the advancement of novel green processing technologies, and among these emerging technologies, plasma technology is considered promising owing to the non-existent/reduced chemical waste associated with its use (Goiana et al., 2023; Esua et al., 2022a). Plasma, which can be naturally occurring or artificial and is generally considered as the fourth state of matter, is fully or partially ionized gas that is produced from high-voltage electrical discharge in a gas (N_2, O_2, Ne, He, Ar, CH_4, H_2, CF_4, NH_3) or gas mixture under ambient and atmospheric pressure (Sudheesh et al., 2019; Okyere et al., 2022a; Aaliya et al., 2022; Goiana et al., 2023). Plasma generation guarantees the production and propagation of numerous reactive chemistries such as ionized chemical species, electrons, excited molecules, and ultraviolet photons, which causes a collision with high kinetic energy that induces etching responsible for modifying biological materials (Okyere et al., 2019; Cubas et al., 2019; Sudheesh et al., 2019; Goiana et al., 2023).

Contingent on the temperature, plasma technology is classified as high-temperature and low-temperature plasma. High-temperature or fusion plasma, also called thermal plasma, is not suitable for processing food, as it is generated by the utilization of high pressures ($\geq 10^5$ Pa), high temperatures (10^6–10^8 K), and high power input (50 MW), with all the particles maintained in local thermodynamic equilibrium (Okyere et al., 2019; Sudheesh et al., 2019; Zhu et al., 2023). On the other hand, low temperature or gas discharge plasma, also called nonthermal plasma (NTP) or cold plasma, is suitable for processing food, as it is generated at lower pressures and much less power input and temperatures close to ambient temperature with the absence of localized thermodynamic equilibrium (Okyere et al., 2019; Sudheesh et al., 2019; Esua et al., 2022b). NTP has been explored for a wide range of food applications, including preservation, microbial decontamination, assisted processing, functionality modification, enzyme inactivation, and degradation of organic and inorganic substances, and has been demonstrated to positive influence on food materials (Sudheesh et al., 2019; Zhang et al., 2022; Esua et al., 2022c; Ali et al., 2023; Obajemihi et al., 2023; Devi et al., 2023).

NTP can be generated at low pressure or atmospheric pressure with numerous reactive oxygen species (ROS) and reactive nitrogen species (RNS), with the process feasibility typically dependent on composition and type of feed gas, configuration of plasma generating equipment, surrounding phase, relative humidity, duration of plasma exposure, voltage, and applied power (Sudheesh et al., 2019; Okyere et al., 2019; Goiana et al., 2023; Devi et al., 2023). The operational feasibility of NTP has been reported to be remarkably enhanced with reduced cost when atmospheric air was employed as the carrier-gas and under atmospheric pressure, in comparison with the use of inert gases like helium and argon under low pressure

DOI: 10.1201/9781032655598-9

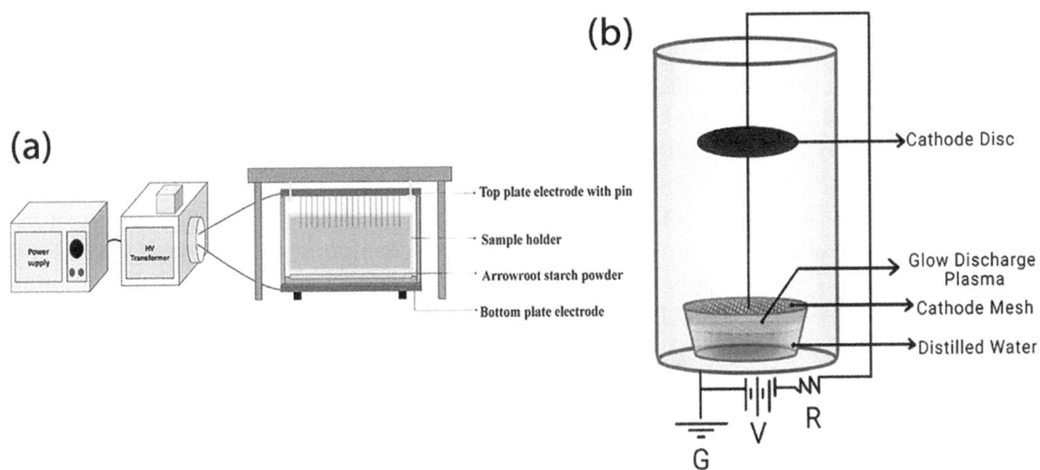

FIGURE 9.1   Schematic representation of (a) direct plasma application and (b) generation of plasma-activated water for indirect plasma application (Aaliya et al., 2022; Devi et al., 2023).

(Devi et al., 2023). NTP can be applied directly or indirectly as plasma-activated water, obtained from discharging NTP either into or directly above water (Figure 9.1) (Aaliya et al., 2022; Devi et al., 2023). Plasma-activated water is acidic and also contains a rich and unique blend of biologically active chemistries of ROS and RNS like H_2O_2, O_3, NO_3^-, and NO_2^-, with high electrical conductivity and oxidation-reduction potential (Esua et al., 2021a; Aaliya et al., 2022).

Therefore, NTP presents promising biopolymer functionality modification from the interaction of the excited electrons, free radicals, and reactive species with the biopolymer, since these interactions can lead to acidification, depolymerization, isomerization, oxidation, crosslinking, etherification, hydrogenation, and hydrolysis (Devi et al., 2023; Goiana et al., 2023). But the modification of starch by NTP is reported to occur due to an increase in surface energy, etching, molecular depolymerization, and crosslinking that alters the hydrophilic or hydrophobic nature to introduce new functional groups to ensure enhanced rheological, physical, and functional properties (Morent et al., 2011; Wongsagonsup et al., 2014; Zhang et al., 2022; Devi et al., 2023). The objective of starch modification is typically dependent upon its application; for instance, minimizing syneresis and enhancing its binding, heat resistance, digestive, and water holding capacity properties ensures direct food preparation applications, while modifying its surface structure and increasing amylopectin content and hydrophobicity would be interesting for its applications in edible films (Goiana et al., 2023).

Modified starch exhibits superior functional and physicochemical properties in comparison with its native unreactive form that shows high retrogradation tendency, low shear force, solubility, paste clarity, swelling, and poor resistance to pH and temperature and plays a significant role in both the food and non-food industries (Sudheesh et al., 2019; Okyere et al., 2019; Bangar et al., 2023). Starch is typically modified using chemical, biological, and physical methods. Chemical modification enhances starch properties by integrating functional groups into the starch molecules and is carried out through phosphorylation, etherification, esterification, acetylation, green solvents, succinylation, and oxidation (Sudheesh et al., 2019; Okyere et al., 2019; Zhang et al., 2022). Even though chemical modification has been reported to result in high efficiency, concerns over food safety, waste management, environmental pollution, and high cost limit its application. Biological modification typically increases yield with few by-products and is carried out through genetic engineering for breeding specific starch varieties or through hydrolyzing starch molecules with enzymes to low dextrose equivalents like maltodextrins but may be relatively expensive (Okyere et al., 2019; Zhang et al., 2022).

Physical modification, on the other hand, has been demonstrated to modify starches successfully and can either be thermal (pre-gelatinization, spray drying, heat treatment, etc.) or nonthermal (NTP, ultrasound treatment, pulsed electric field, etc.) (Wongsagonsup et al., 2014; Sudheesh et al., 2019; Okyere et al., 2019; Zhang et al., 2022). Physical modification methods, especially NTP, are chemical-free and environmentally friendly alternatives that do not generate wastewater or toxic residues and can influence the branching or damage the amylopectin and amylose side chains of starch, thereby affecting the size distribution and shape of the starch granules (Aaliya et al., 2022; Zhang et al., 2022; Goiana et al., 2023). NTP-induced functionality modification in starch is therefore vital for the production of starch for different purposes, either for food or non-food applications. Owing to the growing interests in and promising potential of NTP for modifying starch properties over the conventional chemical and biological methods, this chapter presents an overview on the application of NTP for modifying starch properties, including basic background information of NTP processing, its impact on the structure and functionality of starch, and food applications of the NTP-modified starch. It is hoped that the information presented will provide new insights into the conditions and mechanisms of NTP modification of starch, encourage further interest in the utilization of NTP for creating novel starch functionalities, and ensure industrial scale-up of this green technology for modifying starch properties.

9.2 PRINCIPLES AND GENERATION OF NONTHERMAL PLASMA

Plasma was first described and identified as "radiant matter" in 1879 during observations of oscillations produced during the inhomogeneous dispersal of charges in an ionized gas, and by the middle 1920s, the term "plasma" was introduced by Irving Langmuir (Surowsky et al., 2015; Zhu, 2017). In general, the relative motion between atoms and molecules in solids increases with the application of energy, and this ensures the transition to the liquid state. If the energy is continuously applied, there will be a change to the gaseous state and a further transition to a state of particles that are charged and have equal densities (d'Agostino et al., 2005). This fourth state that contains gas that has undergone partial or total ionization and comprises atoms, photons, ions, and free electrons in the excited stage is called plasma (Thirumdas et al., 2015; Sarangapani et al., 2017; Annapure & Rohit, 2023). Based on the condition of generation, plasma is categorized into equilibrium and non-equilibrium plasma. Thermal or thermodynamic equilibrium plasma displays a high degree of ionization and requires very high pressure levels ($\geq 10^5$ Pa), with the electrons and ions having about the same temperature ($T_{ion} \approx T_{overall} \approx T_{electron}$). Electrons normally lose energy to ions owing to a high frequency of collision to ensure thermodynamic equilibrium (Moreau et al., 2008; Surowsky et al., 2015).

For the non-equilibrium or NTP, ions and electrons are in thermodynamic non-equilibrium; the chemistry is driven by electrons that exhibit low ionization degrees, and electrons and ions display different temperatures. The temperature of ions and neutrals is close to ambient temperature, while that of electrons is usually in the order of 10^4 K (Moreau et al., 2008; Surowsky et al., 2015). NTP is of principal interest in food processing owing to potential processing at low temperatures and can be generated at low pressure or atmospheric pressure using microwave, corona, dielectric barrier, and plasma jet discharges.

9.2.1 Nonthermal Plasma Generation at Low Pressure

The absence of localized thermodynamic equilibrium characterizes the generation of NTP low-power and pressure conditions; thus, it can also be referred to as non-equilibrium plasma (Misra et al., 2016). The non-equilibrium state is sustained due to the rapid and efficient energy transfer from the applied electric field to the electrons, surpassing the energy transfer through electron-heavy particle collisions, causing the gas to remain at low temperatures (Leandro et al., 2023). The basic setup of low-pressure NTP comprises an electromagnetic generator powering unit that can use microwave or radio frequency, alternating current, or direct current; electrodes; pressure gauges; gas controllers; gas feeding tanks; a pumping unit to remove unwanted gases in the system; and a vacuum chamber and may be controlled by a personal computer or microcontroller (Okyere et al., 2022a). It has been suggested that the ability of a low-pressure plasma system to control the composition and level of the critical gas from which it is generated is its main advantage, owing to the fact that operation is carried out under vacuum in a closed vessel, although the vacuum equipment for the plasma setup may be relatively expensive (Choudhury, 2017; Okyere et al., 2022a). The plasma in the closed vessel shows an even distribution, even though processing with low-pressure generated plasmas can only be operated in batches and may not be appropriate for uninterrupted online processing (Okyere et al., 2022a).

The basic low-pressure plasma employed for starch modification includes glow discharge, radio-frequency discharge, and microwave discharge (Figure 9.2). The generated electrons in glow discharge plasma attain the necessary energy from excitation collisions that produce photons that make the glow visible (Vaideki, 2016). Electric current, either in the form of radio frequency (13.56 MHz, 40 kHz), low-frequency (50 Hz) direct current, or alternating current, is applied typically at >100 V over a pair of electrodes through a gas (Okyere et al., 2022a). For radio frequency discharge, the operation is usually in the range of 1–100 MHz, and the plasma is coupled to the power supplied from an electric field either inductively or capacitively, with the inductively coupled radio frequency plasma offering higher processing efficiency in comparison with the conductively coupled radio frequency plasma (Thirumdas et al., 2017a; Wilczek et al., 2020; Okyere et al., 2022a).

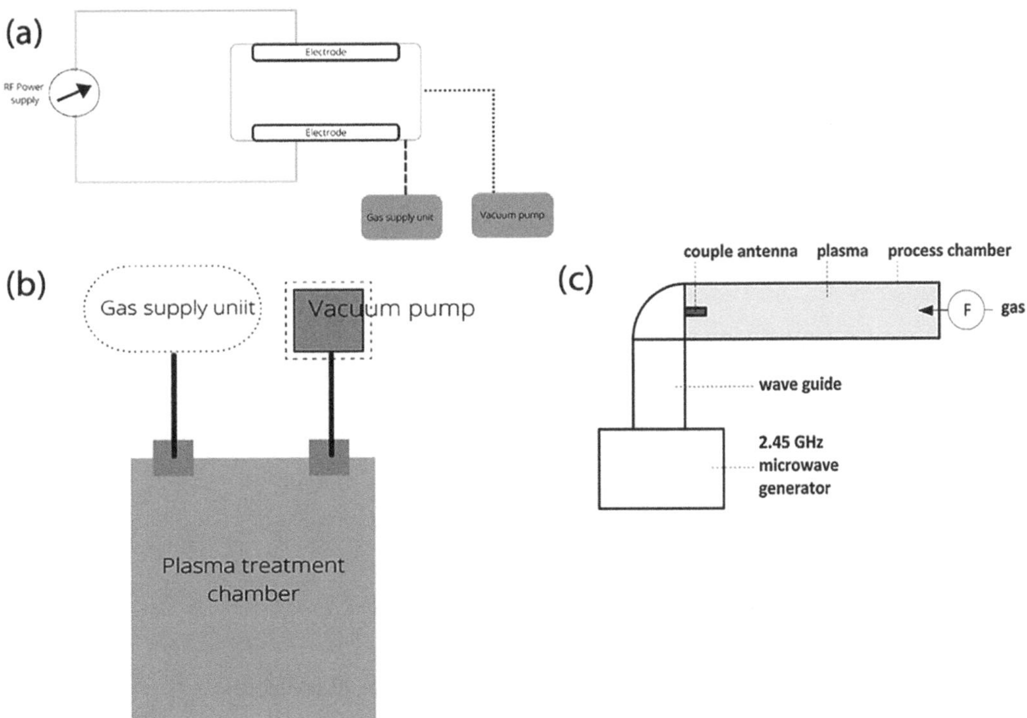

FIGURE 9.2 Schematic representation of low pressure NTP generating systems (a) radio frequency, (b) glow discharge, and (c) microwave discharge (Surowsky et al., 2015; Okyere et al., 2022a).

9.2.2 Nonthermal Plasma Generation at Atmospheric Pressure

Atmospheric pressure-generated NTP can be classified as a distinct form of plasma, whereby the dielectric medium employed is air and is characterized by a low degree of ionization, typically below 1%. This system may be effectively utilized at ambient temperature without heating, decreasing the risk of thermal deterioration in thermosensitive materials (Wu et al., 2019). The majority of NTP treatments used for starch modification are conducted at atmospheric pressure, and this is achieved by utilizing different gas sources, such as argon, oxygen, hydrogen, methane, and ethylene, among others (Leandro et al., 2023; Thirumdas et al., 2017a). Under atmospheric pressure, the system generates NTP characterized by a notably high plasma density, elevated electron temperature, and a substantial presence of active plasma species. The various plasma sources utilized for starch modification at atmospheric pressure include pressure jets, dielectric barrier discharge (DBD), and corona discharge plasma (Figure 9.3) (Okyere et al., 2022a; Ekezie et al., 2017). Plasma jets and DBD are extensively employed in food research. They have gained commercial popularity owing to their simple construction, ease of implementation, and availability of specific configurations.

DBD plasma generation occurs by applying an alternating current, which is produced when two metallic electrodes are separated by a dielectric material at a discharge gap that typically spans 100 mm to a few centimeters (Okyere et al., 2022a). The dielectric layer serves a crucial function in regulating the discharge current to prevent arc transition and promoting the random distribution of streamers on the electrode surface, facilitating a uniform treatment and hindering the occurrence of sparks by impeding the

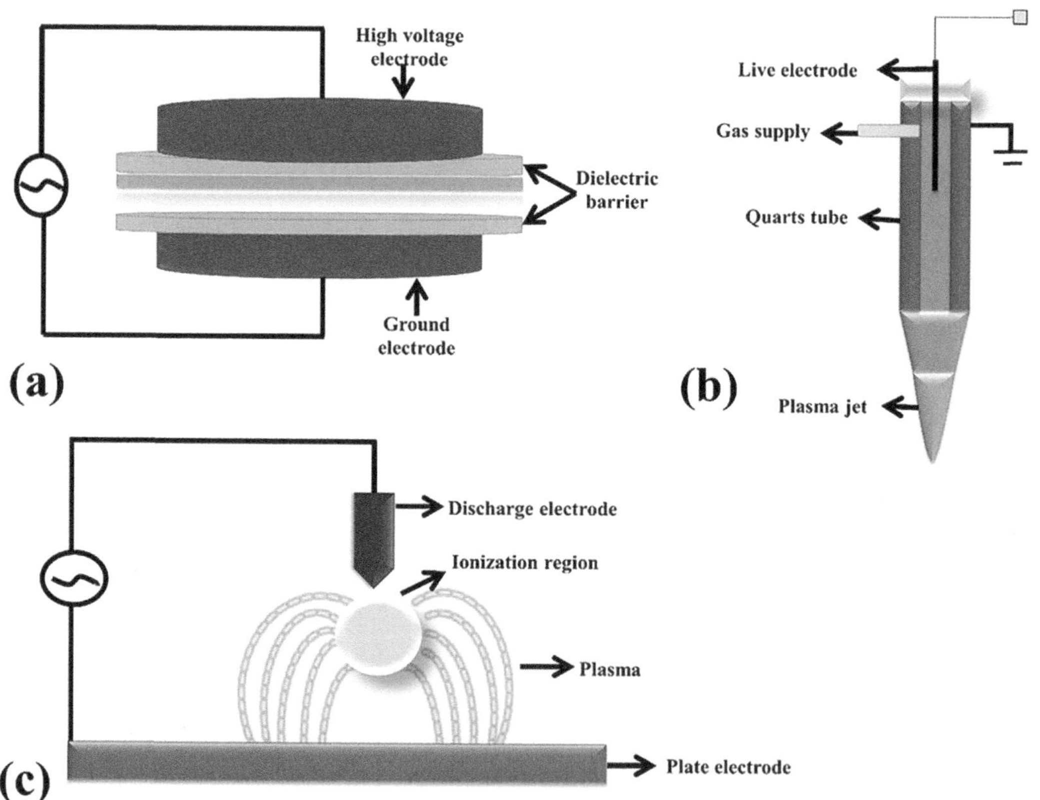

FIGURE 9.3 Schematic representation of atmospheric pressure NTP generating systems (a) dielectric barrier discharge (DBD), (b) plasma jet, (c) corona discharge.

passage of charges (Dong et al., 2017). The gases frequently employed in the DBD plasma system include atmospheric air, nitrogen, helium, and argon (Okyere et al., 2022a). On the other hand, ionization of the gas occurs during plasma jet treatment when a high voltage is applied to the inner electrode in conjunction with a high frequency (usually at 13.56 MHz radio frequency). The ionized gas is then passed through a nozzle onto the material's surface, positioned a few millimeters below the nozzle (Annapure & Rohit, 2023). Also, measuring and controlling the material temperature when utilizing plasma jets is crucial, as temperatures in these systems may often rise, particularly after prolonged exposure (Misra et al., 2016). Corona discharge NTP refers to pulsed discharge plasma generated at or close to atmospheric pressure (Saremnezhad et al., 2021). The application of a potential difference between the electrodes induces the movement of electrons due to the resultant electric fields. The purpose of this electric field is to facilitate the transfer of ionization energy from electrons that are generated randomly to the atoms or molecules of the surrounding gas. Electrodes utilized in corona discharge plasmas frequently exhibit asymmetry, such as a combination of a pointed electrode and a planar electrode (Laroque et al., 2022). Furthermore, corona discharge can manifest as either positive or negative, depending on the polarity of the high voltage supplied to the electrodes. In the case of positive corona discharge, the anode is subjected to high voltage, whereas negative corona discharge occurs when the cathode is exposed to high voltage (Okyere et al., 2022a).

Although plasma treatments conducted at atmospheric pressure may offer advantages such as ease of handling and reduced electrical energy consumption, several limitations hamper their practical application for starch modification. These limitations arise from the fact that for DBD NTP, the reaction is confined solely to the surface of the starch (Wu et al., 2018). Also, it should be noted that laboratory-scale batch methods conducted under atmospheric pressure are limited in their ability to create significant quantities of modified starch. The output of these techniques typically ranges from milligrams to only a few grams per batch (Saremnezhad et al., 2021).

9.3 NONTHERMAL PLASMA CHEMISTRY

9.3.1 Discharge Characteristics of Nonthermal Plasma

Optical emission spectroscopy is typically employed for the qualitative analysis of the discharge characteristics of NTP from the breakdown of the population dynamics of reference spectroscopic levels, and the plasma generated by different devices are distinctive and dependent upon the processing gas (Ekezie et al., 2019; Esua et al., 2021b; Yang et al., 2024). Normally, gases undergo excitation, ionization, and dissociation to form positive and negative species during plasma discharge due to the application of high voltage (Shen et al., 2019; Esua et al., 2021c). When air is employed for generating NTP, O_2 in the air dissociates first into O_2 atoms due to its lower electron energy requirement in comparison with N_2, according to the following equation (Liao et al., 2018; Esua et al., 2021c):

$$e + O_2 \rightarrow O + O + e \tag{1}$$

$$O + O_2 \rightarrow O_3 \tag{2}$$

Continuous discharge will ensure the buildup of the energy needed to dissociate N_2 in the air to nitrogen oxides (NO_x) according to the following equations (Liao et al., 2018; Zhou et al., 2018):

$$e + N_2 \rightarrow N + N + e \tag{3}$$

$$N + O \rightarrow NO \tag{4}$$

$$N + O_2 \rightarrow NO + O \tag{5}$$

$$NO + O \rightarrow NO_2 \tag{6}$$

Air guarantees a sufficient supply of O_2 and N_2, and studies suggest that the emission spectra of air-processed NTP are more complicated when compared with those of inert gases, owing to the generation of more reactive species from the reaction of O_2 and N_2 in the air (Yang et al., 2024). The emission spectra of air-processed NTP from a DBD system are concentrated in the near UV (280–450 nm), and near-infrared and visible regions, as shown in Figure 9.4a (Misra et al., 2015; Kim et al., 2019; Ekezie et al., 2019; Mehr & Koocheki, 2020; Royintarat et al., 2020; Esua et al., 2020; Yang et al., 2024). Dominant in intensity was RNS from the first negative system of N_2^+ ($N_2^+(B^2\Sigma_u^+ - X^2\Sigma_g^+)$) and second positive system vibrational bands of N_2 ($N_2(C^3\Pi_u - B^3\Pi_g)$) at 338–434 nm that indicates low electron excitation energies, in addition to optical transition of CO, ROS like OH, and NO emission at 297–313, while the singlet O_2 transition associated with the quenching of $O(^3P)$ and $O(^5P)$ were identified at 778–850 nm in relatively low intensities (Misra et al., 2015; Mehr & Koocheki, 2020; Esua et al., 2022c, 2022d). Similar emission lines with the first negative system at 391–399 nm and second positive system at 337–382 nm have been reported for air-processed NTP from a plasma jet system (Figure 9.4b), in addition to high-intensity emission of OH radicals at 306–309 nm, excited atomic O_2 species at 650–850 nm, and NO at 223–259 nm (Ekezie et al., 2019; Royintarat et al., 2020; Ali et al., 2021).

When argon is used as the carrier gas, the following equations hold:

$$Ar + e^- \rightarrow Ar^* + e^- \tag{7}$$

$$Ar + e^- \rightarrow Ar^+ + 2e^- \tag{8}$$

$$Ar^* + H_2O \rightarrow Ar + {}^{\cdot}OH + {}^{\cdot}H \tag{9}$$

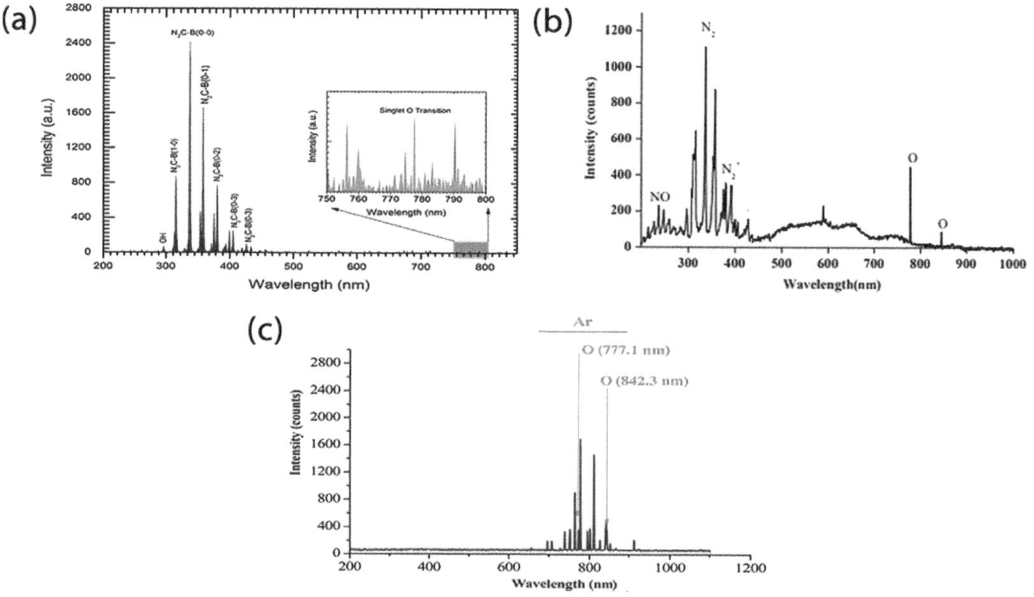

FIGURE 9.4 Optical emission spectra of NTP from (a) atmospheric air processed DBD, (b) atmospheric air processed plasma jet, and (c) argon-oxygen processed plasma jet (Misra et al., 2015; Ekezie et al., 2018, 2019).

The emission spectra of argon-processed NTP from a plasma jet system (Figure 9.4c) comprised mainly the emission lines of argon and the different excitation states since argon is the key component of the carrier gas (Ekezie et al., 2018). There were no visible peaks around the shorter wavelength regions of UV-B and UV-C, but the excited state of argon atoms in the 4p state was visible between 650–850 nm, in addition to strong atomic O_2 emissions at 777.1 and 842.3 nm, and O_2 metastable singlet state at 777.5 nm (Ekezie et al., 2018; Surowsky et al., 2014).

9.3.2 Nonthermal Plasma-Liquid Interactions

During NTP discharge in liquids, the excitation, ionization, and dissociation of gases (Equations 1–9) form negative and positive species that react with the molecules of water to generate numerous unstable chemical species and radicals to increase the acidity of the liquid (Zhou et al., 2018; Cubas et al., 2019; Esua et al., 2021c). When O_2 and N_2 molecules contact the molecules of water, the dissociation of the water molecules occurs, and the ionization of the molecules of water and the rapid reaction with additional molecules of water at the gas–liquid interface acidifies the liquids or solutions according to the following equations (Zhou et al., 2018; Cubas et al., 2019; Ali et al., 2022).

$$e + N_2 \rightarrow N_2^* + e \tag{10}$$

$$N_2^* + 2H_2O \rightarrow 2NO + 4H_2 \tag{11}$$

$$e + H_2O \rightarrow H + OH + e \tag{12}$$

$$e + H_2o \rightarrow H_2o^+ + 2e \tag{13}$$

$$H_2O^+ + H_2O \rightarrow H_3O^+ + OH \tag{14}$$

$$OH + OH \rightarrow H_2O_2 \tag{15}$$

$$H_2O + hv \rightarrow H + OH \tag{16}$$

$$H_2O_2 + hv \rightarrow OH + OH \tag{17}$$

The gas phase–dissociated reactions of N_2 and O_2 form nitrites (NO_2^-) and nitrates (NO_3^-) in the solution, with the actual concentration depending on the pH of the solution, while the concentration of NO_2^- decreases with increasing exposure time of plasma to the solution, and NO_3^- increases owing to disproportionation ($pKa = 3.3$) of NO_2^- into NO_3^- with succeeding reaction in the acidic conditions according to the following equations (Lukes et al., 2014; Yong et al., 2018):

$$NO_{2(aq)} + NO_{2(aq)} + H_2O_{(l)} \rightarrow NO_2^- + NO_3^- + 2H^+ \tag{18}$$

$$NO_{(aq)} + NO_{2(aq)} + H_2O_{(l)} \rightarrow 2NO_2^- + 2H^+ \tag{19}$$

$$3NO_2^- + 3H^+ + H_2O \rightarrow 2NO + NO_3^- + H_3O^+ \tag{20}$$

Under acidic conditions, NO_2^- is typically unstable, but its acid-base equilibrium (HNO_2) may decompose into nitrogen dioxide (NO_2) and nitric oxide (NO) known to possess strong oxidative ability in acidic environments (Lukes et al., 2014).

$$NO_2^- + H^+ \leftrightarrow HNO_2 \tag{21}$$

$$2HNO_2 \rightarrow NO + NO_2 + H_2O \tag{22}$$

Further reactions like the reaction between NO and superoxide anion radicals, NO_2 and OH radical, and NO_{2^-} and H_2O_2 ensures the formation of peroxynitrite as shown in the following.

$$NO_2^- + H_2O_2 \rightarrow ONOOH + H_2O \qquad (23)$$

$$ONOOH \leftrightarrow OH + NO_2 \qquad (24)$$

$$OH + NO_2 \rightarrow ONOO^- + H^+ \qquad (25)$$

$$O_2^- + NO \rightarrow ONOO^- \qquad (26)$$

Peroxynitrites show strong oxidation tendencies and high reactivity towards biological molecules for causing various forms of modifications, either as the protonated (ONOOH) form called peroxynitrous acid, which has a pKa of 6.8, or as peroxynitrite anion (ONOO$^-$) (Lukes et al., 2014; Shen et al., 2019).

9.3.3 Mechanism of Nonthermal Plasma Modifications and Factors Influencing Efficiency

The mechanism of NTP modifications of starch is linked to its high diffusivity and active reactive chemistries that result in surface structuring, depolymerization, and crosslinking/grafting of the amylopectin and amylose side chains of starch (Figure 9.5), owing to ionic bombardment from the synergistic actions of free electrons, charged particles, excited molecules, reactive species, UV photons, and etching (Zhu, 2017; Han et al., 2020; Zhu et al., 2023). Reactive species and charged particles formed during plasma generation can etch the surface of starch and modify starch granules, and these interactions between the plasma chemistries and biopolymers such as starch may result in polar hydrophilic groups being introduced (Morent et al., 2011; Zhu et al., 2023). Molecular crosslinking occurs when there is a breakage in the reduction ends of the chains of two starch polymers (C–OH), such that water is removed and the formation of a new C–O–C bond takes place (Zhu, 2017; Zhu et al., 2023). Polymerization, on the other hand, occurs as a result of the ionization of the molecules of water from ionic bombardment and etching that breaks down the amylopectin and amylose side chains of starch molecules to smaller fragments like maltotetraose, maltotriose, and maltose (Thirumdas et al., 2017b; Zhu et al., 2023). However, the efficacy of NTP modification of starch is contingent upon multiple elements, principally attributed to the unique

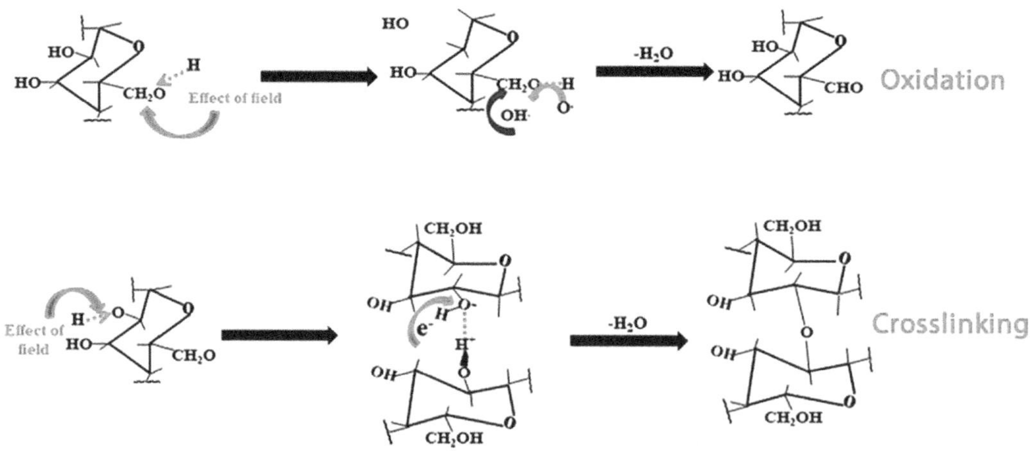

FIGURE 9.5 Schematic representation of the modification process of starch film by DBD NTP (Guo et al., 2022).

features exhibited by different plasmas and the methodologies employed for their generation. The types and quantity of reactive species generated in the discharge and the effectiveness of the treatment method are primarily determined by the processing parameters utilized (Ekezie et al., 2017).

Particularly, the selection of gas substantially impacts the efficiency of NTP generation. For instance, the utilization of specific gases can result in the formation of hydroxyls, ketones, aldehydes, esters, and free radicals, as shown in carbon dioxide and argon gas plasma spectral lines (Okyere et al., 2019). Moreover, the combination of gases, in addition to flow rate and pressure, can significantly facilitate plasma creation with enhanced efficiency, encompassing their impact on electron density and gas temperature (Dong et al., 2017). Additional process parameters that can influence NTP at atmospheric pressure for starch modification may include power, voltage, frequency, treatment duration, electrode arrangement, and the method of coupling employed. In particular, starch granules suffered more severe surface rupture, gradual loss of form, and changes in pasting properties with increasing voltage of NTP generation, and the shallow etching and damage of the starch granules induced by NTP varied depending on the applied voltage (Carvalho et al., 2021; Wu et al., 2018). Also, NTP at high power intensity was also reported to reduce starch viscosity during plasma jet treatment (Wu et al., 2019). Solubility index and swelling properties of rice starch were positively correlated with the increase in power and duration of plasma treatment, while the turbidity of mango seed kernel starch was found to decrease in a voltage and time-dependent manner during atmospheric pressure nonthermal pin-to-plate plasma treatment (Thirumdas et al., 2017a; Kalaivendan et al., 2022).

Furthermore, intrinsic and surrounding factors such as starch type and composition, moisture content, and humidity play a vital role in ensuring the efficacy of starch modification. This is because the cross-linking and depolymerization of starch granules can vary among different types and compositions of starches (Wu et al., 2019). For instance, low-amylose rice starch films were shown to be more susceptible to physical and chemical alterations post-plasma treatment compared with high amylose maize starch films. This finding suggests that the amylose concentration and starch source significantly influence the polymer's interaction with NTP (Pankaj et al., 2017). Also, the presence of moisture in the plasma environment, whether originating from the moisture content of the material or the humidity of the gas phase, can generate various species, such as hydrogen peroxide, superoxide anions, and other reactive oxygen species. These species could potentially influence the properties of starch, as demonstrated by Trinh and Nguyen (2018), who showed that the water solubility index of maize starch increased with decreasing moisture content, while increasing the moisture content led to increases in the swelling factor of maize starch during argon-plasma treatment at atmospheric pressure. Likewise, the level of cross-linking observed in the treated starch under high relative humidity conditions was lower when compared with those treated under low relative humidity conditions (Deeyai et al., 2013).

9.4 MODIFICATION OF STARCH BY NONTHERMAL PLASMA

9.4.1 Effects of Nonthermal Plasma on the Composition of Starch

The chemical composition of starch can be significantly affected by NTP processing, as the treatment decreases the molecular size of starch, which consequently reduces the pH of the starch solution (Lii et al., 2002a, 2002b; Thirumdas et al., 2017a). Notably, a report by Lii et al. (2002b) revealed that the pH of various starches was reduced by around 1.4 to 2.8 after treatment with a low-pressure glow plasma. This reduction in pH could be a result of the interaction of the starch molecules with the reactive species, consequently forming new compounds and introducing new acidic functional groups that are bound to the starch molecules (Khorram et al., 2015). Several reports have indicated that the moisture content of starch can be decreased during treatment with NTP (Lii et al., 2002a, 2002b; Chaiwat et al., 2016). Particularly, the moisture content of cassava starch was reduced from 13.5 to 6.2% with a low-pressure argon plasma treatment applied for 180 min (Chaiwat et al., 2016). Other reports revealed that moisture contents of corn, oat, potato, rice indica, and wheat starches exposed to atmospheric glow plasma treatments for 30 min were decreased from 11.1 to 10.4%, 12.6 to 5.9%, 14.5 to 10.3%, 12.1 to 8.7%, and 9.7 to 5.3%, respectively (Lii et al., 2002a, 2002b). This might be attributed to the release of water molecules bound to the starch, which were induced by the interactions of the reactive species with the starch and water molecules (Chaiwat et al., 2016). Studies by Thirumdas et al. (2017a) revealed that the content of amylose in starch was also reduced by NTP treatment. Specifically, the application of 60 W atmospheric pressure NTP reduced the amylose content of rice starch from 30 to 23%. The reduction in amylose was attributed to the interactions between the amylose and the reactive species.

Reports have indicated that the degree of changes in the chemical composition induced in starch is dependent on the plasma type, experimental conditions, and the type of starch (Zhu, 2017). For example, studies have shown that there was a greater decrease in the pH of rice starch solution (~ 2.6) when compared with maize starch (~ 1.4) when both were treated with a low-pressure oxygen plasma (Lii et al., 2002b). These changes could be due to the differences in the structure and composition of the starch as well as the concentrations and types of reactive species (Lii et al., 2002a, 2002b; Chaiwat et al., 2016). Presently, the understanding of how the type and composition of various starches are susceptible to plasma treatment is still limited. Also, the interactions of other minor components in starch (e.g., lipids, proteins, and phosphorus) during NTP treatment are yet to be fully explored.

9.4.2 Effects of Nonthermal Plasma on the Structure of Starch

9.4.2.1 Molecular Structure

Molecular structure degradation can occur during NTP processing of starch, and literature suggests that the degree of the changes in the range order and molecular weight is dependent on the different types of NTP generating systems and the source of the native starch (Chang et al., 2020; Gao et al., 2019; Sifuentes-Nieves et al., 2020; Wu et al., 2022). To analyze the molecule structure of starch, Fourier transform infrared spectrometry (FTIR), X-ray diffraction (XRD), nuclear magnetic resonance spectrometry (NMR), scanning electron microscopy (SEM), and gel permeation chromatography with multi-angle light scattering detection (GPC-MALS) can be used. The actions of DBD NTP plasma led to the removal of some molecules of water from potato starch, expressed as a decrease in the intensity of the band at 2871 cm^{-1} from FTIR analysis (Guo et al., 2022). The band ratios of 1047/1022 and 1022/1000 cm^{-1} are typically accepted as a measure of starch molecular short-range order. A decrease in this order could be a result of polymerization of starch glycosidic bonds, while an increase in the ratio could result from the reduction in amorphous content due to a breakdown and eventual release of amylose from the amorphous region (Zhou et al., 2019; Yan et al., 2020; Okyere et al., 2022a). In this vein, the short-range molecular order of microcrystalline potato, corn, aria, and banana starches was investigated following atmospheric jet NTP treatment, and while potato, corn, and aria starches presented reductions in the 1047/1022 cm^{-1} ratio, banana starch presented an increase in the 1047/1022 cm^{-1} ratio (Yan et al., 2019, 2020; Zhou et al., 2019; Carvalho et al., 2021). FITR-attenuated total reflectance analysis of granular waxy potato and non-granular waxy rice and maize treated with radio frequency NTP showed the capability of NTP to induce crosslinking in these starches, evident from changes in the internal chain structure and units of the waxy starches (Okyere et al., 2022b). In similar manners, FTIR enabled the detection of crosslinking in starches expressed as an increase in C–O–C linkages in aria and tapioca starches and a reduction in protons of the OH- group in granular and non-granular tapioca starch following NTP treatment (Wongsagonsup et al., 2014; Carvalho et al., 2021).

According to XPS analysis, oxidation of the O–C–O, C–O, and C(C–H) bonds in corn starch by reactive species in NTP led to a decrease in these bonds, while increases were associated with the O=C–OH

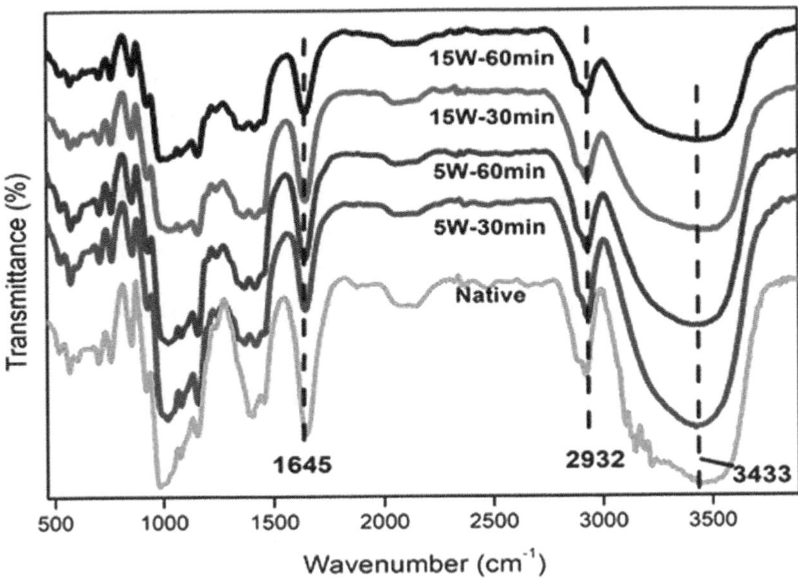

FIGURE 9.6 Fourier transform infrared spectra of native and NTP-treated kithul starch (Sudheesh et al., 2019).

bond (Bie et al., 2016a). Imaging from CLSM showed that the number of the reducing ends of molecules was increased due to the enhanced degradation of molecular chains, demonstrating the capability of NTP for enhancing the molecular chain degradation of starch (Wu et al., 2022). Observation from NMR analysis showed a reduction in the double and single helices of potato starch afterglow discharge NTP treatment, and it was concluded that starch crystallites were converted to amorphous structures following NTP-induced treatment (Zhang et al., 2015). FTIR analysis showed that prolonged treatment time and higher power levels led to the cleavage of hydrogen bonds in kithul starch expressed as a decrease in the intensity of the peaks of the H-O-H bending vibrations (Figure 9.6) (Sudheesh et al., 2019). Reductions in the molecular weight of starches from sources like maize (2.542×10^7 to 8.694×10^6 g/mol Mw) and potato (4.270×10^7–4.716×10^6 g/mol) have been reported after oxygen glow NTP treatment at 245 V using GPC-MALS analysis (Zhang et al., 2014). A decrease in the molecular weight of starch following the degradation of the starch molecules can lead to the separation of macromolecular chains into smaller bits via NTP treatment. For instance, NTP was reported to effectively degrade waxy polymeric starch chains into component units of glucose units, evident in the increase of glucose units around the α-(1,6)-glycosidic linkages (Okyere et al., 2019). In another study, the amylose composition of kithul starch was degraded to glucose units by NTP, which further led to a decrease in the iodine binding capacity of the starch, as starch hydrolysis to form smaller glucose units may limit the iodine binding ability (Sudheesh et al., 2019). The amylopectin and amylose molecular weights of taro starch treated with NTP were progressively decreased, and this was attributed to starch molecule destruction, depolymerization, cross-linking, and oxidation by NTP (Gupta et al., 2023).

9.4.2.2 Granular Morphology

The granular morphology of starch treated with NTP can be investigated through SEM, confocal laser scanning microscopy (CLSM), atomic force microscopy (AFM), and transmission electron microscope (TEM); analysis suggests that the changes in the granular morphology of starch can be influenced by the primary source, treatment intensity, and type of NTP generating system (Zhang et al., 2015; Pankaj et al., 2015; Wu et al., 2018; Banura et al., 2018; Gao et al., 2021). CLSM revealed that NTP not only affected

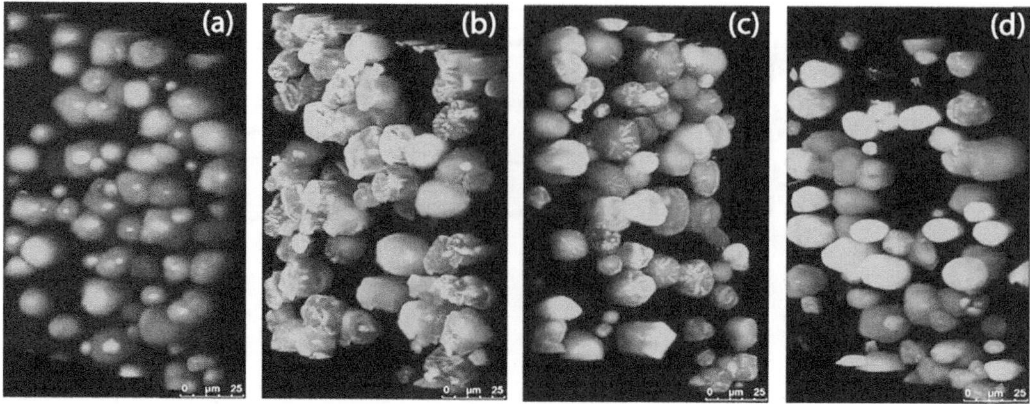

FIGURE 9.7 Three-dimensional CLSM images of maize starch treated by DBD NTP (a) native starch, (b) 75 W—1 min, (c) 75 W—5 min, and (d) 75 W—10 min (Bie et al., 2016a).

starch granule surface but also penetrated maize starch interior partially to cause molecular degradation for increasing the starch molecule reducing end (Figure 9.7) and demonstrated that NTP uses the pinhole structure of maize starch as a pathway for altering the starch granule interior structure (Bie et al., 2016a). The majority of the studies showed that NTP treatment may not produce obvious alterations in the granular morphology of starch, but some research outcomes showed bumps, cracks, or other damage on the starch granular morphology (Gao et al., 2019; Wu et al., 2022). AFM analysis of the surface topography of starch films following high-voltage atmospheric NTP treatment suggested increased surface roughness of the films and demonstrated that films from rice starch with low amylose composition were more susceptible to plasma etching when compared with films from corn starch with high amylose content (Pankaj et al., 2017). There were no observed changes in the shapes of potato, rice, and starch granules following radio frequency NTP treatment at 120 W for 60 min (Okyere et al., 2019). SEM analysis suggests that native kithul starch has an oval shape with smooth surface, and NTP treatment led to the formation of large cavities or fissures owing to partial corrosion fracturing and surface corrosion of the starch that led to molecular depolymerization (Figure 9.8) (Sudheesh et al., 2019), similar to the action of capacitively coupled radio frequency NTP for creating cavities and fissures in rice starch after 10 min treatment at 60 W (Thirumdas et al., 2017a).

Native potato starch presented ellipsoidal-shaped granules and negligible changes in granules following glow discharge NTP using neon gas, but typical visible corrosion-associated destructions were evident when helium gas was employed as the NTP generating gas at various processing times of 30, 45, and 60 min (Figure 9.9) (Zhang et al., 2015). Atmospheric pressure jet NTP led to surface etching of normal and waxy starch granules and corn starch (Zhou et al., 2019; Wu et al., 2019), similar to the observations of fissures and cavities in banana starch with increasing intensity of corona electric discharge NTP treatment (Wu et al., 2018) and cracks on the surface of potato and banana starch following DBD NTP treatment (Yan et al., 2020; Guo et al., 2022). Particle aggregation and fissures were also evident in aria starch granules following DBD NTP treatment at 20 kV (Carvalho et al., 2021). The cavities, fissures, roughness, and deposits on starch surfaces may be attributed to volatilization (etching) from the random bombardments and collisions of ROS and RNS generated during NTP processing on the surface of starch granules (Thirumdas et al., 2017b). However, polarized light microscopy showed that there were no alterations in corn starch Maltese cross structure following atmospheric pressure jet NTP treatment at 80–400 W, just as DBD NTP treatment was not enough to cause significant changes in potato starch Maltese cross structure, suggesting that these treatments could alter starch surface morphology without significantly changing the starch highly ordered structure (Wu et al., 2019; Guo et al., 2022).

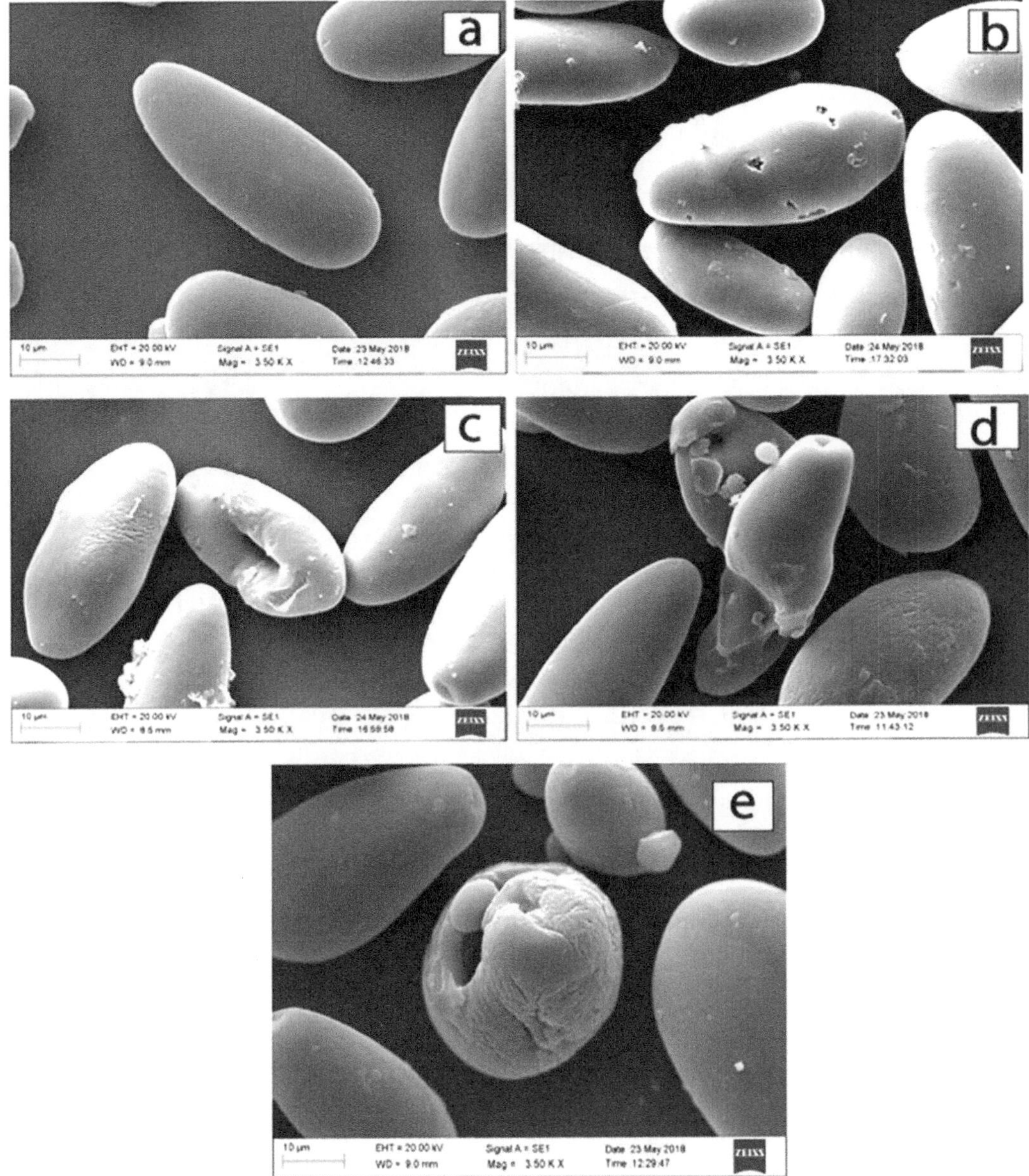

FIGURE 9.8 Scanning electron microscopy of native and NTP-treated kithul starch (a) native, (b) 5 W-30 min, (c) 5 W-60 min, (d) 15 W-30 min, (e) 15 W-60 min (Sudheesh et al., 2019).

9.4.2.3 Crystallinity

The long-range ordered crystallinity of starch can be observed from XRD analysis, and double helices in starch typically crystallize into types A, B, or C allomorphs, with the A-type, B-type, and C-type characteristically representing cereal, tuber, and legume starches, respectively (Okyere et al., 2022a). Strong diffraction peaks were demonstrated by native potato starch at 5.5, 14, 17, 20, 22, and 24° that indicated B-type starch crystals (Figure 9.10), and these peaks were not altered after DBD NTP treatment, indicating that NTP was unable to change potato starch crystal type (Guo et al., 2022). It has been suggested

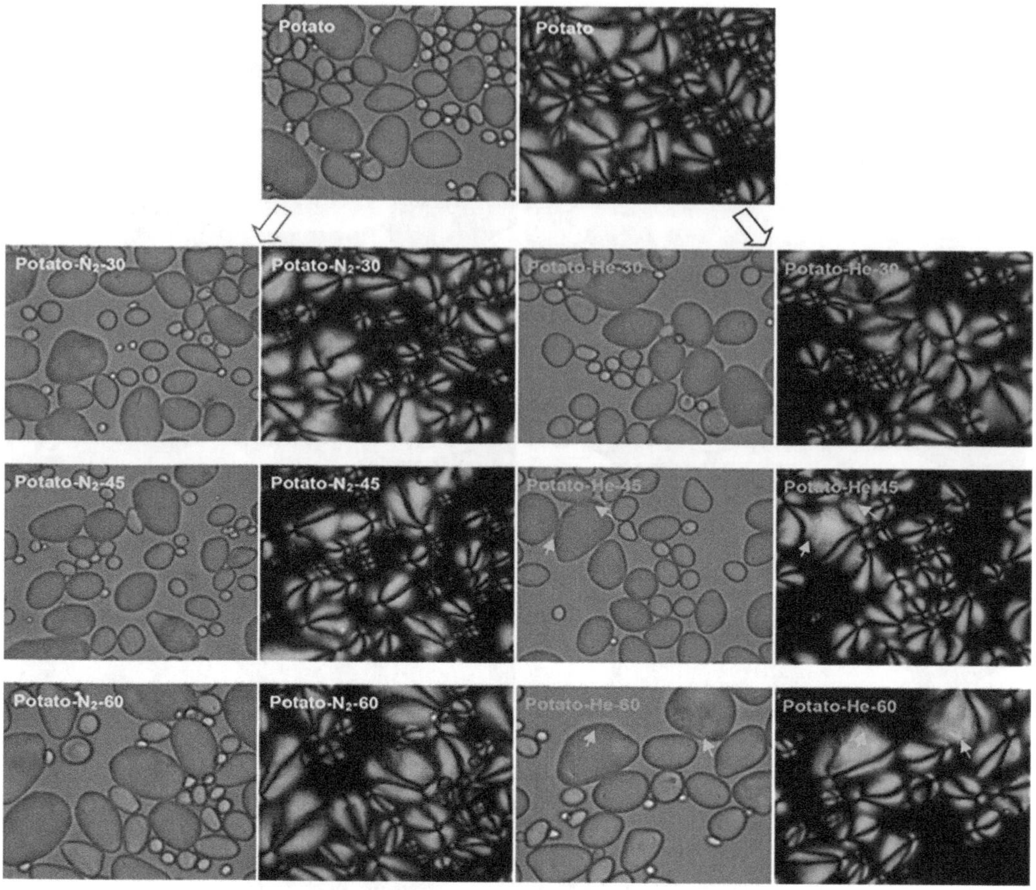

FIGURE 9.9 Micrographs of potato starch treated with NTP using nitrogen (N_2) and helium (He) as the generating gas at exposure times of 30, 45, and 60 min (Zhang et al., 2015).

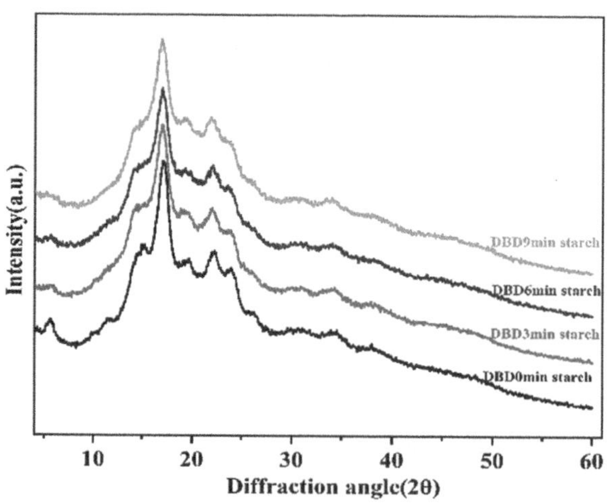

FIGURE 9.10 X-ray diffraction patterns of native and DBD NTP-treated potato starch (Guo et al., 2022).

that the degree of change in starch crystallinity following NTP treatment may be associated with the starch source, granular structure, and treatment intensity of NTP. For instance, the crystallinity of waxy potato starch reduced from 21.6 to 16.1% after CO_2-Ar gas radio frequency NTP treatment at 120 W for 60 min, while starch from maize and waxy rice presented stable crystallinity under the same conditions (Okyere et al., 2019). Radiofrequency NTP at 60 W for 20 min presented minor alterations in the A-type pattern of corn starch crystals (Banura et al., 2018) in comparison with a decrease from 38.48 to 33.79% observed in potato starch relative crystallinity following helium glow NTP treatment (Zhang et al., 2015), and decrease to 42.68% from 45.78% reported for cassava starch crystallinity after treatment with oxygen glow NTP (Bie et al., 2016b). In other studies, the relative crystallinity of potato starch was reported to decrease without a concomitant change in crystalline structure, while normal and waxy maize starches presented decreases in relative crystallinity from 40.1–35.7% and 46.7–42.0%, respectively, following atmospheric pressure jet NTP (Zhou et al., 2019; Yan et al., 2019).

Likewise, the relative crystallinity of banana starch decreased from 21.82 to 17.15% following DBD NTP treatment, in addition to alteration in the crystalline structure to A-type from C-type (Yan et al., 2020). The results, however contrasted a decrease in the relative crystallinity of rice starch without a change in the A-type crystalline structure, as reported by Sun et al. (2022). Starch molecule reorganization disruption that can possibly reduce the perfect long-range crystalline structure of starch and starch chain depolymerization induced by NTP could be the reason for the observed decreases in the relative crystalline structure of starch (Wongsagonsup et al., 2014). It has also been reported that the ability of NTP to cause decreases in the crystallinity of starch may be attributed to the destruction of the starch hydrogen bonding network by the activities of reactive species generated during treatment (Bie et al., 2016b). Overall, NTP treatment can reorganize the double helices of starch structure into an imperfect crystalline structure since the starch model backbone ensures flexibility that allows the reorganization of the longer chain segments (Okyere et al., 2022b). But condensation reactions and ether bond formation induced by reactive species in NTP can also lead to dehydroxylation of starch amorphous regions to cause an increase in the relative crystallinity of starch owing to molecular rearrangement to a more impeccable crystalline structure, as observed by Wu et al. (2018) during corona discharge NTP treatment of banana starch.

9.4.3 Effects of Nonthermal Plasma on the Physicochemical Properties of Starch

9.4.3.1 Swelling and Solubility

The application of NTP can improve the water resistance and swelling behavior of starch from modifications in the amylopectin and amylose content, granule exterior features, size, and structure of the starch. The type of NTP generating system, the source of starch, and the treatment parameters, such as power input and exposure time, could all contribute to the increase or decrease in the solubility and swelling performance of starch. The solubility and swelling properties of starch derived from rice increased from 5.12 ± 0.30 to 6.92 ± 0.71% and 9.56 ± 0.0.74 to 12.26 ± 0.08 g/g, respectively, following NTP treatment at power levels of 40 and 60 W for a period of 5 to 10 min (Thirumdas et al., 2017a). It was believed that the incomplete disintegration of the grains and molecular degradation that led to the formation of smaller fragments contributed to the improved solubility and swelling property. Potato starch treated with DBD NTP formed small fragments as the swelling temperature increased owing to depolymerization or the decomposition of starch, thus increasing solubility (Guo et al., 2022). With the increase in NTP power level from 60–100 W, the solubility of starch derived from bananas increased significantly from 1.35 to 15.05 g/100 g[-1], but a reverse trend was observed for the swelling behavior (Yan et al., 2020).

The swelling power of tapioca starch decreased with increasing voltage application and exposure time, which was consistent with the reducing molecular weight of amylopectin (Figure 9.11a), and

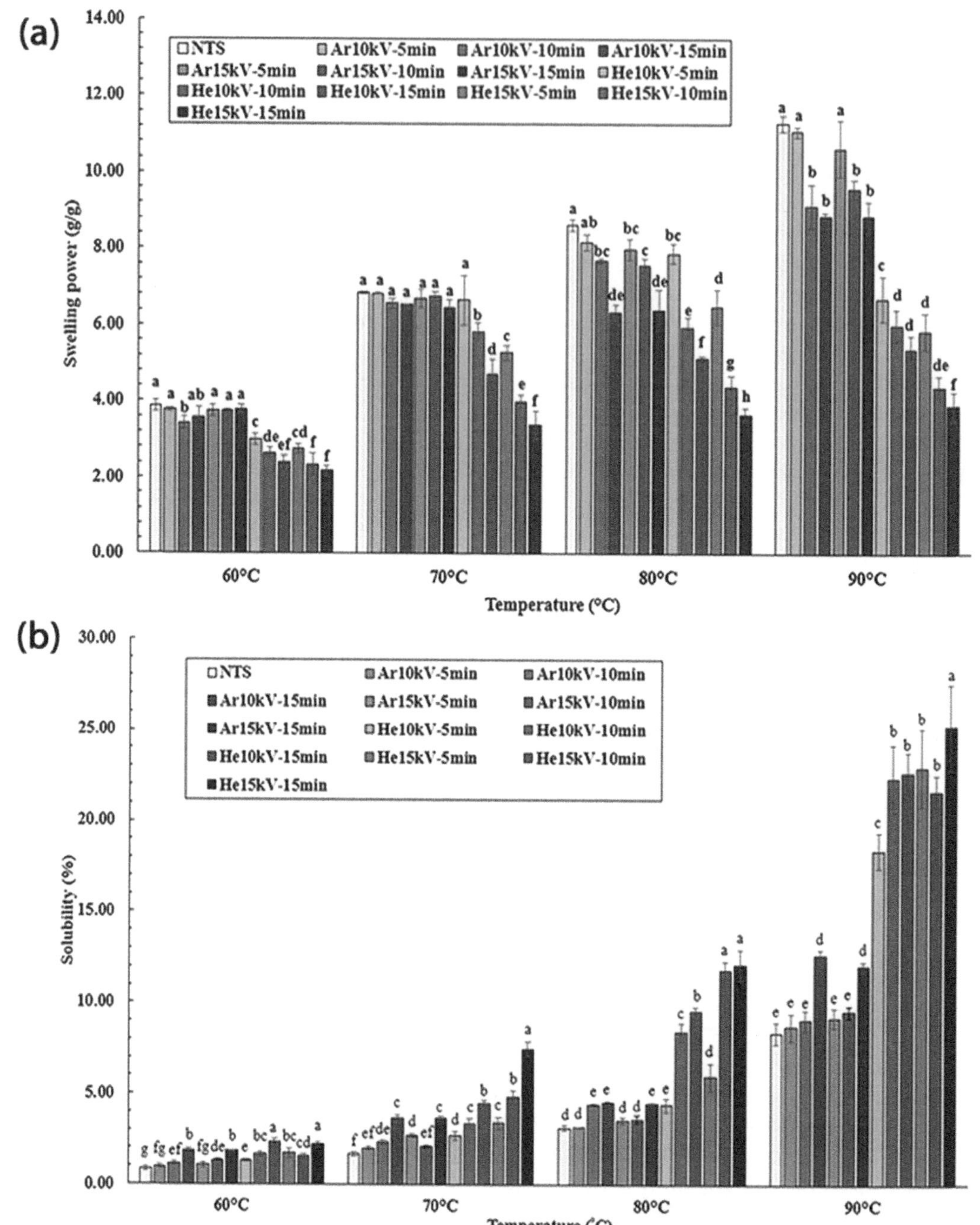

FIGURE 9.11   (a) Swelling power and (b) solubility of native and NTP modified tapioca starch at different temperatures (Sarangapani et al., 2022).

helium-generated NTP presented significantly lower (3.0–8.0 g/g) swelling power in comparison with argon-generated NTP (3.8–11.0 g/g) at the same exposure time and voltage level (Srangsomjit et al., 2022). In contrast, solubility values of treated starch using argon-generated NTP were significantly higher at all temperatures when compared to the native starch (Figure 9.11b), especially increases from 0.97 to 1.87% and 1.05 to 1.83% following treatment time of 5 to 15 min for voltage values of 10 and 15 kV, respectively

(Srangsomjit et al., 2022). Helium-generated NTP was, however, demonstrated to be more effective for modifying the solubility owing to significantly higher values at 80 and 90°C in comparison with argon-generated NTP at the same treatment voltage (Srangsomjit et al., 2022). The decreasing trend observed for swelling properties was attributed to molecular degradation and destabilization of starch and water molecular interactions from the reduction in the three-dimensional structures that affected amylopectin during the swelling process (Yan et al., 2020). Molecular degradation and granular corrosion by NTP could cause further opening of the granular structure of starch and better water adsorption, which may lead to an increase in solubility of starch. The DBD and low-pressure NTP treatments were reported to have relatively increased the swelling force of potato and rice starches, respectively, owing to the interaction of crystalline and amorphous areas in the starch chains (Sarangapani et al., 2016). The crosslinking among starch molecules reduced the water solubility of starch derived from corn after the application of electrical discharge NTP (Nemtanu & Minea, 2006).

9.4.3.2 Rheological Properties

The rheology of starch encompasses the shear thinning, elasticity, and viscosity of flow and deformation performance of starch gel. Various factors, including the treatment conditions, source of native starch, and composition, during NTP treatment can influence it. The increases in the storage and loss modulus of starch are dependent on the degree of crosslinking reactions during NTP treatment, as depolymerization causes poor gel structure (Wu et al., 2022). Following rice starch NTP treatment, Guo et al. (2019) reported hard gel with a high modulus due to the formation of cross-linked networks in starch paste. Kithul starch experienced soft gel performance as indicated by lower loss modulus and storage modulus, owing to molecular depolymerization of starch induced by high-energy neutral nitrogen atoms during glow discharge NTP treatment for 30–60 min and 5–15 W (Sudheesh et al., 2019). Granular tapioca starch subjected to 50 W atmospheric pressure jet NTP treatment ensured the formation of stronger gels, attributed to the crosslinking of starch chains, but the structure of the starch gels was weakened with increasing plasma intensity to 100 W, owing to depolymerization (Wongsagonsup et al., 2014). Native and NTP-treated corn starch presented a parabolic rheological profile with increasing shear rate, which became less apparent with increasing exposure time (Figure 9.12a), indicating the non-Newtonian fluid nature and characteristic pseudo-plastic fluid features of corn starch paste that were debilitated by NTP treatment (Bie et al., 2016a). The coefficient K data for corn starch paste suggested the tendency of corn starch to change from pseudo-plastic to Newtonian fluid, with the viscosity decreasing as the starch concentration increased, and NTP treatment made corn starch shear thinning less apparent (Figure 9.12b). The crosslinking of starch side chains from the actions of reactive species induced by NTP treatment may have prevented the leaching of starch granules into solution to reduce viscosity (Okyere et al., 2022a).

The storage modulus of cooked rice starch paste was observed to decrease from 61 to 35.7 Pa at 90°C following NTP treatment at 60 W for 10 min, and the exception of 60 W treatment for 5 min that presented increases in both moduli, variations in power (40–60 W) and exposure time (5–10 min) led to significant decreases in the loss and storage modulus (Thirumdas et al., 2017a). The increase was attributed to the formation of a more crosslinked network following 60 W treatment for 5 min when compared with the other treatments that resulted in more rigid gel formation expressed as higher moduli. Increased peak viscosity of 4228–4542 cP was also reported for processing variable range in the study when compared to 3113 cP observed for the native starch, and this was linked to starch molecules crosslinking induced by NTP oxidation. The loss modulus (12.29 ± 0.13–17.82 Pa) and storage modulus (57.27 ± 0.24–63.91 ± 0.65 Pa) of potato starch after DBD NTP treatment was relatively higher when the temperature was 90°C (Guo et al., 2022). The moduli increased continuously with an increasing cooling period, possibly due to the interactions between reactive species, substrate surface, and excited nitrogen species generated in the plasma. Different trends in the modulus were, however evident following 30 min cooling at 10°C, as the storage modulus increased while the loss modulus values decreased, proving the formation of a gel via hydrogen bonding of amylose molecules.

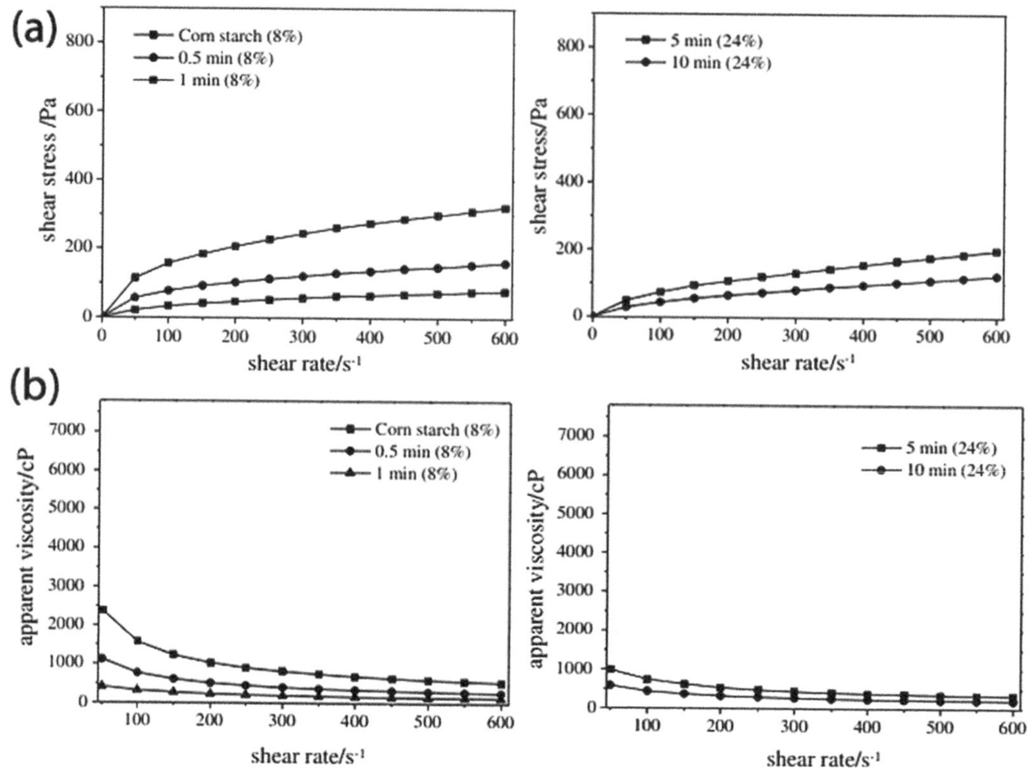

FIGURE 9.12   (a) Rheological profiles and (b) apparent viscosity of native and NTP-treated corn starches at different exposure times (Bie et al., 2016a).

9.4.3.3 Thermal Properties

Heating of starch in water leads to the disruption of molecular order in the granules, which leads to solubilization, viscosity development, melting of native crystallite, loss of birefringence, and swelling of granules, and this occurrence is called gelatinization of starch (Okyere et al., 2022a). A differential scanning calorimeter (DSC) is often employed to study the gelatinization behavior of starch, and DSC thermographs of rice starch following NTP treatment (Figure 9.13) showed that there no differences in the gelatinization temperatures of onset (T_o), peak (T_p), and conclusion (T_c) temperatures between native and treated starch, with the exception of 60 W treatment for 10 min, even though T_p of the native starch decreased from 73.5 to 72.6°C during 60 W treatment for 5 min (Thirumdas et al., 2017a). A similar reduction in gelatinization temperature and enthalpy of gelatinization (ΔH) has been reported for potato starch following helium and nitrogen glow NTP treatment (Zhang et al., 2015) and corn starch following radio frequency NTP treatment (Banura et al., 2018), while an increase in enthalpy and T_p of tapioca starch was observed after atmospheric argon jet and radiofrequency NTP treatment (Wongsagonsup et al., 2014; Banura et al., 2018). Okyere et al. (2019) also reported similar increases in ΔH for potato, rice, and corn starches following radio frequency NTP treatment.

Starches that are highly crosslinked are normally characterized by high ΔH and gelatinization temperatures owing to starch crystallite stability, while decreased gelatinization temperatures may be due to degradation in starch supramolecular lamellar structure (Bie et al., 2016b; Okyere et al., 2022a). Higher gelatinization temperature typically translates to more energy requirements for initiating starch gelatinization, and the decrease reported in these studies was probably due to the changes in amylopectin or

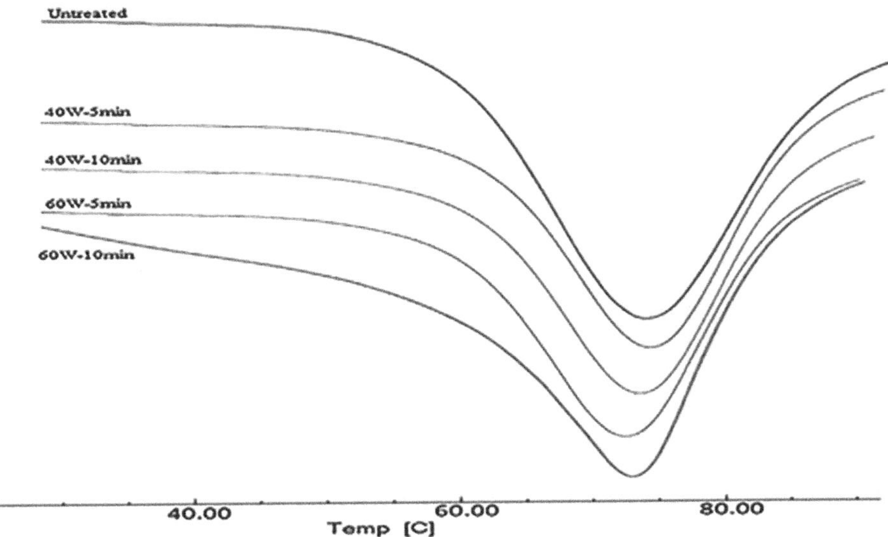

FIGURE 9.13 Differential scanning colorimeter thermographs of native and NTP-treated rice starch (Thirumdas et al., 2017a).

amylose ratio or depolymerization of the starch granules by the actions reactive species produced by NTP (Thirumdas et al., 2017a). Glow discharge atmospheric air NTP treatment of kithul starch led to a decrease in the gelatinization temperatures under power values of 5–15 W from molecular depolymerization of the starch that ensured the crystallites remained stable (Sudheesh et al., 2019). In contrast, the tendency of the active species of NTP to destroy starch crystallites was evident following a reduction in T_o, whereas the increase in T_c led to the formation of more crystallites, following the reordering of amylose chains (Sifuentes-Nieves et al., 2020). The relative increase in the gelatinization temperature was also reported when amylose chains were up to 70% in maize starch. It was suggested that oxygen glow NTP was more effective for modifying the thermal properties of cassava starch in comparison with helium glow NTP when a decrease in the ΔH of the starch was observed (Bie et al., 2016b). Thus, the capability of NTP to cause changes in the gelatinization characteristics of starch will basically be dependent on exposure time, nature/type, power/pressure of the generating system, gas/gases utilized, and the starch source.

9.4.3.4 Retrogradation

Generally, starch retrogradation is a spontaneous process that occurs after the cooked starch is cooled and the amylose and amylopectin chains align. After the initial starch is heated and dissolved in water, the original crystalline structure of the amylose and amylopectin molecules is broken, and the solution becomes viscous after hydration. Thereafter, the linear parts and chains of amylopectin and amylose molecules recrystallize to form parallel alignment of hydrogen bonds (Chang et al., 2021), and this leads to the formation of more viscous gel after cooling (0–4°C). However, retrograded starch is indigestible, which might not be favorable for some applications like reducing the shelf life of food. The effects of NTP treatment on starch retrogradation can be studied through light transmittance and clarity (Zhu, 2017). Depending on the intensity, treatment of starch via NTP has been reported to affect the retrogradation of starch. Light transmittance study of cassava starch paste decreased following low-pressure argon NTP treatment and later increased with increasing treatment time (Chaiwat et al., 2016). Crosslinking and molecular degradation of starch molecules have been reported to have negative effects on starch retro-gradation; thus, the degree of these changes depends largely on the combination of these reactions (Zhu,

2017). For the case of starch derived from tapioca, Srangsomjit et al. (2022) reported that the tendency for starch retrogradation reduced owing to lower setback viscosity of the starch after treatment with argon and helium NTP at 10 and 15 kV. The fragmentation of the linear chain of amylopectin and amylose following NTP treatment reduced the parallel alignment that could lead to retrogradation. Retrogradation plays a critical role in the production of resistance starch as it can affect the quality of starchy products, but there is a paucity of information on the effects of NTP on starch retrogradation properties.

9.5 APPLICATIONS OF NONTHERMAL PLASMA-MODIFIED STARCH

9.5.1 Surface Property Modification of Starch-Based Thermoplastic Films and Bio-Composites

The NTP treatment has been used for surface property modification of starch-based thermoplastic films and bio-composites, and several studies have been reported on various surface property modifications. The surface oxygen concentration in maize starch film was increased when treated with atmospheric DBD plasma (Pankaj et al., 2015). In another study, sulfur and fluoride, induced by a low-pressure glow discharge SF_6 NTP, were added to the maize starch film's surface after treatment (Bastos et al., 2009). There was a decrease in the water adsorption capacity, while the hydrophobicity and water contact angle properties of the maize film increased with the treatment (Bastos et al., 2009). Other studies have also corroborated results regarding the increase in the surface roughness of starch films following NTP treatment (Andrade et al., 2005; Pankaj et al., 2015). In polymorph films, NTP treatment decreased the maximum degradation temperature while its effect on the thermal degradation profile was negligible (Pankaj et al., 2015). Most studies on how NTP treatments modify starch based composite material properties were similar to those reported for starch-based films (Szymanowski et a., 2005; Han et al., 2011; Arolkar et al., 2015). Studies by Han et al. (2011) revealed that the surface hydrophobicity was increased by CF_4 and SF_6 NTP treatment, and this increase was attributed to the incorporation of new fluorine. In another study, the corrosion reactions induced by the plasma treatment increased both the free energy and surface roughness (Arolkar et al., 2015). Starch granules, modified by radio frequency NTP treatment, improved the polyethylene matrix compatibility and increased its hydrophobicity, elongation, and tensile strength (Szymanowski et al., 2005).

9.5.2 Modification of Starch-Based Artificial Tissues as Engineering Scaffolds

Over the years, there has been an increasing research focus on applying starch-based artificial tissues as engineering scaffolds in the medical field. Studies have shown that the compatibility of starch-based bio-composites, including cells and proteins, increases with NTP treatments (Alves et al., 2007). Properties of the starch-based bio-composites, such as surface roughness, energy, and hydrophobicity, were altered by NTP treatment and resulted in better proliferation and adhesion of the cells (Alves et al., 2007). NTP outperformed other ultra-violet or chemical-induced surface modification approaches, as it provided a better cell adhesion effect and a more homogenized surface for grafting, hence confirming the effectiveness of NTP treatment as an ecofriendly/green approach for modifying the surfaces of biomaterials. It is worth noting that desirable surface property modification of starch composite/film can be achieved by varying NTP treatment conditions as well as the type and composition of the starch/composite.

9.5.3 Sterilization of the Solid Media of Starch

The NTP treatment has recently been used for deactivating various microorganisms in contact with food surfaces and sterilizing biomaterial for medical applications. Specifically, NTP has been widely used as a sterilizing technology due to its ability to preserve the quality attributes and nutrients of the treated food, as well as its safety, non-pollution, and fast sterilization. To address the safety concerns and challenges associated with the pollution of cereals by various microorganisms during storage, harvest, and growth, DBD NTP was employed by Los et al. (2018) for treating fungi and bacteria inoculated on barley's surface and significantly reduced the microorganisms. Another study effectively used corona electrical discharge NTP to disinfect grain pollutants such as yeast, aerobic bacteria, and mold with minimal effects on the germination rate of germinated brown rice (Park et al., 2020). Similarly, DBD NTP treatment was successfully employed to degrade the mycotoxin content in oat powder, and although more toxins were degraded with extension in treatment time, the toxins were not completely degraded (Kis et al., 2020). Li et al. (2022) used NTP treatment to reduce and control the initial microorganisms in wheat grains and improve the texture of wheat noodles while their blackening rate was greatly inhibited. The amount of mold and aerobic bacteria on the surface of fresh noodles was reduced after the application of DBD NTP for 45 min, thereby reducing infection from foodborne diseases, inhibiting mold growth during storage, and prolonging the shelf life of noodles (Chen et al., 2019). Despite the sterilization potential of NTP treatment on food products including cereals and cereal products, it has been limited to the shallow surfaces without deep penetration into the food media. Grounding the starch to powder can used to increase the surface exposure and penetration depth during treatment. Also, improving the reaction modes and intensities of the NTP treatment might enhance a homogenous treatment of whole samples in future research.

9.5.4 Mechanical Strength Enhancement of Packaging Materials for Food and Non-Food Purposes

Recently, NTP has been employed as a novel approach for enhancing polysaccharide-based films by uniformly changing their surfaces (Desmet et al., 2009). Regarding the modification of starch-based films for food/non-food packaging, two techniques, which involve either direct application of NTP on the starch-based/composite films or modification of the starch before adding it to the films, have been employed. In addition, NTP has been used to improve several properties of starch-based films. Corn starch treated with DBD NTP resulted in uniform and smooth film surfaces and increased viscosity and solubility of the film-forming solution (Guo et al., 2022). Starch, treated with hexamethyldisiloxane (HMDSO) radio frequency NTP, positively impacted the rigidity of the film (Sifuentes-Nieves et al., 2020). Reports have also revealed that starch-based film could be made durable and soft with better hydrophobicity and rheological properties with HMDSO radio frequency NTP treatment (Hernandez-Perez et al., 2021). In another study, starch-based films directly modified using DBD NTP increased the stiffness of the film, hydrophobicity, and tensile strength (Goiana et al., 2021). Conversely, properties such as water vapor permeability, oxygen permeability, solubility, and moisture absorption all decreased significantly when starch-based films were treated with different strengths of NTP (Sudheesh et al., 2020). The hydrophilicity and tensile properties of the starch/chitosan-based film increased with no significant change observed in the water vapor transmittance after applying argon and atmospheric NTP (Sheikhi et al., 2021). In other studies, properties including tensile strength, elastic modulus, thermal stability, mechanical properties, delamination resistance, interface adhesion, and adhesion resistance of various starch-based composite films were improved with the application of various forms of NTP treatment (Florez et al., 2019; Heidemann et al., 2019; Wiącek et al., 2017). Further research is needed to promote practical applications in food packaging.

9.5.5 Modification of Edible Starch for Improving Functionality and Processing

The modification of starch by NTP has equally found applications in improving the functionality, cooking time, storage period, germination rate, and processing of cereals/starchy foods. For example, a more compact non-covalent complex was formed during NTP DBD modification of quercetin and tartary buckwheat starch that improved the starch thermal stability, while the surface morphology of mango kernel starch was significantly affected by NTP treatment to improve hydration capacity and reduce viscosity, thereby promoting the application of NTP-modified starch in starch-related functional food production (Gao et al., 2021; Kalaivendan et al., 2022). NTP-induced modification from DBD plasma led to cracks on the surfaces of rice starch that ensured the reduction in the hardness, cooking time and enhancement of the textural properties and marketability of the rice, in addition to stabilizing rice color and reducing lipid oxidation during storage (Akasapu et al., 2020; Liu et al., 2021). The NTP-modified starch has also found application in improving the germination rate of rice, as seedling growth rate enhancement and 9.72% improvement in germination rate were reported, attributed to NTP-induced etching of the rice outermost seed coat that changed the surface to smooth from the irregular rough shape before NTP treatment (Billah et al., 2021). NTP-induced modifications in starch-containing foods like fruits, grains and their by-products can improve processes like drying, esterification, and succinylation. For instance, surface etching of starch-containing jujube slices by NTP led to the generation of large cavities in the starch surface that promoted water migration to improve the drying process (Figure 9.14) (Bao et al., 2021). Barley starch that was treated by NTP DBD enhanced the substitution rate of citric acid during citrate esterification of the starch, while reducing the enzyme digestibility and improving the digestion resistance of the starch (Shen et al., 2021). In the same vein, atmospheric pressure-generated NTP jet treatment improved cassava starch succinylation efficiency with increasing treatment duration to reach a value of 39.23% during esterification with octenyl succinic anhydride (Ji et al., 2022). This was attributed to the destruction of the starch double helix structure that enhanced the reaction between the starch and octenyl succinic anhydride, that was also responsible for improved emulsifying capability and even dispersion of the starch at the oil-water interface. Despite the demonstrated potential of NTP as a viable catalytic technique for modifying starch, research on its application for modifying other types of starch, including oxidized starch, are limited.

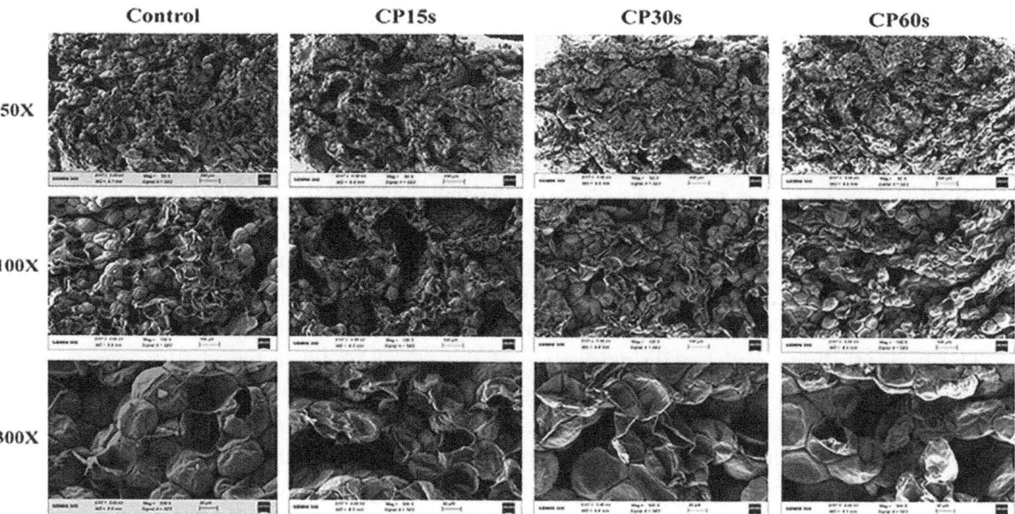

FIGURE 9.14 Microstructural modification on the surface of jujube slices following NTP pretreatments (Bao et al., 2021).

9.6 CONCLUSIONS AND OUTLOOK

This chapter summarized the nonthermal plasma modification of starch for creating new functionalities of starch. Numerous reactive oxygen and nitrogen species, ions, radicals, and electrons generated during NTP treatment interact with the starch to induce chemical changes. The changes include occasional damage to the starch integral structure, etching, holes, and numerous cracks on starch particle surfaces; the reorganization of the amylopectin chain length of starch; changes in the amorphous region and crystalline form of starch; and new functional group formation in the starch molecules. The outcome is the stabilization of starch by functionalization and crosslinking or destabilization by granular corrosion and molecular degradation that may have no effect or increase or decrease a range of properties like enzyme susceptibility, swelling and solubility, retrogradation, degree of crystallinity, viscosity, molecular weight, rheological properties, lamellae granular structure, and thermal properties. The actions of NTP modification of starch are therefore dependent on the two opposing reactions/forces of crosslinking and depolymerization, but the starch composition and type, plasma composition and type, plasma generation method, power input, and exposure time may influence the overall outcomes. Since NTP-modified starch typically presents improved stability during thermal processing, altered surface morphology, enhanced hydrophobicity, improved water absorption capacity, and improved biocompatibility of starch-based biocomposites with cells and proteins, it can find applications as ingredients for different food processing, biocomposite and thermoplastic films for food packaging, and engineering scaffolds in medical uses. When compared with other methods for modifying starch, NTP presents promising potential owing to the advantages of ecofriendliness, simplicity of operation, and no chemical residue; some gray areas need to be addressed to improve the understanding of NTP modification of starch. For instance, the starch-NTP interaction chemistry still remains largely unclear, which has led to a poor understanding of the mechanism of NTP modification of starch. Elaborate studies are required to confirm some opposing results from different studies and properties like retrogradation, where there is a paucity of literature, in addition to proper comparative studies on the influence of different NTP-generating systems and hurdle technologies on the modification process. The indirect NTP application in the form of plasma-activated water has not been fully explored. The indirect method presents options of off-site generation and storageability for future use. More so, the influence of moisture content and water in the starch on NTP modification, especially with the indirect method, needs to be properly studied, and the limit to the degree of starch modification by NTP, especially dextrinization, also needs to be explored through prolonged exposure time and high power levels, as this would provide a basis for waterless dextrinization practice. In addition, NTP treatment could be seen as adding nitrates and nitrites to food, which could become a challenging health issue, as nitrites have the potential to form carcinogenic N-nitroso compounds. Therefore, the toxicity of starch modified by NTP, especially the new elements and functional groups added to the starch, needs to be studied for food applications.

REFERENCES

Aaliya, B., Sunooj, K. V., Navaf, M., Akhila, P. P., Sudheesh, C., Sabu, S., Sasidharan, A., Sinha, S. K., & George, J. (2022). Influence of plasma-activated water on the morphological, functional, and digestibility characteristics of hydrothermally modified non-conventional talipot starch. *Food Hydrocolloids, 130*, 107709.

Akasapu, K., Ojah, N., Gupta, A. K., Choudhury, A. J., & Mishra, P. (2020). An innovative approach for iron fortification of rice using cold plasma. *Food Research International, 136*, 109599.

Ali, M., Cheng, J.-H., & Sun, D-W. (2021). Effects of dielectric barrier discharge cold plasma treatments on degradation of anilazine fungicide and quality of tomato (*Lycopersicon esculentum* Mill) juice. *International Journal of Food Science & Technology, 56*(1), 69–75.

Ali, M., Cheng, J.-H., Tazeddinova, D., Aadil, R. M., Zeng, X.-A., Goksen, G., Lorenzo, J. M., Esua, O. J., & Manzoor, M. F. (2023). Effect of plasma-activated water and buffer solution combined with ultrasound on fungicide degradation and quality of cherry tomato during storage. *Ultrasonics Sonochemistry*, *97*, 106461.

Ali, M., Sun, D.-W., Cheng, J.-H., & Esua, O. J. (2022). Effects of combined treatment of plasma activated liquid and ultrasound for degradation of chlorothalonil fungicide residues in tomato. *Food Chemistry*, *371*, 131162.

Alves, C. M., Yang, Y., Carnes, D. L., Ong, J. L., Sylvia, V. L., Dean, D. D., Agrawal, C. M., & Reis, R. L. (2007). Modulating bone cells response onto starch-based biomaterials by surface plasma treatment and protein adsorption. *Biomaterials*, *28*, 307–315.

Andrade, C. T., Simão, R. A., Thiré, R. M. S. M., & Achete, C. A. (2005). Surface modification of maize starch films by low-pressure glow 1-butene plasma. *Carbohydrate Polymers*, *61*(4), 407–413.

Annapure, U. S., & Rohit, T. (2023). Cold plasma treatment of starch. In *Starch: Advances in modifications, technologies and applications* (pp. 337–359). Springer International Publishing.

Arolkar, G. A., Salgo, M. J., Kelkar-Mane, V., & Deshmukh, R. R. (2015). The study of air-plasma treatment on corn starch/poly (ε-caprolactone) films. *Polymer Degradation and Stability*, *120*, 262–272.

Bangar, S. P., Dhull, S. B., Manzoor, M., Chandak, A., & Esua, O. J. (2023). Functionality and applications of non-conventional starches from different sources. *Starch—Starke*, 2300073. https://doi.org/10.1002/star.202300073

Banura, S., Thirumdas, R., Kaur, A., Deshmukh, R. R., & Annapure, U. S. (2018). Modification of starch using low pressure radio frequency air plasma. *LWT—Food Science and Technology*, *89*, 719–724.

Bao, T., Hao, X., Shishir, M. R. I., Karim, N., & Chen, W. (2021). Cold plasma: An emerging pretreatment technology for the drying of jujube slices. *Food Chemistry*, *337*(866), 127783.

Bastos, D. C., Santos, A. E. F., da Silva, M. L. V. J., & Simao, R. A. (2009). Hydrophobic corn starch thermoplastic films produced by plasma treatment. *Ultramicroscopy*, *109*(8), 1089–1093.

Bie, P., Li, X., Xie, F., Chen, L., Zhang, B., & Li, L., (2016b). Supramolecular structure and thermal behaviour of cassava starch treated by oxygen and helium glow-plasmas. *Innovative Food Science and Emerging Technologies*, *34*, 336–343.

Bie, P., Pu, H., Zhang, B., Su, J., Chen, L., & Li, X. (2016a). Structural characteristics and rheological properties of plasma-treated starch. *Innovative Food Science and Emerging Technology*, *34*, 196–204.

Billah, M., Karmakar, S., Mina, F. B., Haque, M. N., Rashid, M. M., Hasan, M. F., Acharjee, U. K., & Talukder, M. R. (2021). Investigation of mechanisms involved in seed germination enhancement, enzymatic activity and seedling growth of rice (Oryza Sativa L.) using LPDBD (Ar+Air) plasma. *Archives of Biochemistry and Biophysics*, *698*, 108726.

Carvalho, A. P. M. G., Barros, D. R., da Silva, L. S., Sanches, E. A., da Costa Pinto, C., de Souza, S. M., Clerici, M. T. P. S., Rodrigues, S., Fernandes, F. A. N., & Campelo, P. H. (2021). Dielectric barrier atmospheric cold plasma applied to the modification of Aria (*Goeppertia allouia*) starch: Effect of plasma generation voltage. *International Journal of Biological Macromolecules*, *182*, 1618–1627.

Chaiwat, W., Wongsagonsup, R., Tangpanichyanon, N., Jariyaporn, T., Deeyai, P., Suphantharika, M., Fuongfuchat, A., Nisoa, M., & Dangtip, S. (2016). Argon plasma treatment of tapioca starch using a semi-continuous downer reactor. *Food and Bioprocess Technology*, *9*(7), 1125–1134.

Chang, R., Lu, H., Tian, Y., Li, H., Wang, J., & Jin, Z. (2020). Structural modification and functional improvement of starch nanoparticles using vacuum cold plasma. *International Journal of Biological Macromolecules*, *145*, 197–206.

Chang, Q., Zheng, B., Zhang, Y., & Zeng, H. (2021). A comprehensive review of the factors influencing the formation of retrograded starch. *International Journal of Biological Macromolecules*, *186*, 163–173.

Chen, Y., Chen, G., Wei, R., Zhang, Y., Li, S., & Chen, Y. (2019). Quality characteristics of fresh wet noodles treated with nonthermal plasma sterilization. *Food Chemistry*, *297*(29), 124900.

Choudhury, A. K. R. (2017). Various eco-friendly finishes. In *Principles of textile finishing* (pp. 467–525). Elsevier.

Cubas, A. L. V., de Medeiros Machado, M., dos Santos, J. R., Zanco, J. J., Ribeiro, D. H. B., Anderson Andre, A. S., Debacher, N. A., & Moecke, E. H. S. (2019). Effect of chemical species generated by different geometries of air and argon non-thermal plasma reactors on bacteria inactivation in water. *Separation and Purification Technology*, *222*, 68–74.

d'Agostino, R., Favia, P., Oehr, C., & Wertheimer, M. R. (2005). Low-temperature plasma processing of materials: Past, present, and future. *Plasma Processes and Polymers*, *2*(1), 7–15.

Deeyai, P., Suphantharika, M., Wongsagonsup, R., & Dangtip, S. (2013). Characterization of modified tapioca starch in atmospheric argon plasma under diverse humidity by FTIR spectroscopy. *Chinese Physics Letters*, *30*(1), 018103.

Desmet, T., Morent, R., De Geyter, N., Leys, C., Schacht, E., & Dubruel, P. (2009). Nonthermal plasma technology as a versatile strategy for polymeric biomaterials surface modification: A review. *Biomacromolecules*, *10*(9), 2351–2378.

Devi, E., Kalaivendan, R. G. T., Eazhumalai, G., & Annapure, U. S. (2023). Impact of atmospheric pressure pin-to-plate cold plasma on the functionality of arrowroot starch. *Journal of Agriculture and Food Research, 12*, 100567.

Dong, S., Gao, A., Zhao, Y., Li, Y. T., & Chen, Y. (2017). Characterization of physicochemical and structural properties of atmospheric cold plasma (ACP) modified zein. *Food and Bioproducts Processing, 106*, 65–74.

Ekezie, F.-G. C., Cheng, J.-H., & Sun, D.-W. (2018). Effects of mild oxidative and structural modifications induced by argon plasma on physicochemical properties of actomyosin from king prawn (*Litopenaeus vannamei*). *Journal of Agriculture and Food Chemistry, 66*, 13285–13294.

Ekezie, F.-G. C., Cheng, J.-H., & Sun, D.-W. (2019). Effects of atmospheric pressure plasma jet on the conformation and physicochemical properties of myofibrillar proteins from king prawn (*Litopenaeus vannamei*). *Food Chemistry, 276*, 147–156.

Ekezie, F. G. C., Sun, D. W., & Cheng, J. H. (2017). A review on recent advances in cold plasma technology for the food industry: Current applications and future trends. *Trends in Food Science & Technology, 69*, 46–58.

Esua, O. J., Cheng, J.-H., & Sun, D.-W. (2020). Antimicrobial activities of plasma-functionalized liquids against foodborne pathogens on grass carp (*Ctenopharyngodon idella*). *Applied Microbiology and Biotechnology, 104*, 9581–9594.

Esua, O. J., Cheng, J.-H., & Sun, D.-W. (2021a). Novel technique for treating grass carp (*Ctenopharyngodon idella*) by combining plasma functionalized liquids and ultrasound: Effects on bacterial inactivation and quality attributes. *Ultrasonics Sonochemistry, 76*, 105660.

Esua, O. J., Cheng, J.-H., & Sun, D.-W. (2021b). Optimisation of treatment conditions for reducing *Shewanella putrefaciens* and *Salmonella* Typhimurium on grass carp treated by thermoultrasound-assisted plasma functionalized buffer. *Ultrasonics Sonochemistry, 76*, 105609.

Esua, O. J., Cheng, J.-H., & Sun, D.-W. (2021c). Functionalization of water as a nonthermal approach for ensuring safety and quality of meat and seafood products. *Critical Reviews in Food Science and Nutrition, 61*(3), 431–449.

Esua, O. J., Sun, D.-W., Ajani, C. K., Cheng, J.-H., & Keener, K. M. (2022c). Modelling of inactivation kinetics of *Escherichia coli* and *Listeria monocytogenes* on grass carp treated by combining ultrasound with plasma functionalized buffer. *Ultrasonics Sonochemistry, 88*, 106086.

Esua, O. J., Sun, D.-W., Cheng, J.-H., & Li, J.-L. (2022a). Evaluation of storage quality of vacuum-packaged silver Pomfret (*Pampus argenteus*) treated with combined ultrasound and plasma functionalized liquids hurdle technology. *Food Chemistry, 391*, 133237.

Esua, O. J., Sun, D.-W., Cheng, J.-H., Wang, H., & Chen, C. (2022b). Hybridising plasma functionalized water and ultrasound pretreatment for enzymatic protein hydrolysis of *Larimichthys polyactis*: Parametric screening and optimization. *Food Chemistry, 385*, 132677.

Esua, O. J., Sun, D.-W., Cheng, J.-H., Wang, H., & Lv, M. (2022d). Functional and bioactive properties of *Larimichthys polyactis* protein hydrolysates as influenced by plasma functionalized water-ultrasound hybid treatments and enzyme types. *Ultrasonics Sonochemistry, 86*, 106023.

Florez, J. P., Fazeli, M., & Simão, R. A. (2019). Preparation and characterization of thermoplastic starch composite reinforced by plasma-treated poly (hydroxybutyrate) PHB. *International Journal of Biological Macromolecules, 123*, 609–621.

Gao, S., Liu, H., Sun, L., Cao, J., Yang, J., Lu, M., & Wang, M. (2021). Rheological, thermal and in vitro digestibility properties on complex of plasma modified Tartary buckwheat starches with quercetin. *Food Hydrocolloids, 110*, 106209.

Gao, S., Liu, H., Sun, L., Liu, N., Wang, J., Huang, Y., Wang, F., Cao, J., Fan, R., Zhang, X., & Wang, M. (2019). The effects of dielectric barrier discharge plasma on physicochemical and digestion properties of starch. *International Journal of Biological Macromolecules, 138*, 819–830.

Goiana, M. L., de Brito, E. S., Alves Filho, E. G., Miguel, E. de C., Fernandes, F. A. N., Azeredo, H. M. C. de, & Rosa, M. de F. (2021). Corn starch based films treated by dielectric barrier discharge plasma. *International Journal of Biological Macromolecules, 183*, 2009–2016.

Goiana, M. L., & Fernandes, F. A. N. (2023). Influence of dielectric barrier discharge plasma treatment on corn starch properties. *Processes, 11*, 1966.

Guo, Y., Xu, T., Li, N., Cheng, Q., Qiao, D., Zhang, B., Zhao, S., Huang, Q., & Lin, Q. (2019). Supramolecular structure and pasting/digestion behaviors of rice starches following concurrent microwave and heat moisture treatment. *International Journal of Biological Macromolecules, 135*, 437–444.

Guo, Z., Gou, Q., Yang, L., Yu, Q. L., & Han, L. (2022). Dielectric barrier discharge plasma: A green method to change structure of potato starch and improve physicochemical properties of potato starch films. *Food Chemistry, 370*, 130992.

Gupta, R. K., Guha, P., & Srivastav, P. P. (2023). Effect of high voltage dielectric barrier discharge (DBD) atmospheric cold plasma treatment on physicochemical and functional properties of taro (*Colocasia esculenta*) starch. *International Journal of Biological Macromolecules, 253*, 126772.

Han, Y., Manolach, S. O., Denes, F., & Rowell, R. M. (2011). Cold plasma treatment on starch foam reinforced with wood fiber for its surface hydrophobicity. *Carbohydrate Polymers, 86*(2), 1031–1037.

Han, Z., Shi, R., & Sun, D.-W. (2020). Effects of novel physical processing techniques on the multi-structures of starch. *Trends in Food Science & Technology, 97*, 126–135.

Heidemann, H. M., Dotto, M. E. R., Laurindo, J. B., Carciofi, B. A. M., & Costa, C. (2019). Cold plasma treatment to improve the adhesion of cassava starch films onto PCL and PLA surface. *Colloids and Surfaces A: Physicochemical and Engineering Aspects, 580*, 123739.

Hernandez-Perez, P., Flores-Silva, P. C., Velazquez, G., Hern, E., Mendez-Montealvo, G., & Sifuentes-Nieves, I. (2021). Rheological performance of film-forming solutions made from plasma-modified starches with different amylose/amylopectin content. *Carbohydrate Polymers, 255*, 117349.

Ji, S., Xu, T., Huang, W., Gao, S., & Zhong, Y. (2022). Atmospheric pressure plasma jet pretreatment to facilitate cassava starch modification with octenyl succinic anhydride. *Food Chemistry, 370*, 130922.

Kalaivendan, R. G. T., Mishra, A., Eazhumalai, G., & Annapure, U. S. (2022). Effect of atmospheric pressure non-thermal pin to plate plasma on the functional, rheological, thermal, and morphological properties of mango seed kernel starch. *International Journal of Biological Macromolecules, 196*, 63–71.

Khorram, S., Zakerhamidi, M. S., & Karimzadeh, Z. (2015). Polarity functions' characterization and the mechanism of starch modification by DC glow discharge plasma. *Carbohydrate Polymers, 127*, 72–78.

Kim, J., Lee, T., Puligundla, P., Mok, C., (2019). Afterglow dielectric-barrier discharge air plasma (ADDAP) for the inactivation of *Escherichia coli* O157:H7. *LWT—Food Science & Technology, 105*, 177–181.

Kis, M., Milosevic, S., Vuli, A., Herceg, Z., Vuku, T., & Pleadin, J. (2020). Efficacy of low pressure DBD plasma in the reduction of T-2 and HT-2 toxin in oat flour. *Food Chemistry, 316*, 126372.

Laroque, D. A., Seó, S. T., Valencia, G. A., Laurindo, J. B., & Carciofi, B. A. M. (2022). Cold plasma in food processing: Design, mechanisms, and application. *Journal of Food Engineering, 312*, 110748.

Leandro, G. C., Laroque, D. A., Monteiro, A. R., Carciofi, B. A. M., & Valencia, G. A. (2023). Current status and perspectives of starch powders modified by cold plasma: A review. *Journal of Polymers and the Environment, 32*. https://doi.org/10.1007/s10924-023-03027-1

Liao, X., Su, Y., Liu, D., Chen, S., Hu, Y., Ye, X., Wang, J., & Ding, T. (2018). Application of atmospheric cold plasma-activated water (PAW) ice for preservation of shrimps (*Metapenaeus ensis*). *Food Control, 94*, 307–314.

Li, X., Wen, Y., Zhang, J., Ma, D., Zhang, J., An, Y., Song, X., Ren, X., & Zhang, W. (2022). Effects of non-thermal plasma treating wheat kernel on the physicochemical properties of wheat flour and the quality of fresh wet noodles. *International Journal of Food Science & Technology, 57*(3), 1544–1553.

Lii, C. Y., Liao, C. D., Stobinski, L., & Tomasik, P. (2002a). Effects of hydrogen, oxygen, and ammonia low-pressure glow plasma on granular starches. *Carbohydrate Polymers, 49*(4), 449–456.

Lii, C. Y, Liao, C. D, Stobinski, L., & Tomasik, P. (2002b). Exposure of granular starches to low-pressure glow ethylene plasma. *European Polymer Journal, 38*(8), 1601–1606.

Liu, Q., Wu, H., Luo, J., Liu, J., Zhao, S., Hu, Q., & Ding, C. (2021). Effect of dielectric barrier discharge cold plasma treatments on flavor fingerprints of brown rice. *Food Chemistry, 352*(3), 129402.

Los, A., Ziuzina, D., Akkermans, S., Boehm, D., Cullen, P. J., Impe, J. V., & Bourke, P. (2018). Improving microbiological safety and quality characteristics of wheat and barley by high voltage atmospheric cold plasma closed processing. *Food Research International, 106*, 509–521.

Lukes, P., Dolezalova, E., Sisrova, I., & Clupek, M. (2014). Aqueous-phase chemistry and bactericidal effects from an air discharge plasma in contact with water: Evidence for the formation of peroxynitrite through a pseudo-second-order post-discharge reaction of H_2O_2 and HNO_2. *Plasma Sources Science and Technology, 23*, 015019.

Mehr, H. M., & Koocheki, A. (2020). Effect of atmospheric cold plasma on structure, interfacial and emulsifying properties of grass pea (*Lathyrus sativus* L.) protein isolate. *Food Hydrocolloids, 106*, 105899.

Misra, N. N., Kaur, S., Tiwari, B. K., Kaur, A., Singh, N., & Cullen, P. (2015). Atmospheric pressure cold plasma (ACP) treatment of wheat flour. *Food Hydrocolloids, 44*, 115–121.

Misra, N. N., Pankaj, S. K., Segat, A., & Ishikawa, K. (2016). Cold plasma interactions with enzymes in foods and model systems. *Trends in Food Science and Technology, 55*, 39–47.

Moreau, M., Orange, N., & Feuilloley, M. G. J. (2008). Non-thermal plasma technologies: New tools for bio-decontamination. *Biotechnology Advances, 26*(6), 610–617.

Morent, R., De Geyter, N., Desmet, T., Dubruel, P., & Leys, C. (2011). Plasma surface modification of biodegradable polymers: A review. *Plasma Processes and Polymers, 8*(3), 171–190.

Nemtanu, M. R., & Minea, R. (2006). Functional properties of corn starch treated with corona electrical discharges. *Macromolecular Symposia, 245–246*, 525–528.

Obajemihi, O. I., Esua, O. J., Cheng, J.-H., & Sun, D.-W. (2023). Effects of pretreatments using plasma functionalized water, osmodehydration and their combination on hot air drying efficiency and quality of tomato (*Solanum lycopersicum* L.) slices. *Food Chemistry, 406*, 134995.

Okyere, A. Y., Bertoft, E., & Annor, G. A. (2019). Modification of cereal and tuber waxy starches with radio frequency cold plasma and its effects on waxy starch properties. *Carbohydrate Polymers, 223*, 115075.

Okyere, A. Y., Boakye, P. G., Bertoft, E., & Annor, G. A. (2022b). Structural characterization and enzymatic hydrolysis of radio frequency cold plasma treated starches. *Journal of Food Science, 87*(2), 686–698.

Okyere, A. Y., Rajendran, S., & Annor, G. A. (2022a). Cold plasma technologies: Their effect on starch properties and industrial scale-up for starch modification. *Current Research in Food Science, 5*, 451–463.

Pankaj, S. K., Bueno-Ferrer, C., Misra, N. N., O'Neill, L., Tiwari, B. K., Bourke, P., & Cullen, P. J. (2015). Dielectric barrier discharge atmospheric air plasma treatment of high amylose corn starch films. *LWT—Food Science and Technology, 63*(2), 1076–1082.

Pankaj, S. K., Wan, Z., De Leon, J. E., Mosher, C., Colonna, W., & Keener, K. M. (2017). High-voltage atmospheric cold plasma treatment of different types of starch films. *Starch—Starke, 69*(11–12), 1700009.

Park, H., Puligundla, P., & Mok, C. (2020). Cold plasma decontamination of brown rice: Impact on biochemical and sensory qualities of their corresponding seedlings and aqueous tea infusions. *LWT—Food Science and Technology, 131*, 109508.

Royintarat, T., Choi, E. H., Boonyawan, D., Seesuriyachan, P., & Wattanutchariya, W. (2020). Chemical-free and synergistic interaction of ultrasound combined with plasma-activated water (PAW) to enhance microbial inactivation in chicken meat and skin. *Scientific Reports, 10*, 1559.

Sarangapani, C., O'Toole, G., Cullen, P. J., & Bourke, P. (2017). Atmospheric cold plasma dissipation efficiency of agrochemicals on blueberries. *Innovative Food Science and Emerging Technologies, 44*, 235–241.

Sarangapani, C., Thirumdas, R., Devi, Y., Trimukhe, A., Deshmukh, R. R., & Annapure, U. S. (2016). Effect of low-pressure plasma on physico-chemical and functional properties of parboiled rice flour. *LWT—Food Science and Technology, 69*, 482–489.

Saremnezhad, S., Soltani, M., Faraji, A., & Hayaloglu, A. A. (2021). Chemical changes of food constituents during cold plasma processing: A review. *Food Research International, 147*, 110552.

Sheikhi, Z., Mirmoghtadaie, L., Abdolmaleki, K., Reza, M., Mehdi, K., Ehsan, F., Babak, M., & Shojaee-aliabadi, S. (2021). Characterization of physicochemical and antimicrobial properties of plasma-treated starch/chitosan composite film. *Packaging Technology and Science, December, 34*(7), 385–392.

Shen, H., Ge, X., Zhang, B., Su, C., Zhang, Q., Jiang, H., Zhang, G., & Li, W. (2021). Understanding the multi-scale structure, physicochemical properties and in vitro digestibility of citrate naked barley starch induced by non-thermal plasma. *Food & Function, 12*(17), 8169–8180.

Shen, J., Zhang, H., Xu, Z., Zhang, Z., Cheng, C., Ni, G., Lan, Y., Meng, Y., Xi, W., & Chu, P. K. (2019). Preferential production of reactive species and bactericidal efficacy of gas-liquid plasma discharge. *Chemical Engineering Journal, 362*, 402–412.

Sifuentes-Nieves, I., Velazquez, G., Flores-Silva, P. C., Hernández-Hernández, E., Neira-Velázquez, G., Gallardo-Vega, C., & Mendez-Montealvo, G. (2020). HMDSO plasma treatment as alternative to modify structural properties of granular starch. *International Journal of Biological Macromolecules, 144*, 682–689.

Srangsomjit, N., Bovornratanaraks, T., Chotineeranat, S., & Anuntagool, J. (2022). Solid-state modification of tapioca starch using atmospheric nonthermal dielectric barrier discharge argon and helium plasma. *Food Research International, 162*, 111961.

Sudheesh, C., Sunooj, K. V., Sasidharan, A., Sabu, S., Basheer, A., Navaf, M., Raghavender, C., Sinha, S. K., & George, J. (2020). Energetic neutral N_2 atoms treatment on the kithul (*Caryota urens*) starch biodegradable film: Physico-chemical characterization. *Food Hydrocolloids, 103*, 105650.

Sudheesh, C., Sunooj, K. V., Sinha, S. K., George, J., Kumar, S., Murugesan, P., Arumugam, S., Kumar, K. A., Kumar, V. A. S. (2019). Impact of energetic neutral nitrogen atoms created by glow discharge air plasma on the physico-chemical and rheological properties of kithul starch. *Food Chemistry, 294*, 194–202.

Sun, X., Saleh, A. S. M., Sun, Z., Ge, X., Shen, H., Zhang, Q., Yu, X., Yuan, L., & Li, W. (2022). Modification of multi-scale structure, physicochemical properties, and digestibility of rice starch via microwave and cold plasma treatments. *LWT—Food Science and Technology, 153*, 112483.

Surowsky, B., Frohling, A., Gottschalk, N., Schluter, O., & Knorr, D. (2014). Impact of cold plasma on *Citrobacter freundii* in apple juice: Inactivation kinetics and mechanisms. *International Journal of Food Microbiology, 174*, 63–71.

Surowsky, B., Schluter, O., & Knorr, D. (2015). Interactions of non-thermal atmospheric pressure plasma with solid and liquid food systems: A review. *Food Engineering Reviews, 7*(2), 82–108.

Szymanowski, H., Kaczmarek, M., Gazicki-Lipman, M., Klimek, L., & Woźniak, B. (2005). New biodegradable material based on RF plasma modified starch. *Surface and Coatings Technology, 200*(1–4), 539–543.

Thirumdas, R., Kadam, D., & Annapure, U. S. (2017b). Cold plasma: An alternative technology for the starch modification. *Food Biophysics, 12,* 129–139.

Thirumdas, R., Sarangapani, C., & Annapure, U. S. (2015). Cold plasma: A novel non-thermal technology for food processing. *Food Biophysics, 10,* 1–11.

Thirumdas, R., Trimukhe, A., Deshmukh, R. R., Annapure, U. S. (2017a). Functional and rheological properties of cold plasma treated rice starch. *Carbohydrate Polymers, 157,* 1723–1731.

Trinh, K. S., & Nguyen, T. L. (2018). Structural, functional properties and in vitro digestibility of maize starch under heat-moisture and atmospheric-cold plasma treatments. *Vietnam Journal of Science and Technology, 56*(6), 751–760.

Vaideki, K. (2016). Plasma technology for antimicrobial textiles. In *Antimicrobial textiles* (pp. 73–86). Elsevier.

Wiącek, A. E., Jurak, M., Gozdecka, A., & Worzakowska, M. (2017). Interfacial properties of PET and PET/starch polymers developed by air plasma processing. *Colloids and Surfaces A: Physicochemical and Engineering Aspects, 532,* 323–331.

Wilczek, S., Schulze, J., Brinkmann, R. P., Donko, Z., Trieschmann, J., & Mussenbrock, T. (2020). Electron dynamics in low pressure capacitively coupled radio frequency discharges. *Journal of Applied Physics, 127,* 181101.

Wongsagonsup, R., Deeyai, P., Chaiwat, W., Horrungsiwat, S., Leejariensuk, K., Suphantharika, M., Fuongfuchat, A., & Dangtip, S. (2014). Modification of tapioca starch by non-chemical route using jet atmospheric argon plasma. *Carbohydrate Polymers, 102*(1), 790–798.

Wu, T.-Y., Chang, C.-R., Chang, T.-J., Chang, Y.-J., Liew, Y., & Chau, C.-F. (2019). Changes in physicochemical properties of corn starch upon modifications by atmospheric pressure plasma jet. *Food Chemistry, 283,* 46–51.

Wu, T. Y., Sun, N. N., & Chau, C. F. (2018). Application of corona electrical discharge plasma on modifying the physicochemical properties of banana starch indigenous to Taiwan. *Journal of Food and Drug Analysis, 26*(1), 244–251.

Wu, Z., Qiao, D., Zhao, S., Lin, Q., Zhang, B., & Xie, F. (2022). Nonthermal physical modification of starch: An overview of recent research into structure and property alterations. *International Journal of Biological Macromolecules, 203,* 153–175.

Yan, S. L., Chen, G. Y., Hou, Y. J., & Chen, Y. (2020). Improved solubility of banana starch by dielectric barrier discharge plasma treatment. *International Journal of Food Science and Technology, 55*(2), 641–648.

Yan, Y., Zhou, Y., Shi, M., Liu, H., & Liu, Y. (2019). Influence of atmospheric pressure plasma jet on the structure of microcrystalline starch with different relative crystallinity. *International Journal of Food Science & Technology, 54,* 567–575.

Yang, X., Cheng, J.-H., & Sun, D.-W. (2024). Enhancing microorganism inactivation performance through optimization of plate-to-plate dielectric barrier discharge cold plasma reactors. *Food Control, 157,* 110164.

Yong, H. I., Park, J., Kim, H.-J., Jung, S., Park, S., Lee, H. J., Choe, W., & Jo, C. (2018). An innovative curing process with plasma-treated water for production of loin ham and for its quality and safety. *Plasma Processes and Polymers, 15,* 1700050.

Zhang, B., Chen, L., Li, X., Li, L., & Zhang, H. (2015). Understanding the multi-scale structure and functional properties of starch modulated by glow-plasma: A structure functionality relationship. *Food Hydrocolloids, 50,* 228–236.

Zhang, B., Xiong, S., Li, X., Li, L., Xie, F., & Chen, L. (2014). Effect of oxygen glow plasma on supramolecular and molecular structures of starch and related mechanism. *Food Hydrocolloids, 37,* 69–76.

Zhang, K., Zhang, Z., Zhao, M., Milosavljevic, M., Cullen, P. J., Scally, L., Da-Wen Sun, D.-W., & Tiwari, B. K. (2022). Low-pressure plasma modification of the rheological properties of tapioca starch. *Food Hydrocolloids, 125,* 107380.

Zhou, R., Zhou, R., Prasad, K., Fang, Z., Speight, R., Bazaka, K., & Ostrikov, K. K. (2018). Cold atmospheric plasma activated water as a prospective disinfectant: The crucial role of peroxynitrite. *Green Chemistry, 20,* 5276–5284.

Zhou, Y., Yan, Y., Shi, M., & Liu, Y. (2019). Effect of an atmospheric pressure plasma jet on the structure and physicochemical properties of waxy and normal maize starch. *Polymers, 11*(1), 8.

Zhu, F. (2017). Plasma modification of starch. *Food Chemistry, 232,* 476–486.

Zhu, Q., Yao, S., Wu, Z., Li, D., Ding, T., Liu, D., & Xu, E. (2023). Hierarchical structural modification of starch via non-thermal plasma: A state-of-the-art review. *Carbohydrate Polymers, 311,* 120747.

Milling

10

Basheer Aaliya, Sneh Punia Bangar, and K.V. Sunooj

10.1 INTRODUCTION

Milling is a mechanical operation done to reduce the size of solid materials to micro- or nano-sized particles with significant characteristic changes. Milling of grains like rice, maize, barley, and wheat with disc, stone, hammer, roller, and ultra-fine mill involves fragmentation of grains owing to the compressive and shear forces, changing the morphological and functional properties of major components like starch and protein in them (Bangar et al., 2023). The shortcomings of native starch limit their practical applications, as often it is modified by physical, chemical, or enzymatical methods to enhance its functional attributes. Among the physical modifications, milling is considered one of the non-thermal modifications to alter the starch characteristics by friction, collision, shear, impingement, and other forces (Wang et al., 2020). Milling inflicts the application of stress-strain characteristics and produces mechanically activated starch or milling-damaged starch by breaking the starch granules into smaller particles (Joy et al., 2022).

There has been much consideration of ball milling or rod milling as potential and effective techniques that break the chemical bonds of the bulk material through the kinetic energy of the impacted balls by the mechanochemical pulverization process and generate smaller particles with new surfaces and properties that can meet specific application requirements (Joy et al., 2022). Ball milling is gaining attention owing to its simplicity, lower effluent generation, safety, and cost-efficiency compared to other physical modifications. Starches have been milled with various ball mills, that is, planetary, vibratory, attrition, and low-energy tumbling, which mainly use temperature and pressure as variables suitable for in-situ processing and optimization. Although milling is considered a non-thermal process, both mechanical and thermal energies are involved in changing the molecular or granular structures of starch, as the temperature of the rolling ball usually ameliorates with milling time (BeMiller, 2017). Ball milling depolymerizes starch and alters the granular, molecular, and crystalline structures and amylose-amylopectin ratio of starch. The decreased size, enlarged surface area, and enhanced active sites of the starch molecules aid in chemical reactions and change the physicochemical attributes of starch. Ball-milled modified starches are studied for varied applications such as nanoparticles, emulsification, fat replacers, and food products (Luciana Carla González et al., 2020).

10.2 MECHANISM IN BALL MILLING

The starch particles undergo either stress, aggregation, or agglomeration, allowing the granules to interact and alter various physicochemical attributes under different mechanical forces (Bangar et al., 2023). Ball milling shows different friction, shear, and impact on starch granules. In a high-energy mill, starch

DOI: 10.1201/9781032655598-10

is deposited in a hollow cylindrical container with the milling balls fabricated from rubber, ceramic, or stainless steel. An alternative synchronized centrifugal force is developed when the container and the milling balls rotate in opposite directions. A frictional force on the starch is created as the milling balls roll inside the container. The breakdown of the lattice structure and particle size reduction of starch occurs as the kinetic energy produced with the motion of milling balls acts over the starch granules, breaking the chemical bonds of molecules. In order to obtain a specified size reduction with little or no damage to starch particles, it is necessary to optimize the operating parameters of the ball mills in terms of milling speed, time, energy and load, ball-to-powder ratio, and humectants. Also, the friction amongst the starch granules generates heat and modifies their functional attributes (Martínez-Bustos et al., 2007).

The mechanism of ball milling can be looked at as different stages of milling. In the first stage, the impact force of milling balls flattens the starch particles, giving a compressive force, followed by an intermediate stage, where the particles are ground to a microscopic level. A final stage of comminution produces homogeneous mass at the microscopic level, where further dispersion improvement cannot be accomplished. Starch particles milled with a ball mill have greatly deformed metastable structures without visible lamellae under optical microscopy. Thus, as a result of the physical agitation by milling ball, macro-sized starch granules are converted to microscopic fragments with an average size decrease. The fragmentation and size reduction of starch granules is determined by the source and composition of starch and amylopectin branching, which in due course develops morphological, crystalline, and molecular damage (Diop et al., 2012). The degree of starch damage can be elevated by increasing milling intensity, which includes increasing milling energy and milling frequency, minimizing the distance between the grinding medium, increasing jet gas pressure, and adjusting rotor speed (Q. Wang et al., 2020).

10.3 TYPES OF BALL MILLING

Based on the mode of operation, ball mills are divided into direct and indirect milling. In direct milling, the kinetic energy is directly applied to the starch through any shafts. They include attritor, pan, and roll mills. In indirect mills, kinetic energy is transferred to the milling body frame and then transferred to milling balls, and the starch gets charged through friction. Planetary, vibratory, and tumbler mills are indirect mills. Also, based on the different motions created for the generation of momentum in the grinding balls to work on starch granules with different milling forces, ball mills are widely categorized as planetary mills, vibratory mills, tumbler mills, and attrition mills (Figure 10.1).

10.3.1 Planetary Ball Mills

Planetary ball mills are uncomplicated but effective ball mills consisting of two or more jars rotating at unique angles on a rotating sun wheel (Broseghini et al., 2016). A hermetically sealed stainless steel or Teflon jar is used. The planetary ball mills have a superimposed rotational motion as the sun wheel and the grinding jar rotate in opposite directions, producing a non-uniform acceleration inside the chamber by the combined action of two centrifugal fields (S. K. Singh et al., 2018). The generation of centrifugal and Coriolis forces gives 100 times greater kinetic energy to the milling balls than gravitational force. The morphology and particle size of the starch are mainly determined by the speed and duration of milling and the ball-to-powder ratio (Juarez-Arellano et al., 2021).

10.3.2 Tumbler Ball Mills

Gravitational force impinges upon starch granules in the tumbler ball mill. During milling, milling balls rotate axially against the wall owing to friction, rolling over and impacting the granules. The starch

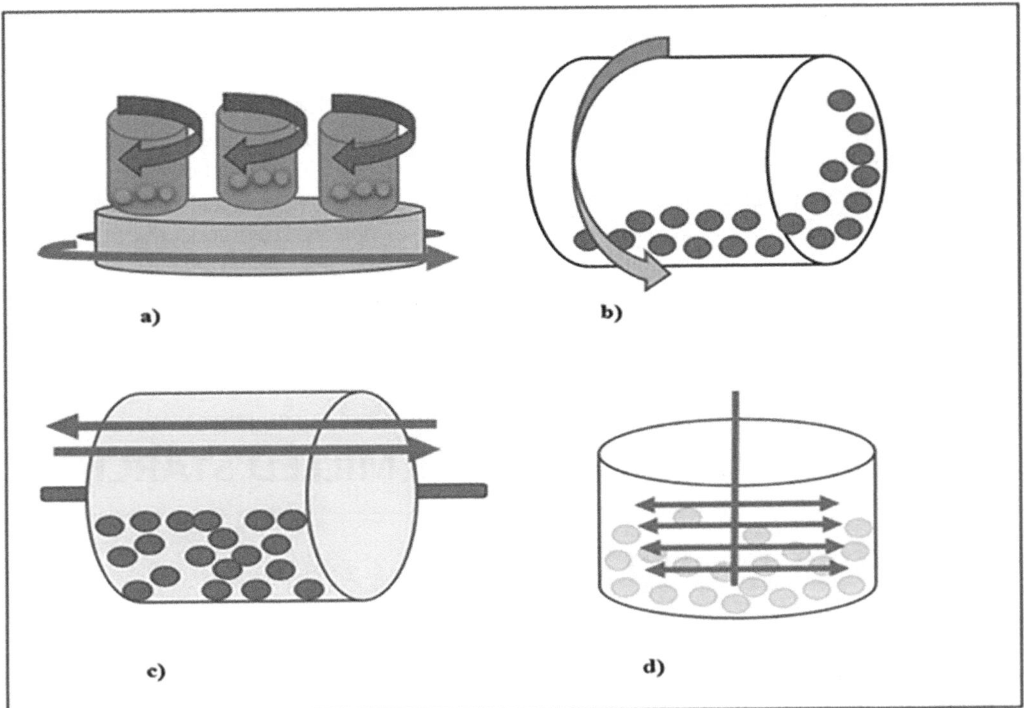

FIGURE 10.1   Types of ball milling: (a) planetary, (b) tumbling, (c) vibratory, and (d) attrition ball mills.

granules experience multiple collisions with attrition and impact forces as they get trapped between the walls of the container and milling balls. The diameter of the milling chamber governs the effectiveness of tumbler milling, and the fall height for the ball increases with an increase in mill diameter. Optimization of the rotational speed of the vessel is necessary to equilibrate the centrifugal force with gravitational force. The bond ball mill is a common type of tumbler mill. There are more than 300 40-mm balls in a small-scale mill, which measures 30 cm in length and diameter. A small-scale mill operates with 50% powder filling and 20% mill filling (Tavares, 2017).

10.3.3 Vibratory Mills

Vibratory mills consist of milling balls and a cylindrical vessel that oscillates at a very high frequency with low amplitude. The mass of the milling medium and vibration amplitude determine the milling efficiency of vibratory mills (Kumar et al., 2020). The fine grinding of starch is possible in vibratory mills as the vibration amplifies the shock-erasing effect because of the variation in frictional components and strength. The increased dynamic interactions between starch particles and cyclic loads increase the rate of starch granule destruction. In general, a 20-mm vibratory amplitude can achieve a 20–30-mm trajectory of the milling balls. Starch granules are subjected to significantly greater force in milling balls than in planetary mills when vibrated (Sitotaw et al., 2021).

10.3.4 Attrition Mills

An attrition mill comprises a stationary vertical cylindrical vessel and a rotating vertical shaft connected to multiple horizontal arms. High-speed rotating arms cause the starch granules and milling balls to tumble together. The collision of milling balls, balls, impellers, and balls and vessel walls create frictional, shear, and impact forces. In comparison to planetary or vibratory mills, attrition mills consume less power, as the milling balls are rotated directly rather than vibrating the entire vessel. However, as a consequence of the intemperate wear on the milling bodies, attrition milling can sometimes lead to sample contamination. The effectiveness of the attrition mill is determined by the milling speed, size and number of milling balls, and temperature of milling (Sitotaw et al., 2021). Furthermore, on the basis of wettability, ball mills are categorized as wet and dry mills. Surface-active media are added to wet mills in order to hinder the development of aggregates, while powder-to-powder friction occurs in dry mills (Joy et al., 2022). In wet milling, starch granule integrity is maintained since water absorbs the heat created upon the milling, thereby impeding the temperature increase and preventing starch damage compared to dry milling (N. Ahmad & Rajab, 2018).

10.4 CHARACTERIZATION OF MILLED STARCHES

10.4.1 Structure of Milled Starch

10.4.1.1 Granular Structure

Generally, native starch granules are flat, slick, and have concave or convex shapes that are often clubbed together or with protein or fiber bodies before any treatment. The process of ball milling makes the starch granules rough, deformed, and distorted with cracks, fissures, pores, or grooves on the surfaces (Wang et al., 2020). In general, cereal starches from rice, wheat, and maize and legume starches from lentils, navy beans, and cowpeas are more resistant to milling treatment than tuber starches from potatoes and cassava. The resistance of cereal starches is due to their smaller granular size, while the resistance of legume starches is because of their greater amylose content and resistance to swelling (Bangar et al., 2023). In a study by Ahmad et al. (2020), native starches from horse chestnuts, lotus stems, and water chestnuts were round or oval with smooth and porous surfaces. Upon ball milling, the starch granules fragmented with cracks on the surfaces. However, some fragmented granules clubbed together upon modification. The agglutination of the powdered starch is due to the frictional force and energy produced upon milling operation. Similarly, ball milling potato starch with oval shapes for 1 hour changed the starch to irregularly shaped granules with rough surfaces (Lv et al., 2019). Ball milling of maize starch for 5 h broke down the granules into smaller peeled-off fractions and made the granules rough from a smooth polyhedral and uniform surface (Amaya-Llano et al., 2011). There has been a change in the shapes of the pearl millet and proso millet starch granules from large, oval, and polyhedral ones to elliptical and irregular ones (Figure 10.2) (Jhan et al., 2020). Additionally, ball milling of rice starch damaged the granules by the impact and frictional forces of the milling balls (Z. Zhang et al., 2010).

The granular modification of ball-milled starch altered the enzymatic hydrolysis of starches. Dhital et al. (2011), in their study, they found that ball milling can be a pretreatment before enzymolysis of starch, as the milling treatment creates fissures and grooves and enhances the enzyme propagation into the inside of the granules. Also, the increased surface area per unit mass with milling operation enhanced the rate of enzymatic hydrolysis. Apart from the alteration in granular shapes, size reduction was observed in maize, water chestnut, lotus stem, and horse chestnut starches (He et al., 2014; Ahmad et al., 2020). The mechanical exertion of ball milling can produce nanoparticles by the attrition and collision of the granules with the milling balls. Various mechanical actions such as friction-shear, compression, attrition, impingement, and impaction lead to coarse granular surfaces. The morphological changes caused by ball milling alter the physicochemical as well as functional attributes of starch (Bangar et al., 2023).

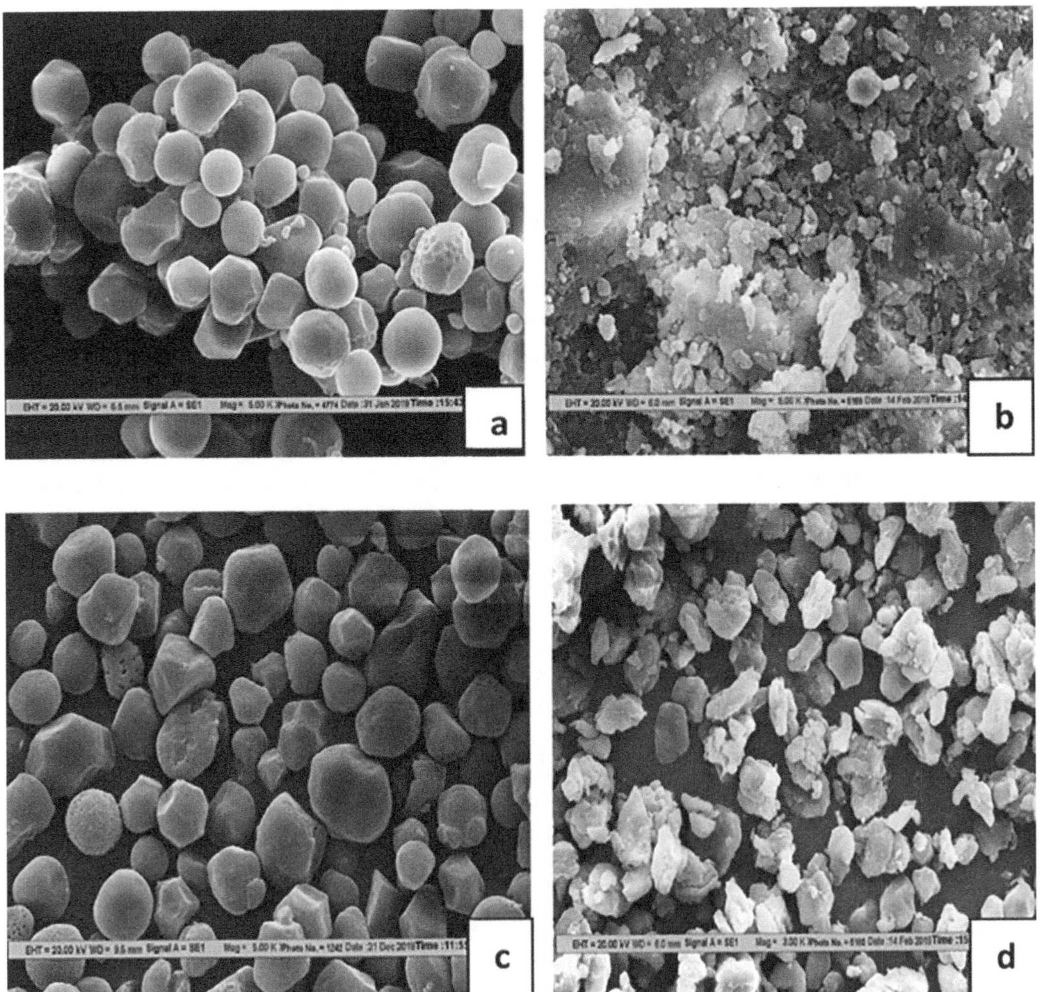

FIGURE 10.2 Scanning electron micrographs of (a) native proso millet starch, (b) ball-milled proso millet starch, (c) native pearl millet starch, and (d) ball-milled pearl millet starch.

10.4.1.2 Particle Size

The particle size, particle size distribution, surface area, and span index of starch are influenced by different ball-milling techniques. The particle size distribution of native starch is bimodal, and ball milling changes it to a monomodal distribution (González et al., 2018). Particle size distribution is acknowledged to decrease significantly with starch damage. A reduction in mean particle size was observed in wheat and cassava starches upon ball-milling treatment (Ren et al., 2010; El-Porai et al., 2013). On the contrary, ball milling does not necessarily decrease starch granule size, as they agglomerate due to excessive milling damage. Damaged starches from maize, wheat, and mung bean showed an increase in average particle size after ball milling compared to their native counterparts (Zhang et al., 2019; Yu et al., 2015). This dissimilarity arises due to varying differences in the degree of damage to starch's interior structures. Ball-milled starches agglomerate or adhere to other fragments depending on the treatment conditions and starch composition and may have a loosened structure, contributing to an increase in particle size (Yu et al., 2015). The lower amylopectin-amylose ratio sequels an increment in the particle size of ball-milled starches. This can be attributed to the ability of amorphous amylose to shield the disruption of crystalline amylopectin by external forces, increasing the particle size (Lu Wang et al., 2018).

10.4.1.3 Crystalline Structure

The X-ray diffraction (XRD) pattern investigates the crystalline characteristics and amorphism of semi-crystalline starches. Ball-milling treatment destroys the crystalline regions of both A-type and B-type starches, decreasing the crystallinity degree compared to native starch. The ball milling of pearl and proso millet did not change the XRD pattern but reduced the crystallinity (Jhan et al., 2020). A reduction in crystallinity was observed in different sources of starch, namely maize, rice, cassava, and carrot (Lu, Xiao, et al., 2018; Moraes et al., 2013; Z. Zhang et al., 2010). The reduction in crystallinity is associated with the disruption in the spatial arrangement of amylose and amylopectin, making the crystalline portion susceptible to mechanical forces upon starch destruction by milling (Kim & Kim, 2014). Milling splits its double helices, resulting in the loss of double helices and, subsequently, the loss of starch crystallinity. The loss of starch crystallinity by ball milling is followed by the loss of its double helices, specifically by the splitting of double helices (Li et al., 2014). Meanwhile, an A-type crystalline pattern is more susceptible to milling damage than a B-type starch. This is because of the ability of B-type crystals to form more hydrogen bonds, as they possess 36 water molecules between helices compared to A-type starches, which contain only 8 inter-helical water molecules. Hence, the compact hydrogen bond network of B-type starches makes the crystalline lamellae stronger and withstand the milling operation compared to A-type starches (Tan et al., 2015).

In a study on maize starch, the XRD pattern showed peaks at 18° and 22° that indicate the semi-crystalline disposition of the starch possessing amorphous and crystalline domains. However, upon ball milling, crystalline regions decreased, and amorphous regions increased (He et al., 2014). In comparison with other modification treatments, such as radiation, ultrasonication, or heat moisture treatment, ball milling has been found to transform semicrystalline starch more effectively into amorphous starch. The shape and size of the granules, amylose/amylopectin ratio, and the nature of amylopectin govern the crystallinity of various starches. Since amylopectin, especially its branch points, is vulnerable to the milling process, the disruption of the amylopectin component reduces the crystallinity of ball-milled starches along the subsequent augmentation of amylose content and amorphous domains (Bangar et al., 2023). Ball milling of potato starch made the diffraction pattern less acute or partly removed it (Lv et al., 2019). The destruction of the inter- and intra-molecular hydrogen bonds of starch molecules and disruption of the crystal portion caused the change in diffraction pattern. The ball-milled potato starch became weak and allowed the easy entry of water into the starch granules. Also, a weak structure of granules is susceptible and supports chemical modification or enzymatic hydrolysis.

10.4.1.4 Molecular Structure

Sequential confirmation and formation form the intricate starch structure. It involves arranging amylose and amylopectin molecules and linking these molecules together. Ball milling affected the molecular weight of the starch upon damage. For instance, the average molecular weight of cassava and maize starches was seen to reduce with ball-milling treatment, which can be ascribed to the alteration in the molecular chain and increase in amylose content at the expense of amylopectin conversion (Liu et al., 2019; Shi et al., 2015). However, no remarkable characteristic difference was noticed in ball-milled wheat starch (Wu et al., 2018). This contradiction in the results is due to the difference in botanical sources and in the treatment process.

Chain length is an additional salient factor of the molecular structural attributes. A significant decrease in longer chain percentage with a polymerization degree of 100–10,000 and an increase in shorter chains with a polymerization degree of 0–100 was seen in wheat starch with ball-milling treatment (Wu et al., 2018). Also, an increase in apparent amylose content in the ball milling of carrot and cassava starch signifies that α-(1, 6) glucosidic linkages endure more attacks, causing a change in chain length (Moraes et al., 2013).

10.4.2 Amylose Content

When starch is subjected to ball milling, the amylose content increases proportionally to milling time due to the debranching of amylopectin at 1,6 linkages and the unchaining of amylose molecules (Cavallini & Franco, 2010). For instance, regarding cassava starch and maize starch, the amylose content increased with milling time as glycosidic linkages of amylopectin were degraded and the long branch chains of amylopectin released. Additionally, cassava starch had a comparatively greater increase in amylose content than maize starch because of the fragility of cassava starch granules (Z. Q. Huang et al., 2007, 2008). A similar result was revealed in sorghum and finger millet starches due to the shear force of milling that easily breaks the breaks amylopectin into smaller linear amylose units (Jhan et al., 2020). This shows the potentiality of ball-milling treatment to change the amylopectin-amylose ratio of starch and change the techno-functional attributes for varied applications in food.

10.4.3 Fourier Transform Infrared Spectroscopy Analysis

Fourier transform infrared spectroscopy (FT-IR) spectral analysis of ball-milled starches indicates that the band representing the hydroxyl stretching decreases the intensity upon treatment. For instance, it was observed that peak intensity between 3600 and 3000 cm^{-1} of hydroxyl stretching associated with inter- and intramolecular H-bonding diminished for ball-milled potato starch (Figure 10.3) (Lv et al., 2019). Also, the reduction in peak intensity of the hydroxyl group correlates with the disruption of steeply ordered crystalline regions of starch upon crystalline analysis. Similarly, milling damage made alterations in the vibrations of chemical bonds of mung bean starch while subsiding its crystallinity (Zhang et al., 2019).

10.4.4 Functional Properties

10.4.4.1 Water Absorption Capacity

Water absorption capacity (WAC) is the capacity of starch to link with water molecules under specific conditions. The damage of starch upon milling enhances the WAC. The corn starch increases its WAC upon an augmentation in ball milling time by the loss of a highly ordered structure and an increase in available surface area (Niemann & Whistler, 1992). A similar result was observed in the starches from lotus stem, water chestnut, horse chestnut, Peruvian carrot, and cassava (Moraes et al., 2013;

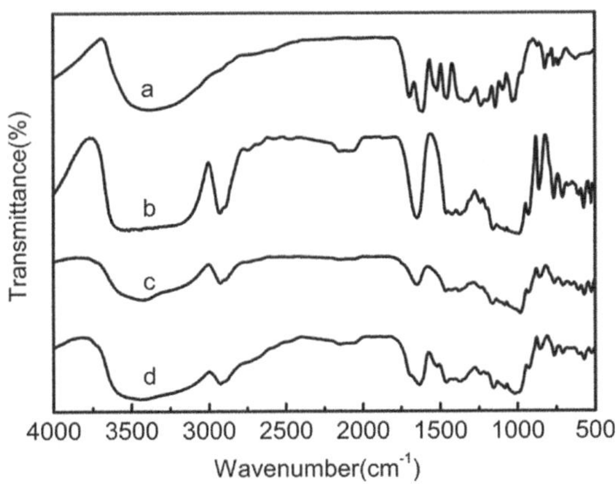

FIGURE 10.3 FT-IR spectra of (a) tea polyphenols, (b) native potato starch, (c) ball-milled starch, and (d) ball-milled starch with tea polyphenols.

M. Ahmad et al., 2020). The molecular structure and distribution, the amount of amorphous or crystalline regions, and the granular size distribution determine the WAC of starch. In ball milling, the amylose and amylopectin are disrupted, revealing the amylose fractions with numerous binding sites available for the absorption of water.

10.4.4.2 Swelling Power

Milling exposes the structural elements of starch, causing an increase in swelling power by increasing water absorption. The augmentation in the swelling power of starch is due to the milling-induced H-bonding damage, exposing more hydroxyl groups inside double helices to interact with surrounding water molecules by forming H-bonds, thus increasing hydration and water retention capacity. Also, milling breaks the starch granules into smaller-sized fragments, enlarges the surface area, and creates surface cracks, channels, or holes in the granules, instituting free spaces for water absorption and aiding in elevating swelling power (E. Li et al., 2014). However, excessive milling collapses the semi-crystalline lamellae and causes a decline in swelling power. In a study on ball-milled Thai rice starch, the swelling power increased owing to an increment in free hydroxyl groups and facile entry of water molecules inside the damaged starch granules (Limpongsa & Jaipakdee, 2020). The swelling power increases as the damaged structure of starch facilitates water retention through hydrogen bonds that are formed with water molecules. Furthermore, the conversion of semi-crystalline form into an amorphous form upon milling with the loss of birefringence, reduction in XRD peaks, and double helix content in amylopectin supports the increment in swelling power of ball-milled starch granules. Hence, the starches that have undergone ball-milling modification can be extensively used as a food ingredient that requires viscous and gel-like consistency.

Ball milling is also studied to elevate the cold water swelling of starch granules. Juarez-Arellano et al. (2021) studied the increase in cold-water swelling of ball-milled potato starch in comparison to native potato starch due to the increment in surface area by the fragmentation of starch molecules. The swelling power was further enhanced by the extension of amorphous rings in the starch molecules and the development of hydrogen bonds between starch and water molecules as a result of crystalline portion reduction. However, ball milling at higher energy (more than 4.4 J/g) did not increase the swelling power due to the complete loss of granular structure.

10.4.4.3 Solubility

The solubility of starches increases with ball-milling treatment due to depolymerization, and it increases with milling severity. Starches from potatoes, cassava, wheat, maize, and carrots were studied to have increased solubility upon ball milling (Dhital et al., 2010; Moraes et al., 2013). In a study, the cold water solubility of Thai rice starch was increased to about 5.6 times by ball-milling treatment for 45 min at 450 rpm with an interval of 40 min in between every 5 min of treatment (Fang et al., 2020). Similarly, ball-milled potato starches show an increased cold-water solubility with 2.2 and 11 J/g ball-milling treatment to 54% and 65%, respectively (Juarez-Arellano et al., 2021). Ball milling rearranges the molecular structure of starch and develops smaller fragments due to the degradation of amylopectin molecules (Z. Zhang et al., 2010). Also, the physical damage to the starch structure is also responsible for the increased solubility.

10.4.4.4 Thermal Properties

Gelatinization of ball-milled starch is the transformation of an ordered semi-crystalline native starch to a disordered amorphous format when starch is heated in excess water. Onset temperature (T_o), peak temperature (T_p), conclusion temperature (T_c), and gelatinization enthalpy (ΔH) are the gelatinization parameters accessed by DSC. Milling-damaged starches have decreased gelatinization temperatures and ΔH compared to native starches. The gelatinization temperature is an indication of crystalline heat stability. Also, ΔH indicates the degree of molecular disorder and heat energy required for melting the amylopectin-based crystals (Akhila et al., 2022). In most cases, ball milling lowers the gelatinization temperatures and ΔH due to the increased disorder in crystalline structure by amylose or amylopectin degradation, which decreases the energy required for gelatinization (Q. Wang et al., 2020). The smaller granules showed greater gelatinization transition temperatures than the larger granules (Chang Liu et al., 2015). The ΔH of ball-milled rice starch shows a positive correlation with its crystalline degree, indicating the destruction of the crystalline area fastens gelatinization tendency (González et al., 2018). In a study with potato starch, ball milling treatment significantly decreased the ΔH and gelatinization transition temperature of the milled starch, revealing an increment in amorphism and reduction in crystallinity. The decreased gelatinization parameters indicate structural defects, poor packing, and a lower degree of stability of residual amylopectin crystallites of ball-milled potato starch compared to the native potato starch (Lv et al., 2019). A similar trend in gelatinization parameters was depicted in rice starch, buckwheat starch, maize starch, and mung bean starch, with an escalation in milling intensity and time contrary to their native counterparts (Shi et al., 2015; Yu et al., 2015; Kim & Kim, 2014; K. Zhang et al., 2019). However, the variation in the gelatinization values among the starches is due to the starch individuality, quantity of amylopectin and amylose in the granules, time and temperature of the treatment, amount of water used in the treatment, and the presence of infinitesimal proteins and lipids in starch.

10.4.4.5 Pasting Properties

The pasting of starch encompasses the swelling of granules through heat, the transition from an ordered to a disordered format in surplus water with the release of molecular components, and, ultimately, the complete rupture of starch granules. The pasting parameters identified by a rapid visco-analyzer (RVA) are pasting temperature (PT), pasting time (Pt), peak viscosity (PV), trough viscosity (TV), breakdown viscosity (BV), final viscosity (FV), and setback viscosity (SV). Upon ball milling, these parameters are determined by the degree of granular starch damage and the presence of short-chained amylopectin in the granules. Ball milling reduces the pasting viscosity and temperature of starch paste from corn, wheat, and rice (Tan et al., 2015; González et al., 2018). Also, viscosity is found to decrease with an enhancement in the degree of damaged starch (Chong Liu et al., 2015). Ball-milled starch has broken granular surfaces, smaller particle sizes, and disrupted crystalline structures that enhance the permeation of water into granules. These factors gradually diminish thermal stability and decrease PT. Ball-milled starch granules with fragile surfaces exhibit lower resistance to shear when heated with excess water, resulting in a reduced pasting viscosity (Tan et al., 2015). Furthermore, the amylose/amylopectin ratio, loss of crystallinity, and reduced gelatinization degree have an influential role in decreasing the pasting viscosity of ball-milled starches (González et al., 2018).

Rice starch upon ball milling showed reduced pasting properties with the progress of mechanical activation (Z. Zhang et al., 2010), but an inversely proportional relation was observed between the pasting parameters and milling energy (González et al., 2018). The pasting viscosity decreased with the increment in milling time and energy, along with an enhancement in solubility and swelling power. The PT was found to decrease initially. However, with the amelioration in milling energy and time, the PT increased because of agglutination as the small granular particles adhered together to form larger particles. In another study with waxy rice starch and waxy maize starch, the pasting temperature decreased with ball milling treatment, owing to the disruption of granular, lamellar, and crystalline structures due to the impingement, impaction, or collision actions by ball milling (Cancan Liu et al., 2021). The disoriented particles with higher entropy and lower enthalpy exhibit lower resistance to heat, and granules quickly rupture when completely swollen, decreasing pasting viscosity with an increase in milling time and energy (Bangar et al., 2023). A similar downturn in pasting properties was observed in ball-milled rice and corn starches (W. Zhang et al., 2015; González et al., 2018). Nevertheless, an augmentation in PT was observed in some ball-milled starches. The ball-milled starch granules hydrate, swell up, and leach out amylose and amylopectin more rapidly than native starch, increasing the viscosity of damaged starches (Barrera et al., 2013). Additionally, the aggregation of granules of ball-milled damaged starch prevents the development of a dispersion system and accounts for the elevation in PT.

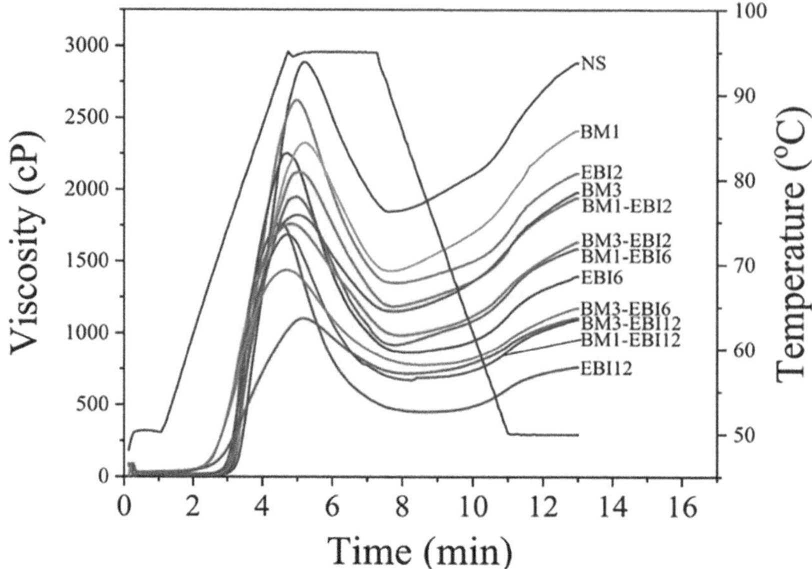

FIGURE 10.4   RVA profile of native corn starch and corn starch modified by single and dual treatment by ball milling and electron beam irradiation.

Ball milling has been synergistically employed with other starch modification techniques to alter the viscosity of starch pastes. Ball milling with electron beam irradiation (2, 6, and 12 kGy doses) decreased the molecular weight of amylopectin, crystallinity, gelatinization temperature, and pasting viscosities of corn starch (Figure 10.4). Ball milling as a pretreatment damaged starch granules and subsequently made the granules vulnerable to electron beam irradiation, leading to the destruction of crystalline structure, chain depolymerization, and lowered viscosity (Shen et al., 2022). In addition, corn starch viscosity was altered by a ball mill-assisted Fenton-like oxidation process using hydrogen peroxide. The apparent viscosity showed a significant reduction in oxidized corn starch paste with a greater degree of oxidative substitution (J. Li et al., 2020). This oxidized corn starch can be used for the warp sizing of high-density or high-count textile fabrics to meet the necessity of advancement in sizing processes in textile industries.

10.4.4.6 Rheological Properties

The granular surface and rheological attributes of milled starch are affected by differences in porosities caused by variations in milling time and milling energy. Hence, it is necessary to understand the pore size, pore volume, and surface area of ball-milled starches. The storage modulus (G′) is the amount of energy stored in the starch paste, the loss modulus (G″) is the amount of energy dissipated, and tan δ is the measure of viscoelastic dominance of a material that represents the ratio of G″ to G′ (Aaliya et al., 2021). The G′, G″, and tan δ of milled starches are related to the granular porosity, and the G′ and G″ increase with the increment in temperature. For instance, the rheological parameters of nanoparticles fabricated from horse chestnut, water chestnut, and lotus stem starches showed an increase in viscosity and G′ and G″ values with ball-milling treatment (Figure 10.5) (M. Ahmad et al., 2020). The study signifies that the size, shape, and dispersion of granules; granular interaction; and amylose content affect the rheological properties of milled starches. The enhancement in G′ upon ball milling is attributed to the granular extension of starch that increases the surface structure and pores. When the temperature is raised further, residual crystallites in the ball-milled starch are depleted, resulting in an increase in amorphous regions and reduced G′ values (Ashogbon, 2021). The destructive nature of ball milling on starch granules and the development

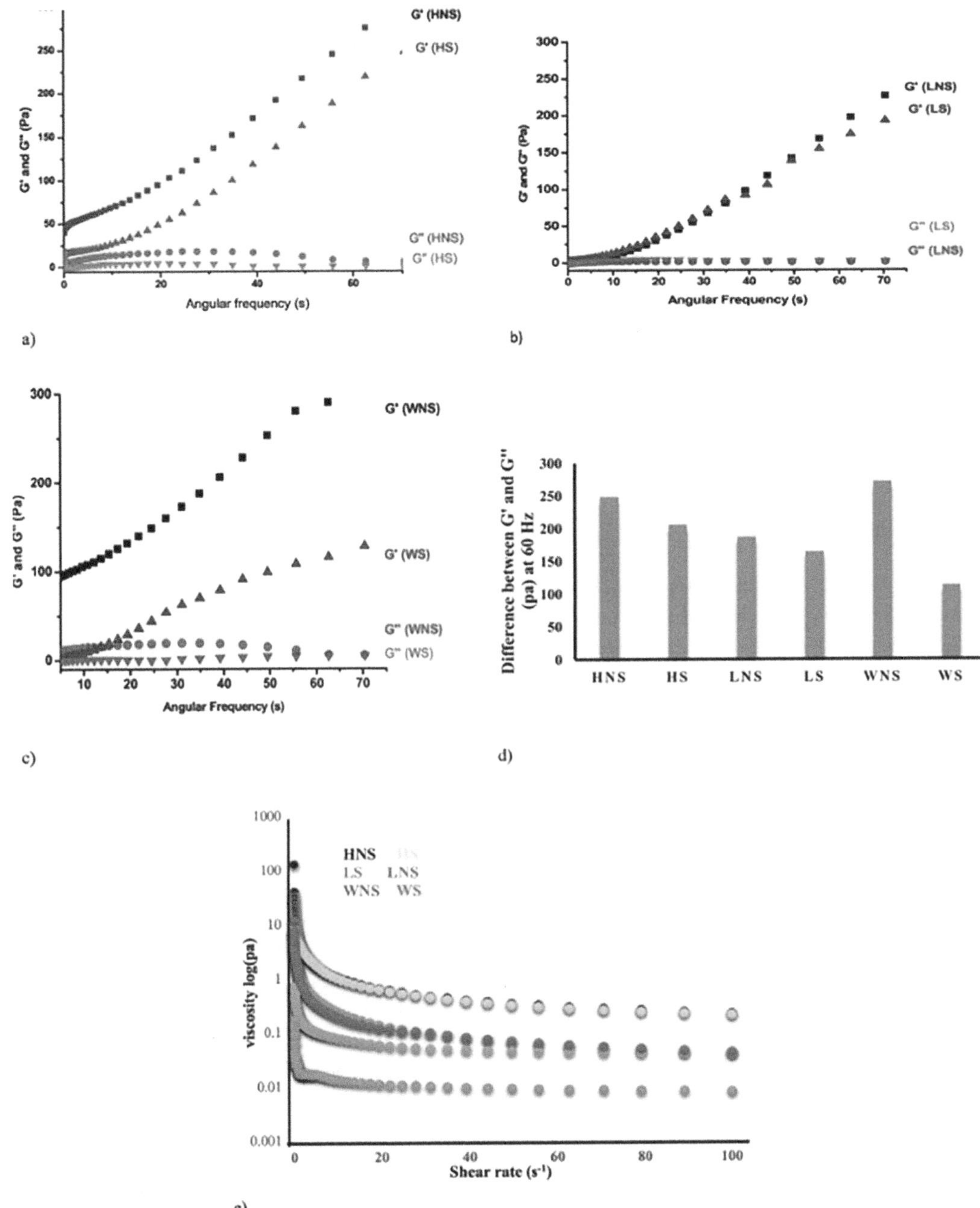

FIGURE 10.5   Rheological attributes of native and ball-milled horse chestnut, water chestnut, and lotus stem starch, respectively. (a–c) storage modulus and loss modulus of starch against the frequency sweep, (d) difference in storage modulus and loss modulus at constant frequency (60 Hz), and (e) apparent viscosity of starch pastes against shear rate (0.01–100 s⁻¹).

of surface pores due to the mechanical actions show that milled starch exhibits adsorptive potential and mucoadhesive attributes in a mucoadhesive polymer (T. Y. Liu et al., 2011; Soe et al., 2020).

The evaluation of the rheological attributes of native and nano-sized millet starches demonstrates that G'' over G' demonstrates more elastic behavior obeying Hooke's law (Jhan et al., 2020). This indicates permanent flow was interrupted by the presence of sparse cross-linkages in the elastic material until the exertion of a robust force. In another study, ball-milled proso millet and pearl millet starches showed enhanced visco-elastic characteristics compared to their native counterparts, showing their traditionally weak gel formation. The developed elastic gel endures small deformations but breaks down with an application of strong force. The implications of milling on starches might be of importance to the polymer sectors. Also, pearl millet starch has an increased viscoelastic nature compared to proso millet starch. This is because of the small hydrodynamic diameter of pearl millet starch that directly influences the rheological aspects of starch, as they can effortlessly create a particulate association and exhibit a more elastic nature. In another study by Huang et al. (2007), the mechanically activated cassava starch developed by ball milling showed a decreasing viscosity at a persistent shear rate with the increase in activation time. The enhancement of fluidity by the reduction in viscosity of the gels indicates weakened inter- and intramolecular bonds that hold together the starch chains, resulting in increased chain mobility. The reduced viscosity also reflects the increased amorphism and loss of crystalline region by mechanical activation, lowering the rigidity by loosening the gel structure. Ball-milling treatment showed its significance in modifying starch for the preparation of viscous or fluid foods. Ball-milled starch suspension exhibited a pseudoplastic or Newtonian flow pattern, which is governed by milling energy and concentration of starch (González et al., 2021).

10.4.4.7 Starch Retrogradation

Retrogradation of starch is a post-gelatinization process observed in starch gels formed by starch gelatinization. Ball milling enhances the retrogradation tendency of the starches, as it increases the amylose content compared to native starch. Upon ball milling of native starch, the amylopectin breaks into low-molecular-weight fragments (mostly amylose), reducing the molecular size with increasing milling time, thereby increasing amylose content (Barrera et al., 2012). For instance, maize and pea starches, when subjected to the ultracentrifugation process, showed that attrition force caused fragmentation of amylopectin molecules and produced fragmented amylopectin and amylose-like material (Tester et al., 2006). Therefore, milled starches increase retrogradation kinetics, which can be linked to a reduction in molecular weight of amylopectin molecules. The kinetics of long-term retrogradation is determined by amylopectin fractions, while amylose molecules usually govern short-term retrogradation kinetics to orient and precipitate (Barrera et al., 2012). The retrogradation of starch produces resistant starches, which possess undeniable health benefits. The high-amylose retrograded starch developed by ball milling is suitable for the preparation of noodles and fermented products. However, in a study with rice starch paste, a milling energy of ~4 kJ/g with 4% starch concentration minimized the syneresis, and a milling energy of ~1 kJ/g with 8% starch concentration eliminated the syneresis tendency (González et al., 2021).

The increase in the damaged starch content increases the degree of amylopectin retrogradation and rate of crystallization. Also, the botanical source, starch composition, milling time, grinding vessel, and intensity of milling influence the duration of retrogradation kinetics. For example, from a comparative study with ball-milled maize in stainless steel and ceramic pots of uniform size and capacity, it was studied that the cold-water solubility was potentially higher in ceramic pots compared to stainless steel pots, indicating the starch milled in the latter is more resistant to ball milling than the former. In addition, maize starch ball milled in ceramic pots had a higher syneresis tendency than starch ball milled in stainless steel pots. The ball-milling treatment depolymerized the starch and led to the loss of crystalline order, causing a reduction in gelatinization temperatures and ΔH and increasing the exudation of water from the ball-milled gelatinized maize gel (He et al., 2014).

10.4.4.8 Starch Transparency

The transparency of milled starches is mostly related to cold water solubility. In a study, the transparency of maize shows a direct proportionality to its cold water solubility due to disruption of crystallinity and transformation to an amorphous phase (He et al., 2014). This can be ascribed to the distortion and shattering of the ordered structure, symmetry, and lowered entropy of the orderly branched amylopectins by the physical mechanisms of ball milling. Ball-milled maize starch finds its potential application in packaging films owing to its molecular chain flexibility and transparency by its enhanced amorphous nature. The collision, shearing, smashing, and attrition between the balls of the ball mill and starch destroy crystallinity and convert them to an amorphous nature. The transparency of maize and cassava starches with varying milling times and milling temperatures was investigated, and it was revealed that both balled milled starches had improved transparency. This can be ascribed to the destroyed crystalline structure and the transformation from a polycrystalline state to an amorphous state. Also, ball-milled cassava starch had higher transparency than maize starch due to the large granular size, lower amylose content, and increased frailty compared to maize starch granules (Z. Q. Huang et al., 2008).

10.4.4.9 In Vitro Digestibility

Digestibility is a crucial aspect to indicate the nutritive value of starches. On the basis of the rate of digestibility, starch is categorized into rapidly digestible starch (RDS), slowly digestible starch (SDS), and resistant starch (RS). The enzymes of the human gastrointestine cannot digest RS, while SDS and RDS can be degraded in the small intestine within 2 h and 20 min, respectively. A moderate increase in the level of digestible starch is beneficial for the daily energy requirement of humans, especially for those who are suffering from malnutrition (Q. Wang et al., 2020).

When starch granules are milled, their fragmentation creates a relatively large surface area that facilitates enzyme hydrolysis (E. Li et al., 2014). This was exemplified in rice, wheat, maize, and potato starches (Dhital et al., 2010; Yu et al., 2015; Fu et al., 2016). The digestibility rate increases with the augmentation in the degree of damage of starch particles (Fu et al., 2016). These differences in digestibility are mostly related to the structural changes caused by the damaged starches. Milling by reducing the granule size, enhances the surface area, and creates channels or spaces in the interior of the granules for the penetration and interaction of enzymes with starch (Almeida et al., 2019). Also, the destruction of the crystalline structure during the milling operation encourages the reach of enzymes inside the starch granules due to the formation of a flexible amorphous state (Lv et al., 2019). However, the complexation of starch with fatty acid decreased the digestibility rate of starch, and the modification of starch granules with milling was found to reduce the digestibility rate. In a study, the correlation between starch digestibility and granular structure based on the intermolecular interaction in starch was explored by the complexation of corn starch with lauric acid under ball milling treatment (Hao et al., 2024). Upon density functional theory calculation, it was shown that hydrophobic intermolecular interactions between corn starch and lauric acid were made of hydrogen bonds. Ball milling destroyed the structure of corn starch and reduced the apparent viscosity, and the development of the short-range ordered structure of starch-fatty acid complex inhibited the enzymatic hydrolysis of starch.

10.5 APPLICATION OF MILLED STARCHES

Ball-milled starch applications have been experimented with in varied food and non-food systems to explore improvements in functionality. Ball-milled starches are used in the development of nanocrystals, emulsification, fat replacers, encapsulation, composites, drug delivery, cationic flocculants, noodles, ethanol, and biolatex.

10.5.1 Nanocrystals

Starch nanocrystals are crystalline platelets with dimensions in the nanometer range (typically ranging from 10 to 100 nanometers) generated from the destruction of the semi-crystalline structure of starch granules by acid hydrolysis, enzymatic treatment, or mechanical methods (George et al., 2021). The development of nanoparticles or nanocrystals by milling can be advantageous for food and non-food applications due to their high surface-to-area ratio, high surface reactivity, and increased storage stability (N. Li et al., 2017). Also, ball milling combined with acid hydrolysis produces starch nanocrystals with higher yields in less duration (Dai et al., 2018). The combined treatment damages the starch granule without disturbing the crystalline structure with 30 min of milling. Starch nanocrystals were developed with a yield of 19.3 wt.% through a combined treatment of ball milling and sulfuric acid hydrolysis over three days, a process that was faster than using acid hydrolysis alone to produce similarly shaped nanocrystals. Ball mill–assisted hydrolysis is more economical and time-saving, as it needs no specific instruments or further purification for mass production when compared to ultrasound-assisted or enzymatic pretreatment acid hydrolysis methods.

10.5.2 Emulsification

Ball-milled starch shows good emulsification capacity and enhanced emulsion stability. Taro ball-milled starch esterified with octenyl succinic anhydride (OSA) was studied to understand the suitability of the stability mechanism of Pickering emulsions (Cancan Liu et al., 2018). The esterified ball-milled starch showed an increase in emulsion viscosity by a decrease in surface tension, which resulted in higher emulsification capacity and stability. A physical barrier was formed as the starch particles adsorbed at the interfacial space between oil and water, and the upper and middle phases of the emulsion were stable and uniform up to a storage period of 30 days. Thus, esterification of ball-milled starch with OSA is a viable and efficient way to prepare preferable starch-based emulsifiers. The ability of ball-milled maize starch to form an emulsion and its stabilization were investigated, and it was found that emulsion stability increased by increasing the treatment time owing to the accessibility of more starch particles in the interface produced by mechanical damage during milling. Milled high-amylose maize starch particles possess greater stability compared to milled normal maize starch particles (Lu, Wang, et al., 2018).

Ball-milled starch-stabilized Pickering emulsions are promising delivery systems for administering curcumin, as they maintain its activity and protect it from harsh gastric conditions. (Figure 10.6). The curcumin-loaded micelles of the digested emulsion showed greater in vitro cytotoxicity and functional cellular uptake (Lu et al., 2019). Milled starch-stabilized Pickering emulsions can deliver lipophilic bioactive compounds like curcumin to develop novel functional foods. In another study, media-milled purple sweet potato fragments were used for the preparation of a sustainable food-grade Pickering emulsion. The synergic action of the two prime components, starch and cellulose, and their interactivity with intrinsic phenolics contributed to the emulsifying ability of purple sweet potato particles (Q. Huang et al., 2022).

10.5.3 Fat Replacer

Modified starches are used to mimic the texture and mouthfeel of fat in certain food products. Ball-milled corn starch was utilized as a fat replacer in cream. The addition of ball-milled corn starch increased the cream overrun in the second stage of whipping (2–4 min), while it decreased after 5 min of whipping because of the formation of larger aggregates by the coalescence of small particles. Ball-milled starch reduces water mobility and increases cream viscosity, and a 15% low-fat replacement level of ball-milled starch exhibited a positive effect on cream quality with better foaming and storage stability (C. Wang et al., 2013). Researchers are exploring various healthier fat replacers with specific formulations without compromising the sensory qualities of the final product.

10.5.4 Encapsulation

Starch, along with other materials like maltodextrins and hydrocolloids, is used as a protective coating or shell for enclosing active ingredients to protect them from breakdown throughout processing or storage.

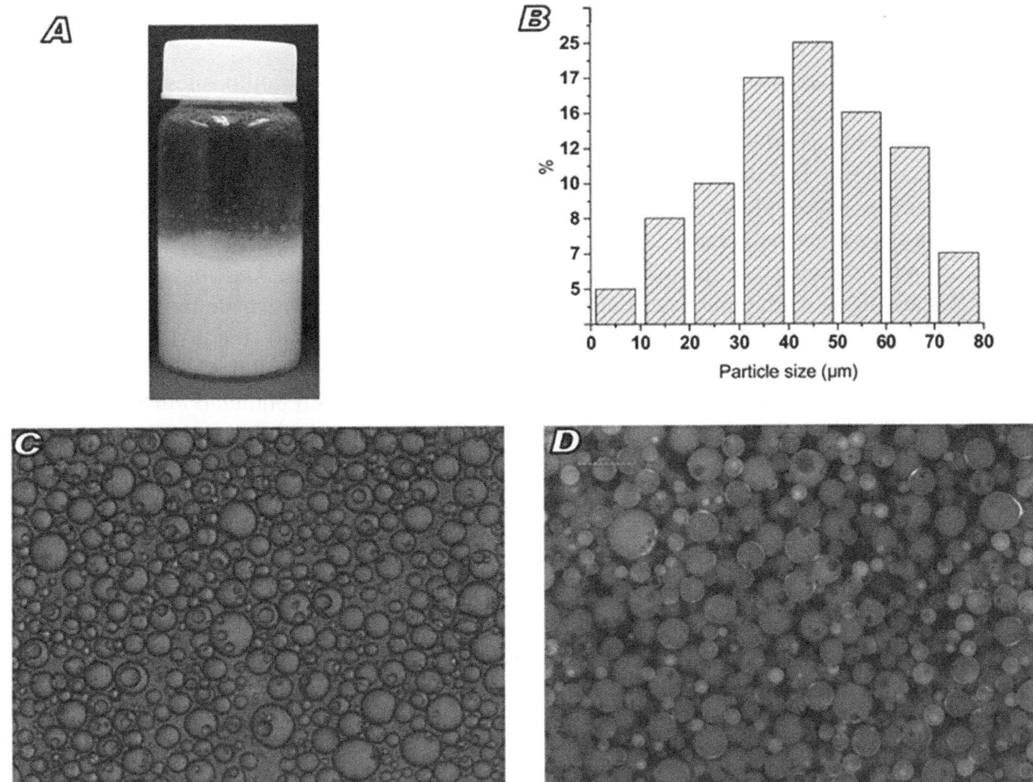

FIGURE 10.6 (a) Digital photograph, (b) droplet size distribution, (c) microscopic, and (d) fluorescence images of curcumin encapsulated in emulsion stabilized by milled maize starch particles after 2 months' storage.

Ball-milled starch has good encapsulation efficiency, which was supported by the enhanced encapsulation and retentivity of β-carotene by milled starches. Gelatine, along with ball-milled rice starch, was found to have four times more encapsulation capacity than native rice starch (González et al., 2020). Similarly, wet and dry ball-milled amaranth starch was used for the encapsulation of β-carotene (Roa et al., 2017). The encapsulated β-carotene content rose from 0.33 to 2.46 mg/g compared to encapsulation by native starch, exhibiting greater encapsulation efficiency and stability. The degradation of amylopectin into smaller-size molecules by milling sped the development of inclusion complexes of β-carotene within starch matrices, thus enhancing encapsulation efficiency.

10.5.5 Drug Delivery

Starch is a widely accepted pharmaceutical excipient in conventional dosage forms as well as drug delivery systems. There have been studies into the potential for milled starches to be used in the delivery of drugs. Planetary ball-milled glutinous rice starch was characterized by its physicochemical, functional, and mucoadhesive properties as a mucoadhesive tablet for its buccal delivery activity. The ball-milling technique was effective in expanding its applicability as a mucoadhesive polymer for the delivery of drugs by altering the starch properties with minimal processing and production costs (Soe et al., 2020). The research concerning the application of milled starches as an active pharmaceutical ingredient for mucoadhesive delivery systems should be further explored for pharmaceutical developments.

10.5.6 Composites

Ball-milling treatment has been adopted for the preparation of various biocomposites. In a study, tapioca starch was used to prepare modified thermoplastic starch using shear-and-heat milling machining along with cellulose nanofiber as filler and glycerol as the plasticizer (Thongmeepech et al., 2023). The modified thermoplastic starch film had decreased crystallinity, several active hydrogen bonds showing Vh crystalline structure, and increased glass transition temperature. The study demonstrated that the shear-and-heat milling machining method is capable of controlling the mechanical properties of glycerol-plasticized starch by improving the interface interaction between cellulose nanofibers. Also, in another recent study, a blend of thermoplastic starch and esterified nanofibrillated cellulose was made by following consecutive steps of ball milling at 500 rpm, dextrinization at 70°C for 1 h, and drying at 105°C for 8 h (Kong et al., 2023). The prepared thermoplastic starch/esterified nanofibrillated cellulose blend was considered as a reinforcing filler for the development of poly(butylene adipate co-terephthalate)/thermoplastic starch composite films. The film showed excellent transparency, foldability, writability, and UV-blocking ability. The findings display concrete data to accelerate the expansion of the application of milling operations to develop high-performance starch-based films.

10.5.7 Noodles

Ball-milled flour was investigated for the preparation of noodles (Oh et al., 1985). The noodles had increased breaking stress, as the flour exhibited high water absorption due to the effect of ball-milled starch. After 8, 16, and 24 h of ball milling, the average water absorption increased by 4.5, 10, and 15%, and breaking stress rose by 5, 9, and 16% causing increased damage of starch in ball-milled flour. The noodles developed showed a comparatively low firmness, indicated by lower surface firmness and cutting stress. Further assessment of the impact of ball-milled starch on the final quality of noodles, like taste and texture, is important, as the functional and cooking properties of noodles may be affected by changes in starch structure upon ball milling.

10.5.8 Bioethanol Production

Ethanol can be produced from starch by the process of fermentation. The starch undergoes enzymatic hydrolysis into simple sugars, where yeast ferments the sugars and produces ethanol and carbon dioxide. Ethanol production can be enhanced by a carbonaceous catalyst bioprocess involving a consecutive catalytic mixed-milling and thermohydrolysis process of cassava starch. An increased concentration of ethanol 15.41 g/L (0.43 g g^{-1} TRS) was obtained after 96 h by the fermentation of cassava starch by mixed-milling/thermohydrolysis hydrolysate at 160°C. An optimized condition of catalytic mixed milling and thermohydrolysis led to the establishment of a robust method for the synthesis of bioethanol with higher yield from the starchy feedstock by an energy-efficient and cost-effective conversion (Intaramas et al., 2019).

10.5.9 Cationic Starch Flocculants

Cationic starch flocculants are water-soluble polymers developed from a modified starch that carries a positive charge. These cationic starch derivatives are applied for the treatment of water by the process called flocculation. Maize starch was activated by ball-milling treatment with 2,3-epoxypropyl trimethyl ammonium chlorides by the destruction of crystal structure and chemical bonding for the synthesis of cationic starch flocculants. The substitution degree and reaction efficiency of mechanically activated starch flocculants were notably higher than native starch flocculants. Thus, the cationic starch flocculants activated by ball milling showed exceptionally good flocculation activity for water treatment (Su et al., 2016).

10.5.10 Biolatex

Starch-based biolatex is an extensively researched topic for the partial replacement of petroleum-based binders in coating paper. In a comparative study on the impact of ultrasonic and ball-milled sago starch suspension on sago starch-incorporated natural rubber latex films, it was revealed that the ball-milling technique showed superior dispersion than ultrasonication due to the smaller particle size of the starch fragments. Hence, milled sago starch–added latex films had greater mechanical properties (Suki et al., 2016). Ball-milling pretreatment destroyed the starch structure by polymerization and enhanced the accessibility and uniformity of α-amylase hydrolysis. The enhanced hydrolysis of starch granules helps in the successive steps of oxidation and cross-linking for the synthesis of starch biolatex. Upon paper coating application, the substitution of 25% petroleum-based latex with starch bio-latex showed excellent adhesive properties with the improved crosslinking structures of biolatex (L. Liu et al., 2020). The development of starch-based biolatex with a ball-milling technique is a cost-efficient and environment-friendly approach with potential in paper coating applications over traditional methods.

10.6 RESEARCH GAPS AND FUTURE RECOMMENDATIONS

Recently, many studies have explored variations in the structure, physiochemical, and functional properties of milled starches. However, there are areas where more focus should be provided concerning the mechanism and functionality of ball-milled starch. The influence of the molecular composition of ball-milled starch should be explored in depth. The pasting mechanism should be studied to understand the impact of molecular chain length on the pasting viscosity of milled starches. As observed in previous studies, starch with single helices and more amylose needs a longer duration to be hydrolyzed by amylolytic enzymes. Hence, more confirmation is required to correlate the alteration of crystalline/molecular orders of balled milled starch in enzymatic digestibility. More research is required to understand the short-range order and changes in functional group intensities by adopting methods like nuclear magnetic resonance spectrometry. Extensive research is needed to ascertain the effect of ball-mill–induced functionality changes in varied starch-based food systems for broadening their application on an industrial scale.

The evolution of research for the progress and broadening of the application of ball milling as a starch modification technique should consider the following future recommendations.

- Optimization of process parameters: Optimizing milling parameters, including milling speed, milling time, and ball-to-starch ratio, in order to optimize efficiency and achieve desired material properties.
- Tailoring starch for distinct applications: Investigation of the potential of ball-milled starch for specific applications in the polymer industry, such as in the development of biodegradable polymers, coatings, or other functional materials.
- Combining processing techniques: Exploration of the synergic impact of combining ball milling with other processing techniques to accomplish unique material properties. For instance, combining ball milling with extrusion or chemical modifications.

- Functionalization and composite materials: Investigation of the possibility of functionalizing ball-milled starch or incorporating it into composite materials to improve specific properties, such as thermal stability, mechanical strength, or barrier properties.
- Scale-up studies: Conduct studies to assess the scalability of the ball-milling process for industrial applications, considering factors like cost, energy efficiency, and large-scale production feasibility.
- Collaborative research: Encourage interdisciplinary collaboration between research in chemistry, materials science, and engineering for the exploration of diverse perspectives and applications for ball-milled starch materials.
- Regulatory considerations: Address regulatory considerations related to the use of ball-milled starch in polymer applications, ensuring compliance with safety and environmental standards.

10.7 CONCLUSION

In conclusion, the ball milling of starch has profound implications for a range of applications, as it is a versatile and efficient technique. As a result of the mechanical forces developed during milling, the starch undergoes structural modifications, which improve its solubility, surface area, and rheological properties. In addition to tailoring starch properties for specific industrial uses, ball milling has contributed to the development of sustainable and cost-effective processes in the food, pharmaceutical, and materials science sectors. With an in-depth understanding of the intricate relationship between milling parameters and starch modifications, ball-milled starch-based materials can be optimized and innovated. Ball milling of starch is expected to be a pivotal technology in the future for a wide range of industries.

REFERENCES

Aaliya, B., Sunooj, K. V., Rajkumar, C. B. S., Navaf, M., Akhila, P. P., Sudheesh, C., George, J., & Lackner, M. (2021). Effect of thermal pretreatments on phosphorylation of Corypha umbraculifera L. stem pith starch: A comparative study using dry-heat, heat-moisture and autoclave treatments. *Polymers, 13*(21), 3855. https://doi.org/10.3390/polym13213855

Ahmad, M., Gani, A., Masoodi, F. A., & Rizvi, S. H. (2020). Influence of ball milling on the production of starch nanoparticles and its effect on structural, thermal and functional properties. *International Journal of Biological Macromolecules, 151*, 85–91. https://doi.org/10.1016/j.ijbiomac.2020.02.139

Ahmad, N., & Rajab, A. (2018). Textural and cooking qualities of dry Laksa noodle made from semi-wet and wet MR253 flours. *Cereal Chemistry, 95*(6), 872–880. https://doi.org/10.1002/cche.10105

Akhila, P. P., Sunooj, K. V., Aaliya, B., Navaf, M., Sudheesh, C., Yadav, D. N., Khan, M. A., Mir, S. A., & George, J. (2022). Morphological, physicochemical, functional, pasting, thermal properties and digestibility of hausa potato (Plectranthus rotundifolius) flour and starch. *Applied Food Research, 2*(2), 100193. https://doi.org/10.1016/j.afres.2022.100193

Almeida, R. L. J., dos Santos Pereira, T., de Andrade Freire, V., Santiago, Â. M., Oliveira, H. M. L., de Sousa Conrado, L., & de Gusmão, R. P. (2019). Influence of enzymatic hydrolysis on the properties of red rice starch. *International Journal of Biological Macromolecules, 141*, 1210–1219. https://doi.org/10.1016/j.ijbiomac.2019.09.072

Amaya-Llano, S. L., Martínez-Bustos, F., Alegría, A. L. M., & de Jesús Zazueta-Morales, J. (2011). Comparative studies on some physico-chemical, thermal, morphological, and pasting properties of acid-thinned jicama and maize starches. *Food and Bioprocess Technology, 4*(1), 48–60. https://doi.org/10.1007/s11947-008-0153-z

Ashogbon, A. O. (2021). The recent development in the syntheses, properties, and applications of triple modification of various starches. *Starch/Staerke, 73*(3–4), 1–10. https://doi.org/10.1002/star.202000125

Bangar, S. P., Singh, A., Ashogbon, A. O., & Bobade, H. (2023). Ball-milling: A sustainable and green approach for starch modification. *International Journal of Biological Macromolecules*, *237*, 124069. https://doi.org/10.1016/j.ijbiomac.2023.124069

Barrera, G. N., Bustos, M. C., Iturriaga, L., Flores, S. K., León, A. E., & Ribotta, P. D. (2013). Effect of damaged starch on the rheological properties of wheat starch suspensions. *Journal of Food Engineering*, *116*(1), 233–239. https://doi.org/10.1016/j.jfoodeng.2012.11.020

Barrera, G. N., León, A. E., & Ribotta, P. D. (2012). Effect of damaged starch on wheat starch thermal behavior. *Starch/Staerke*, *64*(10), 786–793. https://doi.org/10.1002/star.201200022

BeMiller, J. N. (2017). Physical modification of starch. In *Starch in food: Structure, function and applications* (pp. 223–253). Elsevier Ltd. https://doi.org/10.1016/B978-0-08-100868-3.00005-6

Broseghini, M., Gelisio, L., D'Incau, M., Azanza Ricardo, C. L., Pugno, N. M., & Scardi, P. (2016). Modeling of the planetary ball-milling process: The case study of ceramic powders. *Journal of the European Ceramic Society*, *36*(9), 2205–2212. https://doi.org/10.1016/j.jeurceramsoc.2015.09.032

Cavallini, C. M., & Franco, C. M. L. (2010). Effect of acid-ethanol treatment followed by ball milling on structural and physicochemical characteristics of cassava starch. *Starch/Staerke*, *62*(5), 236–245. https://doi.org/10.1002/star.200900231

Dai, L., Li, C., Zhang, J., & Cheng, F. (2018). Preparation and characterization of starch nanocrystals combining ball milling with acid hydrolysis. *Carbohydrate Polymers*, *180*, 122–127. https://doi.org/10.1016/j.carbpol.2017.10.015

Dhital, S., Shrestha, A. K., Flanagan, B. M., Hasjim, J., & Gidley, M. J. (2011). Cryo-milling of starch granules leads to differential effects on molecular size and conformation. *Carbohydrate Polymers*, *84*(3), 1133–1140. https://doi.org/10.1016/j.carbpol.2011.01.002

Dhital, S., Shrestha, A. K., & Gidley, M. J. (2010). Effect of cryo-milling on starches: Functionality and digestibility. *Food Hydrocolloids*, *24*(2–3), 152–163. https://doi.org/10.1016/j.foodhyd.2009.08.013

Diop, C. I. K., Li, H. L., Chen, P., & Xie, B. J. (2012). Properties of maize starch modified by ball milling in ethanol medium and low field NMR determination of the water molecular mobility in their gels. *Journal of Cereal Science*, *56*(2), 321–331. https://doi.org/10.1016/j.jcs.2012.03.002

El-Porai, E. S., Salama, A. E., Sharaf, A. M., Hegazy, A. I., & Gadallah, M. G. E. (2013). Effect of different milling processes on Egyptian wheat flour properties and pan bread quality. *Annals of Agricultural Sciences*, *58*(1), 51–59. https://doi.org/10.1016/j.aoas.2013.01.008

Fang, K., He, W., Jiang, Y., Li, K., & Li, J. (2020). Preparation, characterization and physicochemical properties of cassava starch-ferulic acid complexes by mechanical activation. *International Journal of Biological Macromolecules*, *160*, 482–488. https://doi.org/10.1016/j.ijbiomac.2020.05.213

Fu, Z., Luo, S. J., Liu, W., Liu, C. M., & Zhan, L. jing. (2016). Structural changes induced by high speed jet on in vitro digestibility and hydroxypropylation of rice starch. *International Journal of Food Science and Technology*, *51*(4), 1034–1040. https://doi.org/10.1111/ijfs.13046

George, J., Nair, S. G., Kumar, R., Semwal, A. D., Sudheesh, C., Basheer, A., & Sunooj, K. V. (2021). A new insight into the effect of starch nanocrystals in the retrogradation properties of starch. *Food Hydrocolloids for Health*, *1*, 100009. https://doi.org/10.1016/j.fhfh.2021.100009

González, L. C., Loubes, M. A., Bertotto, M. M., Baeza, R. I., & Tolaba, M. P. (2021). Flow behavior and syneresis of ball milled rice starch and their correlations with starch structure. *Carbohydrate Polymer Technologies and Applications*, *2*, 100168. https://doi.org/10.1016/j.carpta.2021.100168

González, L. C., Loubes, M. A., & Tolaba, M. P. (2018). Incidence of milling energy on dry-milling attributes of rice starch modified by planetary ball milling. *Food Hydrocolloids*, *82*, 155–163. https://doi.org/10.1016/j.foodhyd.2018.03.051

González, L. C., Loubes, M. A., Santagapita, P. R., & Tolaba, M. P. (2020). Co-joined starch modification and β-carotene dispersion in situ by planetary ball milling. *Starch/Staerke*, *72*(11–12), 1–11. https://doi.org/10.1002/star.202000007

Hao, Z., Xu, H., Yu, Y., Gu, Z., Wang, Y., Li, C., Xiao, Y., Liu, Y., Liu, K., Zheng, M., Du, Y., Zhou, Y., & Yu, Z. (2024). Insights into ball milling treatment promotes the formation of starch-lipid complexes and the relation between multi-scale structure and in vitro digestibility based on intermolecular interactions. *Food Hydrocolloids*, *146*, 109277. https://doi.org/10.1016/j.foodhyd.2023.109277

He, S., Qin, Y., Walid, E., Li, L., Cui, J., & Ma, Y. (2014). Effect of ball-milling on the physicochemical properties of maize starch. *Biotechnology Reports*, *3*, 54–59. https://doi.org/10.1016/j.btre.2014.06.004

Huang, Q., Huang, Q., Wang, Y., & Lu, X. (2022). Development of wet media milled purple sweet potato particle-stabilized pickering emulsions: The synergistic role of bioactives, starch and cellulose. *LWT*, *155*, 112964. https://doi.org/10.1016/j.lwt.2021.112964

Huang, Z. Q., Lu, J. P., Li, X. H., & Tong, Z. F. (2007). Effect of mechanical activation on physico-chemical properties and structure of cassava starch. *Carbohydrate Polymers*, *68*(1), 128–135. https://doi.org/10.1016/j.carbpol.2006.07.017

Huang, Z. Q., Xie, X. L., Chen, Y., Lu, J. P., & Tong, Z. F. (2008). Ball-milling treatment effect on physicochemical properties and features for cassava and maize starches. *Comptes Rendus Chimie*, *11*(1–2), 73–79. https://doi.org/10.1016/j.crci.2007.04.008

Intaramas, K., Sakdaronnarong, C., Liu, C. G., Mehmood, M. A., Jonglertjunya, W., & Laosiripojana, N. (2019). Sequential catalytic-mixed-milling and thermohydrolysis of cassava starch improved ethanol fermentation. *Food and Bioproducts Processing*, *114*, 72–84. https://doi.org/10.1016/j.fbp.2018.11.011

Jhan, F., Shah, A., Gani, A., Ahmad, M., & Noor, N. (2020). Nano-reduction of starch from underutilised millets: Effect on structural, thermal, morphological and nutraceutical properties. *International Journal of Biological Macromolecules*, *159*, 1113–1121. https://doi.org/10.1016/j.ijbiomac.2020.05.020

Joy, J., Krishnamoorthy, A., Tanna, A., Kamathe, V., Nagar, R., & Srinivasan, S. (2022). Recent developments on the synthesis of nanocomposite materials via ball milling approach for energy storage applications. *Applied Sciences (Switzerland)*, *12*(18). https://doi.org/10.3390/app12189312

Juarez-Arellano, E. A., Urzua-Valenzuela, M., Peña-Rico, M. A., Aparicio-Saguilan, A., Valera-Zaragoza, M., Huerta-Heredia, A. A., & Navarro-Mtz, A. K. (2021). Planetary ball-mill as a versatile tool to controlled potato starch modification to broaden its industrial applications. *Food Research International*, *140*, 109870. https://doi.org/10.1016/j.foodres.2020.109870

Kim, S. K., & Kim, H. Y. (2014). Pasting properties of colored rice starch by ball-mill treatment. *Journal of Crop Science and Biotechnology*, *17*(3), 161–166. https://doi.org/10.1007/s12892-014-0046-9

Kong, Y., Qian, S., Zhang, Z., & Tian, J. (2023). The impact of esterified nanofibrillated cellulose content on the properties of thermoplastic starch/PBAT biocomposite films through ball-milling. *International Journal of Biological Macromolecules*, *253*, 127462. https://doi.org/10.1016/j.ijbiomac.2023.127462

Kumar, M., Xiong, X., Wan, Z., Sun, Y., Tsang, D. C. W., Gupta, J., Gao, B., Cao, X., Tang, J., & Ok, Y. S. (2020). Ball milling as a mechanochemical technology for fabrication of novel biochar nanomaterials. *Bioresource Technology*, *312*, 123613. https://doi.org/10.1016/j.biortech.2020.123613

Li, E., Dhital, S., & Hasjim, J. (2014). Effects of grain milling on starch structures and flour/starch properties. *Starch/Staerke*, *66*(1–2), 15–27. https://doi.org/10.1002/star.201200224

Li, J., Zhou, M., Cheng, F., Lin, Y., Shi, L., & Zhu, P. X. (2020). Preparation of oxidized corn starch with high degree of oxidation by fenton-like oxidation assisted with ball milling. *Materials Today Communications*, *22*, 100793. https://doi.org/10.1016/j.mtcomm.2019.100793

Li, N., Niu, M., Zhang, B., Zhao, S., Xiong, S., & Xie, F. (2017). Effects of concurrent ball milling and octenyl succinylation on structure and physicochemical properties of starch. *Carbohydrate Polymers*, *155*, 109–116. https://doi.org/10.1016/j.carbpol.2016.08.063

Limpongsa, E., & Jaipakdee, N. (2020). Physical modification of Thai rice starch and its application as orodispersible film former. *Carbohydrate Polymers*, *239*, 116206. https://doi.org/10.1016/j.carbpol.2020.116206

Liu, Cancan, An, F., He, H., He, D., Wang, Y., & Song, H. (2018). Pickering emulsions stabilized by compound modified areca taro (Colocasia esculenta (L.) Schott) starch with ball-milling and OSA. *Colloids and Surfaces A: Physicochemical and Engineering Aspects*, *556*(15), 185–194. https://doi.org/10.1016/j.colsurfa.2018.08.032

Liu, Cancan, Jiang, Y., Liu, J., Li, K., & Li, J. (2021). Insights into the multiscale structure and pasting properties of ball-milled waxy maize and waxy rice starches. *International Journal of Biological Macromolecules*, *168*, 205–214. https://doi.org/10.1016/j.ijbiomac.2020.12.048

Liu, Chang, Wang, S., Copeland, L., & Wang, S. (2015). Physicochemical properties and in vitro digestibility of starches from field peas grown in China. *LWT*, *64*(2), 829–836. https://doi.org/10.1016/j.lwt.2015.06.060

Liu, Chong, Hong, J., & Zheng, X. (2015). Effect of heat-moisture treatment on morphological, structural and functional characteristics of ball-milled wheat starches. *Starch/Staerke*, *67*, 1–9. https://doi.org/10.1002/star.201500141

Liu, L., An, X., Zhang, H., Lu, Z., Nie, S., Cao, H., Xu, Q., & Liu, H. (2020). Ball milling pretreatment facilitating α-amylase hydrolysis for production of starch-based bio-latex with high performance. *Carbohydrate Polymers*, *242*, 116384. https://doi.org/10.1016/j.carbpol.2020.116384

Liu, R., Sun, W., Zhang, Y., Huang, Z., Hu, H., Zhao, M., & Li, W. (2019). Development of a novel model dough based on mechanically activated cassava starch and gluten protein: Application in bread. *Food Chemistry*, *300*, 125196. https://doi.org/10.1016/j.foodchem.2019.125196

Liu, T. Y., Ma, Y., Yu, S. F., Shi, J., & Xue, S. (2011). The effect of ball milling treatment on structure and porosity of maize starch granule. *Innovative Food Science and Emerging Technologies*, *12*(4), 586–593. https://doi.org/10.1016/j.ifset.2011.06.009

Lu, X., Li, C., & Huang, Q. (2019). Combining in vitro digestion model with cell culture model: Assessment of encapsulation and delivery of curcumin in milled starch particle stabilized Pickering emulsions. *International Journal of Biological Macromolecules*, *139*, 917–924. https://doi.org/10.1016/j.ijbiomac.2019.08.078

Lu, X., Wang, Y., Li, Y., & Huang, Q. (2018). Assembly of Pickering emulsions using milled starch particles with different amylose/amylopectin ratios. *Food Hydrocolloids*, *84*, 47–57. https://doi.org/10.1016/j.foodhyd.2018.05.045

Lu, X., Xiao, J., & Huang, Q. (2018). Pickering emulsions stabilized by media-milled starch particles. *Food Research International*, *105*(March 2017), 140–149. https://doi.org/10.1016/j.foodres.2017.11.006

Lv, Y., Zhang, L., Li, M., He, X., Hao, L., & Dai, Y. (2019). Physicochemical properties and digestibility of potato starch treated by ball milling with tea polyphenols. *International Journal of Biological Macromolecules*, *129*, 207–213. https://doi.org/10.1016/j.ijbiomac.2019.02.028

Martínez-Bustos, F., López-Soto, M., San Martin-Martinez, E., Zazueta-Morales, J. J., & Velez-Medina, J. J. (2007). Effects of high energy milling on some functional properties of jicama starch (Pachyrrhizus erosus L. Urban) and cassava starch (Manihot esculenta Crantz). *Journal of Food Engineering*, *78*, 1212–1220.

Moraes, J., Alves, F. S., & Franco, C. M. L. (2013). Effect of ball milling on structural and physicochemical characteristics of cassava and Peruvian carrot starches. *Starch/Staerke*, *65*(3–4), 200–209. https://doi.org/10.1002/star.201200059

Niemann, C., & Whistler, R. L. (1992). Effect of acid hydrolysis and ball milling on porous corn starch. *Starch-Stärke*, *44*(11), 409–414. https://doi.org/10.1002/star.19920441103

Oh, N. H., Seib, P., Ward, B., & Deyoe, C. W. (1985). Influence of flour protein, extraction rate, particle size, and starch damage on the quality characteristics of dry noodles. *Cereal Chemistry*, *62*(6), 441–446.

Ren, G. Y., Li, D., Wang, L. J., Özkan, N., & Mao, Z. H. (2010). Morphological properties and thermoanalysis of micronized cassava starch. *Carbohydrate Polymers*, *79*(1), 101–105. https://doi.org/10.1016/j.carbpol.2009.07.031

Roa, D. F., Buera, M. P., Tolaba, M. P., & Santagapita, P. R. (2017). Encapsulation and stabilization of β-carotene in amaranth matrices obtained by dry and wet assisted ball milling. *Food and Bioprocess Technology*, *10*(3), 512–521. https://doi.org/10.1007/s11947-016-1830-y

Shen, H., Yu, J., Bai, J., Ge, X., Liang, W., Ospankulova, G., Muratkhan, M., Zhang, G., & Li, W. (2022). Electron beam irradiation regulates the structure and functionality of ball-milled corn starch: The related mechanism. *Carbohydrate Polymers*, *297*, 120016. https://doi.org/10.1016/j.carbpol.2022.120016

Shi, L., Cheng, F., Zhu, P. X., & Lin, Y. (2015). Physicochemical changes of maize starch treated by ball milling with limited water content. *Starch/Staerke*, *67*(9–10), 772–779. https://doi.org/10.1002/star.201500026

Singh, S. K., Lillard, J. W., & Singh, R. (2018). Reversal of drug resistance by planetary ball milled (PBM) nanoparticle loaded with resveratrol and docetaxel in prostate cancer. *Cancer Letters*, *427*, 49–62. https://doi.org/10.1016/j.canlet.2018.04.017

Sitotaw, Y. W., Habtu, N. G., Gebreyohannes, A. Y., Nunes, S. P., & Van Gerven, T. (2021). Ball milling as an important pretreatment technique in lignocellulose biorefineries: A review. *Biomass Conversion and Biorefinery*, 1–24. https://doi.org/10.1007/s13399-021-01800-7

Soe, M. T., Chitropas, P., Pongjanyakul, T., Limpongsa, E., & Jaipakdee, N. (2020). Thai glutinous rice starch modified by ball milling and its application as a mucoadhesive polymer. *Carbohydrate Polymers*, *232*, 115812. https://doi.org/10.1016/j.carbpol.2019.115812

Su, Y., Du, H., Huo, Y., Xu, Y., Wang, J., Wang, L., Zhao, S., & Xiong, S. (2016). Characterization of cationic starch flocculants synthesized by dry process with ball milling activating method. *International Journal of Biological Macromolecules*, *87*, 34–40. https://doi.org/10.1016/j.ijbiomac.2015.11.093

Suki, F. M. M., Azura, A. R., & Azahari, B. (2016). Effect of ball milled and ultrasonic sago starch dispersion on sago starch filled natural rubber latex (SSNRL) films. *Procedia Chemistry*, *19*, 782–787. https://doi.org/10.1016/j.proche.2016.03.085

Tan, X., Zhang, B., Chen, L., Li, X., Li, L., & Xie, F. (2015). Effect of planetary ball-milling on multi-scale structures and pasting properties of waxy and high-amylose cornstarches. *Innovative Food Science and Emerging Technologies*, *30*, 198–207. https://doi.org/10.1016/j.ifset.2015.03.013

Tavares, L. M. (2017). A review of advanced ball mill modelling. *KONA Powder and Particle Journal*, *2017*(34), 106–124. https://doi.org/10.14356/kona.2017015

Tester, R. F., Patel, T., & Harding, S. E. (2006). Damaged starch characterisation by ultracentrifugation. *Carbohydrate Research*, *341*(1), 130–137. https://doi.org/10.1016/j.carres.2005.10.018

Thongmeepech, A., Koda, T., & Nishioka, A. (2023). Effect of shear and heat milling of starch on the properties of tapioca starch/cellulose nanofiber composites. *Carbohydrate Polymers*, *306*, 120618. https://doi.org/10.1016/j.carbpol.2023.120618

Wang, C., He, X. W., Huang, Q., Fu, X., & Liu, S. (2013). Physicochemical properties and application of micronized cornstarch in low fat cream. *Journal of Food Engineering*, *116*(4), 881–888. https://doi.org/10.1016/j.jfoodeng.2013.01.025

Wang, Q., Li, L., & Zheng, X. (2020). A review of milling damaged starch: Generation, measurement, functionality and its effect on starch-based food systems. *Food Chemistry, 315*, 126267. https://doi.org/10.1016/j.foodchem.2020.126267

Wu, F., Li, J., Yang, N., Chen, Y., Jin, Y., & Xu, X. (2018). The roles of starch structures in the pasting properties of wheat starch with different degrees of damage. *Starch-Stärke, 1*(2), 1–31. https://doi.org/10.23959/sfdorj-1000010

Yu, J., Wang, S., Wang, J., Li, C., Xin, Q., Huang, W., Zhang, Y., He, Z., & Wang, S. (2015). Effect of laboratory milling on properties of starches isolated from different flour millstreams of hard and soft wheat. *Food Chemistry, 172*, 504–514. https://doi.org/10.1016/j.foodchem.2014.09.070

Zhang, K., Dai, Y., Hou, H., Li, X., Dong, H., Wang, W., & Zhang, H. (2019). Influences of grinding on structures and properties of mung bean starch and quality of acetylated starch. *Food Chemistry, 294*(61), 285–292. https://doi.org/10.1016/j.foodchem.2019.05.055

Zhang, W., Ding, W., Ndeurumi, K. H., Wang, Z., & Feng, Y. (2015). Effect of wet ball milling on physicochemical properties and crosslinking reaction performance of corn starch. *Starch/Staerke, 67*(11–12), 958–963. https://doi.org/10.1002/star.201400251

Zhang, Z., Zhao, S., & Xiong, S. (2010). Morphology and physicochemical properties of mechanically activated rice starch. *Carbohydrate Polymers, 79*(2), 341–348. https://doi.org/10.1016/j.carbpol.2009.08.016

Acid Hydrolysis

11

Rafael Audino Zambelli, Matheus Calixto Saraiva,
Andressa Barbosa Barroso,
and Mariana Lopes dos Anjos

11.1 INTRODUCTION

Natural starch comprises two primary components: amylose and amylopectin. These components contribute to the complex semi-crystalline structure exhibited by starch granules, which are characterized by a relatively high density of approximately 1.5 g/mL. Notably, this density surpasses that of typical plant fibers, which generally fall within the density range of 1.2–1.4 g/mL (Pfister et al., 2020; Bertoft, 2017).

Acid hydrolysis of native starch is a crucial modification method employed in the starch industry to produce starches with enhanced properties, making it a versatile ingredient with wide-ranging applications. This process involves the controlled use of acidic substances to selectively break down the starch molecules, resulting in a range of desirable changes (Compart et al., 2022); it is of significant importance to the food industry due to the wide range of functional benefits it offers. It is extensively utilized in various industries, including food, paper, textiles, and more, due to its ability to improve viscosity, conversion extent, and other key characteristics of starch (Wang & Copeland, 2015).

Acid hydrolysis leads to a reduction in the molecular weight of starch molecules, breaking them into shorter chains and individual glucose units. This reduction in molecular weight results in decreased viscosity, making the modified starch more suitable as a thickening agent in various food products. This improved control over texture and consistency is particularly valuable in the food industry (He et al., 2023; Zambelli et al., 2018) and has been demonstrated as a highly effective method for enhancing our understanding of the intricate structure of starch granules. This approach is particularly valuable because it enables the isolation and examination of the resistant crystalline segments within the granules. This, in turn, facilitates the estimation of two critical components of starch: the easily amorphous portion and semi-crystalline fraction (Silveira et al., 2019).

This enhanced insight into the structural changes brought about by acid hydrolysis not only furthers our comprehension of starch granules but also opens up possibilities for tailoring starch properties for specific applications. By manipulating the degree of hydrolysis and targeting specific structural components, researchers and industries can create modified starches with tailored characteristics, optimizing their functionality in various products and processes. This versatility underscores the significance of acid hydrolysis as a pivotal tool in starch research and its diverse industrial applications (Li et al., 2022; Chen et al., 2021).

DOI: 10.1201/9781032655598-11

11.2 ACID HYDROLYSIS AS A MODIFICATION METHOD

Acid hydrolysis is a chemical process that involves the breakdown of complex molecules through the addition of acid and water. When applied to starch, acid hydrolysis modifies the starch molecules by cleaving the glycosidic bonds that link the glucose units. This process results in the production of smaller sugar molecules, primarily glucose, maltose, and maltodextrins (Wang & Copeland, 2015).

It has been demonstrated that initial treatment, whether chemical or physical, of plant residues, enhances their non-dissociation from the starch matrix and optimizes the qualities of the composite (Quilez-Molina et al., 2023), this technique has significant implications for the final structure and properties of the treated material. Configuring a starch modification is known to affect amorphous *lamellae* (Gonzalez & Wang, 2023).

Amylose or amylopectin molecules, which make up starch and have crystalline and amorphous regions in their structure located on the surface of the granules, are attacked together in the first initial stages of this hydrolytic method. Acting through the slow hydrolysis of the crystallization zone (organized and ordered, resulting from the formation of amylose double helices and/or densely branched regions of amylopectin), which are less accessible, and preferably rapid hydrolysis of the amorphous region (disordered, where the molecules of amylose and amylopectin chains are less organized) due to their looser structure within the starch granules. This hydrolysis generates starches with greater crystallinity and reduces the degree of polymerization (Soler et al., 2020; Gonzalez & Wang, 2023). The functional characteristics of starches in different industries are influenced by various factors, expanding their commercial usability and efficiency across a wide range of processing conditions. Acid-hydrolyzed starch, in particular, has been widely used in the food, pharmaceutical, paper, and textile industries, as well as in the production of biodegradable plastics (Soler et al., 2020).

The importance of these modifications is because native starch, that is, unmodified starch found in its raw form in nature in foods such as cereals, roots, and tubers, has an inert structure, which is why it has limited commercial applications, such as low-stability thermal, as it tends to degrade when subjected to high temperatures, promoting a loss of viscosity and undesirable texture in final products; low shear resistance due to considerable sensitivity to processing under high agitation; and difficult dispersion and complete hydration in specific systems such as liquids due to their large granule size (Zhao et al., 2022). Thus, as acid hydrolysis progresses, the functional characteristics of native starch undergo marked transformations. These changes provide advantages for various industrial applications, as the process changes the granular configuration of this polysaccharide, preventing structural problems and modifying or improving its original characteristics or acquiring new capabilities (Wang et al., 2021). Consequently, acid hydrolysis leads to a distinct reaction when heated in water, resulting in pastes of intrinsically lower viscosity, decreased paste viscosity when hot, increased gel firmness, improved water solubility, and superior film-forming capabilities (Soler et al., 2020).

As mentioned, a major influencing factor for better achievement and use of results through the exposure of starch to chemical modification of hydrolysis by acids is attributed to the configuration of the amorphous or crystalline regions of the amylose and amylopectin molecules, which alternate in the state amorphous or collaborate in the formation of the semi-crystalline region through multiple lamellae (Wang et al., 2021).

In amorphous regions, the action of the acid is prioritized because, due to its more disorganized and, consequently, more open structure, it becomes more susceptible to the penetration and action of the acid used, promoting an increase in relative crystallinity (Wang et al., 2022). Other elements are said to influence the hydrolysis process, more specifically its extent and speed, such as the proportion of amylose/amylopectin, the dimensions of the granules, the existence of pores on the surface, and the amount of lipid-amylose complexes (Gonzalez & Wang, 2023).

The choice of acid and its concentration in the acid hydrolysis of starch depends on the desired properties of the modified starch and the specific application it will be used for. The concentration of the acid is a critical factor that influences the rate and extent of hydrolysis. Higher acid concentrations and longer reaction times generally result in more extensive hydrolysis, leading to smaller molecular fragments (Silveira et al., 2019).

Starches that undergo hydrolysis with acids are normally prepared by treating starch granules with dilute aqueous or alcoholic solutions of mineral acid (hydrochloric or sulfuric acid), occurring at temperatures below the gelatinization temperature for long periods. After reaching the desired degree of hydrolysis, the resulting modified starch is recovered by neutralizing the acidic solution with an alkaline solution, followed by washing. In industrial settings, it is typical to treat starch with diluted hydrochloric acid (HCl) or sulfuric acid (H_2SO_4) at temperatures ranging from 25 to 55°C for extended durations (Soler et al., 2020). HCl and H_2SO_4 are the acids most cited in the literature for carrying out this hydrolytic treatment, as they are strong inorganic acids that claim to have a greater hydrolysis effect; however, hydrochloric acid is the most used (Quilez-Molina et al., 2023). In the current literature, different relationships of concentrations and temperatures with HCl are used in the processes for the occurrence of this breakdown due to chemical action, including solutions with HCl (2.2 M) at 36°C, HCl aqueous solution (7.5% w/w) at 25°C, HCl (0.4 mol/L) at 50°C, and HCl (0.36 M) at 40°C (Soler et al., 2020; Li & Hu, 2021; Wang et al., 2022; Gonzalez & Wang, 2023). It is worth mentioning that the associations of the time-temperature binomial are important from a chemical point of view. Given the temperatures mentioned for carrying out the starch breakdown process, it can be seen that this condition is traditionally maintained at low levels, respecting the limit of 55°C.

Studies carried out by Zhao et al. (2022) state that the limited reactivity of acid hydrolysis usually results in an extensive preparation period, generally 5 to 15 days in processing. Addressing time has been one of the main challenges restricting increased production. Being able to correlate the long duration with low temperatures is commonly established. Practices involving high-temperature acid hydrolysis are little explored due to central concerns that the crystalline structure of starch may be significantly compromised during this process. Thus, these scholars concluded that high-temperature acid hydrolysis of granular starch can be achieved if the crystalline structure is improved through an appropriate pretreatment procedure. Subjecting the polysaccharide to pretreatment with heat and humidity subsequently achieved good results for hydrolyzed starch at a temperature of 69°C lasting up to 60 minutes, obtaining results of improvement in the crystalline structure for resistance to high temperatures.

The cleavage effect caused by acid hydrolysis at glycosidic bonds also results in the fragmentation of long starch chains. This, in turn, contributes to an increase in the proportion of shorter chains, significantly affecting the starch structure. Starch, which is initially composed of glucose monomers linked by multiple glycosidic bonds, sees some of these bonds being broken during the acid hydrolysis process. This breaking of glycosidic bonds not only transforms part of the starch into soluble sugars but also results in a reduction in the amount of modified starch. The presence of shorter chains and the transformation of some of these molecules into soluble sugars can have significant implications for the functional properties of starch, such as viscosity, texture, and ability to form gels. This phenomenon is crucial for obtaining products derived from starch, such as glucose syrups and other industrial products. Furthermore, the acid hydrolysis process can also result in a decrease in molecular weight, mainly due to starch degradation caused by the breakdown of α-1,4 glycosidic bonds in linear starch molecules promoted by hydrochloric acid during the hydrolysis process (Liang et al., 2024a; Wang et al., 2022).

The products of acid hydrolysis of starch are mainly smaller sugar molecules, as the glycosidic bonds between the glucose units in the starch are cleaved. The predominant product of starch hydrolysis is glucose. When subjected to acid hydrolysis, these glycosidic bonds are cleaved by the acid, resulting in the breakdown of starch into its constituent glucose molecules. These products are influenced by changes in conditions such as acid concentration, time of occurrence, and starch composition. The acid-catalyzed reaction hydrolyzes the glycosidic bonds, converting the complex structure of starch into simple sugar units (Khlestkin et al., 2018).

The chemical equation for starch hydrolysis can be represented as follows:

$$\text{Starch} + H_2O + \text{Acid} \rightarrow \text{Glucose and other products} \qquad (1)$$

More extensive hydrolysis produces dextrins with different degrees of polymerization, such as malto-dextrin, which contains easily digestible malto-oligosaccharides (Cordenunsi-Lysenko et al., 2018). These dextrins are used to increase soluble solids, control freezing, and prevent crystallization, providing viscosity and stable texture to foods. However, glucose tends to be the most abundant product in starch hydrolysis under acidic conditions (Hofman et al., 2016).

11.3 PROPERTIES OF ACID-HYDROLYZED STARCH

The physicochemical properties of starch are determined by its molecular structure and the complex hierarchical organization of its structure (Gonzalez & Wang, 2023). On the other hand, the properties of acid-hydrolyzed starch vary depending on the degree of hydrolysis, time, temperature, and acid concentration used (Abdorreza et al., 2012). In general, native starch exists as granules and exhibits different morphologies depending on its origin. The semi-crystalline starch granules consist of two molecules: amylopectin, responsible for the crystalline structure, and amylose, responsible for the amorphous region (Chen et al., 2023).

The amylopectin content is a crucial factor influencing the acid hydrolysis of starch, having a directly proportional effect. In hydrolysis processes using temperatures below the gelatinization temperature, amylopectin is hydrolyzed more rapidly than amylose. An important characteristic is that the physicochemical properties of hydrolyzed starch undergo changes but do not destroy its granular structure (Abdorreza et al., 2012). Acid hydrolysis occurs in two stages: an initial rapid phase in the amorphous lamellae and a slower second phase in the crystalline lamellae.

The type of acid used, starch and acid concentrations, starch crystalline structure, and temperature affect the rate and extent of acid hydrolysis. During acid hydrolysis, the amorphous lamellae of starch are transformed into starches with higher crystallinity, and there is a decrease in the degree of polymerization (Gonzalez & Wang, 2023). Starches from different sources have distinct structures, varying not only in the amylopectin-to-amylose ratio but also in the crystallinity of their structure, presenting different conformations of these compounds. Type A crystals, found in cereal starches, are characterized by monoclinic unit cells with eight water molecules per unit cell and a higher proportion of short chains of amylopectin. Type B crystals are found in tuber starches and consist of an open hexagonal unit cell containing 36 water molecules and a higher proportion of long chains of amylopectin. Type C crystals are a mixture of types A and B and are present in legume starches (Gonzalez & Wang, 2023). Acid-hydrolyzed starch has been used for many years in various industries. In the food industry, it has been applied as gelling agent in the production of gummy candies and cheese bread; as a fat substitute for low-fat mayonnaise, margarine, and ice cream; and as resistant starch powder for biscuits with low digestibility (Wang & Copeland, 2015).

11.3.1 Thickening Properties

The use of acid for starch treatment induces alterations in its structure, morphology, and functionality. The degree of modification varies depending on the type of acid utilized, treatment duration, starch type, and its source. Acid hydrolysis causes the semi-crystalline growth rings of starch to gradually decrease in thickness toward the periphery, while the amorphous growth rings have a more uniform width (Wang & Copeland, 2015).

The thickening properties of acid-modified starch are influenced by the structural changes induced by acid hydrolysis. Acid modification alters the molecular composition and arrangement of starch, leading to enhanced thickening capabilities in certain applications (Sun et al., 2019).

This effect in modified starch allows for the creation of a thicker starch paste, proving useful in the production of certain foods. The viscosity of acid-hydrolyzed starch paste increases as the shear rate rises, showing a direct proportionality. Consequently, modified starches behave as non-Newtonian fluids. This is attributed to rotational shear causing starch clusters to disintegrate, increasing the overall starch concentration. Numerous studies demonstrate that starch suspensions exhibit a high shear rate, leading to an increase in viscosity within a short time frame. This phenomenon arises from the formation of shear-induced structures, suggesting that differences in the molecular structures of amylopectin are linked to this behavior (Thakur et al., 2021).

One notable characteristic is the improved viscosity of acid-modified starch in aqueous solutions. The shortened starch chains, coupled with increased water absorption, contribute to the formation of a thicker and more viscous consistency in liquid mixtures. Additionally, acid-modified starches may display shear-thinning behavior, making them suitable for applications requiring easy mixing, pouring, or dispensing (Li et al., 2020).

These starches often demonstrate stability under varying temperatures, making them suitable for food and industrial processes with temperature fluctuations. Their compatibility with acidic ingredients further broadens their utility, allowing for use in a variety of food and beverage formulations. In the food industry, acid-modified starch is commonly employed for its thickening properties in the formulation of sauces, soups, gravies, and other liquid-based products, contributing to the desired consistency. The degree of acid hydrolysis and the choice of acid can be adjusted to tailor the thickening characteristics, providing versatility for different applications (Tian et al., 2022).

11.3.2 Gelling Properties

Hydrolyzed starches offer certain superior properties as an alternative to gelatin, although they do not exhibit the same gel-forming characteristics useful for instant gelation at room temperature (Li et al., 2018). Acid hydrolysis induces granular structural modifications in starch, resulting in different behaviors upon heating in water. This process produces low-viscosity dough, enhances gel strength, improves water solubility, and enhances starch film-forming capabilities (Wang & Copeland, 2015).

The acidification of starch leads to higher viscosity (0.3314 1/s to native starch and 0.1585 1/s to waxy starch) in hot paste for native starches compared to waxy starch. The lower viscosity in waxy starch is attributed to the reduction in molecular weight caused by acid hydrolysis. Native starch, after acidification, tends to form more stable gels (Ulbrich et al., 2020). The highest viscosity in acidified native starch is achieved under milder hydrolysis conditions, such as using 0.3 M HCl for 4 hours. Ulbrich et al. (2020) found that increased gel hardness is related to reduced molecular weight. They suggest that there is no identifiable optimal degree of molecular degradation that provides greater mechanical gel strength, emphasizing the need for individualized testing to determine ideal hydrolysis conditions for each starch type.

Jiang et al. (2018) observed that starch viscosity gradually decreases with increasing acid concentration in hydrolysis. The viscosity of acidified starch can be adjusted within the range of 292 to 85 cP by varying the acid concentration from 0.25 to 1.5 M, respectively. Starch acidified with a lower acid concentration (0.25 M) demonstrated a viscosity of 292 cP, whereas the starch acidified with a higher concentration (1.50 M) showed a reduced viscosity of 85 cP. The reduction in viscosity is attributed to the destruction of the molecular structure of the non-crystalline region of starch due to acid attack, resulting in lower polymerization and viscosity. Starch retrogradation is primarily related to amylose molecules. During cooling, amylose molecules reorganize through hydrogen bonding, leading to increased retrogradation viscosity. Acid hydrolysis likely disrupts the original amylose structure, reducing starch's ability to reform the gel during cooling. The results indicate that the degree of acid hydrolysis affects the adhesive properties of diluted starch/XG acid mixture.

11.3.3 Emulsification Capabilities

Emulsion capacity refers to the ability to generate an emulsion, allowing the creation of a solution with aqueous and lipid phases, substances that naturally do not mix. Emulsions play a crucial role in various products, such as foods, creams, and skin lotions (McClements & Jafari, 2018).

The emulsification properties and structural characteristics of starch can be enhanced through acid or enzymatic hydrolysis. Acid hydrolysis, known for influencing starch viscosity and gelatinization behavior, is employed in the food industry to obtain oil-in-water emulsions (e.g., seasonings) or water-in-oil emulsions (e.g., margarine) (Hong et al., 2017). The starch granule size plays a pivotal role in forming stable emulsions. Smaller particle sizes contribute to more stable emulsions. Starches from quinoa, oats, and rice exhibit good emulsion properties, while sources like potatoes and sugarcane have large granule starches that require treatments like acid hydrolysis to enhance their physicochemical properties. Certain starch types, such as wheat and barley, are considered bimodal, possessing both large granular (Type A) and small granular (Type B) fractions (Saari et al., 2016).

Increasing the amount of starch reduces the size of oil droplets in the emulsion, but there is a limit to this effect. It can be observed that in emulsions, the droplet sizes can range from 64 (with 36 mg of starch added per mL of oil) to 9.9 µm (3600 mg of starch added per mL of oil) (Rayner et al., 2012). According to Saari et al. (2016), the emulsification index (EI) of quinoa, barley, corn, and potato starches increased with acid hydrolysis time compared to untreated starches. In the case of quinoa starch, the EI value for untreated starch was 0.48, while the same starch, after undergoing 5 days of acid hydrolysis treatment, exhibited an increased EI value of 0.64. The EI values for untreated barley, corn, and potato starches were 0.45, 0.37, and 0.43, respectively. However, these values increased after 5 days of acid hydrolysis to 0.55, 0.46, and 0.53, respectively. Overall, there was an approximately 23% increase in IE after 5 days of treatment. Although both starches from different sources increased the emulsification index with hydrolysis, the degree of EI enhancement varied among starches. Bimodal or larger granule starches, such as barley and potato, showed a smaller increase. As mentioned earlier, the starch granule shape influences emulsification capacity. A study by Saari et al. (2016) demonstrated that, despite barley being a bimodal starch, its round shape and smoother surface favored higher emulsification capacity compared to quinoa and oat starches, which have polyhedral granules with sharp tips.

11.3.4 Solubility and Viscosity

Acid hydrolysis primarily occurs in the amorphous regions of starch, corroding the crystalline regions as the reaction progresses. This results in a significant reduction in starch molecular weight, accompanied by the release of free aldehyde groups, leading to a decrease in viscosity (Liang et al., 2024a). Starches modified by acid generally exhibit Newtonian behavior, potentially offering excellent processing properties (Abdorreza et al., 2012). The solubility of starch is a critical factor in determining its processing performance. Hydrolysis-induced degradation improves starch solubility, primarily due to the broken chains and fragments being more soluble during heating. Acid hydrolysis affects the starch molecule by breaking glycosidic linkages, altering its structure. Once the amylopectin structure is disrupted, a complete network cannot form, and the impaired chains tend to dissolve immediately, as they can no longer retain water (Liang et al., 2024a).

Starch has low solubility in water at room temperature. When subjected to acid hydrolysis, starches exhibit a relatively similar solubility rate, slightly higher, with both cases reaching a solubility rate of around 20%. When hydrolysis is combined with another treatment, such as heat treatment, its solubility can increase up to 90%, depending on the heating time and acid concentration used (Jin & Xu, 2020).

Starch dissolution involves two stages. Firstly, small-sized, rapidly moving solvent molecules preferentially diffuse into polymers, leading to polymer swelling with increased volume. Subsequently, polymer chains are dispersed molecularly in the solvent. Therefore, structural characteristics, including chain length, molecular weight/size, amylose content, and relative crystallinity, play a crucial role in the water solubility of rice starch (Jin & Xu, 2020).

11.3.5 Structure and Microstructure Properties

Observing morphology is a crucial aspect to assess the degree of gelatinization on the starch surface. Analyzing starch granules can provide a more intuitive understanding of the effects of surface gelatinization treatment and esterification on them (Liang et al., 2024b). When starch granules are examined through electron microscopy, fissures and deformations on the granule surface can be observed. These imperfections arise when starch granules rupture, and the crystalline network on the surface disappears, a consequence of erosion caused by acid hydrolysis. Native sorghum starch exhibits spherical and irregular polyhedra with relatively smooth and flat surfaces, although it has some fractured granules. When this starch undergoes acid hydrolysis, even though the integrity of the granules is maintained, the reaction leads to the emergence of a rough and porous appearance on the granule surface, indicating the presence of corroded granules due to acid treatment (Liang et al., 2024a). Approximately after 12 days of hydrolysis, the granular structure of starch is severely altered, with each granule adhering to others to form a continuous network structure (Soler et al., 2020).

Acid modification alters the amylose content in starch, changing its structure and influencing its functional properties. However, the extent of modification varies based on both the acid concentration and hydrolysis time. In rice starch subjected to mild acid hydrolysis, the amylose content and the conformation of the branched chain of amylopectin showed minimal variation compared to native starch, indicating insignificant changes in starch morphology. Weak hydrolysis of amylose occurred because hydrolysis predominantly took place at the branching points of amylopectin, resulting in little alteration in amylose content. Despite the inconspicuous modification in starch morphology, electron microscopy reveals a slightly rougher surface as hydrolysis time and/or acid level increase (Gonzalez & Wang, 2023).

Starch typically has a granular structure composed of both amylose and amylopectin molecules. Acid hydrolysis breaks down these structures into smaller components, such as glucose units. The microstructure of starch is intricately linked to its functional properties. Acid hydrolysis can alter the branching patterns in amylopectin, affecting the overall architecture of the starch granule (Bertoft, 2017).

Figure 11.1 presents the electron micrograph of corn starch-modified with lactic acid. It is possible to observe that there are superficial changes in the microstructure of the starch.

FIGURE 11.1   The electron micrograph of corn starch-modified with lactic acid.

11.3.6 Thermal Properties

Thermogravimetric analysis (TGA) is a technique used to study the thermal decomposition of a material as a function of temperature. In the case of acid-modified starch, TGA can provide valuable information about its thermal stability, decomposition temperature, and mass loss (Tian et al., 2011). In this context, the Figure 11.2 presents a typical TGA chart for native and 15% lactic acid modified cassava starch.

In a TGA analysis it is possible to obtain information about the events related to the thermal decomposition of starch. Initially, the first event is related to dehydration, which is characterized by weight loss at temperatures close to 100°C, followed by the loss of volatile compounds between 100°C and 200°C. Subsequently, it is possible to observe a certain stability in weight loss. A rapid decomposition of starches occurs between 280°C and 380°C. However, starch modified with 15% lactic acid presents an earlier decomposition, compared to native starch, due to the structural changes caused by the acid treatment. According to Zuo et al. (2017), this behavior can be explained by the destruction of the double helix structure caused by the oxidation of acid hydrolysis, which decreases the initial starch pyrolysis temperature.

The thermal properties of acid-hydrolyzed starch vary depending on the acid concentration and hydrolysis time. In a study evaluating the impact of acidification on sago palm starch (*Metroxylonsagu*), Abdorreza et al. (2012) observed an increase in gelatinization temperature from 67°C to 72.3°C compared to native starch. The gelatinization temperature rises with prolonged hydrolysis time.

Another investigation by Gonzalez and Wang (2023) explored the combined effects of acid hydrolysis and heat treatment on the thermal properties of corn, potato, and pea starches. Acid hydrolysis resulted in increased gelatinization temperatures, enthalpy of gelatinization, and relative crystallinity. The gelatinization temperature of corn starch ranged from 65.7°C (for native starch) to 84.4°C (acid modification). In the case of potato starch, the gelatinization enthalpy varied from 21.0 to 24.2 J/g, and the angle associated with relative crystallinity increased from 35.5° to 43.1°. However, susceptibility to α-amylase varied with starch type and the degree of hydrolysis. The acid preferentially hydrolyzes amorphous regions, leading to the formation of more linear chains with higher hydrolysis levels. Corn starch exhibited a higher hydrolysis rate, achieving a 2% hydrolysis level in two 2 days, due to its inferior crystalline structure with shorter A-type chains and less organized crystals derived from branching points within the crystalline region. Potato and pea starches showed an increased hydrolysis rate, indicating an interaction between branching point distribution, open structure, and the presence of more water molecules in B-type crystal units. After 4 days of hydrolysis, pea starch exhibited a lower level of hydrolysis, demonstrating a hydrolysis rate of 2%, whereas corn starch, during the same period, exhibited a hydrolysis rate of 3.5%,

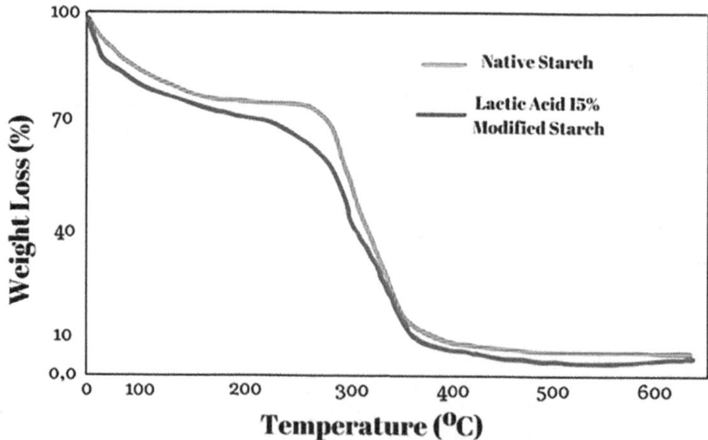

FIGURE 11.2   TGA analysis of native cassava starch and 15% lactic acid modified cassava starch. (Adapted from Zambelli et al., 2018.)

attributed to its higher amylose content compared to corn and potato starches. The presence of crystallized amylose within the granules contributed to the lower hydrolysis rate of pea starch after 4 days. As hydrolysis progressed, the amylose fraction significantly decreased for all studied starches, with potato and pea starches experiencing the most substantial reductions, having less than 1% amylose present after 9% acid hydrolysis. With an increasing degree of hydrolysis, both gelatinization temperature and enthalpy of gelatinization increased, indicating higher energy requirements to break the granular structure and form a gel. For instance, in the case of pea starch with 0% hydrolysis, the gelatinization temperature and enthalpy are 76.2°C and 10.5 J/g, respectively. Conversely, with a 9% hydrolysis rate, these values increase to 84.2°C and 17.4 J/g. Notably, higher acid concentrations in the process made the starch more thermally stable, requiring a higher temperature for gel formation. This stability is advantageous for the industry, providing better control over the storage and processing of products utilizing starch in their formulations. Combining acid hydrolysis with other treatments, such as heat-moisture treatment, has shown synergistic effects in forming thermally stable starch crystals, offering opportunities for further enhancement.

The enthalpy of gelatinization is indicative of the order and extension of the starch double helix, while the gelatinization temperature represents the perfection of these double helices. In general, the enthalpy tends to increase with higher levels of hydrolysis for modified starches. This suggests that as hydrolysis progresses, there is a greater disruption of the starch granular structure, leading to an increase in the enthalpy of gelatinization. The rise in enthalpy may be attributed to the breakdown of crystalline regions and the formation of more amorphous structures, requiring additional energy for gelatinization. Therefore, the combination of elevated enthalpy and changes in gelatinization temperature provides valuable insights into the structural modifications induced by the hydrolysis process in starches (Fonseca et al., 2021).

11.4 OPTIMIZATION OF ACID HYDROLYSIS

Acid hydrolysis is a classical method used to generate crystalline starch particles with micro- or nano-scale dimensions (Angellier et al., 2004; Kumari et al., 2022). Although the efficiency of this starch acidolysis can be enhanced by increasing temperature or acid concentration, it sometimes fails to meet practical requirements (Babu et al., 2015).

An example is the low reactivity of acid hydrolysis, resulting in a prolonged nanoparticle preparation period (5–15 days) (Li et al., 2018). Time consumption has been a major issue limiting larger-scale production (Dai et al., 2019). Hence, the low efficiency and extended reaction time of the hydrolysis process remain challenges to be addressed. In response, the industry seeks optimization methods for starch hydrolysis under mild conditions, such as diluted acid media, lower temperatures (Wang et al., 2022), or higher temperatures, employing different acids combined with the response surface methodology (Angellier et al., 2004).

However, combined methods are typically utilized; thus, along with acid hydrolysis, techniques like ultrasonication (Kim & Huber, 2013), heat-moisture treatment (HMT) (Xing et al., 2017), hydrothermal treatment (Kariuki et al., 2022), succinylation (Mehbob et al., 2015), annealing (Nakazawa & Wang, 2003), enzymes (Lecorre et al., 2011), electric field treatment (Li et al., 2022), and ball-milling pretreatment (Dai et al., 2018), among others, are employed.

The use of two methods proves efficient in achieving functionalities that cannot be attained through a single method (Chavez-Esquivel et al., 2022). Generally, dual modification of starch involves a two-step process, either homogeneous or heterogeneous. On the one hand, the stage is homogenous when methods of the same type are employed, such as dual physical/physical, chemical/chemical, and enzymatic/enzymatic modifications. On the other hand, the stage is heterogeneous when different types are used, for example, physical/chemical, chemical/physical, physical/enzymatic, or vice versa (Ashogbon, 2021). However, the sequence of applying these methods significantly affects the final properties of starch (Khurshida, 2021). The sequence of method application can either benefit or hinder the internal components of starch, thereby facilitating or limiting the effectiveness of the subsequent method. This is evident in cross-linked starches, which lack optimal attributes for freeze–thaw stability with delayed retrogradation. Thus, a dual modification becomes necessary to address these limitations (Chavez-Esquivel et al., 2022). The combined treatment of starch-based materials or native starches is employed to alter further the final properties of these biomaterials, including swelling capacity, crystallinity percentage, microstructure, and other crucial factors in starch (Wang et al., 2020).

It is well known that starch nanoparticles obtained through classical acid hydrolysis are limited for practical use due to a strong tendency to aggregate, especially in the form of dry powders (Dufresne, 2008). The hydrolysis process typically requires a period longer than 5 days, and the yield of preparation from native starch is low, usually less than 20%. Hence, research is conducted to produce starch nanoparticles by applying physical treatments.

An example is the combined process of acid hydrolysis for a short period (2 days at 40°C) followed by easy ultrasonication (60% amplitude, 3 min), which has shown a reduced preparation time and prevented the aggregation of starch nanoparticles. However, ultrasonication disrupted the crystalline structure of starch, and the recovery yield of crystalline nanoparticles prepared exclusively by acid hydrolysis remained low (less than 30%). At low temperatures (40°C–60°C), acid hydrolysis followed by gentle ultrasonic treatment can induce a breakdown at the midpoint of amylose and amylopectin, producing crystalline nanoparticles from waxy corn starch (Kim et al., 2013).

Studies carried out by Chavez-Esquivel et al. (2022) studied the modification of corn starch by combining citric acid treatment and ultrasonication for 1 to 18 days, with the non-combined method, for comparison, designated as single modification (CSD-SM) and double modification (CSD-DM), respectively. The authors found that the synergistic effect of double modification through the sequential combination of chemical and physical treatments improved hydrolysis advancement by 12%, relative crystallinity by 10%, and double helix promotion by 25% during the 18-day combined treatment.

In general, the morphology of CSD-DM was modified due to acid hydrolysis and ultrasonication, exhibiting a loss in granular structure related to gelation, swelling, and hydrolysis-ultrasonication caused by water and carboxylic acid. The surface erosion of corn starch granules (CSGs) formed pores and channels that facilitated the migration from the continuous phase to the dispersed phase of the granule over time. This phenomenon can be advantageous in starch-based materials with synthetic polymers, enhancing and improving the interfacial adhesion of CSGs (Garg & Jana, 2011; Chavez-Esquivel et al., 2022).

A study using low-temperature acid hydrolysis combined with sonication-assisted treatment demonstrated a high yield (78%) of starch nanoparticles (SNCs) (Kim & Huber, 2013). Similarly, the combined treatment of corn starch using ultrasound and sulfuric acid hydrolysis was able to produce SNCs in 45 minutes with a yield of up to 21.6% and crystallinity below 40% (Amini & Ravazi, 2016).

Ultrasonication increased the nanoparticle yield and prevented starch nanoparticle aggregation during acid hydrolysis of waxy corn starch. Kim and Huber (2013) demonstrated a method for producing crystalline starch nanoparticles from waxy corn starch by combining low-temperature acid hydrolysis and ultrasonication. Under refrigerated conditions, acid hydrolysis was used to obtain starch hydrolysates without disturbing crystallinity. The recovered crystalline hydrolysates, precipitated through centrifugation, were then redispersed in water and subjected to ultrasonic treatment, effectively transforming them into nanoparticles. The study found that even after ultrasound treatment, crystallinity could be fully recovered from native starch to nanoparticles. The authors concluded that the combination of low-temperature acid hydrolysis and ultrasonic treatment is an effective process for the large-scale production of crystalline starch nanoparticles.

There is also thermal-acid treatment, which acts as a pretreatment, combining hydrothermal treatment and acid hydrolysis (Xing et al., 2017). Hydrothermal treatment, in brief, involves heating an aqueous starch suspension, generating a fluid comprising a mixture of porous, gelatinized, and swollen granules with an amylopectin skeleton in an amylose solution. Upon cooling, gel networks form, converting random amylose spirals into a crystalline structure. This combined treatment reduces granular size, increases crystallinity and thermal stability, also presenting a more homogeneous surface and color, with no observed adverse effects on properties (Kariuki et al., 2022).

Although acid hydrolysis is widely used to modify granular starch, it is often conducted at temperatures below 55°C for an extended period to avoid crystallinity loss. However, hydrolysis performed at a relatively high temperature offers advantages such as increased hydrogen ion activity and enhanced swelling of the amorphous starch structure, which can improve reactive dynamics (Wang & Copeland, 2015; Liu et al., 2022). Therefore, the behavior of high-temperature acid hydrolysis (HTAH) was tested on heat and moisture-treated (HMT) starch at 69°C, resulting in improved crystalline structure of the starch due to HMT (Shang et al., 2020).

HMT is a promising physical modification method involving a safe and environmentally friendly hydrothermal process (Xing et al., 2017). During HMT, starch molecules are partially destroyed and undergo chain rearrangement, leading to improved functional characteristics of starch granules, such as higher gelatinization temperature, stable paste viscosity, and restricted swelling (Wang & Zheng, 2021; Ambigaipalan et al., 2014).

11.5 APPLICATIONS OF ACID-HYDROLYZED STARCHES

Acid-hydrolyzed starches have versatile applications across multiple industries. In the food sector, they serve as effective thickeners and gelling agents, enhancing texture and stability in products like sauces and desserts. The pharmaceutical industry benefits from their use as excipients in tablet formulations, providing crucial binding and disintegration properties. Additionally, acid-hydrolyzed starches play a vital role in the paper and textile industries, improving sizing, coating, and adhesion. They are integral components in adhesives and glues, contributing to superior bonding characteristics. These starches also find application in the development of biodegradable films and coatings, offering eco-friendly alternatives for packaging materials. In the chemical industry, they serve as raw materials to produce various chemicals and bio-based polymers (Yazid et al., 2018).

Furthermore, acid-hydrolyzed starches play a role in enhanced oil recovery processes in the oil and gas industry as effective thickeners. Their usage extends to textile sizing, where they enhance fiber binding during weaving. In environmental applications, these starches contribute to wastewater treatment processes by facilitating flocculation and sedimentation. The broad spectrum of applications underscores the adaptability and value of acid-hydrolyzed starches in diverse industrial processes (López et al., 2017).

11.5.1 Packaging Applications

Acid-hydrolyzed starches find notable applications in the field of packaging due to their advantageous properties. These modified starches, resulting from acid hydrolysis, exhibit enhanced functionalities that make them valuable in sustainable and eco-friendly packaging solutions (Zhang et al., 2019).

Some key applications in packaging include:

1. Biodegradable films and coatings.
2. Enhanced barrier properties.
3. Adhesive properties.
4. Sustainable packaging solutions.
5. Reduced environmental impact.
6. Compatibilities with other polymers.

Acid-hydrolyzed starches can be utilized in the development of biodegradable films and coatings for packaging materials. These materials offer a sustainable alternative to traditional packaging, contributing to environmental conservation by reducing plastic waste; in this context, pea starch granules partially hydrolyzed with 2% hydrochloric acid for 0.5 to 2.5 hours were applied in biodegradable film formulations. After treatment, it was found that the tensile strength of films increased significantly, while the elongation at break showed the opposite behavior. In this case, acid hydrolysis modified the amylopectin molecules into low molecular weight amylose and starch molecules, which can provide a more viscous flow in the film-forming solution and reinforce the polymer matrix. As the hydrolysis time increased, the starch molecules became smaller to form a tangle between the polymer chains, which facilitates mobility, with a hydrolysis time of 2.0 hours being ideal for improving the mechanical properties of pea starch–based films (Zhang et al., 2019).

In a study conducted by Mukurumbira et al. (2017), Amadumbe (*Colocasia esculenta*) starch films were innovatively reinforced with Amadumbe starch nanocrystals produced from acid hydrolysis with sulfuric acid (3.00 M). The researchers observed a substantial increase in tensile strength at concentrations up to 2.5%. However, beyond this threshold, the aggregation of nanocrystals had a detrimental effect. Amadumbe starch, characterized by 7% amylose content and small particles, demonstrated a high yield in nanocrystal production. Interestingly, when compared to potato starch films reinforced with Amadumbe starch nanocrystals in the same study, the tensile strength values were significantly lower. This was attributed to the formation of a rigid three-dimensional network between the Amadumbe starch nanocrystals and the native film.

Acid modification with 3.0% lactic acid was carried out on sorghum starch for application in biodegradable films, which showed greater tensile strength and stiffness, as well as lower elongation. The increase in the concentration of sorghum starch in the film-forming formulations led to an increase in thickness and permeability to water vapor. No distinction was observed in the water solubility between films made from native starch and those modified. This lack of difference may be attributed to the mild nature of the starch modifications, which also induce subtle alterations in the thermal properties and crystallinity of the starches (Biduski et al., 2017).

Yam starch was used to produce bioplastic to be used as environmentally friendly packaging. Acid modification through citric acid (0.5, 1 and 1.5%) was applied. It was observed that the addition of citric acid promoted starch hydrolysis through better crystalline rearrangement due to increased molecular mobility. The films developed were colorless, translucent, flexible, and smooth in appearance. However, the swelling capacity was lower in films with higher concentrations of citric acid. This behavior can be explained by the fact that the OH groups of water are occupied by citric acid in the films, leaving smaller groups and OH free to form bonds with the water. In general, the mechanical properties of the films were improved with the incorporation of citric acid, especially in a proportion of 1.0%, which obtained 24.69 MPa in the tensile strength analysis (Poudel et al., 2023).

11.5.2 Bakery Applications

Acid-hydrolyzed starch is a valuable and versatile ingredient in bakery applications, contributing to numerous improvements in the quality and characteristics of baked goods. Through the modification of starch molecular structures, acid hydrolysis enhances the functionality of starch, particularly in terms of texture, moisture retention, and overall product quality. This modification results in increased water absorption, aiding in the retention of moisture during the baking process and contributing to extended freshness and improved softness in the final products (Park et al., 2018; Compart et al., 2023). Beyond its contributions to texture, moisture retention, and overall product quality, this modified starch enhances freeze–thaw stability, making it ideal for frozen doughs and par-baked goods. The improved emulsion stability adds value to fillings, icings, and cream-based components (Woo et al., 2021).

Moreover, acid-hydrolyzed starch serves as an effective thickening and stabilizing agent in bakery formulations, ensuring better consistency and appearance. The modified properties also play a role in extending the shelf life of bakery products, offering benefits such as improved moisture retention and texture preservation over time. Acid-hydrolyzed starch is especially useful in gluten-free and clean label formulations, providing an alternative for individuals with gluten sensitivities and aligning with the demand for natural and recognizable ingredients (Zhang et al., 2016; Horstmann et al., 2017).

The gel formation facilitated by acid hydrolysis proves advantageous in certain bakery applications, contributing to the structure, texture, and mouthfeel of baked goods. The modified starch also positively influences crumb structure, resulting in softer and more desirable textures in various baked items like bread, cakes, and pastries. Furthermore, acid hydrolysis helps mitigate starch retrogradation, reducing the staling process in baked goods and contributing to the maintenance of freshness and quality for a more extended period (Miyazaki et al., 2006).

One of its notable advantages lies in the customization of rheological properties, providing bakers with precise control over viscosity, elasticity, and flow characteristics for tailored textures and mouthfeel. In the realm of gluten-free formulations, acid-hydrolyzed starch becomes a valuable ally, overcoming challenges associated with gluten-free products' texture, structure, and moisture retention (Cappelli et al., 2020).

11.6 CONCLUSION

In conclusion, the process of acid hydrolysis of starch represents a fundamental approach to the controlled modification of this crucial biomolecule. Subjecting starch to acidic conditions leads to the cleavage of glycosidic bonds, resulting in alterations in both surface characteristics and microstructure. These modifications bear significant implications for functional properties, including improvements in solubility, viscosity, and gelation capacity.

The observed microstructural changes have a direct impact on the practical applications of acid-hydrolyzed starch, demonstrating broad utility in sectors such as the food and pharmaceutical industries. Understanding the intricacies of this process is essential for unlocking the full potential of modified starch and developing tailored formulations to meet the specific demands of various industrial applications. Furthermore, acid hydrolysis not only enhances the functional attributes of starch but also opens avenues for innovation in material science, biotechnology, and beyond.

The versatility of acid-modified starch extends its applicability to diverse fields, showcasing its adaptability and potential for addressing evolving industry needs. This transformative process underscores not only the complexity but also the relevance of starch acid hydrolysis in producing biomaterials with finely tuned properties, laying the foundation for advancements in multiple sectors.

REFERENCES

Abdorreza, M. N., Robal, M., Cheng, L. H., Tajul, A. Y., & Karim, A. A. (2012). Physicochemical, thermal, and rheological properties of acid-hydrolyzed sago (Metroxylonsagu) starch. *LWT—Food Science and Technology, 46,* 135–141.

Ambigaipalan, P., Hoover, R., Donner, E., & Liu, Q. (2014). Starch chain interactions within the amorphous and crystalline domains of pulse starches during heat-moisture treatment at different temperatures and their impact on physicochemical properties. *Food Chemistry, 143,* 175–184.

Amini, A. M., & Razavi, S. M. A. (2016). A fast and efficient approach to prepare starch nanocrystals from normal corn starch. *Food Hydrocolloids, 57,* 132–138.

Angellier, H., Choisnard, L., Molina-Boisseau, S., Ozil, P., & Dufresne, A. (2004). Optimization of the preparation of aqueous suspensions of waxy maize starch nanocrystals using a response surface methodology. *Biomacromolecules, 5,* 1545–1551.

Ashogbon, A. O. (2021). Dual modification of various starches: Synthesis, properties and applications. *Food Chemistry, 342,* 128–139.

Babu, A. S., Parimalavalli, R., & Rudra, S. G. (2015). Effect of citric acid concentration and hydrolysis time on physicochemical properties of sweet potato starches. *International Journal of Biological Macromolecules, 80,* 557–565.

Bertoft, E. (2017). Understanding starch structure: Recent progress. *Agronomy*, 56–72.

Biduski, B., da Silva, F. T., da Silva, W. M., El Halal, S. L. M., Pinto, V. Z., Dias, A. R. G., & Zavareze, E. R. (2017). Impact of acid and oxidative modifications, single or dual, of sorghum starch on biodegradable films. *Food Chemistry*, *214*, 53–60.

Cappelli, A., Oliva, N., & Cini, E. (2020). A systematic review of gluten-free dough and bread: Dough rheology, bread characteristics, and improvement strategies. *Applied Sciences*, *10*, 6559–65578.

Chavez-Esquivel, G., Cervantes-Cuevas, H., & Vera-Ramírez, M. A. (2022). Effect of dual modification with citric acid combined with ultrasonication on hydrolysis kinetics, morphology and structure of corn starch dispersions. *International Journal of Biological Macromolecules*, *222*, 1688–1699.

Chen, P., Zhang, Y., Qiao, Q., Tao, X., Liu, P., & Xie, F. (2021). Comparison of the structure and properties of hydroxypropylated acid-hydrolysed maize starches with different amylose/amylopectin contents. *Food Hydrocolloids*, *110*, 92–101.

Chen, Z., Yang, Q., Yang, Y., & Zhong, H. (2023). The effects of high-pressure treatment on the structure, physicochemical properties and digestive property of starch—a review. *International Journal of Biological Macromolecules*, *244*, 376–380.

Compart, J., Singh, A., & Fettke, J. (2022). Customizing starch properties: A review of starch modifications and their applications. *Polymers*, *15*, 3491–3504.

Compart, J., Singh, A., Fetkke, J., & Apriyanto, A. (2023). Customizing starch properties: A review of starch modifications and their applications. *Polymers*, *15*, 3491–3502.

Cordenunsi-Lysenko, B. R., Nascimento, J. R. O., Castro-Alves, V. C., Purgatto, E., Fabi, J. P., & Peroni-Okyta, F. H. G. (2018). The starch is (not) just another brick in the wall: The primary metabolism of sugars during banana ripening. *Frontiers in Plant Science*, *10*, 83–92.

Dai, L., Li, C., Zhang, J., & Cheng, F. (2018). Preparation and characterization of starch nanocrystals combining ball milling with acid hydrolysis. *Carbohydrate Polymers*, *180*, 122–127.

Dai, L., Zhang, J., & Cheng, F. (2019). Succeeded starch nanocrystals preparation combining heat-moisture treatment with acid hydrolysis. *Food Chemistry*, *278*, 350–356.

Dufresne, A. (2008). Polysaccharide nano crystal reinforced nanocomposites. *Canadian Journal of Chemistry*, *86*, 484–494.

Fonseca, L. M., El Halal, S. L. M., Dias, A. R. G., & Zavareze, E. R. (2021). Physical modification of starch by heat-moisture treatment and annealing and their applications: A review. *Carbohydrate Polymers*, *274*, 118–130.

Garg, S., & Jana, A. K. (2011). Characterization and evaluation of acylated starch with different acyl groups and degrees of substitution. *Carbohydrate Polymers*, *83*, 1623–1630.

Gonzalez, A., & Wang, Y.-J. (2023). Effects of acid hydrolysis level prior to heat-moisture treatment on properties of starches with different crystalline polymorphs. *LWT—Food Science and Technology*, *187*, 302–310.

He, R., Li, S., Zhao, G., Zhai, L., Qin, P., & Yang, L. (2023). Starch modification with molecular transformation, physicochemical characteristics, and industrial usability: A state-of-the-art review. *Polymers*, *15*, 2935–2949.

Hofman, D. L., van Buul, V. J., & Brouns, F. J. P. H. (2016). Nutrition, health, and regulatory aspects of digestible maltodextrins. *Critical Reviews in Food Science & Nutrition*, *56*, 2091–2100.

Hong, Y., Li, Z., Gu, Z., Wang, Y., & Pang, Y. (2017). Structure and emulsification properties of octenyl succinic anhydride starch using acid-hydrolyzed method. *Starch*, *69*, 160–172.

Horstmann, S. W., Lynch, K. M., & Arendt, E. K. (2017). Starch characteristics linked to gluten-free products. *Foods*, *6*, 29–36.

Jiang, M., Hong, Y., Gu, Z., Cheng, L., Li, Z., & Li, C. (2018). Effects of acid hydrolysis intensity on the properties of starch/xanthan mixtures. *International Journal of Biological Macromolecules*, *106*, 320–329.

Jin, Q., & Xu, X. (2020). Microstructure, gelatinization and pasting properties of rice starch under acid and heat treatments. *International Journal of Biological Macromolecules*, *149*, 1098–1108.

Kariuki, P. N., Arjunan, Y., Nagarajan, U., & Kanth, S. V. (2022). The combined effect of thermal-acid hydrolysis, periodate oxidation, and iodine species removal on the properties of native tapioca (*Manihot esculenta Crantz*) starch. *International Journal of Macromolecules*, *196*, 107–119.

Khlestkin, V. K., Peltek, S. E., & Kolchanov, N. A. (2018). Review of direct chemical and biochemical transformations of starch. *Carbohydrate Polymers*, *181*, 460–476.

Khurshida, S., Das, M. J., Deka, S. C., & Nandan, S. (2021). Effect of dual modification sequence on physicochemical, pasting, rheological and digestibility properties of cassava starch modified by acetic acid and ultrasound. *International Journal of Macromolecules*, *188*, 649–656.

Kim, J. Y., & Huber, K. C. (2013). Heat-moisture treatment under mildly acidic conditions alters potato starch physicochemical properties and digestibility. *Carbohydrate Polymers*, *98*, 1245–1255.

Kumari, S., Yadav, B. S., & Yadav, R. (2022). Characterization of acid hydrolysis based nano-converted mung bean (*Vigna radiata* L.) starch for morphological, rheological and thermal properties. *International Journal of Biological Macromolecules*, *211*, 450–459.

Lecorre, D., Bras, J., & Dufresne, A. (2011). Evidence of micro-and nanoscaled particles during starch nanocrystals preparation and their isolation. *Biomacromolecules, 12,* 3039–3046.

Li, C., & Hu, Y. (2021). Effects of acid hydrolysis on the evolution of starch fine molecular structures and gelatinization properties. *Food Chemistry, 353,* 129–136.

Li, H., Li, J., & Guo, L. (2020). Rheological and pasting characteristics of wheat starch modified with sequential triple enzymes. *Carbohydrate Polymers, 230,* 667–674.

Li, L., Hong, Y., Gu, Z., Cheng, L., Li, Z., & Li, C. (2018). Synergistic effect of sodium dodecyl sulfate and salts on the gelation properties of acid-hydrolyzed-hydroxypropylated potato starch. *LWT—Food Science and Technology, 93,* 556–562.

Li, M., Wang, J., Wang, F., Wu, M., Wang, R., Strappe, P., Blanchard, C., & Zhou, Z. (2022). Insights into the multiscale structure of wheat starch following acylation: Physicochemical properties and digestion characteristics. *Food Hydrocolloids, 124,* 347–356.

Liang, W., Lin, Q., Zeng, J., Gao, H., Muratkhan, M., & Li, W. (2024a). Understanding the improvement of sorghum starch acid hydrolysis modification by E-beam irradiation: A supramolecular structure perspective. *Food Chemistry, 437,* 137–142.

Liang, W., Zhang, Q., Guo, S., Ge, X., Shen, H., Zeng, J., Gao, H., & Li, W. (2024b). Investigating the influence of $CaCl_2$ induced surface gelatinization of red adzuki bean starch on its citric acid esterification modification: Structure–property related mechanism. *Food Chemistry, 436,* 724–731.

Liu, X., Huang, S., Chao, C., Yu, J., Copeland, L., & Wang, S. (2022). Changes of starch during thermal processing of foods: Current status and future directions. *Trends in Food Science and Technology, 119,* 320–337.

López, O. V., Castillo, L. A., Ninago, M. D., Ciolino, A. E., & Villar, M. A. (2017). Modified starches used as additives in enhanced oil recovery (EOR). *Industrial Application of Renewable Biomass Product, 1,* 227–248.

McClements, D. J., & Jafari, S. M. (2018). Improving emulsion formation, stability and performance using mixed emulsifiers: A review. *Advances in Colloid and Interface Science, 251,* 55–79.

Mehboob, S., Ali, T. M., Alam, F., & Hasnain, A. (2015). Dual modification of native white sorghum (*Sorghum bicolor*) starch via acid hydrolysis and succinylation. *LWT—Food Science and Technology, 64,* 459–467.

Miyazaki, M., Hung, P. V., Maeda, T., & Morita, N. (2006). Recent advances in application of modified starches for breadmaking. *Trends in Food Science & Technology, 17,* 591–599.

Mukurumbira, A. R., Mellem, J. J., & Amonsou, E. O. (2017). Effects of amadumbe starch nanocrystals on the physicochemical properties of starch biocomposite films. *Carbohydrate Polymers, 165,* 142–148.

Nakazawa, Y., & Wang, Y. J. (2003). Acid hydrolysis of native and annealed starches and branch-structure of their Naegeli dextrins. *Carbohydrate Research, 338,* 2871–2882.

Park, S. H., Na, Y., Kim, J., Kang, S. D., & Park, K. H. (2018). Properties and applications of starch modifying enzymes for use in the baking industry. *Food Science and Biotechnology, 27,* 299–312.

Pfister, B., Zeeman, S. C., Rugen, M. D., Field, R. A., Ebenhoh, O., & Raguin, A. (2020). Theoretical and experimental approaches to understand the biosynthesis of starch granules in a physiological context. *Photosynthesis Research, 145,* 55–70.

Poudel, R., Dutta, N., & Karak, N. (2023). A mechanically robust biodegradable bioplastic of citric acid modified plasticized yam starch with anthocyanin as a fish spoilage auto-detecting smart film. *International Journal of Biological Macromolecules, 242,* 125–132.

Quilez-Molina, A. I., Oliveira-Salmazo, L., Amezúa-Arranz, C., Lopez-Gil, A., & Rodríguez-Pérez, M. A. (2023). Evaluation of the acid hydrolysis as pretreatment to enhance the integration and functionality of starch composites filled with rich-in-pectin agri-food waste orange peel. *Industrial Crops and Products, 205,* 117–122.

Rayner, M., Timgren, A., Sjöö, M., & Dejmek, P. (2012). Quinoa starch granules: A candidate for stabilizing food-grade Pickering emulsions. *Journal of the Science of Food and Agriculture, 92,* 1841–1847.

Saari, H., Heravifar, K., Rayner, M., Wahlgren, M., & Sjöö, M. (2016). Preparation and characterization of starch particles for use in Pickering emulsions. *Cereal Chemistry, 93,* 116–124.

Shang, J., Li, L., Zhao, B., Liu, M., & Zheng, X. (2020). Comparative studies on physicochemical properties of total, A- and B-type starch from soft and hard wheat varieties. *International Journal of Biological Macromolecules, 154,* 714–723.

Silveira, N. P., Zucatti, R., Vailatti, A. D., & Leite, D. C. (2019). Acid hydrolysis of regular corn starch under external electric field. *Journal of the Brazilian Chemical Society, 30,* 11–10.

Soler, A., Valenzuela-Díaz, E. D., Velazquez, G., Huerta-Ruelas, J. A., Morales-Sanchez, E., Hernandez-Gama, R., & Mendez-Montealvo, G. (2020). Double helical order and functional properties of acid-hydrolyzed maize starches with different amylose content. *Carbohydrate Research, 490,* 107–115.

Sun, S., Hong, Y., Gu, Z., Cheng, L., Li, Z., & Li, C. (2019). Effects of acid hydrolysis on the structure, physicochemical properties and digestibility of starch-myristic acid complexes. *LWT—Food Science and Technology, 113,* 274–282.

Thakur, Y., Thory, R., Sandhu, K. S., Kaur, M., Sinhmar, A., & Pathera, A. K. (2021). Effect of selected physical and chemical modifications on physicochemical, pasting, and morphological properties of underutilized starch from rice bean (*Vigna umbellata*). *Journal of Food Science and Technology, 58*, 4785–4794.

Tian, S., Xue, X., Wang, X., & Chen, Z. (2022). Preparation of starch-based functional food nano-microcapsule delivery system and its controlled release characteristics. *Frontiers in Nutrition, 9*, 982–992.

Tian, Y., Xu, X., & Jin, Z. (2011). Starch retrogradation studied by thermogravimetric analysis (TGA). *Carbohydrate Polymers, 84*, 1165–1168.

Ulbrich, M., Daler, J. M., &Flöter, E. (2020). Acid hydrolysis of corn starch genotypes. II. Impact on functional properties. *Food Hydrocolloids, 98*, 105–112.

Wang, M., Wu, Y., Liu, Y., & Ouyang, J. (2020). Effect of ultrasonic and microwave dual-treatment on the physicochemical properties of chestnut starch. *Polymers, 12*, 1718–1726.

Wang, Q., Li, L., & Zheng, X. (2021). Recent advances in heat-moisture modified cereal starch: Structure, functionality and its applications in starchy food systems. *Food Chemistry, 15*, 344–351.

Wang, S., & Copeland, L. (2015). Effect of acid hydrolysis on starch structure and functionality: A review. *Critical Reviews in Food Science and Nutrition, 55*, 1081–1097.

Wang, W., Xue, L., Dong, Y., Xia, Z., Liu, X., Chen, G., Yang, N., Song, W., & Du, X. (2022). Application of multistage induced electric field for acid hydrolysis of starch in a continuous-flow reactor. *International Journal of Biological Macromolecules, 221*, 703–713.

Woo, S. H., Kim, J. S., Jeong, H. M., Shin, Y. J., Hong, J. S., Choi, H. D., & Shim, J. H. (2021). Development of freeze–thaw stable starch through enzymatic modification. *Foods, 10*, 2269–2276.

Xing, J. J., Li, D., Wang, L. J., & Adhikari, B. (2017). Multiple endothermic transitions of acid hydrolyzed and heat-moisture treated corn starch. *LWT—Food Science and Technology, 81*, 195–201.

Yazid, N. S. M., Abdullah, N., Muhammad, N., & Peralta, H. M. M. (2018). Application of starch and starch-based products in food industry. *Journal of Science and Technology, 10*, 144–174.

Zambelli, R. A., Galvão, A. M. M. T., Mendonça, L. G., Leão, M. V. S., Carneiro, S. V., Lima, A. C. S., & Lemos, C. A. (2018). Effect of different levels of acetic, citric and lactic acid in the cassava starch modification on physical, rheological, thermal and microstructural properties. *Food Science and Technology Research, 8*, 747–754.

Zhang, H., Hou, H., Liu, P., Wang, W., & Dong, H. (2019). Effects of acid hydrolysis on the physicochemical properties of pea starch and its film forming capacity. *Food Hydrocolloids, 87*, 173–179.

Zhang, L. L., Ren, J. N., Zhang, Y., Li, J. J., Guo, Z. Y., Yang, Z. Y., Pan, S. Y., & Fan, G. (2016). Effects of modified starches on the processing properties of heat-resistant blueberry jam. *LWT—Food Science and Technology, 72*, 447–456.

Zhao, X., Xing, J.,An, N., Li, D., Wang, L., & Wang, Y. (2022). Succeeded high-temperature acid hydrolysis of granular maize starch by introducing heat-moisture pretreatment. *International Journal of Biological Macromolecules, 222*, 2868–2877.

Zuo, Y., Liu, W., Xiao, J., Zhao, X., Zhu, Y., & Wu, Y. (2017). Preparation and characterization of dialdehyde starch by one-step acid hydrolysis and oxidation. *International Journal of Biological Macromolecules, 103*, 1257–1264.

Acetylation/Esterification of Starch

12

Ranjan Kaushik, Ankit Kumar, Neha Rani, Rekha Phogat, and Rakesh Gehlot

12.1 INTRODUCTION

Starch, a vital carbohydrate found in various plant-based foods, boasts a complex molecular structure comprising amylose and amylopectin molecules organized into lamellae. These structures, marked by alternating crystalline and amorphous regions, influence starch functionality and behavior in diverse applications (Subroto, 2020). The abundance of hydroxyl groups on starch molecules enhances structural stability, water interaction, and reactivity, affecting properties like gelatinization and viscosity (Otache et al., 2021). Starch, sourced from grains, tubers, legumes, and millets, varies in composition and structure, impacting its versatility (Cornejo-Ramírez et al., 2018). However, inherent limitations such as brittleness, insolubility, and retrogradation necessitate modification for broader industrial use. Modifying starch optimizes its properties for various applications, addressing shortcomings and enhancing functionality in food and industrial contexts (Azeh et al., 2016).

Among the methods of starch modification, acetylation stands out as one of the most widely employed techniques. Acetylated starch is derived from the esterification process of native starch with compounds like acetic anhydride, vinyl acetate, or acetic acid (Diop et al., 2011b; Han et al., 2013). This process alters the physicochemical properties of starch, imparting desirable attributes such as improved stability, solubility, binding, adhesion, thickening, stabilizing, texturing, and film-forming ability. Factors like acetyl content, starch type, molecular components, and pretreatment method influence the properties of resulting starch acetates. Acetyl content measurement serves as a key characterization method, with a variable degree of substitution achievable. Notably, in food applications, the FDA permits only low degrees of substitution (0.01–0.2) (Bello-Pérez et al., 2000). Currently, a diverse range of acetylated starch preparations is available in the market, serving various purposes across different industries. In the food sector, they function as food additives (E 1420), influencing the consistency and stability of food products. Beyond food, these starches find applications in non-food sectors such as pharmaceuticals, textiles, and the production of biodegradable packaging materials (Golachowski et al., 2015).

12.2 MECHANISM OF ACETYLATION

Chemical modification through acetylation represents a crucial avenue for achieving the transformation. Acetylation is a type of esterification reaction; the mechanism of esterification is based on the esterification reaction. Esterification is classified under condensation reaction, which occurs through the interaction of the hydroxyl groups (-OH) of starch with carboxylic acid (RCOOH) or its derivative, typically facilitated by a catalyst (Bhargava et al., 2022). In this chemical transformation, the overarching representation for esterification is typically expressed as follows:

$$RCOOH + R'OH \rightarrow RCOOR' + H_2O$$

Where RCOOH is an organic compound having a carboxyl group (-COOH) in the molecule, R'OH is a starch molecule, RCOOR represents starch ester, and H_2O is a byproduct of the reaction.

The ester bond (RCOOR) formed during esterification is a covalent bond characterized by its stability and central to the modified structure of the starch. The efficiency of esterification is intricately tied to factors such as the inductive, mesomeric, and steric effects inherent in the structures of the reacting acids and alcohols. Evaluation of its effectiveness further hinges on crucial parameters, including reaction time, molar ratio, temperature, and the presence of a catalyst (Hong, Zeng, Buckow, et al., 2016). Esterification reactions can take various forms based on the reactants involved and the conditions under which the reaction occurs like direct esterification, acid anhydride esterification, acid chloride esterification, transesterification, Fischer esterification, and enzymatic esterification. Each type of esterification reaction has its own set of applications and considerations, making them valuable tools in organic synthesis and various industrial processes (Khan et al., 2021).

On the other hand, modification by acetylation has found extensive application in enhancing the quality and stability of starch. It is a specific type of organic esterification reaction where an acetyl group ($-COCH_3$) is introduced into a molecule. Acetylation often involves the use of acetic anhydride or acetyl chloride as reactants, leading to the acetylation of hydroxyl groups. The theoretical upper limit for the degree of substitution (DS) is three, as acetylation can replace the three available hydroxyl groups on the C-2, C-3, and C-6 positions of the anhydroglucose unit within the starch molecule. Notably, the reactivity of these free hydroxyl groups varies, with the primary hydroxyl on C6 exhibiting higher reactivity, leading to quicker acetylation compared to the secondary hydroxyls on C-2 and C-3, which experience steric hindrance (Eshag Osman et al., 2022). This process leads to the formation of acetylated starch, enhancing characteristics such as paste clarity, stability, resistance to retrogradation, and freeze–thaw stability (Kumoro et al., 2015). The general reaction for acetylation can be represented as follows:

$$\text{Starch (R-OH)} + \text{acetyl groups (CH}_3\text{CO-)} \rightarrow \text{acetylated starch (R-COCH}_3)$$

The introduction of bulky acetyl groups during acetylation leads to the degradation of starch structures and weakening of the interactions between starch molecules, resulting in the disorganization of starch multi-scale ordered structure. This steric hindrance hinders hydrogen bond formation; retards self-reassembly behavior between starch molecules; facilitates water percolation; and leads to increased swelling power and solubility, clarity, and softness of starch-based hydrogels. Moreover, the acetyl groups improve water/oil absorption capacity, emulsifiability, film-forming properties, and colonic fermentability of starch while reducing the susceptibility of starch molecules to enzymes (C. Chi et al., 2023; Punia, 2020). The success of starch acetylation can be evaluated through various parameters and characteristics that are indicative of the modification achieved. Some of the factors and methods to assess the success of starch acetylation are given in the following.

12.3 METHODS TO ASSESS STARCH ACETYLATION

12.3.1 Degree of Substitution

Degree of substitution (DS) refers to the average number of hydroxyl groups on the starch molecule that acetyl groups have replaced. A low degree of substitution indicates that only a small proportion of hydroxyl groups on the starch molecule have been replaced with acetyl groups. The degree of change in the physicochemical properties of acetylated starch is directly proportional to the degree of acetylation of the starch molecules (Garg & Jana, 2011). Starches with low DS values are typically modified to a lesser extent, and this may result in subtle changes to their properties. A high degree of substitution implies that a larger number of hydroxyl groups have been substituted with acetyl groups. Starches with high DS values undergo more extensive modification, leading to significant alterations in their physical and chemical characteristics. The choice of DS in starch acetylation depends on the desired application and the specific functional properties required for the modified starch (Trela et al., 2020). Different DS values can be targeted to achieve the desired balance between modified starch properties and the intended end-use. Food-grade acetylated starches typically have a lower degree of substitution (DS) to comply with the FDA's 2.5 g/100 g maximum acetyl content for food starches (Colussi et al., 2015).

Acetylated starches with a low DS (0.01–0.2) are versatile in various applications, serving as film-forming, binding, adhesion, thickening, stabilizing, and texturing agents, widely employed in foods like baked goods, canned pie fillings, sauces, soups, frozen foods, baby foods, salad dressings, and snack foods. Acetylated starches with intermediate DS (0.2–1.5) and high DS (1.5–3) are of research interest due to their solubility in acetone and chloroform, as well as their thermoplasticity. These starches, as biodegradable carbohydrate polymers, have potential applications in hot melt adhesives, coatings, cigarette filters, biodegradable packaging materials, tablet binders, and other pharmaceutical uses (Bello-Pérez et al., 2010; Luo & Shi, 2012; Trela et al., 2020). The degree of substitution value can be determined through various techniques, such as headspace gas chromatography (HS-GC), infrared spectroscopy or FT-IR, nuclear magnetic resonance (NMR), and titration (Hong, Chen, et al., 2016). Different types of starches vary in degrees of substitution that is achieved through acetylation. The degree of substitution for acetylated corn starch samples ACS-160, ACS-260, and ACS-360 were measured at 0.08, 0.15, and 0.21, respectively (Ayucitra, 2012). During the pulsed electric field (PEF)–assisted acetylation of potato starch, the degree of substitution progressively rose from 0.054 (following a reaction time of 15 minutes) to 0.130 (after a reaction time of 60 minutes) (Hong, Chen, et al., 2016). The acetylation process significantly impacts the rice starch composition. While native starch has no measurable acetic anhydride content (0 g/100g), acetylated starches show increasing levels degree of substitutions of 0.05, 0.08, and 0.10, respectively, with acetic anhydride content of with 5 g/100 g, 10 g/100 g, and 20 g/100 g, respectively (Colussi et al., 2015). The degree of substitution of the acetylated starches increases with the rise in catalyst concentration. Under native conditions, DS of barley starch is measured at 0, while with 11.0% NaOH, DS was 2.14 ± 0.08; with 17.0% NaOH, DS increased to 5.77 ± 0.29, and further increase of NaOH (23.0%) resulted in a DS of 7.54 ± 0.35 (El Halal et al., 2015).

12.3.2 Contact Angle Measurement

The measurement of the contact angle indicates the nature of the surface. The most prominent feature of esterified starches is their increased hydrophobicity, which is determined by contact angle measurement. The reduced hydrophilicity of esters is attributed to the replacement of hydrophilic hydroxyls by the relatively hydrophobic ester groups. Upon adding a drop of distilled water, it rapidly spreads on the starch surface, resulting in the lowest contact angle value. The abundance of OH groups on the native starch surface facilitates hydrogen bond formation with water. Through acetylation, the contact angle increased for the modified substance, signifying significant alterations in the surface polarity of starch due to chemical treatments. The contact angle of acetylated corn starch with a degree of substitution of 2.89 was measured at 68.2 degrees (H. Chi et al., 2008). The contact angle value increased from 34° for native cassava starch to 45° for cassava acetylated starch with a 45% acetate anhydride concentration (Sondari et al., 2018).

12.3.3 Hydrophobicity Test

Native starch and acetylated starch samples can be qualitatively tested for hydrophobicity by placing them into transparent test tubes containing two immiscible liquid phases, petroleum ether (upper phase), and distilled water (lower phase). Depending on their DS, samples distribute preferentially in one of the liquid phases. The hydrophilic native starch settled in the lower aqueous phase, causing it to become opaque. In contrast, acetylated starches tended to stay in the lower region of the upper organic phase and did not penetrate into the main aqueous phase (Tupa et al., 2015).

12.3.4 Molecular Weight Distribution

Starch acetylation can influence the molecular weight distribution of the starch polymer. The changes in the molecular weight of starch are linked with changes in the contents of amylose and amylopectin, as starch depolymerization is prone to occur during the acetylation process. Gel permeation chromatography (GPC) or size exclusion chromatography (SEC) can be employed to analyze changes in molecular weight as the acetyl group is heavier than the hydroxyl group, they will separate according to their bulkiness (Bello-Pérez et al., 2010; Han et al., 2013; J. et al., 2007; Shogren & Biswas, 2010). The polydispersity (Mw/Mn) values for a low degree of acetylated barley starch (ABSLS) and a high degree of acetylated barley starch (ABSHS) are lower compared to native starch, with ABSLS at 1.254 ± 0.06 and ABSHS at 1.466 ± 0.03, while native starch exhibits a polydispersity of 1.339 ± 0.10. However, ABSHS showed a significantly lower Mw of 0.51 ± 10 compared to native starch (Mw = 1.44 ± 0.09). Additionally, ABSHS displayed a smaller gyration radius (Rz) of 125.33 ± 3.23 nm compared to ABSLS (152.23 ± 4.68 nm) and native starch (159.10 ± 7.53 nm), indicating greater compactness in structure due to high degree substitution (Bello-Pérez et al., 2010).

12.4 FACTORS AFFECTING ACETYLATION REACTION

Starch acetates, authorized as food industry additives with the designation E1420, are frequently manufactured using acetic acid and acetic anhydride as reagents for starch acetylation. Moreover, an alternative method involves employing vinyl acetate for acetylation (Ackar et al., 2015). The effectiveness of the reaction is influenced by various factors, including the nature of the reagent, its concentration, pH levels, the inclusion of catalysts, reaction duration, medium used, botanical source, and the specific size and structure of starch granules (Colussi et al., 2015).

12.4.1 Medium of Reaction

The choice of the medium (solvent or reaction environment) in starch acetylation plays a crucial role in influencing the reaction kinetics, degree of substitution, and the properties of the resulting acetylated starch. The medium affects the overall reaction efficiency and the characteristics of the modified starch. Low-DS acetylated starches are commonly obtained through esterification in an aqueous medium, while high-DS acetylated starches are known for their solubility in a non-aqueous medium like acetone or chloroform and are prepared under conditions that favor more extensive acetylation (H. Chi et al., 2008; Luo & Shi, 2018; Tian et al., 2018).

12.4.1.1 Aqueous Medium

Water is commonly used as the medium for starch acetylation with a low degree of substitution. The aqueous medium helps facilitate the reaction between starch and acetic anhydride. This method is often conducted under mild conditions, and the presence of water allows for a controlled reaction. Elevated water content is beneficial for the dissociation, diffusion, and adsorption of the esterifying agent, contributing to a favorable reaction. Nevertheless, exceeding a water-to-starch ratio of 1.06:1 leads to a decrease in reaction efficiency due to the occurrence of unwanted side reactions (Han et al., 2012). The presence of water has been reported to have less impact on the DS due to lack of swelling or gelatinization, thereby restricting better penetration of reagent into starch granules (Vanmarcke et al., 2017; Zarski et al., 2019; Zhang, Cheng, et al., 2019). In industrial settings, a large amount of water is required to prevent mixing issues; however, this leads to the hydrolysis of acetic anhydride to acetic acid, thereby reducing reaction efficiency.

12.4.1.2 Non-Aqueous Medium

Alternative solvents for acetylation reactions encompass pyridine and dimethyl sulfoxide (DMSO). These solvents offer enhanced efficiency compared to water, particularly suitable for high-DS starch modifications. It is imperative to note, however, that the use of pyridine and DMSO may pose environmental and health concerns due to challenges in their separation from starch. Consequently, their application warrants careful consideration and is often reserved for specific high DS starch modifications (Muljana et al., 2010; Shogren & Biswas, 2010). Nowadays, "green" solvents like CO_2, ionic liquids, and supercritical CO_2 are also used to solve this problem (Fan & Picchioni, 2020).

The choice between aqueous and non-aqueous mediums depends on the desired properties of the acetylated starch, the degree of substitution required, and the environmental considerations of the process. Aqueous mediums are often associated with low DS acetylation, while non-aqueous mediums are used to achieve high DS.

12.4.2 Catalysts

The introduction of a suitable catalyst can significantly enhance the modification potentials of starch in an aqueous medium, thereby overcoming limitations and facilitating more effective and controlled reactions (Otache et al., 2021). An alkaline catalyst, such as sodium hydroxide (NaOH), potassium hydroxide (KOH), and calcium hydroxide ($Ca(OH)_2$), Na_2CO_3 is used to enhance the reactivity of acetic anhydride with starch (J. et al., 2007; Y. J. Wang & Wang, 2002). The base activates starch by forming starch alcoxide (ST-O−), which reacts with acetic anhydride to build a starch acetate and NaOAc. $Ca(OH)_2$ catalyzed acetylated starch exhibits a slightly higher pasting temperature and a lower β-amylolysis limit compared to NaOH- or KOH-catalyzed starch (Y. J. Wang & Wang, 2002). Alkaline catalysts favor the formation of C-2, with C-6 and C-3 esters formed in lesser amounts, while acids like H_2SO_4 are not region-selective catalysts (Ačkar et al., 2015).

Organocatalytic reactions: Organocatalytic reactions, specifically employing organic acids or carboxylic acids, present a competitive approach for developing processes that are safer and more environmentally sustainable. The range of organocatalytics suitable for use encompasses L-aspartic, citric, L-tartaric, L-malic, L-lactic, glycolic, fumaric acids, and L-proline, α-hydroxycarboxylic acid. These organic catalysts exhibit the ability to generate increased esterification reaction rates and attain elevated degrees of substitution (Casas et al., 2004; Imre & Vilaplana, 2020; Tupa et al., 2015). Various approaches can be applied to obtain a higher degree of substituted starches. In some cases, high-DS acetylated starches are prepared under high-temperature and high-pressure conditions. This can lead to more extensive acetylation and modification of the starch structure (Shogren, 2003). On the other hand, by using excess acetic anhydride, highly acetylated starch can be prepared. The use of excess acetic anhydride can drive the reaction to higher degrees of substitution (Mehltretter & Mark, 1974)

12.5 VARIOUS STRATEGIES FOR STARCH ESTERIFICATION

12.5.1 Chemical Methods

Acetylated starches are commercially produced by acetylation with acetic acid, acetic anhydride, ketene, vinyl acetate, or a combination of these reagents. Chemically modified starches are kinetically rate controlled (Ayucitra, 2012).

12.5.1.1 Acetic Anhydride

This is one of the most common methods for starch acetylation. Starch is typically dispersed in a solvent such as water or ethanol. Acetic anhydride is added to the starch slurry, and the reaction is carried out at an elevated temperature. Alkali, typically sodium hydroxide (NaOH), is added to control the pH and promote acetylation. The reaction can be stopped by neutralizing the alkali, and the modified starch is then isolated, typically by drying. Acetic anhydride reacts with the hydroxyl groups in the starch molecule, resulting in the formation of acetylated starch and acetic acid as a byproduct. The acetylated starch product ($R–CO–CH_3$) is formed when acetyl groups ($-CO-CH_3$) replace the hydroxyl groups in the starch molecule. (Ayucitra, 2012; Bello-Pérez et al., 2000, 2010). A general reaction for obtaining the acetylated starch by reacting with acetic anhydride is given as follows:

$$R\text{-}OH + (CH_3CO)_2O \rightarrow R\text{-}CO\text{-}CH_3 + CH_3COOH$$

$R–OH$ represents the hydroxyl groups in the starch molecule; $(CH_3CO)_2O$ represents acetic anhydride; $R–CO–CH_3$ represents the acetylated starch product; and CH_3COOH represents acetic acid, which is a byproduct of the reaction.

12.5.1.2 Vinyl Acetate

This method involves the reaction of starch with vinyl acetate in the presence of a catalyst, such as a peroxide initiator. The reaction is typically carried out at an elevated temperature, and the vinyl acetate reacts with the starch hydroxyl groups, introducing acetyl groups. After the reaction is complete, the modified starch is typically washed and dried. C-2 esters are specifically generated through the acetylation process using vinyl acetate (Ačkar et al., 2015; Das et al., 2010; García-Tejeda et al., 2015; Mormann & Al-Higari, 2004; Shogren & Biswas, 2010). A general reaction for obtaining the acetylated starch by reacting with vinyl acetate is given as follows:

$$R\text{-}OH + CH_3CO_2CH=CH_2 \rightarrow R\text{-}CO\text{-}CH_3 + CH_2=CH_2$$

In this reaction, $R–OH$ represents the hydroxyl groups in starch, and acetic anhydride $(CH3CO)_2O$ serves as the acetylating agent. The result is the formation of acetylated starch $R–CO–CH_3$ and acetic acid (CH_3COOH) as a byproduct.

12.5.1.3 Ethanoic Acid/Acetic Acid

In this acetylation process, the ethanoic acid reacts with the hydroxyl groups in the starch molecule, resulting in the formation of acetylated starch and water as a byproduct. The acetylated starch product ($R–CO–CH_3$) is formed when acetyl groups ($-CO-CH_3$) replace the hydroxyl groups in the starch molecule (Shah et al., 2017)

The general reaction for obtaining acetylated starch using ethanoic acid (acetic acid) is represented as follows:

$$R\text{-}OH+CH_3COOH \rightarrow R\text{-}CO\text{-}CH_3+H_2O$$

R-OH represents the hydroxyl groups in the starch molecule; CH_3COOH represents ethanoic acid (acetic acid); R-CO-CH$_3$ represents the acetylated starch product; and H_2O represents water, which is a byproduct of the reaction.

12.5.1.4 Acetate Salts

The acetate salts method represents an alternative approach to starch modification, specifically involving the introduction of acetyl groups using acetate salts. In this method, starch is subjected to a reaction with an acetate salt, commonly sodium acetate, in the presence of an acid catalyst. The acid catalyst plays a crucial role in promoting the acetylation reaction, facilitating the incorporation of acetyl groups into the starch structure. The resulting modified starch is subsequently isolated through appropriate procedures. This method adds versatility to the array of techniques available for starch modification, offering an alternative route to tailor the properties of starch for various industrial applications (Muljana et al., 2010; Ulfa et al., 2019).

12.5.1.5 Ketene

It is another reagent used in the acetylation of starch, offering an alternative method for modifying starch molecules. Similar to other acetylation methods, the reaction with ketene leads to the replacement of hydroxyl groups in starch with acetyl groups. The resulting acetylated starch may exhibit altered physical and chemical properties, depending on the reaction conditions and the degree of substitution (Qiao et al., 2006). The general reaction for obtaining acetylated starch using ketene can be represented as follows:

$$R{-}OH + (CH_2CO) \rightarrow R{-}CO{-}CH_3 + CH_2O$$

Here, $R{-}OH$ represents the hydroxyl groups in the starch molecule, and (CH_2CO) represents ketene. $R{-}CO{-}CH_3$ represents the acetylated starch product, and CH_2O represents formaldehyde, which is a byproduct of the reaction. Many researchers have conducted experiments on acetylation where diverse starch sources are subjected to specific reaction conditions using different acetylating agents, as outlined in Table 12.1. These conditions result in various degrees of substitution and yield different functional properties and inferences to be used for different applications.

12.5.2 Enzymatic Methods

The enzymatic method for starch acetylation involves the utilization of enzymes to catalyze the introduction of acetyl groups onto starch molecules (Table 12.2). Starch, extensively employed in industrial applications, frequently undergoes chemical derivatization to enhance its properties, with acetylation playing a crucial role in improving its stability in solution. To address concerns related to chemical pollution, a promising alternative involves the utilization of enzymes in a biological approach. This approach provides a more controlled and selective modification compared to chemical methods (Alissandratos & Halling, 2012; Chakraborty et al., 2005; Qiao et al., 2006). The method for enzymatic acetylation generally follows various steps:

Selection of Enzyme: Enzymes, particularly acetyltransferases, are chosen for their ability to transfer acetyl groups to specific hydroxyl groups on the starch molecule. The enzyme's specificity allows for controlled acetylation.

Starch Dispersion: Starch is dispersed in a suitable medium, often water, to create a starch slurry or solution. The choice of medium can impact the enzymatic activity and overall reaction efficiency.

Enzymatic reaction: The selected enzyme is added to the starch dispersion. The reaction is typically carried out under controlled conditions of temperature and pH. The enzyme facilitates the transfer of acetyl groups from an acetyl donor to hydroxyl groups on the starch molecules.

TABLE 12.1   Acetylation of Various Starch Sources

REAGENT USED FOR ACETYLATION	TYPE OF STARCH	REACTION CONDITIONS	DS	MAJOR RESULTS	REFERENCES
Acetic acid	Corn and wheat starches	• 20% starch slurry • pH: 8 • Temperature: 50°C • Glacial acetic acid: 1:5, 1:6, 1:7 • Reaction time: 40 to 60 min • Terminated reaction by bringing the pH to 5.5	1.95–2.08	• Decrease of the amylose and swelling power • A decrease in the amount of acetylating agent leads to a more uniform reaction • For wheat and corn starch, the optimal glacial acetic acid concentration and reaction time were 1:6 for 40 min and 1:5 for 60 min, respectively • Suitable for the development of biodegradable materials	Yermekov et al., 2021
Acetic acid	Gadung starch	• 20% starch slurry • pH: 8–10 • Glacial acetic acid • Reaction time: 10 to 60 min • Terminated the reaction by 1M HCl	0.18	• The acetylation process brings about alterations in the morphology and structure of gadung tuber flours, resulting in swelling power and water solubility values that are comparable to those of Korean wheat flour	(Kumoro et al., 2015)
Acetic anhydride	Black gram (*Phaseolus mungo* L. cv. PU-19 and T-9)	• 35% starch slurry • pH: 8.0–8.5 • Add acetic anhydride (4 g for 4% acetylation and 8 g for 8% acetylation)	0.03–0.08	• Reduction in gelatinization temperature • Decrease in retrogradation of starch pastes and gels • Lower setback viscosities, syneresis, freeze–thaw stability, and gel hardness • Recommended to use as thickening and stabilizing agents	Wani et al., 2015
Acetic anhydride	Cassava starch	• 44% starch slurry • pH: 8.0–8.5 • Acetic anhydride: 2–20 g per 100 g of starch • Reaction time: 30 min. • Lower down pH to 4.5 with 0.5 mol/L HCl.	0.04	• Increased solubility and swelling power with improved clarity • Decrease in gelatinization temperature • Recommended for application as a thickening agent	(Trela et al., 2020)

Acetic anhydride	Oat starches (Sabzaar, SKO20, SKO90)	• 20% starch slurry pH of starch slurry: 8.0 • Acetic anhydride: 7.65 g • Reaction time: 5 min • Terminated reaction by adjusting pH to 4.5 with HCl (0.5 mol/L)	0.02–0.05	• Alterations in the external morphology • Reduced crystallinity • Decreased starch viscosity, displaying shear-thinning behavior	(Shah et al., 2017)
Vinyl acetate	Chickpea, yellow pea, cowpea starches	• Reagent: 0.088 mol/mole glucose unit • Aqueous starch suspensions: 38–40% • pH: 9–10 • Temperature: 20–25°C • Reaction time: 1–2 hours • Catalyst and buffer: sodium carbonate	0.068, 0.071, and 0.064 for Chickpea, yellow pea, and cowpea starches, respectively	• More substantial degree of granule swelling and higher peak viscosity	(Huang et al., 2007)
Vinyl acetate	Potato, corn, and wheat starches	• 1.23 mol anhydroglucose units • Potato starch: 200 g • Potassium carbonate: 15.76 g • 2.46 mol vinyl acetate: 212.4 g • Distilled off unreacted vinyl ester precipitated in isopropanol	1.1	• A reaction time of 5 hours was found to be adequate • Due to the immiscibility of vinyl esters with water, the organic phase serves as a reagent reservoir, preventing the loss of vinyl esters through reactions with water	(Mormann & Al-Higari, 2004)
Ethanoic acid	Cassava and potato starch	• Starch: 5 g • Commercial vinegar: 40 mL • Catalyst: 17% NaOH • Mercerize starch with 20 mL of 17% sodium hydroxide for 1 h at room temperature • Stir and heat at 100°C • Reaction time: 30 to 180 min • Precipitating agent: acetone and methanol	Increases with increasing concentration of NaOH concentration from 10–40%	• Low water swellability and solubility • Applications as thickening agents, thickeners, and stabilizers in products like ice cream and yogurts	(Azeh et al., 2014)
Acetic anhydride	Pea starch	• 0.33 mol L-(+)-tartaric acid: 20 g • 0.52 mol acetic anhydride: 50 mL, reaction temperature: 70–80°C initially and after 15 minutes raised to 85, 95, 110, and 135°C • Freeze-dried pea starch: 10 g • Reaction time: 30–240 min	0.03 to 2.8	• Increased hydrophobicity, raising the degradation temperature by approximately 17% and the glass transition temperature by up to 38% for pea starch • Elevated tensile strength and improved water barrier properties	(Vidal et al., 2022)

TABLE 12.2   Enzymatic Acetylation of Starch

ENZYME USED	STARCH USED	INFERENCES	REFERENCES
***Aspergillus oryzae* S2 α-amylase**	Corn starch	• At 5% concentration, hydrolyzed acetylated starch had better results for crust color and cake appearance.	(Sahnoun et al., 2016)
***Escherichia coli* maltose acetyltransferase**	Potato	• Acetyl content is low in starch; hence, it can be used to prepare low acetylated starch. • The enzyme doesn't catalyze the acetylation of starch polymers in an in vitro setting.	(Nazarian Firouzabadi et al., 2007)
MAT and galactoside acetyltransferase	–	• MAT selectively introduces acetyl groups to the C6 position of glucose and the C6 position of the nonreducing end glucosyl moiety in maltose. • MAT and galactoside acetyltransferase, both belonging to the lacA family of acyltransferases. • GAT displays specificity in acetylating galactosyl units, while MAT exhibits specificity for glucosyl units while MAT demonstrates the ability to acetylate maltooligosaccharides.	(Lo Leggio et al., 2003)

Reaction control: Parameters such as reaction time, enzyme concentration, and temperature are carefully controlled to achieve the desired degree of acetylation. Monitoring these factors ensures precision in the modification process.

Termination and isolation: The enzymatic reaction is stopped at the desired point, often by adjusting pH or removing the enzyme. The acetylated starch product is then isolated from the reaction mixture using methods like precipitation, filtration, or other separation techniques.

Characterization: The acetylated starch product undergoes characterization to determine its degree of substitution and other relevant properties. Techniques such as Fourier-transform infrared spectroscopy, nuclear magnetic resonance, and chromatography may be employed for thorough analysis.

However, very little research has been documented on the enzymatic acetylation of starches. Some researchers have made efforts to produce acetylated starches by enzymatic methods like Nazarian Firouzabadi et al. (2007) produced acetylated starch by introducing the *Escherichia coli* maltose acetyltransferase (MAT) enzyme into potato plants. The enzyme exhibits the capability to attach acetyl groups to α-glucan. However, some authors, such as (Sahnoun et al., 2016), have utilized enzymatic methods as pre-treatment to enhance acetylation. The enzymatic hydrolysis with *Aspergillus oryzae* S2 α-amylase and subsequent acetylation with vinyl acetate represents an instance where an enzymatic approach has been employed for starch acetylation. In nearly all instances, enzymatic hydrolysis induces the formation of dents and cracks on the surface of starch granules, which is beneficial for the acetylating reagent as it can easily reach the site of action. Several examples demonstrating this phenomenon include the actions of β-amylase (BA)/transglucosidase (TG), branching enzyme (BE)/transglucosidase, BE/maltogenic α-amylase (MA), and amylomaltase/BE. Additionally, similar effects are observed with α-amylase (AA)/amyloglucosidase (AMG) (Ashogbon, 2021). Lo Leggio et al. (2003) conducted a study on acetyltransferases and found that MAT and GAT, members of the lacA acyltransferase family, display distinct specificities in acetylating sugar units. MAT selectively acetylates the C-6 position of glucose and the C-6 position of the non-reducing end glucosyl moiety in maltose, whereas GAT exhibits specificity in acetylating galactosyl units. It is noteworthy that starch, composed of glucose units, undergoes enzymatic breakdown, yielding maltose, a disaccharide of glucose. These findings suggest the potential utility of these enzymes in acetylating starch. Thus, further research is required on enzymes that can be used in acetylating starch.

Enzymes contribute to the acetylation of both glucose and maltose within the context of starch modification. Achieving an effective reaction system involves balancing the hydrophobic nature of the acyl-donor, the hydrophilicity of the polysaccharide (despite its typically low solubility in water or other solvents), and the enzyme's functionality. In this scenario, two polymers are crucial: the starch and the biocatalyst. If both are in a solid state, the likelihood of a functional system diminishes (Alissandratos & Halling, 2012; Bruno et al., 1995). Traditional approaches, such as stabilizing the enzyme through immobilization to enable its operation in an organic solvent, prove unsuccessful for carbohydrates with more than a few units. The reason is that both the carbohydrate and the catalyst are in a solid state. Therefore, for polysaccharides, a successful approach involves solubilizing at least two (preferably all three) components, the acyl-donor, the enzyme, and the polysaccharide, into a single system. This solubilization strategy is essential for the effective synthesis of polysaccharides in enzymatic processes (Alissandratos & Halling, 2012). The behavior of starch in water, specifically its ability for granules to swell and even undergo gelatinization, renders it a highly promising substrate when compared to other polysaccharides.

12.5.3 Mechanical Methods

The dense crystalline structure of starch granules, particularly the thick outermost shell, poses a challenge to efficient acetylation, limiting the entry of acetylating agents (K. Zhang, Cheng, et al., 2019). Overcoming this limitation is crucial to enhancing the modification efficiency of acetylation. In this scenario, mechanical forces emerge as a promising solution, leveraging their ability to alter the energy landscape of chemical reactions and enable reaction pathways, thereby complementing traditional chemistry (Xiao et al., 2019). When subjected to mechanical action such as external friction, shearing, high pressure, and rolling, starch undergoes substantial structural changes. This mechanical energy input results in the exposure of a greater number of active sites within the starch, facilitating its reaction with acetylating agents (K. Zhang, Cheng, et al., 2019).

12.5.3.1 Reactive Extrusion

Reactive extrusion is a type of chemical process where chemical reactions take place during the mechanical shear and thermal conditions inside an extruder. Within the extruder, the combination of shear forces from the rotating screws and the application of heat promotes the reaction between starch, reactant, and catalysts. The extrusion process not only facilitates the mixing of reactants but also provides a controlled environment for the chemical reactions to take place. The advantages of using a reactive extrusion approach include improved reaction kinetics, efficient mass transfer, and the potential for continuous and scalable production (De Graaf et al., 1998).

12.5.3.2 Microwave Heating

Applying microwave energy to starch in the presence of acetylating agents can induce acetylation. Microwaves have the capability to induce the formation of pores, bumps, and cracks in starch granules, enhancing the efficiency of extraction. An instance of this is observed in the case of potato starch granules, where the application of microwaves before acetylation processing resulted in more noticeable pores compared to granules without prior microwave treatment. Additionally, the acetylation process further increases surface roughness. This method is known for its rapid and selective heating, potentially leading to efficient starch modification (Sánchez-Rivera et al., 2010; K. Zhao, Li, et al., 2018).

12.5.3.3 Ultrasonication

Ultrasonic waves can be employed to enhance the penetration of acetylating agents into starch granules, promoting acetylation. This method is characterized by its non-thermal effects and potential for controlled modification (L. Zhang et al., 2012). Ultrasonic waves, operating at frequencies beyond the audible range, induce cavitation, creating localized microturbulence and shock waves that enhance the penetration of acetylating agents into starch granules. This unique approach offers a distinct advantage over traditional methods by allowing precise control over the acetylation process, resulting in modified starch with tailored properties for diverse applications (Jambrak et al., 2010).

12.5.3.4 Cold Plasma Treatment

Plasma processing follows a basic physical principle wherein electric discharge imparts additional energy to a gas, transforming it into a high-energy plasma state, recognized as the fourth state of matter. This plasma comprises positively and negatively charged ions, as well as free electrons, free radicals, and intermediate, highly reactive atoms, molecules, and UV photons, all maintaining a distinct neutral charge. These active species exhibit a preference for surface diffusion and undergo various physical and chemical alterations (Suri & Singh, 2023). Plasma can be used to modify the surface properties of starch, including acetylation. Plasma treatment introduces reactive species that can facilitate the acetylation process without significantly affecting the bulk starch structure (Okyere et al., 2022).

12.5.3.5 Pulsed Electric Field

Pulsed electric field (PEF) treatment stands out as a physical modification technique that has captured the attention of researchers due to its environmentally friendly nature and various advantages, including cost-effectiveness, profitability, and energy efficiency. PEF processing is predominantly applied to liquid materials by subjecting them to high-intensity electrical pulses, typically exceeding 10 kV cm−1, within a short duration, usually in the order of microseconds. The use of PEF in combination with chemical acetylation appears to be a well-defined and potentially innovative method for starch modification (Niu et al., 2020). PEF treatment as part of the starch acetylation process presents a multifaceted approach with potential advantages in controlled modification. It has been shown that the PEF treatment can enhance starch esterification under appropriate conditions (Z. R. Lin et al., 2012; Zeng et al., 2010). Therefore, acetylation of starch combined with PEF treatment could be a novel technique to improve the DS at the same acetic anhydride content.

12.5.3.6 Grinding

Grinding pre-treatment significantly enhances the acetylation of starch by breaking down starch granules into smaller particles, increasing the surface area for a more efficient and uniform reaction with acetylating agents. The improved reactivity and uniform distribution of reactants resulting from grinding lead to reduced reaction times, enhanced gelatinization, and a homogenous mixture. This process ensures that acetylation is thorough and consistent, contributing to improved product quality and potential efficiency gains in the overall acetylation process (K. Zhang, Dai, et al., 2019). Many researchers have explored acetylation experiments employing diverse physical techniques to enhance starch modification, with specific methodologies detailed in Table 12.3.

12.6 FUNCTIONAL AND RHEOLOGICAL PROPERTIES OF MODIFIED STARCH

Functional characteristics are key attributes of starch or flour that directly impact their practical applications. These characteristics are assessed through various parameters to understand the usability of starch

TABLE 12.3 Techniques for Augmenting Mechanical Acetylation

PHYSICAL TREATMENT	ACETYLATING AGENT	DS RANGE	STARCH	INFERENCES	REFERENCES
Microwave (600 W, 30 s)	Acetic anhydride and glacial acetic acid (1:1)	Treated: 0.073 Untreated: 0.059	Potato Acetylated starch	• A higher degree of substitution, increased infrared absorption peak intensity, and lower peak viscosity, gelatinization enthalpy, and relative crystallinity compared to single acetylation	(K. Zhao, Saleh, et al., 2018)
Microwave (2 min, 100 ºC)	Acetic anhydride with iodine (0.16–2.5 mol%)	0.2–0.57	Corn starch	• Rapid reaction without the need for a solvent, low energy input, and low toxicity of the catalyst	(Biswas et al., 2008)
Microwave (100°C, 2 min)	Acetic acid and acetic anhydride (1:1)	0.25–1.75	Ball-milled corn starch	• With an increase in hydrophobicity and a significant decrease in WSI and WAI, starch granules lost initial crystallinity after ball milling • DS showed a lower value compared to unmilled starch • Greater hydrophobicity and relatively improved plasticity	(Diop et al., 2011a)
Grinding	Acetic anhydride	0.0174–0.0215	Mung bean starch	• Pores and umbilical points, reflecting a loose structure, are present in starch granules • These granules undergo continuous deformation, resembling elastic balls during grinding • The mechanochemical effect of grinding is evident significantly enhances the acetylation reaction efficiency to 81.61% after 12 h • Solubility and swelling power of the starch granules show improvement, showcasing the substantial impact of grinding on the overall qualities of the starch	(K. Zhang, Dai, et al., 2019)
Ultrasound	Acetic acid (25% solution)	0.17–0.28	Cassava starch	• The swelling power and solubility decreased for ultrasonicated starches but increased for acetylated starch and dual-modified starch, where acetylation is performed after ultrasonication • Lower viscosities when ultrasonication is carried out after acetylation	(Khurshida et al., 2021)

(Continued)

TABLE 12.3 *(Continued)* Techniques for Augmenting Mechanical Acetylation

PHYSICAL TREATMENT	ACETYLATING AGENT	DS RANGE	STARCH	INFERENCES	REFERENCES
Ultrasound (25 kHz, 40 kHz, and 25 + 40 kHz)	Acetic anhydride (5%)	25 kHz: 1.22%, 25+40 kHz: 1.37%, and 40 kHz: 1.02%	Wheat starch	• Acetylated starch demonstrated elevated peak and final viscosity, particularly at the 25+40 kHz frequency compared to singular frequencies (25 and 40 kHz) • Moreover, gelatinization parameters and gelatinization enthalpy (ΔHgel) were lower in acetylated starches at 25+40 kHz compared to 25 kHz, 40 kHz, and native starches (NS) • Increases damage to surfaces and small granules in acetylated starches at 25 kHz and 40 kHz. Notably, larger granules are more affected in acetylated starches at 2 + 40 kHz	(Abedi et al., 2024)
Electron beam (EB) irradiation Irradiation doze: 4 and 8 kGy	Acetic anhydride	0.027 to 0.109%	Naked barley starch	• Effective in enhancing the multi-scale structure and properties of results in higher degrees of substitution • Promote acetylation and lead to improved acetylated starch characteristics • The pretreatment brings about a reduction in amylose content, amylopectin molecular weight (Mw), relative crystallinity, and short-range order of acetylated starch while simultaneously increasing amylose Mw	(Ge et al., 2023)
Pulsed electric field (3kV/cm-5 kV/cm)	Acetic anhydride (6%)	0.130	Potato starch	• Increased DS with higher PEF strength, optimum DS = 0.13 • This resulted in lower retrogradation, breakdown, and setback values, slightly higher pasting temperature • PEF treatment demonstrated potential for achieving higher DS with a shorter reaction time	(Hong, Zeng, Brennan, et al., 2016)
Reactive extrusion	Vinyl acetate	0.05 to 0.2	Potato starch	• Optimizing with high shear and temperatures boosts starch gelatinization, enhancing selectivity and degree of substitution during acetylation • Increasing screw speed in extrusion with leakage gaps (>1 mm) positively impacts starch gelatinization, leading to higher DS in acetylation	(De Graaf et al., 1998)

in specific contexts. Parameters such as amylose leaching or solubility, oil absorption capacity (OAC), water absorption capacity (WAC), freeze–thaw stability (FTS), and swelling power (SP) are instrumental in evaluating the functional performance of starch or flour (Debet & Gidley, 2006; Ojogbo et al., 2020). The acetyl group, introduced through acetylation, demonstrates dual properties, functioning as both hydrophilic and hydrophobic. Acetylation at low degrees of substitution tends to make starch hydrophilic, while at high DS, it imparts hydrophobic characteristics. These properties significantly influence the functional attributes of starch, potentially impacting parameters like increased water absorption capacity or oil absorption capacity in modified starches.

12.6.1 Water-Holding Capacity

Starch can absorb and retain water, a crucial attribute in the application of modified starch as a thickening, gelling, or texture-enhancing agent in diverse food products. Several researchers (Mbougueng et al., 2012; Trela et al., 2020; Mathew & Ojo, 2015; Rahim et al., 2017) have independently observed increased water-holding capacity (WHC) in acetylated starches. At temperatures surpassing 75°C, the water binding capacity of native cassava starch increases. Likewise, a similar trend was observed in potato starch between 60 and 90°C following acetylation (Mbougueng et al., 2012). Initially, the acetylated arenga starches exhibited higher water-holding capacity than their native counterparts, ranging from 0.89 to 0.97. Water absorption capacity for native oat starches ranged from 0.77 to 0.86 g/g, while acetylated oat starches showed a range of 1.33 to 1.54 g/g (Shah et al., 2017). Furthermore, within the time frame of 15 to 60 minutes, the WHC of acetylated arenga starches demonstrated a noticeable increase (Rahim et al., 2017). The water absorption capacity of acetylated sweet potato starch was observed to be 119.1%, surpassing that of both unmodified and acetylated sweet yam starches (Mathew & Ojo, 2015).

This enhanced WHC is likely attributed to the presence of hydrophilic substituting groups introduced during acetylation, facilitating water retention. Additionally, it is noteworthy that a higher water absorption capacity in starch is associated with the loss of a greater amount of starch constituent material, while a lower WAC contributes to the formation of a more compact starch structure (Mbougueng et al., 2012).

12.6.2 Oil Absorption Capacity

Starch absorbs and retains oil, providing crucial insights into how ingredients interact with fats and oils in diverse food formulations. Researchers (Shah et al., 2017; Rahim et al., 2022; Olagunju et al., 2020; Olatunde et al., 2017; and others) have observed an increase in OAC after acetylation. The oil absorption capacity of native oat starches ranged from 0.29 to 0.34 g/g, whereas acetylated oat starches exhibited a significantly higher range of 0.57 to 0.87 g/g (Shah et al., 2017). The oil absorption capacity of pigeon pea native starch was 1.21, whereas for pigeon pea acetylated starches, it was found to be 1.54, 1.61, and 1.82, respectively, at degrees of substitution of 0.05, 0.09, and 0.14, indicating a noteworthy increase compared to the native starch (Olagunju et al., 2020). Acetylation of native Bengal gram starch increased its oil absorption capability by 20% (Yadav & Patki, 2015). However, a decrease in OAC of acetylated sweet potato starch (57.8%) compared to the native sweet potato starch was observed, leading to a contradiction in findings (Mathew & Ojo, 2015). The degree of acetylation in the starch molecules is identified as a potential factor responsible for this discrepancy. Lower acetyl content in the starch molecules after acetylation may contribute to minimal oil binding capacity in OAC, influencing the overall oil absorption properties of the modified starch. A higher OAC is particularly valuable in enhancing the flavor and mouthfeel of various food products, such as whipped cream, sausages, chiffon cakes, and various processed desserts (Kaushal et al., 2012).

12.6.3 Swelling Power

It refers to the ability of starch granules to absorb and retain water during the process of gelatinization triggered by heating in a liquid. Gelatinization involves the breaking of hydrogen bonds between starch molecules, allowing granules to absorb water and swell, creating a thickened and viscous mixture. The introduction of acetyl groups through acetylation reduces the bond strength between starch molecules, modifying hydrophilicity and increasing the hydration capacity, thereby enhancing swelling power (Mathew & Ojo, 2015; Ojogbo et al., 2020). Researchers have observed increased swelling power in various modified starches attributed to the leaching of amylose (Ayucitra, 2012; Wickramasinghe et al., 2009). Acetylated starches, such as acetylated water yam starch (AWYS) and the acetylated sweet potato/water yam starch blend (SP/WY-20/80), demonstrate markedly higher swelling capacities (9.8–12.8%) at 50°C and 90°C compared to native starches, unacetylated sweet potato starch (0.6–1.7%), and unacetylated water yam starch (3.8–8.1%) (Mathew & Ojo, 2015). The swelling power of acetylated corn starches increases with higher degrees of substitution from 0.08 to 0.21, ranging from 17.37 to 27.85 g/g, surpassing the swelling power of native corn starch (NCS) at 15.15 g/g. Even the acetylated starch with the lowest DS value (0.0297) exhibited a higher swelling power (16.66 g/g) compared to native corn starch (Ayucitra, 2012). Swelling power increased after acetylation of raw starch, with acetylated new cocoyam starch (NCS) reaching 67 g/g at 95°C compared to 53 g/g for native (NCS) (Lawal, 2004). However, some studies noted a decrease in solubility in acetylated starches. Acetylation using the highest concentrations of acetic anhydride (10 and 20 g/100g) led to a decrease in the swelling power of rice starch (Colussi et al., 2015). The introduction of hydrophobic acetyl groups may impede water intake into starch granules, potentially leading to a reduction in swelling power. The observed impact on swelling power is dependent on the degree of substitution and the polymerization of amylose and amylopectin chains (Colussi et al., 2015).

12.6.4 Solubility

It refers to its capacity to dissolve in water or other solvents, resulting in a homogeneous solution. In its natural state, raw starch exhibits limited solubility in cold water. Numerous studies, including those by (Rahim et al., 2017; Sodhi & Singh, 2005), and (N. Singh et al., 2004), have observed an increase in the solubility of various starches after acetylation. The solubility at 90°C of acetylated corn starches with different degrees of substitution (0.08–0.21) ranged from 9.50% to 17.38%, which is higher compared to native corn starch with a solubility of 8.50% (Ayucitra, 2012). Solubility for native sorghum starch varies from 11.9% to 13.8%, while for acetylated starch, it ranges from 12.5% to 15.7%, with a degree of substitution varying from 1.25% to 6.25% (Sodhi & Singh, 2005). Kufri Chandermukhi and Kufri Sindhuri variety potato starches, after acetylation, demonstrated the most notable increase in solubility, with values of 0.185 and 0.168, respectively (N. Singh et al., 2004). The changes in swelling power and solubility following acetylation may be attributed to the introduction of hydrophilic substituting groups, which retain water molecules and form hydrogen bonds within the starch granules. Conversely, a decrease in starch solubility of native starch decreased from 16.48% to 8.83% following dual modification involving annealing and acetylation of mung bean starch (Sitanggang et al., 2020). Acetylated banana starch also exhibited lower solubility (2.24%) compared to native starch (2.58%) (Reddy et al., 2014). The solubility (S) of sweet potato starch ranged from 1.58% to 5.24% after modification, while native sweet potato starch exhibited a solubility range of 1.5% to 9.5% (Ulfa et al., 2019). It can be linked to lower amylose leaching, a consequence of increased interactions between amylose and amylopectin molecules that prevent amylose from leaching out of the granule. Additionally, the introduction of acetyl groups, especially at the C(6) position, may elevate the molecular weight of starch, posing a challenge for amylose to leach from the starch granule. The solubility characteristics of acetylated starch are influenced by the degree of substitution and the polymerization of amylose and amylopectin chains (Colussi et al., 2015; Lawal, 2004). Improvement in solubility of starch by acetylation in cold water is advantageous in applications where rapid dispersion or dissolution is required, providing a more efficient and convenient solution (Park et al., 2020).

Solubility and swelling power are influenced by various factors, encompassing the amylose-lipid complex, molecular weight, granule size, amylose/amylopectin ratio, distribution of amylopectin chain lengths, amylose-amylopectin chain interactions, the molecular structure of starch granules, and crystal arrangement. The amylose content predominantly affects solubility, whereas the amylopectin component plays a significant role in influencing swelling power (Subroto et al., 2023). After acetylation of native mung bean starch, the amylose content decreased from 26.79% to 24.09%, resulting in an increased amylopectin content from 73.21% to 75.91%. This led to a reduction in both solubility, from 16.48% to 8.83%, and swelling power, from 18.11% to 11.62% (Sitanggang et al., 2020). Moreover, the correlation between solubility, swelling power, and water absorption capacity holds significant implications for the overall functionality of starch. The interconnected nature of these parameters directly influences viscosity and crystallinity. Notably, as solubility, SP, and WAC increase, there is a consistent association with a reduction in the crystallinity of starch (Ulfa et al., 2020; S. Wang et al., 2013).

12.6.5 Crystallinity

It refers to the extent of structural order within the starch molecules after acetylation. Acetylation generally tends to reduce the crystallinity of starch. As starch granules absorb water and swell, the structural integrity of the crystalline regions may be disrupted, leading to a decrease in overall crystallinity. The breaking of hydrogen bonds between starch molecules, which is associated with increased solubility and swelling, contributes to this reduction in crystallinity (H. Chi et al., 2008; Colussi et al., 2015; Rahim et al., 2017; Subroto et al., 2023). X-ray diffraction measurements are performed to examine whether modification altered the crystallinity of starch. The X-ray diffraction indicated that with the acetylation process, crystalline structures of native starch are destroyed, and new structures of acetylated starch form (H. Chi et al., 2008; Shah et al., 2017). The native corn starch exhibited distinct diffraction peaks at 15°, 17°, 18°, and 23° (2θ), characteristic of the typical A-pattern observed in cereal starch, while acetylated starch with a degree of substitution of 0.85 displayed a profile similar to the native starch, but with the emergence of a new peak at 9° (2θ). Acetylated starches with DS values of 1.78 and 2.89 demonstrated typical patterns of acetylated starches, featuring broad peaks at 9° and 20° (2θ) (H. Chi et al., 2008). Both native and acetylated oat starches showed typical A-type X-ray patterns, with peaks at 15° and 23°, a doublet at 17° and 18°, and a minor peak at 20°. However, there was a notable decrease in crystallinity and peak intensities at 15°, 17°, 18°, and 23°, with the exception of 20°, indicating increased amylose-lipid complex formation. Highest intensity was observed in SKO20 (1486), followed by SKO90 (1399) and sabzaar (1249) (Shah et al., 2017).

12.6.6 Viscosity

Acetylated starches typically exhibit altered viscosity and thickening properties compared to native starch (Park et al., 2020). It is a crucial parameter that measures the thickness or resistance to deformation of a starch paste or solution. The introduction of acetyl groups weakens the intermolecular bonds between amylose and amylopectin, facilitating water penetration into the amorphous region and increasing water absorption during gelatinization (Adebowale et al., 2006; Sarkar, 2016). The flow behavior index (n) of acetylated oat starches ranged from 0.24 to 0.74, higher than native starches (0.42 to 0.50), indicating shear-thinning behavior. These values are less than unity, confirming the shear-thinning behavior of oat starch solutions (Shah et al., 2017). Elevated hydration ability during gelatinization leads to increased paste viscosity. However, the impact of acetylation on starch viscosity varies between starch types, influenced by factors such as amylopectin chain distribution, amylose content, granule molecular arrangement, and the degree of acetyl group substitution (Huang et al., 2007; Salcedo Mendoza et al., 2016). Acetylation reduces setback viscosity (SB), indicating improved starch storage stability, particularly at cold temperatures. Acetylated amaranth starch showed significant viscosity enhancements compared to native starch, with higher values across peak viscosity (2542 ± 87b cP), trough viscosity (1428 ± 96 cP), breakdown viscosity (1113 ± 90 cP), final viscosity (1617 ± 90 cP), and setback viscosity (188 ± 6.3 cP). Despite these changes, acetylated starch maintained a consistent pasting temperature (69°C) similar to native starch (Sindhu et al., 2021). Acetylation of both potato and cassava starches leads to a reduction in peak viscosity, breakdown viscosity, and final viscosity, with acetylated starches showing peak viscosities as low as 42.1 RVU and final viscosities as low as 35.2 RVU. Acetylated starches tend to have lower setback values, ranging from approximately 8.5 to 55.4 RVU, indicating less retrogradation during cooling compared to their native counterparts (Wickramasinghe et al., 2009). Acetyl group substitution inhibits starch's intermolecular interactions, minimizing retrogradation tendencies (Subroto et al., 2023). Starch with low viscosity is deemed appropriate for liquid-based food products, where a thinner consistency is desired. On the other hand, starch with high viscosity is considered suitable for use as a thickening agent, especially in applications where a thicker texture or enhanced viscosity is required to achieve the desired product characteristics (Park et al., 2020). Acetylation improves starch resistance to retrogradation, enhancing shelf stability and chemical durability in challenging processing conditions. Research (Ayucitra, 2012; Lawal, 2004; Luchese et al., 2015; Sodhi & Singh, 2005) demonstrates that acetylated starches exhibit reduced retrogradation and syneresis, resulting in improved clarity and transparency. The retrogradation behavior of gels from native and acetylated corn starches was observed through syneresis during storage at 4°C. Acetylated starch gels displayed lower syneresis increments compared to native starch gels, indicating reduced retrogradation at the storage time of 24 hours (Ayucitra, 2012). During a 6-day storage period, the transmittance decreased in native potato starch (34.8% to 24.6%), in native corn starch (2.6% to 1.4%), and in acetylated potato and corn starches consistently remained higher than that of their native counterparts throughout storage of 72 hours at 47°C, due to lower levels of retrogradation, which prevented the aggregation of amylose and amylopectin in the starch pastes (N. Singh et al., 2004). The retrogradation of acetylated sweet potato starch with different degrees of substitution decreased in the order: native, 0.032, 0.123, 0.091, 0.059 (H. L. Lee & Yoo, 2009). The substitution of acetyl groups impedes the association between starch molecules, contributing to these enhancements. This enhanced clarity is valuable in food manufacturing, especially for products like salad dressings and confectionery items. Syneresis in freeze–thawed gels is caused by amylose rearrangement at lower temperatures, excluding water from the gel structure. Retrogradation occurs during the cooling of gelatinized starch. An increase in the percent of separated water signifies a lack of freeze–thaw stability. Additionally, acetylation boosts starch hydration, minimizing syneresis and enhancing stability during frozen storage. This process is pivotal for producing starch with superior hydration, stability, and clarity (Ayucitra, 2012; Park et al., 2020; Wani et al., 2010, 2012).

12.7 THERMAL PROPERTIES OF MODIFIED STARCH

The thermal properties of starch can be analyzed using techniques such as differential scanning calorimetry (DSC) or thermogravimetric analysis (TGA). Changes in the thermal behavior of starch can provide insights into the success of acetylation. DSC is a widely used technique to investigate thermal transitions, such as gelatinization and retrogradation, in starch and other materials; the acetylation process led to a reduction in the amount of heat associated with the gelatinization of starch due to the introduction of acetyl groups into the starch structure through acetylation resulting in a less ordered or modified transition compared to the unmodified counterpart (Bello-Pérez et al., 2010). The initial gelatinization temperature (70.8 ± 0.6a°C) of native starch (NWS) is notably higher than that of acetylated starches AWS-20 (63.9 ± 0.4°C) and AWS-3 (64.2 ± 0.5°C). Similarly, the peak gelatinization temperature (79.5 ± 0.8°C) of NWS is significantly higher compared to AWS-20 (68.9 ± 0.9°C) and AWS-3 (68.6 ± 0.6b°C). Gelatinization temperature (88.4 ± 1.1°C) is highest for NWS, whereas AWS-20 (73.4 ± 0.7°C) and AWS-3 (71.6 ± 1.4b°C) exhibited lower values. Furthermore, the enthalpy of gelatinization (DH) is notably higher in NWS (12.3

± 0.5 J/g) compared to AWS-20 (7.4 ± 0.4 J/g) and AWS-3 (8.7 ± 0.3 J/g). This demonstrates that acetylated starches exhibit lower gelatinization temperatures and enthalpy compared to native starch, indicating alterations in their thermal properties induced by acetylation (Luo & Shi, 2018). Comparing the DSC values of native corn starch with acetylated corn starches at varying degrees of substitution reveals distinct trends. As DS increases from 0.071 to 0.133, the gelatinization temperatures (To, Tp, Tc) consistently decrease. Specifically, for NCS, To, Tp, and Tc are higher compared to ACS, with values decreasing from 64.3°C to 54.7°C, 72.8°C to 63.3°C, and 81.8°C to 72.6°C, respectively. Moreover, the enthalpy of gelatinization (ΔHg) decreases with increasing DS, with NCS exhibiting the highest value (1.97 mJ/mg), followed by ACS (DS = 0.071), ACS (DS = 0.105), and ACS (DS = 0.133), with values of 1.59, 1.54, and 1.25 mJ/mg, respectively (Han et al., 2012). Meanwhile, TGA of both native starch and acetylated starch derivatives with varying degrees of substitution provide insights into their thermal decomposition behavior.

Native starch exhibits distinct weight losses associated with the release of physically adsorbed and hydrogen-bonded water and starch decomposition, while acetylated starches exhibit three weight losses: starch dehydration, condensation of remaining hydroxyl molecules with water as the main product, and an additional high-temperature decomposition peak attributed to acetic acid evolution from acetates decomposition (Tupa et al., 2015). The native sago starch exhibited thermal decomposition at 199.8°C, with the highest decomposition occurring around 392.7°C. However, the modified starch acetate showed a shift in peaks, with the onset of decomposition at 229.6°C and the highest decomposition at 404.8°C. This shifting of peaks after acetylation indicates higher thermal stability, likely attributed to the replacement of hydroxyl groups in starch molecules during the acetylation process (A. V. Singh et al., 2011). Both native and acetylated yellow ginger starches demonstrate a two-stage weight loss pattern in thermal analysis. The native starch exhibited initial minor weight loss due to water loss around 60–100°C, followed by decomposition peaking between 277–353°C, with 85.4% decomposition at 500°C. Similarly, acetylated starches display two-stage weight loss, with decomposition onset occurring at higher temperatures between 200–320°C for starch decomposition and 330–420°C for acetylated starch decomposition. Lower stability was observed in partially substituted samples, evident from their lower onset temperatures. Notably, high-DS acetylated starches show superior thermal stability, with the intensity of the first weight loss nearly disappearing at DS 2.67. This trend suggests that thermal stability increases with higher degrees of substitution (L. Zhang et al., 2009).

12.8 MORPHOLOGICAL PROPERTIES OF ACETYLATED STARCH

Microscopic techniques, such as scanning electron microscopy (SEM) or optical microscopy, are used to examine the morphological changes that occur on the surface of starch granules as a result of the acetylation process. SEM allows for high-resolution imaging of the starch granules, providing detailed information about their size, shape, and surface structure. As the degree of substitution increases, more hydroxyl groups are replaced by acetyl groups, leading to alterations in the inter- and intra-molecular hydrogen bonding within the starch granules. This disruption of hydrogen bonding, combined with the introduction of bulkier acetate groups, may weaken the granular structure of starch, causing surface roughness, deformation, and the appearance of grooves. Additionally, the voluminous nature of acetate groups may further contribute to the roughening of the starch surface. In contrast, native starch, with its original hydrophilic and compact structure, exhibits a smoother surface (Shah et al., 2017; J. Singh et al., 2004; Tupa et al., 2015). The granules of native corn starch exhibit a roughly polygonal shape, ranging from 2 to 15 μm in diameter. Notably, large-sized NCS granules display "pin holes" and equatorial grooves or furrows. Conversely, acetylated corn starch granules show slight fragmentation, indentation, and the formation of cavities. Compared to NCS, ACS granules appear rougher, with some loss of definition at the edges, which tend to fuse and form a gelatinized mass (Han et al., 2012). The smooth surfaces of native corn starch granules transformed into rougher surfaces as the degree of starch substitution increased, reaching DS = 0.36. With further esterification, at DS = 0.50, starch acetate granules continued to exhibit rough surfaces, with some granules showing deformed depressed zones. Notably, in starch acetates with DS = 1.13, single grooves appeared in most granules, and in some instances, the interior of the granules was visible (Tupa et al., 2015).

12.9 STRUCTURAL PROPERTIES OF ACETYLATED STARCHES

Several analysis techniques can confirm acetyl groups in modified starch, and Fourier-transform infrared spectroscopy is one such technique used to identify specific functional groups, including acetyl. The FTIR spectra of corn starch acetate exhibits additional bands compared to the native corn starch FTIR spectra attributed to the introduced ester groups. These include the stretching of the C=O of the ester group (1747 cm^{-1}), CH$_3$ deformation vibration of the ester moiety (1376 cm^{-1}), and C–O–C stretching vibration of the ester moiety (1244 cm^{-1}) (Tupa et al., 2015). Increased absorption between 1251 and 1255 cm^{-1} indicates C-O bond stretching by acetyl groups in acetylated yam starch. Absorbance rise from 1650 to 1750 cm^{-1} suggests OH group folding, reflecting starch structural changes. Signal variations in the 3000–3900 cm^{-1} and 2000–2850 cm^{-1} ranges, respectively, correlate with OH group and CH tension, possibly due to acetyl group incorporation into carbon 6 (Salcedo Mendoza et al., 2016). FTIR spectra of various degrees of substitution of acetylated corn starches reveal new absorption bands at 1754, 1435, 1375, and 1240 cm^{-1}, attributed to carbonyl C=O, CH$_3$ antisymmetric deformation vibration, CH$_3$ symmetry deformation vibration, and carbonyl C–O stretch vibration, respectively (H. Chi et al., 2008). Increasing degrees of substitution cause shifts in peak intensities, suggesting the involvement of hydroxyl groups in the reaction (H. Chi et al., 2008; Diop et al., 2011a; Han et al., 2013; Shah et al., 2017; Shogren & Biswas, 2010; Tupa et al., 2015). The chemical analysis of native starch and its acetylated derivatives using nuclear magnetic resonance (^{1}H NMR, ^{13}C NMR, and ^{13}C–^{1}H COSY) also provides valuable insights into the structural characteristics. The spectra allowed for the assignment of chemical shifts, revealing changes in proton environments during the acetylation process. These findings offer a detailed understanding of the low and high degrees of substitution acetylated starch, contributing to the analysis of their overall structure (Chi et al., 2008). In comparison to the ^{1}H NMR spectrum of native starch, the proton chemical environments are altered in acetylated starch. Notably, characteristic peaks shifted: 5.25 ppm to H-3, 5.18 ppm to H-1, 4.74 ppm to H-2, 4.32–4.25 ppm to H-6,6', and 3.96 ppm to H-4. The assigned ^{1}H chemical shifts of the protons at 3.07–3.66 ppm connecting to the proton at 3.35 ppm are designated to H-4, 3.67 ppm to H-3, 3.30 ppm to H-2, 3.59 ppm to H-5, and 3.07 ppm to H-4 (end group), 3.64 and 3.46 ppm to H-6,60, respectively (H. Chi et al., 2008). When native maize starch reacted solely with acetic anhydride-d6, three equally-sized acetyl peaks between 1.8 and 2.3 ppm (DS 0.015) were observed. However, peracetylation enabled the anhydroglucose unit (AGU) protons to be evenly shifted, resulting in well-resolved spectra. Signals at 2.20, 2.03, and 1.99 ppm are identified as acetyl protons substituted at C-6, C-2, and C-3, respectively (Xu & Shi, 2019).

12.10 TEXTURE ATTRIBUTES OF ACETYLATED STARCH GELS

Acetylated starch gels often display enhanced firmness and stability, providing desirable texture attributes for diverse product formulations. Amaranth starch modified with acetylation exhibits lower hardness

(0.68 ± 0.01 g) and reduced adhesiveness (–1.00 ± 0.09 g/s) compared to native starch (NS) (hardness: 1.10 ± 0.05 g; adhesiveness: –1.90 ± 0.07 g/s). Gumminess and chewiness were also notably lower in Ace-S, indicating a softer texture. Cohesiveness remained relatively consistent across treatments. These findings suggest that acetylation alters the gel texture of amaranth starch, resulting in softer and less adhesive gels compared to native starch (Sindhu et al., 2021). Canna starch modified with acetylation demonstrated decreased hardness (Thai-green: 169–193 g; Thai-purple: 114–222 g) and adhesiveness (Thai-green: 297–382 g·s; Thai-purple: 296–327 g·s) compared to native starch, indicative of a softer texture (Saartrat et al., 2005).

12.11 APPLICATIONS OF ACETYLATED STARCHES

Acetylated starches are widely utilized in food and other industries owing to their modified properties that enhance functionality and provide distinctive advantages. These modified starches find diverse applications in various food products. Following are some common uses of acetylated modified starches/flours across different fields and products.

12.11.1 Fat Replacers

Fat replacers are employed in food formulations to replicate the qualities of fats while reducing overall fat content, contributing to the creation of lower-fat and healthier food products. Adhering to guidelines from health organizations like the American Heart Association and the World Health Organization recommends limiting daily fat intake to less than 20% of total caloric intake (L. Wang et al., 2020). It has been demonstrated that acetylated corn starch can be used as a fat replacer in beef patties. Acetylated corn starch significantly improved redness, moisture retention, thickness, and sensory attributes of beef patties while reducing firmness, cooking loss, diameter reduction rate, and dimensional shrinkage. Patties with 15% acetylated corn starch exhibited a microstructure resembling those with 15% animal fat, confirming its efficacy as a fat replacer in meat products during frozen storage (Eshag Osman et al., 2022). The incorporation of chemically modified starches extends beyond meat products to the preparation of low-fat mayonnaise. The addition of a 10% starch paste, chemically modified by acetylation, achieves an impressive 80% oil replacement in the mayonnaise formulation. Remarkably, when chemically modified chestnut starches were utilized, the resulting low-fat mayonnaise exhibited textural and sensory properties quite comparable to those of full-fat mayonnaise (Ansari et al., 2017). Acetylated arrowroot starch, when utilized as a partial fat replacer at levels of 30% and 50% in mayonnaise, demonstrated good emulsion stability of 72.83–85.85%, which is comparable to native starch. The elastic moduli (G′) of fat-reduced mayonnaises containing acetylated arrowroot starch were also lower than the full-fat version (Park et al., 2020). The incorporation of acetylated maize starch in cheese at concentrations of 0.5%, 1.0%, and 1.5% resulted in significant changes in rheological and textural properties. The gel formed with acetylated maize starch exhibited a noticeable disruption of the casein matrix, characterized by smaller and less uniformly distributed swollen granules. The observed interruption of the casein matrix, combined with the acetylated starch acting as filler material and exhibiting very low whey loss (about 86% starch retained in curd), suggests that acetylated maize starch holds promise as a fat replacer in cheese (Diamantino et al., 2019).

12.11.2 Thickening Agents

It is a substance or compound added to a liquid or mixture to increase its viscosity, thereby making it thicker. While native starch paste generally tends to be more viscous than acetylated starches, the stable viscosity behavior exhibited by acetylated starch, characterized by the absence of rapid changes in response to temperature increases, is attributed to the presence of acetyl groups (Han et al., 2012). The significant water-absorbing capacity of acetylated starch has the potential for utilization as a thickening agent. Acetylated starch demonstrates effective thickening capabilities, particularly up to pH 4.5. Furthermore, their surface activity, particularly that of E 1450, makes them suitable as functional components in food emulsions (Prochaska et al., 2007). Acetylated tapioca starch can be used as a thickening agent in food products. The decrease in the pasting temperature is a notable advantage of acetylation. This characteristic enables the utilization of acetylated starches in scenarios where a thickening agent needs to gelatinize at a lower temperature (Babic et al., 2007). Acetylated rice starches exhibit higher final viscosity in comparison to native starches, making them suitable for use in products that necessitate thickening power after preparation by heating. This includes the production of dehydrated soups, instant soups, broths, and sauces (Colussi et al., 2015).

12.11.3 Stabilizing Agents

Acetylated starches find frequent application as stabilizing agents in a diverse range of food products, including sauces, gravies, soups, and dressings. Their role in these formulations is to enhance the desired texture and viscosity. Acetylation modification provides exceptional stability and resistance to retrogradation and syneresis, as mentioned in Section 12.5.6. The syneresis of purple corn starch gels initiates on the first day. It escalates steadily until the 28th day of refrigerated storage, ranging from 27.58% to 52.84% for native starches, while acetylated starches exhibit reduced syneresis, ranging from 17.08% to 35.43%. Generally, native starches demonstrate higher syneresis compared to their acetylated counterparts. Notably, the highest syneresis is observed in the native starch of dark purple corn (52.84%), suggesting that acetylated starches offer greater stability and potential as stabilizers in gel formulations (Cisse et al., 2023). This attribute positions acetylated starch as a promising option for serving as a stabilizer in products that require storage at low temperatures (Han et al., 2012; Subroto et al., 2023).

12.11.4 Texture Improvement

Bakery products, especially breads, typically face a limited shelf life attributed to a gradual increase in hardness during storage. Acetylated starches, belonging to the category of substituted or stabilized starches, are recognized for their ability to resist retrogradation, especially at lower temperatures. Research studies investigating the utilization of acetylated starches have consistently indicated an enhancement in the texture of bread, coupled with potential cost reduction when substituting 20% of wheat flour with modified starch. This cost-effectiveness stems from the lower tariff of the modified starch compared to the stronger wheat flour (Miyazaki et al., 2006). The addition of 20% acetylated tapioca starch decreases arrival and peak time, as measured by a farinograph, accompanied by a slight reduction in mixing tolerance, with no noticeable change in dough extensibility (Dreher et al., 1984).

The introduction of oil-modified acetylated tapioca starch has been demonstrated to enhance the breaking force, deformation, and elasticity of surimi gel, as it can elevate the water-holding capability of surimi gels. Consequently, this transformation leads to a reduction in water mobility within the gel matrix and, thus, improves the texture of the surimi fish gel (X. Zhao et al., 2022). The incorporation of acetylated corn starch into patties improves various attributes, including increased redness, enhanced moisture retention, increased thickness, and positive sensory characteristics. Additionally, the use of acetylated corn starch led to a reduction in firmness, cooking loss, diameter reduction rate, and dimensional shrinkage of the patties. Notably, the microstructure of the patties containing 15% acetylated corn starch closely resembles that of patties with 15% animal fat. This underscores the role of acetylated starch as a texture enhancer in the formulation of these patties (Eshag Osman et al., 2022). In the manufacturing of reduced-fat meat analogs through 3D printing, the incorporation of acetylated and other biosurfactant-modified ingredients influences the structure, texture, and sensory properties, showcasing the potential for improved eating experiences. This study emphasizes the ongoing role of acetylation in inducing fibrous sensations and molding the overall quality of printed meat analogs (Shahbazi et al., 2021).

12.11.5 Reducing Syneresis in Frozen Foods

In frozen foods, acetylated starches can help reduce syneresis (formation of ice crystals), maintaining the quality of products such as ice creams and frozen desserts. The phenomenon of syneresis in freeze/thawed gels is generally a consequence of heightened molecular association between starch chains at lower temperatures, resulting in the exclusion of water from the gel structure. The significant decrease in syneresis observed after acetylation can be ascribed to the impediment of interchain bonding between starch molecules caused by the presence of incorporated acetyl groups. Acetylated Thai-green and Thai-purple canna starch gels exhibit reduced syneresis compared to their native counterparts across all degrees of substitution ranging from 0.05 to 0.10 over an 8-day storage period (Saartrat et al., 2005). Acetylated starches derived from Kufri Sindhuri and Kufri Chandermukhi varieties display an absence of syneresis even after 120 hours of storage at 4°C (N. Singh et al., 2004). Acetylated starch in custards shows no retrogradation after 14 days, suggesting long-term stability and thus offering potential benefits in enhancing cold storage performance and texture (Shaikh et al., 2017). The acetylated rice starch cultivars, namely PR-106, PR-114, IR-8, and PR-113, demonstrate lower syneresis percentages (ranging from 0.09% to 0.15%) compared to native starch cultivars over a period of 168 hours (Sodhi & Singh, 2005). The water-holding capacity of sausages initially increased and then decreased with an increase in the concentration of acetate starch. Among the samples tested, the sausage containing 5% w/w acetate starch exhibited the highest content of immobilized water (F. L. Zhang et al., 2014).

12.11.6 Improved Mouthfeel

Mouthfeel refers to the tactile sensations and perceptions experienced in the mouth while consuming food. Acetylated starches can be used in dairy products like puddings and custards to improve texture, provide creaminess, and enhance overall mouthfeel (Subroto et al., 2023). Native pigeon pea starch is inadequate for yogurt stabilization due to its poor firmness and high syneresis. Experimental yogurt samples incorporating acetylated pigeon pea starch exhibit superior firmness. Despite the lower viscosity of the modified starch ingredient, the heightened firmness of the yogurt may be attributed to the lower level of syneresis. This is noteworthy as texture (firmness, consistency, and syneresis) plays a significant role in determining yogurt quality, influencing its appearance, mouthfeel, and overall acceptability (Olagunju et al., 2020). Acetylated corn starch concentration can influence mouthfeel scores in multicomponent emulsion-based products, emphasizing the role of starch in enhancing viscosity and firmness. This information underscores the versatile use of acetylated starches in various food products to achieve desirable sensory qualities (Cao et al., 2021). The utilization of acetylated-modified cassava or potato starch in the preparation of glutinous rice dumpling skins results in favorable characteristics. The cooked samples exhibit a delicate mouthfeel, consistent color, and luster, accompanied by good transparency. Crucially, there is no evidence of depression or low weight loss, and the dumplings do not adhere to the teeth. These desirable qualities significantly contribute to widespread consumer acceptance (Chen et al., 2023).

12.11.7 Confectionery and Bakery Foods

Acetylated starches prove effective in oat flour cake applications, contributing to increased batter viscosity, cake volume, and enhanced whiteness of the cake crust (Mirmoghtadaie et al., 2009). Additionally, the incorporation of dual-modified starches in cake formulation, such as acetylated and enzymatically hydrolyzed corn starch, has demonstrated superior effects on various quality attributes compared to their single-modified counterparts or native starch (Sahnoun et al., 2016). At 5% concentration, the modified starch led to reduced hardness, cohesion, adhesion, and chewiness of baked cakes. Moreover, they enhanced elasticity, volume, height, crust color, and overall appearance compared to cakes formulated with native starch (Sahnoun et al., 2016). Acetylated rice starch substituted for fat at a level of 20% increased the thickness of the cookies, resulting in a reduced spread ratio. Fat cannot be entirely substituted with starches in low-moisture products like cookies, as fat plays a crucial role in providing tenderness and facilitating mechanical handling (Y. T. Lee & Puligundla, 2016). The bread crafted from wheat flour, incorporating a 20% substitution of acetylated tapioca starch (ATS), demonstrated increased firmness in contrast to bread prepared with native tapioca starch (Miyazaki et al., 2006).

12.11.8 Instant and Convenience Foods

Acetylated starches can be utilized in instant food formulations, contributing to quick dispersion and improved reconstitution properties in products such as instant soups, sauces, and convenience foods. Starch plays a significant role in influencing food quality, particularly instant foods (Ma et al., 2022). Total starch, large A-type starch, and small B-type starch from wheat flour after acetylation with acetic anhydride using both conventional and pulsed electric field-assisted methods, were incorporated into the noodle-making flour; it was observed that all the esterified starches enhanced the color of the noodles and increased the water absorption rate. In comparison to noodles with native B-type starch, those incorporating PEF-assisted esterified B-type starch exhibited a 12 mm increase in stretching distance (Hong et al., 2020). Esterified cassava starch, when introduced in Japanese white salted noodles, led to improved gelatinization, softening, and increased amylopectin content, resulting in a low-density structure. Noodles boiled for 3 min and left standing for 6 min displayed increased elasticity without sacrificing hardness compared to those boiled for 10 min, emphasizing the versatility and efficiency of esterified tapioca starch in noodle preparation (Eguchi et al., 2014). Acetylated corn starch and wheat flour in a 4:1 ratio were used to form noodles, and the addition of ACS imparted a whiter and cleaner appearance to the noodles, accompanied by a reduction in hardness, elasticity, adhesion, chewiness, and tensile properties, indicating a significant influence of acetylated corn starch modification on the textural and mechanical characteristics of the noodles (D. Lin et al., 2019). Furthermore, acetylated potato starch plays a crucial role in enhancing the textural attributes, particularly hardness and adhesiveness, of instant fried noodles when used in conjunction with sodium carboxymethyl cellulose, providing insights into the optimization of noodle quality. Acetylated potato starch (APS) enhances hardness without affecting cohesiveness in instant noodles. Optimal fat uptake is achieved at 10% APS, influencing textural attributes and offering the potential for improved quality in low-protein soft wheat flour formulations (Choy et al., 2012). Overall, these findings underscore the diverse applications and positive influences of esterified starches in enhancing noodle characteristics across various formulations and starch sources.

Coating potato products with high amylose starch acetate enhances crispiness, strength, and flavor, providing excellent texture even after prolonged storage. Rice cake-like foods are created by blending over 10% starch acetate, carbohydrate sweeteners, and water, heating, shaping, and coating with predominantly amylopectin-rich corn starch. They exhibit stability at temperatures below 10°C and maintain a firm rice-cake texture even after defrosting or heating (Sajilata & Singhal, 2005).

12.11.9 Clear Beverages and Fruit Juices

Acetylated starches can be used in the formulation of clear beverages and fruit juices, providing clarity and stability without affecting taste or mouthfeel. Different concentrations of acetylated potato starch phosphate are used as food thickeners in fresh fruit juices of apples, oranges, and peaches and achieve viscosity levels recommended by the National Dysphagia Diet (NDD) standards. After pasteurization and storage for three months, substantial alterations in viscosity, total phenol content, and antioxidant capacity were observed. Starch-thickened orange juices exhibit higher preference ratings in sensory evaluations (Akçay & Alkan, 2023). The versatile nature of acetylated starches makes them valuable in addressing various formulation challenges and enhancing the overall quality of food products. The specific application of acetylated starch depends on the desired functional attributes of a particular food product. Acetylated starches exhibit various abilities, but they lack resistance to elevated temperatures, acidic environments, and mechanical forces. Consequently, acetylated starches are not well-suited for products that require preservation through sterilization processes (Gałkowska et al., 2023).

12.11.10 Eco-Friendly Biocomposites/Films

Acetylation modification enhances the oxygen barrier properties of starch, making it a suitable material for film formation (Subroto et al., 2023). Cassava starch-based films are created by suspending acetylated starch in distilled water and heating it until complete gelatinization. A plasticizer is then added to the acetylated starch, and the resulting filmogenic solution can be employed for strawberry protection (Franco et al., 2017).

As part of the ongoing efforts to find sustainable alternatives to conventional plastics, the integration of acetylated corn starch, acetylated sugarcane fiber (AcSF), and glycerol in biocomposites has been explored (Fitch-Vargas et al., 2019). This innovative approach, utilizing a single-screw extruder and a central composite rotatable design, investigates the impact of AcSF content (0.0–20.0%) and glycerol content (20.0–30.0%) on the overall characteristics of these biocomposites.

12.11.11 Starch-Based Encapsulation Materials

Starch-based encapsulation materials, primarily composed of starch, are utilized across industries like food and pharmaceuticals. They serve the purpose of enclosing substances to protect, preserve, or control their release, which is especially beneficial for achieving controlled release or protection of active ingredients. Acetylation of normal maize starch significantly enhances encapsulation efficiency and storage stability of purple maize anthocyanins in spray-dried microparticles, as demonstrated by higher anthocyanins retention after 30 days at various water activities (García-Tejeda et al., 2015). Enhancing gallic acid encapsulation, acetylated starch proves effective in comparison to its counterpart, acetylated inulin, during spray-drying. Notably, acetylated inulin shows a slower release pattern. These findings suggest the potential for creating quick-acting functional foods, achieving release within less than 9 hours, such as in the design of dry mixes or instantaneous foods (Robert et al., 2012).

12.11.12 Nanocrystals for Protein Delivery

Nanocrystals for protein delivery utilize nanoscale particles to encapsulate and transport proteins. This approach provides benefits, including increased stability, controlled release, and enhanced bioavailability of proteins. Acetylated broken-rice starches enhance solubility, control drug release, and improve the stability of films, making it a promising material for effective and sustained protein drug delivery applications (Xiao et al., 2019). The acetylation process tailors starch properties, offering potential advantages in biocompatibility and controlled drug release mechanisms. A novel controlled-release formulation utilizing PEGylated starch acetate nanoparticles has also been developed by (Minimol et al., 2013), which demonstrated enhanced encapsulation efficiency and bio-adhesive properties. The conjugation of polyethylene glycol with starch acetate results in self-aggregated nanoparticles suitable for controlled delivery of insulin or other proteins, showing the potential for therapeutic applications in drug delivery. Biodegradable nanoparticles for efficient drug delivery by utilizing acetylated dioscorea starch as a biocompatible and preferably biodegradable polymeric material have also been developed (Paulos et al., 2016). The study successfully fabricated nanoparticles with a particle size below 200 nm and a polydispersity index (PDI) less than 0.2.

12.12 CONCLUSION

The acetylation method for starch modification is a versatile and effective technique that imparts unique properties to starch, expanding its applicability in various industries. We delved into the intricacies of acetylation, exploring the chemical process that involves the introduction of acetyl groups to starch molecules. One of the key advantages of acetylation is its ability to enhance the functional properties of starch, such as improved solubility, increased resistance to retrogradation, and altered gelatinization characteristics. These modifications cater to the diverse needs of industries such as food, textiles, and pharmaceuticals, where modified starches find widespread utility. The versatility of acetylated starch extends to its role as a thickening and stabilizing agent in food products, providing a solution for formulations that require specific texture and shelf-life attributes. Additionally, the modified starch demonstrates potential in textile applications, contributing to the development of fabrics with enhanced properties. While acetylation has proven to be a valuable method for starch modification, it is crucial to consider the balance between modification extent and maintaining the inherent qualities of starch. Careful optimization of the acetylation process is essential to achieve the desired functional improvements without compromising the integrity of the starch structure.

REFERENCES

Abedi, E., Roohi, R., Hashemi, S. M. B., & Kaveh, S. (2024). Investigation of ultrasound-assisted starch acetylation by single- and dual-frequency ultrasound based on rheology modelling, non-isothermal reaction kinetics, and flow/acoustic simulation. *Ultrasonics Sonochemistry, 102*(November 2023), 106737. https://doi.org/10.1016/j.ultsonch.2023.106737

Ačkar, D., Babić, J., Jozinović, A., Miličević, B., Jokić, S., Miličević, R., Rajič, M., & Šubarić, D. (2015). Starch modification by organic acids and their derivatives: A review. *Molecules, 20*(10), 19554–19570. https://doi.org/10.3390/molecules201019554

Adebowale, K. O., Afolabi, T. A., & Olu-Owolabi, B. I. (2006). Functional, physicochemical and retrogradation properties of sword bean (Canavalia gladiata) acetylated and oxidized starches. *Carbohydrate Polymers, 65*(1), 93–101. https://doi.org/10.1016/j.carbpol.2005.12.032

Akçay, B., & Alkan, D. (2023). Designing of texture modified fruit juices using food hydrocolloids: Storage influence on viscosity. *Heliyon, 9*(11).

Alissandratos, A., & Halling, P. J. (2012). Enzymatic acylation of starch. *Bioresource Technology, 115*, 41–47. https://doi.org/10.1016/j.biortech.2011.11.030

Ansari, L., Ali, T. M., & Hasnain, A. (2017). Effect of chemical modifications on morphological and functional characteristics of water-chestnut starches and their utilization as a fat-replacer in low-fat mayonnaise. *Starch/Staerke, 69*(1–2), 1600041. https://doi.org/10.1002/star.201600041

Ashogbon, A. O. (2021). Dual modification of various starches: Synthesis, properties and applications. *Food Chemistry, 342*, 128325. https://doi.org/10.1016/j.foodchem.2020.128325

Ayucitra, A. (2012). Preparation and characterisation of acetylated corn starches. *International Journal of Chemical Engineering and Applications, 3*(3), 156–159. https://doi.org/10.7763/ijcea.2012.v3.178

Azeh, Y., Abubakar, R. M., & Umar, M. T. (2014). Isolation and acetylation of starch using ethanoic acid for industrial uses. *Ilorin Journal of Science, 1*(2), 301–318.

Azeh, Y., Paiko, Y. B., Elele, U. U., Adesina, G. O., Mohammad, M. A., & Mohammed, D. Z. (2016). Isolation of cassava starch and synthesis of acetyl derivatives using vinegar. *Lapai Journal of Applied and Natural Sciences*, *1*(1), 89–103.

Babic, J., Subaric, D., Ackar, D., Kovacevic, D., Pilizota, V., & Kopjar, M. (2007). Preparation and characterization of acetylated tapioca starches. *Deutsche Lebensmittel-Rundschau*, *103*(12), 580–585.

Bello-Pérez, L. A., Agama-Acevedo, E., Zamudio-Flores, P. B., Mendez-Montealvo, G., & Rodriguez-Ambriz, S. L. (2010). Effect of low and high acetylation degree in the morphological, physicochemical and structural characteristics of barley starch. *LWT*, *43*(9), 1434–1440. https://doi.org/10.1016/j.lwt.2010.04.003

Bello-Pérez, L. A., Contreras-Ramos, S. M., Jimenez-Aparicio, A., & Paredes-López, O. (2000). Acetylation and characterization of banana (MUSA paradisiaca) starch. *Acta Científica Venezolana*, *51*(3), 143–149.

Bhargava, A., Shelke, S., Dilkash, M., Chaubal-Durve, N. S., Patil, P. D., Nadar, S. S., Marghade, D., & Tiwari, M. S. (2022). A comprehensive review on catalytic etherification of glycerol to value-added products. *Reviews in Chemical Engineering*, *39*. https://doi.org/10.1515/revce-2021-0074

Biswas, A., Shogren, R. L., Selling, G., Salch, J., Willett, J. L., & Buchanan, C. M. (2008). Rapid and environmentally friendly preparation of starch esters. *Carbohydrate Polymers*, *74*(1), 137–141. https://doi.org/10.1016/j.carbpol.2008.01.013

Bruno, F. F., Akkara, M. J. A., Ayyagari, M., Kaplan, D. L., Gross, R., Swift, G., & Dordick, J. S. (1995). Enzymatic modification of insoluble amylose in organic solvents. *Macromolecules*, *28*(26), 8881–8883. https://doi.org/10.1021/ma00130a028

Cao, C., Wang, C., Yuan, D., Kong, B., Sun, F., & Liu, Q. (2021). Effects of acetylated cassava starch on the physical and rheological properties of multicomponent protein emulsions. *International Journal of Biological Macromolecules*, *183*, 1459–1474. https://doi.org/10.1016/j.ijbiomac.2021.05.134

Casas, J., Persson, P. V., Iversen, T., & Córdova, A. (2004). Direct organocatalytic ring-opening polymerizations of lactones. *Advanced Synthesis and Catalysis*, *346*(9–10), 1087–1089. https://doi.org/10.1002/adsc.200404082

Chakraborty, S., Sahoo, B., Teraoka, I., Miller, L. M., & Gross, R. A. (2005). Enzyme-catalyzed regioselective modification of starch nanoparticles. *Macromolecules*, *38*(1), 61–68. https://doi.org/10.1021/ma048842w

Chen, L., Tan, H., Feng, R., Ma, L., Zhang, Y., Yi, H., Yin, L., Liu, W., Hu, L., & Zhu, W. (2023). Effect of modified starches on the quality of skins of glutinous rice dumplings. *International Journal of Biological Macromolecules*, *253*, 127139. https://doi.org/10.1016/j.ijbiomac.2023.127139

Chi, C., Lian, S., Zou, Y., Chen, B., He, Y., Zheng, M., Zhao, Y., & Wang, H. (2023, June). Preparation, multi-scale structures, and functionalities of acetylated starch: An updated review. *International Journal of Biological Macromolecules*, *249*, 126142. https://doi.org/10.1016/j.ijbiomac.2023.126142

Chi, H., Xu, K., Wu, X., Chen, Q., Xue, D., Song, C., Zhang, W., & Wang, P. (2008). Effect of acetylation on the properties of corn starch. *Food Chemistry*, *106*(3), 923–928. https://doi.org/10.1016/j.foodchem.2007.07.002

Choy, A. L., May, B. K., & Small, D. M. (2012). The effects of acetylated potato starch and sodium carboxymethyl cellulose on the quality of instant fried noodles. *Food Hydrocolloids*, *26*(1), 2–8.

Cisse, M., Koffi, D. M., & Akaffou, F. A. (2023). Functional properties of acetylation-modified starches of three purple maize cultivars from Côte d'Ivoire. *European Journal of Nutrition & Food Safety*, *15*(10), 63–72. https://doi.org/10.9734/ejnfs/2023/v15i101346

Colussi, R., El Halal, S. L. M., Pinto, V. Z., Bartz, J., Gutkoski, L. C., Zavareze, E. D. R., & Dias, A. R. G. (2015). Acetylation of rice starch in an aqueous medium for use in food. *LWT*, *62*(2), 1076–1082. https://doi.org/10.1016/j.lwt.2015.01.053

Cornejo-Ramírez, Y. I., Martínez-Cruz, O., Del Toro-Sánchez, C. L., Wong-Corral, F. J., Borboa-Flores, J., & Cinco-Moroyoqui, F. J. (2018). The structural characteristics of starches and their functional properties. *CYTA—Journal of Food*, *16*(1), 1003–1017. https://doi.org/10.1080/19476337.2018.1518343

Das, A. B., Singh, G., Singh, S., & Riar, C. S. (2010). Effect of acetylation and dual modification on physicochemical, rheological and morphological characteristics of sweet potato (Ipomoea batatas) starch. *Carbohydrate Polymers*, *80*(3), 725–732.

Debet, M. R., & Gidley, M. J. (2006). Three classes of starch granule swelling: Influence of surface proteins and lipids. *Carbohydrate Polymers*, *64*(3), 452–465. https://doi.org/10.1016/j.carbpol.2005.12.011

DeGraaf, R. A., Broekroelofs, A., & Janssen, L. P. B. M. (1998). The acetylation of starch by reactive extrusion. *Starch/Staerke*, *50*(5), 198–205. https://doi.org/10.1002/(SICI)1521-379X(199805)50:5<198::AID-STAR198>3.0.CO;2-O

Diamantino, V. R., Costa, M. S., Taboga, S. R., Vilamaior, P. S. L., Franco, C. M. L., & Penna, A. L. B. (2019). Starch as a potential fat replacer for application in cheese: Behaviour of different starches in casein/starch mixtures and in the casein matrix. *International Dairy Journal*, *89*, 129–138. https://doi.org/10.1016/j.idairyj.2018.08.015

Diop, C. I. K., Li, H. L., Xie, B. J., & Shi, J. (2011a). Combinatorial effects of mechanical activation and chemical stimulation on the microwave assisted acetylation of corn (Zea mays) starch. *Starch/Staerke*, *63*(2), 96–105. https://doi.org/10.1002/star.201000106

Diop, C. I. K., Li, H. L., Xie, B. J., & Shi, J. (2011b). Effects of acetic acid/acetic anhydride ratios on the properties of corn starch acetates. *Food Chemistry*, *126*(4), 1662–1669. https://doi.org/10.1016/j.foodchem.2010.12.050

Dreher, M. L., Dreher, C. J., & Berry, J. W. (1984). Starch digestibility of foods: A nutritional perspective. *CRC Critical Reviews in Food Science and Nutrition*, *20*(1), 47–71. https://doi.org/10.1080/10408398409527383

Eguchi, S., Kitamoto, N., Nishinari, K., & Yoshimura, M. (2014). Effects of esterified tapioca starch on the physical and thermal properties of Japanese white salted noodles prepared partly by residual heat. *Food Hydrocolloids*, *35*, 198–208. https://doi.org/10.1016/j.foodhyd.2013.05.012

El Halal, S. L. M., Colussi, R., Pinto, V. Z., Bartz, J., Radunz, M., Carreño, N. L. V., & Rosa Zavareze, E. (2015). Structure, morphology and functionality of acetylated and oxidised barley starches. *Food Chemistry*, *168*, 247–256.

Eshag Osman, M. F., Mohamed, A. A., Mohamed Ahmed, I. A., Alamri, M. S., Al Juhaimi, F. Y., Hussain, S., Ibraheem, M. A., & Qasem, A. A. (2022, April). Acetylated corn starch as a fat replacer: Effect on physiochemical, textural, and sensory attributes of beef patties during frozen storage. *Food Chemistry*, *388*, 132988. https://doi.org/10.1016/j.foodchem.2022.132988

Fan, Y., & Picchioni, F. (2020). Modification of starch: A review on the application of "green" solvents and controlled functionalization. *Carbohydrate Polymers*, *241*. https://doi.org/10.1016/j.carbpol.2020.116350

Fitch-Vargas, P. R., Camacho-Hernández, I. L., Martínez-Bustos, F., Islas-Rubio, A. R., Carrillo-Cañedo, K. I., Calderón-Castro, A., & Aguilar-Palazuelos, E. (2019). Mechanical, physical and microstructural properties of acetylated starch-based biocomposites reinforced with acetylated sugarcane fiber. *Carbohydrate Polymers*, *219*, 378–386.

Franco, M. J., Martin, A. A., Bonfim, L. F., Caetano, J., Linde, G. A., & Dragunski, D. C. (2017). Effect of plasticizer and modified starch on biodegradable films for strawberry protection. *Journal of Food Processing and Preservation*, *41*(4), 1–9. https://doi.org/10.1111/jfpp.13063

Gałkowska, D., Kapuśniak, K., & Juszczak, L. (2023). Chemically modified starches as food additives. *Molecules*, *28*(22), 7543.

García-Tejeda, Y. V., Salinas-Moreno, Y., & Martínez-Bustos, F. (2015, October). Acetylation of normal and waxy maize starches as encapsulating agents for maize anthocyanins microencapsulation. *Food and Bioproducts Processing*, *94*, 717–726. https://doi.org/10.1016/j.fbp.2014.10.003

Garg, S., & Jana, A. K. (2011). Characterization and evaluation of acylated starch with different acyl groups and degrees of substitution. *Carbohydrate Polymers*, *83*(4), 1623–1630. https://doi.org/10.1016/j.carbpol.2010.10.015

Ge, X., Hu, Y., Shen, H., Liang, W., Sun, Z., Zhang, X., Ospankulova, G., Muratkhan, M., Kh, K. Z., & Li, W. (2023). Electron beam irradiation application for improving the multiscale structure and enhancing physicochemical and digestive properties of acetylated naked barley. *Food Chemistry*, *404*, 134674. https://doi.org/10.1016/j.foodchem.2022.134674

Golachowski, A., Zięba, T., Kapelko-Zeberska, M., Drozdz, W., Gryszkin, A., & Grzechac, M. (2015). Current research addressing starch acetylation. *Food Chemistry*, *176*, 350–356. https://doi.org/10.1016/j.foodchem.2014.12.060

Han, F., Gao, C., Liu, M., Huang, F., & Zhang, B. (2013). Synthesis, optimization and characterization of acetylated corn starch with the high degree of substitution. *International Journal of Biological Macromolecules*, *59*, 372–376. https://doi.org/10.1016/j.ijbiomac.2013.04.080

Han, F., Liu, M., Gong, H., Lü, S., Ni, B., & Zhang, B. (2012). Synthesis, characterization and functional properties of low substituted acetylated corn starch. *International Journal of Biological Macromolecules*, *50*(4), 1026–1034. https://doi.org/10.1016/j.ijbiomac.2012.02.030

Hong, J., Chen, R., Zeng, X. A., & Han, Z. (2016). Effect of pulsed electric fields assisted acetylation on morphological, structural and functional characteristics of potato starch. *Food Chemistry*, *192*, 15–24. https://doi.org/10.1016/j.foodchem.2015.06.058

Hong, J., Li, C., An, D., Liu, C., Li, L., Han, Z., Zeng, X. A., Zheng, X., & Cai, M. (2020). Differences in the rheological properties of esterified total, A-type, and B-type wheat starches and their effects on the quality of noodles. *Journal of Food Processing and Preservation*, *44*(3), 14342. https://doi.org/10.1111/jfpp.14342

Hong, J., Zeng, X. A., Brennan, C. S., Brennan, M., & Han, Z. (2016). Recent advances in techniques for starch esters and the applications: A review. *Foods*, *5*(3), 1–15. https://doi.org/10.3390/foods5030050

Hong, J., Zeng, X. A., Buckow, R., Han, Z., & Wang, M. S. (2016). Nanostructure, morphology and functionality of cassava starch after pulsed electric fields assisted acetylation. *Food Hydrocolloids*, *54*, 139–150. https://doi.org/10.1016/j.foodhyd.2015.09.025

Huang, J., Schols, H. A., Jin, Z., Sulmann, E., & Voragen, A. G. (2007). Characterization of differently sized granule fractions of yellow pea, cowpea and chickpea starches after modification with acetic anhydride and vinyl acetate. *Carbohydrate Polymers*, *67*(1), 11–20.

Imre, B., & Vilaplana, F. (2020). Organocatalytic esterification of corn starches towards enhanced thermal stability and moisture resistance. *Green Chemistry*, *22*(15), 5017–5031. https://doi.org/10.1039/d0gc00681e

Jambrak, A. R., Herceg, Z., Šubarić, D., Babić, J., Brnčić, M., Brnčić, S. R., Bosiljkov, T., Čvek, D., Tripalo, B., & Gelo, J. (2010). Ultrasound effect on physical properties of corn starch. *Carbohydrate Polymers*, *79*(1), 91–100. https://doi.org/10.1016/j.carbpol.2009.07.051

Kaushal, P., Kumar, V., & Sharma, H. K. (2012). Comparative study of physicochemical, functional, antinutritional and pasting properties of taro (Colocasia esculenta), rice (Oryza sativa) flour, pigeonpea (Cajanus cajan) flour and their blends. *LWT*, *48*(1), 59–68. https://doi.org/10.1016/j.lwt.2012.02.028

Khan, Z., Javed, F., Shamair, Z., Hafeez, A., Fazal, T., Aslam, A., Zimmerman, W. B., & Rehman, F. (2021). Current developments in esterification reaction: A review on process and parameters. *Journal of Industrial and Engineering Chemistry*, *103*(xxxx), 80–101. https://doi.org/10.1016/j.jiec.2021.07.018

Khurshida, S., Das, M. J., Deka, S. C., & Sit, N. (2021). Effect of dual modification sequence on physicochemical, pasting, rheological and digestibility properties of cassava starch modified by acetic acid and ultrasound. *International Journal of Biological Macromolecules*, *188*, 649–656. https://doi.org/10.1016/j.ijbiomac.2021.08.062

Kumoro, A. C., Amalia, R., Budiyati, C. S., Retnowati, D. S., & Ratnawati, R. (2015). Preparation and characterization of physicochemical properties of glacial acetic acid modified Gadung (Diocorea hispida Dennst) flours. *Journal of Food Science and Technology*, *52*(10), 6615–6622. https://doi.org/10.1007/s13197-015-1723-5

Lawal, O. S. (2004). Composition, physicochemical properties and retrogradation characteristics of native, oxidised, acetylated and acid-thinned new cocoyam (Xanthosoma sagittifolium) starch. *Food Chemistry*, *87*(2), 205–218. https://doi.org/10.1016/j.foodchem.2003.11.013

Lee, H. L., & Yoo, B. (2009). Dynamic rheological and thermal properties of acetylated sweet potato starch. *Starch/Staerke*, *61*(7), 407–413. https://doi.org/10.1002/star.200800109

Lee, Y. T., & Puligundla, P. (2016). Characteristics of reduced-fat muffins and cookies with native and modified rice starches. *Emirates Journal of Food and Agriculture*, *28*(5), 311–316. https://doi.org/10.9755/ejfa.2015-05-227

Lin, D., Zhou, W., Yang, Z., Zhong, Y., Xing, B., Wu, Z., Chen, H., Wu, D., Zhang, Q., Qin, W., & Li, S. (2019). Study on physicochemical properties, digestive properties and application of acetylated starch in noodles. *International Journal of Biological Macromolecules*, *128*, 948–956. https://doi.org/10.1016/j.ijbiomac.2019.01.176

Lin, Z. R., Zeng, X. A., Yu, S. J., & Sun, D. W. (2012). Enhancement of ethanolacetic acid esterification under room temperature and non-catalytic condition via pulsed electric field application. *Food and Bioprocess Technology*, *5*(7), 2637–2645.

Lo Leggio, L., Dal Degan, F., Poulsen, P., Andersen, S. M., & Larsen, S. (2003). The structure and specificity of Escherichia coli maltose acetyltransferase give new insight into the lacA family of acyltransferases. *Biochemistry*, *42*(18), 5225–5235. https://doi.org/10.1021/bi0271446

Luchese, C. L., Frick, J. M., Patzer, V. L., Spada, J. C., & Tessaro, I. C. (2015). Synthesis and characterization of biofilms using native and modified pinhão starch. *Food Hydrocolloids*, *45*, 203–210. https://doi.org/10.1016/j.foodhyd.2014.11.015

Luo, Z. G., & Shi, Y. C. (2012). Preparation of acetylated waxy, normal, and high-amylose maize starches with intermediate degrees of substitution in aqueous solution and their properties. *Journal of Agricultural and Food Chemistry*, *60*(37), 9468–9475. https://doi.org/10.1021/jf301178c

Luo, Z. G., & Shi, Y. C. (2018). Distribution of acetyl groups in acetylated waxy maize starches prepared in aqueous solution with two different alkaline concentrations. *Food Hydrocolloids*, *79*, 491–497. https://doi.org/10.1016/j.foodhyd.2018.01.015

Ma, R., Jin, Z., Wang, F., & Tian, Y. (2022). Contribution of starch to the flavor of rice-based instant foods. *Critical Reviews in Food Science and Nutrition*, *62*(31), 8577–8588. https://doi.org/10.1080/10408398.2021.1931021

Mathew, K. B., & Ojo, J. O. (2015). Influence of acetylation on the physicochemical properties of composited starches from sweet potato (Ipomoea batatas L.) and water yam (Dioscorea alata L.). *African Journal of Biotechnology*, *14*(51), 3340–3349. https://doi.org/10.5897/ajb2015.14881

Mbougueng, P. D., Tenin, D., Scher, J., & Tchiégang, C. (2012). Influence of acetylation on physicochemical, functional and thermal properties of potato and cassava starches. *Journal of Food Engineering*, *108*(2), 320–326. https://doi.org/10.1016/j.jfoodeng.2011.08.006

Mehltretter, C. L., & Mark, A. M. (1974). Process for making starch triacetates. *US Patent*, *795*(670).

Minimol, P. F., Paul, W., & Sharma, C. P. (2013). PEGylated starch acetate nanoparticles and its potential use for oral insulin delivery. *Carbohydrate Polymers*, *95*(1), 1–8. https://doi.org/10.1016/j.carbpol.2013.02.021

Mirmoghtadaie, L., Kadivar, M., & Shahedi, M. (2009). Effect of modified oat starch and protein on batter properties and quality of cake. *Cereal Chemistry*, *86*(6), 685–691. https://doi.org/10.1094/CCHEM-86-6-0685

Miyazaki, M., Van Hung, P., Maeda, T., & Morita, N. (2006). Recent advances in application of modified starches for breadmaking. *Trends in Food Science and Technology*, *17*(11), 591–599. https://doi.org/10.1016/j.tifs.2006.05.002

Mormann, W., & Al-Higari, M. (2004). Acylation of starch with vinyl acetate in water. *Starch/Staerke*, *56*(3–4), 118–121. https://doi.org/10.1002/star.200300238

Muljana, H., Picchioni, F., Heeres, H. J., & Janssen, L. P. B. M. (2010). Green starch conversions: Studies on starch acetylation in densified CO2. *Carbohydrate Polymers*, *82*(3), 653–662. https://doi.org/10.1016/j.carbpol.2010.05.032

Nazarian Firouzabadi, F., Vincken, J. P., Ji, Q., Suurs, L. C. J. M., & Visser, R. G. F. (2007). Expression of an engineered granule-bound Escherichia coli maltose acetyltransferase in wild-type and amf potato plants. *Plant Biotechnology Journal*, *5*(1), 134–145. https://doi.org/10.1111/j.1467-7652.2006.00227.x

Niu, D., Zeng, X.-A., Ren, E.-F., Xu, F.-Y., Li, J., Wang, M.-S., & Wang, R. (2020). Review of the application of pulsed electric fields (PEF) technology for food processing in China. *Food Research International*, *137*, 109715. https://doi.org/10.1016/j.foodres.2020.109715

Ojogbo, E., Ogunsona, E. O., & Mekonnen, T. H. (2020). Chemical and physical modifications of starch for renewable polymeric materials. *Materials Today Sustainability*, *7–8*, 100028. https://doi.org/10.1016/j.mtsust.2019.100028

Okyere, A. Y., Rajendran, S., & Annor, G. A. (2022, January). Cold plasma technologies: Their effect on starch properties and industrial scale-up for starch modification. *Current Research in Food Science*, *5*, 451–463. https://doi.org/10.1016/j.crfs.2022.02.007

Olagunju, A. I., Omoba, O. S., Enujiugha, V. N., Wiens, R. A., Gough, K. M., & Aluko, R. E. (2020). Influence of acetylation on physicochemical and morphological characteristics of pigeon pea starch. *Food Hydrocolloids*, *100*(October 2019), 105424. https://doi.org/10.1016/j.foodhyd.2019.105424

Olatunde, G. O., Arogundade, L. K., & Orija, O. I. (2017). Chemical, functional and pasting properties of banana and plantain starches modified by pre-gelatinization, oxidation and acetylation. *Cogent Food and Agriculture*, *3*(1), 1–12. https://doi.org/10.1080/23311932.2017.1283079

Otache, M. A., Duru, R. U., Achugasim, O., & Abayeh, O. J. (2021). Advances in the modification of starch via esterification for enhanced properties. *Journal of Polymers and the Environment*, *29*(5), 1365–1379. Springer. https://doi.org/10.1007/s10924-020-02006-0

Park, J. J., Olawuyi, I. F., & Lee, W. Y. (2020). Characteristics of low-fat mayonnaise using different modified arrowroot starches as fat replacer. *International Journal of Biological Macromolecules*, *153*, 215–223. https://doi.org/10.1016/j.ijbiomac.2020.02.331

Paulos, G., Mrestani, Y., Heyroth, F., Gebre-Mariam, T., & Neubert, R. H. H. (2016). Fabrication of acetylated dioscorea starch nanoparticles: Optimization of formulation and process variables. *Journal of Drug Delivery Science and Technology*, *31*, 83–92. https://doi.org/10.1016/j.jddst.2015.11.009

Prochaska, K., Kedziora, P., Le Thanh, J., & Lewandowicz, G. (2007). Surface activity of commercial food grade modified starches. *Colloids and Surfaces B: Biointerfaces*, *60*(2), 187–194. https://doi.org/10.1016/j.colsurfb.2007.06.005

Punia, S. (2020). Barley starch modifications: Physical, chemical and enzymatic—a review. *International Journal of Biological Macromolecules*, *144*, 578–585. https://doi.org/10.1016/j.ijbiomac.2019.12.088

Qiao, L., Gu, Q. M., & Cheng, H. N. (2006). Enzyme-catalyzed synthesis of hydrophobically modified starch. *Carbohydrate Polymers*, *66*(1), 135–140. https://doi.org/10.1016/j.carbpol.2006.02.033

Rahim, A., Kadir, S., & Jusman. (2017). The influence degree of substitution on the physicochemical properties of acetylated arenga starches. *International Food Research Journal*, *24*(1), 102–107.

Rahim, A., Mahfudz, M., Kadir, S., & Rostiati, A. (2022). Effect of pH and acetic anhydride concentration on physicochemical characteristics of acetylated sago starch. *IOP Conference Series and Earth Environmental Science*, *1107*, 12124.

Reddy, C. K., Haripriya, S., & Suriya, M. (2014). Effect of acetylationon morphology, pasting and functional properties of starch from banana (Musa Aab). *Indian Journal of Scientific Research and Technology*, *2*(6), 31–36.

Robert, P., García, P., Reyes, N., Chávez, J., & Santos, J. (2012). Acetylated starch and inulin as encapsulating agents of gallic acid and their release behaviour in a hydrophilic system. *Food Chemistry*, *134*(1), 1–8.

Saartrat, S., Puttanlek, C., Rungsardthong, V., & Uttapap, D. (2005). Paste and gel properties of low-substituted acetylated canna starches. *Carbohydrate Polymers*, *61*(2), 211–221. https://doi.org/10.1016/j.carbpol.2005.05.024

Sahnoun, M., Ismail, N., & Kammoun, R. (2016). Enzymatically hydrolysed, acetylated and dually modified corn starch: Physico-chemical, rheological and nutritional properties and effects on cake quality. *Journal of Food Science and Technology*, *53*(1), 481–490. https://doi.org/10.1007/s13197-015-1984-z

Sajilata, M. G., & Singhal, R. S. (2005). Specialty starches for snack foods. *Carbohydrate Polymers*, *59*(2), 131–151.

Salcedo Mendoza, J., Hernández RuyDíaz, J., & Fernández Quintero, A. (2016). Effect of the acetylation process on native starches of yam (Dioscorea spp.). *Revista Facultad Nacional de Agronomia Medellin, 69*(2), 7997–8006. https://doi.org/10.15446/rfna.v69n2.59144

Sánchez-Rivera, M. M., Flores-Ramírez, I., Zamudio-Flores, P. B., González-Soto, R. A., Rodríguez-Ambríz, S. L., & Bello-Pérez, L. A. (2010). Acetylation of banana (Musa paradisiaca L.) and maize (Zea mays L.) starches using a microwave heating procedure and iodine as catalyst: Partial characterization. *Starch/Staerke, 62*(3–4), 155–164. https://doi.org/10.1002/star.200900209

Sarkar, S. (2016). Influence of acetylation and heat-moisture treatment on physio-chemical, pasting and morphological properties of buckwheat (Fagopyrum esculentum) starch. *Asian Journal of Dairy and Food Research, 35*(4), 298–303. https://doi.org/10.18805/ajdfr.v35i4.6628

Shah, A., Masoodi, F. A., Gani, A., & Ashwar, B. A. (2017). Physicochemical, rheological and structural characterization of acetylated oat starches. *LWT, 80*, 19–26. https://doi.org/10.1016/j.lwt.2017.01.072

Shahbazi, M., Jäger, H., Chen, J., & Ettelaie, R. (2021). Construction of 3D printed reduced-fat meat analogue by emulsion gels. Part II: Printing performance, thermal, tribological, and dynamic sensory characterization of printed objects. *Food Hydrocolloids, 121*, 107054. https://doi.org/10.1016/j.foodhyd.2021.107054

Shaikh, M., Ali, T. M., & Hasnain, A. (2017). Utilization of chemically modified pearl millet starches in preparation of custards with improved cold storage stability. *International Journal of Biological Macromolecules, 104*, 360–366. https://doi.org/10.1016/j.ijbiomac.2017.05.183

Shogren, R. L. (2003). Rapid preparation of starch esters by high temperature/pressure reaction. *Carbohydrate Polymers, 52*(3), 319–326.

Shogren, R. L., & Biswas, A. (2010). Acetylation of starch with vinyl acetate in imidazolium ionic liquids and characterization of acetate distribution. *Carbohydrate Polymers, 81*(1), 149–151. https://doi.org/10.1016/j.carbpol.2010.01.045

Sindhu, R., Devi, A., & Khatkar, B. S. (2021, March). Morphology, structure and functionality of acetylated, oxidized and heat moisture treated amaranth starches. *Food Hydrocolloids, 118*, 106800. https://doi.org/10.1016/j.foodhyd.2021.106800

Singh, A. V., Nath, L. K., & Guha, M. (2011). Synthesis and characterization of highly acetylated sago starch. *Starch/Staerke, 63*(9), 523–527. https://doi.org/10.1002/star.201100012

Singh, J., Kaur, L., & Singh, N. (2004). Effect of acetylation on some properties of corn and potato starches. *Starch/Staerke, 56*(12), 586–601. https://doi.org/10.1002/star.200400293

Singh, N., Chawla, D., & Singh, J. (2004). Influence of acetic anhydride on physicochemical, morphological and thermal properties of corn and potato starch. *Food Chemistry, 86*(4), 601–608. https://doi.org/10.1016/j.foodchem.2003.10.008

Sitanggang, A., Sani, P., & Mastuti, T. (2020). *Modification of Mung Bean starch by annealing treatment and acetylation* (pp. 10–19). SciTePress. https://doi.org/10.5220/0009977100100019

Sodhi, N. S., & Singh, N. (2005). Characteristics of acetylated starches prepared using starches separated from different rice cultivars. *Journal of Food Engineering, 70*(1), 117–127. https://doi.org/10.1016/j.jfoodeng.2004.09.018

Sondari, D., Aspiyanto, Amanda, A. S., Triwulandari, E., Ghozali, M., Septiyanti, M., & Iltizam, I. (2018). Characterization edible coating made from native and modification cassava starch. *AIP Conference Proceedings, 2049*. https://doi.org/10.1063/1.5082514

Subroto, E. (2020). Review on the analysis methods of starch, amylose, amylopectin food and agricultural products. *International Journal of Emerging Trends in Engineering Research, 8*(7), 3519–3524. https://doi.org/10.30534/ijeter/2020/103872020

Subroto, E., Cahyana, Y., Indiarto, R., & Rahmah, T. A. (2023). Modification of starches and flours by acetylation and its dual modifications: A review of impact on physicochemical properties and their applications. *Polymers, 15*(14), 2990. https://doi.org/10.3390/polym15142990

Suri, S., & Singh, A. (2023). Modification of starch by novel and traditional ways: Influence on the structure and functional properties. *Sustainable Food Technology, 1*(3), 348–362. https://doi.org/10.1039/d2fb00043a

Tian, S., Chen, Y., Chen, Z., Yang, Y., & Wang, Y. (2018). Preparation and characteristics of starch esters and its effects on dough physicochemical properties. *Journal of Food Quality, 2018*, 1–7. https://doi.org/10.1155/2018/1395978

Trela, V. D., Ramallo, A. L., & Albani, O. A. (2020). Synthesis and characterization of acetylated cassava starch with different degrees of substitution. *Brazilian Archives of Biology and Technology, 63*, 1–13. https://doi.org/10.1590/1678-4324-2020180292

Tupa, M. V., Ávila Ramírez, J. A., Vázquez, A., & Foresti, M. L. (2015). Organocatalytic acetylation of starch: Effect of reaction conditions on DS and characterisation of esterified granules. *Food Chemistry, 170*, 295–302. https://doi.org/10.1016/j.foodchem.2014.08.062

Ulfa, G. M., Putri, W. D. R., Fibrianto, K., Prihatiningtyas, R., & Widjanarko, S. B. (2020). The influence of temperature in swelling power, solubility, and water binding capacity of pregelatinised sweet potato starch. *IOP Conference Series: Earth and Environmental Science, 475*(1), 12036. https://doi.org/10.1088/1755-1315/475/1/012036

Ulfa, G. M., Putri, W. D. R., & Widjanarko, S. B. (2019, July). The influence of sodium acetate anhydrous in swelling power, solubility, and water binding capacity of acetylated sweet potato starch. *AIP Conference Proceedings, 2120.* https://doi.org/10.1063/1.5115697

Vanmarcke, A., Leroy, L., Stoclet, G., Duchatel-Crépy, L., Lefebvre, J. M., Joly, N., & Gaucher, V. (2017). Influence of fatty chain length and starch composition on structure and properties of fully substituted fatty acid starch esters. *Carbohydrate Polymers, 164,* 249–257. https://doi.org/10.1016/j.carbpol.2017.02.013

Vidal, N. P., Bai, W., Geng, M., & Martinez, M. M. (2022). Organocatalytic acetylation of pea starch: Effect of alkanoyl and tartaryl groups on starch acetate performance. *Carbohydrate Polymers, 294*(231), 119780. https://doi.org/10.1016/j.carbpol.2022.119780

Wang, L., Wang, H., Zhang, B., Popkin, B. M., & Du, S. (2020). Elevated fat intake increases body weight and the risk of overweight and obesity among chinese adults: 1991–2015 trends. *Nutrients, 12*(11), 1–13. https://doi.org/10.3390/nu12113272

Wang, S., Jin, F., & Yu, J. (2013). Pea starch annealing: New insights. *Food and Bioprocess Technology, 6*(12), 3564–3575. https://doi.org/10.1007/s11947-012-1010-7

Wang, Y. J., & Wang, L. (2002). Characterization of acetylated waxy maize starches prepared under catalysis by different alkali and alkaline-earth hydroxides. *Starch/Staerke, 54*(1), 25–30. https://doi.org/10.1002/1521-379X(200201)54:1<25::AID-STAR25>3.0.CO;2-T

Wani, I. A., Sogi, D. S., & Gill, B. S. (2012). Physicochemical properties of acetylated starches from some indian kidney bean (Phaseolus vulgaris L.) cultivars. *International Journal of Food Science and Technology, 47*(9), 1993–1999. https://doi.org/10.1111/j.1365-2621.2012.03062.x

Wani, I. A., Sogi, D. S., & Gill, B. S. (2015). Physico-chemical properties of acetylated starches from Indian black gram (Phaseolus mungo L.) cultivars. *Journal of Food Science and Technology, 52*(7), 4078–4089.

Wani, I. A., Sogi, D. S., Wani, A. A., Gill, B. S., & Shivhare, U. S. (2010). Physico-chemical properties of starches from Indian kidney bean (Phaseolus vulgaris) cultivars. *International Journal of Food Science and Technology, 45*(10), 2176–2185. https://doi.org/10.1111/j.1365-2621.2010.02379.x

Wickramasinghe, H. A. M., Yamamoto, K., Yamauchi, H., & Noda, T. (2009). Effect of low level of starch acetylation on physicochemical properties of potato starch. *Food Science and Biotechnology, 18*(1), 118–123.

Xiao, H., Yang, F., Lin, Q., Zhang, Q., Tang, W., Zhang, L., Xu, D., & Liu, G. Q. (2019). Preparation and properties of hydrophobic films based on acetylated broken-rice starch nanocrystals for slow protein delivery. *International Journal of Biological Macromolecules, 138,* 556–564. https://doi.org/10.1016/j.ijbiomac.2019.07.121

Xu, J., & Shi, Y. C. (2019, May). Position of acetyl groups on anhydroglucose unit in acetylated starches with intermediate degrees of substitution. *Carbohydrate Polymers, 220,* 118–125. https://doi.org/10.1016/j.carbpol.2019.05.059

Yadav, D. K., & Patki, P. E. (2015). Effect of acetyl esterification on physicochemical properties of chick pea (Cicer arietinum L.) starch. *Journal of Food Science and Technology, 52*(7), 4176–4185. https://doi.org/10.1007/s13197-014-1388-5

Yermekov, Y. Y., Toimbayeva, D. B., Kamanova, S. G., Murat, L. A., Muratkhan, M., Saduakhasova, S. A., Aidarkhanova, G. S., & Ospankulova, G. K. (2021). Investigation of the effect of acetylation on the physicochemical properties of grain starches. *Bulletin of the Karaganda University Biology. Medicine Geography Series, 104*(4), 22–30.

Zarski, A., Bajer, K., Zarska, S., & Kapusniak, J. (2019). From high oleic vegetable oils to hydrophobic starch derivatives: I. Development and structural studies. *Carbohydrate Polymers, 214,* 124–130. https://doi.org/10.1016/j.carbpol.2019.03.034

Zeng, X. A., Han, Z., & Zi, Z. H. (2010). Effects of pulsed electric field treatments on quality of peanut oil. *Food Control, 21*(5), 611–614. https://doi.org/10.1016/j.foodcont.2009.09.004

Zhang, F. L., Liang, Y., Tan, C. P., Lu, Y. M., & Cui, B. (2014). Research on the water-holding capacity of pork sausage with acetate cassava starch. *Starch/Staerke, 66*(11–12), 1033–1040. https://doi.org/10.1002/star.201400006

Zhang, K., Cheng, F., Zhang, K., Hu, J., Xu, C., Lin, Y., Zhou, M., & Zhu, P. (2019). Synthesis of long-chain fatty acid starch esters in aqueous medium and its characterization. *European Polymer Journal, 119,* 136–147. https://doi.org/10.1016/j.eurpolymj.2019.07.021

Zhang, K., Dai, Y., Hou, H., Li, X., Dong, H., Wang, W., & Zhang, H. (2019). Influences of grinding on structures and properties of mung bean starch and quality of acetylated starch. *Food Chemistry, 294*(61), 285–292. https://doi.org/10.1016/j.foodchem.2019.05.055

Zhang, L., Xie, W., Zhao, X., Liu, Y., & Gao, W. (2009). Study on the morphology, crystalline structure and thermal properties of yellow ginger starch acetates with different degrees of substitution. *Thermochimica Acta, 495*(1–2), 57–62. https://doi.org/10.1016/j.tca.2009.05.019

Zhang, L., Zuo, B., Wu, P., Wang, Y., & Gao, W. (2012). Ultrasound effects on the acetylation of dioscorea starch isolated from Dioscorea zingiberensis C. H. Wright. *Chemical Engineering and Processing: Process Intensification, 54*, 29–36. https://doi.org/10.1016/j.cep.2012.01.005

Zhao, K., Li, B., Xu, M., Jing, L., Gou, M., Yu, Z., Zheng, J., & Li, W. (2018). Microwave pretreated esterification improved the substitution degree, structural and physicochemical properties of potato starch esters. *LWT, 90*(September 2017), 116–123. https://doi.org/10.1016/j.lwt.2017.12.021

Zhao, K., Saleh, A. S. M., Li, B., Wu, H., liu, Y., Zhang, G., & Li, W. (2018). Effects of conventional and microwave pretreatment acetylation on structural and physicochemical properties of wheat starch. *International Journal of Food Science and Technology, 53*(11), 2515–2524. https://doi.org/10.1111/ijfs.13845

Zhao, X., Zeng, L., Huang, Q., Zhang, B., Zhang, J., & Wen, X. (2022). Structure and physicochemical properties of cross-linked and acetylated tapioca starches affected by oil modification. *Food Chemistry, 386*, 132848. https://doi.org/10.1016/j.foodchem.2022.132848

Oxidation of Starch

13

Plachikkattu Parambil Akhila,
Sneh Punia Bangar, and K.V. Sunooj

13.1 INTRODUCTION

Oxidation is widely employed to modify starches characterized by low viscosity at increased solid concentrations. Oxidation of starch is conducted in a mild to moderately alkaline condition with suitable oxidizing reagents. In the process of starch oxidation, OH groups, particularly those located at positions C-2, C-3, and C-6, transform into C=O or COOH groups. Hence, the quantity of C=O and COOH groups present in oxidized starch serves as an indicator of the level of oxidation (Zhang et al., 2012). Oxidized starch has exhibited superior film-forming characteristics and contributes to the enhancement of strength and printability of the paper. Furthermore, oxidized starch has been extensively employed in textile and laundry finishing sectors (Zhang et al., 2012). Currently, hypochlorite oxidation stands as the predominant method for industrially producing oxidized starches (Kuakpetoon & Wang, 2001). Permanganate, persulfate, periodate, and peroxide are being explored as alternative oxidizing agents in laboratory experiments (Zhang et al., 2012). Ozone is perceived as an innovative chemical treatment and is classified as a clean-label ingredient (Maniglia et al., 2021). Although oxidized starch is mostly employed in the paper and textile industries, its applications in the food sectors are expanding owing to its attributes, such as lower viscosity, higher paste clarity, increased stability, improved binding, and film-forming characteristics (Vanier et al., 2017). The food sector uses oxidized starch as coatings and sealants in confectionery, emulsifiers, dough conditioners, gum Arabic substitutes, and binding agents in batters (Kuakpetoon & Wang, 2001). The current chapter focuses on the chemical modification of starch by oxidation, its mechanism, various oxidizing agents, the physicochemical characteristics of modified starch, and its application in the industry.

13.2 STARCH OXIDATION

The practice of oxidation has been widespread since the early 1800s, establishing oxidized starch as the most widely used method of chemical modification (J. Li et al., 2023). Starch oxidation is achieved by subjecting the starch molecule to a suitable oxidizing agent under carefully regulated pH and temperature. Common oxidizing reagents employed in the production of oxidized starch consist of permanganate, hydrogen peroxide, sodium hypochlorite, ozone, periodate, nitrogen dioxide, and chromic acid (Singh et al., 2016; Salimi et al., 2023). Studies regarding major oxidizing agents such as NaOCl, H_2O_2, and O_3 are given in Table 13.1. Among the mentioned oxidants, sodium hypochlorite (NaOCl) stands out as the accepted and oldest oxidizing agent for the commercial production of oxidized starches (Chong et al.,

DOI: 10.1201/9781032655598-13

TABLE 13.1 Studies on the Effect of Oxidation of Starch Using NaOCl, H_2O_2, and O_3

STARCH SOURCE	REAGENT USED AND CONDITIONS	MAJOR FINDINGS	REFERENCE
Corn	Sodium hypochlorite 3% active chlorine 130 min	0.85% COOH, surface fissures, increased RC, reduced Mw, decreased gelatinization temperature.	Y. Li et al. (2023)
Jackfruit seed	Sodium hypochlorite 1% active chlorine 30 min	No change in morphology and crystalline pattern, decreased crystallinity and swelling power, decreased pasting characteristics, lower syneresis, and improved paste clarity.	Naknaen et al. (2017)
Potato	Sodium hypochlorite active chlorine 0.5, 1.0, and 1.5%	Biodegradable films lower tensile strength, water solubility, and water vapor permeability.	Fonseca et al. (2015)
Jackfruit seed	Hydrogen peroxide 0.5 to 2.0%	Rough granular structure, small pores, and the edges lost definition completely. A peak at 1760 cm^{-1} indicates C=O.	Tung et al. (2021)
Cassava	Hydrogen peroxide Catalyst:0.5% $CuSO_4$	Reduced the intrinsic viscosity and thermal stability, increased light transmittance, and reaction time reduced to 1 h from 72 h.	Y. R. Zhang et al. (2012)
Wheat	Sodium hypochlorite or hydrogen peroxide 2, 5, and 10% oxidant	Reduced polymer size, reduction in pasting characteristics, H_2O_2: higher gelatinization temperatures, NaOCl decreased gelatinization temperatures.	Chapagai et al. (2021)
Cassava	Sodium hypochlorite and hydrogen peroxide 3% oxidant 30 and 300 min	NaOCl favored carboxyl group formation. H_2O_2 showed more carbonyl groups, decreased apparent viscosity, increased gelatinization temperatures following H_2O_2 oxidation, and decreased NaOCl oxidation.	Sangseethong et al., (2010)
Potato	Ozone 15, 30, 45, and 60 min	No change in RC; increase in the carbonyl, carboxyl, and reducing sugar contents; decrease in the pH, apparent amylose content, and molecular size.	Castanha et al. (2017)
Arrowroot	Ozone 13 parts per million (ppm) applied repeatedly (10–30 cycles)	Granules become rougher and fissured; decreased crystallinity; increased swelling volume, solubility, water absorption capacity, and gel strength; decreased syneresis and ΔT and increased ΔH.	Handarini et al. (2020)
Tartary buckwheat	Ozone 0, 2.5, 7.5, 15, and 20 min	Reduced granule smoothness and integrity; enhanced short-range and long-range order structure; increased apparent viscosity, swelling power, and solubility; decreased ΔH and viscoelasticity.	Hu et al. (2022)
Corn	Ozone 10 min	Biodegradable film with increased tensile strength; elongation at break; and water contact angle by 276%, 72.5%, and 154% and decreased water vapor permeability by 34.7%.	Huang et al. (2024)

2013). Nevertheless, its use poses environmental challenges due to the persistence of chlorine, potentially resulting in water and air pollution (Salimi et al., 2023; Ojogbo et al., 2020).

Generally, starch oxidation encompasses the following mechanisms:

- Oxidation of starch: The OH group present in C-2, C-3, and C-6 positions in the glucose ring is oxidized to the C=O group and subsequently to more stable COOH functional groups
- Depolymerization of starch: Breakdown of α-1,4 glycosidic linkages of polymer chains in the starch.

The introduction of these voluminous COOH groups leads to a reduced tendency for retrogradation in the oxidized starch gel. The number of C=O and COOH groups present in starch governs the degree of modification (Singh et al., 2016). The degree of modification is influenced by both intrinsic properties, such as the origin, molecular structure, and crystallinity of the starch, as well as reaction conditions, including the concentration of oxidants, type of catalyst, pH, and temperature (Aaliya, Sunooj, Navaf, et al., 2022). A less intense form of oxidation characterized by the concentration <0.1% COOH groups is termed bleaching (Y. F. Chen et al., 2017). Oxidized starches exhibit decreased viscosity, reduced gelatinization temperatures, decreased gelatinization enthalpy, and lower retrogradation rates (Singh et al., 2016). Additional properties of oxidized starch include improved clarity, enhanced binding properties, and film-forming capabilities, making it suitable for applications in batters, breading, and fillings in confectionery.

13.3 OXIDIZING AGENTS

13.3.1 Sodium Hypochlorite (NaOCl)

13.3.1.1 Method of Preparation

Sodium hypochlorite is extensively utilized in the industrial synthesis of oxidized starch, owing to its readily available nature, and has the potential to alter the characteristics of starch efficiently (J. Li et al., 2023). Modification of starch is done by suspending the starch in distilled water and adding the NaOCl solution drop by drop at a temperature below gelatinization (25–50°C) and controlled pH with continuous stirring (Rostamabadi et al., 2022). Starch oxidation with NaOCl is typically carried out under a mild alkaline environment. It is attributed to the significant impact of pH on the chemical kinetics between NaOCl and starch, where an accelerated reaction rate is observed at a pH near 7, while a reduced rate occurs at pH 10 (Dimri et al., 2023). Maintaining a mild alkaline pH facilitates the formation of more COOH groups, which in turn can impede starch retrogradation and contribute to the stabilized viscosity of the starch dispersion (Dimri et al., 2023). The FDA of the US permits a maximum hypochlorite level equal to 0.25 moles of active chlorine per mole of α-D-glucose unit (Xie et al., 2005).

13.3.1.2 Mechanism

Figure 13.1 illustrates the primary reactions involved in NaOCl starch oxidation, which include the breakage of the bonds between the glucose units and the oxidation of OH groups to C=O and COOH groups. According to Richardson and Gorton, starch molecules depolymerize as a result of the breaking of glycosidic linkages during oxidation, which provides an environment that is conducive to the formation of C=O and COOH groups (Zhou et al., 2016). The oxidation process occurs randomly at various sites, including primary OH group (C-6), secondary OH group (C-2, C-3, and C-4), glycol groups (breakage of bond between C-2–C-3), and reactive terminal aldehydes (Xie et al., 2005). The rate of increase in COOH content surpassed that of the C=O content, suggesting that OH groups on the glucopyranose ring were first

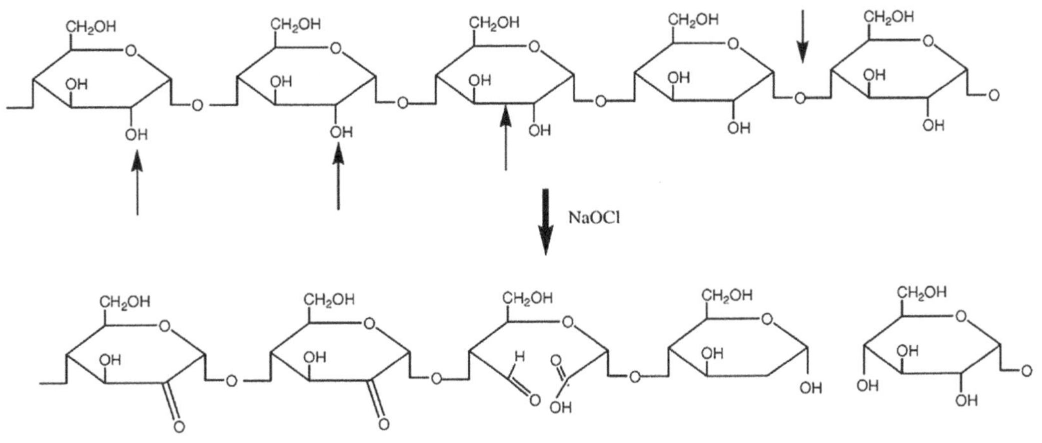

FIGURE 13.1   NaOCl oxidation of starch.

oxidized to C=O groups and subsequently to COOH groups, constituting the primary stable compound (Zhou et al., 2016). The level of the oxidation process is contingent on multiple factors, including reaction time, temperature, pH, concentration of NaOCl, source, and the structural arrangement of the starch. The proportion of C=O and COOH groups within starch significantly influences the physicochemical characteristics of modified starch (Navaf et al., 2023). Starch subjected to oxidation with NaOCl displays increased resistance to amylase activity, the capacity to bind with calcium ions, along with polyelectrolyte traits and enhanced stability at higher temperatures (Dimri et al., 2023).

13.3.1.3 Physico-Chemical and Functional Characteristics

The properties of oxidized starch highly rely on the concentration of the oxidant, level of oxidation, pH, temperature, time, and source of starch. The alteration of OH groups in starch is influenced by its botanical origin as well as other impurities, particularly the protein content (Navaf et al., 2021). Impurities in the starch hinder the oxidation of starch, thereby reducing the effectiveness of the reaction (Lewicka et al., 2015). The dilution of active chlorine influences the levels of C=O and COOH content in modified starches. The impact of bean starch oxidation at various concentrations of active chlorine on the crystalline, morphological, physico-chemical, and pasting characteristics of the starch was examined by Vanier et al. (2012). NaOCl oxidized starch at 1% and 1.5% active chlorine showed enhanced C=O and COOH content, solubility, and whiteness while decreasing relative crystallinity, gel hardness, swelling power, and pasting compared to native starch. A gradual rise in the C=O and COOH content was observed with an increment in the amount of NaOCl in maize starch (Spier et al., 2013). Jackfruit seed starch was oxidized with NaOCl by Naknaen et al. (2017). The findings also supported the previous studies that the characteristics of the hypochlorite-oxidized starch are contingent upon the concentration of active chlorine (Naknaen et al., 2017).

Acidic conditions are typically less favored due to the relatively low oxidation rate and the generation of undesirable chlorine by-products. The COOH functional groups were the predominant outcome of NaOCl treatment, and their prominence was more noticeable under alkaline conditions. Remarkably, Chapagai et al. (2021) found that oxidation with hypochlorite under acidic conditions results in the formation of a comparable amount of C=O group and a lesser amount of COOH in an alkaline environment. Therefore, NaOCl under acidic conditions could be beneficial for the gentle oxidation of starch, resulting in the production of a small amount of C=O functional groups (Chapagai et al., 2021). A 10% NaOCl-modified corn starch exhibited significantly higher COOH content compared to 2% and 5% NaOCl. Potato starch showed a significantly higher COOH group followed by corn, rice, cassava, and wheat at low concentrations (2%) of NaOCl, showing that the source of starch influences the oxidation efficiency

(Chapagai et al., 2021). The impact of NaOCl on molecular structure varied depending on pH. A large size fraction showed a significant reduction percentage with an increase in the concentration of chlorine irrespective of the pH condition. At 10% concentration of NaOCl, a large size fraction showed a 100% reduction in molecular weight under alkaline conditions. At a 5% concentration of NaOCl, the highest depolymerization efficiency exhibited in acidic pH, neutral followed by alkaline condition (Chapagai et al., 2021).

The granular structure of the oxidized starches was unchanged, with cracks and pores on the surface. Additionally, the Maltase cross and the crystalline type were maintained following the modification (Figure 13.2) (Zhou et al., 2016). Oxidation of starch at 1.5% active chlorine exhibited damage on

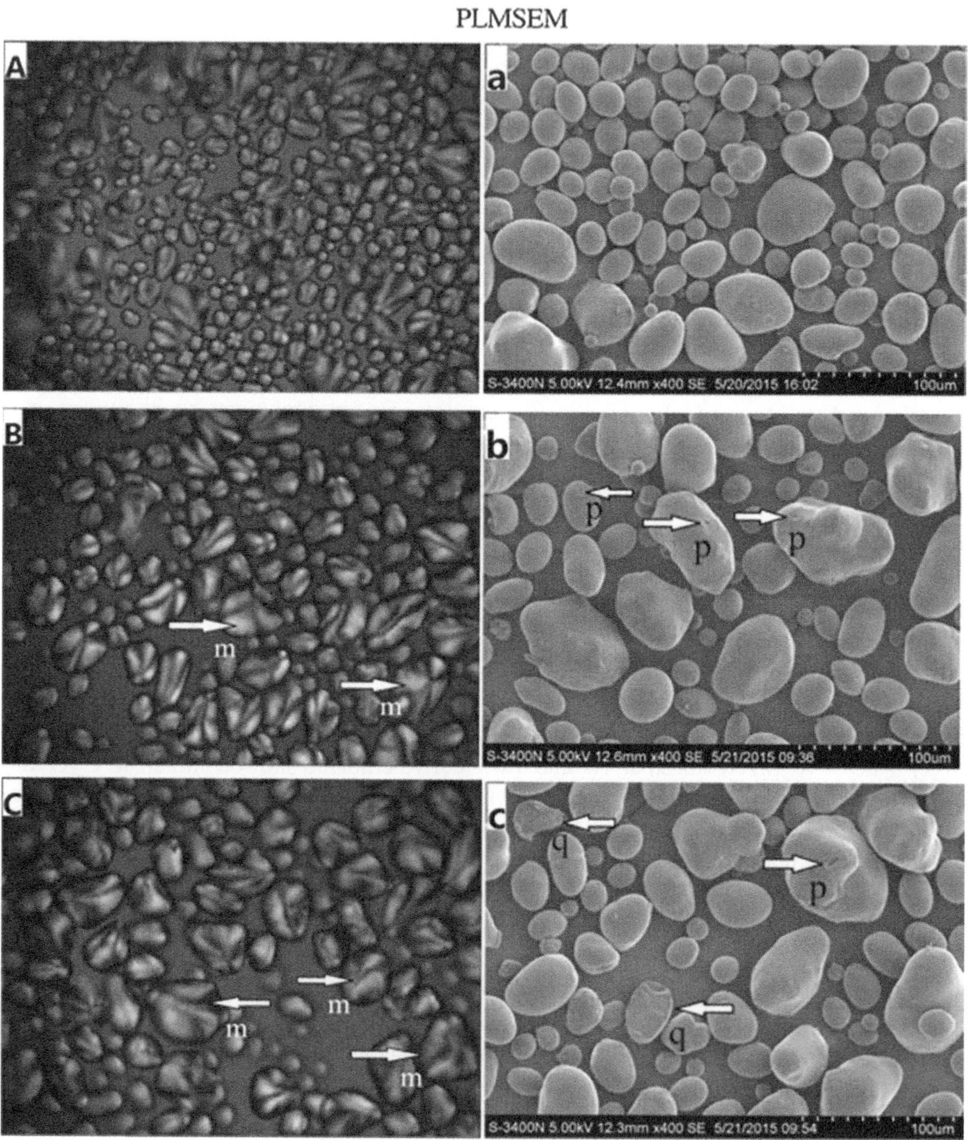

FIGURE 13.2 Morphologies of potato starch samples under polarized light microscopy (PLM) (objective and eyepiece magnifications are 40× and 10×, respectively) and scanning electron microscopy (SEM) (magnification 400×). Native starch (A, a), starch oxidized with 2 g active chlorine/100 g starch (B, b), and with 4 g active chlorine/100 g starch (C, c), m: polarization cross weakened, p: pores, q: crack (Zhou et al., 2016).

the surface of the granule, and their surface appeared rougher than native starch (Vanier et al., 2012). Nevertheless, in another study, SEM data demonstrated that both native and modified *Solanum tuberosum* starch granules displayed a combination of oval and spherical shapes. The morphology of modified starch granules remained unaffected by oxidation, regardless of the concentration of the oxidizing agent (Fonseca et al., 2015). The impact of various concentrations of NaOCl on the properties of *Solanum tuberosum* starch was examined by another study by Zhou et al. (2016). The findings demonstrated that a higher concentration (4%) led to the damage to starch granules, generating cracks, and holes on the granular surface. NaOCl oxidized starches exhibited a similar crystalline pattern with native starch, and a slight change in the relative crystallinity was observed. This implies that the oxidation effect on crystalline regions is constrained, supporting the idea that the primary site of modification is within the amorphous regions (Chapagai et al., 2021). The oxidation process with NaOCl induces the depolymerization of amylose and amylopectin chains within starch, ultimately causing a decrease in both viscosity and retrogradation of the starch (Zhou et al., 2016). The oxidation of *Solanum tuberosum* starches using NaOCl induced alterations in paste characteristics, with the most pronounced changes observed in the oxidized starch with incrementing concentration of the oxidizing agent (Fonseca et al., 2015). Oxidation at a concentration of 0.5%, and 1% NaOCl, potato starch showed an enhanced peak viscosity, whereas 1.5% NaOCl showed reduced viscosity than the unmodified potato starch (Figure 13.3). The elevation in viscosity observed with a lower degree of oxidation might be attributed to the partial breakdown of the starch, which in turn facilitates swelling (Fonseca et al., 2015). Pasting characteristics of oxidized macuna starch reported a significant decrease and lower setback viscosity due to the restriction of the substitute group for the realignment of the starch polymers after the cooling. The study observed a decrease in the cold paste viscosity and hot paste viscosity after the modification of starch. This reduction of viscosity at high concentrations can be ascribed to the change in the molecular and morphology of the starches, the appearance of COOH and C=O groups, and the high degree of depolymerization. Furthermore, the breakdown viscosity of oxidized velvet bean starch increased, suggesting a decrease in its resistance to shear and heat (Adebowale & Lawal, 2003). The reduction in viscosity with hypochlorite can be ascribed to the oxidative breakage of amylose and amylopectin chain, leading to the formation of starch with a smaller molecular size (Sangseethong et al., 2010). In another study, oxidation at low chlorine concentrations (≤1.0%) showed

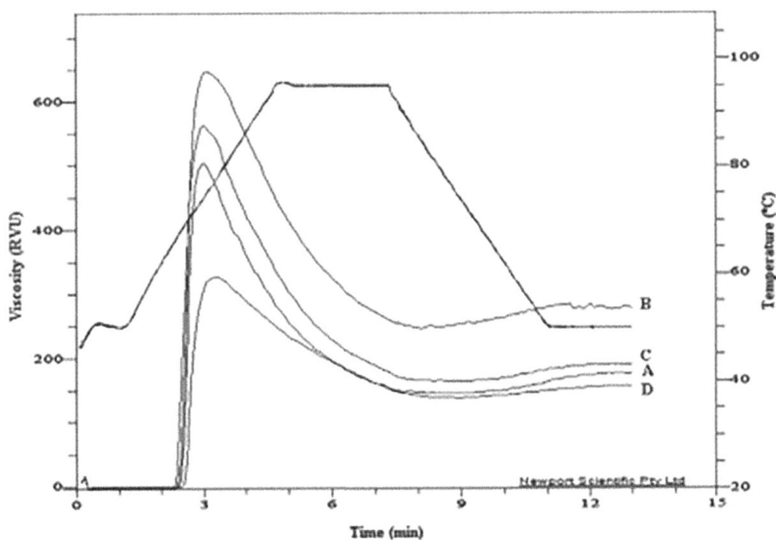

FIGURE 13.3   Pasting characteristics of native NaOCl oxidized potato: native starch (A); starch oxidized with 0.5 g/100 g of active chlorine (B); starch oxidized with 1.0 g/100 g of active chlorine (C); starch oxidized with 1.5 g/100 g of active chlorine (D) (Fonseca et al., 2015).

opposite trends such as significantly elevated peak and final viscosities and low breakdown viscosity. The gel hardness of the starch paste exhibited greater strength at mild oxidation. This can be due to the hemiacetal crosslink formed by the COOH group causing the increased integrity of the starch molecule (Sukhija et al., 2016). A consistent trend was noticed in the *Phaseolus vulgaris* starch oxidized with 10% NaOCl by Vanier et al. (2012).

Digestibility decreased after the starch modification. The resistant starch (RS) and slowly digestible starch (SDS) contents notably increased, accompanied by decreased levels of the rapidly digestible starch (RDS) content (Zhou et al., 2016). The syneresis and retrogradation of the oxidized starches showed a decreasing trend owing to the occurrence of a bulky COOH group produced by the oxidation (Ashogbon & Akintayo, 2014). It was found that the oxidation enhanced polymer chain debranching, leading to a reduced retrogradation of starch by Dias et al. (2011). The C=O and COOH groups formed during the modification process are larger than the OH groups. This spatial difference prevents the molecules from coming closer together, thus inhibiting retrogradation. Oxidized cassava and maize starch revealed a significant reduction in gelatinization parameters and gelatinization enthalpy (ΔH) than the native starches (Sandhu et al., 2008; Sangseethong., 2010). The gelatinization characteristics can primarily be linked to the diminishing strength of intermolecular bonds by the COOH groups in the oxidized starches (Sandhu et al., 2008). The thermal characteristics of vacuum-assisted oxidized canna, tapioca, and maize starches were investigated in a recent study. Native starches had a higher ΔH, followed by oxidized and vacuum-oxidized starches. This indicates that hypochlorite oxidation results in partial damage to the crystalline part of the starch causing reduced ΔH. Maize and canna starches subjected to vacuum-assisted oxidation displayed low onset temperature (T_o), peak temperature (T_p), and conclusion temperature (T_c) in contrast to their native counterparts. Contrastingly, vacuum-assisted oxidized tapioca starch exhibited an opposite trend, displaying an increment in gelatinization temperature compared to its native form. It is indicated that T_o, T_p, and T_c were elevated due to degradation in the amorphous area of the starch granule and subsequent reduction of their destabilizing effect in the crystalline area (Q. Zhang et al., 2018). The change in the thermal properties of starches sourced from multiple origins is potentially ascribed to variations in starch composition. This may comprise factors like the ratio of amylose to amylopectin, the presence of remnant proteins and lipids, the structural arrangement of amylopectin, and the morphology of the granule (Garg & Jana, 2011).

13.3.1.4 Applications

NaOCl-oxidized starch is employed as a food additive and is identified by the code name E1404. The food industry is experiencing a growing trend in utilizing oxidized starch as a food additive, driven by its attributes such as effective binding properties, enhanced stability, and reduced viscosity (Chakraborty et al., 2022). The food industry has employed oxidized starches for their ability to adhere and provide coating properties (Purcell et al., 2014). Oxidized starches act as binding agents in foods, used in batters for meat prior to frying (Xie et al., 2005). The oxidized starches in batters provide improved stability and binding properties without introducing excess viscosity (Purcell et al., 2014). The research examined the use of oxidized starch as a batter for bake-only chicken nuggets. Seven different starches were taken for the preparation of oxidized starches, which were subsequently examined for their thermal and pasting properties. This study focused on exploring the change in textural and sensory attributes of steam-baked coated nuggets by varying the source of starch (Purcell et al., 2014).

In the food sector, oxidized starch finds application due to its remarkable functional attributes and film-forming capabilities (Haq et al., 2019). The hypochlorite oxidized starch film showed enhanced flexibility and improved transparency compared to native film (Vanier et al., 2017). Biodegradable films were synthesized from modified *Solanum tuberosum* starches with 0.5%, 1.0%, and 1.5% active chlorine. Biodegradable film formed from 1.5% active chlorine oxidized potato starch showed reduced water vapor permeability and solubility in contrast to native *Solanum tuberosum* starch film (Fonseca et al., 2015). The conformational changes that occur after oxidation can induce free space between starch molecules. This property is crucial for facilitating interactions with plasticizers during the formation of starch films

(Fonseca et al., 2015). The decreased solubility observed in films made from oxidized starch is ascribed to the presence of the C=O and COOH functional groups. These alterations impact intramolecular bonding and interactions among amylose molecules, contributing to the reduced water absorbance of the oxidized starch (Haq et al., 2019).

Emulsification, sealing, coating, batter binding, and dough conditioning are often carried out using hypochlorite-oxidized starches (Chakraborty et al., 2022). It is used as a thickening agent in sauces, puddings, jellies, and creams, while also serving as a stabilizer in cake fillings and in the production of ketchup (Dimri et al., 2023). It is also used as a substitute for gum Arabic and as a gelation agent (Vanier et al., 2017). The oxidation treatment enhanced both the thermal and cold paste stabilities of potato starch, contributing to the improved stability of thick soup (Zhou et al., 2016). The thickening, binding, and stabilizing properties of oxidized starch paste find application in the adhesive industry and packaging industry (Zhou et al., 2016). Kuakpetoon and Wang (2001) examined the variables affecting the structural and adhesion characteristics of oxidized starch prepared from three sources such as potato, rice, and maize with varying concentrations of oxidizing agent (0.8% and 2% of active chlorine). Oxidized rice starch modified with 0.8% active chlorine showed greater adhesion nature and oxidized corn starch displayed the lowest adhesion properties irrespective of the active chlorine percentage. The study highlighted that the adhesion capability of oxidized starch was primarily affected by factors such as the surface area and the concentration of COOH function groups. Additionally, the surface varies with the change in shape and size of the starch granule (Kuakpetoon & Wang, 2001).

Physio-chemical properties of oxidized starches are helpful in pharmaceutical carriers for the enhancement of regulated discharge of active ingredients (Haq et al., 2019). The functional characteristics of 400 L-NF oxidized corn starch are well-known in the formulation of pharmaceutical and nutraceutical oral doses. This non-ionic powder, renowned for its outstanding super disintegrant properties, proves valuable in the formulation of hard capsules, granules, blends, pellets, and swallowable tablets. Additionally, it serves as a versatile option as a filler or binder (Lemos et al., 2021). Furthermore, paper manufacturers utilize oxidized starch to improve the tear resilience of paper (Dimri et al., 2023).

13.3.2 Hydrogen Peroxide

Hydrogen peroxide proves to be a highly promising oxidant, primarily due to its environmentally friendly attributes, including safety in use and the generation of non-toxic wastes (Lewicka et al., 2015). H_2O_2 possesses a substantial amount of active oxygen (47%) and a higher oxidation potential (1.77 V), rendering it an efficient and cost-effective oxidant. In addition, H_2O_2 exhibits limited reactivity toward many organic functional groups and behaves more like a nucleophile in the presence of electrophilic compounds, lacking strong oxidizing properties (Lemos et al., 2021). To mitigate these undesirable nucleophilic characteristics, H_2O_2 is activated through the presence of metal catalysts such as Cu^{2+}, Fe^{2+}, Fe^{3+}, Ti^{3+}, Co^{2+}, and W^{4+}. This activation facilitates the augmentation of the oxidizing properties of H_2O_2. It's crucial to acknowledge that H_2O_2 oxidized starch with metal catalysts cannot be employed in applications such as food, pharmaceuticals, and cosmetics. This is because the metal ions used in the modification have the ability to bind strongly with the oxidized starch (Dimri et al., 2023).

13.3.2.1 Method of Preparation

The use of H_2O_2 in conventional settings is less common than that of sodium hypochlorite. Nevertheless, numerous studies have shown the application of H_2O_2 for the oxidation of starches (Vanier et al., 2017). Oxidized starches are produced by adding drop by drop of H_2O_2 into the starch suspension under controlled temperature and pH with suitable metal catalysts (e.g., copper sulfate, ferrous sulfate) (Liu et al., 2014). H_2O_2 does not produce hazardous byproducts because it breaks down quickly into oxygen and water. The environmental impact of H_2O_2 is therefore low (Vanier et al., 2017).

13.3.2.2 *Mechanism*

The extensive mechanisms of hydrogen peroxide-induced starch oxidation in the presence of a metal catalyst are illustrated in Figure 13.4. The oxidation of starch with H_2O_2 mediated by metal catalysts can follow either a radical or ionic mechanism. The radical mechanism predominates in reactions catalyzed with iron and copper metal ions, a chemical reaction referred to as Fenton's reaction (Lewicka et al., 2015). Chemical reaction of H_2O_2 involves decomposition and the formation of free $^{\bullet}OH$ radicals, and the recombination of free radicals led to H_2O_2 synthesis (Tolvanen et al., 2009).

The chemical reaction of H_2O_2 follows:

Decomposition:

$$2H_2O_2 \rightarrow 2\,H_2O + O_2 \qquad \text{(Equ. 1)}$$

Formation of $^{\bullet}OH$ radicals

$$Fe^{2+} + H_2O_2 \rightarrow Fe^{3+} + {^{\bullet}OH} + OH^- \qquad \text{(Equ. 2)}$$

$$Fe^{3+} + H_2O_2 \rightarrow Fe^{2+} + {^{\bullet}OOH} + H^+ \qquad \text{(Equ. 3)}$$

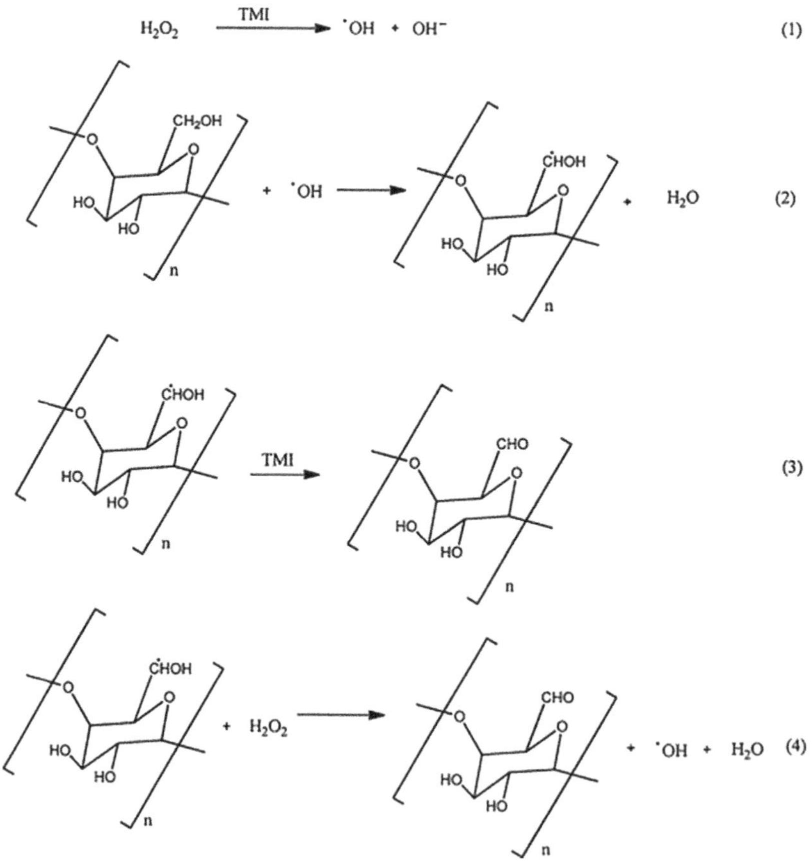

FIGURE 13.4 H_2O_2 oxidation of starch (TMI represents transition metal ion).

Recombination

$$\bullet OH + \bullet OH \rightarrow H_2O_2 \qquad\qquad\qquad\qquad\qquad \text{(Equ. 4)}$$

These $\bullet OH$ radicals readily react with starch by removing hydrogens from C—H groups of glucopyranose rings resulting in $R\bullet CHOH$ radicals. These highly reactive free radicals further react with H_2O_2 or metal ions, causing the formation of C=O or oxidizes OH groups in glucose units to C=O and COOH functional group (Kumoro et al., 2015). Vanier et al. (2017) reported that these radicals further undergo rearrangement, leading to the breakage of glycosidic bonds and the synthesis of a C=O group. Carbohydrates with a free or potentially free C=O group can react further in alkaline conditions, some of which lead to the development of a COOH group (Sangseethong et al., 2010). It should also be emphasized that the depolymerization of polysaccharide chains takes place during the starch oxidation with H_2O_2.

13.3.2.3 Physico-Chemical and Functional Properties

The granular surface of the starch remains unchanged for a short duration, and as time increases surface roughness and fissures can be observed in H_2O_2 oxidized starch (Figure 13.5) (Sangseethong et al., 2010). Oxidation by H_2O_2 in starches results in the formation of a higher amount of C=O groups than the COOH functional group. After the H_2O_2 reaction, a comparatively low amount of COOH groups was seen, which was probably caused by the dropwise addition of peroxide, which prevented the further oxidation of initially generated C=O groups to COOH groups (Chapagai et al., 2021).

The rise in H_2O_2 concentration led to an increment in C=O content in both alkaline and acidic environments, respectively. C=O contents of H_2O_2 oxidized starch at concentrations of 2% and 5% under an alkaline environment showed 3.6 times and 1.3 times greater than under an acidic environment (Chapagai et al., 2021). A pH range of 8.4 and 10 was found to be ideal for the production of C=O and COOH groups, according to Tolvanen et al. (2009). Corn starch was oxidized at varying concentrations of H_2O_2, yielding a degree of oxidation spanning from 0.096 to 0.554. Rice starch was modified using 2.7% H_2O_2, with UV irradiation serving as the photo-initiator in a study conducted by El-Sheikh et al. (2010). The ideal condition for the peroxide modification was observed while treating the starch suspension at a temperature between 60 and 70°C. The effect of catalysts with varying concentrations on the oxidation rate was studied by Y. R. Zhang et al. (2012). 0.5% $CuSO_4$ was found to be an optimum concentration for the oxidation and the reaction time was reduced from 72 to 1 h. The authors also mentioned the possibility of adverse impacts from high catalyst concentrations, such as inadequate oxidation for catalysis and weak catalytic response. Furthermore, the starch may turn an undesirable coloration due to high metal concentrations in the oxidation procedure. Depolymerization during oxidation results in increased low molecular weight starch fraction, increased gelatinization temperature, and decreased viscosity with firm gel (Sangseethong et al., 2010). Typical stretching vibration of the C=O group was observed at the peak at 1760 cm^{-1} after the starch oxidation (Liu et al., 2014; Tung et al., 2021). The molecular weight of starch was noted to decrease during the oxidation process. Oxidized corn starch using Cu^{2+}, and Fe^{3+} catalysts at varied temperatures (70–98°C) and for different periods of time (1–4 h) are synthesized by Yu et al. (2017). An increase in the degree of oxidation resulted in a substantial depolymerization of starch causing a significant reduction in particle size and molecular weight (Yu et al., 2017). Additionally, reduced thermal stability was observed after the modification of starch. DSC results revealed that the oxidized starch did not exhibit a glass transition temperature, and XRD analysis indicated that the oxidation process resulted in the partial destruction of the crystalline structure of the starch granule. A rapid method for the synthesis of oxidized starch with a high degree of oxidation was successfully established by Y. R. Zhang et al. (2012). A higher aldehyde group content is associated with increased water resistance and mechanical properties (S. D. Zhang et al., 2009).

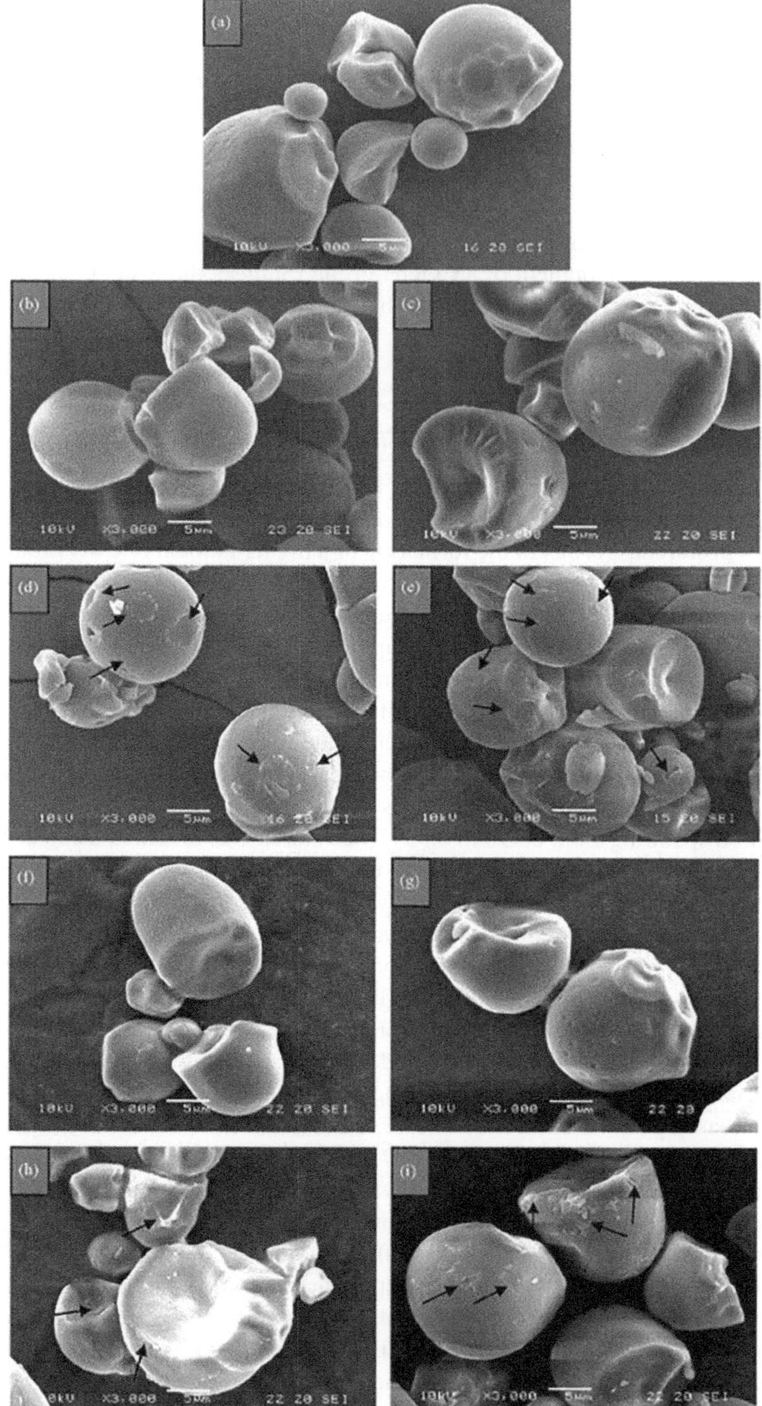

FIGURE 13.5   SEM image of native (a) and oxidized cassava starches prepared with hypochlorite (b–e) and peroxide (H_2O_2) (f–i) at various reaction times: 30 min (b and f), 60 min (c and g), 120 min (d and h) and 300 min (e and i) (Sangseethong et al., 2010).

13.3.2.4 Applications

Hydrogen peroxide–oxidized starch finds extensive application as a surface sizing agent in the paper industry (Aaliya, Sunooj, John, et al., 2022). Oxidized starch can be utilized for the production of thermoplastic starch with versatile properties, offering potential applications as an eco-friendly material (S. D. Zhang et al., 2009). Moreover, the decrease in crystallinity, accompanied by the degradation of hydrogen bonds, results in an increase in hydrophobicity and an enhancement of thermoplastic properties of H_2O_2–oxidized starch (Lewicka et al., 2015). Oxidized starch appears to be a promising candidate for developing a ligand for Zr complexes, given its diverse content of coordinating groups and molecular size. Furthermore, in comparison to synthetic organic ligands, starch-based ligands are presumed to be more sustainable, environmentally friendly, and economical, due to the biodegradability, renewability, and low cost of starch (Yu et al., 2017). The food industry uses oxidized starches to make whipped cream, dough powder mixtures, and puddings (Pietrzyk et al., 2012). The oxidation with H_2O_2 resulted in a reduction in the hardness of biscuits developed from cassava starch, attributed to the promotion of further expansion facilitated by oxidation (Dias et al., 2011). Oxidized starch has potential applications in the development of biodegradable packaging films (Tanetrungroj & Prachayawarakorn, 2018). An oxidized thermoplastic starch film has been prepared and investigated by S. D. Zhang et al. (2009). The findings suggest that the oxidation with H_2O_2 resulted in a decrease in viscosity, crystallinity, and thermal stability of the modified starch sample (S. D. Zhang et al., 2009). H_2O_2–oxidized starch exhibits potential applications in various industries, including food, textile, paper, plastics, and laundry finishing (Y. R. Zhang et al., 2012). Maize starch was oxidized with H_2O_2, and then a gold chloride solution was gradually added, along with either sodium carbonate or sodium hydroxide at 25°C. The addition of a base resulted in a change in the color of the reaction mixture to blue, signifying the development of a biocompatible nanogold (Emam et al., 2017).

13.3.3 Ozone

Ozone is an exceptionally potent oxidizing agent, recognized as a "green" and environment-friendly technique. It can swiftly decompose into oxygen, leaving no remnants in the environment or the food, thus aligning with the global demand for sustainability. O_3 can be produced by exposing air or oxygen to a high-energy source, leading to the conversion of O_2 into O_3 (Castanha et al., 2017). O_3, characterized by its blue color and pungent smell, is a potent oxidant with a higher redox potential (2.07 V). Its widespread use in the food industry is attributed to implicit advantages relative to other oxidizing reagents (Pandiselvam et al., 2019).

13.3.3.1 Method of Preparation

The application of O_3 to modify starch has gained increasing interest. The initial report on the use of O_3 in the modification of starch dates back to 1964 (Pandiselvam et al., 2019). In contrast to NaOCl and H_2O_2, O_3 stands out as a cleaner and more potent oxidizing agent, necessitating fewer downstream purification processes (Dimri et al., 2023). A corona discharge method was used to generate the gaseous O_3 by passing oxygen into an O_3 generator. Oxygen was exposed to an intense electric field between the two electrodes, leading to the dissociation of oxygen molecules into atoms and, consequently, the generation of O_3 (Handarini et al., 2020). Figure 13.6 illustrates the O_3 oxidation in starch (Castanha et al., 2017). Starch oxidation by O_3 can be done either in its gaseous phase or as an aqueous form (Maniglia et al., 2021). The majority of studies have concentrated on investigating the effects of O_3 on powdered starch (Klein et al., 2014).

The ozonation of powdered starch was carried out by placing the starch powder in a rotating reaction vessel that was attached to an O_3 generator. O_3 was flushed to the chamber at a controlled amount at a particular period of time. Rotation of the reaction vessel continued to maintain the uniformity of the O_3 gas during the chemical reaction (Oladebeye et al., 2013). Subsequently, the rotation was halted, and fresh

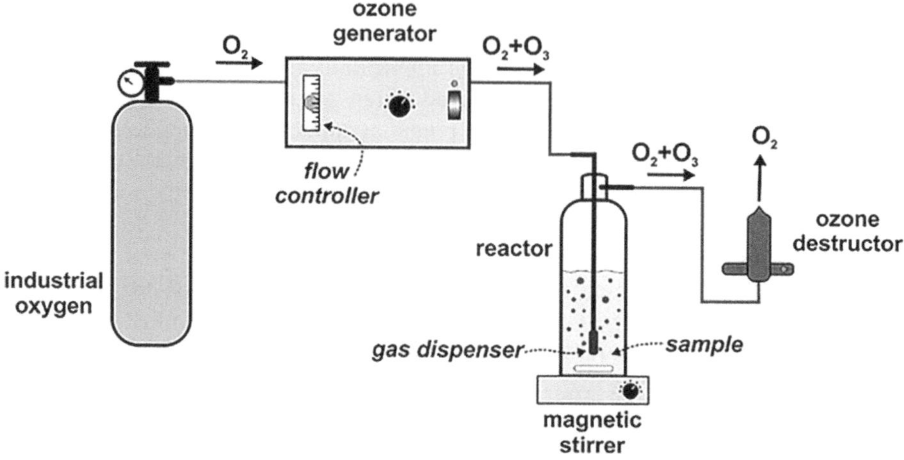

FIGURE 13.6　Schematic representation of O_3 oxidation in starch.

air was introduced into the reaction vessel for 1 min to purge the residual gas (Handarini et al., 2020). The studies of Çatal and Ibanoğlu (2014) and Klein et al. (2014) have provided insights into the changes in starch properties resulting from exposure to O_3 in aqueous solutions. Starch suspension using distilled water was made and placed inside a glass ball reactor under controlled temperature and pH. The reactor was directly connected to a corona-type ozone generator. O_3 gas is bubbled through the aqueous solution to get maximum solubility for a designated time period. Following the reaction, the starch suspension was rinsed with DW and dried (Çatal & Ibanoğlu, 2014; Klein et al., 2014). In the process of oxidation, O_3 generates CO_2, H_2O, inorganic ions, and non-toxic byproducts (Dimri et al., 2023). The ozonation process is affected by various factors, such as the concentration of O_3, contact area and size of bubbles, treatment duration, properties of the reactor, pH, and temperature of the system (Maniglia et al., 2021). O_3 oxidizes starch molecules with remarkable efficiency without a catalyst at room temperature. The utilization of O_3 exhibits higher recovery rates compared to chemical agents, and it entails lower purification costs. O_3 has been applied and its effectiveness has been assessed across a diverse array of starch sources, including potato, rice, wheat, corn, cassava, and sago (Pandiselvam et al., 2019).

13.3.3.2 Mechanism

The mechanism of O_3-starch oxidation is interpreted by many researchers in different ways. This section follows the major possible mechanism of oxidation. The chemical reaction between starch and O_3 can occur either through a direct reaction or via an indirect reaction, as illustrated in Figure 13.7. The direct method entails the reaction between molecular O_3 and starch granule in an acidic medium; however, it is rarely employed due to prolonged reaction time. Conversely, the indirect method in which the O_3 molecule undergoes breakdown prior to the interaction with starch granules in an alkaline environment. Hydroxyl radicals (•OH) are formed as a product during the O_3 reaction mechanism, and they engage in a series of reaction processes involving the elimination of the hydrogen ion, electron transfer, and addition of ozone radicals (Dimri et al., 2023). In a prior investigation, the impact of ozonation on the structural and physico-chemical characteristics of *Solanum tuberosum* starch was assessed and the possible mechanism of oxidation is illustrated in Figure 13.8 (Castanha et al., 2017). The substitution of OH with C=O and COOH functional groups, along with the cleavage of glycosidic linkage in the starch chain, led to a reduction in their sizes, predominantly in the amorphous area of the granules evidenced by the study.

Studies have shown the importance of pH during the oxidation process. Oxidation occurring in the pH range of 6.5 to 9.5 promotes the development of crosslinking between starch chains (Klein et al., 2014). Recently, there has been systematic research on the alteration in the structure of waxy *Oryza*

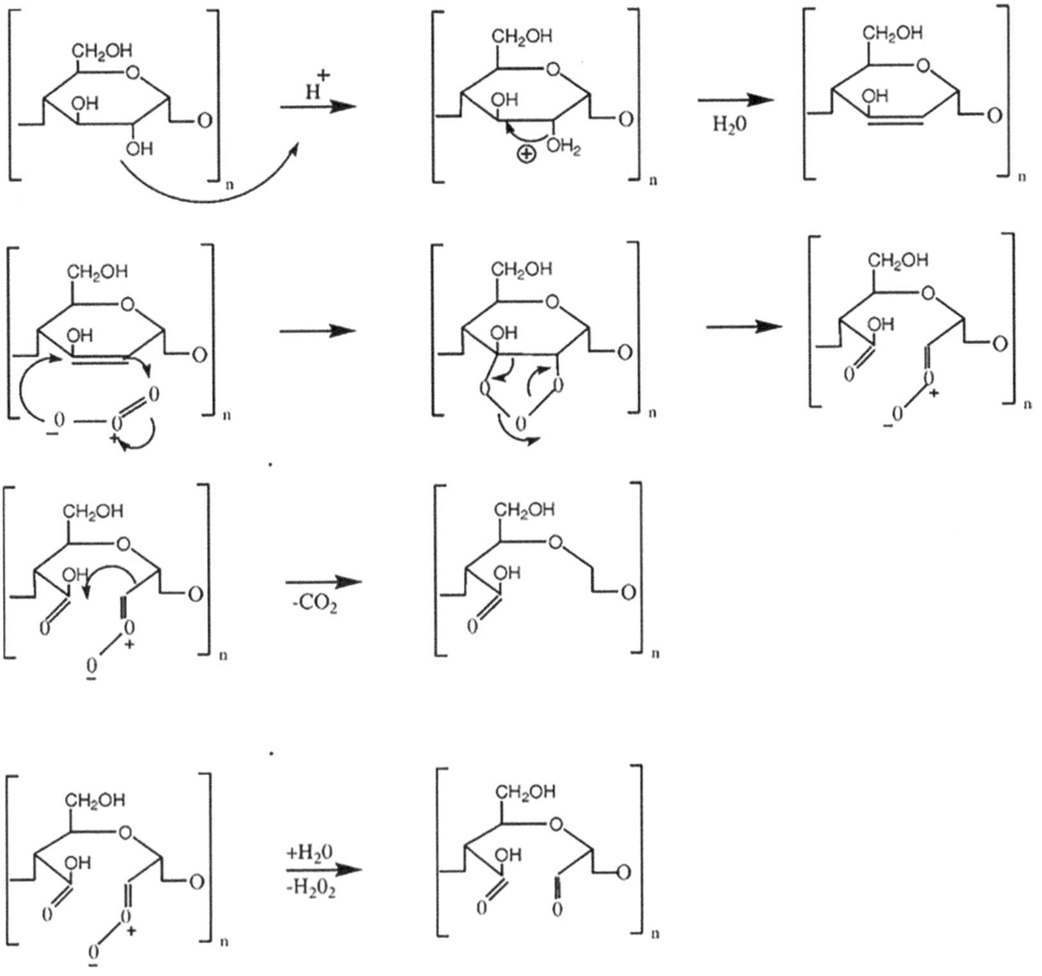

FIGURE 13.7 O_3 oxidation of starch.

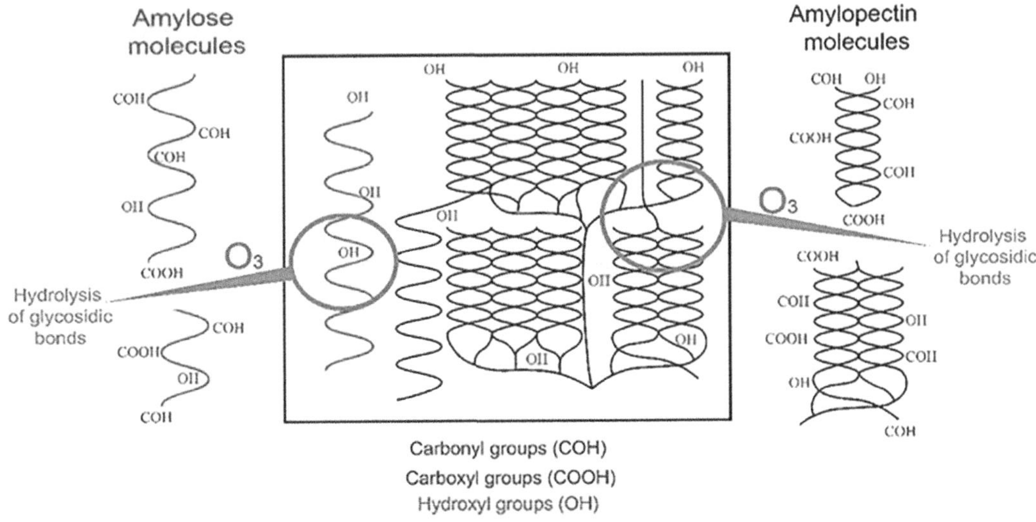

FIGURE 13.8   Illustration of the mechanism of ozone oxidation in starch.

sativa starch following ozone modification (J. Li et al., 2023). The observation indicated that O_3 initially targeted the amorphous region of the granules and subsequently entered the crystalline region. The study demonstrated that O_3 oxidation resulted in ring cleavage between C2 and C3 of the glucopyranose unit through the NMR study. These broken chains in turn form a hemiacetal group by the rearrangement and bonding between the polymer. Among these structures, the primary configurations were intramolecular acetals and intermolecular hemiacetal compounds formed between C6 and C2, as well as C6 and C3 (Figure 13.9).

13.3.3.3 Physico-Chemical and Functional Properties

Ozone, as a potent oxidant, influences the functional groups, granule morphology, thermal properties, swelling, and solubility of native starch (Dimri et al., 2023). Studies have shown that the properties of oxidized starch after 10 min of ozonation are comparable to hypochlorite oxidized starch (Oladebeye et al., 2013). The influence of O_3 treatment on starch characteristics is associated with the conditions of the O_3 reactor (e.g., ozone concentration, pH, temperature, treatment duration, size, and contact area of ozone bubbles), as well as a source of starch (Rostamabadi et al., 2022). The impact of treatment time on the physicochemical characteristics of the potato starch was studied by Castanha et al. (2017). Noticeable rises in the C=O and COOH content and increased reducing-sugar concentration, accompanied by a reduced amylose content, decreased pH, and lowered molecular size were shown with an increase in the ozonation time. This indicates that O_3 processing led to molecular-level modifications in the potato starch. The occurrence of COOH groups in the oxidized starches elevates the acidity of the samples. Indeed, the pH of the samples reduced with the progression of the ozonation process, aligning with the rise in COOH concentration (Castanha et al., 2017).

A study on the impact of ozonation with varying pH was conducted in *Manihot esculenta* starch by Klein et al. (2014). Pasting parameters and the crystallinity of the modified starches were primarily affected by pH, without changing the granular morphology of the starch. Higher C=O and COOH content were observed at a pH of 9.5, and oxidation at a pH of 6.5–9.5 resulted in the crosslinking of depolymerized starch molecules. Modification of *Manihot esculenta* starch granules at a pH range of 3.5 showed increased shear resistance (Klein et al., 2014). Physicochemical characteristics of O_3 oxidized cassava starch at varying pH (7–10) with different periods of time (0–240 min) were investigated by Pudjihastuti et al. (2018). The authors noticed that a greater amount of COOH content, reduced swelling index, and increased solubility were obtained by the ozonation at pH 10.

In the examination of the ozonation process applied to banana flour and starch, it was observed that the starch exhibited a greater quantity of COOH (0.0235%) in comparison to the flour (0.0038%) (Cahyana et al., 2018). The authors showed that the non-starch content also affects the effectiveness of oxidation through the ozonation process. Following O_3 modification, the surface of starch particles became rough, displaying a distinct crushing phenomenon, while the crystallinity of the starch remained unchanged (Huang et al., 2024). The granules of oxidized rice starch exhibited a polygonal shape with multiple depressions on their surfaces, attributed to damage caused by O_3 (J. Li et al., 2023). In contrast, a study performed by Castanha et al. (2017) revealed that the modified samples exhibited increased roughness and heterogeneity with prolonged processing time. Granules with irregular shapes became more common after 30 minutes of ozonation, and some fissures and pores could even be seen on the granules' surface (Figure 13.10) (Castanha et al., 2017). Nevertheless, the morphology of the arracacha starch granules remained unchanged by the ozonation process. Maltese crosses present on the granule surface also showed no alterations, suggesting that the ozonation procedure was unable to significantly alter the granule morphology of arracacha starch (Lima et al., 2020).

The swelling index and solubility of starch increased as the ozonation period was extended (Huang et al., 2024). The existence of an electronegatively active bulkier COOH group induces electrostatic repulsion between the molecules leading to the expansion in their chain conformation and consequently increasing their solubility. This results in starches with enhanced stability and reduced retrogradation of the modified starch samples (Castanha et al., 2017). The gel firmness of potato starch increased with

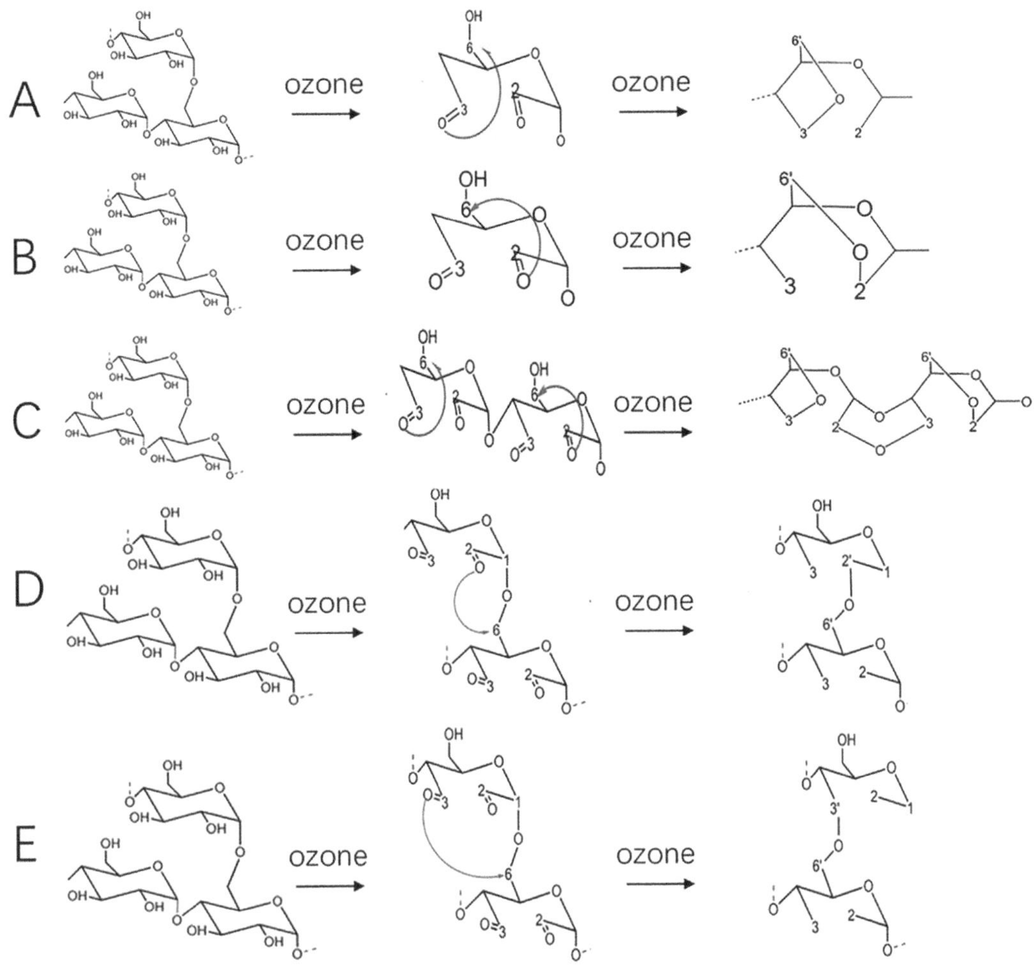

FIGURE 13.9 Illustration of the possible method of formation of intramolecular acetals and intermolecular hemiacetal compounds during ozone oxidation in starch.

an extended duration of ozonation, as reported by Castanha et al. (2017). Extended ozonation results in the depolymerization of the starch and increased mobility of the amylose chains. This in turn results in the increased reassociation of the amylose molecule subsequently enhancing firmness. Ozone-oxidized potato starch for 15 and 30 min exhibited a higher degree of gel firmness compared to the native starch (Castanha et al., 2019). The authors detected that the potato starch ozonated for 15 min achieved a gel firmness value of 0.82 mJ at 65°C, which is comparable with the native starch at 95°C.

The FT-IR and XRD results indicated that a shorter duration of O_3 oxidation on the starch (up to 7.5 min) improved the intermolecular crosslinking between the starch chains thereby increasing the ordered arrangement within the granules. Conversely, ozonation of 15 min or more resulted in the degradation of the short and long-range ordered structure of starch, and thereby slight decrease in the peak intensity can be observed through the FT-IR spectra of Tartary buckwheat starch (Hu, Li, Cheng, Fan, Ma, Hu, et al., 2022). O_3-modified starch exhibited a peak at 1740 cm^{-1}, associated with the stretching vibration peak of C=O, which confirms the modification and indicates that the COOH group is formed after the modification. The crystalline pattern of all the samples was maintained after the modification and the relative crystallinity of the samples showed a mere decrease with an increase in the duration of ozonation. The RC of

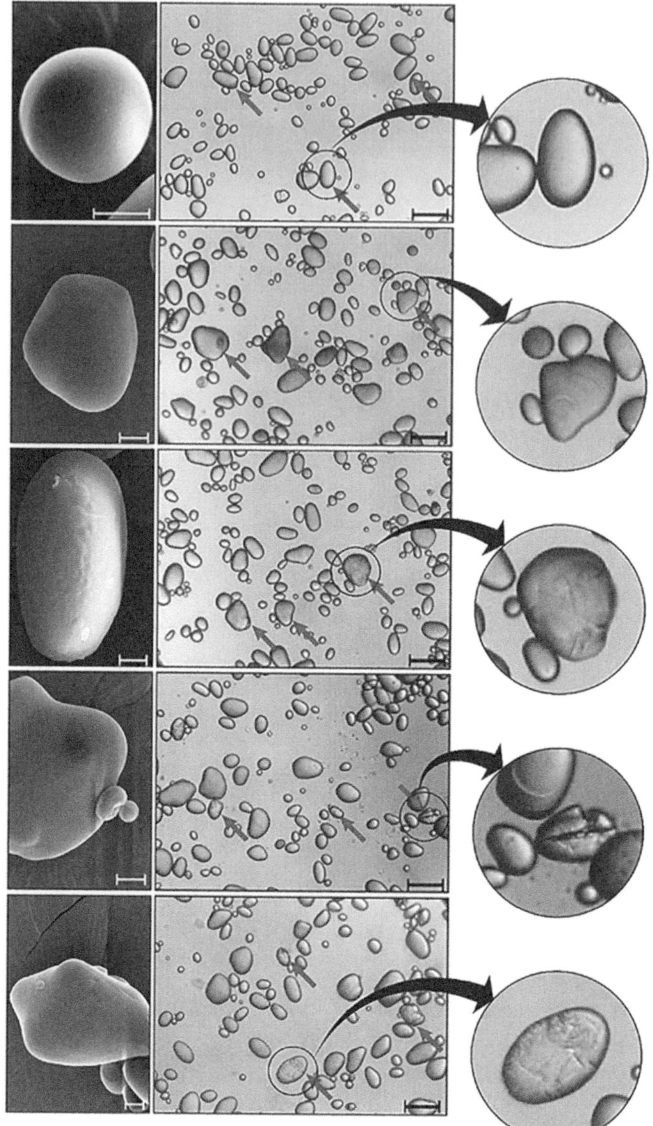

FIGURE 13.10 Microscopic images of the potato starch. From top to bottom: 0, 15, 30, 45, and 60 min samples. Images from the first column are scanning electronic microscopy images. The second column shows images from light microscopy. Further details are shown on the third column (without scale) (Castanha et al., 2017).

the 60 min ozonized starch sample showed an 11% decrease contrasted to unmodified *Manihot esculenta* starch. The alterations in the RC were ascribed to the impact of the oxidizing agent on the amylopectin chains (Lima et al., 2021). The RC of *Manihot esculenta* starch decreased following the oxidation with O_3 for 60 min as reported by Klein et al. (2014), regardless of the pH of the suspension. The authors asserted the reduced RC indicates that ozonation affects both the amorphous and the crystalline region of starch granules.

Ozonation induced the breakdown of polymer chains, resulting in a fragmented starch chain with lower molecular weights (Huang et al., 2024). The gel permeation chromatography analysis conducted by Castanha et al. (2017) revealed a progressive decrease in the high molecular-weight fractions with prolonged ozonation time. The peaks in the first fraction for total carbohydrates (TCH) were indistinguishable, while those in the second fraction increased after 60 min of ozonation (Castanha et al., 2017). Ozonation was found to reduce long-branching compounds and increase smaller molecules in *Manihot esculenta* starch (Lima et al., 2020). Hence, the molecular size distribution indicates that the ozone modification had a significant effect on the depolymerization of polymer chains in cassava starch. Nevertheless, Oladebeye et al. (2013) noticed a rise in the molecular weight after the O_3 oxidation of cocoyam and yam cultivars. A similar trend was observed in another study in which tapioca starch was ozonated for a period of 10 min (Chan et al., 2011).

The flow behavior of the starch is affected by O_3 oxidation, with the majority of the oxidized starch characterized by a decrease in the paste viscosity. A consistent result was observed in maize, sago, and tapioca starch endured to O_3 oxidation in research by Chan et al. (2011). The peak apparent viscosity of the ozonated potato starch samples exhibited a declining trend with an increasing period of processing. This implies that the modified starch granules' ability to retain their integrity during stirring and heating is reduced, resulting in a faster rupture rate in contrast to the unmodified samples (Castanha et al., 2017). Nevertheless, cassava starch exhibited increased peak viscosity and decreased cooking stability after the oxidation of starch (Amorim et al., 2011). The O_3-oxidized potato starch showed a slight increase in pasting temperatures with increased time of ozonization. The rise in pasting temperature can be attributed to molecular depolymerization, as smaller-sized molecules typically demand more energy for gelatinization compared to larger-sized molecules (Castanha et al., 2017). There are reports suggesting that smaller molecules may undergo recombination when subjected to heat, necessitating high temperatures to disrupt the structure of starch (Huang et al., 2024). Enhancements in the thermal characteristics and reductions in the ΔH in O_3-oxidized potato starch were observed by Ibanoğlu et al. (2018). However, when comparing the O_3-modified starches of sago, corn, and tapioca to the unmodified starches, no alterations were seen in the gelatinization characteristics and enthalpy.

13.3.3.4 Applications

O_3 is considered a clean and potent oxidant that leaves no residues behind, in contrast to the hypochlorite oxidation process, which generates significant amounts of salts (Kaur et al., 2012). O_3 oxidized starches were found to be effective as thickening agents. Hydrogels that are 250–300% stronger than native starch can be produced with the modified starches and resistance to acidic conditions has been achieved. Moreover, adjusting the process conditions allows for the production of gels with desired strength. Indeed, the enhanced strength of these hydrogels contributes to improved performance in 3D printing applications (Akhila et al., 2023). Furthermore, the varied molecular interactions of ozonated starches led to improved oven expansion (Maniglia et al., 2021). Conversely, the ozonated starch samples exhibited higher light transmittance compared to the native starch, along with greater stability over the period of storage. This suggests promising applications in the paper, textile, and food industries (Lima et al., 2020). The partial disruption of the glycosidic linkage and the introduction of C=O and COOH groups let oxidized starch with enhanced film-forming capacity. Furthermore, ozonation alters the charge and chemical affinity of starch granules, influencing molecular interactions and thereby modifying the morphology and film-forming properties of thin films (Huang et al., 2024). The cast films prepared from the modified starch by O_3 (10 min) showed a significant improvement in properties, with a 276% increment in tensile strength, a 72.5% increment in elongation at break, a 154% increase in water contact angle, and a 34.7% decrease in water vapor permeability (Huang et al., 2024). In summary, O_3 emerges as a non-thermal and environmentally friendly innovation that imparts potential characteristics upon modified starches (Maniglia et al., 2021).

13.3.4 Other Reagents

Permanganate, persulfate, and periodate have been explored as oxidizing agents in experimental investigations. The molecular structure and properties of the oxidized starch are influenced by the number of oxidation mechanisms. Despite their chemical efficiency, these oxidations result in large quantities of inorganic waste, like chlorinated products (Y. R. Zhang et al., 2012). Dialdehyde starch, a significant form of oxidized starch, can be synthesized using the highly effective oxidant periodic acid. It selectively oxidizes the adjacent OH groups in the starch molecule (Q. Chen et al., 2015). The periodate oxidation process occurs subsequent to starch gelatinization, as both intra- and inter-molecular H bonds break during gelatinization, enhancing the accessibility of chemical reactions within the starch (Dimri et al., 2023). Cleavage of 1,2-glycol structures occurs during the oxidation with periodate and results in two aldehydes per glucose unit. This oxidation reaction is collectively called a selective oxidative diol-scission reaction (Vanier et al., 2017). Periodate-oxidized starch has distinctive properties because it produces extremely reactive dialdehyde groups. It finds applications across industries, including, textiles, paper, proteins, drugs, and biodegradable packaging. Moreover, periodate-modified starch decreases viscosity and imparts improved water solubility. Nevertheless, the industrial utilization of periodate is constrained by its high cost (Dimri et al., 2023). Sodium periodate oxidation can produce dialdehyde starch, which can be used in the leather industry as a crosslinker and to facilitate tanning. Due to the crosslinking nature of sodium periodate, Moreno et al. (2017) synthesized biodegradable active films with improved film strength and barrier capacity and reduced water vapor and oxygen permeability.

13.4 DUAL MODIFICATION

Single modifications have been employed to enhance diverse attributes of starch, encompassing physico-chemical and rheological characteristics, and digestibility (Kumari & Sit, 2023). However, there are instances where a single modification may not fulfill the demands of industrial applications. Therefore, the formation of dual modifications becomes necessary, encompassing combinations of physical and chemical or physical-physical or chemical-chemical methods. Dual modifications, such as crosslinking/oxidation or acetylation/oxidation, have been extensively studied. These methods find wide applications in the food industry as emulsifiers and in the non-food industry as adsorbents (Q. Chen et al., 2015). It was reported that the combination of crosslinking with oxidation enhances the efficiency of the modification and improves the physicochemical characteristics. The dual-modified rice starch with epichlorohydrin and NaOCl observed reduced swelling index and solubility, along with increased paste clarity compared to a single modified sample. Furthermore, crosslinked-oxidized rice starch demonstrated the least tendency to retrograde and the highest resistance to shear compared to other modified starches (Tanetrungroj & Prachayawarakorn, 2018). Elephant foot yam starch was subject to dual modification using NaOCl and STMP. The modification enhanced the water binding capacity and color value (whiteness) of the starches. The dual-modified starch exhibited a decrease in viscosity profile and higher ΔH (Sukhija et al., 2016). An improved paste clarity and reduced retrogradation tendency to the crosslinked-oxidized maize starch prepared using H_2O_2 and STMP were reported by Liu et al. (2014).

Dual-modified starches have also found applications in the preparation of biodegradable packaging. Ibanoğlu et al. (2018) found that using ultrasonication prior to ozonation can yield potato starch with a notable reduction in both the T_O and ΔH. The dual-modified samples showed a lower viscosity at high solid content, enhanced paste clarity, superior film-forming capabilities, and improved binding properties, making them more suitable for applications in the food industry (Ibanoğlu et al., 2018). The impact of dual modification using ultrasound (40 kHz) and ozone (20 g/h) on physico-chemical characteristics and biodegradable film properties of corn starch was evaluated by Huang et al. (2024). The dual-modified starch films exhibited increased light transmittance, reduced, water vapor transmittance, and moisture absorption rate in comparison to the native and single-modified corn starch films (Huang et al., 2024). Tanetrungroj and Prachayawarakorn (2018) synthesized a packaging film from cassava starch using H_2O_2 as an oxidizing agent and boric acid as a crosslinking agent. Biodegradable films synthesized from dual modification exhibited enhancements in strength and extensibility, along with reduced swelling and water vapor permeability (WVP), and a higher thermal degradation temperature. The biodegradable film was prepared with dual-modified Sorghum starch. The study demonstrated that acid modification followed by oxidation of sorghum starch increment the WVP of the films. Films based on dual-modified starch can be employed, where rigid packaging is necessary (Boukhelkhal & Moulai-Mostefa, 2017).

13.5 CHALLENGES AND CONCLUSION

The oxidation of starch using sodium hypochlorite, hydrogen peroxide, and ozone represents a versatile and intriguing area of study within the realm of chemical modification of starch. Each oxidizing agent brings its unique characteristics to the starch, leading to varied applications. There has been a long history of studies and the production of sodium hypochlorite-oxidized starches. Several literature studies have investigated sodium hypochlorite-oxidized starches in relation to the starch source and reaction conditions. Hydrogen peroxide, another oxidant, introduces an alternative pathway for starch oxidation. Its milder reactivity compared to NaOCl may offer a more environmentally safe oxidizing agent compared to hypochlorite. However, mostly oxidation occurs in the presence of a metal catalyst or sometimes by UV light. A recent advancement involves the utilization of ozone in either the gaseous or aqueous phase for starch modification. While research has provided insights into the potential use of ozone for the oxidation of starch, the practical application of ozone in industrial settings for the preparation of modified starch is still unexplored. Therefore, future research needs to concentrate on achieving reproducibility of research results on an industrial scale.

REFERENCES

Aaliya, B., Sunooj, K. V., John, N. E., Navaf, M., Akhila, P. P., Sudheesh, C., Sabu, S., Sasidharan, A., Mir, S. A., & George, J. (2022). Impact of microwave irradiation on chemically modified talipot starches: A characterization study on heterogeneous dual modifications. *International Journal of Biological Macromolecules, 209,* 1943–1955. https://doi.org/10.1016/j.ijbiomac.2022.04.172

Aaliya, B., Sunooj, K. V., Navaf, M., Akhila, P. P., Sudheesh, C., Sabu, S., Sasidharan, A., Sinha, S. K., & George, J. (2022). Influence of plasma-activated water on the morphological, functional, and digestibility characteristics of hydrothermally modified non-conventional talipot starch. *Food Hydrocolloids, 130,* 107709. https://doi.org/10.1016/j.foodhyd.2022.107709

Adebowale, K. O., & Lawal, O. S. (2003). Functional properties and retrogradation behaviour of native and chemically modified starch of mucuna bean (Mucuna pruriens). *Journal of the Science of Food and Agriculture, 83*(15), 1541–1546. https://doi.org/10.1002/jsfa.1569

Akhila, P. P., Aaliya, B., Navaf, M., Sudheer, K. P., Mir, S. A., & Sunooj, K. V. (2023). 3D food printing. In *Cereal-based food products* (pp. 313–341). Springer International Publishing. https://doi.org/10.1201/9781003221210-14

Amorim, E. O. C., Doval, V. C., & Cristianini, M. (2011). Effect of ozonation on the sensory characteristics and pasting properties of cassava starch. *Procedia Food Science, 1,* 914–919. https://doi.org/10.1016/j.profoo.2011.09.138

Ashogbon, A. O., & Akintayo, E. T. (2014). Recent trend in the physical and chemical modification of starches from different botanical sources: A review. *Starch/Staerke, 66*(1–2), 41–57. https://doi.org/10.1002/star.201300106

Boukhelkhal, M., & Moulai-Mostefa, N. (2017). Physicochemical characterization of starch isolated from soft acorns of holm oak (Quercus ilex subsp. ballota (Desf.) Samp.) grown in Algeria. *Journal of Food Measurement and Characterization, 11*(4), 1995–2005. https://doi.org/10.1007/s11694-017-9582-6

Cahyana, Y., Titipanillah, R., Mardawati, E., Sukarminah, E., Rialita, T., Andoyo, R., Djali, M., Hanidah, I. I., Setiasih, I. S., & Handarini, K. (2018). Non-starch contents affect the susceptibility of banana starch and flour to ozonation. *Journal of Food Science and Technology, 55*(5), 1726–1733. https://doi.org/10.1007/s13197-018-3085-2

Castanha, N., Matta Junior, M. D. D., & Augusto, P. E. D. (2017). Potato starch modification using the ozone technology. *Food Hydrocolloids, 66,* 343–356. https://doi.org/10.1016/j.foodhyd.2016.12.001

Castanha, N., Santos, D. N. E., Cunha, R. L., & Augusto, P. E. D. (2019). Properties and possible applications of ozone-modified potato starch. *Food Research International, 116*, 1192–1201. https://doi.org/10.1016/j.foodres.2018.09.064

Çatal, H., & Ibanoğlu, Ş. (2014). Effect of aqueous ozonation on the pasting, flow and gelatinization properties of wheat starch. *LWT, 59*(1), 577–582. https://doi.org/10.1016/j.lwt.2014.04.025

Chakraborty, I., N, P., Mal, S. S., Paul, U. C., Rahman, M. H., & Mazumder, N. (2022). An insight into the gelatinization properties influencing the modified starches used in food industry: A review. *Food and Bioprocess Technology, 15*(6), 1195–1223. https://doi.org/10.1007/s11947-022-02761-z

Chan, H. T., Leh, C. P., Bhat, R., Senan, C., Williams, P. A., & Karim, A. A. (2011). Molecular structure, rheological and thermal characteristics of ozone-oxidized starch. *Food Chemistry, 126*(3), 1019–1024. https://doi.org/10.1016/j.foodchem.2010.11.113

Chapagai, M. K., Fletcher, B., Witt, T., Dhital, S., Flanagan, B. M., & Gidley, M. J. (2021). Multiple length scale structure-property relationships of wheat starch oxidized by sodium hypochlorite or hydrogen peroxide. *Carbohydrate Polymer Technologies and Applications, 2*, 100147. https://doi.org/10.1016/j.carpta.2021.100147

Chen, Q., Yu, H., Wang, L., Ul Abdin, Z., Chen, Y., Wang, J., Zhou, W., Yang, X., Khan, R. U., Zhang, H., & Chen, X. (2015). Recent progress in chemical modification of starch and its applications. *RSC Advances, 5*(83), 67459–67474. https://doi.org/10.1039/c5ra10849g

Chen, Y. F., Kaur, L., & Singh, J. (2017). Chemical modification of starch. In *Starch in food: Structure, function and applications* (pp. 283–321). Elsevier Ltd. https://doi.org/10.1016/B978-0-08-100868-3.00007-X

Chong, W. T., Uthumporn, U., Karim, A. A., & Cheng, L. H. (2013). The influence of ultrasound on the degree of oxidation of hypochlorite-oxidized corn starch. *LWT, 50*, 439–443. https://doi.org/10.1016/j.lwt.2012.08.024

Dias, A. R. G., Zavareze, E. D. R., Helbig, E., Moura, F. A. D., Vargas, C. G., & Ciacco, C. F. (2011). Oxidation of fermented cassava starch using hydrogen peroxide. *Carbohydrate Polymers, 86*(1), 185–191. https://doi.org/10.1016/j.carbpol.2011.04.026

Dimri, S., Aditi, Bist, Y., & Singh, S. (2023). Oxidation of starch. *Starch: Advances in Modifications, Technologies and Applications*, 55–82. https://doi.org/10.1007/978-3-031-35843-2_3

El-Sheikh, M. A., Ramadan, M. A., & El-Shafie, A. (2010). Photo-oxidation of rice starch. Part I: Using hydrogen peroxide. *Carbohydrate Polymers, 80*(1), 266–269. https://doi.org/10.1016/j.carbpol.2009.11.023

Emam, H. E., Zahran, M. K., & Ahmed, H. B. (2017). Generation of biocompatible nanogold using H2O2–starch and their catalytic/antimicrobial activities. *European Polymer Journal, 90*, 354–367. https://doi.org/10.1016/j.eurpolymj.2017.03.034

Fonseca, L. M., Gonçalves, J. R., El Halal, S. L. M., Pinto, V. Z., Dias, A. R. G., Jacques, A. C., & Zavareze, E. D. R. (2015). Oxidation of potato starch with different sodium hypochlorite concentrations and its effect on biodegradable films. *LWT, 60*(2), 714–720. https://doi.org/10.1016/j.lwt.2014.10.052

Garg, S., & Jana, A. K. (2011). Characterization and evaluation of acylated starch with different acyl groups and degrees of substitution. *Carbohydrate Polymers, 83*(4), 1623–1630. https://doi.org/10.1016/j.carbpol.2010.10.015

Handarini, K., Hamdani, J. S., Cahyana, Y., & Setiasih, I. S. (2020). Gaseous ozonation at low concentration modifies functional, pasting, and thermal properties of arrowroot starch (Maranta arundinaceae). *Starch/Staerke, 72*, 1–7. https://doi.org/10.1002/star.201900106

Haq, F., Yu, H., Wang, L., Teng, L., Haroon, M., Khan, R. U., Mehmood, S., Bilal-Ul-Amin, Ullah, R. S., Khan, A., & Nazir, A. (2019). Advances in chemical modifications of starches and their applications. *Carbohydrate Research, 476*, 12–35. https://doi.org/10.1016/j.carres.2019.02.007

Hu, J., Li, X., Cheng, Z., Fan, X., Ma, Z., Hu, X., Wu, G., & Xing, Y. (2022). Modified tartary buckwheat (Fagopyrum tataricum Gaertn.) starch by gaseous ozone: Structural, physicochemical and in vitro digestible properties. *Food Hydrocolloids, 125*, 107365. https://doi.org/10.1016/j.foodhyd.2021.107365

Huang, X., Chen, L., & Liu, Y. (2024). Effects of ultrasonic and ozone modification on the morphology, mechanical, thermal and barrier properties of corn starch films. *Food Hydrocolloids, 147*, 109376. https://doi.org/10.1016/j.foodhyd.2023.109376

Ibanoğlu, Ş., Özaslan, Z. T., & Ibanoğlu, E. (2018). Effects of ultrasonication and aqueous ozonation on gelatinization and flow properties of potato starch. *Ozone: Science and Engineering, 40*(2), 105–112. https://doi.org/10.1080/01919512.2017.1398634

Kaur, B., Ariffin, F., Bhat, R., & Karim, A. A. (2012). Progress in starch modification in the last decade. *Food Hydrocolloids, 26*(2), 398–404. https://doi.org/10.1016/j.foodhyd.2011.02.016

Klein, B., Vanier, N. L., Moomand, K., Pinto, V. Z., Colussi, R., Da Rosa Zavareze, E., & Dias, A. R. G. (2014). Ozone oxidation of cassava starch in aqueous solution at different pH. *Food Chemistry, 155*, 167–173. https://doi.org/10.1016/j.foodchem.2014.01.058

Kuakpetoon, D., & Wang, Y. J. (2001). Characterization of different starches oxidized by hypochlorite. *Starch/Staerke, 53*(5), 211–218. https://doi.org/10.1002/1521-379X(200105)53:5<211::AID-STAR211>3.0.CO;2-M

Kumari, B., & Sit, N. (2023). Comprehensive review on single and dual modification of starch: Methods, properties and applications. *International Journal of Biological Macromolecules, 253,* 126952. https://doi.org/10.1016/j.ijbiomac.2023.126952

Kumoro, A. C., Ratnawati, R., & Retnowati, D. S. (2015). A simplified kinetics model of natural and iron complex catalysed hydrogen peroxide oxidation of starch Andri. *Asia-Pacific Journal of Chemical Engineering, 52*(27), 9351–9358. https://doi.org/10.1002/apj

Lemos, P. V. F., Marcelino, H. R., Cardoso, L. G., Souza, C. O. D., & Druzian, J. I. (2021). Starch chemical modifications applied to drug delivery systems: From fundamentals to FDA-approved raw materials. *International Journal of Biological Macromolecules, 184,* 218–234. https://doi.org/10.1016/j.ijbiomac.2021.06.077

Lewicka, K., Siemion, P., & Kurcok, P. (2015). Chemical modifications of starch: Microwave effect. *International Journal of Polymer Science, 2015.* https://doi.org/10.1155/2015/867697

Li, J., Du, M., Din, Z. ud, Xu, P., Chen, L., Chen, X., Wang, Y., Cao, Y., Zhuang, K., Cai, J., Lyu, Q., Chang, X., & Ding, W. (2023). Multi-scale structure characterization of ozone oxidized waxy rice starch. *Carbohydrate Polymers, 307,* 120624. https://doi.org/10.1016/j.carbpol.2023.120624

Li, Y., Wang, J. H., Wang, E. C., Tang, Z. S., Han, Y., Luo, X. E., Zeng, X. A., Woo, M. W., & Han, Z. (2023). The microstructure and thermal properties of pulsed electric field pretreated oxidized starch. *International Journal of Biological Macromolecules, 235,* 123721. https://doi.org/10.1016/j.ijbiomac.2023.123721

Lima, D. C., Castanha, N., Maniglia, B. C., Matta Junior, M. D., La Fuente, C. I. A., & Augusto, P. E. D. (2021). Ozone processing of cassava starch. *Ozone: Science and Engineering, 43*(1), 60–77. https://doi.org/10.1080/0 1919512.2020.1756218

Lima, D. C., Villar, J., Castanha, N., Maniglia, B. C., Matta Junior, M. D., & Duarte Augusto, P. E. (2020, May). Ozone modification of arracacha starch: Effect on structure and functional properties. *Food Hydrocolloids, 108.* https://doi.org/10.1016/j.foodhyd.2020.106066

Liu, J., Wang, B., Lin, L., Zhang, J., Liu, W., Xie, J., & Ding, Y. (2014). Functional, physicochemical properties and structure of cross-linked oxidized maize starch. *Food Hydrocolloids, 36,* 45–52. https://doi.org/10.1016/j.foodhyd.2013.08.013

Maniglia, B. C., Castanha, N., Rojas, M. L., & Augusto, P. E. (2021). Emerging technologies to enhance starch performance. *Current Opinion in Food Science, 37,* 26–36. https://doi.org/10.1016/j.cofs.2020.09.003

Moreno, O., Cárdenas, J., Atarés, L., & Chiralt, A. (2017). Influence of starch oxidation on the functionality of starch-gelatin based active films. *Carbohydrate Polymers, 178,* 147–158. https://doi.org/10.1016/j.carbpol.2017.08.128

Naknaen, P., Tobkaew, W., & Chaichaleom, S. (2017). Properties of jackfruit seed starch oxidized with different levels of sodium hypochlorite. *International Journal of Food Properties, 20*(5), 979–996. https://doi.org/10.1080/10 942912.2016.1191868

Navaf, M., Sunooj, K. V., Gyati, R., Aaliya, B., Sudheesh, C., Akhila, P. P., Akhila, V., Sabu, S., Sasidharan, A., Mir, S. A., & George, J. (2023). Improving the structural, functional, and rheological properties of nonconventional stem pith starch from Corypha umbraculifera, by different chemical methods: A characterization study. *Journal of Food Measurement and Characterization, 17*(2), 1921–1931. https://doi.org/10.1007/s11694-022-01761-z

Navaf, M., Sunooj, K. V., Krishna, N. U., Aaliya, B., Sudheesh, C., Akhila, P. P., Sabu, S., Sasidharan, A., Mir, S. A., & George, J. (2021). Effect of different hydrothermal treatments on pasting, textural, and rheological properties of single and dual modified corypha umbraculifera L. starch. *Starch—Stärke,* 2100236. https://doi.org/10.1002/star.202100236

Ojogbo, E., Ogunsona, E. O., & Mekonnen, T. H. (2020). Chemical and physical modifications of starch for renewable polymeric materials. *Materials Today Sustainability, 7–8,* 100028. https://doi.org/10.1016/j.mtsust.2019.100028

Oladebeye, A. O., Oshodi, A. A., Amoo, I. A., & Karim, A. A. (2013). Functional, thermal and molecular behaviours of ozone-oxidised cocoyam and yam starches. *Food Chemistry, 141*(2), 1416–1423. https://doi.org/10.1016/j.foodchem.2013.04.080

Pandiselvam, R., Manikantan, M. R., Divya, V., Ashokkumar, C., Kaavya, R., Kothakota, A., & Ramesh, S. V. (2019). Ozone: An advanced oxidation technology for starch modification. *Ozone: Science and Engineering, 41*(6), 491–507. https://doi.org/10.1080/01919512.2019.1577128

Pietrzyk, S., Juszczak, L., Fortuna, T., Łabanowska, M., Bidzińska, E., & Błoniarczyk, K. (2012). The influence of Cu(II) ions on physicochemical properties of potato starch oxidised by hydrogen peroxide. *Starch/Staerke, 64*(4), 272–280. https://doi.org/10.1002/star.201100090

Pudjihastuti, I., Handayani, N., & Sumardiono, S. (2018). Effect of pH on physicochemical properties of cassava starch modification using ozone. *MATEC Web of Conferences, 156,* 2–5. https://doi.org/10.1051/matecconf/201815601027

Purcell, S., Wang, Y.-J., & Seo, H.-S. (2014). Application of oxidized starch in bake-only chicken nuggets. *Journal of Food Science, 79*(5), C810–C815. https://doi.org/10.1111/1750-3841.12466

Rostamabadi, H., Rohit, T., Karaca, A. C., Nowacka, M., Colussi, R., Feksa Frasson, S., Aaliya, B., Valiyapeediyekkal Sunooj, K., & Falsafi, S. R. (2022). How non-thermal processing treatments affect physicochemical and structural attributes of tuber and root starches? *Trends in Food Science and Technology, 128,* 217–237. https://doi.org/10.1016/j.tifs.2022.08.009

Salimi, M., Channab, B. eddine, El Idrissi, A., Zahouily, M., & Motamedi, E. (2023). A comprehensive review on starch: Structure, modification, and applications in slow/controlled-release fertilizers in agriculture. *Carbohydrate Polymers, 322,* 121326. https://doi.org/10.1016/j.carbpol.2023.121326

Sandhu, K. S., Kaur, M., Singh, N., & Lim, S. T. (2008). A comparison of native and oxidized normal and waxy corn starches: Physicochemical, thermal, morphological and pasting properties. *LWT, 41*(6), 1000–1010. https://doi.org/10.1016/j.lwt.2007.07.012

Sangseethong, K., Termvejsayanon, N., & Sriroth, K. (2010). Characterization of physicochemical properties of hypochlorite- and peroxide-oxidized cassava starches. *Carbohydrate Polymers, 82*(2), 446–453. https://doi.org/10.1016/j.carbpol.2010.05.003

Singh, J., Colussi, R., McCarthy, O. J., & Kaur, L. (2016). Potato starch and its modification. In *Advances in potato chemistry and technology* (2nd ed., pp. 195–247). Elsevier Inc. https://doi.org/10.1016/B978-0-12-800002-1.00008-X

Spier, F., Zavareze, E. da R., Silva, R. M., Elias, M. C., & Dias, A. R. G. (2013). Effect of alkali and oxidative treatments on the physicochemical, pasting, thermal and morphological properties of corn starch. *Journal of the Science of Food and Agriculture, 93*(9), 2331–2337. https://doi.org/10.1002/jsfa.6049

Sukhija, S., Singh, S., & Riar, C. S. (2016). Effect of oxidation, cross-linking and dual modification on physicochemical, crystallinity, morphological, pasting and thermal characteristics of elephant foot yam (Amorphophallus paeoniifolius) starch. *Food Hydrocolloids, 55,* 56–64. https://doi.org/10.1016/j.foodhyd.2015.11.003

Tanetrungroj, Y., & Prachayawarakorn, J. (2018). Effect of dual modification on properties of biodegradable crosslinked-oxidized starch and oxidized-crosslinked starch films. *International Journal of Biological Macromolecules, 120,* 1240–1246. https://doi.org/10.1016/j.ijbiomac.2018.08.137

Tolvanen, P., Mäki-Arvela, P., Sorokin, A. B., Salmi, T., & Murzin, D. Y. (2009). Kinetics of starch oxidation using hydrogen peroxide as an environmentally friendly oxidant and an iron complex as a catalyst. *Chemical Engineering Journal, 154*(1–3), 52–59. https://doi.org/10.1016/j.cej.2009.02.001

Tung, N. T., Thuy, L. T. H., Luong, N. T., Van Khoi, N., Ha, P. T. T., & Thang, N. H. (2021). The molecular structural transformation of jackfruit seed starch in hydrogen peroxide oxidation condition. *Journal of the Indian Chemical Society, 98*(11), 100192. https://doi.org/10.1016/j.jics.2021.100192

Vanier, N. L., Da Rosa Zavareze, E., Pinto, V. Z., Klein, B., Botelho, F. T., Dias, A. R. G., & Elias, M. C. (2012). Physicochemical, crystallinity, pasting and morphological properties of bean starch oxidised by different concentrations of sodium hypochlorite. *Food Chemistry, 131*(4), 1255–1262. https://doi.org/10.1016/j.foodchem.2011.09.114

Vanier, N. L., El Halal, S. L. M., Dias, A. R. G., & da Rosa Zavareze, E. (2017). Molecular structure, functionality and applications of oxidized starches: A review. *Food Chemistry, 221,* 1546–1559. https://doi.org/10.1016/j.foodchem.2016.10.138

Xie, S. X., Liu, Q., & Cui, S. W. (2005). Starch modification and applications. *Food Carbohydrates: Chemistry, Physical Properties, and Applications,* 357–405. https://doi.org/10.1201/9780203485286.ch8

Yu, Y., Wang, Y. N., Ding, W., Zhou, J., & Shi, B. (2017). Preparation of highly-oxidized starch using hydrogen peroxide and its application as a novel ligand for zirconium tanning of leather. *Carbohydrate Polymers, 174,* 823–829. https://doi.org/10.1016/j.carbpol.2017.06.114

Zhang, Q., Zhang, S., Deng, L., Wu, C., & Zhong, G. (2018). Effect of vacuum treatment on the characteristics of oxidized starches prepared using a green method. *Starch/Staerke, 70,* 1–10. https://doi.org/10.1002/star.201700216

Zhang, S. D., Zhang, Y. R., Wang, X. L., & Wang, Y. Z. (2009). High carbonyl content oxidized starch prepared by hydrogen peroxide and its thermoplastic application. *Starch/Staerke, 61*(11), 646–655. https://doi.org/10.1002/star.200900130

Zhang, Y. R., Wang, X. L., Zhao, G. M., & Wang, Y. Z. (2012). Preparation and properties of oxidized starch with high degree of oxidation. *Carbohydrate Polymers, 87*(4), 2554–2562. https://doi.org/10.1016/j.carbpol.2011.11.036

Zhou, F., Liu, Q., Zhang, H., Chen, Q., & Kong, B. (2016). Potato starch oxidation induced by sodium hypochlorite and its effect on functional properties and digestibility. *International Journal of Biological Macromolecules, 84,* 410–417. https://doi.org/10.1016/j.ijbiomac.2015.12.050

Starch Grafting

14

Afreen Sultana, Min Jae Shin, and
Simoneth Paulina Jimenez

14.1 INTRODUCTION

Chemical modification is one of the modification methods in which starch is treated with chemical reagents via crosslinking, esterification, grafting, etherification, or condensing reaction (X. Wang et al., 2020). Due to the presence of abundant hydroxyl groups, a wide range of chemical modifications are possible when compared to physical and enzymatic modifications (Suri & Singh, 2023). Among these, grafting is a popular method of chemical modification of starch because it is flexible, allows for simple control over the desired properties of the final product, and facilitates the creation of molecules with improvement of existing properties or introducing completely new functionalities (Noordergraaf et al., 2024). Grafting is a chemical modification in which a monomer or a polymer is covalently attached to the main chain of the polymer. The grafted polymer can be represented as poly(x)-*graft*-poly(y), where x is the primary chain and y is the side chain (Athawale & Rathi, 1999). The resulting grafted starch has the properties of starch as well as the attached polymer/monomer (X. Wang et al., 2020). Graft copolymerization on starch is particularly intriguing since it enables the development of specialized polymer structures. Therefore, the produced novel polymeric molecules come in a variety of forms, which can significantly affect how well they work overall (Sarder et al., 2022). The chemical modification that occurs during the grafting of starch often needs the addition of catalysts or initiators, which help in the grafting process. The reaction can be started using a chemical agent (such as ferrous ammonium sulfate, benzoyl peroxide, ammonium persulfate), enzymes (such as horseradish peroxidase), or radiation (López et al., 2019; Meimoun et al., 2018). Starch grafting can be performed chemically using free radical generators, and in some investigations, metal carbonyl, strong bases, and Lewis acids are also used. Enzymes are sometimes chosen over chemical agents due to their renewable nature and ability to overcome the limitations and difficulties of chemical procedures, especially when it comes to the ecological damage brought by chemical catalysts. The process of starch grafting with radiation uses radiation, such as microwaves, plasma, and irradiation, to start polymerization events and produce graft copolymers with starch as the main component. The absorption of radiation causes the starch to get excited, which in turn produces reactive free radicals (Choudhary et al., 2020; Shokri et al., 2022). The investigators may customize the modified starch to meet unique requirements by adjusting the grafting reagent and reaction parameters, which can improve the starch's tensile properties, thermal resistance, and sustainability.

Grafting starch can be performed via two routes: grafting onto and grafting from. These pathways explain the different approaches used for attaching functional structures or monomers to the starch framework, as shown in Figure 14.1.

In grafting onto, the polymer side chain, which contains many functional groups, and the backbone, which contains just one functional group, are individually synthesized. Then the pre-synthesized functional group is covalently bonded to the reactive sites of the starch main chain (Clauss et al., 2021). In

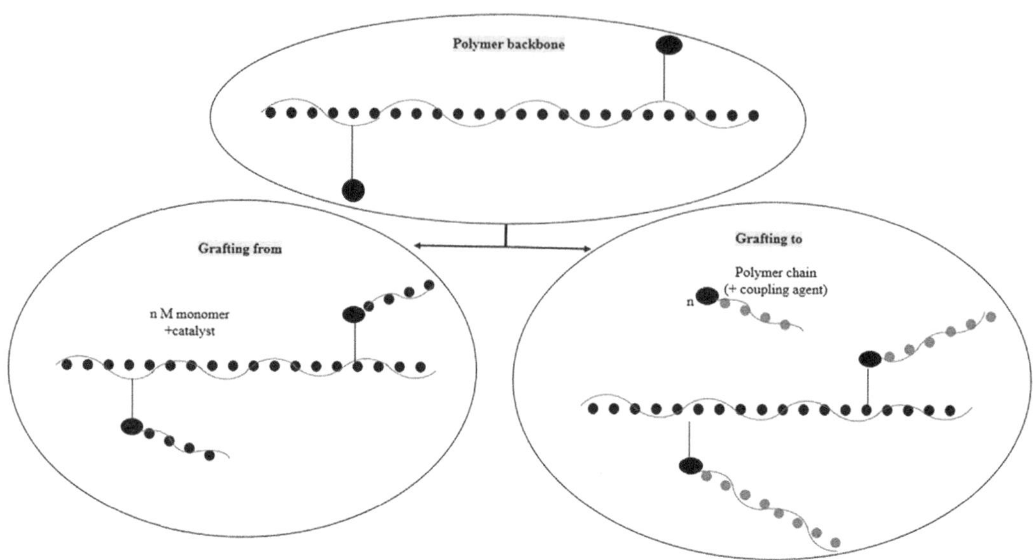

FIGURE 14.1 The two different ways for grafting starch.

order to bind the grafting reagent to the starch's backbone in this process, initiators are not needed because the reagent is already in functional form. Because the characteristics of the grafting agent are predefined, grafting onto provides more control regarding the composition of the grafted chain (Abousalman-Rezvani et al., 2020). It's a simpler way of binding already assembled components to the starch. When using the grafting from approach, the grafting reagent itself is directly attached to the starch backbone, usually because the starch molecule has reactive receptors. In this method, an initiator (chemical, enzyme, or radiation) is needed to trigger the polymerization reaction to attach the functional group to the starch backbone. It is easier to manage the density and create a more uniform distribution when using the grafting from approach (Verheyen et al., 2017). Grafting includes several critical factors to understand the copolymer system, such as graft size, graft spacing, monomer conversion, grafting efficiency, grafting percentage, and un-grafted polymer percentage, also known as homo-polymerization. Graft size can be defined as the average molecular weight of the grafts. Graft spacing is the average population of anhydro-glycose units between two graft attachments. Monomer conversion percentage can be determined as the weight of polymer with respect to the weight of monomer × 100%. Un-grafted polymer percentage can be determined as the weight of homopolymer with respect to the weight of monomer charged × 100%. Grafting efficiency can be determined as the weight of grafted polymer with respect to the weight of monomer charged × 100%. The grafting percentage can be calculated using the following formula (Athawale & Rathi, 2007; Noordergraaf et al., 2018):

$$Grafting\ percentage(\%) = \frac{Weight\ of\ grafted\ polymer}{Actual\ weight\ of\ starch} \times 100$$

To get over the drawbacks of native starch, functional reagents must be included in starch grafting. Examples of its applications in packaging and material science include improving the native starch's thermal stability (Alfuhaid et al., 2023), enhancing mechanical strength (Xu et al., 2017), and lowering its hydrophilicity (Zuo et al., 2019). Starch grafting is beneficial in the food and medicinal fields to modify attributes such as gelation property and viscosity. Starch grafting has a wide range of applications, including drug delivery, packaging, and the food industry. For example, it can be used in the development of pH-sensitive magnetic hydrogels for wound healing for the controlled release of drugs (Nezami et al., 2020).

Using tin(II) 2-ethylhexanoate as an initiator, starch-graft-poly(lactic acid) packaging film is synthesized, which showed a 42.9% reduction in water vapor permeability (Trinh et al., 2022). In the food industry, starch grafting can be used in the development of hydrogels for products such as noodles to improve the swelling property (Baloochestanzadeh et al., 2021).

14.2 TYPES OF STARCH GRAFTING

Starch grafting can be categorized into four types based on their mechanism: radical polymerization, ring-opening polymerization, polycondensation, and grafting onto/to.

14.2.1 Starch Grafting by Radical Polymerization

Using free radicals to start the process of adding polymer chains to the starch molecule is known as starch grafting by radical polymerization. In order to produce free radicals, a radical initiator like peroxide is typically utilized. Because of their extreme reactivity, these free radicals aid in the start of a process by withdrawing hydrogen from starch, which produces reactive sites for further polymerization (P. Purohit et al., 2022). Controlled radical polymerization offers more accuracy than free radical polymerization because it can better regulate the distribution of chain lengths and any side reactions. It facilitates the management and steady expansion of polymer chains, resulting in molecular weights that are more under control (H. S. Wang et al., 2023). The following table represents some advantages and disadvantages of starch grafting by the two radical polymerization methods (Table 14.1).

14.2.1.1 Free Radical Polymerization

The three fundamental steps of free radical polymerization are stage I (initiation), stage II (propagation), and stage III (termination) (Pirman et al., 2021). A hydrogen molecule is delivered to the initiator at stage I of the reaction, which starts the process of radical production on the polymeric chain of starch. This

TABLE 14.1 Difference between Free Radical Polymerization and Controlled Radical Polymerization for Starch Grafting

TYPE OF RADICAL POLYMERIZATION	ADVANTAGES	DISADVANTAGES	REFERENCES
Free radical polymerization	—Most versatile and easily implemented chain-growth polymerization method —Compatible with a wide range of monomers and solvents, including water —Reactive radicals tolerate many functionalities	—Uncontrolled molecular weight distribution through unavoidable fast termination reactions —Unexpected side reactions and uncontrolled propagation —Less predictable product compositions	(Poli, 2019)
Controlled radical polymerization	—Better control and less randomness than free radical polymerization —Increased control of molecular weight and composition —More suitable for high-value industries	—Compatible with a limited number of monomers only —Additional synthetic step for proper initiation —Expensive, not easily available agents	(Noordergraaf et al., 2018)

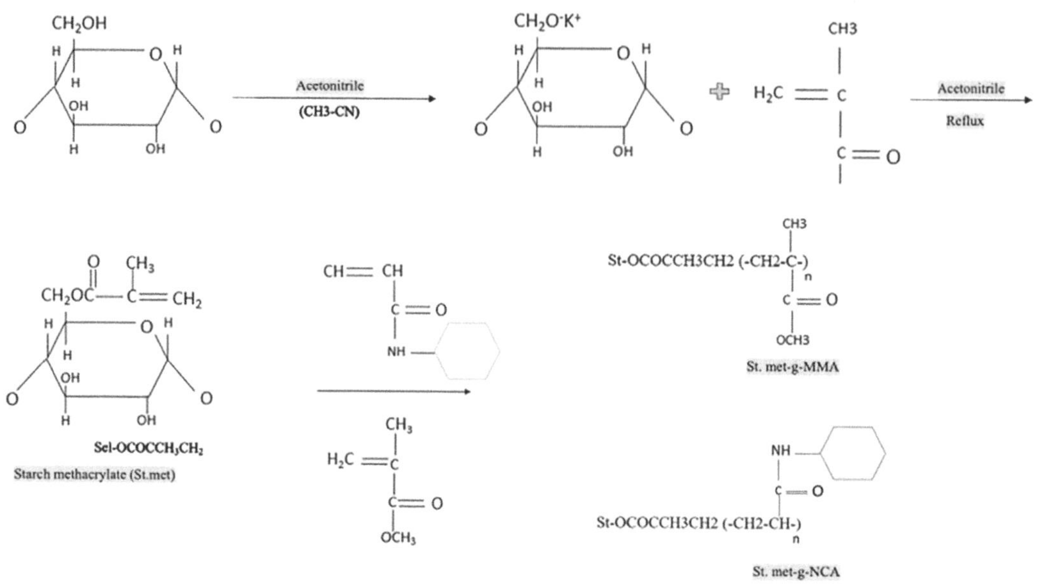

FIGURE 14.2   Mechanism explaining free radical polymerization technique to form starch methacrylate.

starch radical joins a monomer to create a monomer radical that is grafted with starch. These radicals continue to establish bonds with additional monomers until termination (stage III) is reached as a result of disproportionation or combination (Ojogbo, Ogunsona et al., 2020). The mechanism of free radical polymerization can be better understood with the following example (Figure 14.2).

In Figure 14.2, it can be observed that the hydroxy group of native starch was esterified to form starch methacrylate. Then N-cyclohexyl acrylamide and methyl methacrylate monomers were grafted with starch methacrylate. In this mechanism, hydrogen in the primary hydroxy group in starch is replaced by potassium ion by reacting starch with potassium tert-butoxide. Further, it is treated with (acetonitrile-diluted) methacryloyl chloride, resulting in the formation of starch methacrylate. With the help of N-cyclohexyl acrylamide and methyl methacrylate monomer, starch methacrylate was grafted (Çankaya, 2016). Some of the commonly used initiators (cerium complexes, benzoyl, potassium or ammonium, potassium permanganate, and enzymes) and co-initiators (such as ferrous ammonium sulfate and hydrogen peroxide) in free radical polymerization are depicted in Table 14.2.

14.2.1.2 Controlled Radical Polymerization

Controlled radical polymerization includes three types: nitroxide-mediated polymerization, atom transfer radical polymerization (ATRP), and reversible addition-fragmentation chain transfer (Parkatzidis et al., 2020).

14.2.1.2.1 Nitroxide-Mediated Polymerization

This controlled radical polymerization technique involves a nitroxide initiator to form a polymer without the need for air-free transfers, thiol compounds, or metal catalysts (Lamontagne & Lessard, 2020). This mechanism involves the cleavage of an alkoxyamine bond to create a radical that can add monomers or react with free nitroxide radicals to reversibly deactivate propagating radical chains (Zeinali et al., 2023). This mechanism can be better understood through an example (Figure 14.3).

In an investigation (Figure 14.3), cold-water-soluble starch was treated with 4-vinylbenzyl chloride, followed by a reaction with BlocBuilder MA. Then three different copolymers, acrylic, acrylate, and methacrylate, were grafted from the formed macroinitiator. Results revealed that cold water starch grafted

TABLE 14.2   Recent Studies Involving Starch Grafting Polymerization via Radical Polymerization, Ring-Opening Polymerization, and Polycondensation

STARCH GRAFTING USING CHEMICALS AS INITIATORS VIA FREE RADICAL POLYMERIZATION

STARCH	INITIATOR	MONOMER	GE (%)	GP (%)	REFERENCES
Cassava starch	Potassium persulfate	Methacrylic acid	–	89.7	(Weerapoprasit & Prachayawarakorn, 2019a)
Cassava starch	Potassium persulfate	Methacrylamide	36.9	85.6	(Weerapoprasit & Prachayawarakorn, 2019b)
Corn starch	Ceric ammonium nitrate	Acrylic acid	–	–	(Czarnecka & Nowaczyk, 2020)
Soluble starch	Ceric ammonium nitrate	Acrylic acid	–	–	(Czarnecka & Nowaczyk, 2020)
Maize starch	Ceric ammonium nitrate	Methacrylic acid	16.6	–	(Güler et al., 2020)
Potato starch	Ceric ammonium nitrate	Methacrylic acid	41.4	–	(Güler et al., 2020)
Rice starch	Ceric ammonium nitrate	Methacrylic acid	14.3	–	(Güler et al., 2020)
Wheat starch	Ceric ammonium nitrate	Methacrylic acid	43.6	–	(Güler et al., 2020)
Potato starch	Potassium persulfate	Methacrylic acid	67	46	(Worzakowska, 2020)
Corn starch	Potassium persulfate	Acrylamide	52	36.1	(Bai et al., 2021)
Maize starch	Potassium persulfate	Diallyl amine	–	56	(Saleh et al., 2021)
Corn starch	Ceric ammonium nitrate	Acrylamide	–	–	(Tajbakhsh et al., 2022)
Maize starch	Potassium persulfate	Maleamic acid	116.63	93.04	(Alfuhaid et al., 2023)
Soluble starch	Ceric ammonium nitrate	Acrylamide	71	95	(Y. Luo et al., 2023)
Potato starch	Potassium persulfate	Acrylic acid	–	56.5	(Worzakowska, 2023)
Starch Grafting Using Chemicals as Co-Initiator Free Radical Polymerization					
Corn starch	Ferrous ammonium sulfate and potassium persulfate	Oleic acid	–	–	(Mittal et al., 2020)
Cassava starch	Ferrous ammonium sulfate hexahydrate and hydrogen peroxide	Methyl methacrylate and butyl acrylate	–	283	(Rodrigues et al., 2020)
Corn starch	Ferrous ammonium sulfate and hydrogen peroxide	Sodium allyl sulfonate	50.6	97.9	(W. Li et al., 2021)
Corn starch	Ferrous ammonium sulfate and potassium persulfate	Oleic acid	–	–	(Mittal et al., 2021)
Cassava starch	Ferrous ammonium sulfate hexahydrate and hydrogen peroxide	Methyl methacrylate and butyl acrylate	–	207	(Rodrigues et al., 2021)
Corn starch	Ceric ammonium nitrate and ammonium persulfate	Acrylamide	44.72	93.18	(Jiang et al., 2022)

(Continued)

TABLE 14.2 *(Continued)*　Recent Studies Involving Starch Grafting Polymerization via Radical Polymerization, Ring-Opening Polymerization, and Polycondensation

STARCH	INITIATOR	MONOMER	GE (%)	GP (%)	REFERENCES
Starch Grafting Using Methods Such as Ring-Opening Polymerization					
Cassava starch	Potassium tert-butoxide	Lactide	–	–	(Noivoil & Yoksan, 2020)
Tapioca starch	Epichlorohydrin	Triethylamine	–	–	(Sarmah & Karak, 2020)
Corn starch	Potassium tert-butoxide	Lactide	–	–	(Dong et al., 2021)
Corn starch	Phosphoric acid	ε-caprolactone	–	68–73	(Meimoun et al., 2021)
Corn starch	Tin(II) 2-ethylhexanoate	Lactide	–	–	(Trinh et al., 2022)
Starch Grafting Using Methods Such as Polycondensation					
Corn starch	Stannous octoate	Lactic acid	0.12–12.45	–	(Zuo et al., 2019)
Corn starch	Dibutylin dilaurate	Hexamethylene diamine	17.37	–	(Ojogbo, Ward, et al., 2020)
Wheat starch	2,5-bis(tert-butylperoxy)-2,5-dimethylhexane	Lactic acid	16.11–24.21	–	(Alikarami et al., 2022)
Potato starch	Sodium hypophosphite monohydrate	Octadecenylsuccinic anhydride	10.9–13.0	–	(Baniasadi et al., 2023)
Sugar palm starch	Stannous octoate	Lactic acid	–	–	(Triwulandari et al., 2023)

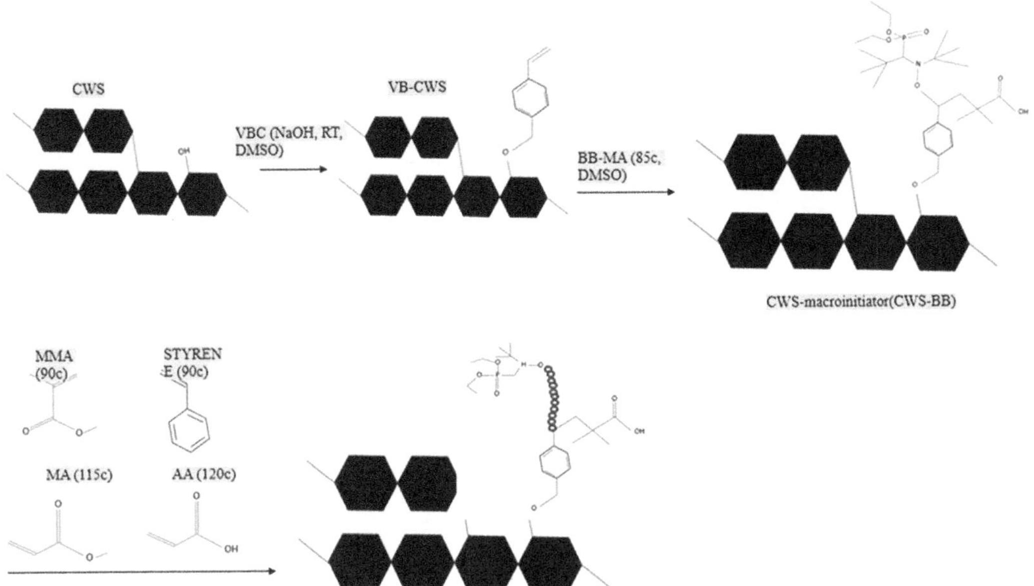

FIGURE 14.3   Mechanism explaining controlled radical polymerization technique involving nitroxide initiator.

with methacrylate at 0.11 M (D = 1.36) grafting syntheses were the most well controlled showing linear reaction and relatively low dispersities (<1.5) for the methyl methacrylate-co-styrene grafts (Fritz et al., 2020).

14.2.1.2.2 Atom Transfer Radical Polymerization

In ATRP, the reaction takes place between the metal complex (at low oxidation state) and the dormant chain (capped with halogen) (Dworakowska et al., 2022). As a result, a metal complex with a high degree of oxidation is formed, along with a coordinated halide ligand. This ligand behaves as a deactivation element and reverts the radical to its dormant state (Matyjaszewski, 2018). The quantity of developing macroradicals is minimal, and termination chemical reactions are reduced as a result of the ATRP equilibrium being substantially tilted in the direction of dormant species (Lorandi et al., 2022). One of the examples to explain this mechanism can be done by functionalization of nano-starch (Figure 14.4).

In Figure 14.4 it is observed that, first, the surface area of starch was increased by acid-catalyzed hydrolyzation reaction. The hydroxy group of nano-starch is replaced by one of the halogen atoms of 2-bromoisobutyryl bromide (acts as an initiator). The other halogen which was at the end of 2-bromoisobutyryl bromide was used to graft dimethylaminoethyl methacrylate (acts as a monomer), as shown in the figure below (Su et al., 2020). ATRP has the advantage that no homopolymer is formed in this process, but the removal of residual (such as copper) is challenging (Dworakowska et al., 2022; Matyjaszewski & Xia, 2001).

14.2.1.2.3 Reversible Addition-Fragmentation Chain Transfer

A reversible chain transfer agent, a vinyl-type monomer, and a method for producing reactive initiator radicals are the three basic components of a reversible chain transfer system (Thum et al., 2020). The incorporation of radicals into the reversible chain transfer agent starts a reaction in which the chances of removing an initiating radical or a polymer chain are equal. A quick pre-equilibrium is obtained when a chain transfer agent has captured all radical species. This pre-equilibrium inhibits termination events, controls the concentration of reactive radicals in the steady state, and encourages controlled chain development by degenerative chain transfer (Perrier, 2017). Due to its inert carbon atom, which is challenging

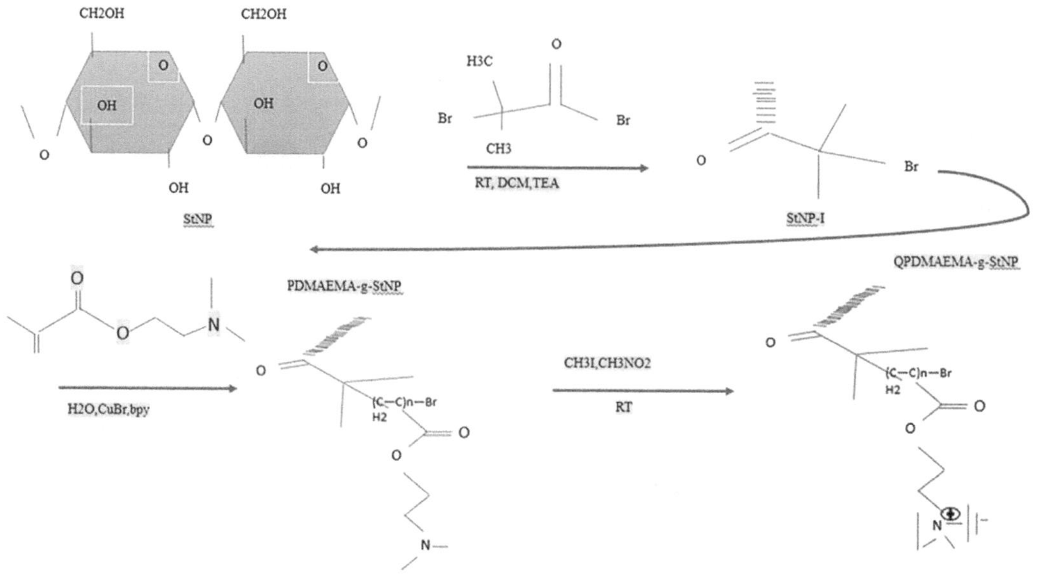

FIGURE 14.4　Mechanism explaining functionalization of nano-starch via ATRP technique.

to modify, this technique lacks applications where degradable properties are required. This challenge can be overcome by step growth polymerization, which involves the incorporation of desired functions into the backbone (Tanaka et al., 2021)

14.2.2　Starch Grafting Using Ring-Opening Polymerization

In ring-opening polymerization, the carbon–carbon double bonds in the cyclic olefin rings are broken, and the resultant ends that are open are connected to create polymer chains (Plummer et al., 2023). An initiator, which might be anionic or cationic, usually performs the ring opening of a cyclic monomer to produce an ionic propagating species (V. B. Purohit et al., 2022). Ring-opening polymerization of starch is initiated by the hydroxyl group in the presence of a catalyst, and this method is used for polymerizing cyclic esters. Such as oligo (lactic acid)-grafted starch can be prepared by reacting potassium tert-butoxide with starch, followed by a reaction with lactide, and the reaction is terminated by adding an aqueous HCl solution. Oligo (lactic acid)-grafted starch with a DS of 1.2 and DP of 2.0 was synthesized (Noivoil & Yoksan, 2020). An example of ring-opening polymerization in-situ involving caprolactone helps clarify the process (Figure 14.5).

As shown in Figure 14.5, a study that was carried out in an open atmosphere uses tetra(phenylethynyl) tin as a catalyst and ultrasonication to disperse guar gum (co-initiator) in a liquid monomer (caprolactone). Hydrophobicity was decreased, which is a desirable attribute for repelling biomaterials, and it was also observed that crystallinity increased with the rise in guar gum concentration. Fourier transform infrared spectroscopy confirmed the effective grafting of poly-caprolactone on guar gum (El Assimi et al., 2019). The following table represents some more examples of ring-opening polymerization (Table 14.2).

14.2.3　Starch Grafting via Polycondensation

Starch grafting through polycondensation involves the reaction between the bifunctional or polyfunctional monomers, resulting in the release of small compounds such as water molecules (Cummings et al., 2019).

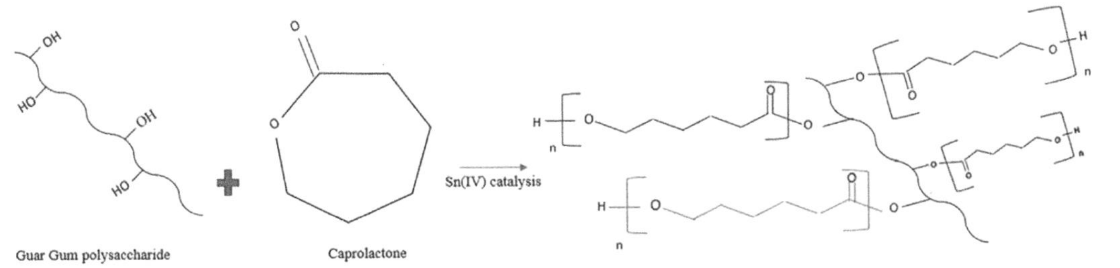

Guar Gum polysaccharide

Caprolactone

Sn(IV) catalysis

FIGURE 14.5   Representation of ring-opening polymerization using guar gum polysaccharide as a co-initiator.

In order to increase reactivity, starch must first be activated in this process before polycondensation. Activated starch and monomers (for example, methyl acrylate, maleic anhydride, and lactic acid) consist of functional groups that react to produce a molecule that has a higher molecular weight (Zuo et al., 2019). For instance, in a study, thermoplastic was developed using gelatinized starch and furfural through the process of polycondensation under acidic conditions. The developed plastic was smooth, homogeneous, without cracks, and showed a glass transition temperature of 129°C, suggesting the scope of polycondensation in starch grafting for polymer development (J. Zhang et al., 2022). Some similar examples are represented in Table 14.2.

14.2.4 Grafting Onto/to

Grafting onto/to takes place between a functional group of macromolecules, such as starch or cellulose, and a molecule with a reactive end. It can be performed via coupling agents, activated esters or acids, click chemistry, and amidation (Yadav et al., 2023). In recent research, acrylonitrile monomers were grafted onto starch nanoparticles using the microwave initiation method to avoid the use of a toxic initiator. The grafting yield was recorded as 51.6%, thermal gravimetric analysis showed a weight loss of 43.5% between the temperature range of 200–300°C for starch nanoparticles, and modified starch showed a weight loss of 35% between 300–400°C. These results suggest that the grafting of acrylonitrile onto starch nanoparticles enhanced thermal stability (Mostafa & EL-Sanabary, 2020). In the esterification process, maleic anhydride acts as a coupling agent to help graft polyethylene glycol onto starch. Maleic anhydride first interacts with the hydroxyl end groups of polyether in the first phase. Following this step, the carboxylic acids that are produced interact subsequently with the hydroxyl groups. This novel grafting technique improves starch's solubility and modifies its structural properties (Verheyen et al., 2017). One benefit of grafting onto/to is that it allows you control over the characteristics of the final polymer since it gives you the ability to characterize the molecules or polymer chain prior to grafting (Yadav et al., 2023). However, it is limited in its ability to achieve high grafting densities due to steric repulsions between the grafted chains (Z. Wang et al., 2021).

14.3 FACTORS AFFECTING STARCH GRAFT POLYMERIZATION

The effectiveness and properties of starch graft polymerization might vary depending on several parameters. Here are some further details on some of the critical factors affecting starch grafting polymerization:

14.3.1 Starch Composition

The efficiency of grafting is significantly influenced by the ratio of amylose to amylopectin. High amylose content shows a lower grafting ratio; the reaction occurs slowly in terms of its rheological characteristics (such as flow and deformation) (Bao et al., 2019; H. Luo et al., 2021). In comparison to high amylose–content starch, starch with a high amylopectin content is more advantageous for grafting functional groups like epoxy resins (Tratnik et al., 2020). The amylose content of the native starch may have more influence on the viscosity of grafted starches (L. Wang et al., 2020). On the other hand, according to NMR data for waxy starch, functional groups like the (acrylamide group) predominantly grafted onto C6, and the length of the grafted segment decreased as amylopectin levels increased generally. The grafting ratio and efficiency were shown to improve with increasing amylose content under the same reaction conditions (Zou et al., 2012).

14.3.2 Reaction Temperature

Temperature is one of the factors affecting the rate of polymerization; an increase in temperature is directly proportional to the rate of polymerization. The presence of additives such as Lewis acid, aluminum compounds, water, and strong nucleophiles have a strong impact on ATRP by either promoting or terminating the reaction (Dworakowska et al., 2022; Matyjaszewski & Xia, 2001). In one of the studies, starch was grafted using three different monomers: maleic anhydride, lactic acid, and methyl acrylate. Results showed that with the increase in temperature up to 80°C, the grafting ratio increased, but on further increasing the temperature, the grafting ratio decreased because of the evaporation of molten monomers (Zuo et al., 2019).

14.3.3 Reaction Time

Studies show that grafting percentage increases with an increase in time up to reaching a peak; after that, it decreases before becoming stable. The decrease in grafting percentage might be due to exhaustion of monomers. It's also possible that the increased yield of graft copolymers is increasing viscosity, which would impede reaction collision and lead to reduced grafting ratios (Alfuhaid et al., 2023). It was reported in result discussion of starch grafted with carbonyl monomers that with increase in time from 0.5 to 2 hours, the grafting ratio increased, but on further increasing the time, the reaction became stable due to saturation (Zuo et al., 2019).

14.3.4 Type of Solvent

Different solvents have been used for different monomers such as alcohol, benzene, water, and many others. Choosing the appropriate solvent is crucial because it can impact the structure of the catalyst and it is also needed in certain situations when the polymer is insoluble in its monomer to improve the solubility (Dworakowska et al., 2022). When choosing a solvent, one must take into account a variety of aspects, including compatibility, reaction kinetics, solvent removal and recycling, and environmental effect. Different starches have different reaction abilities, and it was reported in a study that a catalyst decomposes around ten times faster in tetrahydrofuran (THF) and N,N-dimethylformamide than in toluene (Blosch et al., 2022). Contrarily, research has revealed that the quality of the solvent affects the grafting of polymers onto surfaces. It was discovered that poor-quality solvent (6%), as opposed to excellent-quality solvent (14%), exhibited less of an impact on the distortion of molecular weight (Michalek et al., 2019). The way grafted polymer chains behave, their conformity, and their dynamic properties are influenced by the solvent quality. The morphology of end-grafted polymer chains may change from a stretched regime to a more collapsed regime when the purity of the solvent degrades. A nonuniform polymer canopy may also emerge at this time (Kim & Thérien-Aubin, 2020). Solvent quantity also affects the grafting percentage; an increase in solvent concentration can lead to an increase in grafting percentage (Mohy Eldin et al., 2016). Graft frequency is increased when a solvent with radical chain transfer capabilities is utilized, which may promote the development of new grafts due to transferred radical activity (Noordergraaf et al., 2018).

14.3.5 Concentration of Monomers

The concentration of monomers has an effect on grafting efficiency, grafting yield, and grafting percentage (Abd El-Ghany et al., 2019; Misra et al., 2020). The effect of monomer concentration varies depending on the reaction's other components. Differential structural configurations of functional moieties or polymer chains result from variations in monomer concentration. When polymer chains are present in high quantities, they take on a brush-like shape with considerable spatial flexibility, but at low concentrations, they interlace to create a complex structure. The grafted membrane surface's physio-chemical properties are significantly impacted by this concentration-related orientation (Suresh et al., 2021). For example, in one study, monomer concentration was in the range of 0.10–0.75 mol/L, and it was reported that on increasing the concentration up to 0.5 mol/L, maximum grafting efficiency and grafting percentage were achieved, and on further increasing monomer concentration, a decrease was observed (Abd El-Ghany et al., 2019). On the contrary, in a study related to grafting acrylonitrile onto starch nanoparticles, it was reported that increasing the concentration up to 1 ml of acrylonitrile increased the graft yield percentage, and on further increase in concentration, the graft yield became constant (Mostafa & EL-Sanabary, 2020). Decreasing monomer content improves reaction control, and it can be demonstrated by reduced grafted polymer dispersity and increased reaction kinetic linearity (Fritz et al., 2020). The ratio of monomer to polymer also plays a critical role in grafting percentage. High monomer concentration will reduce the grafting percentage (Fan & Picchioni, 2020).

14.4 MODIFICATIONS IN GRAFTED STARCH PROPERTIES

When compared to unmodified starch, the characteristics of grafted starch might alter in a number of ways as a result of the transformation. The particular qualities that alter will depend on elements like the nature of the graft chains, the grafting procedure, starch type, and the intended use. Some of these typical changes possible are discussed in the following.

14.4.1 Morphology

Granules of starch vary in size from over 100 μm to less than 1 μm, and size plays a critical role in applicability in food and non-food applications (M. Li et al., 2023). Grafting can have an effect on the morphology of starch by potentially causing surface damage and increasing granule roughness. Oval-shaped banana starch granules ranging between 7 and 50 μm, when grafted with polyethylene monoalcohol, showed cracks in the granules, suggesting that the chemical interactions involved in grafting of starch might lead to changes in morphology (Ramírez-Hernández et al., 2020). According to a study, maize starch's spherical form become flaky when it was oxidized, and when it was further grafted with ethylene glycol diglycidyl ether, it took on a needle-like structure (Hao et al., 2021). In another recent study, graft polymerization was performed between cassava starch and glycidyl methacrylate using ceric ammonium nitrate as an initiator. Scanning electron microscopy showed the spherical shape of cassava starch was

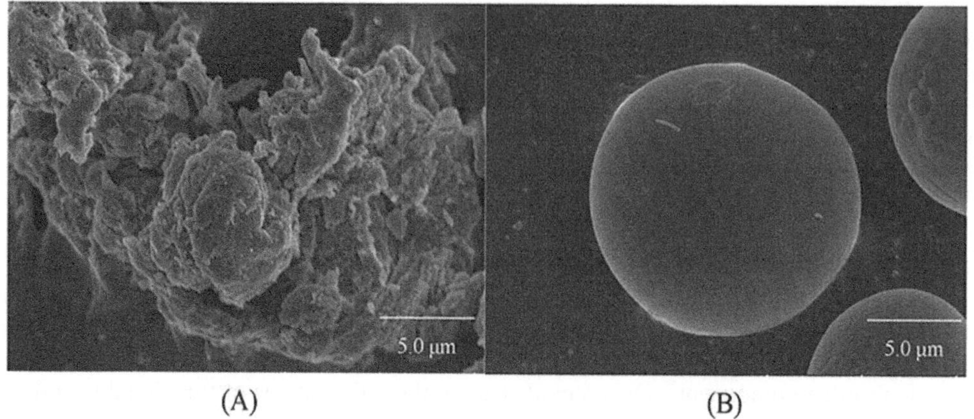

(A)   (B)

FIGURE 14.6   Scanning electron microscopy image of soluble starch (A) and polymer after starch grafting (B).

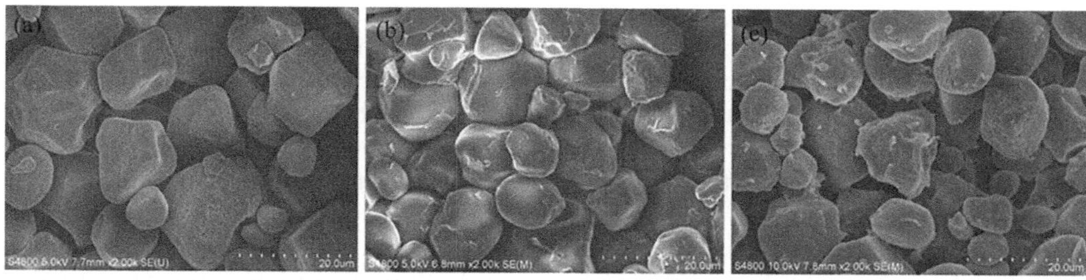

FIGURE 14.7   Scanning electron microscopy image of corn starch (a), starch macroinitiator (b), and polymer after starch grafting (c) (W. Li et al., 2023).

broken due to grafting (X. Chen et al., 2022). As shown in Figure 14.6., the smooth spherical shape of soluble starch was changed to a peak-like surface with some holes and wrinkles due to graft polymerization of starch with acrylamide-allyl trimethyl ammonium chloride (Y. Luo et al., 2023b).

On the contrary, there are some studies which suggest that grafting causes very minute changes on the morphology of starch. In an investigation, food-grade corn starch was treated to form a starch macroinitiator and then further grafted with acrylic acid and butyl acrylate. The polygonal morphology of starch was changed to granules with some damage on the surface. It was stated that the granular form of starch is retained if the reaction occurs below gelatinization temperature, and therefore very slight damage was observed, which was only on the surface (Figure 14.7) (W. Li et al., 2023).

Based on these investigations, it can be concluded that depending on the starch source and modification properties, starch grafting may result in either little or major alteration to the morphological properties of starch. Grafting can change the smooth spherical shape of starch to a peak-like surface with wrinkles and holes (Y. Luo et al., 2023), and in some cases, it might only cause slight changes on the surface of granules without changing the granular shape of starch (W. Li et al., 2023).

14.4.2 Crystallinity

Starch crystallinity and its molecular structure can be analyzed by X-ray diffraction analysis. Native starch can have three different forms of patterns: A (example: cereal starch), B (example: tuber starch), and C (example: legume starch), depending on the botanical source. Any enzymatic, physical, and chemical alterations to starch are reflected in its crystallinity pattern (Govindaraju et al., 2021). The V-type starch pattern is observed when amylose present in starch undergoes complexation (Sarkar et al., 2021). The crystallinity pattern can change due to starch grafting; for example, in a study, the V-type structure of starch nanoparticles showing a sharp peak at 13.5° and 20.8 disappeared when it was grafted with acrylonitrile (Mostafa & EL-Sanabary, 2020). In some cases, there might be observed decrease in peak intensity to indicate a weakening of the starch's crystalline structure or the merging of two peaks to indicate a shift in crystallinity (Y. Zhang et al., 2021). Graft polymerization between starch and copolymers (acrylonitrile and acrylamide) was performed by via reversible addition–fragmentation chain transfer. The diffraction peaks of starch was recorded at 2θ of around 15°, 17°, 18°, and 23°, but for grafted starch, a broad peak was recorded at 18°, indicating a decrease in crystallinity. It was revealed that a higher number of side chains causes destruction of the structural regularity of starch crystallinity, resulting in lower crystallinity (Cai et al., 2020). Due to decreasing hydrogen connections between the molecules, which increase thermoplasticity, there is a possibility that the grafted structure will have less crystallinity when compared to native starch (Zuo et al., 2019). Results from certain research indicate that after undergoing spinning conditions (thermal setting and drawing-stretching multiples), the crystallinity of grafted starch fibers synthesized via wet-spinning process can be enhanced (Chang et al., 2021). However, there are a few investigations which showed no significant changes to the crystalline structure of starch. For example, a study where starch nanocrystal was grafted with polyethyleneimine showed the invariant A-type crystal structure. Starch nanoparticles and grafted starch were found to have somewhat similar relative crystallinity, measuring 40.11% and 38.23%, respectively (Q. J. Chen et al., 2021). Thus, it could be stated that structural changes of starch granules which affect the crystallinity depend on the structure and composition of their components (Cruz-Benítez et al., 2022).

14.4.3 Thermal Stability

Grafting helps in improving the thermal stability of starch significantly. Increased resistance to thermal degradation of grafted starch may be related to the monomer's character, particularly if it has an aromatic nature and the capacity to produce carbonaceous residues when heated (Abd El-Ghany et al., 2019). When a monomer is grafted with starch, the grafted polymer may undergo chemolysis at a temperature greater than that of native starch. This phenomenon results in a maximum weight loss at a higher temperature than native starch, which affects thermal stability (Liu, Guo, Wei, et al., 2023). In a study, native wheat starch, soluble wheat starch, waxy maize starch, and cationic tapioca starch were grafted with acrylamide monomers using twin screw extrusion technology via free radical copolymerization. According to the thermogravimetrical analysis data, around 60% of weight loss occurred at 325°C, 412°C, 467°C, 414°C, and 430°C for native wheat starch, wheat starch grafted copolymer, cationic tapioca starch grafted copolymer, waxy maize starch grafted copolymer, and soluble wheat starch grafted copolymer, respectively. This suggests that the thermal stability of starch can be increased by grafting (Siyamak, Laycock, et al., 2020). Likewise, a different study found that weight loss of grafted starch (approximately 76%) was lower than native starch (approximately 91%), suggesting an increase in the thermal stability of starch (Alfuhaid et al., 2023).

In some circumstances, grafting can have a negative impact on a grafted starch's thermal properties, for example, causing grafted starch to be less thermally stable than native starch. A study that used oligolactic acid to graft cassava starch revealed that the native starch had a onset of decomposition temperature of 223°C, while the grafted starch had 306°C, suggesting that a reduction in decomposition temperature could happen because the packed starch structure was destroyed (Noivoil & Yoksan, 2020).

14.4.4 Hydrophobicity

Starch grafting can be used to improve the hydrophobicity characteristics of starch by replacing the hydroxy group with hydrophobic functional groups (X. Wang et al., 2020). For example, in a recent investigation, the hydroxy group of an anhydro glucose unit was replaced by carbonyl groups using three different monomers (methyl acrylate, maleic anhydride, and lactic acid). The resulting grafted starches were compared, and results showed that the contact angle of native starch was 37°, whereas grafted starch showed an increase to 40°, 55°, and 61° for methyl acrylate, maleic anhydride, and lactic acid, respectively. The findings of the contact angle were consistent with the water absorption data, which demonstrated a lower rate of water absorption for grafted starch relative to native starch (Zuo et al., 2019). Grafted starch can also be used as a compatibilizer in blending starch with synthetic polymers (such as linear low-density polyethylene). In an investigation, stearic acid–grafted starch was used as a compatibilizer for preparation of linear low-density polyethylene/thermoplastic films. The result indicated that grafted starch helped in decreasing the phase separation of the developed polymer film. Hence the water vapor permeability of the developed film was reduced from 13×10^{-5} µg m/m² h Pa (films without compatibilizer) to 10×10^{-5} µg m/m² h Pa (films with compatibilizer) (Khanoonkon et al., 2016). Films developed using 5% of starch grafted with poly(ε-caprolactone) with maleic anhydride and glycidyl methacrylate blended with 40% of PLA recorded a 67% reduction in water vapor permeability compared to native starch films (Collazo-Bigliardi et al., 2019). In a study, oligo (lactic acid)-grafted starch was incorporated into a polylactic acid and thermoplastic starch blend to develop a film, resulting in decreased barrier properties. Grafted starch was shown to increase the distribution of thermoplastic starch within the film matrix and decrease the quantity of free hydroxyl groups accessible for interaction with water molecules (Noivoil & Yoksan, 2020). Thus, these investigations suggest that incorporation of starch-grafted polymer helps in improving the hydrophobicity of starch.

14.4.5 Molecular Weight

Starch's molecular weight is impacted by grafting, which can be examined by measuring the pasting viscosity of the starch. A study revealed that pasting viscosity was increased to 27,100 mPa.s (grafted starch) from 16,500 mPa.s (native starch), which indicates that starch's molecular weight increases as a result of graft polymerization (Zuo et al., 2019). The effective grafting of branches onto starch may be the cause of the increased molecular weight of the grafted starch (Liu, Guo, Ren, et al., 2023). Additionally, under the same conditions (such as the same quantity of initiator), the dosage of the monomer might also directly affect the rise in molecular weight of the grafts (Noordergraaf et al., 2018).

14.4.6 Selective Adsorption

The selective adsorption capacity of starch can be enhanced by starch grafting. Grafted starch may be employed for the adsorption of compounds such as cationic dyes, bisphenol, and heavy metals because of its hierarchical porosity and high hydrophilic properties. Adsorption of cationic dyes such as methylene blue and crystal violet were 1428.6 and 2500.0 mg/g, respectively, by starch-grafted hydrogel nanocomposites (Moharrami & Motamedi, 2020). Similar to this, starch-grafted styrene hydrophilic porous resin shows impressive removal capacity toward bisphenol, regardless of the various aqueous conditions (Yuan et al., 2022). In another study, hydrogels made of starch grafted with itaconic acid were created, and the findings showed that at pH 8, they were capable of adsorbing heavy metals, including 1.11 mg/g of hexavalent chromium and 14.13 mg/g of trivalent chromium (Solis-Ceballos et al., 2023). These findings demonstrated that grafting of starch aids in altering its chemical and structural features, hence improving the starch's capacity for selective adsorption.

14.5 APPLICATIONS IN THE FOOD INDUSTRY

In the food industry sector, starch grafting polymerization can be employed in a number of ways for modifying the characteristics of starch to satisfy certain functional or technical needs. There is limited research performed in the field of modification of starch via grafting technique for food processing applications; mostly it is used in agricultural, drug delivery, electrical engineering, and bioplastic (food packaging) (Compart et al., 2023). For instance, in a study, starch grafting was used in the development of hydrogels, which are substances that, when physically and chemically crosslinked, generate three-dimensional networks and have the capacity to inflate and capture vast amounts of water without dissolving (Baloochestanzadeh et al., 2021). In the food industry, hydrogels are used in the preparation of noodles and carriers for active substances (Cui et al., 2022). But in terms of development of packaging material, starch grafting is widely studied. Bioplastics made from starch have demonstrated superior qualities that are equivalent to those of commercial packaging materials but lack some desired properties, which can be overcome by grafting (Iqbal et al., 2021). Starch grafting has shown potential in improvement in enhancing barrier properties, thermal resistivity, biodegradability, and electrical conductivity of packaging films, which are further shown in Table 14.3. A substantial body of research is being done on novel starch morphologies, such as nanoparticles and nanocrystals, since they enhance surface area and generate more stable emulsions (Campelo et al., 2020). Some of the recent research using nano-starch grafting techniques in the development of packaging materials is also mentioned in Table 14.3.

14.6 SUSTAINABILITY AND FUTURE SCOPE

Starch grafting is not used in the food processing industry due to toxicity concerns but in the future can be used in enhancement of a wide range of properties such as viscosity, stability, and shelf life of foods. It can also be used to improve the gel-forming and thickening characteristics of starch to be used in soups and sauces. Some of the ways to overcome the drawbacks of conventional starch-grafting techniques (such as poor graft yield rates, limited control over how homopolymers grow, reduced repeatability, and low chances of commercial synthesis) are discussed in the following.

14.6.1 Sources of Monomers or Polymers

Starch used in starch grafting should be considered to be taken from non-conventional sources; for example, in the case of starch, it can be isolated from millets, roots, or seeds rather than conventional sources (which are staple foods for the majority of the population) (Tagliapietra et al., 2021). The science community has turned its attention to the investigation of sustainable monomers and polymers generated from sources that are renewable, mainly due to concerns about negative ecological impact and carbon footprint (Getzler & Mathers, 2022). According to a report, globally around 31.9 million metric tons of polymer garbage contaminated coastal regions in 2010, highlighting the growing problem of polymer pollution. It consequently becomes essential to move away from the use of conventional fossil fuel feedstocks for monomers that result in slow-degrading polymers (Law & Narayan, 2021). For example, in recent research, grafting was performed via reversible addition-fragmentation chain transfer to develop elastomers using lignin, chitin, and plant oil (Z. Wang et al., 2023). Various sustainable substances such as fatty acids, polysaccharides such as chitosan, proteins such as zein, and terpenes can be used as raw materials to develop a variety of sustainable goods and composite materials (Techie-Menson et al., 2021).

14.6.2 Reagents

In starch-grafting modification, one of the most critical factors is selecting the appropriate reagent (solvent or catalyst), as it helps in achieving a higher degree of substitution. Starch modification such as starch grafting involves the use of corrosive and hazardous chemicals. Organic solvents such as dimethyl

TABLE 14.3   Applications of Starch Grafting Modification Techniques in the Development of Packaging Films/Coatings

STARCH	POLYMER	POLYMERIZATION	RESULTS	REFERENCES
Cassava starch	Polymethacryl acid/starch	Chemicals as initiator	Comparing starch films grafted by MAA at a high percentage of grafting to starch films with native starch and grafted starch by MAA at a low percentage of grafting, the former showed improvements in extendibility, WVP, water uptake, and thermal degradation temperature.	(Weerapoprasit & Prachayawarakorn, 2019a)
Cassava starch	Polymetha-crylamide/starch	Chemicals as initiator	When compared to native starch film, the stresses at maximum load of grafted starch films grafted with methacrylamide (MAM) were better. Additionally, compared to grafted starch films with a low percentage of grafting, those grafted with a high percentage of MAM showed reduced water uptake, WVP, and swelling properties as well as increased extendibility.	(Weerapoprasit & Prachayawarakorn, 2019b)
Corn starch	Polyvinyl alcohol/ starch	Chemicals as co-initiator	The hydrophilicity of the films as measured by water uptake (%) and water contact angle (CA) similarly decreased after polyvinyl alcohol (PVA)/starch was included.	(Mittal et al., 2020)
Corn starch	Polyvinyl alcohol/ starch	Chemicals as co-initiator	After starch and starch nanocrystals (SNCs) were added to the PVA matrix, the films' water absorption (%) decreased. Furthermore, when comparing the PVA/grafted starch nanocomposite film to PVA/starch and PVA/starch nanocomposite films, respectively, there was a 59% and 35% decrease in the water uptake (%). Lower water absorption into the matrix resulted from the SNCs' increased crystallinity and excellent dispersion into the PVA/S matrix.	(Mittal et al., 2021)
Commercial starch	Polymethacrylate/ starch	Free radical polymerization	With 0.75 g of starch and 3 ml of monomer, grafting efficiency was attained, resulting in a biodegradable polymer that showed 31.45% breakdown in 60 days. Biodegradability of polymer which added Ni nanoparticles was recorded as approximately 20%. This study concluded that the polymer nanocomposite is highly stable and has potential applications as a food packaging material.	(Iqbal et al., 2022)
Corn starch	Starch-graft-poly(lactic acid)	Ring-opening polymerization	The produced starch/PLA films demonstrated superior oxygen barrier capabilities and enhanced tensile performances because of the good distribution of PLA phases in the grafted starch matrix. However, it was also shown that the crystalline morphology of PLA structures was diminished by the bridging between grafted starch and PLA phases via ex-situ compatibilization.	(Trinh et al., 2022)

sulfoxide are the most commonly used, but it has the risk of toxicity and is non-ecofriendly. The solvent used in starch grafting can be replaced by sustainable solvents such as water, ionic liquids, and supercritical carbon dioxide (Sarder et al., 2022). To replace organic solvents, other techniques such as radiation or extrusion have also been investigated. Starch nanoparticles were used in a recent study to graft copolymerize with acrylonitrile using microwave irradiation without the need of hazardous chemical initiators. It was revealed that on increasing the microwave power up to 500 watts, the grafting yield percentage was increased. Higher copolymer graft yields were reported as the consequence of a rapid interaction between these free radical sites on starch nanoparticles and acrylonitrile through a typical free radical reaction process (Mostafa & EL-Sanabary, 2020). Extrusion (via twin-screw extruder) can also be used for quick and inexpensive technique for starch grafting using monomers (such as acrylamide and 2-acrylamido-2-methylpropane sulfonic acid) (Siyamak, Luckman, et al., 2020). The recent research emphasis is shifting towards creating sustainable polymers that use natural molecules such as enzymes and their immobilized analogs that are easily broken down by the ecosystem via the process of hydrolysis, microbial activity, or photolysis (Sarder et al., 2022). If the finished product has applications in the food business, catalysts with zero or low toxicity and cheap cost must be considered. For high grafting efficiency, it is important to extensively investigate catalyst stability. It is advisable to select catalysts that are active in low-energy visible light to reduce the likelihood of unfavorable reactions brought on by higher-energy UV light (Olson et al., 2020).

14.6.3 Cost Effectiveness and Amount of Power Consumed

The cost of starch grafting varies depending on the ingredients used and sometimes may be subject to experimental settings. For example, although polyhydroxyalkanoates are hydrophobic and biodegradable, the process of synthesizing them from microorganisms is expensive. So customization of polyhydroxyalkanoates is required to get a low-cost replacement (Z. Li et al., 2016). In order to achieve sustainability, polymerization processes with minimal energy consumption during manufacture and composting or recycling must be preferred. Life cycle assessment is a crucial technique that permits evaluation of these many ends and the accompanying use of assets for a better understanding (De Hoe et al., 2022).

14.6.4 Management of Residue

The remaining products/byproducts (such as catalysts and solvents) from starch grafting polymerization are often waste materials which could be harmful (Sanchez-Salvador et al., 2021). The remains of starch-grafting polymerization should be recovered and may be recycled in an effort to limit waste. One of the most frequent issues in graft polymerization is catalyst residue, which may become hazardous and is often formed when the catalysts co-precipitate in the polymers (Noordergraaf et al., 2018). Because of this, catalyst residue for the majority of applications needs to be kept to a low level for precautionary purposes. To overcome this challenge, strategies such as surface-initiated photo ATRP utilizing organic semiconductor graphitic carbon nitride catalysts in polymer brushes is one of the most effective techniques to reduce catalyst leftovers (Suresh et al., 2021).

14.6.5 Ethics and Regulations

Consumers in the current digital age are more interested in eating foods that are good for them, but they are also curious about technology that doesn't negatively impact the environment or their wellness (Wansink & Chandon, 2014). Since the final output of starch production is a human-related product, the raw materials and their dose must be viewed as falling within an acceptable range by regulatory organizations. Regulatory bodies such as the Food and Drug Administration, the US Department of Agriculture, and the Centers for Disease Control and Prevention should be taken into consideration even though there is no special regulating agency for starch grafting polymerization (Wypij et al., 2022). The European Commission and the United States Food and Drug Administration have created laws that address the migration of nanomaterials into food products. Generally, clients need a manufacturer's certification that the product is created in accordance with the FDA guideline paper in order to avoid legal liabilities (Paidari et al., 2021). Because polymerization processes are exothermic and auto-accelerating in nature, they are highly susceptible to reaction runaways, which have resulted in several severe accidents in the past. The primary concern, in these terms, is the monomer's heat instability. Because of their thermal instability, monomers—the building units of polymers—can spontaneously react or decompose when exposed to temperature. During storage or polymerization, this thermal instability might uncontrollably accelerate the process faster than intended or tolerable levels (Zhao et al., 2019).

With the ability to overcome obstacles such as maximizing grafting efficiency, preserving functionality, and guaranteeing cost-effectiveness, starch grafting has a wide range of uses in the food industry. Future applications of starch grafting in the food sector show great potential for resolving issues with functionality, sustainability, and health.

14.7 CONCLUSIONS

This chapter briefly covered the current state of research focused on starch grafting and its applications in the food and packaging industries. Grafting is a polymerization method to develop a polymer made up of one or a number of types of blocks (acts as side chains) attached to the primary chain; these side chains have different configurational or constitutional properties from the primary polymer chain. Each grafting technique exhibits unique properties depending on the type of main chain polymer (starch), copolymer, and initiator used. Starch grafting can be closely monitored by having control over crucial factors such as reaction time, the temperature of the reaction, and the composition of starch and selecting the right solvent and monomer concentration. Starch grafting helps modify various characteristics of native starch, such as morphology, crystalline properties, thermal stability, absorption capability, and molecular weight. In the food processing business, starch grafting can used for improving a variety of food qualities, including viscosity, stability, and shelf life. Additionally, it can be utilized to enhance the starch's ability to thicken and form gel when added to soups and sauces. Starch grafting helps in performing blending with other polymers to improve the hydrophobicity and mechanical strength of packaging material. Recent developments in nanotechnology have made it possible to create nanostarch-based packaging materials by starch grafting. In order to accomplish sustained starch-grafting polymerization, some of the parameters to be taken into account are highlighted in this chapter related to raw materials, end product residues, energy consumption, ethics, and consideration of consumer demands.

REFERENCES

Abd El-Ghany, N. A., Abdel Aziz, M. S., Abdel-Aziz, M. M., & Mahmoud, Z. (2019). Antimicrobial and swelling behaviors of novel biodegradable corn starch grafted/poly(4-acrylamidobenzoic acid) copolymers. *International Journal of Biological Macromolecules, 134*, 912–920. https://doi.org/10.1016/J.IJBIOMAC.2019.05.078

Abousalman-Rezvani, Z., Eskandari, P., Roghani-Mamaqani, H., & Salami-Kalajahi, M. (2020). Functionalization of carbon nanotubes by combination of controlled radical polymerization and "grafting to" method. *Advances in Colloid and Interface Science, 278*, 102126. https://doi.org/10.1016/J.CIS.2020.102126

Alfuhaid, L., Al-Abbad, E., Alshammari, S., Alotaibi, A., Malek, N., & Al-Ghamdi, A. (2023). Preparation and characterization of a renewable starch-g-(MA-DETA) copolymer and its adjustment for dye removal applications. *Polymers, 15*(5). https://doi.org/10.3390/POLYM15051197

Alikarami, N., Abrisham, M., Huang, X., Panahi-Sarmad, M., Zhang, K., Dong, K., & Xiao, X. (2022). Compatibilization of PLA grafted maleic anhydrate through blending of thermoplastic starch (TPS) and nanoclay nanocomposites for the reduction of gas permeability. *International Journal of Smart and Nano Materials, 13*(1). https://doi.org/10.1080/19475411.2022.2051639

Athawale, V. D., & Rathi, S. C. (1999). Graft polymerization: Starch as a model substrate. *Journal of Macromolecular Science—Polymer Reviews, 39*(3), 445–480. https://doi.org/10.1081/MC-100101424

Athawale, V. D., & Rathi, S. C. (2007). Graft polymerization: Starch as a model substrate. *Journal of Macromolecular Science—Polymer Reviews, 39*(3), 445–480. https://doi.org/10.1081/MC-100101424

Bai, X., Zhang, X., Xu, Y., & Yong, X. (2021). Synthesis and characterization of sodium carboxymethyl starch-graft acrylamide/1-vinyl-2-pyrrolidone copolymers via central composite design and using as filtration loss agent in drilling muds. *Starch/Staerke, 73*(3–4). https://doi.org/10.1002/star.202000151

Baloochestanzadeh, S., Hassanajili, S., & Escrochi, M. (2021). Rheological properties and swelling behavior of nano-composite preformed particle gels based on starch-graft-polyacrylamide loaded with nanosilica. *Rheologica Acta, 60*(10), 571–585. https://doi.org/10.1007/S00397-021-01287-Z/FIGURES/10

Baniasadi, H., Madani, Z., Mohan, M., Vaara, M., Lipponen, S., Vapaavuori, J., & Seppälä, J. V. (2023). Heat-induced actuator fibers: Starch-containing biopolyamide composites for functional textiles. *ACS Applied Materials & Interfaces, 15*(41), 48584–48600. https://doi.org/10.1021/acsami.3c08774

Bao, X., Yu, L., Shen, S., Simon, G. P., Liu, H., & Chen, L. (2019). How rheological behaviors of concentrated starch affect graft copolymerization of acrylamide and resultant hydrogel. *Carbohydrate Polymers, 219*, 395–404. https://doi.org/10.1016/J.CARBPOL.2019.05.034

Blosch, S. E., Alaboalirat, M., Eades, C. B., Scannelli, S. J., & Matson, J. B. (2022). Solvent effects in grafting-through ring-opening metathesis polymerization. *Macromolecules, 55*(9), 3522–3532. https://doi.org/10.1021/ACS.MACROMOL.2C00254/SUPPL_FILE/MA2C00254_SI_001.PDF

Cai, S., Gu, S., Li, X., Wan, S., Chen, S., & He, X. (2020). Controlled grafting modification of starch and UCST-type thermosensitive behavior in water. *Colloid and Polymer Science, 298*(8), 1053–1061. https://doi.org/10.1007/S00396-020-04670-Z/FIGURES/6

Campelo, P. H., Sant'Ana, A. S., & Pedrosa Silva Clerici, M. T. (2020). Starch nanoparticles: Production methods, structure, and properties for food applications. *Current Opinion in Food Science, 33*, 136–140. https://doi.org/10.1016/J.COFS.2020.04.007

Çankaya, N. (2016). Synthesis of graft copolymers onto starch and its semiconducting properties. *Results in Physics, 6*, 538–542. https://doi.org/10.1016/J.RINP.2016.08.010

Chang, L., Wang, F., Guo, Y., Li, J., Gong, Y., & Shi, Q. (2021). Green preparation of thermochromic starch-based fibers through a wet-spinning process. *ACS Applied Polymer Materials, 3*(1), 436–444. https://doi.org/10.1021/ACSAPM.0C01194/ASSET/IMAGES/LARGE/AP0C01194_0005.JPEG

Chen, Q. J., Zhao, Y. L., Xie, Q. H., Liang, C. Y., & Zong, Z. Y. (2021). Polyethyleneimine grafted starch nanocrystals as a novel biosorbent for efficient removal of methyl blue dye. *Carbohydrate Polymers, 273*, 118579. https://doi.org/10.1016/J.CARBPOL.2021.118579

Chen, X., Sun, C., Wang, Q., Tan, H., & Zhang, Y. (2022). Preparation of glycidyl methacrylate grafted starch adhesive to apply in high-performance and environment-friendly plywood. *International Journal of Biological Macromolecules, 194*, 954–961. https://doi.org/10.1016/J.IJBIOMAC.2021.11.152

Choudhary, S., Sharma, K., Sharma, V., & Kumar, V. (2020). Grafting polymers. *Reactive and Functional Polymers Volume Two: Modification Reactions, Compatibility and Blends, 2*, 199–243. https://doi.org/10.1007/978-3-030-45135-6_8/COVER

Clauss, Z. S., Wardzala, C. L., Schlirf, A. E., Wright, N. S., Saini, S. S., Onoa, B., Bustamante, C., & Kramer, J. R. (2021). Tunable, biodegradable grafting-from glycopolypeptide bottlebrush polymers. *Nature Communications, 12*(1), 1–11. https://doi.org/10.1038/s41467-021-26808-5

Collazo-Bigliardi, S., Ortega-Toro, R., & Chiralt, A. (2019). Using grafted poly(ε-caprolactone) for the compatibilization of thermoplastic starch-polylactic acid blends. *Reactive and Functional Polymers, 142*, 25–35. https://doi.org/10.1016/J.REACTFUNCTPOLYM.2019.05.013

Compart, J., Singh, A., Fettke, J., & Apriyanto, A. (2023). Customizing starch properties: A review of starch modifications and their applications. *Polymers, 15*(16). https://doi.org/10.3390/POLYM15163491

Cruz-Benítez, M. M., Castro-Rosas, J., González-Morones, P., Santos-López, E. M., Villagómez-Ibarra, J. R., Tlahuextl-Romero, A. M., Fonseca-Florido, H. A., & Gómez-Aldapa, C. A. (2022). Biocomposites based on starch with multi-functionalized graphene oxide: Effect of graft composition and concentration. *Polymer Composites, 43*(1), 267–281. https://doi.org/10.1002/PC.26372

Cui, C., Jia, Y., Sun, Q., Yu, M., Ji, N., Dai, L., Wang, Y., Qin, Y., Xiong, L., & Sun, Q. (2022). Recent advances in the preparation, characterization, and food application of starch-based hydrogels. *Carbohydrate Polymers, 291*, 119624. https://doi.org/10.1016/J.CARBPOL.2022.119624

Cummings, S., Zhang, Y., Smeets, N., Cunningham, M., & Dubé, M. A. (2019). On the use of starch in emulsion polymerizations. *Processes*, 7(3), 140. https://doi.org/10.3390/PR7030140

Czarnecka, E., & Nowaczyk, J. (2020). Semi-natural superabsorbents based on starch-g-poly(acrylic acid): Modification, synthesis and application. *Polymers*, 12(8). https://doi.org/10.3390/polym12081794

De Hoe, G. X., Şucu, T., & Shaver, M. P. (2022). Sustainability and polyesters: Beyond metals and monomers to function and fate. *Accounts of Chemical Research*, 55(11), 1514–1523. https://doi.org/10.1021/ACS.ACCOUNTS.2C00134/ASSET/IMAGES/LARGE/AR2C00134_0006.JPEG

Dong, X., Wu, Z., Wang, Y., Li, T., Yuan, H., Zhang, X., Ma, P., Chen, M., & Dong, W. (2021). Design of degradable core-shell starch nanoparticles by radical ring-opening polymerization of 2-methylene-1,3-dioxepane and their toughening of poly (lactic acid). *Composites Communications*, 27. https://doi.org/10.1016/j.coco.2021.100808

Dworakowska, S., Lorandi, F., Gorczyński, A., & Matyjaszewski, K. (2022). Toward green atom transfer radical polymerization: Current status and future challenges. *Advanced Science*, 9(19), 2106076. https://doi.org/10.1002/ADVS.202106076

El Assimi, T., Roko Blažic, El Kadib, A., Raihane, M., Beniazza, R., Luinstra, G. A., Vidović, E., & Lahcini, M. (2019). Synthesis of poly(ε-caprolactone)-grafted guar gum by surface-initiated ring-opening polymerization. *Carbohydrate Polymers*, 220, 95–102. https://doi.org/10.1016/J.CARBPOL.2019.05.049

Fan, Y., & Picchioni, F. (2020). Modification of starch: A review on the application of "green" solvents and controlled functionalization. *Carbohydrate Polymers*, 241, 116350. https://doi.org/10.1016/J.CARBPOL.2020.116350

Fritz, A. T., Cazotti, J. C., Garcia-Valdez, O., Smeets, N. M. B., Dubé, M. A., & Cunningham, M. F. (2020). Graft modification of cold water-soluble starch via nitroxide-mediated polymerisation. *Polymer Chemistry*, 11(25), 4180–4191. https://doi.org/10.1039/D0PY00239A

Getzler, Y. D. Y. L., & Mathers, R. T. (2022). Sustainable polymers: Our evolving understanding. *Accounts of Chemical Research*, 55(14), 1869–1878. https://doi.org/10.1021/ACS.ACCOUNTS.2C00194/SUPPL_FILE/AR2C00194_SI_002.XLSX

Govindaraju, I., Chakraborty, I., Baruah, V. J., Sarmah, B., Mahato, K. K., & Mazumder, N. (2021). Structure and morphological properties of starch macromolecule using biophysical techniques. *Starch—Stärke*, 73(1–2), 2000030. https://doi.org/10.1002/STAR.202000030

Güler, M. A., Gök, M. K., & Özgümüş, S. (2020). Effects of the starch types and the grafting conditions on the in vitro mucoadhesiveness of the starch-graft-poly(methacrylic acid) hydrogels. *Starch/Staerke*, 72(7–8). https://doi.org/10.1002/star.201900266

Hao, D., Wang, X., Liu, X., Su, R., Duan, Z., & Dang, X. (2021). Chrome-free tanning agent based on epoxy-modified dialdehyde starch towards sustainable leather making. *Green Chemistry*, 23(23), 9693–9703. https://doi.org/10.1039/D1GC03472C

Iqbal, S., Nadeem, S., Bano, R., Bahadur, A., Ahmad, Z., Javed, M., AL-Anazy, M. M., Qasier, A. A., Laref, A., Shoaib, M., Liu, G., & Qayyum, M. A. (2021). Green synthesis of biodegradable terpolymer modified starch nanocomposite with carbon nanoparticles for food packaging application. *Journal of Applied Polymer Science*, 138(25), 50604. https://doi.org/10.1002/APP.50604

Iqbal, S., Nadeem, S., Javed, M., Alsaab, H. O., Awwad, N. S., Ibrahium, H. A., & Mohyuddin, A. (2022). Controlled preparation of grafted starch modified with Ni nanoparticles for biodegradable polymer nanocomposites and its application in food packaging. *Microscopy Research and Technique*, 85(6), 2331–2337. https://doi.org/10.1002/JEMT.24089

Jiang, T., Chen, F., Duan, Q., Bao, X., Jiang, S., Liu, H., Chen, L., & Yu, L. (2022). Designing and application of reactive extrusion with twice initiations for graft copolymerization of acrylamide on starch. *European Polymer Journal*, 165. https://doi.org/10.1016/j.eurpolymj.2022.111008

Khanoonkon, N., Yoksan, R., & Ogale, A. A. (2016). Effect of stearic acid-grafted starch compatibilizer on properties of linear low density polyethylene/thermoplastic starch blown film. *Carbohydrate Polymers*, 137, 165–173. https://doi.org/10.1016/J.CARBPOL.2015.10.038

Kim, Y. G., & Thérien-Aubin, H. (2020). Impact of the solvent quality on the local dynamics of soft and swollen polymer nanoparticles functionalized with polymer chains. *Macromolecules*, 53(17), 7561–7569. https://doi.org/10.1021/ACS.MACROMOL.0C00346/SUPPL_FILE/MA0C00346_SI_001.PDF

Lamontagne, H. R., & Lessard, B. H. (2020). Nitroxide-mediated polymerization: A versatile tool for the engineering of next generation materials. *ACS Applied Polymer Materials*, 2(12), 5327–5344. https://doi.org/10.1021/ACSAPM.0C00888/ASSET/IMAGES/LARGE/AP0C00888_0010.JPEG

Law, K. L., & Narayan, R. (2021). Reducing environmental plastic pollution by designing polymer materials for managed end-of-life. *Nature Reviews Materials*, 7(2), 104–116. https://doi.org/10.1038/s41578-021-00382-0

Li, M., Daygon, V. D., Solah, V., & Dhital, S. (2023). Starch granule size: Does it matter? *Critical Reviews in Food Science and Nutrition*, 63(19), 3683–3703. https://doi.org/10.1080/10408398.2021.1992607

Li, W., Wu, L., Zhu, Z., Zhang, Z., Liu, Q., Lu, Y., & Ke, H. (2021). Incorporation of poly(sodium allyl sulfonate) branches on corn starch chains for enhancing its sizing properties: Viscosity stability, adhesion, film properties and desizability. *International Journal of Biological Macromolecules, 166*. https://doi.org/10.1016/j.ijbiomac.2020.11.025

Li, W., Zhang, Z., Liu, Z., & Tao, X. (2023). Introducing the grafted poly(acrylic acid-co-butyl acrylate) branches onto biological corn starch macromolecule for imparting it with superior sizing properties. *International Journal of Biological Macromolecules, 253*, 126957. https://doi.org/10.1016/J.IJBIOMAC.2023.126957

Li, Z., Yang, J., & Loh, X. J. (2016). Polyhydroxyalkanoates: Opening doors for a sustainable future. *NPG Asia Materials, 8*(4), e265. https://doi.org/10.1038/am.2016.48

Liu, X., Guo, Q., Ren, S., Guo, J., Wei, C., Chang, J., & Shen, B. (2023). Synthesis of starch-based flocculant by multi-component grafting copolymerization and its application in oily wastewater treatment. *Journal of Applied Polymer Science, 140*(4), e53356. https://doi.org/10.1002/APP.53356

Liu, X., Guo, Q., Wei, C., Chang, J., Guo, J., Ren, S., & Shen, B. (2023). Preparation and kinetics of thermal analysis of inorganic–organic composite modified carboxymethyl starch and its applicability as flocculant for oily wastewater treatment. *Journal of Polymers and the Environment, 31*(5), 1839–1852. https://doi.org/10.1007/S10924-022-02713-W/FIGURES/9

López, O. V., Ninago, M. D., Lencina, M. M. S., Ciolino, A. E., Villar, M. A., & Andreucetti, N. A. (2019). Starch/poly(ε-caprolactone) graft copolymers synthetized by γ-radiation and their application as compatibilizer in polymer blends. *Journal of Polymers and the Environment, 27*(12), 2906–2914. https://doi.org/10.1007/S10924-019-01568-Y/TABLES/2

Lorandi, F., Fantin, M., & Matyjaszewski, K. (2022). Atom transfer radical polymerization: A mechanistic perspective. *Journal of the American Chemical Society, 144*(34), 15413–15430. https://doi.org/10.1021/JACS.2C05364/ASSET/IMAGES/LARGE/JA2C05364_0017.JPEG

Luo, H., Dong, F., Wang, Q., Li, Y., & Xiong, Y. (2021). Construction of porous starch-based hydrogel via regulating the ratio of amylopectin/amylose for enhanced water-retention. *Molecules, 26*(13). https://doi.org/10.3390/MOLECULES26133999

Luo, Y., Lin, J., Wang, X., Yeerken, A., & Ma, H. (2023a). Study on flocculation performance of new cationic starch-based sludge wastewater flocculant. *Starch/Staerke, 75*(5–6). https://doi.org/10.1002/star.202200226

Luo, Y., Lin, J., Wang, X., Yeerken, A., & Ma, H. (2023b). Study on flocculation performance of new cationic starch-based sludge wastewater flocculant. *Starch/Stärke, 75*(5–6), 2200226. https://doi.org/10.1002/STAR.202200226

Matyjaszewski, K. (2018). Advanced materials by atom transfer radical polymerization. *Advanced Materials, 30*(23), 1706441. https://doi.org/10.1002/ADMA.201706441

Matyjaszewski, K., & Xia, J. (2001). Atom transfer radical polymerization. *Chemical Reviews, 101*(9), 2921–2990. https://doi.org/10.1021/CR940534G/ASSET/IMAGES/MEDIUM/CR940534GF00015.GIF

Meimoun, J., Wiatz, V., Saint-Loup, R., Parcq, J., David, A., Stoclet, G., Gaucher, V., Favrelle-Huret, A., Bonnet, F., & Zinck, P. (2021). A one pot one step combined radical and ring-opening route for the dual functionalization of starch in aqueous medium. *Carbohydrate Polymers, 254*. https://doi.org/10.1016/j.carbpol.2020.117399

Meimoun, J., Wiatz, V., Saint-Loup, R., Parcq, J., Favrelle, A., Bonnet, F., & Zinck, P. (2018). Modification of starch by graft copolymerization. *Starch—Stärke, 70*(1–2), 1600351. https://doi.org/10.1002/STAR.201600351

Michalek, L., Mundsinger, K., Barner, L., & Barner-Kowollik, C. (2019). Quantifying solvent effects on polymer surface grafting. *ACS Macro Letters, 8*(7), 800–805. https://doi.org/10.1021/ACSMACROLETT.9B00336/SUPPL_FILE/MZ9B00336_SI_001.PDF

Misra, N., Rawat, S., Goel, N. K., Shelkar, S. A., & Kumar, V. (2020). Radiation grafted cellulose fabric as reusable anionic adsorbent: A novel strategy for potential large-scale dye wastewater remediation. *Carbohydrate Polymers, 249*, 116902. https://doi.org/10.1016/J.CARBPOL.2020.116902

Mittal, A., Garg, S., & Bajpai, S. (2020). Fabrication and characteristics of poly (vinyl alcohol)-starch-cellulosic material based biodegradable composite film for packaging application. *Materials Today: Proceedings, 21*. https://doi.org/10.1016/j.matpr.2019.11.210

Mittal, A., Garg, S., Premi, A., & Giri, A. S. (2021). Synthesis of polyvinyl alcohol/modified starch-based biodegradable nanocomposite films reinforced with starch nanocrystals for packaging applications. *Polymers and Polymer Composites, 29*(5). https://doi.org/10.1177/09673911120922429

Moharrami, P., & Motamedi, E. (2020). Application of cellulose nanocrystals prepared from agricultural wastes for synthesis of starch-based hydrogel nanocomposites: Efficient and selective nano-adsorbent for removal of cationic dyes from water. *Bioresource Technology, 313*, 123661. https://doi.org/10.1016/J.BIORTECH.2020.123661

Mohy Eldin, M. S., Gouda, M. H., Abu-Saied, M. A., El-Shazly, Y. M. S., & Farag, H. A. (2016). Development of grafted cotton fabrics ions exchanger for dye removal applications: Methylene blue model. *Desalination and Water Treatment, 57*(46), 22049–22060. https://doi.org/10.1080/19443994.2015.1128363

Mostafa, K. M., & EL-Sanabary, A. A. (2020). Green and efficient tool for grafting acrylonitrile onto starch nanoparticles using microwave irradiation. *Journal of Polymer Research*, *27*(4), 1–10. https://doi.org/10.1007/S10965-020-02069-6/TABLES/2

Nezami, S., Sadeghi, M., & Mohajerani, H. (2020). A novel pH-sensitive and magnetic starch-based nanocomposite hydrogel as a controlled drug delivery system for wound healing. *Polymer Degradation and Stability*, *179*, 109255. https://doi.org/10.1016/J.POLYMDEGRADSTAB.2020.109255

Noivoil, N., & Yoksan, R. (2020). Oligo(lactic acid)-grafted starch: A compatibilizer for poly(lactic acid)/thermoplastic starch blend. *International Journal of Biological Macromolecules*, *160*, 506–517. https://doi.org/10.1016/J.IJBIOMAC.2020.05.178

Noordergraaf, I. W., Fourie, T. K., & Raffa, P. (2018). Free-Radical graft polymerization onto starch as a tool to tune properties in relation to potential applications: A review. *Processes*, *6*(4), 31. https://doi.org/10.3390/PR6040031

Noordergraaf, I. W., Witono, J. R., & Heeres, H. J. (2024). Grafting starch with acrylic acid and Fenton's initiator: The selectivity challenge. *Polymers*, *16*(2). https://doi.org/10.3390/POLYM16020255/S1

Ojogbo, E., Ogunsona, E. O., & Mekonnen, T. H. (2020). Chemical and physical modifications of starch for renewable polymeric materials. *Materials Today Sustainability*, *7–8*, 100028. https://doi.org/10.1016/J.MTSUST.2019.100028

Ojogbo, E., Ward, V., & Mekonnen, T. H. (2020). Functionalized starch microparticles for contact-active antimicrobial polymer surfaces. *Carbohydrate Polymers*, *229*. https://doi.org/10.1016/j.carbpol.2019.115422

Olson, R. A., Korpusik, A. B., & Sumerlin, B. S. (2020). Enlightening advances in polymer bioconjugate chemistry: Light-based techniques for grafting to and from biomacromolecules. *Chemical Science*, *11*(20), 5142–5156. https://doi.org/10.1039/D0SC01544J

Paidari, S., Tahergorabi, R., Anari, E. S., Nafchi, A. M., Zamindar, N., & Goli, M. (2021). Migration of various nanoparticles into food samples: A review. *Foods*, *10*(9), 2114. https://doi.org/10.3390/FOODS10092114

Parkatzidis, K., Wang, H. S., Truong, N. P., & Anastasaki, A. (2020). Recent developments and future challenges in controlled radical polymerization: A 2020 update. *Chem*, *6*. https://doi.org/10.1016/j.chempr.2020.06.014

Perrier, S. (2017). 50th anniversary perspective: RAFT polymerization—a user guide. *Macromolecules*, *50*(19), 7433–7447. https://doi.org/10.1021/ACS.MACROMOL.7B00767/ASSET/IMAGES/LARGE/MA-2017-00767N_0008.JPEG

Pirman, T., Ocepek, M., & Likozar, B. (2021). Radical polymerization of acrylates, methacrylates, and styrene: Biobased approaches, mechanism, kinetics, secondary reactions, and modeling. *Industrial and Engineering Chemistry Research*, *60*(26), 9347–9367. https://doi.org/10.1021/ACS.IECR.1C01649/ASSET/IMAGES/LARGE/IE1C01649_0007.JPEG

Plummer, C. M., Li, L., & Chen, Y. (2023). Ring-opening polymerization for the goal of chemically recyclable polymers. *Macromolecules*, *56*(3), 731–750. https://doi.org/10.1021/ACS.MACROMOL.2C01694/ASSET/IMAGES/LARGE/MA2C01694_0001.JPEG

Poli, R. (2019). Contribution of computations to metal-mediated radical polymerization. In *Computational quantum chemistry: Insights into polymerization reactions*. Elsevier. https://doi.org/10.1016/B978-0-12-815983-5.00007-6

Purohit, P., Bhatt, A., Mittal, R. K., Abdellattif, M. H., & Farghaly, T. A. (2022). Polymer Grafting and its chemical reactions. *Frontiers in Bioengineering and Biotechnology*, *10*. https://doi.org/10.3389/FBIOE.2022.1044927

Purohit, V. B., Pięta, M., Pietrasik, J., & Plummer, C. M. (2022). Recent advances in the ring-opening polymerization of sulfur-containing monomers. *Polymer Chemistry*, *13*(34), 4858–4878. https://doi.org/10.1039/D2PY00831A

Ramírez-Hernández, A., Hernández-Mota, C. E., Páramo-Calderón, D. E., González-García, G., Báez-García, E., Rangel-Porras, G., Vargas-Torres, A., & Aparicio-Saguilán, A. (2020). Thermal, morphological and structural characterization of a copolymer of starch and polyethylene. *Carbohydrate Research*, *488*, 107907. https://doi.org/10.1016/J.CARRES.2020.107907

Rodrigues, L. D. A., Guerrini, L. M., & Oliveira, M. P. (2021). Thermal and mechanical properties of cationic starch-graft-poly(butyl acrylate-co-methyl methacrylate) latex film obtained by semi-continuous emulsion polymerization for adhesive application. *Journal of Thermal Analysis and Calorimetry*, *146*(1). https://doi.org/10.1007/s10973-020-09915-1

Rodrigues, L. D. A., Hurtado, C. R., Macedo, E. F., Tada, D. B., Guerrini, L. M., & Oliveira, M. P. (2020). Colloidal properties and cytotoxicity of enzymatically hydrolyzed cationic starch-graft-poly(butyl acrylate-co-methyl methacrylate) latex by surfactant-free emulsion polymerization for paper coating application. *Progress in Organic Coatings*, *145*. https://doi.org/10.1016/j.porgcoat.2020.105693

Rui-He, Wang, X. L., Wang, Y. Z., Yang, K. K., Zeng, J. B., & Ding, S. D. (2006). A study on grafting poly(1,4-dioxan-2-one) onto starch via 2,4-tolylene diisocyanate. *Carbohydrate Polymers*, *65*(1), 28–34. https://doi.org/10.1016/J.CARBPOL.2005.11.031

Saleh, M., Abdel-Naby, A., Al-Ghamdi, A., & Al-Shahrani, N. (2021). Graft copolymerization of Diallylamine onto starch for water treatment use characterization, removal of Cu (II) cations and antibacterial activity. *Journal of Polymer Research, 28*(6). https://doi.org/10.1007/s10965-021-02558-2

Sanchez-Salvador, J. L., Balea, A., Monte, M. C., Negro, C., & Blanco, A. (2021). Chitosan grafted/cross-linked with biodegradable polymers: A review. *International Journal of Biological Macromolecules, 178*, 325–343. https://doi.org/10.1016/J.IJBIOMAC.2021.02.200

Sarder, R., Piner, E., Rios, D. C., Chacon, L., Artner, M. A., Barrios, N., & Argyropoulos, D. (2022). Copolymers of starch, a sustainable template for biomedical applications: A review. *Carbohydrate Polymers, 278*, 118973. https://doi.org/10.1016/J.CARBPOL.2021.118973

Sarkar, A., Biswas, D. R., Datta, S. C., Dwivedi, B. S., Bhattacharyya, R., Kumar, R., Bandyopadhyay, K. K., Saha, M., Chawla, G., Saha, J. K., & Patra, A. K. (2021). Preparation of novel biodegradable starch/poly(vinyl alcohol)/bentonite grafted polymeric films for fertilizer encapsulation. *Carbohydrate Polymers, 259*, 117679. https://doi.org/10.1016/J.CARBPOL.2021.117679

Sarmah, D., & Karak, N. (2020). Double network hydrophobic starch based amphoteric hydrogel as an effective adsorbent for both cationic and anionic dyes. *Carbohydrate Polymers, 242*. https://doi.org/10.1016/j.carbpol.2020.116320

Shokri, Z., Seidi, F., Saeb, M. R., Jin, Y., Li, C., & Xiao, H. (2022). Elucidating the impact of enzymatic modifications on the structure, properties, and applications of cellulose, chitosan, starch and their derivatives: A review. *Materials Today Chemistry, 24*, 100780. https://doi.org/10.1016/J.MTCHEM.2022.100780

Siyamak, S., Laycock, B., & Luckman, P. (2020). Synthesis of starch graft-copolymers via reactive extrusion: Process development and structural analysis. *Carbohydrate Polymers, 227*, 115066. https://doi.org/10.1016/J.CARBPOL.2019.115066

Siyamak, S., Luckman, P., & Laycock, B. (2020). Rapid and solvent-free synthesis of pH-responsive graft-copolymers based on wheat starch and their properties as potential ammonium sorbents. *International Journal of Biological Macromolecules, 149*, 477–486. https://doi.org/10.1016/J.IJBIOMAC.2020.01.202

Solis-Ceballos, A., Roy, R., Golsztajn, A., Tavares, J. R., & Dumont, M. J. (2023). Selective adsorption of Cr(III) over Cr(VI) by starch-graft-itaconic acid hydrogels. *Journal of Hazardous Materials Advances, 10*, 100255. https://doi.org/10.1016/J.HAZADV.2023.100255

Su, Q., Wang, Y., Zhao, X., Wang, H., Wang, Z., Wang, N., & Zhang, H. (2020). Functionalized nano-starch prepared by surface-initiated atom transfer radical polymerization and quaternization. *Carbohydrate Polymers, 229*, 115390. https://doi.org/10.1016/J.CARBPOL.2019.115390

Suresh, D., Goh, P. S., Ismail, A. F., & Hilal, N. (2021). Surface design of liquid separation membrane through graft polymerization: A state of the art review. *Membranes, 11*(11). https://doi.org/10.3390/MEMBRANES11110832

Suri, S., & Singh, A. (2023). Modification of starch by novel and traditional ways: Influence on the structure and functional properties. *Sustainable Food Technology, 1*(3), 348–362. https://doi.org/10.1039/D2FB00043A

Tagliapietra, B. L., Felisberto, M. H. F., Sanches, E. A., Campelo, P. H., & Clerici, M. T. P. S. (2021). Non-conventional starch sources. *Current Opinion in Food Science, 39*, 93–102. https://doi.org/10.1016/J.COFS.2020.11.011

Tajbakhsh, S. F., Mohmmadipour, R., & Janani, H. (2022). One-pot production of a graft copolymer of cationic starch and cationic polyacrylamide applicable as flocculant for wastewater treatment. *Journal of Macromolecular Science, Part A: Pure and Applied Chemistry, 59*(10). https://doi.org/10.1080/10601325.2022.2112516

Tanaka, J., Archer, N. E., Grant, M. J., & You, W. (2021). Reversible-addition fragmentation chain transfer step-growth polymerization. *Journal of the American Chemical Society, 143*(39), 15918–15923. https://doi.org/10.1021/JACS.1C07553/SUPPL_FILE/JA1C07553_SI_001.PDF

Techie-Menson, R., Rono, C. K., Etale, A., Mehlana, G., Darkwa, J., & Makhubela, B. C. E. (2021). New bio-based sustainable polymers and polymer composites based on methacrylate derivatives of furfural, solketal and lactic acid. *Materials Today Communications, 28*, 102721. https://doi.org/10.1016/J.MTCOMM.2021.102721

Thum, M. D., Wolf, S., & Falvey, D. E. (2020). State-dependent photochemical and photophysical behavior of dithiolate ester and trithiocarbonate reversible addition-fragmentation chain transfer polymerization agents. *Journal of Physical Chemistry A, 124*(21), 4211–4222. https://doi.org/10.1021/ACS.JPCA.0C02678/ASSET/IMAGES/LARGE/JP0C02678_0004.JPEG

Tratnik, N., Kuo, P. Y., Tanguy, N. R., Gnanasekar, P., & Yan, N. (2020). Biobased epoxidized starch wood adhesives: Effect of amylopectin and amylose content on adhesion properties. *ACS Sustainable Chemistry and Engineering, 8*(49), 17997–18005. https://doi.org/10.1021/ACSSUSCHEMENG.0C05716/ASSET/IMAGES/LARGE/SC0C05716_0004.JPEG

Trinh, B. M., Tadele, D. T., & Mekonnen, T. H. (2022). Robust and high barrier thermoplastic starch—PLA blend films using starch-graft-poly(lactic acid) as a compatibilizer. *Materials Advances, 3*(15), 6208–6221. https://doi.org/10.1039/D2MA00501H

Triwulandari, E., Ghozali, M., Restu, W. K., Septiyanti, M., Sampora, Y., Sondari, D., Devi, Y. A., Mawarni, R. S., Meliana, Y., & Chalid, M. (2023). Molecular weight distribution of lactic acid oligomer from the polycondensation without catalyst and its application for the starch modification. *Journal of Polymers and the Environment*, *32*, 1–15. https://doi.org/10.1007/s10924-023-03092-6

Verheyen, L., Leysen, P., Van Den Eede, M. P., Ceunen, W., Hardeman, T., & Koeckelberghs, G. (2017). Advances in the controlled polymerization of conjugated polymers. *Polymer*, *108*, 521–546. https://doi.org/10.1016/J.POLYMER.2016.09.085

Wang, H. S., Parkatzidis, K., Junkers, T., Truong, N. P., & Anastasaki, A. (2023). Controlled radical depolymerization: Structural differentiation and molecular weight control. *Chem*, *10*. https://doi.org/10.1016/J.CHEMPR.2023.09.027

Wang, L., Zhang, X., Xu, J., Wang, Q., & Fan, X. (2020). How starch-g-poly(acrylamide) molecular structure effect sizing properties. *International Journal of Biological Macromolecules*, *144*, 403–409. https://doi.org/10.1016/J.IJBIOMAC.2019.12.143

Wang, X., Huang, L., Zhang, C., Deng, Y., Xie, P., Liu, L., & Cheng, J. (2020). Research advances in chemical modifications of starch for hydrophobicity and its applications: A review. *Carbohydrate Polymers*, *240*, 116292. https://doi.org/10.1016/J.CARBPOL.2020.116292

Wang, Z., Bockstaller, M. R., & Matyjaszewski, K. (2021). Synthesis and applications of ZnO/polymer nanohybrids. *ACS Materials Letters*, *3*(5), 599–621. https://doi.org/10.1021/ACSMATERIALSLETT.1C00145/ASSET/IMAGES/LARGE/TZ1C00145_0012.JPEG

Wang, Z., Tang, P., Chen, S., Xing, Y., Yin, C., Feng, J., & Jiang, F. (2023). Fully biobased sustainable elastomers derived from chitin, lignin, and plant oil via grafting strategy and Schiff-base chemistry. *Carbohydrate Polymers*, *305*, 120577. https://doi.org/10.1016/J.CARBPOL.2023.120577

Wansink, B., & Chandon, P. (2014). Slim by design: Redirecting the accidental drivers of mindless overeating. *Journal of Consumer Psychology*, *24*(3), 413–431. https://doi.org/10.1016/J.JCPS.2014.03.006

Weerapoprasit, C., & Prachayawarakorn, J. (2019a). Characterization and properties of biodegradable thermoplastic grafted starch films by different contents of methacrylic acid. *International Journal of Biological Macromolecules*, *123*. https://doi.org/10.1016/j.ijbiomac.2018.11.083

Weerapoprasit, C., & Prachayawarakorn, J. (2019b). Effects of polymethacrylamide-grafted branch on mechanical performances, hydrophilicity, and biodegradability of thermoplastic starch film. *Starch/Staerke*, *71*(11–12). https://doi.org/10.1002/star.201900068

Worzakowska, M. (2020). The preparation, physicochemical and thermal properties of the high moisture, solvent and chemical resistant starch-g-poly(geranyl methacrylate) copolymers. *Journal of Thermal Analysis and Calorimetry*, *140*(1). https://doi.org/10.1007/s10973-019-08801-9

Worzakowska, M. (2023). Experimental studies on the preparation and properties of starch-graft-poly(hexyl acrylate) copolymers. *Journal of Applied Polymer Science*, *140*(28). https://doi.org/10.1002/app.54029

Wypij, M., Trzcińska-Wencel, J., Golińska, P., Avila-Quezada, G. D., Ingle, A. P., & Rai, M. (2022). The strategic applications of natural polymer nanocomposites in food packaging and agriculture: Chances, challenges, and consumers' perception. *Frontiers in Chemistry*, *10*. https://doi.org/10.3389/FCHEM.2022.1106230

Xu, P., Zeng, Q., Cao, Y., Ma, P., Dong, W., & Chen, M. (2017). Interfacial modification on polyhydroxyalkanoates/starch blend by grafting in-situ. *Carbohydrate Polymers*, *174*, 716–722. https://doi.org/10.1016/J.CARBPOL.2017.06.048

Yadav, C., Lee, J. M., Mohanty, P., Li, X., & Jang, W. D. (2023). Graft onto approaches for nanocellulose-based advanced functional materials. *Nanoscale*, *15*(37), 15108–15145. https://doi.org/10.1039/D3NR03087C

Yuan, W., Zhou, L., Zhang, Z., Ying, Y., Fan, W., Chai, K., Zhao, Z., Tan, Z., Shen, F., & Ji, H. (2022). Synergistic dual-functionalities of starch-grafted-styrene hydrophilic porous resin for efficiently removing bisphenols from wastewater. *Chemical Engineering Journal*, *429*, 132350. https://doi.org/10.1016/J.CEJ.2021.132350

Zeinali, E., Marien, Y. W., George, S. R., Cunningham, M. F., D'Hooge, D. R., & Van Steenberge, P. H. M. (2023). How phase transfer increases the number of kinetic regimes from three to seven in nitroxide mediated polymerization of n-butyl acrylate in aqueous miniemulsion. *Chemical Engineering Journal*, *470*, 144162. https://doi.org/10.1016/J.CEJ.2023.144162

Zhang, J., Liu, B., Zhou, Y., Essawy, H., Liang, J., Zhou, X., & Du, G. (2022). Preparation and characterization of a bio-based rigid plastic based on gelatinized starch cross-linked with furfural and formaldehyde. *Industrial Crops and Products*, *186*, 115246. https://doi.org/10.1016/J.INDCROP.2022.115246

Zhang, Y., Guo, Z., Chen, X., Ma, Y., & Tan, H. (2021). Synthesis of grafting itaconic acid to starch-based wood adhesive for curing at room temperature. *Journal of Polymers and the Environment*, *29*(3), 685–693. https://doi.org/10.1007/S10924-020-01912-7/FIGURES/8

Zhao, L., Zhu, W., Papadaki, M. I., Mannan, M. S., & Akbulut, M. (2019). Probing into styrene polymerization runaway hazards: Effects of the monomer mass fraction. *ACS Omega*, *4*(5), 8136. https://doi.org/10.1021/ACSOMEGA.9B00004

Zou, W., Yu, L., Liu, X., Chen, L., Zhang, X., Qiao, D., & Zhang, R. (2012). Effects of amylose/amylopectin ratio on starch-based superabsorbent polymers. *Carbohydrate Polymers*, *87*(2), 1583–1588. https://doi.org/10.1016/J.CARBPOL.2011.09.060

Zuo, Y., He, X., Li, P., Li, W., & Wu, Y. (2019). Preparation and characterization of hydrophobically grafted starches by in situ solid phase polymerization. *Polymers*, *11*(1), 72. https://doi.org/10.3390/POLYM11010072

Enzymatic Modifications
Properties and Applications

15

Guifang Huang, Long Chen, and Ming Miao

15.1 INTRODUCTION

Natural starch has some limitations, such as insolubility in water, easy dehydration and condensation, high gelatinization temperature, easy retrogradation after gelatinization, poor acid resistance, and weak shear resistance, which greatly limit its application in industry (Ruidi, et al., 2023; Wang, et al., 2017). Therefore, in order to make full use of the excellent properties of starch, overcome the shortcomings of starch, and broaden its application range, it is necessary to modify starch (Punia, 2020). The modification of natural starch is to change the structure distribution of its particles by various specific methods, so that it has new and meet the requirements of the characteristics (Zhai et al., 2022). Starch modification can improve the competitive advantage in new products, ensure the consistency and stability of products, and comprehensively improve the quality of products (Kaur et al., 2012). Researchers have carried out a lot of work on starch modification. At present, the modification methods generally include physical, chemical, biological enzymes, transgenic methods, and the combination of these modification methods (Compart et al., 2023). Physical modification mainly changes the crystallinity and reactivity of natural starch, including ultrasonic methods, high pressure or extrusion, ionization, and heat treatment. This method has the advantages of convenience and being harmless to the environment (Zhong et al., 2022). However, the operating conditions are limited to a specific physical environment, resulting in high equipment costs and low production capacity, which makes it difficult to meet large-scale production needs (Punia, 2020). Chemical modification changes the molecular structure of natural starch by a series of chemical treatment methods, such as changing the molecular weight and introducing groups on the molecular chain. Common techniques include oxidation, etherification, esterification, acidification, and grafting (Amaraweera et al., 2021; Compart et al., 2023). However, the chemical substances used in the chemical modification process will produce wastewater, pollute the environment, and may have potential hazards to human health (Rajan & Abraham, 2006). Enzyme modification has recently attracted extensive attention from researchers because it can provide starches with different functional properties (Ashim & Nandan, 2017).

Compared with other starch modification methods, enzymatic modification has many advantages. Enzymatic modification adopts enzyme preparation without introducing chemical composition, which is in line with the concept of a "clean label" (Dura & Rosell, 2016; Miao et al., 2018, 2023). Enzyme preparations can be directly added to food as ingredients and are classified as processing aids rather than being shown in the ingredients table. For example, in bread production, amylase directly acts on

DOI: 10.1201/9781032655598-15

starch granules to achieve in-situ modification of wheat starch (Zhai et al., 2022). Enzymatic modification conditions are mild, and the starch granule morphology can be retained with modification (Liu et al., 2022). Enzymes can also act on specific parts of starch granules, which are more selective than chemical modification, such as the production of porous starch by enzymatic methods (Dura et al., 2014; Wang et al., 2017). However, the hydrolysis of starch glycosidic bonds in the acid hydrolysis process of chemical modification is random, which is not conducive to the conversion of starch to the target product, and it is difficult to recover the acid (Xu et al., 2021). The specificity, selectivity, and activation energy reduction of enzyme treatments can reduce the occurrence of side reactions, promote the conversion of substrates to net products, and prevent the formation of harmful products (Dura & Rosell, 2016). In addition, from the perspective of the environment and consumers, enzyme treatment has the advantages of energy saving, environmental protection, high yield, safety, and health (Dura & Rosell, 2016).

Commonly used starch-modifying enzymes are glycosyl hydrolases and glycosyltransferases, including α-amylase, β-amylase, glucoamylase, starch invertase, debranching enzyme, and α-glucan transferase, which can cause changes in the molecular structure, pasting properties, rheological properties, and digestive properties of natural starch (Bangar et al., 2022; Miao et al., 2018). Glycosyl hydrolases can catalyze the hydrolysis of glycosidic bonds to produce dextrin or free monosaccharides (Dura et al., 2014). Glycosyltransferase can catalyze the transfer of glycosyl from one glycoside (donor) to another glycoside (receptor) and form new glycosidic bonds (Dura et al., 2014). These starch-modifying enzymes are mostly isolated from microorganisms (such as *Bacillus subtilis, Aspergillus niger, Bacillus magaterium*), and a few are derived from plants (such as malt, Ramie leaf) (Bangar et al., 2022).

In recent years, a great deal of research has been conducted by scholars on the enzymatic modification technology of starch (Miao et al., 2018, 2023). Enzymatic modification induces structural transformation of natural starch, which can be used to produce modified starch and its derivatives with different physicochemical and processing characteristics. It has been shown that the application of enzymatically modified starch can broaden the application areas and scope of starch and further enhance the utilization value. Therefore, this chapter introduces commonly used enzymes for starch modification, focusing on the effects of enzyme modification on the main physicochemical and processing properties of starch, as well as the main applications of enzyme-modified starch in industrial production.

15.2 TYPES AND EFFECT MECHANISMS OF ENZYMES USED FOR ENZYMATIC MODIFICATIONS

Enzyme-modified starch includes single enzyme, composite enzyme modification, and enzyme modification combined with other modification methods (Table 15.1). This section mainly introduces several common enzymes in enzyme modification, including starch hydrolase (maltogenic amylase, β-amylase, and glucoamylase) and glycosyltransferase (branching enzyme, amylosucrase, and cyclodextrin glycosyltransferase).

15.2.1 Glycosyl Hydrolases

15.2.1.1 Maltogenic Amylase

Maltose amylase (MA, dextran 1,4-α-maltohydrolase, EC 3.2.1.133) has a unique activity to catalyze the hydrolysis of α-1,4 and α-1,6 glycosidic bonds compared to conventional α-amylases. See Figure 15.1A. It has a wide range of hydrolyzable substrates, including straight chain starch, branched chain starch, and maltose (Miao et al., 2014). This enzyme mainly focuses on external stoichiometric hydrolysis but also has some internal hydrolysis (Ruan et al., 2021). Maltose amylase is homologous to cyclic glycosyltransferases

TABLE 15.1   Common Enzyme Modification Methods and Their Effects on Starch Properties

MODIFICATION METHODS	TYPES OF ENZYMES/OTHER METHODS	TYPES OF STARCH	PROPERTY CHANGE	REFERENCES
Single	Maltogenic amylase	Wheat starch	Increase in short chains, decrease in medium chains, increase in resistant starch content reduces starch digestion	(Korompokis et al., 2021)
	β-amylase	Rice starch	Decreases the ratio of crystalline to amorphous regions in starch, increases resistant starch, and slows digestion rate	(Ye et al., 2018)
	Glucoamylase	Maize starch	Starch surface generates deep pores with higher peak viscosity, final viscosity, and average particle size for high adsorption capacity	(Han et al., 2021)
	1,4-α-glucan branching enzyme	Corn starch	Increased glucan content and proportion of short chains, forming more cluster structures and inhibiting the digestion of corn starch	(Ren et al., 2020)
	Amylosucrase	Waxy corn starch	Modified starch has extended branched chains, more long-range double helix structures, showing better thermal stability, and a significant increase in resistant starch content	(Zhang et al., 2017b)
	Cyclodextrin glycosyltransferase	Corn starch	The molecular weight of starch decreases and the depolymerization effect of debranched samples is obvious, which can inhibit starch aging	(Ji et al., 2020)
Compound	α-amylase and glucoamylase	Corn starch	Elevated pasting temperature and excellent adsorption capacity of maize porous starch prepared by α-amylase and glucoamylase	(Zhang et al., 2012)
	α-amylase and glucoamylase	Potato starch	α-amylase and glucoamylase lead to reduced digestibility of potato pasteurized starch containing higher phosphorus content	(Absar et al., 2009)
	Glycosyltransferase and branching enzyme	Potato, corn, wheat, and sweet potato starches	Abundant macropores are generated on the surface of starch granules, producing more shorter-branched and less straight-chain starch	(Guo et al., 2020)
	Amylosucrase and pullulanase	Amylopectin from maize	Sequential modification of AS and pullulan amylase produces 200–400 nm starch particles, and changes in molecular and crystal structure reduce starch digestibility	(Zhang et al., 2019)
	Glycogen branching enzyme and amylosucrase	Sweet potato starch	Increase in the number of short chains in BE-modified starch, increase in the proportion of long side chains by AS treatment, and sequential treatment with combined enzymes produce a new BEAS-modified starch with slow-digesting properties	(Jo et al., 2016)

Enzyme modification combined with other methods	α-D-glucan branching enzyme (GBE) and cyclodextrin glucosyltransferase (CGTase)	Corn starch	Double enzyme modification increases the content of α-1,6 glycosidic bonds, the proportion of short chains increases, and straight-chain starch-cyclodextrin V complexes appear	(Li et al., 2019b)
	Glucoamylase and moderate electric field	Corn starch	Low-intensity electric field induces electrophoretic effect and polarization effect to increase the activity of glucoamylase, medium-intensity electric field enhances the hydrolysis of corn starch by glucoamylase, and high-intensity electric field can directly destroy the structure of starch granules	(Li et al., 2022)
	Glucoamylase and ultrasound	Potato starch	Glucoamylase and ultrasound have a synergistic effect on potato starch hydrolysis, with glucoamylase treatment increasing starch solubility by about 20%, and further assisted ultrasound treatment doubling solubility	(Wang et al., 2017)
	Amylosucrase and hydrothermal treatment	Waxy corn starch	Hydrothermal treatment promotes the formation of longer double helices in sucrose amylase-modified starch, improves starch crystallinity, and facilitates the enhancement of resistant starch content	(Wang et al., 2023d)
	Repeated heat-moisture treatment (HMT) and compound enzymes hydrolysis	Porous wheat starch	Double modification resulted in more surface pores of starch, and the increase of specific surface area and total pore volume improved the adsorption capacity of porous starch	(Xie et al., 2019)
	Extrusion technology combined with enzymatic hydrolysis	Corn starch	Combined modification resulted in increased crystallinity and gelatinization temperature of porous starch, more uniform and extensive pore structure, and enhanced adsorption capacity	(Wu et al., 2020)
	Improved extrusion cooking technology (IECT) and enzymatic modification	Corn starch	The gelatinization temperature of starch decreased, the proportion of bound water decreased, and the degree of solubility and molecular structure disorder increased	(Wang et al., 2023c)

and is a retention enzyme that hydrolyzes starch to maltose and oligosaccharides (Miao et al., 2014). It is not limited in its action by the formation of non-reducing ends by the endo-type substrate and is able to degrade β-limit dextrin.

15.2.1.2 β-Amylase

β-amylase is an exonuclease that sequentially cleaves spaced α-1,4 glycosidic bonds from the non-reducing end of the starch chain, and the hydrolysis product is β-maltose (Gui et al., 2021a). The hydrolysis of β-amylase stops when it encounters the branching point of the starch chain, and it cannot hydrolyze the α-1,6 glycosidic bond of the branched starch, and large molecules of β-limit dextrin will remain (Guo et al., 2022). β-amylase is commonly used as a saccharification step in maltose production and has a good effect on pasting starch (Das & Kayastha, 2019).

15.2.1.3 Glucoamylase

Glucoamylase (E.C. 3.2.1.3, glucoamylase 1,4-α-glucosidase) has a broad spectrum of hydrolysis of α-glycosidic bonds, specifically hydrolyzing α-1,4 and α-1,6 glycosidic bonds. As a fungal extra-acting enzyme, it generally hydrolyzes D-glucose from the non-reducing end of the starch chain, and is widely used in the study of the internal structure of starch granules. Glucoamylase undergoes three steps of diffusion, adsorption, and catalysis to exert hydrolysis (Jiang et al., 2011).

15.2.1.4 Branching Enzymes

Starch branching enzyme (BE, EC 2.4.1.18) (Figure 15.1B) is the only enzyme capable of adding branching linkages to branched starch. It hydrolyzes the α-1,4 glycosidic bond, transferring oligosaccharides from the (1,4)-α-D-glucan chain to the acceptor molecule, adding branches to the linear glucan (Wang et al., 2023a). Branched chain starch accounts for 65–85% of the starch composition, is the key component of starch, and has a great influence on the structure and properties of starch (Li et al., 2020b).

15.2.1.5 Amylosucrase

Amylosucrase (AS, EC 2.4.1.4) (Figure 15.1C) belongs to the family of glycoside hydrolases and is an enzyme that uses sucrose as a substrate to release free fructose for the synthesis of glucan polymers containing only α-1,4 glycosidic bonds (Kim et al., 2014). In particular, amylosucrase catalyzes the elongation of branched chains with non-reducing ends in the presence of an acceptor (such as starch and glycogen) (Kim et al., 2013). In contrast to α-glucan phosphorylase, this enzyme synthesizes polysaccharides without the need for expensive precursors such as adenosine diphosphate (ADP) or uridine diphosphate (UDP) and utilizes only the energy released during the cleavage of the glycosidic bond in sucrose to synthesize the glycosidic bond in the polymer (Zhang et al., 2018). Therefore, sucrose amylase is a simple, effective, and convenient tool for the industrial synthesis of polysaccharide polymers for glycosyl transfer. Based on its catalytic properties, AS has been widely used for the production of resistant starch and slowly digestible starch.

15.2.1.6 Cyclodextrin Glycosyltransferase

Cyclodextrin glycosyltransferase (CGTase, EC 2.4.1.1) (Figure 15.1D) belongs to the α-amylase family of endonucleases (Dura et al., 2016). CGTase catalyzes four main reaction types: three types of transglycosylation reactions (disproportionation, cyclization, and coupling) and minor hydrolysis (Dura et al., 2016). Cyclization is the characteristic reaction of CGTase, which is an intramolecular transglycosylation reaction based on the transfer of O-4 or C-4 from the non-reducing terminus of a straight-chain malt oligosaccharide to the reducing terminus of the same straight chain to form a cyclodextrin (Benavent-Gil et al., 2021). The coupling reaction, which is the reverse of the cyclization reaction, opens the ring

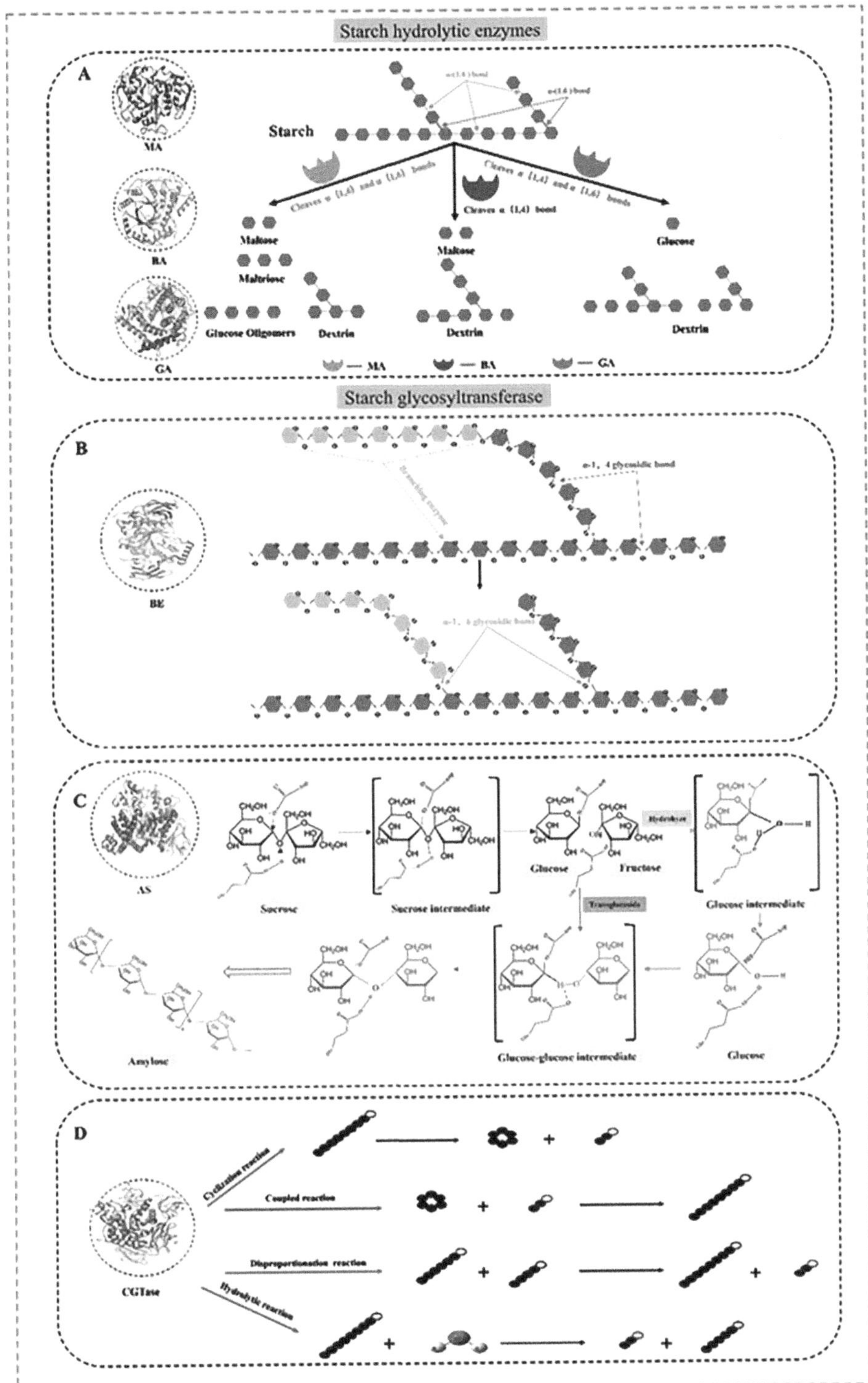

FIGURE 15.1   Mechanism of action of different enzymes: (A) maltogenic amylase (MA), β-amylase (BA) and glucoamylase (GA); (B) branching enzyme (BE); (C) amylosucrase (AS); (D) cyclodextrin glycosyltransferase (CGTase).

of the cyclodextrin and then transfers it to the straight-chain malt oligosaccharides. Disproportionation is the process of cutting off a straight-chain malt oligosaccharide and then transferring one segment to another straight-chain receptor (Charoenlap et al., 2004). In the modification of starch, the most widely used method is the use of CGTase to catalyze the internal α-glycosidic bond breaking to produce oligosaccharides, which are then cyclized into a cyclic structure of six to eight sugar groups corresponding to α-, β-, and γ-cyclodextrins, respectively (Dura & Rosell, 2016). Cyclodextrins have been widely studied due to their ability to reduce glycemic index, retain flavor, prevent oxidative degradation, and sequester cholesterol (Dura & Rosell, 2016).

15.3 EFFECT OF ENZYMATIC MODIFICATIONS ON STARCH STRUCTURES

15.3.1 Molecular Structure

Starch is a polymer with a semi-crystalline structure (Figure 15.2A), which is composed of a crystalline region composed of amylopectin (AP) and an amorphous region composed of amylose (AM) (Li et al., 2020a). A lipophilic core is present on the inner surface of AM, and when AM is complexed with complexing agents such as phospholipids, monoglycerides, and fatty acids, molecules containing AM-lipid complexes with a left-handed single helix structure are formed (Li, 2023). The different branched chains of the AP molecule are named A, B, and C chains, and the arrangement of these chains affects the effective stacking and physicochemical properties of the starch molecule (Li & Gong, 2021). Changes in the starch molecular structure are mainly related to molecular weight (Mw), chain length distribution, and functional group transformation (Lei et al., 2024).

Maltose amylase induced the molecular weight of amylopectin in corn starch to shift to low Mw, and the degree of polymerization (DP) peak of amylose decreased sharply. Similarly, maltose amylase resulted in significant degradation of the outer chain (A and B_1 chain) of corn starch, and the short chain (DP < 9) level increased from 23.7% to 44.2%. However, maltose amylase had little effect on the inner chain length, and it only changed from exogenous to inner in the later stage (Miao et al., 2014). In order to further observe the fine structure of the remaining waxy corn starch after hydrolysis of maltose amylase, isoamylase was used for debranching treatment. The results showed that maltose amylase could shorten the outer chain length (CL) of amylopectin and had a good anti-aging effect even under 20% starch hydrolysis conditions (Grewal et al., 2015). Maltose amylase mainly reduced the content of long chains in pea starch, increased the content of short chains, and had little effect on the content of medium chains. This may be due to the formation of non-reducing D-glucose residues after the cleavage of α-1,4 glycosidic bonds by maltose amylase and the formation of α-1,6 chains after transfer. This change in the fine structure of pea starch affects the digestibility of starch to a large extent (Zheng et al., 2022). Yifan Gui and colleagues compared the effects of commercial or endogenous enzymes on the structure of corn starch. Corn starch was modified by one or two enzymes of commercial α-amylase (AA), β-amylase (BA), and endogenous maltase (MS), and then pullulanase (PUL) was added to it for sequential modification. The results showed that the sequential triple commercial enzyme treatment (AA and BA→PUL) led to more reductions in medium and long chains, while MA/PUL-MS treatment led to more reductions in short and medium long chains. It may be due to the restrictive dextrinase contained in the endogenous maltose amylase to further debranch the short-chain branches. However, the highest amylose content (45.56%) was detected in MA/PUL-MS, which may be attributed to the synergistic treatment of multiple enzymes (α-amylase, β-amylase, and limit dextrinase) in MA to produce more linear long-chain dextrin. The results in the X-ray diffraction pattern can also be demonstrated. MA/PUL-MS contains the most obvious V-shaped scattering peak of amylose-lipid complex (Guo et al., 2022). Compared with natural rice starch and rice starch modified by pullulanase alone, BA/TG/PUL modified rice starch had more short chains, and the chain length distribution was a more uniform "regular triangle". Interestingly, the maximum peak

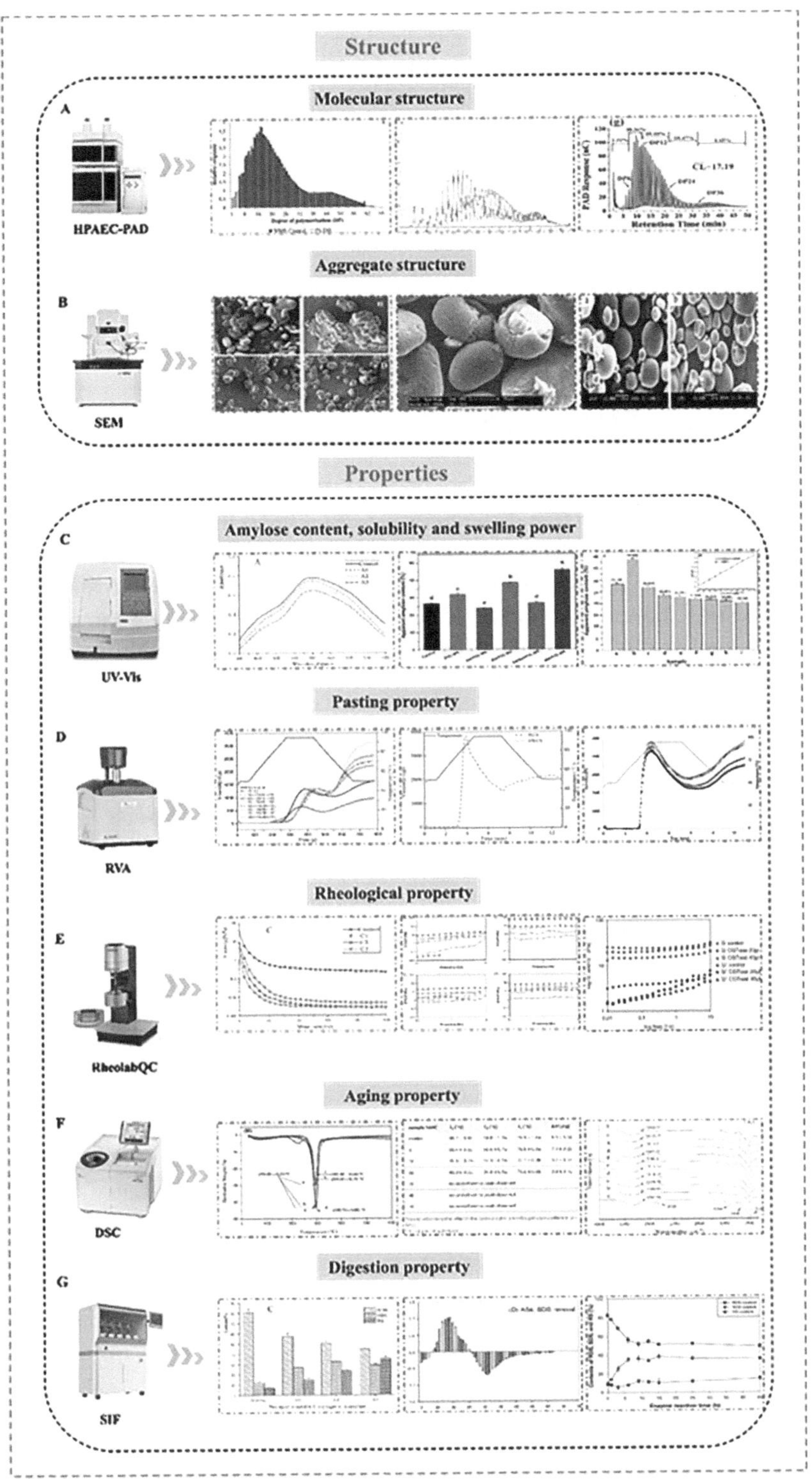

FIGURE 15.2 Effects of enzymatic modification on the structure and properties of starch. (A) Molecular structure (Grewal et al., 2015; Li et al., 2019a; Zheng et al., 2022); (B) aggregate structure (Chen et al., 2011; Das & Kayastha, 2019; Jiang et al., 2011); (C) amylose content, solubility and swelling power (Guo et al., 2022; Li et al., 2019a; Zheng et al., 2022); (D) pasting property (Dura & Rosell, 2016; Li et al., 2020b; Zhang et al., 2017a); (E) rheological properties (Gujral & Rosell, 2004; Shin et al., 2010; Zheng et al., 2022); (F) aging properties (Grewal et al., 2015; Guo et al., 2022; Wang et al., 2023a); (G) digestion properties (Kim et al., 2014, 2017; Zheng et al., 2022).

of BA/TG/PUL-modified rice starch shifted to DP 8–11, and the content was positively correlated with TG treatment time (Li et al., 2019a). Wang and coworkers showed that there was a difference between the average chain lengths of MA→BE and BE→MA-modified sweet potato starch. This was attributed to the different catalytic activities of MA and BE and was also related to the mechanism by which MA hydrolyzes long chains from the outside in. Notably, BE-induced accumulation of short chains of branched starch from the outside inhibits the reaction process of MA (Wang et al., 2023a). For amylosucrase-modified glutinous rice starch, once the reaction started, the consumption of sucrose and the release of fructose occurred together. The shorter chains (DP ≤ 12 and DP 13–24) were gradually extended to those of DP 25–36 as the reaction time increased, but the reaction rate was lower than that of the earlier stages after 6h (Kim et al., 2014).

15.3.2 Aggregate Structure

Enzyme-modified starch differs significantly from natural starch in terms of morphological characteristics (Figure 15.2B). β-amylase occurs in two different ways when breaking down starch granules: centripetal corrosion from the surface towards the center of the granule and centrifugal corrosion that occurs at the periphery of the granule. Das and Kayastha compared the hydrolysis of potato starch and maize starch by β-amylase. The results showed that potato starch hydrolyzed by β-amylase showed scratches, exposed lamellar structures, and displayed rough surfaces and even craters, exhibiting centrifugal corrosion. However, corn starch hydrolyzed by β-amylase had fewer cracks and holes, showing centripetal corrosion (Das & Kayastha, 2019). Jiang and colleagues investigated the characteristics of three different crystalline starches after hydrolysis by glucoamylase. The results showed that type A, B, and C starches exhibited different hydrolysis mechanisms. Glucoamylase penetrated into starch granules through natural pores, leading to severe degradation of type-A starch granules and even formation of cavities. Glucose amylase corroded type B starch by attacking the cracks formed on the surface as well as the small cracks inside, but most of the starch granules were structurally intact and did not produce significant changes. Glucoamylase corroded by attacking the interior of type-C starch and then attacking the exterior of the starch granules, with some cracks and depressions appearing on the surface (Jiang et al., 2011). The surface pores of the porous corn starch synergistically prepared by α-amylase and glucoamylase under optimal reaction conditions extended from the surface to the interior, and the specific surface area of the granules increased with good adsorption capacity (Zhang et al., 2012). Hydrolysis of cassava starch with a mixture of the two enzymes resulted in a similar trend of morphological changes in the granules, but the enzyme corrosion took place mainly in a stratified manner on the surface of the granules (Chen et al., 2011).

The porous starches produced by glycosyltransferases and branched enzymes showed better adsorption compared to the combination of α-amylase and glucoamylase, mainly in terms of increased number of pores and pore size (>50 nm). The rice starch and barley starch samples treated with amylosucrase did not have a complete granular structure, and only a continuous irregular matrix structure was observed, which was attributed to the pre-gelatinization of the starch (Jo et al., 2016). The surface of CGTase-treated corn starch appeared to have an inhomogeneous porous structure, with even crevices and large areas of random erosion over time (Dura et al., 2016). It is possible that CGTase migrates through the channel and hydrolyzes starch from the inside out (Dura & Rosell, 2016).

15.4 EFFECT OF ENZYMATIC MODIFICATIONS ON STARCH PROPERTIES

15.4.1 Amylose Content, Solubility, and Swelling Power

Amylose (Figure 15.2C) is a linear component of starch granules with limited branches. Starch can be classified according to the content of amylose: the content of amylose in conventional starch is about 30%, the content of amylose in high amylose ≥ 70%, and the content of amylose in waxy starch is extremely low (Zhou et al., 2024). After the starch granules are heated and expanded in water, the amylose is first dissolved, which enhances the solubility, which causes the viscosity and fluidity of the starch system to change. Because amylose has a specific affinity for iodine, iodine binding analysis is often used to determine and evaluate the content and purity of amylose (Khan et al., 2023).

A decrease of absorbance was observed in the hydrolysis of pea starch by maltogenic amylase, which was attributed to the destruction of the linear structure of amylose by enzymatic hydrolysis, and the decrease of double helix content led to the decrease of the number of complexable iodine molecules (Zheng et al., 2022). Enzymatic treatment produces rectilinear starch with shorter chain lengths and low molecular sugars, and the formation of rectilinear starch-lipid complexes reduces the content of water-soluble rectilinear starch, which together increase starch solubility (Guo et al., 2022). The apparent straight-chain starch content of BA→TG→PUL-treated rice starch was reduced to 20.10%. TG cleaves the α-1,4 glycosidic bond of straight-chain starch to form branched glucan by transglycosylation, and branched glucan after debranching by pullulanase generates many short straight chains, which leads to the decrease of the content of straight-chain starch that can be complexed with iodine (Li et al., 2019a). The swelling power of AS-treated rice starch decreased, which was attributed to the reduction of α-glucan short chains, which formed a more stable crystal structure and hindered the penetration of water into rice starch (Jung et al., 2020). Similarly, CGTase-treated corn starch also showed lower swelling capacity, which may be related to the physical barriers of some bound water molecules and the hydrophobic pinhole inner wall (Dura & Rosell, 2016).

15.4.2 Pasting Properties

Starch will change its pasting properties (Figure 15.2D) during thermal processing, which is mainly reflected by the pasting process. It mainly includes particle expansion, AM leaching, the three-dimensional network structure of leaching components, and interaction between particle residues and leaching components. Starch granules are heated in water to absorb water and swell, hydrogen bonds between and within molecules are broken, and starch hydration molecules are diffused (Tong et al., 2023). The pasting properties of starch are related to the fine structure of starch, which is usually measured by a rapid viscosity analyzer (RVA) (Gui et al., 2021b).

The pasting temperature of PUL-modified rice starch was reduced compared to natural starch. This was attributed to the disruption of the double helix and hydrogen bond breaking during debranching. However, the pasting temperature of BA/TG/PUL-modified rice starch was significantly higher due to the increase in the ordered arrangement of the starch, and the narrower range of pasting temperatures also indicated a more homogeneous crystal formation and greater thermal stability of the starch (Li et al., 2019a). The gelatinization viscosity of BA/TG/PUL-modified wheat starch increased significantly, which was attributed to the fact that the decrease of amylose content enhanced the hydration ability of starch, and more DP 6–12 linear short chains contributed to the formation of network and increased the final viscosity during cooling (Li et al., 2020b). Significant differences in the viscosity profiles of AS-modified glutinous corn starch (WCS) and acid-diluted glutinous rice starch (AWCS) were observed. The peak viscosity of WCS reached more than 3500 cP, whereas the viscosity of AWCS was maintained almost unchanged at 17 cP, which was attributed to the hydrolysis of glycosidic bonds of branched starch in the amorphous region (Zhang et al., 2017a). CGTase has been widely used in the production of cyclodextrins. RVA was used to study the pasting characteristics of CGTase-treated corn starch at subgelatinization temperature. It was readily observed that pH had a significant effect on starch viscosity change, and CGTase at pH 6.0 induced more hydrolysis of straight-chain starch than CGTase at pH 4.0, resulting in lower peak viscosity of modified starch (Dura & Rosell, 2016). It has also been explained that the hydrolysis products (CD and oligosaccharides) obtained from hydrolysis of maize starch by CGTase affect the molecular rearrangement, leading to differences in the pasting characteristics of the modified starches (Dura et al., 2016). Studies of CGTase addition to wheat starch confirmed that the presence of cyclodextrins disrupts straight-chain starch-lipid complexes and cyclodextrin-lipid inclusions to increase the peak viscosity of modified wheat starch pastes (Gujral & Rosell, 2004).

15.4.3 Rheological Properties

The rheological characterization of starch (Figure 15.2E) is usually studied with starch pasting systems to characterize the change of starch and viscosity with shear rate, which is closely related to the molecular structure of starch. Shorter chain molecules are more resistant to shear, which may be determined by the fact that the high rate of migratory reforming of short chains is greater than the rate of intermolecular entanglement destruction. Steady-state shear rheology measurements were performed on maltose amylase-treated pea starch, and the starch paste showed a tendency toward shear thinning characteristic of pseudoplastic fluids. The results showed that the starch paste was a weakly gelatinous system containing swollen starch granules that were fully hydrated. In particular, the viscosity of the maltose amylase-treated starch paste was substantially reduced, which was attributed to the disruption of the interconnected network structure by enzymatic hydrolysis (Zheng et al., 2022). Shin and colleagues utilized AS-modified various starches to investigate the relationship between changes in the physicochemical properties of the modified starches and the SDS and RS contents. The results showed that the G' values of AS-modified starches at a concentration of 2% were comparable to those of 6% of the original starches, indicating that the elongated branched chains of AS-modified starches have more contactable points and a stronger tendency to gel (Shin et al., 2010). CGTase-treated wheat starch showed overall solid-elastic behavior, but G" was more variable than G" at low frequencies, which was attributed to the hydrolysis of starch by CCTase (Gujral & Rosell, 2004).

The debranching of the side chains induced by the attack of the α-1,6 glycosidic bond by purulanase, as well as the creation of shorter linear chains, leads to a reduction in the entanglement between the chains during the pasting process and a dramatic decrease in the apparent viscosity (Guo et al., 2022). The storage modulus (G') reflects the elastic behavior of the gel, and the loss modulus (G") represents the viscous behavior. The G' and G" and of starch treated by MA→BE and BE→MA were lower than those of natural sweet potato starch, which was attributed to the higher degree of branching of starch caused by MA and BE, which hindered the formation of hydrogen bonds and molecular rearrangement and made it difficult to form a gel system. However, when treated with MA alone for 3 h, it was found that the increase of medium- and long-chain content in sweet potato starch led to the increase of G' and G", which was beneficial to the formation of a gelatinized starch gel network (Wang et al., 2023a). Li et al. (2020b) explored the effect of sequential enzyme BA→TG→PUL treatment on the viscoelasticity and gel strength of wheat starch. BA→TG→PUL-modified wheat starch enhanced the steady-state shear fluidity, which was mainly manifested in two aspects. On the one hand, BA/TG/PUL-modified starch is more likely to expand and become larger, and the gap between the particles is reduced, resulting in an increase in apparent viscosity. On the other hand, more short straight chains of DP 6–12 in BA/TG/PUL-modified starch are easy to move and aggregate, and the formation of ordered molecular structure can increase the shear resistance of starch paste. It should be noted that the shear strength of BA/TG/PUL-modified starch showed a certain time dependence. The results of dynamic rheological properties showed that the loss tangent angles of BA→TG→PUL-modified starch were less than 1, indicating that the gel system exhibited solid-like behavior and increased stability. This is attributed to the increase and entanglement of short linear chains.

15.4.4 Aging Properties

Generally, starch aging (Figure 15.2F) is unfavorable and can cause huge economic losses in the food industry. Many studies have shown that starch aging is due to the formation of a tight double helix structure in the crystalline region (Guo et al., 2022). The chain length of amylopectin is an important factor affecting starch aging. Maltose amylase is an anti-aging enzyme, which is widely used as a preservative for baked products to improve softness and prolong shelf life. Maltose amylase isolated from *Bacillus licheniformis* R-53 can be used in bread production to improve the operation characteristics of dough. In the production stage, maltose amylase hydrolyzes starch to produce maltose, which supplies the substrate for maltose enzyme in yeast to produce glucose, promotes dough fermentation, and increases the specific volume of bread. During storage, maltose amylase further showed internal hydrolysis, cutting amylopectin (mainly destroying α-1,4 glycosidic bonds) to generate maltose and low molecular weight dextrin. These small molecular weight sugars can interfere with the entanglement and recrystallization of starch and protein and delay the retrogradation of bread (Ruan et al., 2021). Similar results were obtained in a study by Navneet Grewal et al. Under the action of maltose amylase, the retrogradation enthalpy of corn starch hydrolysate decreased significantly, and the shortened outer chain and the increase of small molecule dextrin interfered with the recrystallization of starch (Grewal et al., 2015). FTIR spectroscopy results showed that the hydrogen bond strength of sweet potato starch decreased after MA and BE treatment, which indicated that the destruction of hydrogen bonds by enzyme treatment inhibited the molecular rearrangement of sweet potato starch and delayed starch aging (Wang et al., 2023a). CGTase has been widely used to retard the preservation of dough products, mainly relying on the hydrolytic destruction of straight-chain starch and branched-chain starch by CGTase, which generates small-molecule polysaccharides, cyclodextrin-lipids, and cyclodextrin-protein complexes that impede water migration, delaying the loss of water and aging of bread (Yang et al., 2022).

15.4.5 Digestion Properties

The digestibility of starch (Figure 15.2G) has an important impact on human health. Starch is hydrolyzed and converted to glucose by amylase in the mouth and small intestine, causing a rise in postprandial blood glucose (Zheng et al., 2022). Failure to maintain glucose homeostasis can lead to common diseases such as obesity, diabetes, cardiovascular and cerebrovascular diseases, and some cancers (Kim et al., 2014). Studies have shown that starch with low glycemic index can be slowly digested and absorbed by the gastrointestinal tract, which can reduce the incidence of chronic diseases to a certain extent (Dura & Rosell, 2016). Digestibility is an important characteristic of starch, which has a high degree of correlation with blood glucose response (Zhang et al., 2019). According to different enzymatic hydrolysis rates, the composition of natural starch can be divided into three parts: rapidly digestible starch (RDS), slowly digestible starch (SDS), and resistant starch (RS) (Li et al., 2019a). The proportion of starch with three different digestion rates in different types and sources of starch is different. However, enzymatic modification can affect the digestibility of starch.

The difference in digestibility of starch is the result of a combination of many factors (such as starch type, crystallinity, particle pore and shape, and particle size), but it mainly depends on changing the ratio of amylose to amylopectin or the ratio of long amylopectin to short amylopectin to change the digestibility of natural starch. Pea starch is the remaining by-product after protein extraction from peas. The digestibility of natural pea starch was originally lower than that of cereal starch, and the content of SDS and RS increased after malt amylase treatment, resulting in a further decrease in digestibility (Zheng et al., 2022). Miao and colleagues used maltose amylase to regulate the digestion rate of corn starch. The results of Englyst assay showed that the content of SDS in corn starch treated with 6 h enzyme increased from 11.1% to 19.6%, which was attributed to the fact that the transglycosylation of maltose amylase increased the content of amylopectin oligosaccharides. Branched oligosaccharides mainly contain α-1,4 glycosidic bonds with slow hydrolysis rate, which have potential as prebiotics (Miao et al., 2014). However, the hydrolysis time and the amount of enzyme have a great influence on the degree of hydrolysis and the product of starch. The SDS content (38.5%) obtained under 60.00 ppm maltose amylase treatment was higher than other levels (Zheng et al., 2022). Hui Li and his colleagues used sequential enzymes (BA→TG→PUL) to modify rice starch, hoping to reduce the glycemic index of rice starch. The results showed that the RDS level of BA/TG/PUL-modified starch was lower than that of natural starch, and the RS level increased significantly. However, the content of SDS and RS began to decrease when the TG treatment time reached 24 h. Therefore, controlling the TG treatment time can effectively slow down the digestion rate of rice starch (Li et al., 2019a). AS modified starch mainly prolongs the outer chain of amylopectin and forms intermolecular double helix between amylopectin. The perfect crystal structure leads to high SDS or RS content, which is conducive to slow digestion (Kim et al., 2014). Kim and his colleagues prepared modified starch with different amylopectin length distribution by amylosucrase, and their digestibility was studied in detail. The results showed that different amylopectin lengths determined the number and structure of RDS and SDS by affecting the crystal arrangement, thus affecting the degree of hydrolysis and digestion of starch (Kim et al., 2017). Further studies showed that for a given source of starch, the content of RS in AS modified starch was not related to the content of AM but determined by the structural change of AP (Kim et al., 2013). A V-type structure was detected in CGTase- and GBE-modified corn starch. Corn starch is first catalyzed by CGTase to produce cyclodextrins, and then the addition of GBE increases the branching points of cyclodextrins, which is conducive to the capture and binding of cyclodextrins by amylopectin clusters to form complexes. The formation of a V-type crystal structure reduced the digestibility of modified corn starch (Li et al., 2019b).

15.5 APPLICATIONS

The emergence of amylase modification technology has significantly improved the performance defects of natural starch, making it more suitable for practical applications. It has become one of the low-cost raw materials for improving processing technology, improving product performance and reducing production costs in the food, pharmaceutical, and material industries (Figure 15.3).

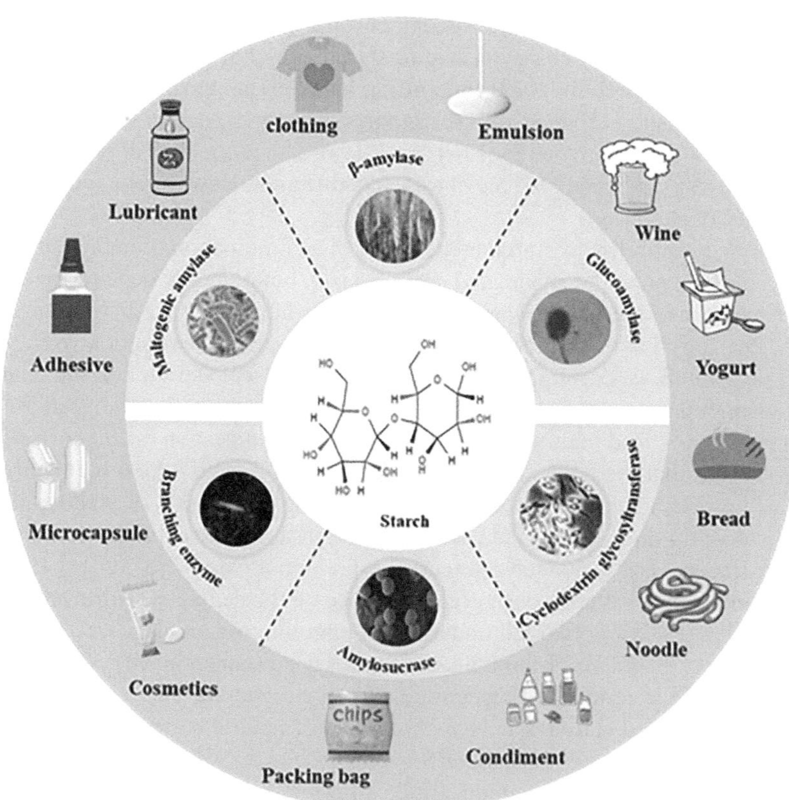

FIGURE 15.3 The main application of enzyme-modified starch in food, medicine, and material industry.

15.5.1 Food Industry

Starch is a major source of calorie intake for humans and an important additive in processed foods. Natural starch is weak in shear resistance, poor in heat resistance, susceptible to thermal degradation, and prone to spoilage and has limited application in some industrial foods (Xu et al., 2021). Enzyme-modified starch has many applications in the food industry, and the properties of starch obtained from starch granules modified by different enzymes and their combinations vary. According to the different catalytic effects of enzymes that can affect the digestive and regenerative properties of starch, slow-digesting starch, resistant starch, and cyclodextrins can be prepared and applied to the production of food products with blood glucose–regulating effects. Raw doughs obtained by replacing wheat flour with enzyme-modified resistant corn starch showed "strain hardening" and increased expansion, and breads baked with doughs of this formulation showed a lower level of aging after storage (Altuna et al., 2016). Potato starch modified by *Streptococcus thermophilus* GtfB enzyme exhibits glucose slow-release properties and can be used as a carbohydrate to help control postprandial blood glucose (Li et al., 2021). Gluten-free breads are usually characterized by a smaller specific volume and poorer texture compared to wheat breads. The hydrolytic and cyclization activity introduced by the addition of cyclodextrin glycosyltransferase releases cyclodextrins from starch and improves the texture of rice bread, which can be used as an improver for gluten-free bread (Gujral et al., 2003).

Enzyme-modified starch can be used not only in pasta products, meat products, sauces, and other food products but also in thickeners, adhesives, stabilizers, and other food additives by changing the viscosity of the starch after the enzyme action. 4-alpha-glucan transferase-modified maize and rice starch forms a gel that shows a discontinuous, partially dense, and fibrous structure, which is brittle, stiffer, and more flexible than the natural gel. It can be used as a substitute for gelatin in the production of jellies and dairy products (Do et al., 2012). Octenyl succinic anhydride (OSA) starch emulsifiers have been successfully applied to stabilize oil-in-water (O/W) emulsion systems in foodstuffs to resist enzymatic degradation by α-amylase. Li Wang and his colleagues modified glutinous corn starch with OSA after debranching by prulanase, and the resulting modified resistant starch has great potential to be used as a food emulsifier (Wang et al., 2023b). Enzyme-modified porous starch can be used as a fat replacer and as an adsorbent for food ingredients. Excessive intake of fat can adversely affect human health, and the study of fat substitutes is particularly important. The bound water content, viscosity, and consistency of ice cream mixtures were increased after reducing the branched chains of glutinous corn starch using debranching amylase and then adding it to ice cream to partially replace fat. More interestingly, the interaction between the shorter straight-chain molecules improved the melt resistance of the ice cream (Jin et al., 2024). An increase in the content of straight-chain starch in enzyme-modified starches favors the formation of V-complexes with guest molecules. In addition, the gaps and pores in the granules also offer the possibility of encapsulating food components as effective carriers of fatty acids, flavors, probiotics, and phenolic compounds, which show great potential for enhancing the nutritional and sensory properties of food products (Guo et al., 2021).

15.5.2 Non-Food Industry

In addition to applications in the food industry, modified starch is also widely used in other industrial fields. Starch is widely used in biomedical applications due to its excellent biocompatibility, high biodegradability, and inexpensive and non-toxic properties. Starch in the composition of pharmaceutical formulations is mainly used as a solid dosage form with different functions such as diluent, binder, disintegrant, sliding agent, anti-adhesive, and lubricant. Antidigestive starch is widely used in pharmaceutical applications to develop new products. Pullulanase increases the resistant starch content of corn starch by 20-fold, and the slow digestibility and high rectilinear amylose content give it the potential to be used as a safe encapsulant material for delivery of sensitive drugs to the colon (Khan et al., 2023). Gu and colleagues sequentially digested highly branched corn starch (HBCS) and encapsulated ascorbic acid (AA) with it using α-amylase and glycogen branching enzyme. The results showed that HBCS-AA inclusion bodies were detected in differential scanning calorimetry, and the embedded AA had enhanced photostability and thermal stability, which could be used as an effective delivery material (Gu et al., 2021). Debranched starches prepared by branched amylases have a high modulus of elasticity and can be used as direct compression tablet binders for the preparation of carriers for active agents that can be released instantaneously or delayed. It is also of interest to note that the results of experiments with branched amylases on branched maize starch with different degrees of debranching indicate that modified starches that can form strong gel networks can be used as superdisintegrants in tablets to delay the release of drugs (Liu et al., 2017).

In terms of environmental protection, it can be used as an adsorbent in the treatment of industrial wastewater, and it can also be used to prepare starch film with enzyme-modified starch to partially replace plastic products to reduce the pollution of the environment (Hu et al., 2015). In other aspects, enzyme-modified starch can be used to change the viscosity of starch, which can be applied to the textile industry, and it can also be used as an emulsifier to make products that meet green standards. In other aspects, enzyme-modified starch can be used to change the viscosity of starch, which can be applied to the textile industry and can also be used as an emulsifier in the cosmetic industry to produce cosmetics in line with the green concept (Boualis et al., 2023).

15.6 SUMMARY AND FUTURE PROSPECTS

Natural starch has been widely used in production and life. With the continuous development of science and technology, modified starch has become the main research object. Through enzymatic modification, starch still maintains a molecular hierarchy and even particle structure but obtains unique properties. This can not only improve the quality of related products but also reduce the use of chemical reagents in the modification process (Miao et al., 2018, 2023). It is a kind of green starch modification method, which is in line with the "clean label" concept of the current food industry. On the one hand, using the specificity of the enzyme, considering the combination of physical, chemical, and other modification methods, modified starch with good specific properties is manufactured and applied to low-calorie foods, emulsion stabilizers, drug delivery, materials, and other fields. On the other hand, although enzymes have relatively high safety, due to the scarcity of starch resources and the variability of the environment, in the process of finding and extracting enzymes from plants or microorganisms, we should also pay attention to the protection of consumer health and natural environment.

ACKNOWLEDGMENTS

This study was financially supported by the National Natural Science Foundation of China (No. 32101990, No. 32372475, No. 32130084), National Key Research and Development Program of China (No. 2022YFD2100602), the Interdisciplinary Integration Innovation Project (SFST2023-KY-02), and the Collaborative Innovation Center of Food Safety and Quality Control in Jiangsu Province, Jiangnan University (2022–1–1).

REFERENCES

Absar, N., Zaidul, I. S. M., Takigawa, S., Hashimoto, N., Matsuura-Endo, C., Yamauchi, H., & Noda, T. (2009). Enzymatic hydrolysis of potato starches containing different amounts of phosphorus. *Food Chemistry, 112*(1), 57–62. https://doi.org/10.1016/j.foodchem.2008.05.045.

Altuna, L., Ribotta, P. D., & Tadini, C. C. (2016). Effect of a combination of enzymes on the fundamental rheological behavior of bread dough enriched with resistant starch. *LWT-Food Science and Technology, 73*, 267–273. https://doi.org/10.1016/j.lwt.2016.06.010.

Amaraweera, S. M., Gunathilake, C., Gunawardene, O. H. P., Fernando, N. M. L., Wanninayaka, D. B., Dassanayake, R. S., Rajapaksha, S. M., Manamperi, A., Fernando, C. A. N., Kulatunga, A. K., & Manipura, A. (2021). Development of starch-based materials using current modification techniques and their applications: A review. *Molecules, 26*(22), 30. https://doi.org/10.3390/molecules26226880

Ashim, D., & Nandan, S. (2017). Modification of foxtail millet starch by combining physical, chemical and enzymatic methods. *International Journal of Biological Macromolecules, 95*, 314–320. https://doi.org/10.1016/j.ijbiomac.2016.11.067.

Bangar, S. P., Ashogbon, A. O., Singh, A., Chaudhary, V., & Whiteside, W. S. (2022). Enzymatic modification of starch: A green approach for starch applications. *Carbohydrate Polymers, 287*, 27. https://doi.org/10.1016/j.carbpol.2022.119265.

Benavent-Gil, Y., Rosell, C. M., & Gilbert, E. P. (2021). Understanding CGTase action through the relationship between starch structure and cyclodextrin formation. *Food Hydrocolloids, 112*, 13. https://doi.org/10.1016/j.foodhyd.2020.106316.

Boualis, H., Wu, X. D., Wang, B. Y., Li, Q., Liu, M. W., Zhang, L., Lyu, M., & Wang, S. J. (2023). Dextranase production using marine microbacterium sp. XD05 and its application. *Marine Drugs, 21*(10), 12. https://doi.org/10.3390/md21100528

Charoenlap, N., Dharmsthiti, S., Sirisansaneeyakul, S., & Lertsiri, S. (2004). Optimization of cyclodextrin production from sago starch. *Bioresource Technology, 92*(1), 49–54. https://doi.org/10.1016/j.biortech.2003.07.007

Chen, Y. S., Huang, S. R., Tang, Z. F., Chen, X. W., & Zhang, Z. F. (2011). Structural changes of cassava starch granules hydrolyzed by a mixture of α-amylase and glucoamylase. *Carbohydrate Polymers, 85*(1), 272–275. https://doi.org/10.1016/j.carbpol.2011.01.047.

Compart, J., Singh, A., Fettke, J., & Apriyanto, A. (2023). Customizing starch properties: A review of starch modifications and their applications. *Polymers, 15*(16), 20. https://doi.org/10.3390/polym15163491

Das, R., & Kayastha, A. M. (2019). Enzymatic hydrolysis of native granular starches by a new β-amylase from peanut (*Arachis hypogaea*). *Food Chemistry, 276*, 583–590. https://doi.org/10.1016/j.foodchem.2018.10.058

Do, H. V., Lee, E. J., Park, J. H., Park, K. H., Shim, J. Y., Mun, S., & Kim, Y. R. (2012). Structural and physicochemical properties of starch gels prepared from partially modified starches using *Thermus aquaticus* 4-α-glucanotransferase. *Carbohydrate Polymers, 87*(4), 2455–2463. https://doi.org/10.1016/j.carbpol.2011.11.021

Dura, A., Blaszczak, W., & Rosell, C. M. (2014). Functionality of porous starch obtained by amylase or amyloglucosidase treatments. *Carbohydrate Polymers, 101*, 837–845. https://doi.org/10.1016/j.carbpol.2013.10.013

Dura, A., & Rosell, C. M. (2016). Physico-chemical properties of corn starch modified with cyclodextrin glycosyltransferase. *International Journal of Biological Macromolecules, 87*, 466–472. https://doi.org/10.1016/j.ijbiomac.2016.03.012

Dura, A., Yokoyama, W., & Rosell, C. M. (2016). Glycemic response to corn starch modified with cyclodextrin glycosyltransferase and its relationship to physical properties. *Plant Foods for Human Nutrition, 71*(3), 252–258. https://doi.org/10.1007/s11130-016-0553-6

Grewal, N., Faubion, J., Feng, G. H., Kaufman, R. C., Wilson, J. D., & Shi, Y. C. (2015). Structure of waxy maize starch hydrolyzed by maltogenic α-amylase in relation to its retrogradation. *Journal of Agricultural and Food Chemistry, 63*(16), 4196–4201. https://doi.org/10.1021/jf506215s

Gu, Z. X., Chen, B. C., & Tian, Y. Q. (2021). Highly branched corn starch: Preparation, encapsulation, and release of ascorbic acid. *Food Chemistry, 343*, 9. https://doi.org/10.1016/j.foodchem.2020.128485

Gui, Y. F., Zou, F. X., Li, J. H., Tang, J., Guo, L., & Cui, B. (2021a). Corn starch modification during endogenous malt amylases: The impact of synergistic hydrolysis time of α-amylase and β-amylase and limit dextrinase. *International Journal of Biological Macromolecules, 190*, 819–826. https://doi.org/10.1016/j.ijbiomac.2021.09.052.

Gui, Y. F., Zou, F. X., Li, J. H., Zhu, Y., Guo, L., & Cui, B. (2021b). The structural and functional properties of corn starch treated with endogenous malt amylases. *Food Hydrocolloids, 117*, 10. https://doi.org/10.1016/j.foodhyd.2021.106722

Gujral, H. S., Guardiola, I., Carbonell, J. V., & Rosell, C. M. (2003). Effect of cyclodextrinase on dough rheology and bread quality from rice flour. *Journal of Agricultural and Food Chemistry, 51*(13), 3814–3818. https://doi.org/10.1021/jf034112w

Gujral, H. S., & Rosell, C. M. (2004). Modification of pasting properties of wheat starch by cyclodextrin glycosyltransferase. *Journal of the Science of Food and Agriculture, 84*(13), 1685–1690. https://doi.org/10.1002/jsfa.1861

Guo, L., Cui, B., Gui, Y. F., Wei, X. Y., Yang, N., Zou, F. X., Lu, L., Liu, P. F., & Fang, Y. S. (2022). Comparison of structural and functional properties of maize starch produced with commercial or endogenous enzymes. *International Journal of Biological Macromolecules, 209*, 2213–2225. https://doi.org/10.1016/j.ijbiomac.2022.04.202

Guo, L., Li, J. H., Gui, Y. F., Zhu, Y., Yu, B., Tan, C. P., Fang, Y. S., & Cui, B. (2020). Porous starches modified with double enzymes: Structure and adsorption properties. *International Journal of Biological Macromolecules, 164*, 1758–1765. https://doi.org/10.1016/j.ijbiomac.2020.07.323

Guo, Y. B., Qiao, D. L., Zhao, S. M., Zhang, B. J., & Xie, F. W. (2021). Starch-based materials encapsulating food ingredients: Recent advances in fabrication methods and applications. *Carbohydrate Polymers, 270*, 24. https://doi.org/10.1016/j.carbpol.2021.118358

Han, X. Y., Wen, H. L., Luo, Y., Yang, J., Xiao, W. H., Ji, X. Y., & Xie, J. H. (2021). Effects of? Amylase and glucoamylase on the characterization and function of maize porous starches. *Food Hydrocolloids, 116*, 8. https://doi.org/10.1016/j.foodhyd.2021.106661

Hu, J., Tian, T., & Xiao, Z. B. (2015). Preparation of cross-linked porous starch and its adsorption for chromium (VI) in tannery wastewater. *Polymers for Advanced Technologies, 26*(10), 1259–1266. https://doi.org/10.1002/pat.3561

Ji, H. Y., Bai, Y. X., Li, X. X., Zheng, D. N., Shen, Y., & Jin, Z. Y. (2020). Structural and property characterization of corn starch modified by cyclodextrin glycosyltransferase and specific cyclodextrinase. *Carbohydrate Polymers, 237*, 8. https://doi.org/10.1016/j.carbpol.2020.116137

Jiang, Q. Q., Gao, W. Y., Li, X., & Zhang, J. Z. (2011). Characteristics of native and enzymatically hydrolyzed *Zea mays* L., *Fritillaria ussuriensis* Maxim. and *Dioscorea opposita* Thunb. starches. *Food Hydrocolloids, 25*(3), 521–528. https://doi.org/10.1016/j.foodhyd.2010.08.003

Jin, Y. Z., Gu, Z. B., Cheng, L., Li, C. M., Li, Z. F., & Hong, Y. (2024). Physicochemical characterization of debranched waxy rice starches and their effect on the quality of low-fat ice cream mixtures. *Food Bioscience, 57*, 9. https://doi.org/10.1016/j.fbio.2023.103485

Jo, A. R., Kim, H. R., Choi, S. J., Lee, J. S., Chung, M. N., Han, S. K., Park, C. S., & Moon, T. W. (2016). Preparation of slowly digestible sweet potato Daeyumi starch by dual enzyme modification. *Carbohydrate Polymers, 143*, 164–171. https://doi.org/10.1016/j.carbpol.2016.02.021

Jung, H. T., Park, C. S., Shim, Y. E., Shin, H., Baik, M. Y., Kim, H. S., Yoo, S. H., Seo, D. H., & Lee, B. H. (2020). Enzymatically elongated rice starches by amylosucrase from *Deinococcus geothermalis* lead to slow down the glucose generation rate at the mammalian α-glucosidase level. *International Journal of Biological Macromolecules, 149*, 767–772. https://doi.org/10.1016/j.ijbiomac.2020.01.266

Kaur, B., Ariffin, F., Bhat, R., & Karim, A. A. (2012). Progress in starch modification in the last decade. *Food Hydrocolloids, 26*(2), 398–404. https://doi.org/10.1016/j.foodhyd.2011.02.016

Khan, A., Siddiqui, S., Rahman, U. U., Belduz, A. O., Shah, A. A., Badshah, M., Hasan, F., & Khan, S. (2023). Enzymatic modification of maize flour improves its functional properties, digestion resistibility, and antioxidant potential. *Journal of Food Measurement and Characterization, 16*. https://doi.org/10.1007/s11694-023-02072-7

Kim, B. K., Kim, H. I., Moon, T. W., & Choi, S. J. (2014). Branch chain elongation by amylosucrase: Production of waxy corn starch with a slow digestion property. *Food Chemistry, 152*, 113–120. https://doi.org/10.1016/j.foodchem.2013.11.145

Kim, B. S., Kim, H. S., Hong, J. S., Huber, K. C., Shim, J. H., & Yoo, S. H. (2013). Effects of amylosucrase treatment on molecular structure and digestion resistance of pre-gelatinised rice and barley starches. *Food Chemistry, 138*(2–3), 966–975. https://doi.org/10.1016/j.foodchem.2012.11.028

Kim, H. R., Choi, S. J., Park, C. S., & Moon, T. W. (2017). Kinetic studies of *in vitro* digestion of amylosucrase-modified waxy corn starches based on branch chain length distributions. *Food Hydrocolloids, 65*, 46–56. https://doi.org/10.1016/j.foodhyd.2016.10.038

Korompokis, K., Deleu, L. J., De Brier, N., & Delcour, J. A. (2021). Investigation of starch functionality and digestibility in white wheat bread produced from a recipe containing added maltogenic amylase or amylomaltase. *Food Chemistry, 362*, 8. https://doi.org/10.1016/j.foodchem.2021.130203

Lei, X. Q., Xu, J. Y., Han, H., Zhang, X. L., Li, Y. H., Wang, S., Li, Y. L., & Ren, Y. M. (2024). Fine molecular structure and digestibility changes of potato starch irradiated with electron beam and X-ray. *Food Chemistry, 439*, 10. https://doi.org/10.1016/j.foodchem.2023.138192

Li, C. (2023). Starch fine molecular structures: The basis for designer rice with slower digestibility and desirable texture properties. *Carbohydrate Polymers, 299*, 12. https://doi.org/10.1016/j.carbpol.2022.120217

Li, C., & Gong, B. (2021). Relations between rice starch fine molecular and lamellar/crystalline structures. *Food Chemistry, 353*, 7. https://doi.org/10.1016/j.foodchem.2021.129467

Li, C., Wu, A., Yu, W. W., Hu, Y. M., Li, E. P., Zhang, C. Q., & Liu, Q. Q. (2020a). Parameterizing starch chain-length distributions for structure-property relations. *Carbohydrate Polymers, 241*, 12. https://doi.org/10.1016/j.carbpol.2020.116390

Li, D., Fu, X. X., Mu, S. Y., Fei, T., Zhao, Y. K., Fu, J. C., Lee, B. H., Ma, Y. L., Zhao, J., Hou, J. M., Li, X. L., & Li, Z. Y. (2021). Potato starch modified by *Streptococcus thermophilus* GtfB enzyme has low viscoelastic and slowly digestible properties. *International Journal of Biological Macromolecules, 183*, 1248–1256. https://doi.org/10.1016/j.ijbiomac.2021.05.032

Li, D. D., Wu, Z. Z., Wang, P., Xu, E. B., Cui, B., Han, Y. B., & Tao, Y. (2022). Effect of moderate electric field on glucoamylase-catalyzed hydrolysis of corn starch: Roles of electrophoretic and polarization effects. *Food Hydrocolloids, 122*, 10. https://doi.org/10.1016/j.foodhyd.2021.107120

Li, H., Li, J. H., & Guo, L. (2020b). Rheological and pasting characteristics of wheat starch modified with sequential triple enzymes. *Carbohydrate Polymers, 230*, 9. https://doi.org/10.1016/j.carbpol.2019.115667

Li, H., Li, J. H., Xiao, Y., Cui, B., Fang, Y. S., & Guo, L. (2019a). *In vitro* digestibility of rice starch granules modified by β-amylase, transglucosidase and pullulanase. *International Journal of Biological Macromolecules, 136*, 1228–1236. https://doi.org/10.1016/j.ijbiomac.2019.06.111

Li, Y., Li, C. M., Gu, Z. B., Cheng, L., Hong, Y., & Li, Z. F. (2019b). Digestion properties of corn starch modified by α-D-glucan branching enzyme and cyclodextrin glycosyltransferase. *Food Hydrocolloids, 89*, 534–541. https://doi.org/10.1016/j.foodhyd.2018.11.025

Liu, G. D., Gu, Z. B. A., Hong, Y., Cheng, L., & Li, C. M. (2017). Structure, functionality and applications of debranched starch: A review. *Trends in Food Science & Technology, 63*, 70–79. https://doi.org/10.1016/j.tifs.2017.03.004

Liu, Y. J., Jiang, F., Du, C. W., Li, M. Q., Leng, Z. F., Yu, X. Z., & Du, S. K. (2022). Optimization of corn resistant starch preparation by dual enzymatic modification using response surface methodology and its physicochemical characterization. *Foods, 11*(15), 12. https://doi.org/10.3390/foods11152223

Miao, M., & BeMiller, J. N. (2023). Enzymatic approaches for structuring starch to improve functionality. *Annual Review of Food Science and Technology, 14*, 271–295. https://doi.org/10.1146/annurev-food-072122-023510

Miao, M., Jiang, B., Jin, Z., & BeMiller, J. N. (2018). Microbial starch-converting enzymes: Recent insights and perspectives. *Comprehensive Reviews in Food Science and Food Safety, 17*(5), 1238–1260. https://doi.org/10.1111/1541-4337.12381

Miao, M., Xiong, S. S., Ye, F., Jiang, B., Cui, S. W., & Zhang, T. (2014). Development of maize starch with a slow digestion property using maltogenic α-amylase. *Carbohydrate Polymers, 103*, 164–169. https://doi.org/10.1016/j.carbpol.2013.12.041

Punia, S. (2020). Barley starch modifications: Physical, chemical and enzymatic—a review. *International Journal of Biological Macromolecules, 144*, 578–585. https://doi.org/10.1016/j.ijbiomac.2019.12.088

Rajan, A., & Abraham, T. E. (2006). Enzymatic modification of cassava starch by bacterial lipase. *Bioprocess and Biosystems Engineering, 29*(1), 65–71. https://doi.org/10.1007/s00449-006-0060-5

Ren, J. Y., Chen, S. D., Li, C. M., Gu, Z. B., Cheng, L., Hong, Y., & Li, Z. F. (2020). A two-stage modification method using 1,4-α-glucan branching enzyme lowers the *in vitro* digestibility of corn starch. *Food Chemistry, 305*, 8. https://doi.org/10.1016/j.foodchem.2019.125441

Ruan, Y. Q., Xu, Y., Zhang, W. C., & Zhang, R. Z. (2021). A new maltogenic amylase from *Bacillus licheniformis* R-53 significantly improves bread quality and extends shelf life. *Food Chemistry, 344*, 9. https://doi.org/10.1016/j.foodchem.2020.128599

Ruidi, H., Songnan, L., Gongqi, Z., Ligong, Z., Peng, Q., & Liping, Y. (2023). Starch modification with molecular transformation, physicochemical characteristics, and industrial usability: A state-of-the-art review. *Polymers, 15*(13), 2935–2935. https://doi.org/10.3390/polym15132935

Shin, H. J., Choi, S. J., Park, C. S., & Moon, T. W. (2010). Preparation of starches with low glycaemic response using amylosucrase and their physicochemical properties. *Carbohydrate Polymers, 82*(2), 489–497. https://doi.org/10.1016/j.carbpol.2010.05.017.

Tong, C., Ma, Z., Chen, H. J., & Gao, H. Y. (2023). Toward an understanding of potato starch structure, function, biosynthesis, and applications. *Food Frontiers, 4*(3), 980–1000. https://doi.org/10.1002/fft2.223

Wang, D. L., Ma, X. B., Yan, L. F., Chantapakul, T., Wang, W. J., Ding, T., Ye, X. Q., & Liu, D. H. (2017). Ultrasound assisted enzymatic hydrolysis of starch catalyzed by glucoamylase: Investigation on starch properties and degradation kinetics. *Carbohydrate Polymers, 175*, 47–54. https://doi.org/10.1016/j.carbpol.2017.06.093

Wang, D. Y., Mi, T. T., Gao, W., Yu, B., Yuan, C., Cui, B., Liu, X., & Liu, P. F. (2023a). Effect of modification by maltogenic amylase and branching enzyme on the structural and physicochemical properties of sweet potato starch. *International Journal of Biological Macromolecules, 239*, 10. https://doi.org/10.1016/j.ijbiomac.2023.124234

Wang, L., Zhu, S. P., Chen, Y., Karthik, P., & Chen, J. S. (2023b). Fabrication and characterization of O/W emulsion stabilized by Octenyl Succinic Anhydride (OSA) modified resistant starch. *Food Hydrocolloids, 141*, 11. https://doi.org/10.1016/j.foodhyd.2023.108750

Wang, N., Dai, J. Q., Miao, D., Li, C., Yang, X. Y., Shu, Q. X., Zhang, Y., Dai, Y. Y., Hou, H. X., & Xu, S. B. (2023c). Influence of enzymatic modification on the basis of improved extrusion cooking technology (IECT) on the structure and properties of corn starch. *International Journal of Biological Macromolecules, 253*, 9. https://doi.org/10.1016/j.ijbiomac.2023.127274

Wang, R., Rui, P. X., Wang, T., Feng, W., Chen, Z. X., Luo, X. H., & Zhang, H. (2023d). Resistant starch formation mechanism of amylosucrase-modified starches with crystalline structure enhanced by hydrothermal treatment. *Food Chemistry, 414*, 9. https://doi.org/10.1016/j.foodchem.2023.135703

Wu, W. Q., Jiao, A. Q., Xu, E. B., Chen, Y., & Jin, Z. Y. (2020). Effects of extrusion technology combined with enzymatic hydrolysis on the structural and physicochemical properties of porous corn starch. *Food and Bioprocess Technology, 13*(3), 442–451. https://doi.org/10.1007/s11947-020-02404-1

Xie, Y., Li, M. N., Chen, H. Q., & Zhang, B. (2019). Effects of the combination of repeated heat-moisture treatment and compound enzymes hydrolysis on the structural and physicochemical properties of porous wheat starch. *Food Chemistry, 274*, 351–359. https://doi.org/10.1016/j.foodchem.2018.09.034

Xu, T., Li, X. X., Ji, S. Y., Zhong, Y. H., Simal-Gandara, J., Capanoglu, E., Xiao, J. B., & Lu, B. Y. (2021). Starch modification with phenolics: Methods, physicochemical property alteration, and mechanisms of glycaemic control. *Trends in Food Science & Technology, 111*, 12–26. https://doi.org/10.1016/j.tifs.2021.02.023

Yang, L., Cai, J., Qian, H., Li, Y., Zhang, H., Qi, X., Wang, L., & Cao, G. (2022). Effect of cyclodextrin glucosyltransferase extracted from *Bacillus xiaoxiensis* on wheat dough and bread properties. *Frontiers in Nutrition, 9*. https://doi.org/10.3389/fnut.2022.1026678

Ye, J. P., Liu, C. M., Luo, S. J., Hu, X. T., & McClements, D. J. (2018). Modification of the digestibility of extruded rice starch by enzyme treatment (β-amylolysis): An *in vitro* study. *Food Research International, 111*, 590–596. https://doi.org/10.1016/j.foodres.2018.06.002

Zhai, Y. T., Li, X. X., Bai, Y. X., Jin, Z. Y., & Svensson, B. (2022). Maltogenic a-amylase hydrolysis of wheat starch granules: Mechanism and relation to starch retrogradation. *Food Hydrocolloids, 124*, 9. https://doi.org/10.1016/j.foodhyd.2021.107256

Zhang, B., Cui, D. P., Liu, M. Z., Gong, H. H., Huang, Y. J., & Han, F. (2012). Corn porous starch: Preparation, characterization and adsorption property. *International Journal of Biological Macromolecules, 50*(1), 250–256. https://doi.org/10.1016/j.ijbiomac.2011.11.002

Zhang, H., Chen, Z. X., Zhou, X., He, J., Wang, T., Luo, X. H., Wang, L., & Wang, R. (2018). Anti-digestion properties of amylosucrase modified waxy corn starch. *International Journal of Biological Macromolecules, 109*, 383–388. https://doi.org/10.1016/j.ijbiomac.2017.12.106

Zhang, H., Wang, R., Chen, Z. X., & Zhong, Q. X. (2019). Enzymatically modified starch with low digestibility produced from amylopectin by sequential amylosucrase and pullulanase treatments. *Food Hydrocolloids, 95*, 195–202. https://doi.org/10.1016/j.foodhyd.2019.04.036

Zhang, H., Zhou, X., He, J., Wang, T., Luo, X. H., Wang, L., Wang, R., & Chen, Z. X. (2017a). Impact of amylosucrase modification on the structural and physicochemical properties of native and acid-thinned waxy corn starch. *Food Chemistry, 220*, 413–419. https://doi.org/10.1016/j.foodchem.2016.10.030

Zhang, H., Zhou, X., Wang, T., He, J., Yue, M., Luo, X. H., Wang, L., Wang, R., & Chen, Z. X. (2017b). Enzymatically modified waxy corn starch with amylosucrase: The effect of branch chain elongation on structural and physicochemical properties. *Food Hydrocolloids, 63*, 518–524. https://doi.org/10.1016/j.foodhyd.2016.09.043

Zheng, Y., Zhang, H. X., Li, R. M., Zhu, X. L., & Zhu, M. W. (2022). Effects of maltogenic α-amylase treatment on the proportion of slowly digestible starch and the structural properties of pea starch. *Food Bioscience, 48*, 8. https://doi.org/10.1016/j.fbio.2022.101810

Zhong, Y. Y., Xu, J. C., Liu, X. X., Ding, L., Svensson, B., Herburger, K., Guo, K., Pang, C. F., & Blennow, A. (2022). Recent advances in *enzyme* biotechnology on modifying gelatinized and granular starch. *Trends in Food Science & Technology, 123*, 343–354. https://doi.org/10.1016/j.tifs.2022.03.019

Zhou, X., He, S., & Jin, Z. Y. (2024). Impact of amylose content on the formation of V-type granular starch. *Food Hydrocolloids, 146*, 9. https://doi.org/10.1016/j.foodhyd.2023.109257

Index

Note: Page numbers in *italics* indicate a figure and page numbers in **bold** indicate a table on the corresponding page.